Springer Series in Solid and Structural Mechanics

Volume 10

Series Editors

Michel Frémond, Rome, Italy
Franco Maceri, Department of Civil Engineering and Computer Science,
University of Rome "Tor Vergata", Rome, Italy

The Springer Series in Solid and Structural Mechanics (SSSSM) publishes new developments and advances dealing with any aspect of mechanics of materials and structures, with a high quality. It features original works dealing with mechanical, mathematical, numerical and experimental analysis of structures and structural materials, both taken in the broadest sense. The series covers multi-scale, multi-field and multiple-media problems, including static and dynamic interaction. It also illustrates advanced and innovative applications to structural problems from science and engineering, including aerospace, civil, materials, mechanical engineering and living materials and structures. Within the scope of the series are monographs, lectures notes, references, textbooks and selected contributions from specialized conferences and workshops.

More information about this series at http://www.springer.com/series/10616

Vincenzo Vullo

Gears

Volume 1: Geometric and Kinematic Design

 Springer

Vincenzo Vullo
University of Rome "Tor Vergata"
Rome, Italy

ISSN 2195-3511   ISSN 2195-352X   (electronic)
Springer Series in Solid and Structural Mechanics
ISBN 978-3-030-36504-2   ISBN 978-3-030-36502-8   (eBook)
https://doi.org/10.1007/978-3-030-36502-8

This Springer imprint is published by the registered company Springer Nature Switzerland AG
The registered company address is: Gewerbestrasse 11, 6330 Cham, Switzerland

Aphorism

Ορῶν δέ σε, καθάπερ ἐγω, σπουδαῖον καὶ φιλοσοφίας προεστῶ τα ἀξιο λόγως καὶ τὴν ἐν τοῖς μαθήμασιν κατὰ τὸ ὑποπίπτον εωρίαν τετιμηκότα ἐδοκίμασα γράψαι σοι καὶ εἰς τὸ αὐτὸ βιβλίον ἐξορίσαι τρόπου τινὸς ἰδιότητα, καθ᾽ ὄν οι παρεχόμενον ἔσται λαμβάνειν ἀφορμὰς εἰς τὸ δύνασθαί τινα τῶν ἐν τοῖς μαθήμασι θεωρεῖν διὰ τῶν ηχανικῶν. τοῦτο δὲ πέπεισμαι χρήσιμον εἶναι οὐδὲνῆσσον καὶ εἰς τὴν ἀπόδειξιν αὐτῶν τῶν θεωρημάτων. αἱ γάρ τινα τῶν πρότερον μοι φανέντων μηχανικῶς ὕστερον γεωμετρικῶς ἀπεδείχθη διὰ τὸ χωρὶς ποδείξεως εἶναι τὴν διὰ τούτο τοῦ τρόπου θεωρίαν· ἑτοιμότερον γάρ ἐστι προλαβόντα διὰ τοῦ τρόπου γνῶσίν τινα τῶν ητημάτων πορίσασθαι τὴν ἀπόδειξιν μᾶλλον ἢ μηδενὸς ἐγνωσμένου ζητεῖν.

Archimedes, The Method, Introduction
(*Translation in Vol. 3, Sect. 3.2*)

*To my wife Maria Giovanna,
my sons Luca and Alberto,
my nephew Nicolò
and my students*

Preface

Gears and gear power transmissions used in most mechanical applications and, especially, in the automotive and aerospace industry constitute one of the classical topics of Machine Design Theory and Methodology. In the opinion of many scholars, including the author of this monothematic textbook, the gears are, by far, one of the most complicated but fascinating subjects of the mechanical engineering, because it is extremely polyhedral, as it calls into question a remarkable variety of disciplines, of which the main ones are: plane and space geometry; various branches of theoretical and applied mechanics, such as kinematics, dynamics, elasto-hydrodynamics, vibrations and noise; continuum mechanics and machine design theory and methodology; static and dynamic material strength; structural optimization; surface contact and lubrication theories and tribology; materials science and metallurgy; cutting processes and, more generally, technological processes of manufacture; maintenance, etc.

Despite the fact that, from a historical perspective (see *Gears—Vol. 3: A Concise History*), the gears constitute, after the potter's wheel, the oldest mechanism invented by Homo Sapiens Sapiens, we can say that a unified scientific theory of the gears, able to consider all the aforementioned disciplines, amalgamating them in a "*unicum*" does not yet exist at present. Currently, according to the most widespread thinking that is shared by historians of science and scientists, a science, to be such, must have at least the following essential features:

- All that science claims should not involve real objects, but specific theoretical entities.
- The theory on which a science is based must have a strictly deductive structure, i.e., it must be characterized by a few fundamental statements (called axioms or postulates or principles), and by a unified method, universally accepted, to deduce from them an unlimited number of consequential properties.
- Its applications to real objects must be based on rules of correspondence between the theoretical entities and real objects.

Well, a unified and comprehensive science of the gears, which meets all these requirements with reference to multiplicity of the disciplines mentioned above, does not exist and, perhaps, it will not exist for a long time yet. The way to achieve this important goal is still long and fraught with difficulties, since it is necessary to agree the different scientific principles that form the bases of the various disciplines involved. Based on the current state of knowledge, we can only say that there are different scientific theories of the gears, as many as the disciplines that contribute to defining a gear design.

Gear design scholars and experts are well aware that, with a design based on the fundamental statements related to only one of the disciplines involved, there is a reasonable certainty of not meeting the postulates related to other disciplines. The difficulties to make a theoretically perfect design of a gear power transmission, i.e., respectful of the countless design requirements related to the various disciplines involved, are actually insurmountable, so the designer must fall back to a compromise design (the so-called design optimized, but often very far from the theoretical one), able to reconcile as best as possible the different and almost always conflicting requirements, dictated by these disciplines.

The scholars of vaunted credit, who claim to know everything and to be able to speak or write about any subject and who boast or attribute themselves knowledge in any field, consider the gears as a synonym of obsolescence, a symbol of the past or, when they are benevolent, a nineteenth-century old stuff. This way of thinking of those we have benevolently referred to as scholars of vaunted credit (no one can therefore accuse this author of not being kindly gentlemen) implies at least the ignorance of the historical evidence that the theory of the gears, considered as mechanisms, is much older, having it characterized the birth of science, in the first Hellenism (see Gears—Vol. 3, Chap. 3). But the unjustified conviction of these so-called scholars hides a far more serious ignorance. In fact, they prove not to know that the most significant contributions for the calculation of the load carrying capacity of the gears were brought gradually in the entire twentieth century (with the exception of Lewis, 1982—see references in Gears—Vol. 2, Chap. 3) and that in this beginning of the third millennium new and equally significant contributions of high scientific value appeared and continued to appear at the horizon (see, e.g., Vol. 2, Chaps. 10 and 11).

On the contrary, the true scholars and experts of gear power transmission systems are well aware that the gears were yesterday, continue to be today and for a long time yet will continue to be an ongoing scientific and technological challenge. Even today, surely it is worth investing significant financial resources, in terms of manpower, tools and means, in R&S activities on this important area, as all the technologically advanced countries continue to do. This depends on the fact that the gears are a very complex multidisciplinary field, as few in mechanical engineering, and knowledge still be acquired is numerous. We can affirm, without the fear of being denied, that gears are the result of ancient knowledge in continuous updating.

Knowledge still to be acquired concerns not only those specific to each of the disciplines involved (geometry and kinematics; static and dynamic loads, including those due to impact; friction and efficiency; dynamic response and noise emission;

static and fatigue tooth root strength; contact stresses and surface fatigue durability; nucleation of fractures of any kind and their propagation until breakage; full film, mixed and boundary lubrication; scuffing and wear; materials and heat treatments; new materials; cutting processes and other manufacturing processes, etc.), but also those arising from their mutual interactions. Other important challenges are those related to the new fields of application (see, e.g., those related to the helicopter and aerospace industry and wind power generators), which require the introduction and design of new types of gear drives that are not reflected in the current technological landscape.

Researchers working in this field are well aware of being in front of an important goal: to develop a comprehensive and unified scientific theory of the gears. Really, at present, a scientific theory of gears having a general horizon, capable of considering and balancing all the aspects involved in their design (i.e., geometric and kinematic aspects, mechanical strength aspects and technological and production aspects) does not yet exist. Only attempts of partial scientific theories of gearing exist, which are limited only to the first of the three aforementioned aspects. At the moment, partial scientific theories concerning the other two aspects mentioned above cannot unfortunately be reported, as non-existent.

Among the partial scientific theories concerning the geometric and kinematic aspects, it is however worth mentioning the laudable attempt by Radzevich (2013, 2018: see references in this volume, Chaps. 1 and 11) who, for the first time, tried to develop a scientific theory of gearing. In reality, Radzevich is heavily indebted to Litvin (1994: see references in this volume, Chap. 12) and his followers, inasmuch as he treasures the substantial contributions on the subject in terms of differential geometry brought by them, presenting already known knowledge concerning the above aspects in the systematic form of a scientific theory. However, Radzevich has the great merit of having set up a general scientific theory of gearing that, although limited to geometric and kinematic aspects, is capable of treating all types of gears, including hypoid gears that include, as special cases, all other types of gears.

Notoriously, gears constitute a classic multidisciplinary machine design topic. In this multidisciplinary framework, numerous and different areas can be identified, which however are between them interdependent, since the variation of a parameter or quantity relative to a given area, inexorably, affects a parameter or quantity of other areas involved. Usually, three areas (it is to be noted that each of these areas is composed of two or more disciplines) are introduced, which identify three characteristic aspects of the gears. They are as follows:

- Geometric–kinematic area, which considers the gears as ideal theoretical entities, and that, in these ideal conditions, tries to optimize their geometric and kinematic quantities.
- Stress analysis area, which aims to stretch as much as possible the lifetime of the gear, considered as an actual structural machine element, working under real operating conditions.

- Technological-productive area that, in relation to the material used and its chemical and metallurgical properties, including heat treatment, or in relation to the development of new and more efficient materials, has as its main objective to make the final product in accordance with the requirements imposed by the designer.

Notoriously, a good design of gears or gear power transmission systems should consider, in a balanced manner, all aspects and specific issues of the three afore-mentioned areas. In other words, it should try to balance optimally the pros and cons of different design choices to be made. Usually, textbooks on gears address one or two of these areas, with some mention to the third area. To my knowledge, a textbook which in an equal manner addresses the subjects of all three of these areas does not exist today. Only gear handbooks (and not all) discuss all the issues concerning these three areas as well as those of the various disciplines that char-acterize each area. However, in most cases, the different problems are discussed with a sectorial vision, without enlarging the horizons for evaluating the conse-quences of the decisions pertaining to them on the issues relating to the other areas.

This textbook also favors the first two of the aforementioned areas, but, in the discussion of the typical topics of the disciplines of each of these two areas, special attention is paid to issues of third area, namely that concerning the technological-productive problems of the gears, with a special reference to those for cutting of the gears. Issues concerning this third area would deserve at least the same attention and the same space as those reserved to the first two areas, but the limitations of a usual textbook generally do not allow a discussion of such a broad horizon. To try to satisfy at least partially this need, this author has considered it appropriate to discuss the most salient aspects of gear cutting in the framework of related topics concerning the other two areas; this is done whenever these aspects can significantly influence the gear design, so they cannot be overlooked by the designer.

This textbook fits into the long trail of gear monographs. It starts with the textbook entitled *Théorie Géométrique Des Engrenages Destinés a Transmetter Le Mouvement De Rotation Entre Deux Axes Situés Ou Non Situés Dans En Même Plan* by Théodore Olivier (1842: see references in *Gears*—Vol. 3, Chap. 5), which is likely to be considered the first monograph ever written and published on the gears. Other textbooks and monographs on this subject have gradually written and published that, however, are not specifically mentioned here, as the main ones are referred to several times in the references of this monothematic textbook. All these textbooks and monographs collect and synthesize, sometimes in remarkable fash-ion, the theoretical bases of the gears, that is, the geometric and kinematic concepts as well as practical experiences and technologies to manufacture them. However, with the exception of the aforementioned textbooks of Radzevich and Litvin, none of these monographs has for its subject a real "*Theory of Gearing*," i.e., a scientific theory of the gears. In fact, none of the other textbooks and monographs is really in line with the deepest and fullest meaning of scientific theory that, as we said before, is based on a set of postulates from which we can deduce the entire body of knowledge of a given area concerning the gears.

Even this monograph on the gears is substantially aligned with those published previously. However, unlike the latter, this monograph favors the basic concepts of the calculation of the gear strength, inclusive of the rating of their load carrying capacity in terms of tooth root fatigue strength, surface durability (pitting), micropitting, tooth flank fatigue fracture, tooth interior fatigue fracture, scuffing, wear, etc. However, these calculations cannot be performed accurately without in-depth knowledge of the geometric–kinematic aspects of the mating tooth flank surfaces as well as those concerning the tooth cutting processes. In this perspective, a large part of this textbook (the first volume) is dedicated to the fundamental geometric–kinematic aspects, while the second volume of the same textbook is reserved to the basic concepts of strength and durability design, which constitute the basis of international standards on the topic. The technological aspects of the tooth cutting are not however discussed in a specific part of this monograph, comparable by extension and deepening with the parties reserved to the two aforementioned aspects; they are just called up and described here and there, in the framework of the first two aspects, just enough because the designer has a picture as complete and comprehensive as possible to be fulfilled for a good gear design.

This textbook mainly collects the lectures held by the author, in the four universities where his academic career took place, namely in chronological order: Polytechnic of Turin, first as an assistant professor and later as an associate professor, from 1971 to 1980; University of Naples "Federico II", as a full professor, from 1980 to 1986; University of Rome, "Tor Vergata", as a full professor from 1986 to 2013 and as a full professor retired from 2014 to 2017; Cusano Telematic University of Rome, as a contract professor from 2018 to date.

However, it is noteworthy that this textbook takes into account the lectures held in these four universities, in the framework of many courses of Bachelor's degree, Master's degree and Ph.D./D.Phil. terminal degree in Mechanical Engineering Science, but also lectures for the training of researchers and research technicians, specialized in gear design, which the author held for Costamasnaga S.p.A. The textbook also collects and builds on many years' experience (over fifteen years) gained from the author at this company, as a consultant, especially for industrial research programs and projects of technological innovation regarding special and custom-made gear systems.

This monothematic textbook has the twofold aim of meeting the needs of university education and those of the engineering profession. In the first area, the goal is to provide a link between the matters covered in the most basic textbooks on gears intended for students in three-year first-cycle degree programs and those addressed in the more advanced monographs on gear theory to be used in second-cycle and third-cycle or doctoral programs. In the second area, the purpose is to provide practitioners and professional engineers working in research and industry with an advanced knowledge on the subject that can serve as a basis for designing gear systems efficient and technologically sophisticated, or for developing new and innovative applications of the gears.

Anyone who has ever worked with the analysis and design of gears is well aware that their actual engineering applications, with no exception, are characterized by the fact that a gear or a gear train is part of a complex mechanical system, which must be analyzed in its entirety, that is considering all the possible interactions with other mechanical parts or mechanical units that make up the entire system. These interactions between the components or groups of the system enhance the resulting stress and strain states, including those that arise under dynamic operating conditions, such as impact loading, vibration, high- and low-cycle fatigue and so forth. The higher the transmission error, the greater the deviations between the theoretical conditions and real operating conditions.

The analysis of such complex systems requires the use of sophisticated and refined mathematical models. To save time and costs for the calculations to be made, these mathematical models can use simultaneously numerical models, based both on the discretization of the continuum, such as finite element method (FEM) and boundary element method (BEM), and on the discretization of the governing differential equations, such as finite differences (FD) and analytical models, such as step-by-step integration methods of the differential equations governing the dynamic behavior of the system to be examined and modal analysis methods for the study of the system's response to the dynamic loads applied to it.

All the problems presented in this textbook on the gears are approached and solved preferably using, whenever possible, theoretical methods, and attention is mainly focused on analytical and methodological aspects. The analytical definition of the tooth flank surface geometry, including that of the tip and fillet with all or part of the rim of the gear members to be analyzed, has another great advantage compared to traditional methods that do not use the differential analytical geometry. In fact, the equations that analytically define the aforementioned surfaces including their intentional modifications allow to automatically generate refined FE models or BE models, with which to perform the contact analysis as well as the stress analysis of the gear pair or gear drive under consideration, thus avoiding the loss of accuracy due to the development of solid models by computer-aided design (CAD) computer programs.

Some topic is discussed using different methods, especially when each of the considered methods allows to examine it from different points of view that allow to better evaluate significant design aspects. In this framework, also some method that may appear a little dated is described, especially when it allows us to better understand the physical phenomenon underlying the topic under discussion. Moreover, this choice has also the advantage of keeping alive the historical memory of the pioneers who have traced new paths that have facilitated the task of developing the most sophisticated current calculation methods.

When necessary for the understanding of the phenomena of interest, hints are made to methods of experimental analysis, and to numerical methods. However, generally, these are not extended to the whole mechanical system, but only limited to the single gear pair. This is because the attention wants to be called on the understanding of the specific phenomena, which are proper of the same gear pair, without the risk that they are more or less substantially altered or overshadowed by the above-mentioned interaction effects with other components and groups of the mechanical system to be examined.

The analytical and numerical solutions proposed here are formulated so as to be of interest not only to academics, but also to the designer who deals with actual engineering problems concerning the gears. This is because such solutions, though sophisticated and complex (see, e.g., the solution of the matrix equations that define the contact surfaces between the mating tooth flanks), have become immediately usable by practitioners in the gear field thanks to today's computers, which can readily solve demanding equations. For the reader, moreover, these solutions provide the grounding needed to achieve a full understanding of the requirements laid out in the standards applying in this area. In most cases, the analytical relationships developed for use in the design analysis and/or in the response analysis (the latter analysis, also called verification analysis, is used mainly by the standards) are also presented in graphical form. This gives the reader an immediate grasp of the underlying physical phenomena that these formulae explain and clarifies exactly which major quantities must be borne in mind by the designer.

In dealing with certain topics, including for example the geometric–kinematic aspects, fully developed exercises have been included to draw the reader's attention to the problems that are of greatest interest to the designer, as well as to clarify the calculation procedure. The author has not considered it necessary to provide such exercises in cases where the problems could be solved immediately with the relationships presented in each chapter, as it was felt that they would add nothing to an understanding of the text.

Each topic is addressed from a theoretical standpoint, but in such a way as not to lose sight of the physical phenomena that characterize various types of gears gradually examined, up to hypoid gears, which are those with a more complex geometry. The study of the gears proceeds in steps, starting from the geometric–kinematic aspects, which are related to the cutting conditions of the teeth, and gradually coming to the analysis of the loads and, subsequently, to the analysis of stress and strain states, which allow to evaluate the tooth bending strength, surface durability in terms of macropitting (pitting) and micropitting, scuffing load capacity, tooth flank fracture load capacity, tooth interior fatigue fracture load capacity, wear strength and service lifetime even under variable loading. The material is thus organized so that the knowledge gained in the beginning chapters provides the grounding needed to understand the topics covered in the chapters that follow.

This monothematic textbook, entitled *Gears*, is also intended for the students in the course on Machine Design Theory and Methodology (Progettazione Meccanica e Costruzione di Macchine) at the Università degli Studi di Roma "Tor Vergata". It consists of the following three volumes:

- *Gears—Vol. 1: Geometric and Kinematic Design*. This volume is organized in 14 chapters, which tackle the geometric and kinematic design of various types of gears most commonly used in practical applications, also considering the problems concerning their cutting processes.

- *Gears—Vol. 2: Analysis of Load Carrying Capacity and Strength Design.* This volume is organized in 11 chapters and an Annex. Eleven chapters address the main problems related to the strength and load carrying capacity of almost all the gears (with the only exception of the worm gears), providing the theoretical bases for a better understanding of the calculation relationships formulated by the current ISO standards. A brief outline of the same problems related to worm gears is made in the Annex.
- *Gears—Vol. 3: A Concise History.* This volume is organized in five short chapters, which summarize the main stages of the development of the gears, and the gradual acquisition of knowledge inherent to them, since the down of the history of *Homo Sapiens Sapiens* to this day.

In Volume 1, the aspects of geometric–kinematic design of involute profile gears are discussed. The cylindrical spur and helical gears are first considered, determining their main geometric quantities in light of interference and undercut problems, as well as the related kinematic parameters. Particular attention is paid to the profile shift of these types of gears either generated by rack-type cutter or by pinion-rack cutter. Among other things, profile-shifted toothing allows to obtain teeth shapes capable of greater strength and more balanced specific sliding, as well as to reduce the number of teeth below the minimum one to avoid the operating interference or undercut. These very important aspects of geometric–kinematic design of cylindrical spur and helical gears are then generalized and extended to the other examined types of gears most commonly used in practical applications, such as straight bevel gears; crossed helical gears; worm gears; spiral bevel and hypoid gears. Ordinary gear trains, planetary gear trains and face gear drives are also discussed.

For the discussion of the simplest geometry gears, which are the cylindrical spur and helical gears and, in some ways, the crossed helical gears, the traditional methods are mainly used, which make use of descriptive geometry and analytical procedures often accompanied by empirical criteria. Instead, for the discussion of the gears with the most complex geometry, which are the worm gears (crossed helical gears can be approached in the same way), face gears and spiral and hypoid gears (the latter are notoriously the most general gear configuration of a gear from which it is possible to obtain, as special cases, all the previous types of gears), differential geometry methods and matrix methods are mainly used. These methods are the advantage of allowing not only a more accurate definition of the instantaneous area of contact during the meshing cycle, but also an immediate and automatic generation of the finite element models through the use of the equations of the teeth surfaces, including the teeth and fillet geometry, portion of the rim and tooth profile modification geometry. In this way, losses of accuracy due to the development of solid models with the use of CAD computer programs are avoided.

However, even the most traditional vector methods are used as well as even the more traditional mixed methods, based on descriptive geometry and on the plane and spherical trigonometry. The discussion of the topics on the basis of these methods is done not only because, as we have said before, the memory of the latter is not lost, but also and above all to demonstrate that the results obtained with their use, although approximate, are not far from those obtained with the aforementioned differential geometry and matrix methods, and therefore, have full design validity. These stunning results obtained with these traditional mixed methods give credit to their creators, who also knew the differential geometry and matrix methods equally well, but could not use them because in their time the current calculation tools were not available.

About the formulae, diagrams and figures taken or derived from the ISO standards, it should be pointed out that, in the form in which they are discussed and presented here, they do not guarantee the accuracy of the results obtained. They are to be considered as important points of reference for the calculations concerning the determination of the geometric and kinematic parameters and the assessment of the load carrying capacity of the gears as well as a clear demonstration of the usefulness of the theoretical bases previously discussed. The guarantee of reliability of the results with formulae, diagrams and figures drawn from ISO standards is in any case given by the use of the original ISO standards, which moreover almost always add to the standard specification, technical specification or technical report concerning a certain topic an equally standard specification, technical specification or technical report regarding interesting and clarifying examples of calculation for the same topic, to which the user can directly draw.

The first two volumes of this monothematic textbook draw their origin from the continuous and incessant stimulation of the numerous students of the author in the three public universities where he taught. The thirst for knowledge led them to treasure the following famous maxim of Aristotle, quoted by Diogenes Laërtius "τῆς παιδείας τὰς μὲν ῥίζας εἶναι πικράς, τὸν δὲ καρπὸν γλυκύν", i.e., "*The roots of education are bitter but the fruit is sweet.*" It has stimulated the author's treatment of unusual and even little difficult subjects, which the students have always shown to appreciable for their growth in view of their inclusion in the world of profession or research. The author's greatest satisfaction is to feel their gratitude when, for some occasional reason, they come back to visit him. They have never let the author miss their attachment and affection, and they still do not miss him. The deepest and heartfelt thanks of the author go to them, who represent the most authentic, sincere and vital component of the university.

Last but not least, the author wishes to express his affectionate, warm and heartfelt thanks to Dr. Ing. Alberto Maria Vullo, his son, who in addition to showing enthusiasm and meticulous care in the drafting of the drawings and graphs of this textbook has validly collaborated in the preparation of its format. The author then wishes to express his equally affectionate thanks to his wife, for her great willingness to transfer in print format the textbook manuscript and formulae as well as for her uncommon patience shown in having the author taken away from the family needs, given the commitment made in almost four-year work. Finally, the author's thanks are also due to his first-born son, Dr. Ing. Luca Vullo, who even

from a distance has been able to give a small but valuable contribution. These
heartfelt thanks are a small gesture of gratitude from the author to ask for an ocean
of excuses for having stolen time and affection from the whole family, to which,
however, he has the pleasure of expressing the deepest movement of his soul.

Rome, Italy   Vincenzo Vullo

Contents

Symbols, Notations and Units

d'	Operating or working center distance (mm)
a_0	Nominal or reference center distance (mm)
a_0	Sum or difference of the radii of base circles (mm)
a_c	Instantaneous value of the cutting center distance (mm)
a_{min}	Minimum center distance (mm)
a_v	Center distance of virtual cylindrical gear (mm)
a	Center distance (mm)
a	Hypoid offset (mm)
$\boldsymbol{a}$	Shortest distance vector
A	Lower end point of the path of contact (–)
A^*	Point on the pitch circle of pinion corresponding to the lower end point of path of contact, A (–)
$b_{1,2}$	Face width of pinion or wheel (mm)
$b_{e1,2}$	Face width from calculation point to outside of pinion or wheel (mm)
b_{eH}	Effective face width (mm)
$b_{i1,2}$	Face width from calculation point to inside of pinion or wheel (mm)
b_v	Active face width or face width of virtual cylindrical gear (mm)
b_{vir}	Virtual face width (mm)
b	Face width (mm)
b	Moment or lever arm (mm)
$\boldsymbol{b}$	Position vector
b	Tooth width coordinate for straight bevel gear, starting from the back cone (mm)
B	Lower point of single pair tooth contact (–)
$c_{1,2}$	Bottom clearance or clearance between pinion and wheel or between wheel and pinion (mm or percent of m)
c_{ham}	Mean addendum factor of wheel (–)
c_{be2}	Face width factor (–)
c	Clearance (mm or percent of m)

c	Feed rate of cutter (m/s)
C	Pitch point (–)
d_0	Nominal diameter (mm)
$d_{1,2}$	Pitch or reference diameter of pinion or wheel (mm)
$d_{a1,2}$	Addendum, tip or outside diameter of pinion or wheel (mm)
d_{ae}	Diameter of outside circle (mm)
$d_{ae1,2}$	Outside diameter of pinion or wheel (mm)
d_b	Base diameter (mm)
$d_{b1,2}$	Base diameter of pinion or wheel (mm)
d_{cf}	Profile control diameter (mm)
d_e	Diameter of circle through outer point of single pair tooth contact or outer pitch diameter (mm)
$d_{e1,2}$	Outer pitch diameter of pinion or wheel (mm)
$d_{f1,2}$	Root diameter of pinion or wheel (mm)
d_{fe}	Diameter of root circle (mm)
d_m	Diameter of mean pitch circle, mean pitch diameter or diameter at mid-face width (mm)
$d_{m1,2}$	Mean pitch diameter of pinion or wheel (mm)
d_v	Reference diameter of virtual cylindrical gear (mm)
d_{va}	Tip diameter of virtual cylindrical gear (mm)
d_{vb}	Base diameter of virtual cylindrical gear (mm)
d	Reference diameter of a gear wheel (mm)
D	Upper point of single pair tooth contact (–)
e_0	Nominal space width (mm)
$e_{1,2}$	Transverse space width of pinion or wheel (mm)
$e_1 = e_{a2}$	Arc of approach (mm)
$e_2 = e_{a1}$	Arc of recess (mm)
e_n	Normal space width (mm)
$e_{i(i=1,2,3)}$	Unit vectors
e	Evolute (–)
e	Total arc of contact (mm)
e	Transverse space width (mm)
E^*	Point on the pitch circle of pinion corresponding to the upper end point of path of contact E (–)
E	Distance between the center of curvature of rounded tip corner and axis of symmetry of the tooth (mm)
E	Upper end point of the path of contact (–)
$f_{H\alpha}$	Profile slope deviation (μm)
$f_{H\beta}$	Helix slope deviation (μm)
$f_{f\alpha}$	Profile form deviation (μm)
$f_{f\beta}$	Helix form deviation (μm)
f_i''	Tooth-to-tooth radial composite deviation (μm)
f_{is}	Tooth-to-tooth single flank composite deviation (μm)
f_{isT}	Tooth-to-tooth single flank composite tolerance (μm)

f_p	Single pitch deviation (μm)
f_{pb}	Transverse base pitch deviation (μm)
f_{pi}	Individual single pitch deviation (μm)
f_{s2}	Meshing equation between shaper and face gear
f_α	Total profile deviation (μm)
$f_{\alpha lim}$	Influence factor of limit pressure angle (–)
f	Tooth deformation or deviation (μm)
F_μ	Friction force (N)
F_a	Axial component of tooth force or axial load (N)
F_{ax}	Axial force (N)
F_i''	Total radial composite deviation (μm)
F_{is}	Total single flank composite deviation (μm)
F_{isT}	Total single flank composite deviation (μm)
$F_{mt1,2}$	Tangential force at mean diameter of pinion or wheel (N)
F_n	Nominal normal force or tooth force (N)
F_{nt}	Nominal transverse tangential load (N)
F_p	Total cumulative pitch deviation or total index deviation (μm)
F_{pi}	Individual cumulative deviation (μm)
F_r	Radial component of tooth force or radial load (N)
F_r	Total run out deviation (μm)
F_{rad}	Radial force (N)
F_t	Nominal transverse tangential force at reference cylinder per mesh or transmitted load (N)
F_β	Total helix deviation (μm)
F	Force or load and instantaneous force or load (N)
F	Composite and cumulative deviation (μm)
g_P	Algebraic value of distance between the point of contact and instantaneous center of rotation (mm)
g_a	Recess path of contact of pinion or wheel (mm)
$g_{a1} = g_{f2}$	Length of path of recess (mm)
$g_{a2} = g_{f1}$	Length of path of approach (mm)
g_f	Approach path of contact of pinion or wheel (mm)
g_α	Length of path of contact (mm)
g	Path of contact (mm)
$h_{1,2}$	Tooth depth of pinion or wheel (mm)
h_a	Addendum (mm)
$h_{a1,2}$	Addendum of pinion or wheel (mm)
$h_{a1,2max}$	Maximum addendum of pinion or wheel (mm)
h_{aP}	Addendum of basic rack of cylindrical gears (mm)
$h_{ae1,2}$	Outer addendum of pinion or wheel (mm)
$h_{am1,2}$	Mean addendum of pinion or wheel (mm)
$h_{amc1,2}$	Mean chordal addendum of pinion or wheel (mm)
$h_{e1,2}$	Outer whole tooth depth of pinion or wheel (mm)
h_f	Dedendum (mm)

$h_{f1,2}$	Dedendum of pinion or wheel (mm)
h_{fP}	Dedendum of basic rack of cylindrical gears (mm)
$h_{fe1,2}$	Outer dedendum of pinion or wheel (mm)
$h_{fi1,2}$	Inner dedendum of pinion or wheel (mm)
$h_{fm1,2}$	Mean dedendum of pinion or wheel (mm)
h_m	Mean whole tooth depth (mm)
h_{mw}	Mean working tooth depth (mm)
h_{t1}	Pinion whole tooth depth (mm)
h_w	Operating or working tooth depth (mm)
h	Tooth depth from root circle to tip circle (mm)
H	Mounting distance (mm)
H	Surface hardness (MPa)
i_T	Torque conversion factor (–)
i_n	Speed conversion factor or transmission function (–)
i_0	Characteristic ratio (–)
i	Involute or involute curve (–)
i	Transmission ratio (–)
$\boldsymbol{i}$	Unit vector
j_{en}	Outer normal backlash (mm)
j_{et}	Outer transverse backlash (mm)
j_{mn}	Mean normal backlash or normal backlash at calculation point (mm)
j_{mt}	Mean transverse backlash or transverse backlash at calculation point (mm)
j_n	Normal backlash (mm)
j_t	Transverse or circumferential backlash (mm)
$\boldsymbol{j}$	Unit vector
$\bar{k}_0$	Addendum factor of rack-type cutter (–)
k_M	Material factor (–)
k'	Dimensionless constant (–)
$k_{1,2}$	Addendum factor of pinion or wheel (–)
$k_{1,2\max}$	Maximum addendum factor of pinion or wheel (–)
k_1	Dimensionless factor (–)
k_2	Dimensionless factor (–)
k_{hap}	Basic crown gear addendum factor related to m_{mn} (–)
k_{hfp}	Basic crown gear dedendum factor related to m_{mn} (–)
k_c	Clearance factor (–)
k_d	Depth factor (–)
k_n	Addendum factor in normal section (–)
k_t	Circular thickness factor (–)
k	Dimensionless constant (–)
k	Distance of tracing point of epicycloids or hypocycloids from the center (mm)
$\boldsymbol{k}$	Unit vector
K_Σ	Cumulative factor (–)

K_g	Sliding factor (–)
K_z	Dimensionless factor (–)
K	Dimensionless wear coefficient (–)
K	Truncation coefficient (–)
$l_{1,2}$	Apex distance of pinion or wheel of a hypoid pair from origin of fixed coordinate system (mm)
l_{max}	Maximum length of instantaneous line of contact or action (mm)
l_t	Total length of line of contact or action (mm)
l	Chordal tooth space in transverse mid-plane of a double-helical gear wheel (mm)
l	Distance along wheel axis between crossing point and intersection of contact normal (mm)
l	Generation straight line or any straight line (–)
l	Instantaneous line of contact (–)
l	Length of instantaneous line of contact or action (mm)
L	Chordal tooth thickness in transverse mid-plane of a double-helical gear wheel (mm)
m' or m_w	Operating or working module (mm)
m_0	Nominal module (mm)
m_b	Base module (mm)
m_{en}	Normal module on the back cone (mm)
m_{et}	Outer transverse module (mm)
m_{mn}	Mean normal module or normal module of hypoid gear at mid-face width (mm)
m_{mt}	Mean transverse module or transverse module in the mean cone (mm)
m_n	Normal module (mm)
m	Dimensional parameter in expression of unit normal to a helicoid ruled surface (mm)
m or m_t	Module or transverse module (mm)
M_{01}	Rotational matrix about z_0-axis
M_{12}	Inverse matrix of M_{21} $\left(M_{12} = M_{21}^{-1}\right)$
M_{1a}	Coordinate transformation matrix from auxiliary coordinate system O_a to fixed coordinate system O_1
M_{21}	Coordinate transformation matrix for transition from coordinate system O_1 to coordinate system O_2
M_{2m}	Rotational matrix about z_2-axis with the unit vector
M_{2n}	Rotational matrix about z_n-axis
M_{2s}	Coordinate transformation matrix from coordinate system O_s to coordinate system O_2
M_{m0}	Translational matrix from coordinate system O_0 to coordinate system O_m
M_{mn}	Translation matrix in transition from coordinate system O_n to coordinate system O_m

M_{nm}	Rotational matrix about z_m-axis
M_{ns}	Rotational matrix about z_n-axis with the unit vector
M	Matrix
$n_{1,2}$	Rotational speed of pinion or wheel (s^{-1} or min^{-1})
n_i	Rotational speed of the input shaft or first driving shaft (s^{-1} or min^{-1})
n_o	Rotational speed of the output shaft or last driven shaft (s^{-1} or min^{-1})
$n_{i(i=1,2)}$	Unit normal in coordinate system, i, rigidly connected to pinion or wheel
n_s	Unit normal vector to tooth flank surface of shaper
n	Common normal of contact between two mating profiles or line of action (–)
n	Integer (–)
n	Number of adjacent pitches (–)
$\boldsymbol{n}$	Unit normal vector or simply unit normal
$N_{i(i=1,2)}$	Normal in coordinate system, i, rigidly connected to pinion or wheel
N_s	Normal vector to tooth flank surface of shaper
N	Normal vector or simply normal
$O_i(x_i, y_i, z_i)$	Coordinate systems, with $i = 0, 1, 2, a, b, m, n, s$
p' or p_w	Operating or working circular pitch (mm)
p_0	Nominal pitch (mm)
p_1, p_2	Involutes
p_{a0}	Hob axial pitch (mm)
p_b	Base pitch (mm)
p_{bn}	Normal base pitch (mm)
p_{bt}	Transverse base pitch (mm)
p_f, p_m	Fixed and moving centrodes or evolute curves (–)
p_n	Normal pitch (mm)
p_{n0}	Normal pitch of hob (mm)
p_z	Lead (mm)
p or p_t	Pitch, circular pitch, transverse circular pitch or transverse pitch (mm)
p	Dimensional screw parameter (mm)
$P_{1,2}$	Input and output power (kW)
P_b	Power dissipated in bearings (kW)
P_d	Total power loss (kW)
P_i	Input power (kW)
P_o	Output power (kW)
P_v	Power dissipated for all other losses (kW)
P_z	Power dissipated at contact between mating teeth (kW)
P	Transverse diametral pitch (number of teeth per inch of diameter pitch)
q_a	Arc of approach (mm)

q_r	Arc of recess (mm)
q	Arc of action (mm)
$\bar{r}_P$	Dimensionless radius at a current point on fillet profile (–)
r_0	Nominal radius (mm)
r_0	Pitch radius of hob (mm)
$r_{1,2}$	Pitch radius of pinion or wheel (mm)
$r'_{1,2}$ or $r_{w1,2}$	Radius of operating or working pitch circle of pinion or wheel (mm)
$r^*_{1,2}$	Shortest distance of Mozzi's axis from the driving or driven wheel axis of a hyperboloid gear pair, or radius of driving or driven wheel of hyperboloid gear members at their throat cross section (mm)
r_I	Distance of center of curvature of rounded tip corner of the pinion-type cutter from rotation axis (mm)
r_L	Radius of effective addendum circle of a pinion-type cutter (mm)
r_P	Radial distance of a current point on fillet profile from pole (mm)
r_a	Radius of addendum circle (mm)
r_a	Radius of pitch circle of annulus gear (mm)
r_b	Base radius (mm)
r_b	Moment arm of normal to profile (mm)
r_{b0}	Radius of the base circle of the pinion-type cutter (mm)
$r_{b1,2}$	Base radius of pinion or wheel (mm)
r_{bs}	Radius of shaper base circle (mm)
r_c	Radius of limit circle (mm)
r_{c0}	Cutter radius (mm)
r_f	Radius of dedendum circle (mm)
r_f	Radius of fillet circle (mm)
r_{fil}	Radius of fillet circle in the general condition in which cutting pitch line does not coincide with datum line (mm)
r_i	Individual radial measurement (μm)
r_{l1}	Radius of limit circle of pinion (mm)
r_{l2}	Radius of limit circle of annulus (mm)
r_p	Radius of pitch circle of planet gear (mm)
r_s	Radius of pitch circle of sun gear (mm)
r_s	Radius of pitch cylinder of shaper (mm)
r_{th}	Radius of worm wheel face or throat form radius (mm)
r_v	Radius of osculating circle or virtual pitch circle (mm)
$\boldsymbol{r}_{u,v}$	Partial derivative of position vector r with respect to curvilinear coordinates (u, v)
$\boldsymbol{r}_\alpha$	First derivative of position vector r with respect to α
$\boldsymbol{r}_{\alpha\alpha}$	Second derivative of position vector r with respect to α
$\boldsymbol{r}_\vartheta$	First derivative of position vector r with respect to ϑ
$\boldsymbol{r}_{\vartheta\vartheta}$	Second derivative of position vector r with respect to ϑ
r	Reference radius (mm)
$\boldsymbol{r}$	Position vector
R_e	Outer cone distance (mm)

$R_{e1,2}$	Outer cone distance of pinion or wheel (mm)
$R_{i,e}$	Inner or outer radius of generation crow wheel (mm)
R_i	Inner cone distance (mm)
$R_{i1,2}$	Inner cone distance of pinion or wheel (mm)
R_m	Mean cone distance (mm)
$R_{m1,2}$	Mean cone distance of pinion or wheel (mm)
s_0	Nominal tooth thickness or nominal standard value of tooth thickness at standard pitch circle (mm)
$s_{1,2}$	Transverse tooth thickness of pinion or wheel or involute arc on the transverse tooth profile of pinion or wheel (mm)
$s'_{1,2}$	Tooth thickness on the pitch circle of pinion or wheel (mm)
$s_{1,20}$	Tooth thickness on the cutting pitch circle of pinion or wheel (mm)
$s_{a1,2}$	Tooth thickness on tip circle of pinion or wheel (mm)
s_b	Tooth thickness on base circle (mm)
$s_{f1,2}$	Tooth thickness on root circle of pinion or wheel (mm)
s_{mn}	Mean normal circular tooth thickness (mm)
$s_{mn1,2}$	Mean normal circular tooth thickness of pinion or wheel (mm)
$s_{mnc1,2}$	Mean normal chordal tooth thickness of pinion or wheel (mm)
s_n	Normal tooth thickness (mm)
s_t or s	Transverse tooth thickness (mm)
s_w	Wear depth (mm)
s	Face advance (mm or degrees)
s	Transverse tooth thickness (mm)
t_{12}	Instantaneous axis of rotation between pinion and face gear
t_{s1}	Instantaneous axis of rotation between shaper and pinion
t_{s2}	Instantaneous axis of rotation between shaper and face gear
t_{x0}	Crown to crossing point (mm)
$t_{xi1,2}$	Front crown to crossing point of hypoid pinion or wheel (mm)
$t_{xo1,2}$	Pitch cone apex to crown or crown to crossing point of hypoid pinion or wheel (mm)
$t_{z1,2}$	Pitch apex beyond crossing point of hypoid pinion or wheel (mm)
$t_{zF1,2}$	Face apex beyond crossing point of hypoid pinion or wheel (mm)
$t_{zR1,2}$	Root apex beyond crossing point of hypoid pinion or wheel (mm)
$t_{zi1,2}$	Crossing point to inside point along axis of hypoid pinion or wheel (mm)
$t_{zm1,2}$	Crossing point to mean point along axis of hypoid pinion or wheel (mm)
t	Time (s)
t	Unit tangent vector or simply unit tangent
T_0	Output torque (Nm)
$T_{1,2}$	Nominal torque at the pinion or wheel or input and output torque (Nm)
$T_{1,2}$	Points of interference (–)
T_{1e}	Effective driving torque (Nm)

T_{Flim}	Limit torque corresponding to nominal stress number for bending (Nm)
T_i	Input torque (Nm)
T_{max}	Maximum torque (Nm)
T_r	Reaction torque (Nm)
T	Torque or operating torque (Nm)
$\boldsymbol{T}$	Tangent vector or simply tangent
u_a	Equivalent gear ratio (–)
u_s	Coordinate along z_s-axis of a current point on tooth flank surface of shaper
u_v	Gear ratio of virtual cylindrical gear (–)
u	Curvilinear or Gaussian coordinate (mm)
u	Gear ratio (–)
$U_{i(i=1,2)}$	Gaussian coordinate for operating pitch cone of pinion or wheel (mm)
$\boldsymbol{U}_{1,2}$	Unit vector of axis of pinion or wheel of a hypoid gear
U	Face or tooth advance, or offset (mm)
v_0	Instantaneous linear speed of cutting tool (m/s)
v_Σ	Cumulative velocity or sum of velocities in the mean point P (m/s)
$v_{P1,2}$ or $v_{1,2}$	Tangential velocity of pinion or wheel at a given point of contact (m/s)
v_{gm}	Fictitious average sliding velocity (m/s)
$v_{g\alpha 1}$	Profile sliding velocity (m/s)
$v_{g\beta 1}$	Helical sliding velocity (m/s)
$v_{g\gamma 1}$	Total sliding velocity or maximum sliding velocity at tip of pinion (m/s)
v_t	Pitch line velocity (m/s)
v_x	Velocity of point of contact along the direction of gear wheel axis (m/s)
$\boldsymbol{v}^{(12)}$ or $\boldsymbol{v}^{(21)}$	Relative velocity vectors or sliding velocity vectors
$\boldsymbol{v}^{(12)}$	Velocity vector of worm or worm wheel
$\boldsymbol{v}_{1,2}$	Velocity vector at any profile point of pinion or wheel
$\boldsymbol{v}_\Sigma$	Cumulative velocity vector
$\boldsymbol{v}_P$	Absolute velocity vector
$\boldsymbol{v}_g$	Sliding velocity vector
$\boldsymbol{v}_{g\alpha,\beta,\gamma}$	Profile, helical or total sliding velocity vector
$\boldsymbol{v}_i^{(12)}$	Sliding velocity vector in coordinate system O_i
$\boldsymbol{v}_{n1,2}$	Components of vector $\boldsymbol{v}_{1,2}$ along the direction of line of action
$\boldsymbol{v}_r^{(s)}$	Velocity vector component in relative motion over tooth surface, σ_s
$\boldsymbol{v}_r$	Relative velocity vector
$\boldsymbol{v}_s^{(s2)}$	Relative velocity vector between a point on surface σ_s with respect to a point on surface σ_2
$\boldsymbol{v}_w$	Cumulative semi-velocity vector
v	Curvilinear or Gaussian coordinate (mm)

v	Tangential velocity at reference circle or at pitch circle (m/s)
$\boldsymbol{v}$	Velocity vector
W_m	Work done by the driving torque (Nm)
W_{m2}	Wheel mean slot width (mm)
W_μ	Energy or work lost by friction (Nm)
$x_{1,2}$	Profile shift coefficient of pinion or wheel (–)
x_{hm}	Profile shift coefficient (–)
$x_{hm1,2}$	Profile shift coefficient of pinion or wheel (–)
x_{sm}	Thickness modification coefficient (–)
$x_{sm1,2}$	Thickness modification coefficient of pinion or wheel, backlash included (–)
x_{smn}	Theoretical thickness modification coefficient (–)
x	Profile shift coefficient (–)
z_0	Number of equal blade groups of a milling head cutter (–)
$z_{1,2}$	Number of teeth of pinion or wheel (–)
z_1	Number of threads or starts of worm (–)
z_{1min}	Minimum number of teeth of pinion (–)
z^*_{1min}	Minimum number of teeth of pinion with profile shift (–)
z_F	Form number or diameter quotient of worm (–)
z_n	Virtual number of teeth of a helical gear (–)
z_p	Number of teeth of crown gear (–)
z_s	Number of teeth of shaper (–)
z_v	Number of teeth of virtual cylindrical gear or virtual number of teeth (–)
z	Number of teeth
Z_0	Virtual number of teeth of fictitious generation or cutting crown wheel (–)
Z_{Hyp}	Hypoid factor (–)
Z_ε	Contact ratio factor for pitting (–)
Z	Integer (–)
α' (or α_w)	Operating or working pressure angle (°)
α^*	Pressure angle at upper end point of the involute part of tooth profile of the pinion-type cutter (°)
α^*_w	Working displacement angle to avoid rubbing (°)
α_0	Nominal pressure angle (°)
α_I	Angle between straight lines O_0I and O_0T (°)
α_L	Pressure angle at point L (°)
α_P	Pressure angle of basic rack for cylindrical gears (°)
α_{a0}	Axial pressure angle of hob (°)
$\alpha_{a1,2}$	Pressure angle at tip point of pinion or wheel (°)
α_{dC}	Nominal design pressure angle on coast side (°)
α_{dD}	Nominal design pressure angle on drive side (°)
α_{eC}	Effective pressure angle on coast side (°)
α_{eD}	Effective pressure angle on drive side (°)

α_{lim}	Limit pressure angle (°)
α_n	Normal pressure angle (°)
α_{n0}	Normal pressure angle of hob (°)
α_{nC}	Generated pressure angle on coast side (°)
α_{nD}	Generated pressure angle on drive side (°)
α_s	Pressure angle of involute shaper (°)
α_t	Transverse pressure angle (°)
α_t' (or α_{wt})	Transverse pressure angle at the pitch cylinder or transverse working pressure angle (°)
α_{vt}	Transverse pressure angle of virtual cylindrical gear (°)
α_x	Axial pressure angle (°)
α	Difference between the involute roll angle and involute polar angle (° or rad)
α	Pressure angle at reference cylinder (°)
β_0	Helix angle of hob (°)
$\beta_{1,2}$	Angles that define the angular position of pinion and annulus gear with respect to centerline (°)
$\beta_{1,2}^*$	Angle on a plane normal to shortest distance between the projection of Mozzi's axis and that of driving or driven wheel of a hyperboloid gear pair (°)
β_b	Base helix angle, helix angle at base circle (°)
β_{bv}	Helix angle at base circle of virtual cylindrical gear (°)
$\beta_{e1,2}$	Outer spiral angle of pinion or wheel (°)
$\beta_{i,e}$	Inner or outer spiral angle at inner or outer radius of the generation crown wheel (°)
$\beta_{i1,2}$	Inner spiral angle of pinion or wheel (°)
β_m	Mean spiral angle or spiral angle at mid-face width (°)
$\beta_{m1,2}$	Mean spiral angle of pinion or wheel (°)
β	Helix angle at reference or pitch cylinder (°)
β	Spiral angle at a point (°)
γ'	Side rack angle (°)
γ_0	Lead angle of hob (°)
$\gamma_{1,2}$	Angle subtending tooth half-thickness on the base circle of pinion or wheel (°)
γ_b	Base lead angle, lead angle at base cylinder (°)
$\gamma = (\varphi + \vartheta)$	Angle subtended by an arc of base circle (°)
γ	Angle between the normal to tooth force and contact generatrix of operating pitch cones (°)
γ	Angle subtending tooth half-thickness on the base circle (°)
γ	Angular width of a tool (°)
γ	Lead angle at pitch cylinder (°)
γ	Relief angle (°)
$\delta_{1,2}$	Pitch cone angle of pinion or wheel (°)

δ_P	Angular coordinate or polar angle between the radial distance of a current point on the fillet profile and the polar axis (rad)
δ_a	Face angle (°)
$\delta_{a1,2}$	Face angle of pinion or wheel (°)
δ_{a2}	Face angle of blank of hypoid wheel (°)
$\delta_{b1,2}$	Base cone angle of pinion or wheel (°)
δ_f	Root angle (°)
$\delta_{f1,2}$	Root angle of pinion or wheel (°)
δ_s	Angle between the shaper axis and instantaneous axis t_{s2} (°)
δ	Distance of the instantaneous line of application of force from the pitch point (mm)
δ	Pitch angle of bevel gear, reference cone angle (°)
Δ^*	Non-dimensional parameter (–)
Δb_{x1}	Pinion face width increment (mm)
Δg_{xe}	Increment along the pinion axis from calculation point to outside (mm)
Δg_{xi}	Increment along pinion axis from calculation point to inside (mm)
Δ	Non-dimensional parameter (–)
$\Delta \Sigma$	Shaft angle departure from 90° (°)
$\varepsilon_{1,2}$	Addendum contact ratio of pinion or wheel (–)
$\varepsilon_{1,2}$	Thickness factor of pinion or wheel (–)
ε_a	Recess contact ratio (–)
ε_f	Approach contact ratio (–)
$\varepsilon_{n1,2}$	Addendum contact ratio of pinion or wheel in normal section (–)
ε_α	Transverse contact ratio (–)
ε_β	Overlap ratio or face contact ratio(–)
ε_γ	Total contact ratio (–)
ε	Angle (°)
ε	Contact ratio, overlap ratio, relative eccentricity, thickness factor (–)
$\zeta_{1,2}$	Slide-roll ratio on pinion profile or wheel profile (–)
$\zeta_{1,2}$	Specific sliding on pinion profile or wheel profile (–)
ζ_R	Pinion offset angle in root plane (°)
ζ_m	Pinion offset angle in axial plane (°)
ζ_{mp}	Offset angle of pinion and wheel in pitch plane (°)
ζ_o	Pinion offset angle in face plane (°)
ζ	Angle of rotation of plane (x_a, y_a) about z_1-axis (°)
η_1	Second auxiliary angle (°)
η_T	Constance torque factor (–)
η_i	Instantaneous efficiency (%)
η_m	Average efficiency (%)
η_t	Total contact efficiency or total efficiency of a gear unit (%)
η_z	Contact efficiency (%)
η_ω	Isogonality factor (–)
η	Efficiency (%)

η	Wheel offset angle in axial plane (°)
ϑ_a	Addendum angle (°)
$\vartheta_{a1,2}$	Addendum angle of pinion or wheel (°)
$\vartheta_{a1,2}$	Tooth thickness half angle on tip circle of pinion or wheel (° or rad)
$\vartheta_{b1,2}$	Tooth thickness half angle on base circle of pinion or wheel (° or rad)
ϑ_{e0}	Space width half angle on cutting pitch circle (° or rad)
$\vartheta'_{e1,2}$	Space width half angle on operating pitch circle of pinion or wheel (° or rad)
ϑ_f	Dedendum angle (°)
$\vartheta_{f1,2}$	Dedendum angle of pinion or wheel (°)
$\vartheta_{i(i=1,2)}$	Gaussian coordinate for operating pitch cone (°)
ϑ_s	Involute roll angle of involute shaper (°)
ϑ'_s	Space width half angle on the base cylinder of involute shaper (°)
ϑ_{s0}	Tooth thickness half angle on cutting pitch circle (° or rad)
$\vartheta'_{s1,2}$	Tooth thickness half angle on operating pitch circle of pinion or wheel (° or rad)
ϑ	Angle between x_a-axis and position vector of a current point of helicoid generation curve on plane $z_a = 0$
θ	Angle subtended by the arc of base circle or auxiliary angle (°)
ϑ	Involute roll angle (rad)
λ	Constant (–)
λ	Angle (°)
λ	First auxiliary angle (°)
λ	Parameter to calculate reduction of path of contact due to cutting interference or undercut (–)
μ^*	Friction amplification factor (–)
μ_0	Coefficient of friction at a given point of path of contact (–)
μ	Coefficient of sliding friction, coefficient of kinetic friction or coefficient of friction (–)
ν	Lead angle of cutter (°)
$\xi_{1,2}$	Deviation of generation pitch cone angle of pinion or wheel from 90° (°)
ξ_1	Angle of approach (°)
ξ_2	Angle of recess (°)
ξ	Total angle of contact (°)
π	Any plane
Π	Common tangent plane to pseudo-pitch cylinders of a crossed helical gear pair
Π	Product notation
$\rho_{1,2}$	Radii of cylinders of friction (mm)
$\rho_{1,2}$	Radius of curvature at two top edges of shaper (mm)
$\rho_{1,2}$	Radius of curvature of pinion or wheel (mm)
ρ_F	Root fillet radius at point of contact of 30° tangent (mm)

ρ_{Fn} Root fillet radius at point of contact of 30° tangent in normal section (mm)

ρ_{P0} Offset or crown gear to cutter center (mm)

ρ_{aP} Distance between the center of curvature of rounded tip corner and cutter tip line (mm)

ρ_b Radius of epicycloid base circle (mm)

ρ_c Radius of centrode circle (mm)

ρ_{fP} Radius of the rounded corner at the tooth tip of pinion-type cutter (mm)

ρ_{fP} Root fillet radius of basic rack for cylindrical gears (mm)

ρ_{lim} Limit radius of curvature (mm)

$\rho_{m\beta}$ Tooth mean radius of curvature in lengthwise direction (mm)

$\rho_{n,min}$ Minimum principal radius of curvature in the normal plane (mm)

ρ_t Radius of curvature in the transverse plane (mm)

ρ Radius of curvature (mm)

ρ Ratio of pitch radii of pinion and wheel (–)

$\boldsymbol{\rho}$ Position vector

$\sigma_{1,2}$ Tooth flank surface of pinion or wheel

σ_2 Tooth flank surface of face gear

σ_{Flim} Nominal stress number for bending (N/mm^2)

σ_s Tooth flank surface of shaper

σ Curve of pointed teeth for worm gear pairs

σ Surface in three-dimensional space

$\Sigma\vartheta_f$ Sum of dedendum angles (°)

$\Sigma\vartheta_{fC}$ Sum of dedendum angles for constant slot width taper (°)

$\Sigma\vartheta_{fM}$ Sum of dedendum angles for modified slot width taper (°)

$\Sigma\vartheta_{fS}$ Sum of dedendum angles for standard taper (°)

$\Sigma\vartheta_{fU}$ Sum of dedendum angles for uniform depth taper (°)

Σ_s Supplementary angle of shaft angle (°)

Σ Plane tangent to a helicoid surface at any point

Σ Shaft angle, crossing angle of virtual crossed axes helical gear (°)

Σ Summation notation

ΣP Sum of powers (kW)

ΣT Sum of torques (Nm)

τ' Curve for which two branches of lines of contact occur for worm gear pairs

$\tau_{1,2}$ Angular pitch of pinion or wheel (° or rad)

$\tau_{v1,2}$ Angular pitch of virtual pinion or wheel (° or rad)

$\tau_{i(i=1,2)}$ Unit vector

τ Angular pitch (° or rad)

τ Curve of interference for worm gear pairs

$\varphi_{1,2}$ Angular displacement or angle of rotation of the pinion or wheel about their axes (° or rad)

φ_2 Intermediate angle (°)

φ_I	Polar angle at point I (° or rad)
φ_L	Polar angle at point L (° or rad)
φ_R	Angle defining polar axis position (° or rad)
$\varphi_{i,o}$	Angular displacement or angle of rotation of driving or driven member of a gear unit (°)
φ_s	Angle of rotation of shaper about its axis (° or rad)
φ	Angle of friction (°)
φ	Angle subtended by arc of base circle (°)
φ	Angular distance between planet gears (° or rad)
φ	Angular parameter in parametric equations of trochoid/cycloid (°)
φ	Involute polar angle (rad)
ψ	Angle between axes of hob and helical gear wheel to be cut (°)
ψ	Reciprocal of characteristic ratio (–)
ω_0	Angular velocity of head cutter or hob (rad/s)
$\omega_{1,2}$	Angular velocity of pinion or wheel (rad/s)
ω_a	Angular velocity of annulus gear (rad/s)
ω_c	Angular velocity of planet carrier (rad/s)
ω_i	Angular velocity of input shaft (rad/s)
ω_o	Angular velocity of output shaft (rad/s)
ω_p	Angular velocity of planet gear (rad/s)
ω_s	Angular velocity of shaper (rad/s)
ω_s	Angular velocity of sun gear (rad/s)
$\omega^{(1,2)}$	Angular velocity vector of worm or worm wheel
$\omega_{1,2}$	Angular velocity vector of pinion or wheel
ω_r	Relative angular velocity vector
ω	Angular velocity (rad/s)
Ω_0	Angular velocity of generation crown wheel (rad/s)
Ω	Angular velocity of conical hob about the crown wheel axis (rad/s)

Chapter 1
Gears: General Concepts, Definitions and Some Basic Quantities

Abstract In this chapter, the main types of gearing used in practical applications for transferring power between parallel, intersecting and crossed axes are described. The tooth parts and some quantities that characterize the toothing are also described and defined. The first law of gearing to be met to have a constant transmission ratio is also recalled, and the main geometric and kinematic parameters of gear drives, gear pairs, and gear wheels are defined. The geometrical parameters concerning the toothing and its sizing are expressed in terms of transverse module, but indications are given to express them in terms of transverse diametrical pitch. Moreover, the ranges of the reference values of efficiency obtainable with the various types of gears are provided, with indications for reaching the maximum values. Finally, attention is focused on the parameters that define the precision of the teeth and the accuracy grade of the gears, highlighting their influence on the transmission error, under static and dynamic conditions.

1.1 Introduction

The transmission of power between shafts can be accomplished in various ways, such as: cogs, i.e. toothed members or gears; flexible elements, such as chains and belts; mechanical couplings and universal joints; hydrodynamic drivers, such as fluid couplings and torque converters; hydrostatic torque converters; step-less speed change gears; friction wheel drives; etc. With gear drives, chain drives and toothed belt drives, which are characterized by a positive geometric coupling, the transmission ratio is constant and depends on the number of teeth of the wheels. With smooth belt drives, trapezoidal belt drives, and friction wheel drives, the circumferential force is transmitted by the friction force, and there is a sliding that, depending on the load, may vary from 1 to 3%, for which the transmission ratio is not, in the strict sense, constant.

Among the various ways of transmitting mechanical power between two shafts, the gears are notoriously not only the most ancient solution, but also the solution more satisfactory, robust, reliable and durable. Gears that make up the mechanical power transmissions must convert the input power characteristics (torque and angular velocity at the input), adapting their to the output power characteristics (torque and angular velocity at the output).

In the manufacturing industry, gear units are widely used for several reasons: versatility of use, which makes them suitable to meet most of the requirements posed by the vast majority of practical applications, even those with strong environmental requirements (with the exception of non-low acoustic emission, which can still be remedied by means of active and passive design solutions); inexpensiveness from the point of view of the costs of purchase and from that of running costs; robustness and ease of maintenance; positive transmission of power; ability to cover almost all application fields, from small miniature instrument installations to huge powerful gears, like those used in gear drives for turbines, mining skips, astronomical telescopes, cement kilns, etc.; widespread interchangeability thanks to the standardization in size and shape of the gears.

Gear design therefore requires that two closely interdependent aspects be considered and analyzed: the geometric-kinematic aspect, and the aspect of the mechanical strength. Both these issues are very complex and interconnected, but we need to address them simultaneously if we want to assure the performance of assembled drive gear systems. The first aspect involves the deepening of the gearing geometry and kinematics (module, tooth profile, pressure angle, line of action, length of path of contact, contact ratio, tip and root relief, crowning, end relief, profile shift or addendum modification, meshing interference, cutting interference or undercutting, specific sliding, absolute and relative speeds, etc.) in order to achieve predetermined design goals. The second aspect involves the analysis of the static and fatigue strength as well as the scuffing and wear resistance, with the purpose of ensuring the predetermined durability under the given operating conditions. On these two aspects, we refer the reader well to traditional textbooks, such as Buckingham [4], Merritt [26], Dudley [9], Giovannozzi [13], Pollone [34], Henriot [15], Niemann and Winter [29, 30], Townsend [39], Maitra [25], Dooner and Seireg [7], Jelaska [23], Litvin and Fuentes [24], and Radzevich [36].

The two above-mentioned aspects are deeply interconnected, constituting de facto two sides of the same coin. In the description that follows, they are analyzed separately, but only to make immediate and understandable basic concepts. In this context, it should be kept in mind the fact that the choice of any quantity of interest carried out by geometric-kinematic considerations inevitably has an impact on the quantities correlated to the mechanical strength, and vice versa. Therefore, the separate analysis of the two aspects must be continuously revised and governed by a more general point of view, which allows to evaluate the mutual interdependencies, with the aim of being able to make the optimum choice of all the quantities of interest in relation to the design goals to be achieved.

Of course, the gear designer must necessarily take into account the technological and manufacturing aspects, which include not only the cutting technologies and, more generally, the manufacturing process of the gears, but also the choice of materials and heat treatments. These important issues are not directly addressed, but only briefly recalled to memory in the framework of the two design aspects mentioned above, when it is deemed useful for a better understanding of the specifically addressed and analyzed phenomena. For further study on this last point, we refer the reader to traditional textbook, such as Woodbury [41], Galassini [12], Rossi [37], Henriot [16], Micheletti [27], Boothroyd and Knight [3], Björke [2], and Radzevich [35].

However, before addressing the detailed study of the above mentioned two aspects of the gear design, and the various related issues, which together constitute the main objective of this textbook, we consider necessary to premise some general concepts regarding gears and gear drives, also called gear systems, gear units, gearboxes or gear transmissions.

These drive gear systems consist of both *gear reduction units*, i.e. units that reduce the speed, also called *speed reducing gears* or *speed reducing gear trains*, and *gear multiplier units* or *overgears*, i.e. units that increases the speed, also called *speed increasing gears* or *speed increasing gear trains*. These are mechanical systems that accomplish a constant *transmission ratio* between the input shaft, rotating at the *angular velocity*, ω_i, and the output shaft, rotating at the angular velocity ω_0; this transmission ratio is given by:

$$i = \frac{\omega_i}{\omega_0} = \frac{n_i}{n_0} = const, \tag{1.1}$$

where $\omega_i = 2\pi n_i/60$ and $\omega_0 = 2\pi n_0/60$; n_i and n_0 are the rotation speeds of the first driving shaft and respectively of the last driven shaft (ω_i and ω_0 in rad/s, n_i and n_0 in min^{-1}, i.e. rpm).

Figure 1.1 shows schematically a mechanical system where a gear unit, **R**, carries out its design function. In this figure, **M** and **U** represent, in order, the *driving machine* or *prime mover* and the *driven machine*, which are connected to the gear unit, or gearbox, by means of suitable couplings, **G**, and shafts. Generally, as design data, the rotational speed n_0 and starting torque T_0 of the driven machine are know; they depend on the working requirements of this machine.

In this regard, we may be faced with very different cases as: (i), very high and constant rotational speed, and torque also constant, but relatively low, as in the case of a rotary compressor, for reasons of economy; (ii), very low and constant rotational speed, and torque also constant but relatively high, as in the cases of belt and screw conveyors, escalators, moving walkways, rotary kilns and mills for cement, crushers, machine tools, etc.; (iii), low rotational speed and high torque for a vehicle in the start-up phase, and alternately high rotational speed and low torque for the

Fig. 1.1 Diagram of the mechanical system in which the gear drive carries out its design functions

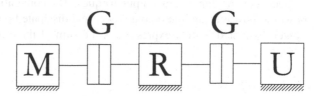

same vehicle during the running regime (we have operating conditions similar to the latter in the rectilinear advancement motion at different speeds, as occur, for example, in some machine tools and rolling carpets).

In most cases, the driving machine, which is characterized by a given rotational speed n_i and by a given torque T_i, is not able to meet the above-mentioned requirements. For example, a three-phase asynchronous electric motor has its own operating rotational speed, which depends on the number of its poles and the frequency of the network; an internal combustion engine, in condition of maximum efficiency, must operate within a narrow range of variability of the rotational speed; small and medium power turbines work well at very high rotational speeds; etc.

Therefore the gear unit, which is interposed between the driving and driven machines, performs the function of changing, according to the fixed transmission ratios, the rotational speed and torque available to the output shaft of the driving machine, adapting their to the needs of the driven machine, represented by the rotational speed and torque that must be available to the input shaft of the latter.

In cases where the driven machine must operate at variable rotational speed, the use of a gear unit with fixed transmission ratio would be unsuitable. In these cases, it is necessary to use variable speed drives, which allow to obtain a continuous variation of the transmission ratio, within a given variability range. To meet the same requirements, a static frequency converter can be used that, through the energy distribution network, supplies the three-phase electric motor (this solution is, however, very expensive). Alternatively, a gear unit with fixed transmission ratio, interposed between the driving machine and the driven machine, and a static frequency converter upstream of the electric motor can be used simultaneously (this solution is cheaper than the previous one). In some cases, the optimal technical solution is just the simultaneous use of a *static frequency converter* and a gear unit; in these cases, regardless of the type of the driving machine, it is certainly preferable to make the continuous variation of the rotational speed on the input shaft, on which a lower torque is applied, and thus obtain higher torque, with a corresponding low rotation speed, by means of a gear drive.

With reference to the Fig. 1.1, we consider separately the gear unit **R**. Since in a mechanical system in steady-state equilibrium the sum ΣP of the mechanical powers that come into play must be equal to zero, we can write the following power balance equation:

$$\Sigma P = P_i + P_0 + P_d = T_i \omega_i + T_0 \omega_0 + P_d = 0, \tag{1.2}$$

where P_i, P_0 and P_d are respectively the input power, output power and total power loss, that is the power dissipated for the various causes of loss, while T_i and T_0 are respectively the input and output torques. By convention, we assume the input power as positive, and the output power and dissipated power as negative. The total power dissipated is then expressed as the sum of three main contribution, i.e. as:

$$P_d = P_z + P_b + P_v, \tag{1.3}$$

where P_z is the power dissipated at the contact between the mating teeth, P_b is the power dissipated in the bearings (rolling bearings, plain bearings, etc.), and P_v is the power dissipated for all other losses (ventilating effect, splash lubrication, contact seals, idling, etc.).

Therefore, the total efficiency of the gear unit is equal to:

$$\eta_t = \frac{P_0}{P_i} = \frac{P_0}{P_0 + P_d} = \frac{P_i - P_d}{P_i} = 1 - \frac{P_d}{P_i}. \tag{1.4}$$

So we get $P_d = P_i(1 - \eta_t)$. If attention is focused only to power dissipated at the contact between the mating teeth, the contact efficiency can be defined by the relation $\eta_z = (P_i - P_z)/P_i$.

In a mechanical system in steady-state equilibrium also the sum of the torques, ΣT, must be equal to zero; therefore, we can write the following torque balance equation:

$$\Sigma T = T_i + T_0 + T_r = 0, \tag{1.5}$$

where, remaining the same the meaning of the quantities already introduced, T_r is the reaction torque applied to the gear unit housing by the structure to which it is connected. In dependence on the functional requirements of the driven machine, it will be possible to have T_0 greater than, less than or equal to T_i (the difference between T_0 and T_i is compensated by the reaction torque T_r), and, consequently, n_0 less than, greater than or equal to n_i.

When the absolute value of the transmission ratio is grater then unity, that is when $|i| > 1, |n_i| > |n_0|$ and $|\omega_i| > |\omega_0|$, we have a *speed reduction ratio*, and the gear unit is a *speed reducing gear drive*. Vice versa, when the absolute value of the transmission ratio is less than unity, that is when $|i| < 1, |n_i| < |n_0|$ and $|\omega_i| < |\omega_0|$, we have a *speed increasing ratio*, and the gear unit is a *speed increasing gear drive*. The ratio $i_n = n_0/n_i$ between the output speed n_0 and the input speed n_i in a gear power train is called the *speed conversion factor*; the ratio $i_T = T_0/T_i$ between the output torque T_0 and the input torque T_i is called the *torque conversion factor*.

A transmission ratio $|i| \neq 1$ should only arise when there is both speed and torque conversion. In the ideal case of *power losses* equal to zero, the product $i_n i_T$ be unitary, and therefore we will have $i_n < 1$ only if $i_T > 1$ and vice versa. In the actual case, which is always characterized by power losses, we have the following relationship that correlates the quantities i_n, i_T and i (for the sign conventions, see below):

$$i_T = \frac{T_0}{T_i} = -\frac{\eta_t n_i}{n_0} = -\frac{\eta_t}{i_n} = -\eta_t i. \tag{1.6}$$

It should be noted that, in the literature concerning this topic, the reciprocal i_n of i, that is $i_n = 1/i = \omega_0/\omega_i = n_0/n_i$, is better known as *transmission function*, and is often used improperly as transmission ratio.

1.2 Gear Units and Gears

Gear units or gear drives, widely used in practical applications, are combinations of gear pairs arranged in a wide variety of ways in order to form a *gear train* or *powertrain*. We can have two types of gear train: a *simple train* (in this type of gear train, one or more idle gears are introduced between the driving and the driven member of the gear system, in such a way that each idle gear is able to reverse the direction a rotation, without changing the transmission ratio), and a *compound train* (in this type of gear train, the intermediate shafts carry at least two separate gear wheels or compound gear wheels having different numbers of teeth, and each stage of reduction provides its own transmission ratio). A powertrain is conceived so as to meet particular requirements, such as: gear ratios, total transmission ratio, direction of rotation, position of the input, intermediate and output shafts, etc.

The gear trains composed of three coaxial members and precisely by two end gear wheels (the *sun gear* and the *ring gear*, also called *annulus gear* or simply *annulus*) with fixed axis and a frame (the *planet carrier*, also called *carrier, spider* or *arm*) free to rotate around to the common axis of these end gear wheels, are called *simple epicyclic gear trains* or *simple planetary gear trains*. The planet carrier supports the movable axis (or axes) of one (or more) gear wheels called *planet gear(s)*, which mesh simultaneously with the sun gear and annulus gear. The connection between this planetary gear train and any external mechanism is accomplished only through these three coaxial members. We have a *compound epicyclic gear train* or *compound planetary gear train* in the two cases of a planetary gear train with compound planet gears (the planet gears themselves are compound, instead of simple gears) or a combination of two or more simple planetary gear trains. All the epicyclic gear trains (simple or compound) are studied with reference to the *basic train*, that is the gear train obtained by helding at rest the planet carrier, while the other two coaxial members rotate. By fixing the planet carrier, the simple epicyclic gear train becomes a simple gear train, while the compound epicyclic gear train becomes a compound gear train.

According to ISO 1122-1:1998 [18], a planetary gear train (also called *planetary gear, epicyclic gear* or *epicyclic gear train*) is a combination of coaxial elements, of which one or more are annulus gears, and one or more are planet carries, which rotate around the common axes and support one or more planet gears which mesh with the annulus gears and one or more sun gears. This definition is quite general, as it includes both the simple planetary gear trains and the compound planetary gear trains.

Whatever its configuration, each gear train can be thought of as made up of pairs of toothed wheels (*gear pair*) which mesh mutually. The term toothed wheel indicates a tooted member which, through the action of teeth which mesh in succession, rotates another or is driven in rotation by another. Generally, the term *gear* indicates an elementary mechanism consisting of two toothed wheels, one which drags the other and vice versa, and rotating around axes in invariable relative position; it is therefore synonymous with gear pair. But it should be noted that, often the term *gear* is used to indicate one of the two toothed wheels of the gear pair, the one with larger number of teeth. In any case, we call *pinion* and *wheel* respectively the smallest and the largest of the toothed wheels of the gear pair. To avoid misunderstandings, for the member with larger number of teeth it would be good to use the term *gear wheel*.

One of the gear design data consists of the specification of the relative position of the shafts connected by the two toothed members of a gear pair. With reference to the relative position of the two shafts (Fig. 1.2), their axes may be parallel (coaxial gears constitute a particular case), intersecting or crossed (or skew), i.e. neither parallel nor intersecting (axes at right angles constitute further particular cases of intersecting and crossed axes). Correspondingly, the gear pair connected to the two shafts is a parallel gear pair (*parallel gears*, Fig. 1.3a), an intersecting gear pair (*intersecting gears*, Fig. 1.3b) or a crossed gear pair (*crossed gears* or *skew gears*, Fig. 1.3c).

Parallel gears are generally *cylindrical gears*, consisting of a pair of cylindrical toothed gear wheels. However, it should be noted that *bevel gears* with parallel axes exist as well (see Niemann and Winter [30]). These cylindrical gears can be *spur gears*, *helical gears* and *double-helical gears*. Spur gears may be *external spur gears* (the term spur gears without qualification implies an external spur gear), *internal spur gears*, and *spur rack-pinion pair*. The term spur gears implies also

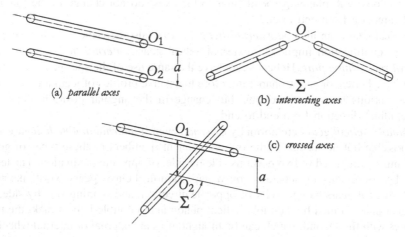

Fig. 1.2 Relative position of the axes: **a** parallel axes; **b** intersecting axes; **c** crossed or skew axes (*a* and *Σ* are respectively the center distance and shaft angle)

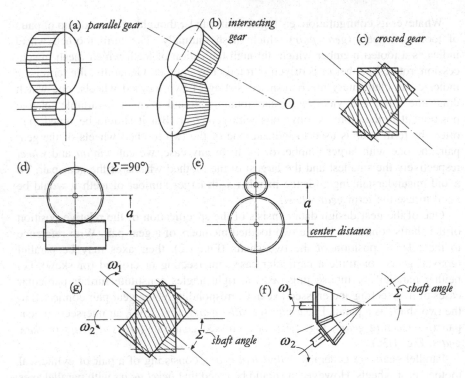

Fig. 1.3 Types of gears and quantities defining the relative position: **a** parallel gear; **b** intersecting gear; **c** crossed gear; **d** worm gear; **e** center distance in a parallel cylindrical gear; **f** shaft angle in an intersecting gear; **g** shaft angle in a crossed gear

that the teeth are parallel to the axes, so that their profiles in the transverse sections (i.e. sections with planes perpendicular to the axes) do not change in the longitudinal direction from end to end.

Helical gears can also be *external helical gears* (also here the term helical gear without qualification implies an external helical gear), *internal helical gears* and *helical rack-pinion pair*. Helical gears have the same use of the spur gears, but their teeth have helical or screw shape; therefore transverse profiles of teeth are the same in the various transverse sections, but change in the angular position along the longitudinal direction from end to end.

Double-helical gears are normally *external*, even if *internal double helical gears* are possible but not usual. In the two component members of these types of gear, toothing is composed of two portions side by side, of opposite inclination. The teeth may be continuous or separated by a gap. Double helical gears work like two single-helical gears having helix of opposite direction, and running side by side. In the gear pairs formed by a double-helical pinion and a double-helical rack, the first meshes with the second, which can be thought of as an external or internal wheel of infinite radius. Note that *triple-helical gears* have been made, but they are not used since, compared to several disadvantages, do not present particular advantages with respect to the double-helical gears.

An intersecting gear consists of a *bevel gear* pair, whose axes intersect at a common apex, O, and form a given shaft angle, Σ (Fig. 1.2). The shafts axes are usually, but not necessarily, perpendicular; the shaft angle may have any value between $0°$ and $180°$, but limitations exist in cutting machines. Each type of bevel gear used to connect two intersecting axes finds a correspondence in the analogous type of cylindrical gear used to connect parallel shafts. Then we can have *external bevel gears*, *internal bevel gears*, and *crown gear-pinion pairs* or *crownwheel-pinion pairs*, the latter corresponding to spur rack-pinion pairs. *Miter gears* or *mitre gears* represent a special case of external bevel gears, characterized by a unitary transmission ratio.

Bevel gears have teeth shaped generally like spur gears, except that the teeth surface are conical surface. Bevel gears may be: *straight bevel gears* (note that the term bevel gear without qualification implies a straight bevel gear); *helical bevel gears* (also called *skew bevel gears*), and *spiral bevel gears*. Correspondingly the toothing is straight, helical and spiral. It may also have *double-helical bevel gears* in which the spiral toothing is composed by two portions side by side, of opposite inclinations; these type of bevel gears have been made and used, but are not common.

Crossed (or skew) gears cover a wide and interesting type of gears, among which we mention specifically:

- *Crossed helical gears*: they are cylindrical gears having helical toothing, but used to connect crossed (or skew) shafts. Therefore the terms helical gears and crossed helical gears indicate the same helical cylindrical gears, but used in different context, the first to connect parallel shafts, and the second to connect skew shafts. These crossed helical gears are also called *crossed-axes helical gears* or *screw gears* (they are also called *spiral gears* but this denomination should be avoided because it can generate misunderstandings). Unlike the helical gears, which give a line contact between the teeth, in the crossed helical gears we have theoretically a point contact, and a longitudinal sliding motion between the teeth.
- *Spiral rack-pinion pairs*: this type of gear combine the features of both helical and crossed helical gears, as it provides a line contact rather than a point contact, but the teeth have a longitudinal sliding motion. We can have different configurations: the pinion in the form of a spur gear, and the rack with teeth inclined to the direction of its motion; the rack in the form of a spur rack, and the pinion with helical teeth; both rack and pinion with inclined teeth. In any case, the pinion drives the rack in a direction not perpendicular to the axis of the same pinion.
- *Worm gears*: this *worm gear pair* (Fig. 1.3d), which consists of a *worm* and a *worm-wheel*, is essentially a screw meshing with a special helical gear wheel. The worm can have one or more threads (any number up to six or more may be used), and its geometry is similar to that a power screw. Worm gears connect

crossed shafts, usually, but not necessarily, perpendicular. Rotation of the worm, which simulates a motion of retrograde linear advancement of a rack, determines the consequent rotation of the worm gear. Generally, the profiles of the worm threads have the same geometry as those of helical and spiral gears, and the geometry of the worm-wheel, also called worm gear, is similar to that of a helical gear, except that the teeth are curved and concave to envelop the worm. The contact between the worm threads and the worm-wheel teeth is a line contact, but between them there is a sliding motion much higher compared to what occurs in other types of gears. Sometimes the worm-wheel or both worm and worm-wheel are modified to better envelop the mating member, and we can have (see ISO 1122-2: 1999(E/F) [19]): a *single enveloping worm gear pair*, where an *enveloping worm* is in meshing with a *cylindrical worm wheel*; a single enveloping worm gear pair, where a *cylindrical worm* is in meshing with an *enveloping worm-wheel*; a *double-enveloping worm gear pair*, where an enveloping worm is in meshing with an enveloping worm-wheel. This last case, which is the most demanding from the technological point of view, allows for a greater area of contact, but cutting processes and mounting techniques more precise are required. The term *globoidal worm gear pair* is reserved for a worm gearing in which both worm and worm-wheel have modified form, and are concave on their respective axial sections.

– *Hypoid gears*: they are spiral bevel gears, the axes of which, however, do not intersect. This type of gear is a special case of *hyperboloid gear*, to which it resembles; it is generated using cutting processes somewhat similar to those of the spiral bevel gears. Hypoid gears are usually used in motor vehicles (cars and trucks), with the axis of the pinion disposed below of the crown wheel. Note that, from the theoretical point of view, hyperboloid gears represent the general case of which cylindrical and bevel gears are particular cases; in these hyperboloid gears, teeth are formed on hyperboloids of revolution.

These type of gears do not exhaust the possible cases, but cover a substantial portion of the most significant practical applications. Gears for special purpose, such as pump gears, timing gears, instrument gears, sprocket wheels, and so on, remain outside of the classification above. These special gears must have specific characteristics and must meet given design requirements. They include *non-circular gears*, which are used for the generation of a prescribed transmission function or as a generating driving mechanism to modify the *displacement function* or the *velocity function*. These types of gears are not discussed in this monographic book.

Now let's go back to the specification of the relative position of the shafts connected by the gear pair, and see its interdependence with respect to the direction of rotation of the shafts, which is another of the design date of the gears. For parallel cylindrical gears, the relative position of the axes is completely defined only by the *center distance* between the shafts (Figs. 1.2a and 1.3e), and for an external gear there is only one possible direction of relative rotation, once fixed the direction of rotation of one of the two members of the gear pair. Since their axes are parallel, the shaft angle Σ, that is the other of the two quantities that, in the general case, define the position of two axes in a three-dimensional space, is equal to zero. Conversely, for intersecting gears, the relative position of the axes is completed defined only by the shaft angle Σ (Figs. 1.2b, e and 1.3f), i.e. the smaller of the two angles of which one of the two axes must rotate to bring it to coincide with the other, so that the two gear members have opposite directions of rotation; having intersecting axes, which are therefore coplanar axes, the center distance is equal to zero. Finally, in the case of crossed gears, which represents the more general case, the relative position of the axes is uniquely fixed by giving simultaneously both the center distance, defined as the minimum distance between the two axes, and the shaft angle (Figs. 1.2c and 1.3d), also defined here as the smaller of the two angles of which one of the two axes must rotate to bring it to be parallel to the other, so that the two gear members have opposite direction of rotation; in this general case, the center distance and the shaft angle are both different from zero.

1.3 Efficiency of the Gears

The designer of gears and gear transmission systems must have a clear and precise knowledge of efficiency achievable with various types of gears that he must design. Among other influences, with the other variables that remain the same, the efficiency of the gears and gear systems is influenced by the value of the transmission ratio actually achieved. The selection of the appropriate gears or the appropriate gear systems for different practical applications can be made, so as accurately and precisely by the designer, after reasoned comparison of the efficiency of the different gear types, taking account of the transmission ratio to be realized.

The analytic determination of efficiency of the gears having involute tooth profiles (these are the gears that interest in this textbook) is a problem that is generally not easy to be solved (see: Panetti [32]; Ferrari and Romiti [11]; Scotto Lavina [38]). We will focus on this issue with more detail in the chapters on the various types of gears used in most current practical applications. Here we want to give the reader a general guideline on the variability ranges of efficiency values achievable with the various types of gears. The Eq. (1.4) shows that the total efficiency η_t of a gear transmission system can be simply calculated as the percentage value of the quotient of the output shaft power divided by the input shaft power, i.e. $\eta_t = (P_0/P_i)100\%$. We have then $P_0 = (P_i - P_d)$, i.e. the output power is the difference between the input power, P_i, and the power total losses, P_d.

The Eq. (1.3) then shows that the power total losses in gear transmission systems mainly depend on the tooth contact friction losses, P_z, bearing losses, P_b, and lubrication churning losses, P_v. Friction losses depend on the gear design and their manufacturing process, gear size, transmission ratio, pressure angle, and coefficient of friction. For the main types of gear pairs usually used in practical applications, a rough estimate of the efficiency range associated with the tooth friction can be made using the guideline data shown in Table 1.1, which also show the usual variability range of the transmission ratio, i, and the usual pitch line velocity, v (in m/s).

Table 1.1 Efficiency range for various types of gear pairs

Type of gear pair	Normal transmission ratio range	Usual pitch line velocity (m/s)	Efficiency range (%)
Cylindrical spur	1:1–6:1	25	98–99
Cylindrical helical	1:1–10:1	50	98–99
Cylindrical double helical	1:1–15:1	150	98–99
Straight bevel	1:1–4:1	20	97–98
Spiral bevel	1:1–4:1	50	98–99
Crossed cylindrical helical	1:1–6:1	30	70–98
Worm	5:1–75:1	30	20–98
Hypoid	10:1–200:1	30	80–95
Cycloid	10:1–100:1	10	75–85

It is to be noted that the values of efficiency due to tooth friction losses shown in Table 1.1 are valid only for single tooth meshes. For gear transmission systems consisting of a given number of gear pairs of the same type, the efficiency related to tooth friction is obtained by multiplying the efficiency of the single gear pair by the number of gear pairs. In the case in which the gear unit is constituted by different gear pairs, the efficiency is obtained by multiplying the various efficiencies of the individual gears pairs.

Also it is to be noted that the theoretical efficiency ranges of the various gear types shown in Table 1.1 do not include bearing and lubricant losses. Furthermore they assume ideal mounting and assembly conditions in regard to axis orientation and center distance. Deviations from these ideal conditions will downgrade the efficiency values.

Bearing losses, P_b, and churning losses, P_v, are almost independent of the type of gear and transmission ratio. The determination of these two losses is not easy, so at least in the initial stage of a new gear drive design, the designer would do well to base the design on the experience gained in similar cases. Very often these two losses are being combined into a single loss, and then they are evaluated globally. In any case they are respectively influenced by the types of bearings used and the tangential velocity of the gears passing through the lubricant contained in the oil sump.

It is not the place to delve into this interesting topic. To get a rough idea of what happens in the rolling bearings, we refer the reader to [31]. Always to have a rough idea about what is happening in the plain bearings (radial and axial hydrodynamic bearings, radial and axial hydrostatic bearings, pneumostatic and aerostatic bearings and magnetic bearings) we refer the reader to Niemann et al. [31], and Chirone and Vullo [6].

Now we return to focus our attention only on friction losses, P_z, which occur at the contact between the mating teeth. From data shown in Table 1.1, we can deduce the following (see also Dubbel [8]):

- Cylindrical spur gears can guarantee much higher efficiencies compared to other type of gears; their efficiency varies between 98 and 99%. Table 1.1 shows the same efficiency values for cylindrical, helical and double helical gears. In this regard, however, it is to be noted that, with the same accuracy grade and the same operating and assembly conditions, the efficiencies of these two types of gears are slightly smaller than the efficiency of cylindrical spur gears. For the cylindrical spur gears, the highest values of efficiency are related to lower gear transmission ratios, while the cylindrical helical and double helical gears can work with very high pitch line velocities and can reach their higher efficiency with maximum transmission ratios up to 10:1.
- Straight bevel gears have efficiencies a little lower than those of cylindrical spur gears and, as for the latter gears, the highest efficiency values are correlated to the lowest values of the transmission ratio. With the same accuracy grade and the same operating and assembly conditions, spiral bevel gears have efficiencies greater than those of the equivalent straight bevel gears; this is due to tooth spiral shape, from which even less noise and vibration are the result.
- The efficiencies of the crossed cylindrical helical gears, worm gears, and hypoid gears are much lower than those of the gears examined above, having the same accuracy grade and the same operating and assembly conditions. In Chaps. 10, 11 and 12, where these gear types are respectively analyzed in detail, we will deepen the subject concerning their efficiencies, and we will provide the information necessary to obtain, if possible, their maximum values, through optimization of the design choices. We will provide also reliable relationships to calculate their values as a function of the quantities on which they depend. However, we must already emphasize the high transmission ratios attainable with hypoid gears (up to 200:1), with efficiencies that can be considered sufficiently satisfactory.
- Although in this textbook we do not speak of cycloidal gears, we considered it appropriate to provide the data relating to these gears, in order to have the terms of comparison with the gears having involute tooth profiles. As Table 1.1 shows, under normal working conditions and with the same accuracy grade and the same operating and assembly conditions, cycloidal gears are characterized by lower efficiencies than those of the gears with involute tooth profiles. In addition, cycloidal gears allow to obtain much higher transmission ratios, and can work in very high efficiencies at relatively high values of the transmission ratio, above 30:1.

Finally it should be stressed the fact that, for the various type of gears taken into consideration, the data summarized in Table 1.1 constitute the guideline values of the efficiencies obtainable for a single gear pair of these gears. In addition, these guideline values only consider the friction losses at the contact between mating gear teeth. In order to obtain sufficiently reliable values of the efficiency of a gear transmission system, which usually is constituted by a multiplicity of gear pairs, which may also be of different types, it is necessary to evaluate the individual losses of each gear pair, which make up the gear unit, as well as the bearing losses and churning losses. It should also be borne in mind that the maximum values of the total efficiency of a gear unit will be obtained under full load (see: Dubbel [8]; Grote and Antonsson [14]), while the ones under partial load, and worse still in the start-up, when the temperature of lubricant inside the oil sump is lower, the total efficiency is considerably lower.

1.4 Basic Law of Mating Gear Teeth

From the geometric-kinematic point of view, the conversion of the characteristics of the input power into those of the output power in a spur gear pair, takes place by means of the rolling without sliding of two imaginary cylindrical surfaces, each of which is rigidly connected to each of the two members of the same gear pair (Fig. 1.4). These cylinders are the *loci* of the instantaneous axis of rotation in relative motion between the two members of the gear pair, called *axodes*, *operating cylinders*, *operating surfaces* or *pitch cylinders*, which are synonymous. The outer surfaces of these cylinders correspond to those of two friction wheels which roll without sliding on one another; this rolling is therefore a pure rolling. As shown in Fig. 1.4, pitch cylinders are tangent along the line *C-C*, which is the *instantaneous axis of rotation* or *pitch axis*, belonging to the common tangent plane. This is the *pitch plane*, which can be thought of as the materialization of a thin sheet of paper; when this sheet is pulled in the direction perpendicular to the pitch axis, it drags in pure rolling the aforementioned two cylinders, one clockwise and the other counterclockwise. This plane is the pitch plane of the rack that meshes well with the two members of the gear pair under consideration.

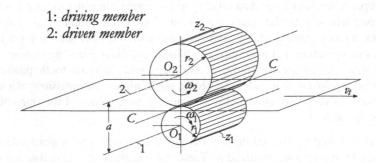

Fig. 1.4 Pitch cylinders and common pitch plane; 1 and 2, axes of pinion and wheel

Quantities shown in Fig. 1.4 are: $r_1=d_1/2$ and $r_2=d_2/2$, the pitch radii and pitch diameters of the two cylinders; ω_1 and ω_2 their angular velocities ($\omega_1=2\pi n_1/60$ and $\omega_2=2\pi n_2/60$, where n_1 and n_2 are the rotational speeds); $v_t=\omega_1 r_1=\omega_2 r_2$, the pitch line velocities or tangential velocities; $a=r_1+r_1 = (d_1+d)/2$, the center distance; z_1 and z_2 their number of teeth. Note that the subscripts 1 and 2 indicate, respectively, quantities referring to the pinion (the smaller wheel) and the wheel (the larger wheel). Note also that for external gears a, z_1 and z_2 are considered positive, whereas for internal gears, a and z_2 have a negative sign, and z_1 has a positive sign. The transverse sections of the two pitch cylinders with a transverse plane, i.e. with a plane normal to their axes, are the *pitch circles* of two flat gear wheels by which the kinematics of their relative motion is described. Using these pitch circles we also study the shape of the teeth profiles which realize a geometric coupling with a predetermined transmission ratio between the two members of the gear pair.

The basic requirement that the geometry of the gear-tooth profile needs to satisfy is the guarantee of a transmission ratio exactly constant. It is noteworthy that a geared mechanical transmission may also be made with a specified variable transmission ratio, but this type of geared drive is not among those considered in this textbook. Notoriously this requirement is satisfied by two *mating profiles* that perform a *conjugated action*. The basic law of conjugated action between two mating profiles states that as the profiles rotate, the common normal to the profiles at the point of contact must always intersect the line of centers at the same point C, called the *pitch point* (Fig. 1.5). This basic law of conjugate action is known as the *first law of gearing*.

From a more general point of view, mating gear teeth acting against each other to produce a rotational motion are similar to cams. The profiles of cams or teeth must be designed so as to get a constant transmission ratio during meshing, and then to have a conjugate action. In theory, it is possible to select any arbitrary profile for one cam or

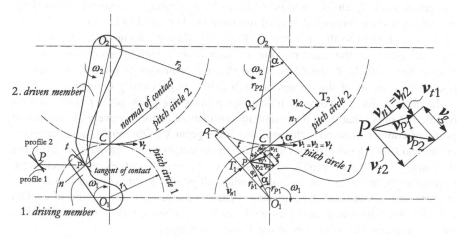

Fig. 1.5 Basic law of conjugate profiles, normal of contact, and velocities

tooth and then to find the correspondent theoretical profile for the meshing cam or tooth which will give a conjugate action. Among the possible pairs of profiles, *involute profiles* are those of greatest interest; in fact, these involute profiles, with few exceptions, are used universally for gear teeth. Our attention is focused on these profiles, even if *cycloidal profiles*, *circular-arc profiles* or *Wildhaber-Novikov profiles* where used in the past and in some specific cases are still used (see: Giovannozzi [13]; Niemann and Winter [29]; Litvin and Fuentes [24]).

Regarding the discussion that follows, we assume the teeth (or cams) to be perfectly shaped, perfectly smooth and absolutely rigid. Of course, these assumptions are unsuitable in practice, due to errors in cutting and finishing operations of the gears, mounting and assembly errors as well as to elastic deformations related to workloads.

Let us now consider two flat mating profiles of two conjugate members 1 and 2, that rotate about their axes O_1 and O_2 (Fig. 1.5), with angular velocities ω_1 and ω_2. When one profile pushes against the other, the point of contact P occurs where the two conjugate profiles are tangent to each other, and the forces exchanged at every instant are direct along the common normal n to the two profiles at point P. The line n is the *line of action*, that is the line representing the direction of action of the forces exchanged between the two mating profiles, under the hypothesis that there are no friction losses.

This line of action will intersect the line of centers O_1O_2 at the same point C. The *transmission ratio* $i = \omega_1/\omega_2$ between the two conjugate members is inversely proportional to their radii to the point C. Circles drawn through point C from each center O_1 and O_2 are the *loci* of the instantaneous center of rotation in relative motion between the two conjugate members; they are called *centrodes* or *pitch circles*, which are synonymous. The radius of each pitch circle is the *pitch radius*, while point C is the *instantaneous center of rotation* or *pitch point*. Therefore, the rotational motion of the two conjugate members about their respective axes O_1 and O_2 can be replaced by a motion cinematically equivalent: the rolling motion without sliding of the pitch circle related to the member 2 to the pitch circle related to the member 1. Obviously, during rolling of the pitch circles on each other, at the current point of contact P a sliding also occurs, unless P coincides with C. It should be noted that the centrodes or pitch circles are the transverse sections of the axodes or pitch cylinders previously defined.

In order to have a constant transmission ratio, the pitch point C must always remain the same, that is all the instantaneous lines of action for every instantaneous point of contact must pass through the same point C. Obviously, manufacturing errors and tooth deformations will cause slight deviations in transmission ratio, but the analysis of tooth profiles is based on theoretical curves that meet these requirements. In the case of involute profiles, all points of contact occur on the same straight-line n, that is all straight-lines normal to the profiles at the point of contact coincide with the line n, and thus we have a transmission with uniform rotational motion and constant direction of the forces exchanged.

Since the angular velocities ω_1 and ω_2 can be expressed as the ratio of the tangential velocity at any point of the two conjugate members 1 and 2 and the distance of this point from the centers of rotation O_1 and O_2 of the same members, with the symbol shown in Fig. 1.5 we have:

$$\omega_1 = \frac{v_{P1}}{r_{P1}} = \frac{v_1}{r_1} = \frac{v_{n1}}{r_{b1}} \tag{1.7}$$

$$\omega_2 = \frac{v_{P2}}{r_{P2}} = \frac{v_2}{r_2} = \frac{v_{n2}}{r_{b2}}. \tag{1.8}$$

By the similarity of the triangles CT_1O_1 and CT_2O_2 of Fig. 1.5, we obtain:

$$\cos\alpha = \frac{r_{b1}}{r_1} = \frac{r_{b2}}{r_2} \tag{1.9}$$

and then

$$v_{n1} = v_1 \cos\alpha \tag{1.10}$$

$$v_{n2} = v_2 \cos\alpha. \tag{1.11}$$

Since the mating profiles must be in contact without backlash and no penetration, vectors v_{n1} and v_{n2} must be equal, i.e. they should have the same direction and the same absolute value. It is worth nothing that vectors are indicated in bold letters. From equality $v_{n1} = v_{n2}$, taking into account the Eqs. (1.10) and (1.11), it follows that also the tangential velocities must be the same on the line of centers, i.e. $v_1 = v_2 = v_t$. This is only possible at the operating pitch point C, where the pitch circles touch and are tangent with each other. Then we will have:

$$\frac{\omega_1}{\omega_2} = \left(\frac{v_t}{r_1}\right) / \left(\frac{v_t}{r_2}\right) = \frac{r_2}{r_1}. \tag{1.12}$$

Normally the transmission ratio $i = \omega_1/\omega_2 = n_1/n_2$ is a constant; furthermore, with a fixed center distance, a, also the pitch point C is a fixed point on the line of centers. In the case of variable transmission ratio, also the position of the pitch point C on the line of centers is variable, in accordance with Eq. (1.12). This occurs, for examples, in the case of elliptical gears as well as of spur gears having oscillating centers of rotation.

We have so far considered two any flat mating profiles, i.e. two profiles having constant transverse section whatever longitudinal position be chosen, as occurs in the cylindrical spur gears. In this case, not only the teeth profiles are identical, but also the point of contact of the teeth will be similarly positioned on all transverse sections for any given angular position. But the concepts described above can be generalized to the case of three-dimensional mating profiles, such as those of the teeth of cylindrical helical gears or screw gears.

In this more general case, all transverse sections are similar in so far as tooth profile is concerned (this profile may be identical with that of spur gear), but change in the longitudinal position of the transverse plane of section, producing a corresponding change in the angular position of the individual tooth sections. The position of the point of contact of the teeth on any transverse section will vary from one end of the gears to the other, and the line of contact is no longer parallel to the teeth axes, but inclined. However, there is no longitudinal sliding, and the motion is transmitted uniformly even if the aforementioned law of the mating profiles is satisfied only for one position of the point of contact corresponding to a given transverse section. As shown in Fig. 1.6, the profiles obtained by sectioning the tooth with transverse planes 1,2,3 and 4 come into contact in succession. The point of contact P moves with velocity $v_x = v_t/\text{tg }\beta$ along the face-width, where β is the helix angle (see Chap. 8). Note that the position of the point of contact P in each transverse section remains constant.

Whatever the geometry of the profile (involute profile, cycloidal profile, circular-arc profile, etc.), we have:

- *Transmission ratio, i*: the transmission ratio of a gear pair is the ratio between the angular velocity of the driving wheel and the angular velocity of the driven wheel, i.e.:

$$i = \frac{\omega_1}{\omega_2} = \frac{n_1}{n_2} = \frac{r_2}{r_1} = -\frac{z_2}{z_1}. \tag{1.13}$$

It should be noted that, from here on, unless we say otherwise, we assume that members 1 and 2 of a gear pair are respectively the driving and driven members. For an external gear pair, the directions of rotation of the pinion and wheel are opposite, whereby, z_1 and z_2 being both positive, the transmission ratio is negative; obviously also the quantities ω_2, n_2 and r_2 are to be considered as negative quantities (conventionally, clockwise rotation is regarded as positive, and counterclockwise rotation as negative). Vice versa, in an internal gear pair, the direction of rotation of both gear wheels is the same, for which the transmission ratio is positive, as z_2 is conventionally negative. More generally, in a gear train, the transmission ratio is the ratio between the angular velocity of the first driving wheel and the angular velocity of the last driven wheel. Therefore, the transmission ratio of a gear train is the product of individual transmission ratios of component gears pairs. However, it should be remembered that, especially in cases where misunderstandings are not possible, only the absolute value of the above mentioned quantities are considered, without taking into account their conventional sign.

Speed reducing gear pairs (or speed reducing gear trains) and speed increasing gear pairs (or speed increasing gear trains) are those for which the angular velocity of the second (or last) driven wheel is respectively less than or greater than that of the first driving wheel. With the exception of special cases (for example,

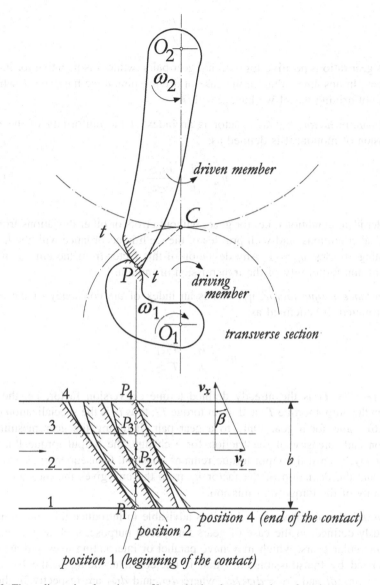

Fig. 1.6 Three-dimensional mating profiles and related transmission of motion

the elliptical gears), the cases of most interest are those for which the transmission ratio is constant.

- *Gear ratio*, u: the gear ratio of a gear pair is the ratio between the number of teeth of the wheel and that of the pinion, i.e.:

$$u = \frac{z_2}{z_1} = \frac{r_2}{r_1}. \tag{1.14}$$

This gear ratio is positive for external gear pairs, while it is negative for internal gear pairs. In absolute value, in the case of driving pinion we have $u = i$, while in the case of driving wheel we have $u = 1/i$.

- *Isogonality factor*, η_ω: this factor is an index of the uniformity of the transmission of motion; it is defined as:

$$\eta_\omega = \frac{i}{u} = \frac{\omega_1 z_1}{\omega_2 z_2}. \tag{1.15}$$

Under ideal conditions, i.e. for gears without errors or other deviations from the theoretical conditions, and with profiles of the teeth in accordance with the laws of the mating profiles, $\eta_\omega = 1$. Any deviation of this factor from the unit is an indication of non-uniformity of the transmission of motion.

- *Constance torque factor*, η_T: this is an index of the constancy of the torque transmitted; it is defined as:

$$\eta_T = \frac{i_T}{u} = \frac{T_2 z_1}{T_1 z_2}, \tag{1.16}$$

where $i_T = T_2/T_1$ is the already defined torque conversion factor, i.e. the ratio between the output torque T_2 and input torque T_1; thus i_T is the specialization of the Eq. (1.6), valid for a gear train, to a gear pair. If during the gear meshing the direction and intensity of the friction force change, the output torque fluctuates accordingly, even if the input torque remains constant. Consequently also i_T fluctuates, and the deviation of the factor η_T from the unit gives the measure of the irregularity of the torque transmission.

- *Transmission function*, $i_n = 1/i$: it is advisable to introduce this function, previously defined, in the case of gears for special purpose, such as for example non-circular gears, which may have parallel or intersecting axes and are characterized by transmission ratio that is not constant. Being in the Eq. (1.13) $\omega_1 = d\varphi_1/dt$ and $\omega_2 = d\varphi_2/dt$, where $d\varphi_1$ and $d\varphi_2$ are respectively the differential angular displacements of the driving and driven members about their axes, this transmission function is given by the following relationship:

$$i_n = \frac{1}{i} = \frac{d\varphi_2}{d\varphi_1}. \tag{1.17}$$

It therefore defines the relationship that correlates the angular position of the driving member with the corresponding angular position of the drive member.

1.5   Tooth Parts and Some Quantities of the Toothing

Figure 1.7 shows the main parts of a tooth having any profile, and some quantities characterizing a cylindrical spur gear. We define axial and transverse planes respectively planes containing the axis and perpendicular to the axis of the gear; instead we define normal planes the planes perpendicular to the axis of the tooth. Obviously transverse and normal planes coincide only in the case of straight teeth, while they are different in the case of curved teeth. We call *blank* the member of gear pair (pinion or wheel) before the cutting of the teeth. *Top land* or *crest* of the tooth is the portion of *tip surface* (also called *addendum surface* or *outside surface*) between the opposite tooth flank surfaces of the same tooth, while the *bottom land* is the portion of *root surface* or *dedendum surface* between the roots of two adjacent teeth. *Rim* is the portion of a gear wheel that provides full support for the tooth roots. *Tooth flanks* are the portions of the surfaces of a tooth between the addendum surface and the dedendum surface.

The *reference surfaces* are imaginary conventional surfaces with respect to which the dimensions of the teeth are defined. Instead, as we have seen in the previous section, the *pitch surfaces* of a given gear pair are the geometrical surfaces described by the instantaneous axis of rotation of the relative motion of the mating gear member, in relation to the gear member under consideration. We also said that pitch surfaces and axodes are synonymous.

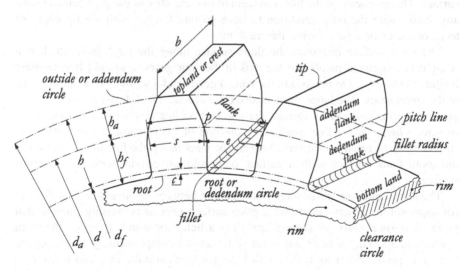

Fig. 1.7 Main parts of a tooth, and some quantities of a spur gear

The *pitch surfaces* of parallel or intersecting gears are those ideal surfaces of revolution, having the same axes of the two gears, which touch each other along a common line of contact and roll on each other without sliding when the two wheels to which they are rigidly connected rotate with angular velocities inversely proportion to their pitch diameters. For parallel cylindrical gears (spur or helical gears) and intersecting gears (straight or curved-teeth bevel gears) these pitch surfaces are respectively cylinders and cones, and are therefore termed *pitch cylinders* and *pitch cones*. Instead, crossed gears have no pitch surfaces understood with the above defined meaning, i.e. ideal surfaces which rotate with the gears and roll together without slide; for each of these crossed gears the term pitch surface has a different meaning, which will be explained case by case in the next chapters.

The pitch surface divides the tooth flank (left flank and right flank) in two portions: the *addendum flank* and *dedendum flank* respectively between the tip surface (or addendum surface) and pitch surface, and between the pitch surface and root surface (or dedendum surface). The *tip* is the edge where the flank meets theoretically the crest; in practice, however, the flank and crest surface are connected by a connecting surface, for which we still use the term *tip*, but with a larger meaning, which indicates the portion of the tooth surface adjacent to that edge. In this respect, often the *tip edges* are rounded or chamfered, in order to facilitate the engagement of the mating gear members. These two types of modification of the tip edge, the first characterized by a given radius of curvature, and the second by a given depth, are called respectively *rounding* and *chamfering*. Instead the *fillet* is the portion of the flank surface that connect the usable flank surface to the root surface. The geometry of the fillets of teeth of one member of the gear pair can have any shape, with the only condition to have no interference with the tip edges of teeth of the mating gear during the meshing.

The pitch surface intersects the flank surface along the *pitch line*, which is a straight line segment parallel to the axis in the spur gears, a straight line segment inclined with respect to the axis in the helical gears, and a segment of a planar curve in the spiral gears. The latter curve is also called *tooth spiral*. Instead, in the case of gear having helical teeth or threads the intersection line above is called *tooth helix* or *thread helix*. We define *tooth trace* the line of intersection of the tooth flank surface with the reference surface. Instead, the *flank line* is the line of intersection of the tooth flank surface with a surface of revolution coaxial with the reference surface.

The helix angle (or *spiral angle)* is the angle between a tangent to a tooth helix (or tooth spiral) at any point and a pitch-surface generatrix passing thought that point. A *normal helix* (or *normal spiral*) is a helix (or spiral) lying on the pitch cylinder and having a helix angle (or spiral angle) complementary to that of the helix (or spiral) which form the tooth helix (or tooth spiral). The *lead angle of a helix* (or *spiral*) is the angle between a tangent to the helix (or spiral) and a transverse plane. The *lead of a helix* (or *spiral)* is the *axial advance* per revolution, i.e. the axial distance between similar points in successive convolution.

The intersections of the pitch surface, tip surface and root surface with a transverse plane, i.e. a plane perpendicular to the gear wheel axis, are respectively the *pitch circle, addendum circle* and *dedendum circle*. *Tooth profiles* are the intersections of the tooth flank surface with any defined surface which also cuts the reference surface, that is the conventional surface with respect to which the dimensions of the teeth are defined (for a cylindrical gear it is the pitch surface of the gear wheel meshing with the basic rack). The intersections of the tooth flank surface with a transverse plane, normal plane (plane perpendicular to the tooth axis) and axial plane (plane containing the gear wheel axis), are respectively the *transverse profile, normal profile* and *axial profile*.

It is also necessary to keep in mind other possible intentional modifications of the tooth shape, among which those described hereunder deserve specific mention. The *tip relief* or *tip-easing* and *root relief* or *root-easing* are intentional modifications of the tooth profile at the expense of the material thickness (a small amount of material is removed from the flank towards the crest or the root), whose purpose is to soften the beginning of the contact between the mating flanks, during the meshing.

The *undercut* is another intentional modification: it concerns the fillet and results in a removal of material (this can be obtained for example with a cutting tool fitted with protrusion), aimed to facilitate any subsequent operation to the cut. It is also important to remember the *barrelling* or *crowning*, that is the modification by which the teeth are slightly and progressively thinned from the centerline towards their ends, in order to avoid concentration of loading at the ends. The *end relief* serves the same purpose: it consist of a progressive reduction of the teeth thickness at their ends, on a narrow portion of the face width.

The main quantities which characterize a toothing and a gear are the following (other quantities will be introduced at the appropriate time):

- *Circular pitch* or *transverse circular pitch, p:* it is the distance between corresponding profiles (both right profiles or both left profiles) of two successive teeth measured along the pitch circle. Indicating with d, z and m, respectively, the *reference diameter* (i.e., the diameter of the *reference circle*), *number of teeth* and *module*, we will have (p and m are given in mm):

$$p = \frac{\pi d}{z} = \pi m. \tag{1.18}$$

This module is the *transverse module*. It is a standardized quantity, since it is impractical to calculate the circular pitch with irrational numbers; it is instead much easier to use a scaling factor that replaces the circular pitch with a regular value represented by the module. The Eq. (1.18) shows that the transverse module is defined as the ratio between the circular pitch and number π or as the ratio between the reference diameter and number of teeth. It represents a very important quantity in the sizing of the toothing and gear; indeed, the transverse module is the magnitude to which are referred other significant quantities, such as the addendum, dedendum, thickness and so on. It should be remembered that the term *module*, without qualifying adjective, indicates the transverse module.

It is noteworthy that, in the English-speaking word, the reference quantity is not the transverse module, but the *transverse diametral pitch, P,* which is defined as the ratio between the number π and transverse pitch, expressed in inches, or as the ratio between the number of teeth and the reference diameter, expressed in inches. Therefore, m, p and P are related by the relationship: $pP = \pi$ or $m = 25.4/P$. It also to be noted that, for the involute toothing, the pitch circle and reference circle may be different. Finally, it is to keep in mind that sometimes we use the *angular pitch* τ (in rad), which is the angle subtended by the circular pitch; thus it is defined as the ratio between the round angle and the number of teeth, i.e. $\tau = 2\pi/z$.

- *Reference diameters*, d_1 and d_2, and *center distance, a*: of course, to have a theoretically correct engagement, the pitches of the pinion and wheel must coincide. Therefore, from Eq. (1.18) we obtain the following relationship.

$$d_1 = 2r_1 = z_1 m = z_1 p/\pi \qquad (1.19)$$

$$d_2 = 2r_2 = z_2 m = z_2 p/\pi \qquad (1.20)$$

$$a = r_1 + r_2 = (d_1 + d_2)/2 = m(z_1 + z_2)/2. \qquad (1.21)$$

Addendum, h_a, dedendum, h_f, and *tooth depth, h*: addendum, dedendum and tooth depth are respectively the radial distance between: the addendum circle and pitch circle; the pitch circle and dedendum circle; the addendum circle and dedendum circle. Usually $h_a \cong m$ and $h_f \cong (1.1 \div 1.3)m$. Of course, $h = h_a + h_f$; note that the notation h without subscript indicates the tooth depth from root circle to tip circle. However, it is to keep in mind that the tooth profile is not all usable for the contact. In this regard we define as *active profile* (it correspond to the *active flank*) the portion of the profile along which the contact is carried out with the profile of the mating gear tooth, while we define as *usable profile* (it corresponds to the *usable flank*) the maximum portion of the tooth profile of a gear, considered individually, which may be used as active profile. In this framework, it is evident that the *operating depth* or *working depth* h_w of the tooth is equal, at most, to the sum of the addenda of the teeth of pinion and wheel; hence, it is given by:

$$h_w = h_{a1} + h_{a2} = [(d_{a1} + d_{a2})/2] - a, \qquad (1.22)$$

where d_{a1} and d_{a2} are the *addendum* (or *outside* or *tip*) *diameters* of the pinion and wheel. For the involute toothing this diameters are given by (Fig. 1.8):

$$d_{a1} = d_1 + 2h_{a1} = 2a - d_{f2} - 2c_1 \qquad (1.23)$$

$$d_{a2} = d_2 + 2h_{a2} = 2a - d_{f1} - 2c_2, \qquad (1.24)$$

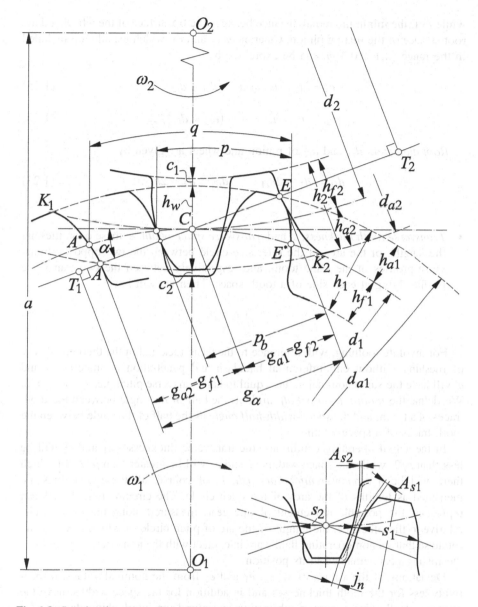

Fig. 1.8 Spur involute gear pair: main geometric quantities

where c_1 and c_2 are the *bottom clearances* or simply *clearances* between pinion and wheel, and between wheel and pinion: in other words, c_1 is the minimum radial distance between the tip surface of the pinion and root surface of the mating wheel, while c_2 is the minimum radial distance between the tip surface of the wheel and the root surface of the mating pinion. Clearances c_1 and c_2, which usually are included in the range $(0.1 \div 0.3)m$, can be expressed by:

$$c_1 = h_1 - h_w = a - (d_{a1} + d_{f2})/2 \tag{1.25}$$

$$c_2 = h_2 - h_w = a - (d_{a2} + d_{f1})/2. \tag{1.26}$$

Root diameters d_{f1} and d_{f2} of pinion and wheel are given by:

$$d_{f1} = d_1 - 2h_{f1} \qquad d_{f2} = d_2 - 2h_{f2}. \tag{1.27}$$

- *Transverse tooth thickness, s,* and *transverse space-width, e*: these quantities are the lengths of the arcs of the reference circle between the two opposite transverse profiles of the same tooth, and respectively between the two transverse profiles lying at each side of a tooth space. Then we will have:

$$s + e = p. \tag{1.28}$$

For involute toothing, with reference to the basic rack and in the theoretical case of meshing without circumferential backlash and penetration of material, s and e will have the same nominal value, equal to half transverse pitch, i.e. $s = e = p/2$. We define the *tooth thickness half angle* as the half of the angle between the tooth traces of a tooth, and the *space-width-half angle* as the half of the angle between tooth traces of a space-width.

In the actual operating conditions, the transverse thicknesses s_1 and s_2 will be less than $p/2$, while the space-widths e_1 and e_2 will be greater than $p/2$, for which there will be a *circumferential backlash*, j_t; of course, all these quantities are measured as lengths of the arcs of the pitch circle. The circumferential backlash represents the possible movement of one gear, measured along the pitch circle, relative to the other, i.e. the length of the arc of pitch circle of which a wheel can rotate so that its non-operating flanks are in contact with the non-operating flanks of the mating gear remained in its position.

Deviations of the values of s_1, s_2, e_1 and e_2 from the nominal values, respectively less for the tooth thicknesses and in addition for the space-widths, have the purpose of allowing a correct lubrication and avoid the interlocking danger of a tooth into the space-width of the other, due to effects of thermal expansions and assembling inaccuracies. This circumferential backlash is given by:

$$j_t = p - (s_1 + s_2). \tag{1.29}$$

Instead, the *normal backlash*, $j_n = j_t \cos \alpha$ (α is the *pressure angle*, as defined below) is the minimum distance between the non-operating flanks of two wheels when the operating flanks are in contact. For helical or spiral gears, j_n is the possible movement of one gear measured in the common pitch plane normal to the tooth helix or spiral, relative to the other.

To avoid the danger of scuffing, which would cause irreparable damage to the teeth flanks, it is necessary that such circumferential or normal backlashes have adequate values. In design, they must be calculate taking into account the tolerances of the machining operation (tooth thicknesses, space-widths, center distance, etc.), mounting tolerances, swelling tendency of some materials (e.g., polymers and plastics), temperature differences between the gear wheels and the box, especially during start up, etc. It is also necessary to take into account the fact that, due to the actual workloads, the space-widths undergo changes caused by the bending and shear deformations and local Hertzian deformations. Therefore, the backlashes that we have under real working conditions are a complex function of the load transmitted.

- *Length of path of contact*, g_α, and *pressure angle*, α: both these quantities are very important. First of all we consider the *line of action*, that is the common normal to two transverse tooth profiles at their point of contact, along which the contact may take place. In the case of involute gears (this is the case that most interests us), the line of action is one of the two common tangents to the base circles, and precisely the one compatible with the directions of rotation of the mating gears. The *plane of action* is the plane containing the lines of action of a parallel cylindrical involute gear pair.

We define *path of contact* that portion of the line of action along which contact actually take place, and *length of path of contact* g_α, its length. Therefore the path of contact is the *locus* of successive points of contact between the mating tooth profiles in a transverse plane. Path of contact is composed of the sum of *approach contact* and *recess contact*: the approach contact refers to contact anywhere along the path of contact between the addendum circle of the driven gear and the pitch point; instead, the recess contact refers to contact between the pitch point and the addendum circle of the driving gear. In the case of unmodified involute profile free from tip-easing, the length of path of contact is that portion of the line of action between the point on it intercepted by addendum circles of the two mating gears. It is noteworthy that, using cutting tool with modified profile, the path of contact will no longer be a straight line. This occurs, for examples, when a symmetrically modified basic rack is used; if the profile of basic rack is unsymmetrically modified to provide for tip-easing, from the theoretical point of view the path of contact ceases at the point at which tip-easing begins.

In this regard, it would be convenient to introduce, in place of the addendum circles, the circles passing through the points where the involute profile ends and intentional modification of the same profile near the tooth tip begins. We could call these circles as *tip relief circles*, with the understanding that tip relief here has a more general meaning, which includes not only the true tip relief, but also any other intentional modification in the gear tooth profile close to its tip. These tip relief circles are the circles that determine, with their intersections with the line of action, the extreme points of the useful portion of this line, and hence the effective path of contact and its length.

Beyond these circles, the teeth extend up to the addendum circles, but with profiles (whether rounding or chamfering or tip relief) which theoretically do not take part to the contact and have the purpose of facilitating the engagement and avoid scratches and scuffing. The portions of the profiles between the tip relief circles and addendum circles can have any geometry, as long as this is compatible with the need to avoid the interference, only if the cut of the teeth is performed by means of milling cutter designed in order to produce the desired geometry of the profiles near to the tips. If instead the cutting of the teeth is performed by envelope cutting (generation by rack-type cutters, pinion-type cutters or shapers and hobs), the geometric shapes of the tip profiles is automatically defined by the shape of fillet curves of the cutting tools, i.e. rack-type cutters, shapers and hobs.

Figure 1.8 refers to the theoretical case in which the involute profile is extended to the addendum circle. It shows that $g_\alpha = g_{f1} + g_{a1} = g_{a2} + g_{f2}$, where $g_{a2} = g_{f1}$ and $g_{a1} = g_{f2}$ are respectively the *length of recess path* and *length of approach path*. The lengths of approach path and recess path are the lengths of those parts of the path of contact along which approach contact and respectively recess contact occur. With driving pinion, in the approach path of contact, a point of the dedendum flank of pinion is in contact with a corresponding point of the addendum flank of the driven wheel, while during the recess path of contact, a point of the addendum flank of pinion is in contact with a corresponding point of the dedendum flank of the wheel. This fact is highlighted by a double subscript, the first of which (a or f) refers to the addendum or dedendum, while the second (1 or 2) refers to driving pinion or driven wheel.

The path of contact starts at the point A, where the dedendum flank of one tooth of the driving pinion meets for the first time the tip of the driven wheel, and ends at the point E, where the tip of the same tooth of the pinion leaves the contact with the dedendum flank of the wheel. Points A and E on the line of action match on the pitch circle of the pinion points A^* and E^*. The arc $\overgroup{A^*E^*}$ is called *transverse arc of action* or simply *arc of action*, and its length q is called *length of the transverse arc of action* or *length of path of rotation*, while the corresponding angle is called *transverse angle of action* and its length is called *length of the transverse angle of action*. It should be noted that the arc of action, q, is the arc measured on the pitch circle, during which two conjugated profiles remain in contact. Therefore it is the arc of pitch circle through which a tooth profile moves from the beginning to the end of contact with the mating profile.

We can write $q = (q_a + q_r)$, i.e. the arc of action is the sum of the *arc of approach*, q_a, and the *arc of recess*, q_r. The arc of approach, q_a, and arc of recess, q_r, are therefore the arcs of the pitch circle through which a tooth profile moves from the beginning of contact to the pitch point and, respectively, from the pitch point until contact ends. The generally used convention is to measure the arc of approach on the driving wheel pitch circle (usually, the pinion), and the arc of recess on the driven wheel pitch circle. However, it is obviously equivalent to measuring these arcs of approach and recess on one or the other of the two pitch circles, as these roll one on the other, without sliding. In order to ensure continuity of transmission of motion, the following condition must be met: $q = (q_a + q_r) \geq p$. In this way a new pair of mating teeth will be in engagement before the previous one leaves the contact.

Figure 1.8 also highlights the active portions of the two profiles, i.e. the active profiles $\widehat{AK_1}$ and $\widehat{EK_2}$. It is noteworthy that the profiles of the dedendum flanks must be shaped so that the tips of the two gear members are in no way therein to collide; in other words, the trajectory of the crest of the mating gear must develop at the outside of the profile of the dedendum flanks. Finally is to be observed that points T_1 and T_2 intercepted on the line of action by the straight lines perpendicular to it conducted for the centers O_1 and O_2 are the points of interference, which cannot be exceeded by the addendum circles of the two mating gears (note that T_1, A, C, E, T_2 are points on line of action; they derive only from geometrical consideration and are related to the meshing evolution from pinion root to pinion tip, regardless of whether pinion or wheel drives).

Pressure angle, α, is the acute angle, measured in a chosen plane, between the common pitch plane and the common normal to the profiles of the two gear teeth at a point of contact. In transverse, normal and axial planes, we have respectively the *transverse pressure angle*, α_t, the *normal pressure angle*, α_n, and the *axial pressure angle*, α_x. Without subscript and when there is no possibility of misunderstanding, α indicates the transverse pressure angle. The value of the pressure angle influences the shape of the tooth profiles as well as the components of the load exchanged between the teeth. When α increases, the tooth profiles tend to flatten out, up to incurring the risk of the pointed tooth, but the tooth thickness at the bottom increases. On the contrary, when α decreases, the tooth profile tends to curve, the thickness in addendum grows and that in dedendum decreases. Moreover, when α increases, the radial component of the force exchanged between the teeth increases well; this component does not participate in the transmission of power, but over-loads bearings.

- *Transverse contact ratio*, ε_α: this is the ratio between the length of path of rotation q and circular pitch p or, what is the same, the ratio between the length of path of contact g_α and base pitch p_b (see the following chapter). If $q > p$ or $g_\alpha > p_b$, before a pair of meshing teeth comes out from the contact, the next pair of teeth is already engaged in the contact. Just so the rolling motion of the pitch circles is a uniform motion. Then the following inequality must be necessarily verified, otherwise irregularities of motion occur:

$$\varepsilon_\alpha = \frac{q}{p} = \frac{g_\alpha}{p_b} > 1. \tag{1.30}$$

It is however necessary to consider that, with $\varepsilon_\alpha > 1$, when the tip of the tooth flank of the driving pinion comes out of contact at the point E, it moves towards the crest of the tooth flank of mating wheel. This actual behavior under load is due to the elastic springback which occurs when the load instantly vanishes. If the teeth were not loaded, the contact at the point A between the tip of the wheel tooth and dedendum flank of the pinion would be a regular contact, i.e. without any impact, provided that the teeth do not show errors. However, since the teeth are loaded, and thus are subjected the bending, we have an inlet impact at the point A, because the edge tip of the tooth flank of the wheel tends to penetrate into the dedendum flank of the pinion, shaving the latter towards its root. Then a relative sliding is added to impact: consequently we have irregular motion, heavy wear of the addendum flank of the wheel and dedendum flank of the pinion, and considerable noise. At least within certain limits, we can remedy this behavior with an adequate tip relief of the teeth of the driven wheel or with a suitable root relief of the teeth of the driving member.

- *Root fillet radius*, ρ_F: it is the radius of the root fillet, i.e. the radius of the portion of the clearance curve joining the flank profile to the dedendum circle. As the profile portion between the tip relief circle and addendum circle, also the root fillet (for involute profiles, the fillet develops from the *fillet circle* until the dedendum circle) can have any geometry, as long as this is compatible with the need to avoid the interference, only if the cut of toothing is performed by means of a milling cutter designed in order to produce the desired geometry of the profile adjacent to the root. When the teeth cutting is instead performed by envelope cutting, the geometric shape of the root fillet is automatically defined by the shape of the tip curve of the cutting tool (for example, the basic rack profiles which can be with and without undercut). In the design of the gears, the study of the root fillet geometry has great importance. Indeed a root fillet well designed strengthens the tooth at its root and decreases considerably the *notch effect* at the interface tooth/rim.

1.6 Precision and Accuracy Grade of the Gears

The load capacity and regular operating conditions of the gears considerably depend on the accuracy of the production process by which the two members of a gear pair are obtained. As for any mechanical member or engineering component, it is practically impossible to produce a gear wheel without manufacture errors, which represent all the deviations of dimensions and geometry of the real gear wheel with respect to those theoretically perfect of the same gear wheel. To ensure good operating conditions for the gears, these deviations must be contained within appropriate limits, which are known as *tolerances* or *allowable errors*.

For economic reasons, these permissible errors must have values as high as possible (i.e., tolerances must be as wide as possible), as the selection of too narrow tolerances in cases where precision is not necessary would make the product too expensive or even unsaleable. On the other end, it is risky to choose too large tolerances, which could compromise the same functionality of the gear to be designed. Therefore, to make a reasoned choice of allowable errors, the gear designer must make an in-depth analysis, taking into account the related consequences to every design choice made by him.

In this regard, the factors to be considered are numerous and differentiated, and often each of them plays a contrasting role than the other. Among this factors, some deserve special attention, such as: the type of gear unit to be designed, and its working conditions; the expected lifetime in hours of work; the maintenance conditions inclusive of the possibility of interchangeability of gear wheels; the available manufacturing process, inclusive of cutting machines and shaving or grinding machines; the control and testing procedures in relation to the available measuring instruments; and above all the techno-economic aspects regarding the economic feasibility and viability of the gear design. It would, therefore, most appropriate that the gear designer made use and treasure of previous knowledge and experiences, when these were available.

The unavoidable manufacturing errors affecting both the body of gear wheels and the teeth are due mainly to: deviation in pitch, tooth profile, and flank line; center distance deviations; deviations from parallelism or misalignment of the axes of the two members of a gear pair with respect to their direction or position theoretically correct; radial and axial *runouts* of blank or body of a gear wheel, which represent the rotation accuracy in the radial and respectively axial directions, and greatly influence the precision attainable by means of the technological process by which the toothing is obtained; etc. This errors may result in: increased operating noise, torsional vibration, and loss of rotational accuracy of the driven gear wheel with respect to the driving gear wheel; uneven distribution of the pressure of contact on the active flanks, with consequent non-uniform wear of the same flank; irregular operating conditions; uneven distribution of the load along the tooth flank line; increased dynamic loads; alterations in the load distribution in the single and double contact areas of path of contact in comparison with the theoretical distribution, and other detrimental effects.

However, it is to be noted that not all the deviations and errors that may take place are equally important to ensure the correct operating conditions of a gear unit. Only the dimensional sizes of a gear wheel which are essential for the aforementioned correct operation of the gear transmission system must be subject to tolerance, since tolerance have a considerable cost, both because it is necessary to provide an adequate process of machining for they can be obtained, and because all the dimensional sizes subject to tolerance must be checked and tested by means of suitable measuring instruments. To give a rough idea of the cost of tolerance, just remember that, for accuracy grades between 5 and 8, the production cost of a gear wheel increases by about $(60 \div 80)\%$ from an accuracy grade to that immediately finer [40].

It is also be noted that deviations of the single dimensional sizes are all important to ensure the proper operating conditions of the gear unit, but their importance may be higher for some quantities, and less for the others. In this regard, for a reasoned judgment on the members of a gear transmission system, it is necessary to check how individual deviations affect together the regularity of motion and load transmission. To this end, appropriate quantities that allow to evaluate this global effect are introduced. The determination of the tolerances to be assigned to the individual dimensional sizes of gears is much more complex compared to what happens for fits or other mechanical members coupled together, as kinematic considerations come into play in the case of the gears, which greatly complicate the problem. The introduction of the comprehensive quantities above seeks to facilitate, at least a little, the solution of this problem. In the absence of specific experience gained in the design of gear units similar to the one of interest, standards codified in this regard may be a valid and reliable guideline for the gear designer.

The ISO Standard provide a system of accuracy for cylindrical (ISO 1328-1: 2013 [20], and ISO 1328-2: 1997 [21]) and bevel gears (ISO 17485: 2006 [22]), and allow to determine the tolerances to be assigned to the dimensional sizes and comprehensive quantities of these gears. We refer the reader directly to these ISO standards that, for obvious reasons of brevity, are summarized below only on their essential directives. Only for particular aspects not expressly covered in the ISO Standards we will briefly refer to other national standards.

These ISO Standards provide 12 *accuracy grades* (or *quality grades*) which, in decreasing order of accuracy, are designed by the numbers from 1 to 12. The accuracy grade 1 and 2 refer to *master gear wheels* (or *reference test gear wheels*), which are produced by means of highly sophisticated machines. The accuracy grades 3 and 4 refer respectively to high-precision gears and medium-high precision gears, used for measuring instruments. The accuracy grade 5 is mainly used in fine mechanics. The accuracy grades from 6 to 12 are those which are interest in practical applications of the mechanical industry. However, it is to be born in mind that, for the production of usual gears, the accuracy grades 6, 7 and 8 are those normally used; they correspond respectively to: high-accuracy gears that work at high speed and high load; normal-accuracy gears that operate at high speed and moderate load or at moderate speed and high load; low-accuracy gears that operate at low speed and low load. The accuracy grades from 9 to 12 relate to gears of progressively decreasing qualitative characteristics, so that the accuracy grade 12 corresponds to coarse-quality and low-speed gears, namely to the poorest quality gears in the considered scale.

Usually, for a pair of mating gears, both members are manufactured with the same accuracy grade. This custom, however, is not a rule to be respected strictly; it may be waived by agreement between the manufacturer and user.

To complete the subject, we consider appropriate to briefly describe below the main individual errors and cumulative errors. We recall once again that the accuracy grade of a gear depends on the tolerance limits for the circular pitch, tooth profile and flank line (this last gives a measure of the tooth alignment). We recall also that the deviations of parameters of interest from their theoretical values can be accurately measured, by means of suitable measuring instruments, but they do not allow to evaluate properly and exhaustively what are their effects on the regularity of the motion of running gears, because the running quality depends mainly on the sum of these errors as well as on the radial and axial runouts.

The main types of individual errors are as follows (here we give symbol and unit of each of them, while in regard to the choice of their values we refer the reader to the aforementioned standards):

- *single pitch deviation* or *pitch error*, f_p(μm): it is the maximum absolute value of all the *individual single pitch deviation*, f_{pi}, which are observed; f_{pi} is the algebraic difference between the actual pitch and the corresponding theoretical pitch in the transverse plane, detected on the measurement circle of the gear wheel. Then $f_p = \max|f_{pi}|$. We consider also the *total cumulative pitch deviation* or *total index deviation*, f_p(μm), defined as the largest algebraic difference between the *individual cumulative deviation* values for a specified flank obtained for all the teeth of a gear, F_{pi}; this last is also called *individual index deviation*, and is defined as the algebraic difference, over a sector of n adjacent pitches, between the actual length and the theoretical length of the relevant arc. Then $F_p = (\max F_{pi} - \min F_{pi})$. The *transverse base pitch deviation* or simply *base pitch deviation*, f_{pb}(μm), is the difference between the actual and the ideal base pitch; it is related to the single pitch deviation by means of relationship $f_{pb} = f_p \cos \alpha$. The base pitch deviation is essential for uniformity of transmission of motion and distribution of load over all tooth pairs in meshing. It is crucial that the base pitches of pinion and mating wheel coincide, while equal base pitch deviations of pinion and mating wheel cancel one another during the engagement.

- *Total profile deviation*, f_α(μm): it is defined as the distance between two facsimiles of the *design profile*, which enclose the *measured profile* over the *profile evaluation range*. The total profile deviation is the result of the *profile form deviation*, $f_{f\alpha}$(μm), and the *profile slope deviation*, $f_{H\alpha}$(μm). Profile form deviation is defined as the distance between two facsimiles of the *mean profile line*, which enclose the measured profile over the profile evaluation range, while the profile slope deviation is defined as the distance between two facsimiles of the design profile, which intersect the extrapolated *mean profile line* at the *profile control diameter*, d_{cf}(mm), and the *tip diameter*, d_a(mm). For further detail on this subject, we refer the reader directly to ISO 1328-1: 2013, in which five interesting examples of profile deviations are shown, namely: profile deviations with unmodified involute; profile deviations with pressure angle modified; profile deviations with profile crowning modification; profile deviations with profile modified with tip relief; and profile deviations with profile modified with tip and root relief. All five examples show the influences of the profile form and profile slope deviations on the total profile deviation.

- *Total helix deviation* or *total flank line deviation*, $F_\beta(\mu m)$: it is defined as the distance between two facsimiles of the *design helix* which enclose the *measured helix* over the *helix evaluation range*. The total helix deviation is the result of the *helix form deviation*, $f_{f\beta}(\mu m)$, and the *helix slope deviation*, $f_{H\beta}(\mu m)$. Helix form deviation is defined as the distance between two facsimiles of the *mean helix line*, which enclose the measured helix over the helix evaluation range, while the helix slope deviation is defined as the distance between two facsimiles of the design helix, which intersect the extrapolated *mean helix line* at the end points of the face-width, $b(mm)$. Again, for the further details on this subject, we refer the reader directly to ISO 1328-1: 2013, in which five other interesting examples of helix deviations are represented, namely: helix deviations with unmodified helix; helix deviations with helix angle modification; helix deviations with helix crowning modification; helix deviations with helix end relief; helix deviations with modified helix angle with end relief. All five examples show the influences of helix form and helix slope deviations on the total helix deviation.
- *Runout*, $F_r(\mu m)$: it should first be noted that the term *runout*, without any specification, indicates the *radial runout*. The runout (or radial runout) is the difference between the maximum and the minimum individual radial measurement, $r_i(\mu m)$; this is the radial distance from the gear wheel axis to the center or other defined location of a probe (cylinder, ball or anvil), which is placed successively in each tooth space. During each measurement, the probe contacts both the right and left flanks at approximately mid-tooth depth. It can also be calculated from pitch measurements. The runout is a measure of the eccentricity of the gear wheel teeth, i.e. the measure of the off-center error due to the fact that the gear wheel is not exactly round.
- *Axial runout*: this deviation is usually measured in the testing stage of a gear wheel, and gives a measure of the *wobble* of the same gear wheel. The measurement is performed by placing a dial gauge, the axis of which is held at a specific distance and parallel to the axis of rotation of the gear wheel under consideration.

For a judgment on the operation of the toothed members of a gear unit it is necessary to check how the various individual deviations jointly affect the regularity of the motion and load transmissions. The overall effect due to two or more individual errors, acting simultaneously, can be verified by means of suitable *composite deviation texts*. These type of tests approximate the operation of the gear examined in conditions of service. The composite deviation tests are of two types: *single flank composite testing* and *double flank composite testing*.

In both the two types of tests, the gear wheel to be tested is rotated through at least one full revolution in close contact with a *master gear wheel* (or *reference gear wheel*) of know accuracy grade. However, with the single flank composite deviation test, the gear wheel pair simulates the actual meshing conditions, including the *backlash*, but with a reduced load; on the contrary, with the double flank composite deviation test, the two gear wheels are made to rotate without backlash, radially pressing them one against the other. It is to be noted that the first type of test, which simulates the actual meshing conditions, allows to evaluate the errors in angular transmission, whereas the second type of test allows to evaluate the deviations in center distance.

With the single flank composite deviation testing, the master gear wheel is rotated at an angular velocity strictly kept constant, and the center distance and alignment are adjusted so as to ensure the design backlash; only the rolling on any one operating flank of the gear wheel to be tested is examined, and any deviation in the angular motion is measured. The conclusions on the operating behavior of a gear wheel obtained with this testing method are more realistic and unique compared to those allowed by the double flank composite deviation testing, because with this second testing method also the deviations of the non-operating flank come into play. However, the equipment and measuring instruments to carry out the checks with the first testing method (i.e. the single flank composite testing) are more complex, sensitive and expensive compared to those of the second testing method.

The single flank composite deviation testing is that used for the evaluation of the *transmission error* of a gear. Transmission error is defined as the deviation of the angular position of the driven gear wheel, for a given angular position of the driving gear wheel, from the position that the driven gear wheel should theoretically have if the gear was geometrically perfect. In even more general terms, the transmission error can be defined as the difference between the actual position of the output gear wheel and the position it would occupy if the gear drive was perfectly conjugated. For gears with theoretically perfect involutes and an infinite stiffness, the rotation of the output gear wheel would be a function of the input rotation and the transmission ratio. Therefore, in these ideal conditions, a constant rotation of the output shaft would also correspond to a constant rotation of the input shaft. Instead, in the actual operating conditions, a motion error of the output gear wheel with respect to the input gear wheel will occur, due to both intentional shape modifications and unintentional modifications, as well as assembly errors and defects in stiffness [17, 39, 33].

The transmission error can be measured statically or dynamically, under load or without load. For a gear pair, this error is measured with an apparatus that analyzes the two component gear wheels to be tested, in their mutual engagement, but the same apparatus is used, as mentioned above, to test the individual gear wheels of the gear pair, each of which run against a master gear wheel, to measure the individual contributions to the transmission error.

Test for evaluation of the transmission error are normally carried out at very light torques, so as to eliminate or minimize the contribution of deflections of the same test apparatus, which may affect the measurement results. Tests of the same type, but under heavy loads, such as those in actual practical applications, are normally carry out in the actual gearbox or in a very rigid special text box, using other types of equipment.

All deviations of the gear pair detected with the above-mentioned testing method are useful to control gear functional characteristics; also nicks and burrs can be detected. Given the above-mentioned test procedures, the results obtained allow to calculate the *total no-load transmission error*, in terms of *total single flank composite deviation*, $F_{is}(\mu m)$, and the *tooth-to-tooth error*, in terms of *tooth-to-tooth single flank composite deviation*, $f_{is}(\mu m)$. In order to have correct operating conditions for the gear in question, these two errors must fall within specific limits, which are respectively defined by the *total single flank composite tolerance*, $F_{isT}(\mu m)$, and the *tooth-to-tooth single flank composite tolerance*, $f_{isT}(\mu m)$. About the calculation method and the values of these two tolerances, to be taken in the design stage, we refer the reader to the afore-mentioned ISO Standard.

It should be remembered that the transmission error, together with the mesh stiffness variation, are considered to be the primary source of excitation of vibrations and noise, which are radiated outwards through the gearbox. The characteristics of low or high vibration and noise level of a gearbox depend on the instantaneous meshing conditions that arise between the various meshing tooth pairs. Under load, but at very low speeds, we have the *static transmission error*, which mainly depends on tooth deflections and manufacturing and assembly errors. Under operating conditions where both loads and speeds are high, the mesh stiffness variation (it is due to changes in the length of contact line and tooth deflections) and the excitation, which is localized at the instantaneous point of contact, generate dynamic mesh forces, which are transmitted to the housing through shafts and bearings; in this case, we are talking about *dynamic transmission error*. Noise radiated by the gearbox is closely related to the vibratory level of the housing.

It is to be noted that the tooth-to-tooth single flank composite deviation is a parameter of great importance in reference to the control of vibration and noise, and smoothness of operation. The primary source of this deviation is due to errors in tooth profiles with respect to their theoretical geometry, which alter the correct conjugacy of mating teeth. Even the modified tooth shape (tip and root relief, profile crowning, helix modification, etc.) can greatly influence the tooth-to-tooth single flank composite deviation, as the gears are tested at low load, while the teeth are designed so as to have the right conjugacy only to the specific design load, which is very high. It is also to be noted that the primary source of the no-load total transmission error is due to the accumulated pitch error, i.e. the already defined total cumulative pitch deviation, F_P.

The double flank composite deviation testing is used to evaluate the *total radial composite deviation*, $F_i''(\mu m)$, and the *tooth-to-tooth radial composite deviation*, $f_i''(\mu m)$. The total radial composite deviation, F_i'', is defined as the difference between the maximum and minimum values of center distance, which occur in the test when the gear wheel to be tested, with its right and left flanks simultaneously in contact with those of the master gear wheel, is rotated through one complete revolution. Instead, the tooth-to-tooth radial composite deviation, f_i'', is defined as the value of the radial composite deviation corresponding to one angular pitch, $\tau = (360°/z)$, during the complete cycle of engagement of all the teeth of the gear wheel to be tested.

These two radial composite deviations are affected by the accuracy of the master gear wheel and the total contact ratio of the gear pair composed by the master gear wheel and the gear wheel to be tested. Both these deviations show the combined effect of different errors, such as profile error, pitch error and variation in tooth thickness. The total radial composite deviation includes, in addition to these deviations, even runout and wobble. In the usual types of test equipment, errors may be recorded in the form of a circular trace or a linear trace; these traces can be processed using statistical methods, such as shown in Murari et al. [28].

Finally, it should be noted that the topic covered in this textbook assumes the reader has the basic knowledge regarding the dimensional tolerances (including ISO system of limits, fits and tolerances), and geometrical tolerances and deviations. These are very important issues which, for reasons of space, cannot be recalled here. Only when the clarity requires a memory recall, a brief mention will be made, in their essential concepts. For basic concepts, we refer the reader to ISO specifications, as regards the general aspects, or to specialized textbooks, as regards particular aspects, such as the computer aided toleracing (see: Björke [1]; Chévalier [5]; ElMaraghy [10]).

References

1. Björke Ö (1978) Computer aided tolerancing. Tapir Publishers, Trondheim
2. Björke Ö (1995) Manufacturing systems theory. Tapir Publishers, Trondheim
3. Boothroyd G, Knight WA (1989) Fundamentals of machining and machine tools, 2nd edn. Marcell Decker, New York
4. Buckingham E (1949) Analytical mechanics of gears. McGraw-Hill Book Company Inc, New York
5. Chévalier A (1983) Manuale del Disegno Tecnico, revised and expanded Italian edition by Chirone E, Vullo V: Società Editrice Internazionale, Torino
6. Chirone E, Vullo V (1984) Cuscinetti a strisciamento. Libreria Editrice Universitaria Levrotto & Bella, Torino
7. Dooner DB, Seireg AA (1995) The kinematic geometry of gearing: a concurrent engineering approach. Wiley, New York
8. Dubbel H (1984) Taschenbuch für den Maschinenbau, 15th edn. Springer, Berlin, Heidelberg
9. Dudley DW (1962) Gear handbook. The design, manufacture, and application of gears. McGraw-Hill Book Company, New York
10. ElMaraghy HA (ed) (1998) Geometric design tolerancing: theories, standards and applications. Chapman & Hall, London
11. Ferrari C, Romiti A (1966) Meccanica Applicata alle Macchine. Unione Tipografica – Editrice Torinese (UTET), Torino
12. Galassini A (1962) Elementi di Tecnologia Meccanica, Macchine Utensili, Principi Funzionali e Costruttivi, loro Impiego e Controllo della loro Precisione, reprint 9th edn. Editore Ulrico Hoepli, Milano
13. Giovannozzi R (1965b) Costruzione di Macchine, vol II, 4th edn. Casa Editrice Prof. Riccardo Pàtron, Bologna
14. Grote K-H, Antonsson EK (eds) (2009) Springer handbook of mechanical engineering, vol 10. Springer Science, New York
15. Henriot G (1979) Traité théorique and pratique des engrenages, vol 1, 6th edn. Bordas, Paris
16. Henriot G (1972) Traité théorique et pratique des engrenages. Fabrication, contrôle, lubrification, traitement thermique, vol 2. Dunod, Paris
17. Houser DR (1986) The root of gear noise-transmission error. Power Transmission Design 86(5):27–30

18. ISO 1122-1: 1998 Vocabulary of gear terms—Part 1: definition related to geometry
19. ISO 1122-2: 1999(E/F) Vocabulary of gear terms—Part 1: definition related to worm gear geometry
20. ISO 1328-1: 2013 Cylindrical gears-ISO system of flank tolerance classification-Part 1: definitions and allowable values of deviations relevant to flank of gear teeth
21. ISO 1328-2: 1997 Cylindrical gear-ISO system of flank tolerance classification-Part 2: definitions and allowable values of deviations relevant to radial composite and runout information
22. ISO 17485: 2006 Bevel gears-ISO system of accuracy
23. Jelaska DT (2012) Gears and gear drives. Wiley, Chichester
24. Litvin FL, Fuentes A (2004) Gear geometry and applied theory, 2nd edn. Cambridge University Press, Cambridge
25. Maitra GM (1994) Handbook of gear design, 2nd edn. Tata McGraw-Hill Publishing Company Ltd, New Delhi
26. Merritt HE (1954) Gears, 3th edn. Sir Isaac Pitman & Sons, Ltd., London
27. Micheletti GF (1977) Tecnologia Meccanica, 2nd edn. Unione Tipografica – Editrice Torinese (UTET), Torino
28. Murari G, Stroppiana B, Vullo V (1981) Approccio statistico per la descrizione di caratteristiche strutturali di superfici lavorate. ATA 5:351–355
29. Niemann G, Winter H (1983) Maschinen-Elemente, Band II: Getriebe allgemein, Zahnradgetriebe-Grundlagen, Stirnradgetriebe. Springer, Berlin, Heidelberg
30. Niemann G, Winter H (1983) Maschinen-Elemente, Band III: Schraubrad-, Kegelrad-, Schnecken-, Ketten-, Rienem-, Reibradgetriebe, Kupplungen, Bremsen, Freiläufe. Springer, Berlin, Heidelberg
31. Niemann G, Winter H, Höln B-R (2005) Maschinenelemente: Konstruktion und Berechung von Verbindungen, Lagern, Wellen, Band 1. - 4. Auflage. Springer, Berlin, Heidelberg
32. Panetti M (1937) Lezioni di Meccanica Applicata alle Macchine, Parte IIᵃ, Ruote – Roteggi – Macchine Funicolari – Cingoli. Arti Grafiche Pozzo, Torino
33. Podzharov E, Syromyatnikov V, Ponce Navarro JP, Ponce Navarro R (2008) Static and dynamic transmission error in spur gears. Open Ind Manuf Eng J 1:37–41
34. Pollone G (1970) Il Veicolo. Libreria Editrice Universitaria Levrotto&Bella, Torino
35. Radzevich SP (2017) Gear cutting tools: science and engineering, 2nd edn. CRC Press, Taylor&Frencis Group, Boca Raton
36. Radzevich SP (2018) Theory of gearing: kinematics, geometry and synthesis, 2nd edn. CRC Press, Taylor&Francis Group, Boca Raton
37. Rossi M (1965) Macchine Utensili Moderne. Editore Ulrico Hoepli, Milano
38. Scotto Lavina G (1990) Riassunto delle Lezioni di Meccanica Applicata alle Macchine: Cinematica Applicata, Dinamica Applicata - Resistenze Passive - Coppie Inferiori, Coppie Superiori Ingranaggi – Flessibili – Freni. Edizioni Scientifiche SIDEREA, Roma
39. Townsend DP (1991) Dudley's gear handbook. McGraw-Hill, New York
40. Vullo V (1983) Calcolo delle Tolleranze con Metodi Probabilistici e Criteri di Minimo Costo. Cooperaria Libraria Universitaria Torinese, Torino
41. Woodbury RW (1958) History of the gear-cutting machines: a historical study in geometry and machines. M.I.T. Technology Press, Cambridge, Massachusetts

Chapter 2
The Geometry of Involute Spur Gears

Abstract In this chapter, the fundamentals of involute spur gears geometry are given. Once the way of generating an involute of base circle has been described, the polar coordinates of any of its current points are given. The parametric equations that describe the position vector as a function of both the involute roll angle and the involute polar angle are also obtained, also defining the tangent and normal vectors as well as their unit vectors. The discussion is then generalized to include the ordinary, extended and shortened involute curves. The fundamental properties of the involute curves are subsequently described, with special reference to their use as gearing teeth profiles. Particular attention is given to the geometric sizing of the gears, in terms of modular sizing, without however neglecting the one based on the diametral pitch. The main quantities defining the teeth geometry are given, particularly those that ensure the appropriate kinematic operation of the gears and those that play a fundamental role for the addendum modification in terms of profile shift. Finally, a reference is made to standard and no-standard basic rack tooth profiles.

2.1 Generation of the Involute and Its Geometry

As we mentioned in the Chap. 1, in order to avoid severe dynamic problems, gear pairs and gear trains must meet the primary requirement of the constancy of angular velocities. This goal is achieved through a conjugate action between the mating gear tooth profiles, which is referred to as the *law of gearing* [3, 10, 29]. According to this law, whatever the position of contact teeth, the common normal to the tooth profiles at their instantaneous point of contact must necessarily pass through a fixed point on the line of centers, which is the pitch point. Two any profiles (or curves) that mesh each other and that satisfy this law of gearing are *conjugate profiles* (or *conjugate curves*).

An almost infinity of curves can be used to meet the law of gearing. However, in practice, with few exceptions (e.g., the clock gears), the curve of almost generalized use is the *involute of the circle*, also known as *evolvent of the circle*. The reasons for the success of this curve with respect to the almost infinity of other possible curves

© Springer Nature Switzerland AG 2020 39
V. Vullo, *Gears*, Springer Series in Solid and Structural Mechanics 10,
https://doi.org/10.1007/978-3-030-36502-8_2

are to be found in the many and significant advantages associated with it, among which the following advantages deserve specific mention: the conjugate action is independent of the variation of center distance; it allows to obtain a high accuracy grade of the gears, since the standard basic rack teeth have straight-sided profiles (as well as those of the cutting tools derived from them), so they can be made as accurately as possible; a single cutting tool can generate gear wheels of a given module, with any number of teeth.

By definition, an involute is the path described by any point on a taut inextensible thin strip as it unwinds from a given curve, called *evolute*. Therefore, given a planar curve (evolute) and a tangent straight-line to it at a given point, the curve generated by this *tracing point* of the tangent straight-line (or any other tracing point on this tangent straight-line), when it rolls without sliding on the given curve, is the involute of it (see Gray [13], McCleary [21], Smirnov [30]) . Any convex curve can be used for this purpose, but in gear design the term involute implies the involute of a circle. The evolute, that is the circle from which the involute of the circle is generated, is called the *base circle*.

With few exceptions, the involute of a circle is, nowadays, the curve characterizing the tooth profiles of cylindrical gears with parallel and crossed axes and the involute worm. With its generalization in three-dimensional space, which leads to the spherical involute, it also forms the basis of the bevel and spiral bevel gears. The knowledge of the geometry of the involute of a circle, which is a planar curve, and of the spatial or spherical involute, which is a three-dimensional curve, is of fundamental importance for the understanding of the gears. In this Chapter, we summarize the basic concepts of the involute of a circle, while in Chap. 9 we will summarize the basic concepts of the spherical involute. For any further details, we refer the reader to Tuplin [31], Merritt [22], Colbourne [6], Phillips [26], Litvin and Fuentes [20], and Dooner [8].

Figure 2.1 shows the involute of a circle with fixed center at point O, base radius r_b and base diameter $d_b = 2r_b$, generated as the trace of the tracing point P_0 of the tangent straight-line t-t, when the latter rolls round the base circle in a clockwise direction without sliding. A symmetrical involute with respect to the radial straight-line OP_0 (not shown) could be generated as the trace of the same tracing point P_0, when the tangent straight-line t-t rotates in a counterclockwise direction. The involute clearly cannot extend inside its base circle, and can be developed as far as desired outside the same base circle. All geometric relationships of the involute derive from the fact that, due to the rolling without sliding of the tangent straight-line t-t round the base circle, whatever the tracing point P, the segment PT of tangent line t-t is equal in length to the corresponding arc $\overset{\frown}{P_0T}$ of the base circle.

It is to be noted that, at every its point P, the involute is normal to the taut inextensible thin strip, which materializes the generating straight-line. In the position of this generating line represented by P_iT_i in Fig. 2.1, the point T_i on the base circle is the instantaneous center of rotation of the generating straight-line. Hence an infinitesimal length of involute arc at point P_i is indistinguishable from a circular arc having its center at point T_i and radius P_iT_i; then P_iT_i is the radius of

Fig. 2.1 Generation of an involute of the base circle having radius r_b

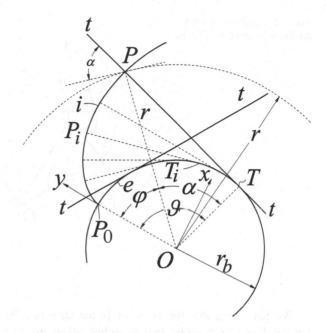

curvature of the involute at point P_i, which varies continuously according to a non-linear function, being zero at point P_0 and a higher value at point P.

Consequently, the generating line P_iT_i is normal to the involute whatever the point P_i and, at the same time, it is always tangent to the base circle: thus the properties from which every normal to an involute is a tangent to the base circle, and the evolute, e, to a regular curve, i, is the envelope of the family of normals P_iT_i to curve i. Also, the evolute, e, to involute curve, i, i.e. the base circle, is the locus of the centers of curvature, T_i. In Fig. 2.1, P_0 is the *origin of the involute* on the base circle, also called *point of regression of the involute*. This point of regression is not a *regular point*, as it is not possible to identify two limiting rays which define the tangent at this point to the planar curve under consideration (i.e., the involute curve), in the sense defined by Zalgaller [33]; it is instead a variety of *singular point*, as only a half-tangent exists at this point, coinciding with the straight-line through this point and the center of the base circle (see also Litvin and Fuentes [20]).

Distinct points marked on the tangent straight-line t-t, when it rolls without sliding round the base circle, describe many involutes, all parallel to each other. Then the perpendicular distance between any two adjacent involutes is a constant, equal to the distance between the corresponding tracing points measured along the generating straight-line t-t. Figure 2.2 shows a family of parallel involute curves corresponding to the operating profiles of an involute spur gear. The distance p_b between two adjacent involutes, measured along the corresponding rolling straight-lines, is called *base pitch*, because it is equal to the distance between the origins of the involutes measured round the base circle.

Fig. 2.2 Family of parallel
involutes having base pitch p_b

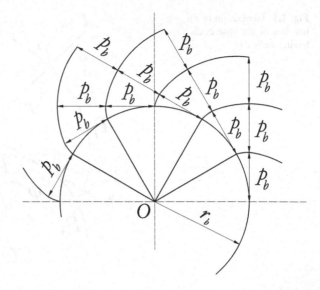

We have seen that the point of initial tangency P_0 (Fig. 2.1) of any rolling straight-line can generate two branches of an involute curve symmetrical with respect to the radial straight-line OP_0, when the generating straight-line rolls without sliding over the base circle clockwise and counterclockwise, respectively. From a geometrical point of view, each of these two branches represents the profiles of the right and left flanks of teeth of an involute spur gear. If the involute is used as profile of the teeth flanks, two families of involute equally spaced along the base circle have to be constructed. In other words, the symmetry of the tooth requires the tracing of as many profiles on the same base circle obtained by unwinding the taut inextensible thin strip in the opposite rotational directions.

From the point of view of the conditions of contact between the mating profiles, which will be discussed soon, it is interesting to consider the correlation between the arcs (and the corresponding angles) of the base circle described by the generating straight-line and the corresponding arcs of the involute described by the tracing point during the involute generation. Figure 2.3 highlights the fact that, at equal arcs of the base circle described by the generating straight-line during its rolling without sliding round the base circle, progressively increasing arcs of the involute correspond.

To define analytically the geometry of the involute, we return to consider the Fig. 2.1, where P is any current point on the involute with origin at point P_0. At point P, the angle of rotation during the rolling motion of the generating straight-line t-t from the initial position (point of tangency with the base circle at P_0) is ϑ. This is the *involute roll angle* through which the taut inextensible thin strip has been unwound (according to [16], it is also the angle whose arc on a circle of unit radius is equal to the tangent of the pressure angle at a given point in an involute to that circle). With ϑ measured in radians, the involute arc, $\widehat{P_0 T}$, and the segment,

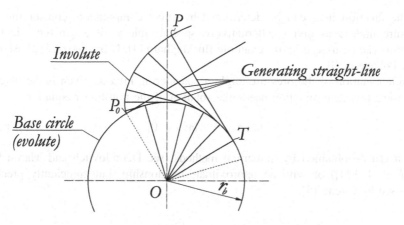

Fig. 2.3 Arcs of base circle and corresponding arcs of involute

$\overline{PT}$, will be both equal to ϑr_b, that is $\widehat{P_0T} = \overline{PT} = \vartheta r_b$. If we denote respectively with φ and α the *involute polar angle* and the difference between the involute roll angle and the involute polar angle, we can write the following relationship:

$$\varphi = \vartheta - \alpha = \frac{\widehat{P_0T}}{r_b} - \alpha = \frac{\overline{PT}}{r_b} - \alpha = \tan\alpha - \alpha, \tag{2.1}$$

where α and φ must be also measured in radians. The quantity:

$$\varphi = \tan\alpha - \alpha = \text{inv}\alpha \tag{2.2}$$

of a given value of α is the *involute function* or simply *involute* of the angle α; it is indicated as $\text{inv}\alpha$. The polar coordinates (r, φ) of the involute of a base circle having radius r_b are given by:

$$r = \overline{OP} = \frac{r_b}{\cos\alpha} - r_b \sec\alpha \tag{2.3}$$

$$\varphi = \text{inv}\alpha, \tag{2.4}$$

where $r = \overline{OP}$ is the *radius vector* to any current point P of the involute curve, and φ is the *involute polar angle* or *involute vectorial angle*, i.e. the angle between the radius vector to the current point under consideration and the radial straight-line through the origin of the involute (see ISO 1122-1 [16]).

With reference to Fig. 2.1, we draw the circle having radius r passing through the current point P on the involute curve. The angle that the generating straight-line, normal to the involute at point P, forms with the tangent to this circle at the same point P is equal to α, and is called the *pressure angle* of the involute on the circle of radius r. Therefore, given an involute of base circle, for each value of r we have the corresponding pressure angle, α.

The function invα can be determined by direct computation, considering the pressure angle α as given. Alternatively, special table available in the scientific literature can be used (see, for example Buckingham [4], Giovannozzi [12], Merritt [22], Pollone [27]).

Determination of the pressure angle α considering invα as given is the inverse operation; this determination needs the solution of the non-linear equation

$$\tan\alpha - \alpha - \text{inv}\alpha = 0, \tag{2.5}$$

which can be obtained by numerical methods (see Démidovitch and Maron [7], Moré et al. [24]) or with an approximate relationship, but sufficiently precise, proposed by Cheng [5].

2.2 Parametric Representation of the Involute Curve

To perform computer aided calculations, it is very useful to have a parametric representation of the involute curve. To this end, it is advisable to introduce a Cartesian reference system $O(x, y)$ like the one shown both in Figs. 2.1 and 2.4, where the y-axis coincides with the straight-line through the point of regression of the involute, P_0. As variable parameters we can assume α (Fig. 2.4a) or ϑ (Fig. 2.4b). The corresponding *position vectors* $r(\alpha)$ or $r(\vartheta)$, drawn from the origin, O, of the reference system to any current point, P, of the involute curve are represented by one of the following two *vector functions* (it should be reiterated that in this textbook vectors and vector quantities are indicated in bold letters):

$$r(\alpha) = x(\alpha)\boldsymbol{i} + y(\alpha)\boldsymbol{j}, \tag{2.6}$$

or

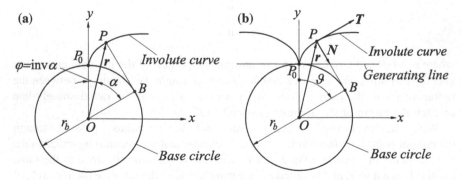

Fig. 2.4 Parameters for parametric representation of an ordinary involute curve: **a**, parameter α; **b**, parameter ϑ

$$r(\vartheta) = x(\vartheta)\boldsymbol{i} + y(\vartheta)\boldsymbol{j}, \tag{2.7}$$

to be used as an alternative. In this equations, $\boldsymbol{i}$ and $\boldsymbol{j}$ are the *unit vectors* of the x-axis and respectively y-axis, while $x(\alpha)$ and $y(\alpha)$ are continuous functions of the variable parameter, α, which can vary within the open interval $(a < \alpha < b)$; in a similar way, $x(\vartheta)$ and $y(\vartheta)$ are continuous functions of the variable parameter, ϑ, which can vary within the open interval $(c < \vartheta < d)$.

We leave aside the cases of the *extended involute curve* and *shortened involute curve*, which are *self-intersecting curves*, and focus our attention on an *ordinary involute curve* or *conventional involute curve*, which does not have *points of self-intersection*. Therefore, this last curve is a *simple curve*, inasmuch as it is characterized by one-to-one correspondence between the current point on the involute curve and the selected parameter, what does not happen for extended and shortened involute curves. In addition, the ordinary involute curve is a *regular curve*, because the following conditions are satisfied:

- Functions $x(\alpha)$ and $y(\alpha)$ as well as functions $x(\vartheta)$ and $y(\vartheta)$ have continuous derivative to the first order at least;
- In terms of variable parameters α, it must be:

$$r_\alpha = \frac{dr(\alpha)}{d\alpha} = x_\alpha \boldsymbol{i} + y_\alpha \boldsymbol{j} \neq 0, \tag{2.8}$$

which is equivalent to the inequality

$$x_\alpha^2 + y_\alpha^2 \neq 0, \tag{2.9}$$

where $x_\alpha = (dx/d\alpha)$ and $y_\alpha - (dy/d\alpha)$.

- In terms of variable parameter ϑ, similarly it must be:

$$r_\theta = \frac{dr(\vartheta)}{d\vartheta} = x_\vartheta \boldsymbol{i} + y_\vartheta \boldsymbol{j} \neq 0, \tag{2.10}$$

which is equivalent to the inequality

$$x_\vartheta^2 + y_\vartheta^2 \neq 0, \tag{2.11}$$

where $x_\vartheta = (dx/d\vartheta)$ and $y_\vartheta = (dy/d\vartheta)$.

We recall then the well-known conditions of existence of singular points of a planar curve, which, as a function of α or ϑ, may be written in the form

$$r_\alpha = 0 \text{ and } r_{\alpha\alpha} = \frac{d^2 r(\alpha)}{d\alpha^2} \neq 0 \tag{2.12}$$

or

$$r_\vartheta = 0 \text{ and } r_{\vartheta\vartheta} = \frac{d^2 r(\vartheta)}{d\vartheta^2} \neq 0. \tag{2.13}$$

The point of regression of an ordinary involute curve can be determined by imposing any of the above pairs of conditions, as it is nothing but a special case of singular point. Furthermore, vector $r_{\alpha\alpha}$ or vector $r_{\vartheta\vartheta}$ determines the direction of the tangent at this point of regression.

The two vector functions $r(\alpha)$ and $r(\vartheta)$ of an ordinary involute curve are given respectively by:

$$r(\alpha) = \frac{r_b}{\cos \alpha} [\sin(\text{inv}\alpha)i + \cos(\text{inv}\alpha)j], \tag{2.14}$$

where $\text{inv}\alpha = (\tan\alpha - \alpha)$ in accordance with Eq. (2.2), with α that varies within the open interval $(-\pi/2 < \alpha < \pi/2)$;

$$r(\vartheta) = r_b[(\sin \vartheta - \vartheta \cos \vartheta)i + (\cos \vartheta + \vartheta \sin \vartheta)j], \tag{2.15}$$

with ϑ which varies within the open interval $(-\infty < \vartheta < \infty)$.

It should be noted that Eq. (2.14) is easily obtainable by means of the relationships obtained in the previous section, while Eq. (2.15) is obtained considering that $\overline{OP} = \overline{OB} + \overline{BP}$ (Fig. 2.4). Furthermore, Eq. (2.15) may be obtained from Eq. (2.14) by changing parameter α to parameter ϑ using the continuous strongly monotonic function $\alpha(\vartheta) = \arctan(\vartheta)$, with $(-\infty < \vartheta < \infty)$. For further details, we refer the reader to Litvin et al. [19], and Litvin and Fuentes [20].

To extend the framework and define in a more detailed way the geometry of an ordinary involute curve, which is a planar curve, we still need to write the equations of the tangent T, the unit tangent t, the normal $N = T \times k$, and the unit normal n; only the tangent T and normal N are shown in Fig. 2.4b, while the unit tangent t and unit normal n are not shown. It is to be noted that N is defined by a cross product where the unit vector, k, of the z-axis appears, which completes the right-hand trihedron of the reference system. To do this, we use the vector function given by Eq. (2.15), which expresses the position vector as a function of the variable parameter, ϑ. The latter is the *involute roll angle*, i.e. the rotation angle of the generating straight-line of the involute curve during its rolling motion over the base circle (Fig. 2.4b), starting from the line OP_0. The reason for using the parameter ϑ, and therefore Eq. (2.15), is that in doing so we get simpler final equations. So, we have:

$$T = r_\vartheta = \frac{dr(\vartheta)}{d\vartheta} = x_\vartheta i + y_\vartheta j = r_b \vartheta (\sin \vartheta i + \cos \vartheta j) = T_x i + T_y j, \tag{2.16}$$

where $T_x = r_b \vartheta \sin \vartheta$ and $T_y = r_b \vartheta \cos \vartheta$;

$$t = \frac{T}{|T|} = \frac{T}{\sqrt{T_x^2 + T_y^2}} = \sin \vartheta i + \cos \vartheta j, \qquad (2.17)$$

where $|T| = \sqrt{T_x^2 + T_y^2}$ is the absolute value of T, with the condition that it is valid provided $\vartheta \neq 0$, i.e. at any current point on the involute curve, with the exception of its point of regression;

$$N = T \times k = \begin{vmatrix} i & j & k \\ T_x & T_y & 0 \\ 0 & 0 & 1 \end{vmatrix} = \begin{bmatrix} T_y \\ -T_x \\ 0 \end{bmatrix} \begin{Bmatrix} i \\ j \\ k \end{Bmatrix} = r_\vartheta \vartheta (\cos \vartheta i - \sin \vartheta j), \quad (2.18)$$

which highlights the fact that the direction of normal N at any current point P coincides with the direction of tangent PB to the base circle (Fig. 2.4b);

$$n = \frac{N}{|N|} = \frac{T_y i - T_x j}{\sqrt{N_x^2 + N_y^2}} = \frac{T_y i - T_x j}{\sqrt{T_x^2 + T_y^2}} = \cos \vartheta i - \sin \vartheta j, \qquad (2.19)$$

where $|N| = \sqrt{N_x^2 + N_y^2} = \sqrt{T_x^2 + T_y^2}$ is the absolute value of N, with the condition that it is valid provided $\vartheta \neq 0$, i.e. at any current point on the involute curve, with the exception of its singular point. This singular point, i.e. the point of regression of the involute curve, is determined by imposing the two conditions $r_\vartheta = dr(\vartheta)/d\vartheta = 0$, and $r_{\vartheta\vartheta} = d^2 r(\vartheta)/d\vartheta^2 \neq 0$, where

$$r_{\vartheta\vartheta} = \frac{d^2 r(\vartheta)}{d\vartheta^2} = r_b [(\sin \vartheta + \vartheta \cos \vartheta)i + (\cos \vartheta - \vartheta \sin \vartheta)j]. \qquad (2.20)$$

From Eq. (2.16), we get the point of regression corresponds to $\vartheta = 0$, while Eq. (2.20) with $\vartheta \neq 0$ gives $r_{\vartheta\vartheta} = r_b j$. Thus, we infer that the half-tangent at point P_0 has the direction of the positive y-axis (see Figs. 2.1 and 2.4).

To solve specials problems regarding the gears, it may be necessary to consider the geometry of extended and shortened involute curves. For example, the extended involute curve comes into play as a trajectory which is drawn by the center of circular arc corresponding to the tip fillet of a rack cutter, during its relative motion with respect to the gear wheel to be cut, when we need to determine the shape of the root fillet of the same gear wheel.

As Fig. 2.4b shows, an ordinary or conventional involute curve is generated as a trajectory of a tracing point of the generating line (P_0 in the figure), during the rolling motion of this line over the base circle. Two branches of an involute curve may be obtained, depending on whether this rolling motion is clockwise or counterclockwise. Now imagine that the tracing point $P_0' \neq P_0$ is not a point of the

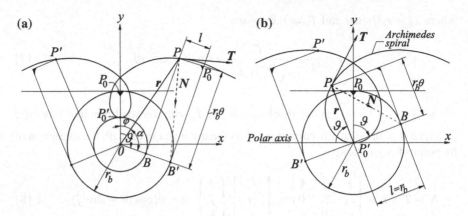

Fig. 2.5 a extended involute curve; **b** Archimedean spiral as special case of an extended involute curve

generating line, but a point that is on the straight-line OP_0, at the distance from P_0, $l = P_0 P_0'$, called *offset*, and rigidly connected with the same generating line. If P_0' is between O and P_0, the generated curve is an extended involute curve (see Fig. 2.5a), while if P_0' is external to P_0, the generated curve is a shortened involute curve.

Using an approach similar to the previous one, and taking ϑ as a variable parameter, we get the following equations:

$$
\begin{aligned}
x(\vartheta) &= (r_b \mp l) \sin \vartheta - r_b \vartheta \cos \vartheta \\
y(\vartheta) &= (r_b \mp l) \cos \vartheta - r_b \vartheta \sin \vartheta,
\end{aligned}
\tag{2.21}
$$

where the upper sign corresponds to the extended involute curve, and the lower sign corresponds to the shortened involute curve. Even in these two cases, two branches of extended and shortened involute curves are obtained, which are symmetric with respect to the y-axis. However, contrary to what happens for an ordinary involute curve, where the point P_0 which is common to its two branches is a singular point, in the cases of extended and shortened involute curves the point P_0', which is common to their two branches, is a regular point.

Taking into account Eqs. (2.21), we can generalize Eq. (2.15), which becomes:

$$
\boldsymbol{r}(\vartheta) = [(r_b \mp l) \sin \vartheta - r_b \vartheta \cos \vartheta]\boldsymbol{i} + [(r_b \mp l) \cos \vartheta + r_b \vartheta \sin \vartheta]\boldsymbol{j}.
\tag{2.22}
$$

For $l = 0$, Eqs. (2.22) and (2.21) are reduced to those of an ordinary involute curve, obtained previously. Another particular case is interesting, that for which $l = r_b$, whereby P_0' coincides with the origin O of the Cartesian reference system (see Fig. 2.5b); in this case, the extended involute curve turns out into an Archimedean spiral, defined by equation $r(\vartheta) = r_b \vartheta$.

The components along the coordinate axes of the tangent T and the normal $N = T \times k$ for extended and shortened involute curves are given by the relationship:

$$T_x = \mp l \cos \vartheta - r_b \vartheta \sin \vartheta$$
$$T_y = \pm l \sin \vartheta + r_b \vartheta \cos \vartheta, \tag{2.23}$$

and respectively by the relationship:

$$N_x = \pm l \sin \vartheta + r_b \vartheta \cos \vartheta$$
$$N_y = \pm l \cos \vartheta - r_b \vartheta \sin \vartheta, \tag{2.24}$$

2.3 Involute Properties and Fundamentals

Let us now consider the involute profile to examine how it satisfies the requirement for the transmission of the uniform motion between two parallel axes by means of an external spur gear. Figure 2.6 shows a spur gear consisting of two gear wheels with fixed centers O_1 and O_2 (these are the traces of the two axes on the drawing plane), and having base circles whose respective radii are $r_{b1} = O_1 T_1$ and $r_{b2} = O_2 T_2$, where T_1 and T_2 are the points of tangency of the straight-line t-t with

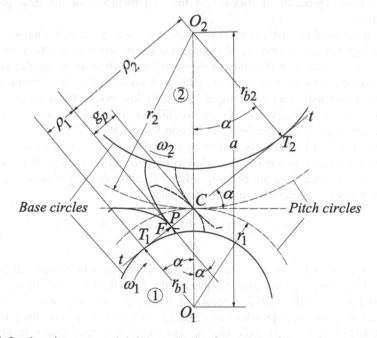

Fig. 2.6 Involutes in contact and their mutual action for an external gear pair

these circles. This straight-line *t-t* is one of the two common tangents to the two base circles, the one compatible with the directions of rotation of the two gear wheels, obviously with the opposite direction relative to one another.

The contact between any two mating profiles during meshing takes place at a given point P, located on the above-mentioned common tangent, which is also the common normal to the two mating profiles at the same point P. When the driving wheel 1 rotates in the clockwise direction, as indicated by arrow, the point of contact P moves from T_1 to T_2. The two base circles undergo the same peripheral displacement, moving as if they were drag by the straight-line *t-t*, which moves parallel to itself along its direction, by rolling on the base circles without sliding. Therefore, the two base circles have equal tangential velocities, and thus their angular velocities ω_1 and ω_2, and their rotational speeds n_1 and n_2 (ω_1 and ω_2 in rad/s, and n_1 and n_2 in rpm) will be inversely proportional to their radii.

Point C, that is the point of intersection of the common tangent *t-t* to the center distance, a, of the gear wheels, is distant from the center O_1 and O_2 of lengths proportional to the radii of the base circles, and therefore we have:

$$\frac{\omega_1}{\omega_2} = \frac{n_1}{n_2} = \frac{r_{b2}}{r_{b1}} = \frac{CO_2}{CO_1}. \tag{2.25}$$

The two circles having radii $r_1 = CO_1$ and $r_2 = CO_2$, supposed rigidly connected with the two wheels, also have the same tangential velocity, and then roll against each other without slide. They are the *pitch circles* of the gear pair under consideration. Direction of rotation of the two members of the gear pair are opposite.

We now imagine that pitch circles represent two cylindrical friction wheels pressed together. In the absence of sliding, rotation of one friction wheel (i.e. pitch circle) causes rotation of the other in the opposite direction, at an angular velocity ratio inversely proportional to their radii or diameters. But the friction wheels transmit notoriously very low torque (and then low power for a given angular velocity), which is closely related to the coefficient of friction. In order to transmit more torque than is possible with the friction wheels alone, we now introduce, in addition to the friction wheels, a driving crossed belt (or a taut inextensible thin strip) that drives the two base circle used as two cylindrical pulleys. If the smaller pulley rotates clockwise, the belt (or strip) will cause the larger pulley to rotate counterclockwise in accordance with the equation:

$$\frac{\omega_1}{\omega_2} = -\frac{r_{b2}}{r_{b1}} = -\frac{d_{b2}}{d_{b1}} = -\frac{r_2}{r_1} = -\frac{d_2}{d_1}, \tag{2.26}$$

where the minus sign indicates that the two pulleys rotates in opposite directions.

In order to transmit even more torque, the third and the last step is to replace the combined action of the friction wheels and driving crossed belt (or strip) with that of teeth having involute profiles that exchange their mutual actions along the tangent straight-line *t-t* to the base circles, and therefore behave kinematically as a

driving belt. We can imagine that these profiles are generated from the tracing point
P, when the belt is cut at this point, and the two branches of the belt are winded and
un-winded on the two base circles. In this way the tracing point, which coincides
with the two ends of the belt which has been cut, describes the two involutes drawn
with solid line in Fig. 2.6. These involutes are thus generated simultaneously by the
tracing point P, which is the point of contact, while the portion of the belt $T_1 T_2$ is
the *generation straight-line*.

This generation straight-line is always tangent to the base circles, and does not
change position, but moves parallel to itself from T_1 to T_2 causing the rolling
without sliding of the two base circles about their axes. The point of contact P also
moves along the generating straight-line, and the latter is always normal to the
involutes at the point of contact. Then the requirements for uniform motion and so
the law of gearing are satisfied.

If the traces O_1 and O_2, of the axes of the two wheels in the drawing plane are
arranged on the same side with respect to the area where the teeth touch, the larger
wheel turns in an *internal gear* or *annulus gear* or *ring gear*, having fixed center
O_2, radius of the base circle r_{b2}, and teeth protruding inward. Both wheels rotate in
the same direction (Fig. 2.7). When the driving wheel having fixed center O_1

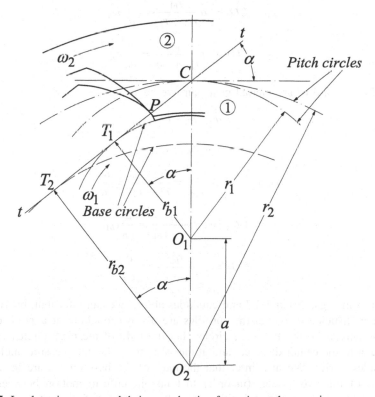

Fig. 2.7 Involutes in contact and their mutual action for an internal gear pair

(the pinion) rotates in the clockwise direction, as shown in Fig. 2.7, the contact of the profiles is still at a point P of the common tangent to the two base circles.

During the rotation, the point of contact moves along this common tangent from T_1 to the right. Pitch circles of the two wheels are still those with fixed centers O_1 and O_2, passing through point C, that is the intersection of this tangent with the line of centers O_1O_2. For this internal gear pair still apply Eqs. (2.25) and (2.26), with the caveat that the radius of the ring gear is to be assumed negative. In this way, the ratio (ω_1/ω_2) in Eq. (2.26) becomes positive, namely the two gear wheels rotate in the same direction. If r_1 and r_2 are the radii of the pitch circles, α is the pressure angle of the toothing with respect to these pitch circles, a is the center distance and a_0 is the sum or the difference of the radii of the base circles, for the case of an external gear pair we have the relations:

$$a = r_1 + r_2 \tag{2.27}$$

$$a_0 = r_{b1} + r_{b2} \tag{2.28}$$

$$\cos \alpha = \frac{r_{b1}}{r_1} = \frac{r_{b2}}{r_2} = \frac{r_{b1} + r_{b2}}{r_1 + r_2} = \frac{a_0}{a} \tag{2.29}$$

$$r_1 = CO_1 = a \frac{r_{b1}}{r_{b1} + r_{b2}} = \frac{a}{a_0} r_{b1} \tag{2.30}$$

$$r_2 = CO_2 = a \frac{r_{b2}}{r_{b1} + r_{b2}} = \frac{a}{a_0} r_{b2}, \tag{2.31}$$

while, for the case of an internal gear pair, we have:

$$a = r_2 - r_1 \tag{2.32}$$

$$a_0 = r_{b2} - r_{b1} \tag{2.33}$$

$$\cos \alpha = \frac{r_{b1}}{r_1} = \frac{r_{b2}}{r_2} = \frac{r_{b2} - r_{b1}}{r_2 - r_1} = \frac{a_0}{a} \tag{2.34}$$

$$r_1 = CO_1 = a \frac{r_{b1}}{r_{b2} - r_{b1}} = \frac{a}{a_0} r_{b1} \tag{2.35}$$

$$r_2 = CO_2 = a \frac{r_{b2}}{r_{b2} - r_{b1}} = \frac{a}{a_0} r_{b2}. \tag{2.36}$$

Both from Figs. 2.6 and 2.7 and equations above, we can infer that, by varying the center distance a, the mating profiles are always touching at a point of the common tangent to the two base circles, but the radii of the pitch circles vary in proportion to the center distance, and at the same time also the pressure angle α of the profiles varies. We also infer that the radii of the base circles are invariable elements of the two wheels (in order to transmit uniform motion between two rotating shafts with angular velocities ω_1 and ω_2, the base circles must satisfy the relationships (2.25 and 2.26), while the radii of the pitch circles and the pressure angle are elements that vary with the center distance of the gear pair.

With reference to Fig. 2.6, suppose that the member of the gear pair having center O_1 is the driving wheel and that it rotates clockwise. During the motion, the mating profiles in mutual contact are sliding over one another, but, if the friction is negligible, the line of application of the force F that they transmit coincides with the common normal, i.e. the tangent straight-line to the base circles. This interior common tangent to the base circles is the *line of action* (more exactly the *transverse line of action*) of the teeth, i.e. the line along which contact may take place. This line meets the line of centers O_1O_2 at the *pitch point C*. Since then for a given center distance, a, the position of the pitch point C is independent of the position of the point of contact P, the profiles would ensure a uniform angular velocity ratio.

The line of action forms with the normal to the line of centers the *pressure angle* α. As far as we said earlier, we infer that, given the radius r_{b1} of the base circle of the wheel 1 and the direction of rotation of the latter, and chosen the radius r_1 of the pitch circle, which must operate coupled to the wheel 2, and the direction of the line of centers, the pitch point C and direction of the line of action are determined (Fig. 2.6). Then also the pressure angle α is determined by means of Eq. (2.29).

Now consider the meshing between any two involutes of two base circles constrained to rotate about their fixed centers O_1 and O_2 (Fig. 2.6). The point of contact, P, between the two involute mating profiles will move along the line of action. If the two base circles, and then the two involutes related to them, rotate uniformly with respect to their centers, the length of the segment T_1P will increase uniformly, while the length of segment T_2P will decrease uniformly, since the length of the line of action, T_1T_2, remains constant.

On the other hand, the ratio (ω_1/ω_2) which is closely linked to radii of the two base circles in accordance with Eq. (2.25), does not change whatever the center distance, a. Moreover, whatever the center distance, the two involute mating profiles will touch however along the line of action; of course, this occurs within the limits related to the radial development of the two involutes or, what is the same, within limits related to the tooth depths of the two mating teeth. The variation of the center distance will determine instead a change of the pressure angle, α, and diameters of the pitch circles, d_1 and d_2.

The line of action is closely related to the radii of the two base circles and center distance. Considering an actual parallel involute gear pair in the three-dimensional space, we can define the *plane of action* as the plane containing the infinity of lines of action related to any transverse section. Then we define the *transverse path of contact*, or simply *path of contact*, as the locus of successive points of contact between mating tooth profiles on any transverse section. For involute gear pair with parallel axes, the transverse path of contact is the portion of the line of action lying between their tip circles. Therefore, the path of contact depends on tooth depths.

Therefore, the line of contact must not be confused, as sometimes happens, with the *path of contact*. The latter is in fact a portion of the line of action along which contact actually takes place. In the case of unmodified involute profiles free from undercutting, the path of contact is the portion of the line of action intercepted between the addendum circles of the two members of the gear pair. However, the profile of the cutting tool (rack-cutter, hob, or shaper) may be symmetrically or unsymmetrically modified, and the path of contact will no longer be that corresponding to the theoretically conditions. For example, in the case in which the profile of the cutting tool is modified in order to obtain the tip-easing (or tip relief), the theoretical path of contact will end at the point at which tip relief begins.

If now, with $\alpha = const$, we grow progressively the radius r_2 of the pitch circle of the wheel 2 (see Figs. 2.6 or 2.7), the radius r_{b2} of the base circle, center distance a, and distance CT_2, i.e. the distance of point C from the point of tangency T_2 of the line of action with the base circle, will increase progressively. But if the radius r_2 becomes infinitely large, the pitch circle of the wheel 2 degenerates into the normal straight-line to O_1C through the point C, while point T_2 moves at point at infinity of the line of action, and base circle is transformed into the line at infinity of the plane. Therefore, the involute tooth profile of this wheel having infinite pitch radius becomes the straight-line normal to the line of action.

Therefore, the *rack*, i.e. the wheel with pitch circle degenerated into a straight-line that has as conjugate the pitch circle chosen for the wheel 1, has teeth with straight flanks, inclined of the angle α (pressure angle) with respect to O_1C, that is with the normal to the *pitch line*. In this way, starting from an external or internal gear pair, we have obtained a *pinion-rack pair*. Therefore, invariable elements of the *involute rack* are the pitch line, the active straight flank of the tooth, and the pressure angle α, that is the angle formed by this straight flank with the normal to the pitch line. Then, given a rack as the one shown in Fig. 2.8, which transmit the motion in the direction indicated by the arrow, and the radius r_1 of the pitch circle of a gear wheel intended to meshing with it, the direction of the line of action, which is normal to the tooth flank at point C, and radius of its base circle $r_{b1} = r_1 \cos \alpha$ are determined. Contrary to what happens for an external gear pair, a variation of the center distance of the pinion-rack pair does not cause any change of the pressure angle, since this is an invariable quantity of the involute rack.

2.4   Characteristic Quantities of the Involute Gears

First it is necessary to examine what might be the maximum length of the path of contact. We have already seen above that, in the case of unmodified involute profiles, the path of contact is the portion of line of action intercepted between the addendum circles of the two members of the gear pair. However, both for external and internal gear pairs, the *addendum circles* of the two gear wheels which make up the gear pair cannot be arbitrarily large, but must satisfy well defined limits.

For an external gear pair, such as that shown in Fig. 2.6, the addendum circles of the pinion and wheel can have a radius at most equal to O_1T_2 and O_2T_1 respectively. In fact, it should be noted that un-conjugate behavior exists if the initial and final engagement between two teeth occurs outside of the two points T_1 and T_2 respectively, where the line of action is tangent to the two base circles. These points

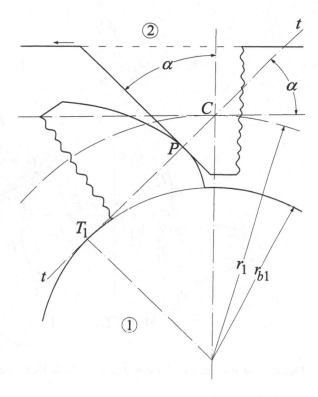

Fig. 2.8 Involutes in contact and their mutual action for rack-pinion pair

are the *limit points* beyond which, as shown below, the so-called *theoretical interference* or *involute interference* occurs.

Figure 2.9 shows the limit condition above, with the addendum circles that pass through the points T_1 and T_2. The profiles p_1 and p_2 of the two teeth in contact at point T_1 are the two involutes generated by the same tracing point T_1 of the line of action, when it rolls without sliding on the two base circles. With the tooth depths shown in figure, whereas the driving wheel 1 (the pinion) rotates clockwise, the contact can extend from point T_1 up to point T_2. When the point of contact P is located in an intermediate position between T_1 and T_2, the centers of curvature of the two mating profiles p_1 and p_2, which coincide with points T_1 and T_2, are located one on one side and the other on the other side with respect to the point P. Therefore, until the point P is in an intermediate position between the points T_1 and T_2, the transmission of motion is kinematically correct.

Suppose now that the addendum circle of the pinion has a radius larger than O_1T_2. When the point of contact P is located beyond the point T_2, the interference occurs. In fact, if we consider a point P on the line of action positioned beyond T_2, the profiles generated from it are those drawn with a dashed line and indicated with p_2' and p_1 (Fig. 2.9). We see immediately that the driving profile p_1 instead of push, should pull the driven profile p_2'. In the passage of the contact through the point T_2, the profile of the driven tooth should change and assume a curvature symmetrical

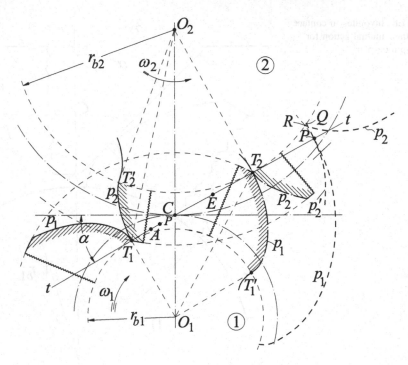

Fig. 2.9 Path of contact of an external gear pair in the limit condition to avoid interference

with respect to that of the profile p_2'. We can also see that at point P, the radius of curvature of the profile p_1 is PT_1, while the radius of curvature of the profile p_2' is PT_2. The two centers of curvature T_1 and T_2 of the two profiles p_1 and p_2' at point P are therefore located on the same side with respect to point P. Thus, the profile p_2' has a curvature in the same direction of that of the profile p_1; therefore, it is nothing other than the fictitious branch of the profile p_2 symmetrical with respect to the radius O_2R, where R is the origin of the two branches p_2' and p_2 of the involute curve.

However, as we can see from Fig. 2.9, the origin R of the involute p_2' on the base circle having radius O_2T_2 is inside the involute p_1. Therefore, the profile p_1 cuts the actual branch of the involute of the profile p_2 at point Q. This is the *theoretical interference* or *involute interference* or *primary interference*. Then teeth cannot properly transmit motion beyond the point T_2.

We can achieve the same conclusions in the case where the addendum circle of the driven wheel has a radius greater than O_2T_1, and the point of contact P is to the left of the point T_1. Therefore, to avoid incorrect contacts due to the theoretical interference, it is necessary that the radii of the addendum circles of the pinion and wheel do not exceed respectively O_1T_2 and O_2T_1. Thus, the maximum possible length of the path of contact is equal to T_1T_2.

When instead we analyze the contact problem between the teeth of an internal gear pair, such as that shown in Fig. 2.7, we deduce that correct contacts cannot occur at points located on the line of action to the left of point T_1, while at first sight limitations would not exist to the extension of the contacts on the line of action to the right of this point. When, however, we will examine more thoroughly the problem of interference in an internal gear pair (see Chap. 5), we will see that, for other reasons, also in this case there are limitations on the length of the path of contact.

In the limit condition to avoid interference shown in Fig. 2.9, we note that, during the rotation of the pinion (the driving wheel 1), while the point of contact between the teeth moves from T_1 to T_2, the point T_1 of the wheel 1 moves in T_1', and the point T_2' of the wheel 2 moves in T_2. The arcs $\overset{\frown}{T_1T_1'}$ and $\overset{\frown}{T_2T_2'}$, equal to each other because both equal to the maximum length of the path of contact T_1T_2, are the *arcs of action* or *transverse arcs of transmission* on the base circles of the two teeth; the angles $\overset{\frown}{T_1O_1T_1'}$ and $\overset{\frown}{T_2O_2T_2'}$ corresponding to these arcs are instead different from each other, as the radii r_{b1} and r_{b2} of the base circles are generally different. Therefore, the arc of action is the arc through which one tooth moves along the base circles (and then along the path of contact), from the beginning until the end of contact with its mating tooth. It is to be noted that those shown in Fig. 2.9 are the maximum possible lengths of arcs of action on the base circles.

If each gear member was fitted with a single tooth, the pinion 1 could transmit the motion to wheel 2 for a maximum angle of rotation equal to the angle $\overset{\frown}{T_1O_1T_1'}$ (it is the *angle of action* or *transverse angle of transmission*), after which, lacking the contact between the same pair of teeth profiles, there would be no more transmission of motion. If the teeth, as always happens, have depths smaller than those shown in Fig. 2.9, being truncated, for example, for the pinion by the addendum circle passing through the point E, and for the wheel by the addendum circle passing through point A, their transverse arcs of transmission on the base circles will have only the length AE, and the transmission of motion can be only carried out for a corresponding rotation to an arc of such length.

It is noteworthy that all the aforementioned arcs and angles above on the base circles have their counterparts on the corresponding arcs and angles on the pitch circles. Therefore, the transverse arcs (angles) of transmission on the pitch circles are the arcs (angles) of rotation of the pitch circles from the beginning until the end of meshing for the same pair of profiles.

Now consider an external gear pair such as that shown in Fig. 2.10, characterized by the addendum circles of the pinion and wheel respectively having radii equal to $r_{a1} = d_{a1}/2 = O_1A$ and $r_{a2} = d_{a2}/2 = O_2E$. The length of the arc of action on the base circles, equal to the length of path of contact $g = g_\alpha = AE$, is uniquely determined. Points A and E indicate the points of contact on the line of action at the beginning and at the end of meshing for the same pair of teeth profiles respectively. We define *meshing cycle* or *cycle of engagement* the total angular displacement that exists between the instant a pair of teeth come into contact until they become separated.

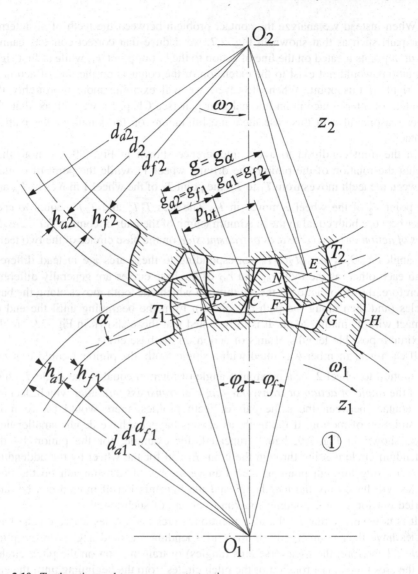

Fig. 2.10 Teeth action and transverse contact ratio

The initial contact will take place when the dedendum flank of the driving tooth comes into contact with the tip of the driven tooth. This occurs at point A where the addendum circle of the driven gear crosses the line of action. If we now depict tooth profiles through point A and, in this position, we draw lines from the intersection of the tooth profiles with the pitch circles to the centers O_1 and O_2, we obtain the *angles of approach* for each gear. From this initial point of contact onwards there is a mixed contact between the meshing profiles, that is a contact composed of a rolling contact combined with a sliding contact. Furthermore, while developing the meshing, the point of contact, which moves on the line of action from A to E, will move along the same profiles, from the root to the tip on the driving profile, and from the tip to the root on the driven profile. We will have pure rolling without sliding only in the instant in which the point of contact passes through the pitch point C. From this instant onwards, the point of contact continues to move over the tooth profiles in the same direction until contact finally ceases when the same point of contact, moving on the line of action, reaches point E. This final point of contact is located where the addendum circle of the driving gear crosses the line of action.

Following a similar procedure as before, that is depicting tooth profiles through point E and drawing lines from the intersection of the tooth profiles with the pitch circles to the centers O_1 and O_2, we obtain the *angles of recess* for each gear. In order not to complicate Fig. 2.10, we have not traced the profiles at the beginning and at the end of the contact, for which the angles of approach and recess therein shown are approximate. In any case the sum of these two angles is called the *angle of action*. These three angles have their corresponding arcs, i.e. *arc of approach*, *arc of recess* and *arc of action* on the pitch circles. Note that the arcs of approach and recess are normally measured on the pitch circle of the driving gear and on the pitch circle of the driven gear, respectively. To be more precise, the arcs of approach is the arc through which any tooth moves along the path of contact, from the beginning of contact with its mating tooth until the contact arrives at the pitch point, while the recess arc is the arc through which the same tooth moves along the path of contact, from time in which the contact coincides with the pitch point until the contact with its mating tooth ceases.

In order to ensure the continuity of motion, the two members of a gear must be characterized by teeth that, on the base circles, follow one another at a distance measured on the same base circles smaller than the length of path of contact AE (Fig. 2.10). In this way, before the end of the contact between one pair of teeth, the next pair is already in contact, and therefore the continuity of motion is guaranteed. Obviously the arcs of action on the base circles (not shown in the figure) have lengths equal to that of the path of contact, while, on the same circles, the distance measured between corresponding flanks (or homologues flanks) of two successive teeth is equal to the length of the segment PN, which is the *transverse base pitch* or simply *base pitch*, $p_{bt} = p_b$, of the toothing. Therefore, we infer that two toothed gear wheels meshing properly between them have teeth of equal base pitch, p_b. Thus, if z_1 and z_2 are the numbers of teeth of the pinion and wheel, we will have:

$$z_1 = \frac{2\pi r_{b1}}{p_b} \qquad z_2 = \frac{2\pi r_{b2}}{p_b}. \tag{2.37}$$

The *circular pitch* (or *transverse pitch* or simply *pitch*), p, is the distance measured on the pitch circle between corresponding profiles of two adjacent teeth, and then it is the sum of the tooth thickness and space width, both measured on the pitch circle. Therefore, the circular pitch is the quotient of the length of circumference of the pitch circle divided by the number of teeth, i.e.:

$$p = \frac{2\pi r_1}{z_1} = \frac{2\pi r_2}{z_2}. \tag{2.38}$$

We define as *angular pitch*, τ, the quotient of the round angle (expressed in degree or radians) divided by the number of teeth, that is:

$$\tau = \frac{360°}{z} = \frac{2\pi}{z}. \tag{2.39}$$

The *transverse module* (or simply *module*), m, is defined as the ratio between the circular pitch in millimeters and the number π, or in equivalent terms as the ratio between the pitch diameter in millimeters and the number of teeth (in other words, number of millimeters of pitch diameter per teeth), i.e.

$$m = \frac{p}{\pi} = \frac{d_1}{z_1} = \frac{d_2}{z_2} = \frac{d}{z}. \tag{2.40}$$

It is noteworthy that, in English units, the *diametral pitch P* (or *transverse diametral pitch*) is usually used; it is defined as the number of teeth per inch of pitch diameter, i.e.:

$$P = \frac{z_1}{d_1} = \frac{z_2}{d_2} = \frac{z}{d}. \tag{2.41}$$

Therefore, since the diametral pitch is the ratio between the number of teeth and pitch diameter in inches, or in equivalent terms the ratio between the number π and the circular pitch in inches, the two indices of gear-tooth size are related by the relationships:

$$pP = \pi \tag{2.42}$$

with p in inch, and P in teeth per inch, and

$$m = \frac{25.4}{P}. \tag{2.43}$$

To avoid misunderstanding, it is good to keep in mind that, with SI units, the word *pitch* without a qualifying adjective, means *circular pitch* or *standard circular pitch*, whereas with Imperial units, *pitch* means *diametral pitch*. Gears are usually made to a standard value of module for SI units or an integral value of diametral pitch for Imperial units. With SI units considered in this textbook, the standard values of modules for cylindrical gears to be used for general engineering and for heavy engineering are the ones shown in [18]; they are comprised between 1 and 50 mm. For other engineering applications (for examples, watches and similar), other ISO standards provide modules less than 1 mm.

We have already seen above that, in the design of a gear pair, it is necessary to ensure the continuity of motion. To do this, the proportions of the involute profiles of the mating teeth must be chosen so that a new pair of mating teeth comes into contact before the previous pair is out of contact. In this regard, the most significant parameter is the *transverse contact ratio*, ε_α (or simply the *contact ratio*, ε, for cylindrical spur gear). This parameter is defined as the quotient of the transverse angle of transmission (or angle of action, i.e. the angle through which the gear rotates from the beginning to the ending of contact on the transverse profile) divided by the angular pitch. If the angle of action be less than the angular pitch, there would be a time interval during which no tooth pair would be in contact.

We can express the transverse contact ratio, ε_α, in various way, of course using homogenous quantities. With reference to Figs. 2.10 and 1.8, we can write:

$$\varepsilon_\alpha = \frac{AE}{p_{bt}} = \frac{g_\alpha}{p_{bt}} = \frac{A^*E^*}{AE} = \frac{q}{p}. \tag{2.44}$$

In order to ensure the continuity of the motion, it is obviously necessary that the tooth profiles be sized so that a second pair of mating teeth come into contact before the first pair is out of contact. In other words, the contact ratio must be greater than unity. However, to have good working condition of the gear pair, it is desirable that this contact ratio is at least equal to $(1.4 \div 1.7)$. It is to remember that the contact ratio represents the average number of teeth in contact as the gears rotate together. A contact ratio of 1 or 2 means that one pair or two pairs of teeth are engaged at all times during meshing, while a contact ratio of 1.5 indicates that two pairs of teeth are engaged for 50% of the time, and only one pair is engaged for the remaining 50% of the time. AGMA (American Gear Manufactures Association) Standards state that for satisfactory performance, the contact ratio should never be less than 1.2. However, in the gear design, under no circumstance the contact ratio should drop below 1.1, when it is calculated with all tolerances at their worst values.

From Eqs. (2.29, 2.37 and 2.38) we can deduce the following relationship that relates the circular pitch, p, base pitch, p_b, and the pressure angle, α:

$$p_b = p \cos \alpha. \tag{2.45}$$

From the previous relationship we infer that between the module on the pitch circle, m, and the base module (i.e. the module on the base circle), m_b, we have a relationship similar to Eq. (2.45), that is:

$$m_b = m \cos \alpha. \tag{2.46}$$

If the center distance and the radii of the base and addendum circles are given, and AE and p_b are respectively the arc of action of the base circle and the base pitch, the arc of action on the pitch circle, q, and circular pitch, p, will be given by the following relationships:

$$q = AE\frac{r_1}{r_{b1}} = AE\frac{r_2}{r_{b2}} \tag{2.47}$$

$$p = p_b\frac{r_1}{r_{b1}} = p_b\frac{r_2}{r_{b2}}. \tag{2.48}$$

Therefore, we will have $\varepsilon_\alpha = q/p$ [see last equality of Eq. (2.44)].

In order to be able to transmit the motion in both direction of rotation of the gear members, the two flanks of each tooth must have involute profiles symmetrical with respect to the axis of the tooth itself, as shown in Fig. 2.10. Between two adjacent teeth of any of the two gear members there must be sufficient space width to the meshing of the mating member. Since the two gear wheels meshing with each other must have not only the same base pitch, but also the same circular pitch, their number of teeth will be given by relationships similar to Eqs. (2.37), i.e.:

$$z_1 = \frac{2\pi r_1}{p} \quad z_2 = \frac{2\pi r_2}{p}. \tag{2.49}$$

The *gear ratio*, u, is the quotient of the number of teeth of the wheel, z_2, divided by the number of teeth of the pinion, z_1. For a single gear pair it coincides with the transmission ratio, i, defined as the quotient of the angular velocity of the driving gear, ω_1, divided by the angular velocity of the driven wheel, ω_2; therefore, for a single gear pair, we have:

$$u = \frac{z_2}{z_1} = i = \frac{\omega_1}{\omega_2} = \frac{n_1}{n_2} = \frac{r_{b2}}{r_{b1}} = \frac{r_2}{r_1}. \tag{2.50}$$

It should be noted that the aforesaid coincidence between the gear ratio, u, and the transmission ratio, i, no longer exists for a train of gears, since in this case the gear ratio applies to the individual gear pairs that compose it, while the transmission ratio regards the whole gear train. In fact, the transmission ratio of the latter is defined as the quotient of the angular velocity ω_i of the first driving wheel, which is the power input wheel, divided by the angular velocity ω_0 of the last driven wheel, which is the power output wheel; it is given by Eq. (1.1).

From Eqs. (2.37) we infer that, for two gear wheels meshing with each other, the condition that they have the same base pitch may be replaced by the one for which the radii (or diameters) of the base circles are proportional to the numbers of teeth. For these meshing wheels it is preferable that the usable tooth depth to be divided into two equal portions, inside and outside the pitch circle. For a given tooth depth, this condition corresponds to the minimum relative sliding between the mating profiles (this topic will be discussed later). Then we need to introduce the *reference circles* in relation to which we define the main quantities concerning the toothing and the gear blank, which include the thicknesses of the teeth, space widths, and pressure angle. It is however necessary to keep in mind that the reference circle (the one shown on the design drawings) is often different from the pitch circle (the one correlated to the actual operating conditions).

Now we denote with the subscript, 0, the quantities referred to this reference circle, named *nominal quantities*. The *nominal tooth thickness*, s_0, of the tooth and the *nominal space width*, e_0, between two adjacent teeth are theoretically equal between them and equal to half the *nominal pitch*, p_0, that is:

$$s_0 = e_0 = \frac{p_0}{2} = \frac{\pi m_0}{2}, \qquad (2.51)$$

where m_0 is the *nominal module*. The *nominal diameter* of this reference circle will be $d_0 = 2r_0 = m_0 z$ (r_0 is the nominal radius), while the *nominal pressure angle* on the same reference circle is α_0. Once these nominal quantities are known, also the diameter d_b of the base circle and the base module m_b can be determined by means of the relationships:

$$d_b = 2r_b = d_0 \cos \alpha_0 \qquad (2.52)$$

$$m_b = m_0 \cos \alpha. \qquad (2.53)$$

These quantities related to the base circle allow to calculate the pressure angle and the module on any circle of the gear wheel. Therefore, the pressure angle and the module vary with the radius of the circle under consideration. It is also worth noting that many of the aforementioned nominal quantities come into play in the generation processes of the teeth (see Woodbury [32], Galassini [11], Rossi [28], Henriot [15], Micheletti [23], Boothroyd and Knight [2], and Björke [1]).

If instead we consider a rack (it is a wheel having infinite radius), a shift of the pitch line parallel to itself corresponds to a change of the radius. However, since in this case the involute is a straight-line (the teeth have trapezoidal shape), the shift of the *pitch line* parallel to itself does not cause any change of the pitch, module and pressure angle, which are therefore invariable quantities of the rack. This shift instead determines a variation of the thickness of the tooth and space width, for which it is necessary to give their nominal values with reference to the *nominal* or *reference pitch line*, the position of which is uniquely determined by its distance from the *dedendum line* (or *root line*) or from the *addendum line* (or *tip line*).

Since in the rack the quantities p, m and α are invariable irrespective of the pitch line choice on it, a toothed gear having z_1 teeth which must mesh with this rack will be tangent to the said pitch line (Fig. 2.11). Therefore, this toothed gear wheel will have a pitch circle and a base circle whose radii are respectively given by:

$$r_1 = m \frac{z_1}{2} \qquad (2.54)$$

$$r_b = r_1 \cos \alpha = \left(m \frac{z_1}{2} \right) \cos \alpha. \qquad (2.55)$$

Module and pressure angle become then the module and the pressure angle of the toothed gear wheel on the pitch circle having radius r_1.

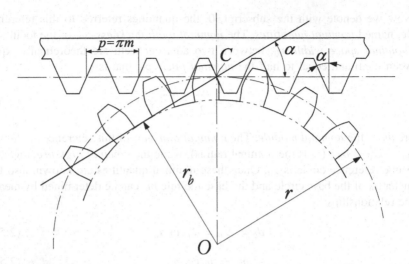

Fig. 2.11 Rack-pinion gear pair

2.5 Gear-Tooth Sizing

The teeth thickness of an involute toothing is variable as function of the radius of the circle under consideration. Conventionally, the *tooth thickness* is measured by the length of the arc *GH* (Fig. 2.10) of the reference circle between the two opposite profiles (left and right) of the same tooth. The space width is also measured by the length of the arc *FG* of the same reference circle between the right profile of a tooth and the left profile of the next tooth (Fig. 2.10). The thickness so defined should not be confused with the *chordal tooth thickness*, which is the one detected by the measuring instruments, and is defined as the minimum distance between two corresponding points (i.e. on the same circle) of the two opposite flanks of the same tooth (see Colbourne [6], Henriot [14], Niemann and Winter [25]).

In the previous section we said that the nominal thickness of the tooth, s_0, and nominal space width between two adjacent teeth, e_0, are equal between them, and equal to half the nominal pitch, p_0 (2.51). In Sect. 1.5 we also said that, in the actual cases, the tooth thickness will be less than half the pitch, while the space width will be greater than half the pitch, and we have given the reasons due to the need to have a circumferential backlash.

With reference to the modification of addendum, understood as profile shifting, which we will analyze in detail in Chaps. 6 and 7, it is necessary to be able to calculate the tooth thickness s' on a given circle, having radius r', when the thickness s of the same tooth on another circle having radius r is known. The two circles having radii r and r' intersect the profile of the tooth left flank at the points P and P' (Fig. 2.12). If we call φ and φ', and α and α' the angles φ and α shown in

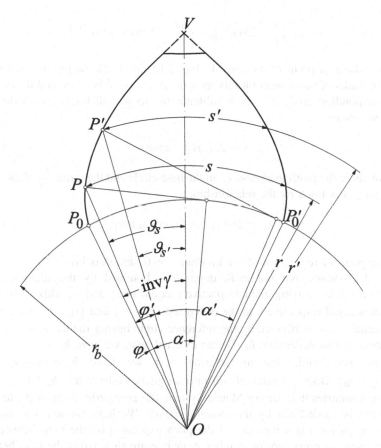

Fig. 2.12 Tooth thicknesses on different circles

Fig. 2.1, which correspond to points P and P' (φ and φ' are the involute polar angles related to point P and P', while $\alpha = (\vartheta - \varphi)$ and $\alpha' = (\vartheta' - \varphi')$, where ϑ and ϑ' are the corresponding involute roll angles), we have:

$$\widehat{POP'} = \varphi' - \varphi = \mathrm{inv}\alpha' - \mathrm{inv}\alpha. \tag{2.56}$$

Since then:

$$OP = r = \frac{r_b}{\cos\alpha} \qquad OP' = r' = \frac{r_b}{\cos\alpha'}, \tag{2.57}$$

where r_b is the radius of the base circle of the gear, we infer that thicknesses s and s' on the circles having radii r and r' are related by the relationship:

$$s' = r'\left(\frac{s}{r} - 2\widehat{POP'}\right) = r'\left[\frac{s}{r} + 2(\text{inv}\alpha - \text{inv}\alpha')\right]. \tag{2.58}$$

If we take as a point P' the apex V (Fig. 2.12) in which the profiles of the two opposite flanks of tooth meet (at this apex the thickness s' is zero), and denote invγ the corresponding angle φ' which subtends the tooth half-thickness on the base circle, we obtain:

$$s = 2r(\text{inv}\gamma - \text{inv}\alpha). \tag{2.59}$$

If we know the tooth thickness s_b on the base circle and the radius r_b of the latter, the angle γ is defined by the relationship:

$$s_b = 2r_b \text{inv}\gamma = 2r_0 \cos\alpha_0 \text{inv}\gamma \tag{2.60}$$

as at the point of regression of the involute $\alpha = 0$, and thus inv$\alpha = 0$.

Fig. 2.13 shows that the tooth depth h is bounded by the addendum and dedendum circles, having radii respectively equal to r_a and r_f. This tooth depth h which is equal respectively to the difference $(r_a - r_f)$ and $(r_f - r_a)$ for external and internal gears, is divided by the reference circle having radius r_0 in two parts: addendum h_a and dedendum h_f. For an external gear, we have $h_a = (r_a - r_0)$ and $h_f = (r_0 - r_f)$, while for an internal gear we have $h_a = (r_0 - r_a)$ and $h_f = (r_f - r_0)$. Both for external and internal gears we have $h = h_a + h_f$.

In the Countries that use the Metric System, the *geometric sizing* of the tooth of spur gears is carried out by the *modular sizing*. We have already said that the module m (in mm) is a measure of the pitch expressed by the ratio between the circular pitch (in mm) and the number π, or in equivalent terms the ratio between the pitch diameter (in mm) and the number of teeth [see Eq. (2.40)]. Nominal standard modules are those standardized by [18]. The nominal sizing most widely used provides a standard addendum equal to the nominal module and a standard dedendum equal to 1.25 times the nominal module, that is: (Fig. 2.13)

$$\begin{aligned} h_a &= m_0 \\ h_f &= 1.25\, m_0 = (5/4)m_0 \\ h &= h_a + h_f = 2.25\, m_0 = (9/4)m_0. \end{aligned} \tag{2.61}$$

The fillet radius ρ at the root of the tooth is included in the range $(0.2m_0 < \rho < 0.4m_0)$, and it is commonly assumed equal to $(m_0/3)$. It is noteworthy that the dedendum can be even greater than the standard value when we want a larger radius of curvature between the flank of the tooth and its root. For external gears, the nominal pitch diameter, and diameters of the addendum and dedendum circles are given respectively by:

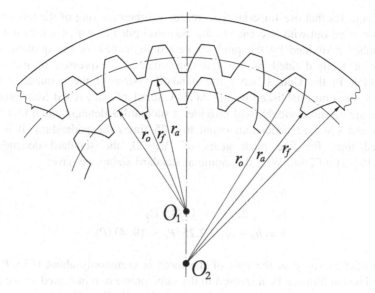

Fig. 2.13 Reference, addendum and dedendum circles

$$d_0 = 2r_0 = m_0 z$$
$$d_a = d_0 + 2m_0 = m_0(z+2)$$
$$d_f = d_0 - 2.5\,m_0 = m_0\left(z - \frac{10}{4}\right),$$

(2.62)

while for internal gears the same quantities are given by:

$$d_0 = 2r_0 = m_0 z$$
$$d_a = d_0 - 2m_0 = m_0(z-2)$$
$$d_f = d_0 + 2.5m_0 = m_0\left(z + \frac{10}{4}\right).$$

(2.63)

The nominal center distance, a_0, of an external gear pair with z_1 and z_2 teeth is given by:

$$a_0 = \frac{z_1 + z_2}{2}m_0,$$

(2.64)

while the same quantity for an internal gear pair with z_1 and z_2 teeth is given by:

$$a_0 = \frac{z_2 - z_1}{2}m_0.$$

(2.65)

In Countries that use Imperial System, the geometric sizing of the teeth of spur gears is carried out with reference to the diametral pitch P, which is the quotient of the number π divided by the pitch expressed in inches, or the quotient of the number of teeth divided by the reference diameter, expressed in inches [see Eq. (2.41)]. In the United States the standard system of the nominal diametral pitches is the one established by the AGMA Standards and ANSI Standards. The nominal sizing most widely used provides a standard addendum equal to $1/P_0$ (in inches) and standard dedendum equal to 1.25 times the addendum. It is to be observed that, for fine pitch gears of $P_0 \geq 20$, the standard dedendum is $[(1.20/P_0) + 0.002 \text{ in}]$. With this nominal standard sizing we have:

$$
\begin{aligned}
h_a &= 1/P_0 \\
h_f &= 1.25/P_0 = (5/4)/P_0 \\
h &= h_a + h_f = 2.25/P_0 = (9/4)/P_0.
\end{aligned}
\tag{2.66}
$$

The fillet radius ρ at the root of the tooth is commonly about $0.35/P_0$. The standard dedendum can be increased in the same proportion indicated above for the modular sizing. For external gears the nominal pitch diameter, and diameters of the addendum and dedendum circle are respectively given by:

$$
\begin{aligned}
d_0 &= 2r_0 = z/P_0 \\
d_a &= d_0 + 2/P_0 = (z+2)/P_0 \\
d_f &= d_0 - 2.5/P_0 = \left(z - \frac{10}{4} \right)/P_0,
\end{aligned}
\tag{2.67}
$$

while for internal gears the same quantities are given by:

$$
\begin{aligned}
d_0 &= 2r_0 = z/P_0 \\
d_a &= d_0 - 2/P_0 = (z-2)/P_0 \\
d_f &= d_0 + 2.5/P_0 = \left(z + \frac{10}{4} \right)/P_0.
\end{aligned}
\tag{2.68}
$$

The nominal center distance of an external gear pair with z_1 and z_2 teeth is given by:

$$
a_0 = \frac{z_1 + z_2}{2P_0},
\tag{2.69}
$$

while the same quantity for an internal gear pair with z_1 and z_2 teeth is given by:

$$
a_0 = \frac{z_2 - z_1}{2P_0}.
\tag{2.70}
$$

From the comparison of Eqs. (2.61–2.65), valid for SI units, with Eqs. (2.66–2.70), valid for Imperial System units, we see immediately that the two sizing systems are perfectly equivalent. If this is so, we can ask what the reason for which one system of sizing is not internationally used. Unfortunately, the large number of existing cutters and designs, based on the two systems using SI and Imperial System, makes it unlikely that any one of them will become the sole standard.

In addition to the two above-mentioned nominal sizings, special sizings are sometimes used. The reason why these special sizings are used can be many, such as for example:

- the necessity to have pinions with few teeth avoiding the cutting interference and, especially, the interference in the working condition between the gears that mesh with each other;
- the reduction of the relative sliding between the teeth;
- the need to reduce the tooth depth in gears very stressed under bending loads; etc.

Therefore, gears are realized with teeth of different depth from that given by the Eqs. (2.61 and 2.66), or with equal depths to those given by these equations, but divided in different proportions between the addendum and dedendum, and with different division of the pitch between the tooth thicknesses and space widths.

An example of special sizing is that represented by the *stub system*, in which the stub teeth have depth smaller than those given by Eqs. (2.61 and 2.66). Both for SI and Imperial System, a standard stub system was the one with pressure angle of 20°, and addendum and dedendum shortened according to the relationships:

$$h_a = 0.8m_0 = 0.8/P_0$$
$$h_f = (1.0 \div 1.1)m_0 = (1.0 \div 1.1)/P_0.$$
(2.71)

Stub teeth have a higher strength than teeth having nominal depth, but the contact ratio is lower. Gears with stub teeth have been produced in the past, and still continue to be produced, having tooth depth different from the one given by Eq. (2.71), and pressure angle different from 20°.

Another example of special sizing is represented by gears with high tooth depth, i.e. with tooth depth deeper than that of a tooth having nominal depth. These gears have the advantage of a higher contact ratio, but the strength of the teeth under bending loads is considerably lower. When, however, the high tooth depth is obtained also choosing an addendum greater than that of the standard basic rack, the risk can appear of exceed the interference limiting points T_1 and T_2 (Fig. 1.8). In fact, for zero-shifted tooth profiles, any tooth addendum that extends beyond these interference points is not only without usefulness, but also interferes with the root fillet area of the mating tooth. This results in a typical undercut tooth, that determines not only a weakening of the tooth to its root, but also the removal of a remarkable part of useful involute profile adjacent to the base circle.

The most commonly used pressure angle, α, with both SI and Imperial System, is $20°$. In the United States, $\alpha = 25°$ is also a standard value. However, the following pressure angles are still commonly used: $\alpha = 14°1/2; \alpha = 15°$; $\alpha = 22°1/2$. In certain cases, for internal gear pairs, a pressure angle of $30°$ is used. The pressure angle of $20°$ commonly used allows to have gear pairs able to solve most of the design problems concerning the gear transmissions.

The actual pressure angles that we have in the working conditions may differ significantly from the nominal pressure angles, as we shall see when we analyze the topic about the addendum modification in terms of profile shift. It is however necessary to keep in mind that, in the case of large values of the pressure angle, it may have drawbacks when the shafts on which gears are mounted are too flexible.

Finally, it must be remembered that even the face width b, i.e. the width of the toothed part of a gear, measured along a generatrix of the reference cylinder, is given as a function of the module or diametral pitch. This face width is not standardized, but it is generally included within the following ranges:

$$9m_0 < b < 14\ m_0$$
$$9/P_0 < b < 14/P_0. \tag{2.72}$$

The wider the face width, the more difficult is to manufacture and mount the gears so that contact is uniformly distributed across the entire face width.

As we already said, the involute tooth profile, analytically defined by Eq. (2.2), cannot extend inside the base circle. On the other hand, if we choose the gear-tooth sizing defined by the relationships (2.61), or the equivalent relationships (2.46) when we use the gear-tooth sizing based on the diametral pitch, we see immediately that only for very high number of teeth the profile is all outside the base circle. Indeed, in the extreme case in which the profile from tip to root is all active, the nominal quantities involved must satisfy the following inequality:

$$(r_0 - r_b) = r_0(1 - \cos \alpha_0) \geq h_f = 1.25m_0 \tag{2.73}$$

From this inequality, taking into account that $m_0 = d_0/z = 2r_0/z$, we obtain:

$$z \geq \frac{5}{2(1 - \cos \alpha_0)}. \tag{2.74}$$

Tooth profile will then be all outside the base circle only if: for $\alpha_0 = 15°, z \geq 74$; for $\alpha_0 = 18°, z \geq 52$; for $\alpha_0 = 20°, z \geq 42$; for $\alpha_0 = 22°, z \geq 35$; for $\alpha_0 = 24°, z \geq 29$; for $\alpha_0 = 26°, z \geq 25$; for $\alpha_0 = 28°, z \geq 22$; for $\alpha_0 = 30°, z \geq 19$. In cases where the correspondences are not fulfilled, for which z is lower than the value given by the ratio to the right side of the inequality (2.74), it is necessary to extend the profile of the tooth inside the base circle. The portion of the tooth profile inside the base circle can have any geometric shape provided that interference will not occur; this shape will be the one related to the profile of the tool head used in the cutting process of the gear.

It is then preferable that the portion of the involute profile in the vicinity of the base circle is not used. This is because the involute profile immediately close to the base circle has a radius of curvature which decrease to zero in correspondence of the same base circle. The involute profile in this region is therefore very difficult to obtain accurately using current manufacturing processes. Furthermore, in this area, the relative sliding, wear of the tooth and local compression contact stress tend to grow strongly.

2.6 Standard Basic Rack Tooth Profile

Large-quantity production of gears usually involves separate machine and cutting tools to fabricate both members of a gear pair. One member (the pinion) may be manufactured using one particular machine tool, with its equally particular cutting tool, while the other member (the wheel) is fabricated using another particular machine tool, with its equally particular cutting tool. Two distinct cutting tools, and then two distinct tool profiles are used to manufacture the two members of the gear pair. This mode of manufacture that uses two distinct profiles to manufacture a gear pair is known as a *complementary rack* (see Dooner [8], Dooner and Seireg [9]).

Obviously, however, it is preferable, as well economically convenient, to conceive and use only one cutting tool (and hence one machine tool) capable of generating both the members of a gear pair. Then the profiles of the two cutting tools must have the same pressure angle or be self-complementary. When a single rack is capable of manufacturing both the members of a gear pair, we have the technology known as *basic rack*. This basic rack is the rack that forms a part of an intermating series. The counterpart of the basic rack serves as the basis of the shape of a *generating cutting tool* (hereinafter also called simply *cutter*) which will produce an intermating series of gears (see Dooner [8], Dooner and Seireg [9], Henriot [15]).

The *standard basic rack tool profile* defines the characteristics common to all cylindrical gears with tooth profiles having involute geometry. It is the profile of the section of the rack used to define the size of the standardized toothings of a system of involute gear wheels. Each gear wheel of the standard system (both external and internal toothed gear wheels) may be considered geometrically generated by the *standard basic rack*, which is characterized by a straight-line profile. This basic rack is a fictitious rack which has, in the section normal to the flanks, the standard basic rack tooth profile, which corresponds to an external gear wheel with number of teeth $z = \infty$ and diameter $d = \infty$.

Figure 2.14 shows the standard basic rack tooth profile according to the [17]. It refers to a theoretical toothing without backlash. The tooth depth h_P (subscript P indicates quantities that relate to the basic rack) is equal to $2.25m$, but the working portion h_{wP} of this profile, equally divided above and below the *datum line P-P* and corresponding to the involute of a circle having an infinite radius, is deep $2m$ (it should be noted that, from here on, we will use the symbol m without any subscript to indicate the module; only when misunderstandings are possible, we will add the

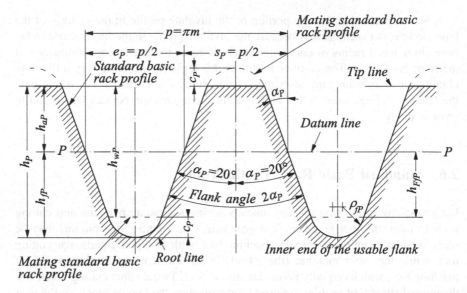

Fig. 2.14 Standard basic rack tooth profile and mating standard basic rack profile according to ISO 53

subscript, specifying its meaning). Addendum h_{aP} and dedendum h_{fP} of this profile are equal to m and $1.25m$ respectively, but the straight portion of the standard basic rack tooth dedendum (i.e. the portion of dedendum really usable) has depth equal to $h_{FfP} = m$. The tooth of the standard basic rack is bounded by the tip line at the top and by the root line at the bottom. Of course, datum line, tip line and root line are parallel lines. The working straight portion of the basic rack tooth profile is interconnected with the root line by the fillet, which has the shape of a circular arc with a radius equal to ρ_{fP}.

The working portions, that is the involute portions of the standard basic rack tooth profiles, are the straight lines passing through the equally-spaced points located symmetrically with respect to the axis of symmetry of the tooth, which bisects the thickness of the tooth. These straight lines are inclined at the pressure angle $\alpha_P = 20°$ with respect to the axis of symmetry of the tooth. On the datum line, the tooth thickness s_P is equal to the space width e_P, and both are equal to one-half the pitch, i.e.: $s_P = e_P = p/2 = \pi m/2$.

The *mating standard rack tooth profile* is the rack tooth profile symmetrical to the standard basic rack tooth profile with respect to the datum line P-P, and displaced by half a pitch relative to it. Figure 2.14 also shows this mating standard rack tooth profile drawn with a dashed line. Furthermore, Fig. 2.14 shows the following main characteristics of the standard basic rack tooth profile having module, m, and pitch $p = \pi m$:

- the flanks are straight for the portion $h_{wP} = (h_{aP} + h_{FfP})$, and are inclined at the pressure angle $\alpha_P = 20°$ to a line normal to the datum line P-P;

- the tip and the root lines are parallel to the datum line and respectively at distances of $h_{aP} = m$ and $h_{fP} = 1.25 \, m$ from it;
- all dimensions of the standard basic rack tooth profile and the mating standard basic rack tooth profile use line P-P as the base datum, and the active tooth depth, h_{wP}, of both profiles is equal to $2 \, h_{aP}$;
- with certain exceptions (see below), the standard value of the *bottom clearance*, c_P, between the standard basis rack tooth and mating standard basic rack tooth is equal to $0.25m$, that is $c_P = \left(h_{fP} - h_{FfP}\right) = 0.25m$ (this clearance is measured along a line normal to the datum line);
- the *fillet radius*, ρ_{FP}, of the standard basic rack is determined by the standard clearance, c_P.

It is obvious that the higher the bottom clearance (and then the greater the ratio c_P/m), the larger can be the *fillet radius of the basic rack*, with corresponding improvement of the bending strength of the tooth. Figure 2.14 shows that the center of the fillet radius is not on the center of the rack space. The maximum fillet radius of the basic rack, $\rho_{fP\,max}$, which instead is centered on the rack space, can be calculated with the following relationships:

$$\rho_{fP\,max} = \frac{c_P}{1 - \sin \alpha_P} \tag{2.75}$$

$$\rho_{fP\,max} = \frac{\left[(\pi m)/4 - h_{fP} \tan \alpha_P\right]}{\tan\left[(90° - \alpha_P)/2\right]}, \tag{2.76}$$

which are valid for a basic rack where $\alpha_P = 20°$, $c_P \leq 0.295m$ and $h_{FfP} = 1 \, m$ (h_{FfP} is the straight portion of the standard basic rack tooth dedendum), and respectively for a basic rack where $\alpha_P = 20°$ and $0.295m < c_P \leq 0.396 \, m$.

It is to be noted that the actual root fillet, which is outside the active profile, can vary depending on various influences such as the profile shift, number of teeth, and manufacturing method. Four types of basic rack tooth profiles are provided, respectively called A, B, C and D. Table 2.1 summarizes the main characteristics of these four types of profiles, to be used as an alternative depending on the practical application requirements. As a guideline, the standard basic rack tooth profile type A is recommended for gears transmitting high torques, while types B and C are

Table 2.1 Basic rack tooth profiles according to ISO 53: 1998 (E)

Symbol	Types of basic rack tooth profiles			
	A	B	C	D
α_P	20°	20°	20°	20°
h_{aP}	1 m	1 m	1 m	1 m
c_P	0.25 m	0.25 m	0.25 m	0.40 m
h_{fP}	1.25 m	1.25 m	1.25 m	1.40 m
ρ_{fP}	0.38 m	0.30 m	0.25 m	0.39 m

recommended for normal service (type C is also applied for manufacturing with some standard hobs). Profile type D is equivalent to a full radius shape for the fillet; the high value of dedendum, $h_{fP} = 1.40\,m$, with the associated fillet radius, $\rho_{fP} = 0.39\,m$, and the bottom clearance, $c_P = 0.40\,m$, permit the finishing tool to work without interference, while maintaining the maximum fillet radius. It is recommended for high precision gears with tooth flanks finished by shaving and grinding, and transmitting high torques. During finishing of these gears care should be taken to avoid notches of the fillet, which would generate stress concentrations.

The standard basic rack tooth profiles described above are without undercut. For gears cut by a *protuberance tool* and finished by shaving or grinding, a basic rack tooth profile with a given *undercut*, U_{FP}, and a given *angle of undercut*, α_{FP}, is used. Figure 2.15 shows the basic rack tooth profile with a given undercut. The specific values of the undercut, U_{FP}, and the angle of undercut, α_{FP}, depend on various influences, the main of which is the method of manufacturing.

Figure 2.16a and b show the normal sections of addendums of two cutting tools, the first with protuberance, and the second without protuberance. They are used for cutting teeth with undercut and, respectively, without undercut. The same figures show the main quantities that define the geometry of these two types of cutting tools. Among the quantities we have not yet introduced, we have: q_{pr}, *protuberance of the tool*; q, *material allowance for finish machining*; $s_{pr} = q_{pr} - q$, *residual fillet undercut*. These three quantities are given in mm. The other quantities that characterize the tooth sizing in the normal section will be defined in next chapters.

The standard profile of the rack type cutter, with which the teeth of an intermating series of gears can be cut, is the profile with which a rack, that is a toothed wheel having infinite diameter, is generated. With the exception of precision gears, in general the tip cylindrical surface of a toothed gear wheel is not manufactured by

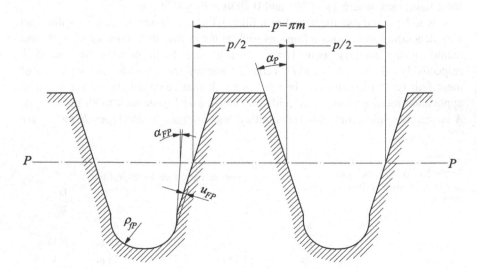

Fig. 2.15 Basic rack tooth profile with given undercut and angle of undercut

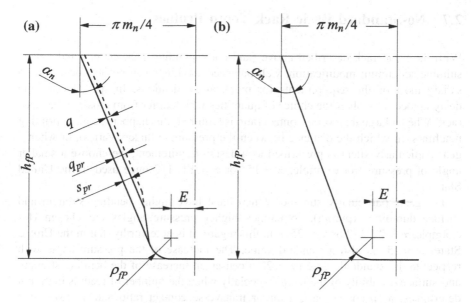

Fig. 2.16 Normal sections of a tooth of two cutting tools: **a** with protuberance, and finishing allowance for each tooth flank; **b** without protuberance

the cutter, but it is commonly manufactured by using another machine tool. In fact, the starting product in gear machining is the *gear blank*, which is obtained by means of the so-called blanking operations carried out using machine tools other than gear cutting machines.

To obtain a suitable circumferential backlash, the standard profile of the rack-type cutter must cut more deeply than the theoretical condition provides. In gashing operations instead, the cutter must cut less than the theoretical condition provides, in order to leave the suitable *finish stock*. In the latter case, a greater diameter of the root circles is obtained, but during the finishing it is necessary to operate in such a way that the root circle is not cut, and that the bottom clearance is not reduced. For this reason, the roughing cutters are characterized by an adequate machining allowance, q, so as to obtain the desired root circle with subsequent machining operations. Even the finishing cutters used to prepare super-finishing machining, such as shaving and grinding, are characterized by allowance which can be less than or greater than the allowance of the roughing cutters. For other considerations in this regard, we refer the reader to Niemann and Winter [25].

2.7   No-Standard Basic Rack Tooth Profiles

With the basic rack described above (and corresponding rack cutter), and with a suitable addendum modification, well-balanced toothings can be obtained, able to satisfy most of the requirements for practical applications. In special case, the design process involves the choice of quantities which differ from those of the basic rack. When a large transverse contact ratio is required (this happens, e.g., in printing machines, in which the distance between the pressure cylinders varies), or when a gear particularly silent is prescribed as a design requirement, we choose a smaller angle of pressure (for example, $\alpha = 15°$ or $\alpha = 14° \ 1/2$ once used in the United States).

For gears particularly stressed, which must have higher bending strength and surface durability (pitting), sometimes higher pressure angles are chosen (for example, $\alpha = 22°1/2$ or $\alpha = 25°$). In this regard, it is noteworthy that in the United States $\alpha = 25°$ is also a standard value. The increase of the pressure angle with respect to the standard value ($\alpha = 20°$) causes an increase of the bending strength and surface durability of the teeth, especially when the number of teeth is high, but determines, as a counterpart, a minor transverse contact ratio and a lower fillet radius at the tooth root, and puts limitations on the addendum modification. Furthermore, with increasing pressure angle, the thickness of the tooth at the top land decreases (for $\alpha \cong 33°$ we have $s = 0$, and then a pointed tooth) and, therefore, the risk of a tooth chipping increases appreciably, particularly when the teeth have been case-hardened.

Figure 2.17 shows the great influence that the pressure angle has on the shape of the profile of the tooth; it relates to a toothed gear wheel with number of teeth $z = 20$ and a given base circle, but the consequences of the change of the pressure angle are quite general. The figure shows that, when the pressure angle increases, assuming values greater than the standard basic rack value ($\alpha_P = 20°$), the tooth thickness at the bottom land, the pitch diameter and, as we will show later, the radial component of the transmitted force and loads on the bearings increase. On the contrary, when the pressure angle decreases, assuming values lower than the standard basic rack value, the portion of the tooth dedendum affected from the undercutting grows. The four profiles A, B, C and D are related respectively to the pressure angles $\alpha = 15°, \alpha = 20°, \alpha = 25°$ and $\alpha = 30°$.

Figure 2.18, which refers to the same toothed gear wheel of the Fig. 2.17, shows instead the change, as a function of pressure angle α, of the ratio ($\rho_\alpha/\rho_{20°}$), where ρ_α and $\rho_{20°}$ are the radii of curvature on the pitch circles related respectively to the pressure angle α and to the standard basic rack pressure angle $\alpha_P = 20°$, and to the same base circle. As we will see in Vol. 2, Chap. 2 regarding the calculation of surface durability (pitting), the radii of curvature are very important, because they control the value of the Hertzian contact stress.

The actual tooth depth may be higher or lower than the tooth depth of standard basic rack $\left(h_P = h_{aP} + h_{fP} = 2.25 \ m \right)$. We can get gears particularly silent with teeth having tooth depth greater than the standard tooth depth. These are the

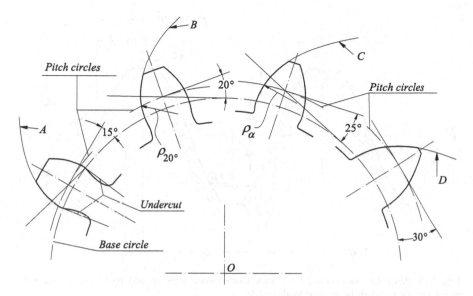

Fig. 2.17 Correlation between the pressure angle and the tooth profile shape, for a given base circle

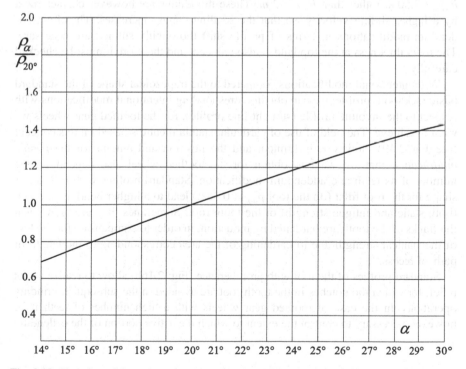

Fig. 2.18 Variation of the ratio $\rho_\alpha/\rho_{20°}$ as a function of the pressure angle α

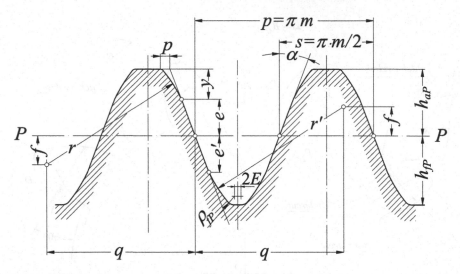

Fig. 2.19 Particular no-standard basic rack tooth profile with tip and root modification with respect to the standard basic rack tooth profile

so-called high depth teeth, that can have working depth $h_{wP} = 2.25\,m$ or even $h_{wP} = 2.50\,m$ rather than $h_{wP} = 2\,m$. These thoothings are however characterized by a higher sliding velocity, so that the scuffing danger increases. Therefore, addendum modifications in terms of profile shift thoroughly studied are necessary. The tooth thickness at the top land is also reduced, and therefore it must be checked carefully.

With intentional modifications compared to the trapezoidal shape of the standard basic rack tooth profile, we can obtain corresponding intentional modifications with respect to the involute profile (straight-line profile), for the toothed gear wheels we want to achieve. The field of use of particular standard cutters based on this concept (see Fig. 2.19), however, is limited, and the position and amount of the *profile modification* that we want to obtain for the toothed wheel also depend on the number of its teeth and addendum modification. Standard profiles with full radius shape for the root fillet (on the tool $\rho_{a0} = 0.38m$) lead to a higher bending strength (both static and fatigue strength) of the tooth root. Sometimes they are used when the flanks of the tooth are hardened by induction, in order to compensate for the less of mechanical strength due to hardening of the area corresponding to the length of path of recess.

Standard profiles of the tool with protuberance (Fig. 2.16b) allow to realize a root relief, for which the notches in the tooth root are avoided in the subsequent grinding operations. In the case of toothed gear wheels with a high number of teeth it is, however, necessary to control the extent to which the active portion of the dedendum flank is shortened by the intentional undercut during the gear rough-machining. It is also necessary to check carefully the problems that can arise for toothed gear wheels with small numbers of teeth, and small addendum modification.

References

1. Björke Ö (1995) Manufacturing systems theory. Tapir Publishers, Trondheim
2. Boothroyd G, Knight WA (1989) Fundamentals of machining and machine tools, 2nd edn. Marcell Decker, New York
3. Buckingham E (1949) Analytical mechanics of gears. McGraw-Hill Book Company Inc, New York
4. Buckingham E (1928) Spur gears. McGraw-Hill, New York
5. Cheng HH (1992) Derivation of explicit solution of the inverse involute function and its application. In: Advancing power transmission into the 21st century, ASME DTC, 1, pp. 161–168
6. Colbourne JR (1987) The geometry of involute gears. Springer-Verlag New York, Inc, New York
7. Démidovitch B, Maron I (1973) Éléments de Calcule Numérique. Éditions MIR, Moscou
8. Dooner DB (2012) Kinematic geometry of gearing, 2nd edn. Wiley & Sons Inc, New York
9. Dooner DB, Seireg AA (1995) The kinematic geometry of gearing: a concurrent engineering approach. Wiley & Sons Inc, New York
10. Ferrari C, Romiti A (1966) Meccanica Applicata alle Macchine. Unione Tipografica – Editrice Torinese (UTET), Torino
11. Galassini A (1962) Elementi di Tecnologia Meccanica, Macchine Utensili, Principi Funzionali e Costruttivi, loro Impiego e Controllo della loro Precisione, reprint 9th edn. Editore Ulrico Hoepli, Milano
12. Giovannozzi R (1965) Costruzione di Macchine, vol II. Casa Editrice Prof. Riccardo Pàtron, Bologna
13. Gray A (1997) Modern differential geometry of curves and surfaces with mathematica, 2nd edn. CRC Press, Boca Raton, FL
14. Henriot G (1979) Traité théorique and pratique des engrenages, vol 1, 6th edn. Bordas, Paris
15. Henriot G (1972) Traité théorique et pratique des engrenages. Fabrication, contrôle, lubrification, traitement thermique, vol 2. Dunod, Paris
16. ISO 1122-1: 1998 (1998) Vocabulary of gear terms- part 1: definitions related to geometry
17. ISO 53: 1998 (E) (1998) Cylindrical gears for general and for heavy engineering – Standard basic rack tooth profile
18. ISO 54: 1996 (1996) Cylindrical gears for general and for heavy engineering – Modules
19. Litvin FL, Demenego A, Vecchiato D (2001) Formation by branches of envelope to parametric famielies of surfaces and curves. Comput Methods Appl Mech Eng 190:4587–4608
20. Litvin FL, Fuentes A (2004) Gear geometry and applied theory, 2nd edn. Cambridge University Press, Cambridge
21. McCleary J (1995) Geometry from a differential viewpoint. Cambridge University Press, Cambridge
22. Merritt HE (1971) Gear engineering. Pitman, London
23. Micheletti GF (1977) Tecnologia meccanica, 2nd edn. Unione Tipografica – Editrice Torinese (UTET), Torino
24. Moré JJ, Garbow BS, Hilstrom KE (1980) User guide for MINIPACK-1, Argonne National Laboratory, Report ANL-80–74

25. Niemann G, Winter H (1983) Maschinen-Elemente, Band II: Getriebe allgemein, Zahnradgetriebe-Grundlagen, Stirnradgetriebe. Springer, Berlin, Heidelberg
26. Phillips J (2003) General spatial involute gearing. Springer Science & Business Media, Berlin
27. Pollone G (1970) Il Veicolo. Libreria Editrice Universitaria Levrotto & Bella, Torino
28. Rossi M (1965) Macchine Utensili Moderne. Ulrico Hoepli, Milano
29. Scotto Lavina G (1990) Riassunto delle Lezioni di Meccanica Applicata alle Macchine: Cinematica Applicata, Dinamica Applicata - Resistenze Passive - Coppie Inferiori, Coppie Superiori (Ingranaggi – Flessibili – Freni). Edizioni Scientifiche SIDEREA, Roma
30. Smirnov S (1970) Cours de Mathématiques Supérieures. Éditions MIR, Tome II, Moscou
31. Tuplin WA (1962) Involute gear geometry. Chatto and Windus, London
32. Woodbury RW (1958) History of the gear-cutting machines: a historical study in geometry and machines. M.I.T. Technology Press, Cambridge, Massachusetts
33. Zalgaller VA (1975) Theory of envelopes. Publishing House Nauka, Moscow (in Russian)

Chapter 3
Characteristic Quantities of Cylindrical Spur Gears and Their Determination

Abstract In this chapter, the minimum number of teeth to avoid interference in the operating conditions of rack-pinion, external and internal spur gear pairs is first determined, and useful considerations on this topic are made. Other important characteristic quantities of these types of gears are then determined, such as the lengths of the path of contact, path of approach and path of recess as well as the lengths of the corresponding arcs and values of the angles correlated to the latter. The relationships for calculating the transverse contact ratio and the instantaneous radii of curvature of the involute mating profiles are obtained, and the generalized laws of gearing are described. Subsequently, the main kinematic quantities related to the rolling and sliding motions of the mating teeth flanks are determined, with particular attention to relative sliding and specific sliding to which wear damage and gear efficiency are correlated. In particular, the relationships that express the instantaneous and average efficiencies as well as those that define the total contact efficiency and effective driven torque along the path of contact are determined. Finally, short notes on the main cutting processes of cylindrical spur gears are given.

3.1 Minimum Number of Teeth to Avoid Interference

We have already seen that, for an external gear pair with unmodified involute profiles (Fig. 2.9), to prevent the *theoretical interference* or *involute interference*, the maximum radii $r_{a1\,max}$ and $r_{a2\,max}$ of the addendum circles of the pinion and gear wheel can be equal to O_1T_2 and O_2T_1 respectively. In these limit conditions, the addendum circles pass through the points of interference T_1 and T_2 previously defined (see Sect. 1.5, and Figs. 1.8 and 2.10).

Now let's go back to examine the external gear pair shown in Fig. 2.10, where the condition of non-interference is met, since the two addendum circles intersect the line of action at points E and A, located internally with respect to the interference points T_2 and T_1. The condition of non-interference can be expressed analytically by the following two inequalities, which must be both satisfied (see Colbourne [13], Ferrari and Romiti [21]).

© Springer Nature Switzerland AG 2020
V. Vullo, *Gears*, Springer Series in Solid and Structural Mechanics 10,
https://doi.org/10.1007/978-3-030-36502-8_3

$$CE \leq CT_2 \qquad CA \leq CT_1. \tag{3.1}$$

By the same Fig. 2.10, however, we deduce immediately that the more restrictive inequality is that related to the addendum circle of the gear wheel having a larger diameter. Therefore, it is sufficient to consider the inequality corresponding to it, i.e. the second of the inequality Eqs. (3.1). Then, since $CT_1 = r_1 \sin \alpha$, the latter can be written in the form:

$$CA \leq r_1 \sin \alpha. \tag{3.2}$$

Thus, from the triangle ACO_2 we have:

$$\overline{AO_2}^2 = \overline{CO_2}^2 + \overline{CA}^2 + 2\overline{CO_2}\,\overline{CA} \sin \alpha, \tag{3.3}$$

from which, by putting $x = \overline{CA}$, we obtain:

$$x^2 + 2r_2 x \sin \alpha - (r_2 + h_{a2})^2 + r_2^2 = 0. \tag{3.4}$$

After simplification, this equation is reduced to equality

$$x(2r_2 \sin \alpha + x) = h_{a2}(2r_2 + h_{a2}), \tag{3.5}$$

that can be written in the form:

$$x\left(\sin \alpha + \frac{x}{2r_2}\right) = h_{a2}\left(1 + \frac{h_{a2}}{2r_2}\right). \tag{3.6}$$

Evidently (see Fig. 2.10) the risk of theoretical interference is increased when the point A is approaching the point T_1 or when the point T_1 approaches the point A. This occurs, respectively, when the gear wheel diameter increases, while that of the pinion remains unchanged, or when the pinion diameter decreases, while that of the wheel remains unchanged. Therefore, the risk of interference is increased when the number of teeth z_2 of the gear wheel increases, and the number of teeth z_1 of the pinion decreases.

3.1.1 Minimum Number of Teeth for Rack-Pinion Pair

This case corresponds to the limit case of an external cylindrical spur gear of infinite pitch radius r_2 ($r_2 \to \infty$), while r_1 remains unchanged. From Eq. (3.6) we can deduce that, when r_2 increases, x also grows. Therefore, for an external gear pair, the maximum value of x is that corresponding to $r_2 \to \infty$, i.e. to the case of the rack. In this case ($r_2 \to \infty$), from Eq. (3.6) we get:

$$x = \frac{h_{a2}}{\sin \alpha} \tag{3.7}$$

and, since $x = \overline{CA}$, from this, taking into account the inequality Eq. (3.2), we obtain:

$$\frac{h_{a2}}{\sin \alpha} \leq r_1 \sin \alpha; \tag{3.8}$$

this is the inequality that should be checked to avoid interference.

In general terms, for the modular sizing (the only one considered from here on), each quantity of the toothing is related to the module m. Therefore, the addendum of the pinion and wheel can be expressed as:

$$h_{a1} = k_1 m \quad h_{a2} = k_2 m, \tag{3.9}$$

where k_1 and k_2 are the *addendum factors*, i.e. the proportionality coefficients that correlate the addendum of the two members of the gear pair to the module. It is to remember that:

- $h_{a1} = h_{a2} = m$, for the usual standard sizing;
- $h_{a1} = h_{a2} = 0.8m$, for the usual stub-sizing;
- $h_{a1} \neq m$ and $h_{a2} \neq h_{a1}$, for particular sizings, that we will discuss in the next chapters.

From the inequality Eq. (3.8), taking account of the second of Eq. (3.9), and recalling that $m = (2r_1/z_1)$, we obtain:

$$z_1 = z_{min} \geq \frac{2k_2}{\sin^2 \alpha}. \tag{3.10}$$

This is the minimum number of teeth of the toothed gear wheel, which must mesh with the rack of the corresponding series. If we assume z_{min} in accordance with the inequality (3.10), we avoid the theoretical interference during the meshing between the wheel under consideration and the rack having the same circular pitch. Therefore, for the considerations made above, we also avoid this type of interference between the same wheel and any other external wheel drawn in accordance with the modular sizing.

Of course, with the same inequality (3.10) we can calculate the minimum number of teeth that can be cut without interference, when instead of considering the meshing conditions between a pinion and a rack, we consider the cutting operation, that is, we replace the rack-type cutter to the rack gear.

3.1.2 Minimum Number of Teeth for an External Cylindrical Spur Gear

Now let's examine how to avoid theoretical interference during the meshing of the two members of an external cylindrical spur gear characterized by a given *gear ratio* $u = (z_2/z_1) = (r_2/r_1)$. As we said above, in this case we find a value of the minimum number of the teeth lower than the value obtained for the rack-pinion pair.

From Fig. 2.10, we infer that, for a given value of the pressure angle α, the maximum value of $x = \overline{CA}$ corresponds to the maximum value of h_{a2}. Therefore, the limit value of h_{a2} to avoid the theoretical interference is the one obtained from Eq. (2.5) when we substitute in it the maximum value $x_{max} = \overline{CA}_{max} = CT_1 = r_1 \sin \alpha$, given by the second of Eq. (3.1). So we have:

$$h_{a2}(2r_2 + h_{a2}) = r_1 \sin \alpha (2r_2 \sin \alpha + r_1 \sin \alpha) = r_1 \sin^2 \alpha (2r_2 + r_1). \qquad (3.11)$$

Introducing in this equation the absolute value $u = |r_2/r_1|$ of the gear ratio, and simplifying it, we get the following algebraic equation of the second degree:

$$\left(\frac{h_{a2}}{r_1}\right)^2 + 2u\left(\frac{h_{a2}}{r_1}\right) - (2u+1)\sin^2 \alpha = 0, \qquad (3.12)$$

from whose solution, taking the positive root, we obtain:

$$\frac{h_{a2}}{r_1} = -u + \sqrt{u^2 + (2u+1)\sin^2 \alpha}. \qquad (3.13)$$

This equation provides the limit value of z_1, as $h_{a2} = k_2 m$ and $r_1 = (mz_1/2)$. Therefore, we obtain the following final relationship:

$$z_1 = z_{min} \geq \frac{2k_2}{\left[-u + \sqrt{u^2 + (2u+1)\sin^2 \alpha}\right]} = 2k_2 \frac{\left[u + \sqrt{u^2 + (2u+1)\sin^2 \alpha}\right]}{(2u+1)\sin^2 \alpha}.$$

$$(3.14)$$

3.1.3 Minimum Number of Teeth for an Internal Cylindrical Spur Gear

In the case of an internal cylindrical spur gear pair, the value of the minimum number of teeth to avoid theoretical interference is greater than that given by

Eq. (3.10). *Mutatis mutandis*, proceeding in a similar way to that described in the previous section, we obtain the following relationship:

$$z_1 = z_{\min} \geq \frac{2k_2}{\left[u - \sqrt{u^2 - (2u - 1)\sin^2 \alpha}\right]} = 2k_2 \frac{\left[u + \sqrt{u^2 - (2u - 1)\sin^2 \alpha}\right]}{(2u - 1)\sin^2 \alpha}.$$

(3.15)

3.2 Considerations on the Minimum Number of Teeth

The chart shown in Fig. 3.1 allows calculating the minimum number of teeth, $z_{1\,\min}$, for external and internal gears without profile shift (see Chap. 6), for different values of the pressure angle, α, and the standard values of k_2 (i.e., $k_2 = 1$), as a function of the ratio $(1/u) = (z_1/z_2)$. In this figure, we preferred to use the ratio $(1/u) = (z_1/z_2)$, instead of the gear ratio, u, that appears in the inequalities Eqs. (3.10, 3.14 and 3.15). This in order to have an absolute value of the input data on the ordinate axis (positive for external gears, and negative for internal gears) included in the range between 0 and 1, instead in the range between 1 and ∞.

The use of the chart is very simple. First, we enter it, on the ordinate axis, with a horizontal line corresponding to the design value of the ratio $(1/u) = (z_1/z_2)$. We then extend this line to the right, until it intersects the curve of the value of the pressure angle referred to the nominal center distance. Finally, we exit with a vertical line up to the abscissa, where we read $z_{1\,\min}$. For sizing other than the usual standard, i.e. for $k_2 \neq 1$, the minimum number of teeth, $z_{1\,\min}$, is obtained by multiplying the value derived of the chart for the value of k_2 actually used.

Both from Eqs. (3.10, 3.14 and 3.15), and from the curves shown in Fig. 3.1, the strong influence of the pressure angle α on the minimum number of teeth $z_{1\min}$ is very clear. Equations (3.14) and (3.15) also show that, for an external or internal gear pair, the minimum number of teeth $z_{1\,\min}$ of the wheel with smaller diameter (the pinion) is proportional to the addendum factor k_2 of the mating gear.

For the reasons that we will describe later, often the gear designer is forced to choose a pinion with a number of teeth, $z_{1\,\min}$, less than that obtained using Eqs. (3.14 or 3.15). To avoid the theoretical interference that would occur in this case, a possible solution is to reduce k_2, multiplying it with the ratio between the desired number of teeth and the minimum number of teeth obtained by using the above equations for the given values of the pressure angle α and gear ratio u. The negative aspect of this solution consists in the fact that the path of contact is reduced, due to the consequent decrease of the addendum. In addition, if we want to keep the same bottom clearance that we have in the usual standard sizing ($c = 0.25m$), once the value of k_2 has been calculated in accordance with the procedure described above, the dedendum must be obtained using the relationship $(h_f/m) = k_2 + 0.25$.

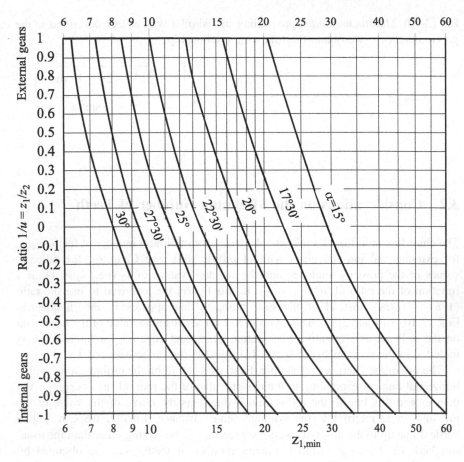

Fig. 3.1 Minimum number of teeth $z_{1\text{min}}$ as a function of the ratio $(1/u) = (z_1/z_2)$, for $k_2 = 1$ and different pressure angles $(\alpha = 15°; \alpha = 17°30'; \alpha = 20°; \alpha = 22°30'; \alpha = 25°; \alpha = 27°30'; \alpha = 30°)$

The problem of determining the minimum number of teeth $z_{1\,\text{min}}$ to prevent the theoretical interference must be considered not only in the operating conditions of the gear, but also during the related cutting operations. From this second point of view, it is noteworthy that the minimum number of teeth $z_{1\text{min}}$ defined by the Eqs. (3.14 and 3.15) can be cut without interference only if the cutting is performed by milling, with form milling cutters (see Sect. 3.11.1).

If instead the teeth are cut with generation process, by means of rack-type cutters, hobs, and pinion-type cutters (Fellows gear shapers or simply Fellows shapers), the minimum numbers of teeth that we can cut without interference are:

- those corresponding to a gear ratio $u = \infty$, given by Eq. (3.10), in the case of cutting with rack-type cutters or hobs;
- those corresponding to a gear ratio u equal to the ratio between the minimum number of teeth, z_{1min}, that can be cut without interference (it is unknown), and the number of teeth of the pinion-type cutter, in the case of cutting with Fellows shaping machine (see Sect. 3.11.2).

In this second case, some attempt is necessary to find $z_{1\,min}$. An approximate initial value of $(z_{1\,min})_i$ is determined using Eqs. (3.14, 3.15) or the curves of Fig. 3.1, according to an approximate initial value of $u_i = (z_2/z_1)_i$ of the gear pair made up of the wheel that must be cut and the pinion-type cutter. With this initial value of $(z_{1\,min})_i$ thus obtained, the new value of u will be calculated, dividing this value of $(z_{1\,min})_i$ by the number of teeth, z_0, of the pinion-type cutter. The iterative procedure, which also leads to fast convergence, is repeated until there are two subsequent values of $z_{1\,min}$ coincident or slightly different within predetermined limits.

About the interference, it should be noted that the cutter tool, having to cut a tooth with dedendum $h_f = 1.25m$, must have an addendum $h_{a0} = 1.25m$ (remember that the subscript 0 applies to quantities related to the tool, and more generally to cutting operations). In practice, however, the tooth of the rack-type cutter (the focus is temporary limited to cutting by rack-type cutter) has an addendum formed by a straight-line portion (and therefore with involute geometry) of tooth depth equal to about the value of the module, while the remaining portion consists of the curve that connects the flank profile to the top land.

Generally, we can assume that, as regards the danger of interference, things are as if the addendum profile of the cutter ended at the beginning of fitting curve of the tip, i.e. that this portion of the addendum profile is not dangerous in respect to the interference. Therefore, when Eq. (3.10) is used, we assume $h_{a0} = m$, although the cutter tool has addendum depth equal to 1.25 m. In the more general case, the value of $k_2 = k_0$ will be evaluated by the quotient of the addendum of the rack-type cutter, decreased of depth of rounded curve near the tip, measured along the normal to the pitch line, divided by the module. Quite similar considerations can be made with regard to the gear cutting with hobs or Fellows shaping cutters.

In some case (see, for example, the cutting by hobs of large gears for marine transmissions), hobs with curved transverse section near the tip and root of the tooth, having radii of curvature more extensive than those corresponding to the rack-type cutter as described above, are used. In these cases, effects of cutting interference of these curved portions of the tool profile with the tooth profile of the gear to be cut cannot be excluded a priori. In these cases, it is necessary to perform a more detailed study of the interaction between the tooth profile of the cutting tool and tooth profile of the wheel, which must be cut, in order to check the effects of the cutting interference. This study can be done using the known methods of Applied Mechanics (see Ferrari and Romiti [21], Levi-Civita and Amaldi [35], Litvin and Fuentes [36], Scotto Lavina [47]); however, it is necessary only in very special cases.

Finally, we must keep in mind that, while no interference can be allowed between the two mating profiles of a gear in its operating conditions, in the generation cutting process of a gear wheel a light interference between the cutter profile and profile generated by it is not only perfectly permissible, but it is indeed often desirable. This is because such a light interference undoubtedly determines a light reduction of the active profile, but in return offers the significant advantage of eliminating the part of involute curve immediately near the base circle, to which, as we will see, different problems are related.

Therefore, in this framework, it is admitted that, in practice, the minimum number of teeth that can be cut by a generation process, using a rack-type cutter, is equal to 5/6 of the theoretical value given by Eq. (3.10). Figure 3.2, valid for

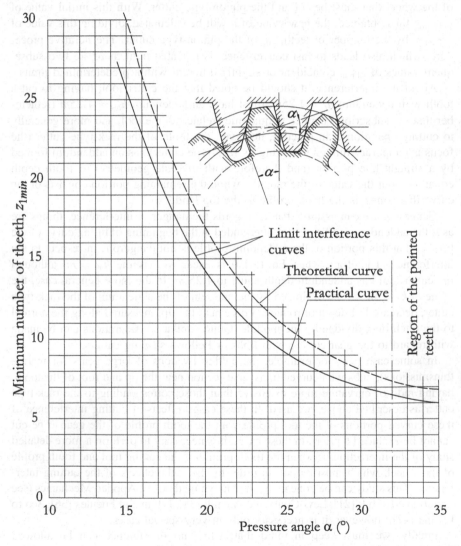

Fig. 3.2 Theoretical and practical number of teeth, z_{1min}, as a function of the pressure angle, α, for cylindrical spur gears cut with rack-type cutter

cylindrical spur gears obtained by cutting with a rack-type cutter, shows the variation of $z_{1\,min}$ as a function of the pressure angle α, both in the cutting theoretical condition and in cutting practical condition.

3.3 Lengths of the Path and Arc of Contact, and Angles of Contact

Once the geometric quantities of the two members of the gear pair under consideration have been determined, it is necessary to assess the length $g = g_\alpha$ of the path of contact (see Figs. 1.8 and 2.10). Considering the case in which the pinion drives the wheel or rack (*mutatis mutandis*, in the same way we can proceed when the pinion is driven by the wheel or rack), the total length of the path of contact is equal to the sum of the lengths of the *path of approach*, AC and *path of recess*, CE i.e., $g_\alpha = (g_{f1} + g_{a1}) = (g_{a2} + g_{f2})$. The three possible cases are those of external gear pairs, pinion-rack pairs and internal gear pairs.

For an external gear pair (Fig. 3.3a), we have:

$$g_{f1} = g_{a2} = AC = AT_2 - CT_2 = \sqrt{\left(r_{a2}^2 - r_{b2}^2\right)} - r_2 \sin \alpha \qquad (3.16)$$

$$g_{a1} = g_{f2} = CE = T_1E - T_1C = \sqrt{\left(r_{a1}^2 - r_{b1}^2\right)} - r_1 \sin \alpha. \qquad (3.17)$$

So, if we consider that $r_{a2} = (r_2 + k_2 m)$, $r_{a1} = (r_1 + k_1 m)$, $r_{b2} = r_2 \cos \alpha$, $r_{b1} = r_1 \cos \alpha$, $r_2 = (mz_2/2)$ and $r_1 = (mz_1/2)$, we get:

$$g_{f1} = g_{a2} = \frac{m}{2} \left(\sqrt{z_2^2 \sin^2 \alpha + 4k_2(z_2 + k_2)} - z_2 \sin \alpha \right) \qquad (3.18)$$

$$g_{a1} = g_{f2} = \frac{m}{2} \left(\sqrt{z_1^2 \sin^2 \alpha + 4k_1(z_1 + k_1)} - z_1 \sin \alpha \right). \qquad (3.19)$$

To the lengths $g_{f1} = g_{a2}$ and $g_{a1} = g_{f2}$, measured along the line of action, the lengths of the *arc of approach*, $e_1 = e_{a2}$, and *arc of recess*, $e_2 = e_{a1}$, correspond. They are given respectively by:

$$e_1 = e_{a2} = \frac{AC}{\cos \alpha} \qquad (3.20)$$

$$e_2 = e_{a1} = \frac{CE}{\cos \alpha}. \qquad (3.21)$$

It should be noted that the symbol, e, used here to indicate arcs of contact, should not be confused with the same symbol, also used to indicate space width; here the

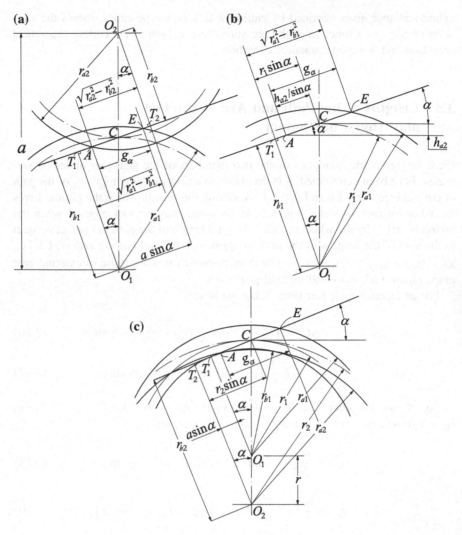

Fig. 3.3 Schematic diagrams for evaluation of lengths $g_{f1} = g_{a2}$, $g_{a1} = g_{f2}$, and $g_\alpha = (g_{f1} + g_{a1})$, for: **a** external gear pair; **b** pinion-rack gear pair; **c** internal gear pair

symbols e, e_1 and e_2 indicate respectively the total arc of contact, arc of approach, and arc of recess. It should also be noted that the most precise double subscript calls only the addendum flanks of the two members of the gear pair. It indicates that, in the path of approach, the addendum flank of the driven wheel is in contact, while in the path of recess the addendum flank of the driving pinion is in contact. For brevity, from here on, we will use only one subscript, as misunderstandings are not possible.

These arcs of approach and recess are the lengths of addendum contact respectively of the wheel and pinion, for which they are measured along the pitch circles of the wheel and, respectively, of the pinion. Usually however, according to the most commonly used convention, the arc of approach and arc of recess are respectively measured on the pitch circles of the driving and driven members of the gear pair under consideration. However, the complete equivalence of measuring these arcs on one or the other of the pitch circles of the two members of the gear pair is obvious, given that they roll over each other without sliding.

The correlated angles of approach and recess are respectively found by dividing the path of approach, AC, and path of recess, CE, by the radius of the base circle of the driving (or driven) wheel; they are expressed in circular measure or radians. In completely equivalent terms, using quantities referred to the pitch circle of the pinion, they are given by the following relationships:

$$\xi_1 = \frac{e_1}{r_1} \tag{3.22}$$

$$\xi_2 = \frac{e_2}{r_1}. \tag{3.23}$$

Thus, the total lengths of the path of contact and arc of contact, and the total angle of contact are given by the following final relationships:

$$
\begin{aligned}
g_\alpha &= g_{f1} + g_{a1} \\
&= \frac{m}{2} \left(\sqrt{z_2^2 \sin^2 \alpha + 4k_2(z_2 + k_2)} \right. \\
&\quad \left. + \sqrt{z_1^2 \sin^2 \alpha + 4k_1(z_1 + k_1)} - (z_1 + z_2)\sin\alpha \right)
\end{aligned} \tag{3.24}
$$

$$
\begin{aligned}
e = e_1 + e_2 &= \frac{g_\alpha}{\cos\alpha} \\
&= \frac{m}{2\cos\alpha} \left(\sqrt{z_2^2 \sin^2 \alpha + 4k_2(z_2 + k_2)} + \sqrt{z_1^2 \sin^2 \alpha + 4k_1(z_1 + k_1)} - (z_1 + z_2)\sin\alpha \right)
\end{aligned} \tag{3.25}
$$

$$
\begin{aligned}
\xi = \xi_1 + \xi_2 &= \frac{1}{z_1 \cos\alpha} \left(\sqrt{z_2^2 \sin^2 \alpha + 4k_2(z_2 + k_2)} \right. \\
&\quad \left. + \sqrt{z_1^2 \sin^2 \alpha + 4k_1(z_1 + k_1)} - (z_1 + z_2)\sin\alpha \right).
\end{aligned} \tag{3.26}
$$

For a pinion-rack pair (Fig. 3.3b), we have:

$$g_{f1} = g_{a2} = AC = (h_{a2}/\sin\alpha) \tag{3.27}$$

$$g_{a1} = g_{f2} = CE = T_1E - T_1C = \sqrt{\left(r_{a1}^2 - r_{b1}^2\right)} - r_1 \sin\alpha; \tag{3.28}$$

therefore, if we consider that $h_{a2} = k_2 m$, $r_{a1} = (r_1 + k_1 m)$, $r_{b1} = r_1 \sin \alpha$, and $r_1 = (m z_1 / 2)$, we get:

$$g_{f1} = g_{a2} = (k_2 m / \sin \alpha) \tag{3.29}$$

$$g_{a1} = g_{f2} = \frac{m}{2} \left(\sqrt{z_1^2 \sin^2 \alpha + 4 k_1 (z_1 + k_1)} - z_1 \sin \alpha \right). \tag{3.30}$$

To the lengths $g_{f1} = g_{a2}$ and $g_{a1} = g_{f2}$ measured along the line of action, the lengths of the arc of approach, e_1, and arc of recess, e_2, correspond; they are given by Eqs. (3.20) and (3.21). For this type of gear pair, these quantities are also the lengths of addendum contact of the rack and pinion, but with the difference that they are both measured along the pitch circle of the pinion. Thus, the related angles of approach and recess are given by:

$$\xi_1 = \frac{e_1}{r_1} = \frac{AC}{r_1 \cos \alpha} \tag{3.31}$$

$$\xi_2 = \frac{e_2}{r_1} = \frac{CE}{r_1 \cos \alpha}. \tag{3.32}$$

Thus, the total lengths of the path of contact and arc of contact, and the total angle of contact are given by:

$$g_\alpha = g_{f1} + g_{a1} = \frac{m}{2} \left(\sqrt{z_1^2 \sin^2 \alpha + 4 k_1 (z_1 + k_1)} - z_1 \sin \alpha + \frac{2 k_2}{\sin \alpha} \right) \tag{3.33}$$

$$e = e_1 + e_2 = \frac{g_\alpha}{\cos \alpha} = \frac{m}{2 \cos \alpha} \left(\sqrt{z_1^2 \sin^2 \alpha + 4 k_1 (z_1 + k_1)} - z_1 \sin \alpha + \frac{2 k_2}{\sin \alpha} \right) \tag{3.34}$$

$$\xi = \xi_1 + \xi_2 = \frac{e}{r_1} = \frac{1}{z_1 \cos \alpha} \left(\sqrt{z_1^2 \sin^2 \alpha + 4 k_1 (z_1 + k_1)} - z_1 \sin \alpha + \frac{2 k_2}{\sin \alpha} \right). \tag{3.35}$$

Lastly, for an internal gear pair (Fig. 3.3c), we have:

$$g_{f1} = g_{a2} = AC = T_2 C - T_2 A = r_2 \sin \alpha - \sqrt{(r_{a2}^2 - r_{b2}^2)} \tag{3.36}$$

$$g_{a1} = g_{f2} = CE = T_2 E - T_2 C = T_2 T_1 + T_1 E - T_2 C$$
$$= a \sin \alpha + \sqrt{(r_{a1}^2 - r_{b1}^2)} - r_2 \sin \alpha. \tag{3.37}$$

So, if we consider that $r_{a2} = (r_2 - k_2 m)$, $r_{a1} = (r_1 + k_1 m)$, $r_{b2} = r_2 \cos \alpha$, $r_{b1} = r_1 \cos \alpha$, $r_2 = (m z_2 / 2)$ and $r_1 = (m z_1 / 2)$, we get:

$$g_{f1} = g_{a2} = \frac{m}{2} \left(-\sqrt{z_2^2 \sin^2 \alpha - 4 k_2 (z_2 - k_2)} + z_2 \sin \alpha \right) \tag{3.38}$$

$$g_{a1} = g_{f2} = \frac{m}{2} \left(\sqrt{z_1^2 \sin^2 \alpha + 4 k_1 (z_1 + k_1)} - z_1 \sin \alpha \right). \tag{3.39}$$

The lengths at the arcs of approach and recess e_1 and e_2 related to the lengths $g_{f1} = g_{a2}$ and $g_{a1} = g_{f2}$ are given by Eqs. (3.20) and (3.21). They are the lengths of addendum contact of the annulus (the ring gear) and pinion, measured along the pitch circles of the annulus and pinion respectively. The related angles of the approach and recess are given, also for this type of gear pair, by Eqs. (3.22 and 3.23).

Thus, the total lengths of the path of contact and arc of contact, and the total angle of contact are as follows:

$$g_\alpha = g_{f1} + g_{a1}$$
$$= \frac{m}{2} \left(-\sqrt{z_2^2 \sin^2 \alpha - 4 k_2 (z_2 - k_2)} \right.$$
$$\left. + \sqrt{z_1^2 \sin^2 \alpha + 4 k_1 (z_1 + k_1)} + (z_2 - z_1) \sin \alpha \right) \tag{3.40}$$

$$e = e_1 + e_2 = \frac{g_\alpha}{\cos \alpha}$$
$$= \frac{m}{2 \cos \alpha} \left(-\sqrt{z_2^2 \sin^2 \alpha - 4 k_2 (z_2 - k_2)} \right.$$
$$\left. + \sqrt{z_1^2 \sin^2 \alpha + 4 k_1 (z_1 + k_1)} + (z_2 - z_1) \sin \alpha \right) \tag{3.41}$$

$$\xi = \xi_1 + \xi_2$$
$$= \frac{1}{z_1 \cos \alpha} \left(-\sqrt{z_2^2 \sin^2 \alpha - 4 k_2 (z_2 - k_2)} \right.$$
$$\left. + \sqrt{z_1^2 \sin^2 \alpha + 4 k_1 (z_1 + k_1)} + (z_2 - z_1) \sin \alpha \right). \tag{3.42}$$

3.4 Transverse Contact Ratio

By definition, the transverse contact ratio ε_α is the ratio between the length of the arc of contact and circular pitch or, in equivalent terms, the ratio between the length of the path of contact and transverse base pitch, i.e.:

$$\varepsilon_\alpha = \frac{e}{p} = \frac{g_\alpha}{p_{bt}}. \tag{3.43}$$

Since $p_{bt} = p \cos \alpha$ and $p = \pi m$, considering the pairs of Eqs. (3.24) and (3.25), (3.33) and (3.34), (3.40) and (3.41), respectively valid for external gear pairs, pinion-rack gear pairs, and internal gear pairs, we get:

- For an external gear pair:

$$
\begin{aligned}
\varepsilon_\alpha = \frac{1}{2\pi \cos \alpha} \Bigg(& \sqrt{z_2^2 \sin^2 \alpha + 4k_2(z_2 + k_2)} \\
& + \sqrt{z_1^2 \sin^2 \alpha + 4k_1(z_1 + k_1)} - (z_1 + z_2) \sin \alpha \Bigg);
\end{aligned} \tag{3.44}
$$

- For an pinion-rack gear pair:

$$\varepsilon_\alpha = \frac{1}{2\pi \cos \alpha} \left(\sqrt{z_1^2 \sin^2 \alpha + 4k_1(z_1 + k_1)} - z_1 \sin \alpha + \frac{2k_2}{\sin \alpha} \right) \tag{3.45}$$

- For an internal gear pair:

$$
\begin{aligned}
\varepsilon_\alpha = \frac{1}{2\pi \cos \alpha} \Bigg(& -\sqrt{z_2^2 \sin^2 \alpha - 4k_2(z_2 - k_2)} \\
& + \sqrt{z_1^2 \sin^2 \alpha + 4k_1(z_1 + k_1)} + (z_2 - z_1) \sin \alpha \Bigg).
\end{aligned} \tag{3.46}
$$

We have already said that the meshing cycle of one tooth pair begins when the teeth first make contact, and ends when the contact is broken. In order the transmission of motion between the two meshing gear wheels is continuous, there must clearly be at least one tooth pair in contact at all times. However, it is well known that a smooth working condition is only possible when the contact between one tooth pair continues until sometime after the contact between the next pair of teeth is started.

Essentially, there must be parts of the meshing cycle during which two pairs of teeth are in contact simultaneously. The transverse contact ratio is then a measure of the amount of this overlap. For example, $\varepsilon_\alpha = 1.4$ means that one pair of teeth is in contact for 60% of the meshing cycle, two pairs of teeth are in meshing for the remaining 40%. Indeed, $\varepsilon_\alpha = 2.2$ means that three pairs of teeth are in contact for 20% of the meshing cycle and two pairs of teeth are in contact for the remaining 80% of the meshing cycle. Thus, in this last case, at least two pairs of teeth are theoretically in contact at all times; whether or not they are actually in contact depends on the tooth stiffness, applied load, and precision of manufacture and assembly.

In general, the greater the transverse contact ratio the smoother and quieter the working condition of the gear pair. From the point of view of the gear design, it is necessary to bear in mind that, for gears with low speed of rotation, it is not appropriate that the value of ε_α falls below $(1.2 \div 1.3)$, while for fast gears the value of ε_α must be greater than $(1.4 \div 1.5)$. In any case, the value of ε_α must not fall below the unit, otherwise the continuity of motion is lost.

If the addenda of the two members of an external gear pair (Fig. 3.3a) assume their maximum values $k_1 m$ and $k_2 m$, compatible with non-interference (in this case, the point A in the figure coincides with the point T_1, while the point E coincides with the point T_2), we will have (see also Giovannozzi [24]):

$$\varepsilon_\alpha = \frac{(r_1 + r_2)\sin\alpha}{p\cos\alpha} = \frac{(z_1 + z_2)\tan\alpha}{2\pi}. \tag{3.47}$$

From this equation, we obtain the following minimum value of the sum of the numbers of teeth of wheel and pinion compatible with the limiting condition of continuity of the transmission of motion ($\varepsilon_\alpha = 1$):

$$(z_1 + z_2)_{\min} = 2\pi\cot\alpha. \tag{3.48}$$

Therefore, in this case, to ensure the continuity of motion, the minimum sum $(z_1 + z_2)_{\min}$ must not fall below: 24 teeth for $\alpha = 15°$; 18 teeth for $\alpha = 20°$; and 11 teeth for $\alpha = 30°$.

Both from Fig. 3.3a and Eqs. (3.18, 3.19, 3.20, 3.21 and 3.44), we deduce that an increase in the addendum factors k_1 and k_2 increases the arc of approach and the arc of recess, and consequently, the transverse contact ratio between external gears in meshing. The limiting values for the arc of approach and arc of recess are when either one of them exceeds the pressure angle α. The transverse contact ratio for an external gear pair also increases when the radius of the pitch circle of the gear members increases. By contrast, an increase of the pressure angle causes a decrease of the transverse contact ratio. Figure 3.4 shows the effect of the numbers of teeth z_1 and z_2 on the transverse contact ratio, for an external gear pair with $k_1 = k_2 = 1$ and $\alpha = 20°$.

Mutatis mutandis, starting from Fig. 3.3b and Eqs. (3.29, 3.30, 3.20, 3.21 and 3.45), entirely similar considerations can be made for the pinion-rack gear pair, as it can be considered the limit case of an external gear pair for $z_2 = \infty$ ($u = z_2/z_1 = \infty$). The same Fig. 3.4 is valid for this gear pair, characterized by $k_1 = k_2 = 1$ and $\alpha = 20°$.

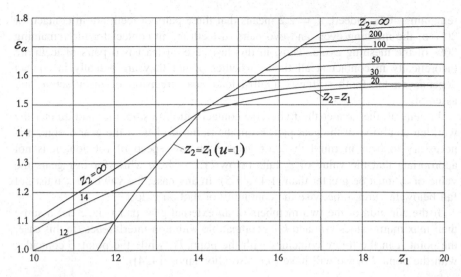

Fig. 3.4 Effect of the numbers of teeth z_1 and z_2 on the transverse contact ratio ε_α, for external gear pairs and pinion-rack gear pairs, with $k_1 = k_2 = 1$ and $\alpha = 20°$

Both from Fig. 3.3c and Eqs. (3.38), (3.39), (3.20), (3.21) and (3.46), we infer that also for an internal gear pair an increase in the addendum factors k_1 and k_2 increases the arc of approach and the arc of recess, and consequently, the transverse contact ratio. We also infer that the transverse contact ratio increases when the radius of the pitch circle of the pinion increases, while it decreases when the radius of the pitch circle of the annulus increases. In addition, in this case, an increase of the pressure angle determines a decrease of the transverse contact ratio.

3.5 Radius of Curvature of Involute Tooth Profiles and Generalized Laws of Gearing

Let consider a current point P between two mating profiles of the external gear pair shown in Fig. 2.6. This point, as Fig. 2.10 shows, can be located between the points A (start of contact) and E (end of contact). The determination of the radii of curvature of the two involute profiles is very simple, because we can make use of one of the special properties of the involute curve [25].

In Sect. 2.1 we found that at every point P the involute of a base circle is normal to the generating line, and in any position of this generating line (see position P_iT_i in Fig. 2.1) its point of tangency with the base circle is the instantaneous center of rotation. Therefore, an infinitesimal length of involute arc at point P_i (Fig. 2.1) is indistinguishable from a circular arc having its center at point T_i and radius P_iT_i; in other words, P_iT_i is the radius of curvature of the involute at point P_i.

For an external gear pair, whatever the position of the point of contact P on the line of action (see Fig. 2.6), the instantaneous centers of curvature of the two conjugate profiles are then the points T_1 and T_2, while the instantaneous radii of curvature are $\rho_1 = PT_1$ for the driving profile, and $\rho_2 = PT_2$ for the driven profile. We can achieve the same result (i.e., T_1 and T_2 instantaneous centers of curvature, and $\rho_1 = PT_1$ and $\rho_2 = PT_2$ instantaneous radii of curvature of the two mating profiles) using the well-known *Euler-Savary equation* for planar motion, as the involutes considered here are planar curves (see Belfiore et al. [5]).

Then, if we denote by g_P the algebraic value of the distance between the point of contact P and the instantaneous center of rotation C (Fig. 2.6), measured along the line of action, the instantaneous radii of curvature of the two profiles will be expressed by the following equations:

$$\rho_1 = PT_1 = r_1 \sin\alpha + g_P \tag{3.49}$$

$$\rho_2 = PT_2 = r_2 \sin\alpha - g_P. \tag{3.50}$$

In these equations, we conventionally assume g_P positive during the recess contact (P between points C and E in Fig. 2.10), and negative during the approach contact (P between points A and C in Fig. 2.10). The distance between the centers of curvature T_1 and T_2 is constant and equal to:

$$T_1 T_2 = \rho_1 + \rho_2 = (r_1 + r_2) \sin\alpha. \tag{3.51}$$

If for an internal gear pair (Fig. 2.7) we repeat, *mutatis mutandis*, the same considerations made for an external gear pair, we still find that the points T_1 and T_2 are the instantaneous centers of curvature, while the instantaneous radii of curvature of the two mating profiles are given by:

$$\rho_1 = PT_1 = r_1 \sin\alpha + g_P \tag{3.52}$$

$$\rho_2 = PT_2 = r_2 \sin\alpha + g_P. \tag{3.53}$$

For a pinion-rack gear pair, the instantaneous radius of curvature ρ_1 is given by Eqs. (3.49 or 3.52), while $\rho_2 = \infty$.

It should be noted that the above procedure, based on the special properties of the involute, is a very simple method for calculating the instantaneous radii of curvature during the meshing between two mating gears. This method, however, is valid only for involute profiles; therefore, it cannot be used for finding the radius of curvature in the fillet. In this case, as we will see, we will make use of the Euler-Savary equation that relates the radii of curvature of two conjugate profiles, and has general validity for planar profiles. However, in the case of involute profiles, the Euler-Savary equation allows to show that the centers of curvature of the two conjugate profiles that are touching in any position of the point of contact on the line of action are always points T_1 and T_2.

In this framework, in addition to the *Euler-Savary equation*, it is necessary to mention the *Arhnold-Kennedy instant center theorem*, which states that if any three bodies have a relative motion to each other, their instantaneous centers lie on a straight line. When this theorem is applied to a gear pair, it implies that the pitch point must always lie on the line that connects the two centers of rotation of the driving and driven members of the gear under consideration. This theorem is a special case of the *vector loop equation*, which defines the relative motion between the three-link 1-dof kinematic chain, constituting the gear pair.

In Sect. 1.4 we defined the first law of gearing, which establishes that a common normal to the tooth profiles at their point of contact must pass through a fixed point on the line of centres, i.e. the pitch point, whatever the position of the contacting teeth. This law of gearing must be generalized, to extend it from planar gearing to spatial gearing, for a given *transmission function*, defined as the relationship between the angular position of the input member of a gear pair and the corresponding angular position of the output member. Therefore, in the most general case of axes in any way arranged in the Euclidean three-dimensional space, for a given position of axes and a given transmission function, the laws of gearing become three, and are formulated as follows:

- A unique relationship exists between the instantaneous displacement of the output member and the instantaneous displacement of the input member (*first law of gearing*).
- A unique relationship exists between the spiral angle and the pressure angle at the contacts between conjugate surfaces in order to provide motion transmission as defined by the first law of gearing (*second law of gearing*). This second law results from applying *Ball's reciprocity relation* to direct contact mechanisms [3, 4].
- The conjugate action requires a unique effective curvature at the contacts which satisfy the second law of gearing (*third law of gearing*).

Knowledge of the effective curvature between two conjugate surfaces in mesh enables the distance between the two surfaces to be determined. These three laws of gearing are equally valid for any direct-contact mechanism. The third law of gearing is the spatial equivalence of the Euler-Savary equation for planar gearing. Several attempts have been made to generalize the Euler-Savary equation for planar gearing in a similar relationship for spatial motion. In this regard, it is worth mentioning the very appreciable work of Disteli [17]. Each of these attempts, however, provide results other than a unique relationship for the effective curvature of two conjugate meshing surfaces [8, 49]. The differences between the Euler-Savary equation for planar gearing and the third law of gearing prove necessary to take into account the non-degenerate relationships that spatial motion exhibits over planar motion.

3.6 Kinematics of Gearing: Rolling and Sliding Motions of the Teeth Flanks

Now we analyze the kinematics of contact between two involute conjugate profiles of an external gear pair. The general validity of the discussion is not compromised, as the considerations and deductions made for the external gear pairs are provided, with the variations of the case, even for the internal gear pairs. The choice of considering the external gear pairs is motivated by the fact that these gears are the most widely used in practical applications.

Here we assume that the meshing gears and mechanical elements for supporting loads, which globally constitute the system for power transmission, are rigid. In fact, all of these system components will deform as a function of the transmitted loads. These deformations, due to deflection of the teeth relative to the gear blanks, bending and torsional displacements of the shafts and gear blanks, deflections of the bearing supports, and compliance of the housing used to support the bearings, will be introduced in a later stage of deepening.

In Fig. 2.3 we highlighted the fact that progressively increasing arcs of the involute correspond to equal arcs of the base circle described by the generating line during its rolling without sliding about the base circle. It follows that the point of contact between two mating involute profiles, associated with their base circles, describes on the two involute curves two arcs of different lengths in the same time interval.

Consequently, during the meshing cycle, the contact between the two mating profiles is not a pure rolling contact, but a contact of rolling and sliding together. Figure 3.5, which refers to two mating involute profiles related to two base circles having the same diameters, show that, whatever the direction of rotation, the lengths of the involute arcs described by the point of contact on the two profiles are not equal.

It is therefore necessary to examine the relative motion between the two mating profiles, to which a continuously variable relative sliding between the two involute curves is associated. Even in the simplest case of planar gears (in this case, the flank surfaces are reduced to their transverse sections, which are the involute profiles), different types of velocities come into play. These velocities are: angular velocity; tangential velocity at pitch circle, also called pitch line velocity; tangential velocity at the base circle; absolute velocity; sliding velocity; entrainment velocity; spinning velocity; etc. The evaluation of these velocities is based on an *idealized center of contact*, which coincides with the *theoretical point of contact*, and neglects the effects of micro-slip (see Cattaneo [10], Johnson [31], Kalker [33], Mindlin [38], Polach [40]).

With these velocities, it is possible to determine the tribological conditions at the contact zone. As matter of fact, the mechanisms of wear as well as the analysis of the rheological phenomena between two surfaces in contact depend on the relative displacements between the two surfaces. Only the magnitudes of these velocities are needed to predict such tribological conditions and related effects during the meshing.

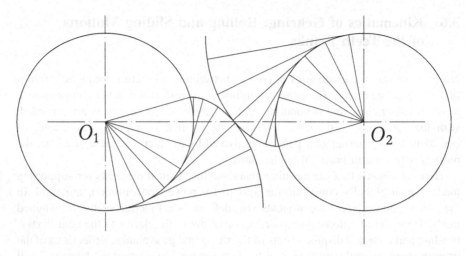

Fig. 3.5 Meshing of two involute profiles related to two base circles having the same diameters

With these premises, we examine in details the kinematics of the external gear pair shown in Fig. 2.10, and considered as a mechanism. It is a three-link mechanism consisting of three elements: the two-toothed gears and the fixed frame. This mechanism, according to the *Grübler's mobility criterion* (see Ambekar [1], Di Benedetto and Pennestrì [16], Hain [26]), has mobility equal to unity, i.e. it is a 1-dof mechanism. In other terms, as one of the two gears rotates, the other gear must rotate according to the *transmission function*, i_n defined by Eq. (1.17).

In order to analyze the kinematic behavior of gears characterized by a non-constant transmission ratio (for example, non-circular gears, elliptical gears, harmonic gears, and similar), the concept of transmission function must also be generalized. For these types of gears, which are not discussed in this monographic textbook, the transmission function is not constant. However, it is always given by Eq. (1.17) and, as we have already said, helps to define the first law of gearing, according to which for a given position of the axes and a given transmission function, a unique relationship exists between the instantaneous displacement of the driven member and the instantaneous displacement of the driving member.

In general terms, the transmission function is the reciprocal of the instantaneous transmission ratio. Therefore, for a gear train it is defined as the quotient of the instantaneous angular displacement of the driven member, $d\varphi_0$, divided by the instantaneous angular displacement of the driving member, $d\varphi_i$, i.e. $i_n = (1/i) = (d\varphi_0/d\varphi_i)$. This function has a constant value for circular gears, while for non-circular gears and similar it is not constant. To generalize the problem, taking into account both circular and non-circular gears, a further constraint equation must be added to the aforementioned first law of gearing. This constraint equation is given by the following relationship:

$$\int_0^{2\pi} i_n d\varphi_i = rational \ function; \tag{3.54}$$

this function must always be a rational function because, otherwise, the driven member could not sustain an indefinite number of cycles with the desired transmission function [22].

To minimize dynamic effects, the transmission function must be as smooth as possible. The actual graph of this function assumes decisive importance in the design of non-circular gears. Since the integrals $\int d\varphi_0$ and $\int i_n d\varphi_i$ define the angular position of the driven member, φ_0, for a given angular position of the driving member, φ_i, it is evident that any discontinuity of the transmission function would result in a not one to one correspondence between these two angular positions, with the risk of having two or more simultaneous angular positions of the driven member for a given angular position of the driving member.

Since the transmission function expresses the change $d\varphi_0$ in the angular position φ_0 of the driven member with respect to the change $d\varphi_i$ in the angular position φ_i of the driving member, it is also known as *velocity ratio*. The derivative $i_n' = d^2\varphi_0/d^2\varphi_i$ is called *acceleration ratio*; it is related to the curvature of the two centrodes. If this derivative is too high, the centrodes become pointed, so their manufacture is difficult. Furthermore, as the required output torque increases as the acceleration ratio increases, the loads acting on the mating teeth will also increase accordingly. The next derivative $i_n'' = d^3\varphi_0/d^3\varphi_i$ is called *jerk ratio*, while the two subsequent derivatives $i_n''' = d^4\varphi_0/d^4\varphi_i$ and $i_n'''' = d^5\varphi_0/d^5\varphi_i$ are called, respectively, *snap ratio* and *crackle ratio*.

The angular velocities ω_1 and ω_2 of the two gear members are given respectively by:

$$\omega_1 = \frac{d\varphi_1}{dt} \qquad \omega_2 = \frac{d\varphi_2}{dt}, \tag{3.55}$$

where dt is the infinitesimal time interval. Therefore, we introduced the time function, without which it is not possible to analyze the inertial effects during the meshing. In fact, up to now we have considered the kinematic geometry of the mating gear as independent of time, t.

During the time interval dt, the pitch circles roll describing arcs respectively equal to $r_1 d\varphi_1 = r_1\omega_1 dt = v_1 dt$ and $r_2 d\varphi_2 = r_2\omega_2 dt = v_2 dt$, and equal to each other, as the *pitch-line velocity v* (or *tangential velocity* at pitch circle) is the same on the two pitch circles, i.e.:

$$v = v_1 = r_1\omega_1 = v_2 = r_2\omega_2. \tag{3.56}$$

In fact, the rolling of pitch circles relative to one another is a pure rolling, i.e. without sliding.

As shown in Fig. 1.5 and, on a larger scale, in Fig. 3.6, the point of contact P is any current point of path of contact in any transverse section. This point can be considered belonging to both the driving and driven profiles. The *absolute velocity vector* v_{P1} of the point P belonging to the driving profile is a vector perpendicular to the line connecting P and the fixed center O_1. Similarly, the absolute velocity v_{P2} of the same point P belonging to the driven profile is a vector perpendicular to the line connecting P and the fixed center O_2. These vectors are respectively given by the two cross products (it is still to be noted that vectors are indicated by bold letters):

$$v_{P1} = \omega_1 \times \overline{O_1P} \quad v_{P2} = \omega_2 \times \overline{O_2P}. \tag{3.57}$$

Obviously, these two vectors are not collinear. They become collinear only when, during meshing, the point P lies on the line connecting the two fixed centers O_1 and O_2, that is when it comes to coincide with the pitch point C.

Vector v_{P1} can be resolved into two components: one of these components, v_{n1}, is parallel to the line of action, while the other component, v_{t1}, is perpendicular to it. Similarly, vector v_{P2} can also be resolved into two components: one of these components, v_{n2}, is parallel to the line of action, while the other component v_{t2} is perpendicular to it (Fig. 3.6).

In order to satisfy the basic law of conjugate profiles, and the profiles themselves remain in meshing, the components v_{n1} and v_{n2} along the line of action must be equal, i.e. $v_{n1} = v_{n2} = v_n$; otherwise, the two profiles would no longer be in contact,

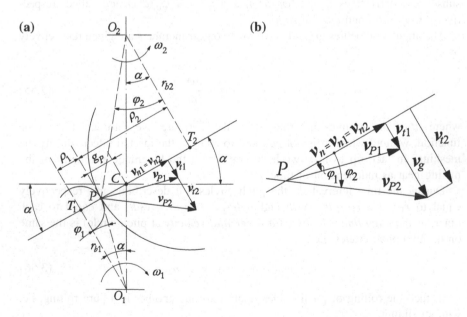

Fig. 3.6 **a** Velocity vectors associated to any position of the current point of contact P along the line of action; **b** Detail on a larger scale

but separated. These velocities can be conceived as the absolute velocities of points of the base circles about their fixed centers. Thus, they are nothing more than the tangential velocities at base circles, so we can write in terms of their absolute values:

$$v_n = v_{n1} = v_{n2} = \omega_1 r_{b1} = \omega_2 r_{b2} = \omega_1 r_1 \cos \alpha = \omega_2 r_2 \cos \alpha. \qquad (3.58)$$

In order to calculate the components v_{t1} and v_{t2}, we can follow two alternatives, both related to the motion of the point P, whether belonging to the driving or driven profiles. According to the first alternative, the motion of point P is regarded as compounded of a motion along the line of action, directed from T_1 to T_2, and a rotation around T_1 and T_2. According the second alternative, kinematically equivalent to the first, the motion of point P is regarded as compounded of a motion along the straight-line tangent to the pitch circle at the pitch point, and a rotation around the pitch point. Here it is convenient to follow the first alternative, as we have already calculated the absolute values of the components v_{n1} and v_{n2}. According to this alternative, taking into account Eqs. (3.49) and (3.50), we have, in terms of their absolute values:

$$v_{t1} = \omega_1 \rho_1 = \omega_1 (r_1 \sin \alpha + g_P) = v \left(\sin \alpha + \frac{g_P}{r_1} \right) \qquad (3.59)$$

$$v_{t2} = \omega_2 \rho_2 = \omega_2 (r_2 \sin \alpha - g_P) = v \left(\sin \alpha - \frac{g_P}{r_2} \right). \qquad (3.60)$$

With pinion driving, the term g_P is regarded as negative during the approach path of contact (contact between the dedendum flank of the pinion and addendum flank of the wheel), and positive during the recess path of contact (contact between the addendum flank of the pinion and dedendum flank of the wheel).

Notoriously, two virtual equivalent cylinders, with the appropriate rolling and sliding motions along their contact line, may represent the conditions of teeth engagement. In the case of spur gears, all transverse sections are similar, and the contact line between the two cylinders is perpendicular to the plane of section, and thus parallel to the axes of the gears. In any transverse section, the two cylinders become two circles, also virtual, touching at point P (this is the intersection of the contact line between the cylinders with the transverse section under consideration), and having centers at fixed points T_1 and T_2, and radii ρ_1 and ρ_2, the latter variable without solution of continuity during the meshing (Fig. 3.6a). So, it is evident that v_{t1} and v_{t2} are nothing more than the instantaneous tangential velocities or absolute velocities at these circles; these velocities are known in the scientific literature as the *rolling velocities*.

As Eqs. (3.59) and (3.60) show, the magnitudes of the absolute values of the rolling velocities v_{t1} and v_{t2} vary linearly as a function of the position of the point of contact P on the line of action, i.e. as a function of g_P (note that, in the case of non-involute profiles, these velocities differ in both magnitude and direction), but

v_{t1} increases and v_{t2} decreases during meshing. These velocities have different magnitudes, but they are collinear, as their straight line of application is the common tangent to the mating profiles at point of contact, that is the perpendicular straight line at point P to the line of action; however, they become equal only when, during the meshing, the point of contact P comes to coincide with the pitch point C.

Equation (3.59) shows that the rolling velocity v_{t1} is always positive (it would be equal to zero only in the case were the point of contact P were to coincide with the point of interference T_1 of the pinion), and acts outwards of the pinion, with the pinion driving. In addition, the rolling velocity v_{t2} given by Eq. (3.60) is positive (this would also be equal to zero in the case in which the point of contact P were to coincide with the point of interference T_2 of the wheel), and acts radially inwards over the wheel teeth.

During recess (g_P positive), with the pinion driving, the absolute value of v_{t1} is greater than the one of v_{t2}, and the *rolling velocity ratio*, defined as the ratio of the smaller to the larger rolling velocity, is given by:

$$\frac{v_{t2}}{v_{t1}} = \frac{r_2 \sin \alpha - g_P}{u(r_1 \sin \alpha + g_P)}. \tag{3.61}$$

Conversely, during approach (g_P negative), the absolute value of v_{t1} is less than the one of v_{t2}, and thus the rolling velocity ratio is given by:

$$\frac{v_{t1}}{v_{t2}} = \frac{u(r_1 \sin \alpha + g_P)}{r_2 \sin \alpha - g_P}. \tag{3.62}$$

We define as *cumulative velocity vector*, v_Σ, the vector sum:

$$v_\Sigma = v_{t1} + v_{t2} = v\left[2 \sin \alpha + g_P\left(\frac{1}{r_1} + \frac{1}{r_2}\right)\right] = v\left[2 \sin \alpha + \frac{g_P}{r_1}\left(1 + \frac{1}{u}\right)\right], \tag{3.63}$$

to be used with the sign convention described above. This cumulative velocity is a very important quantity, because it is related to the pressure that is generated within the lubricant film between the flanks of mating teeth, and therefore to the actually gear capacity to withstand the working loads. From Eq. (3.63) we deduce the following dimensionless ratio, K_Σ, between the absolute values of the velocities, v_Σ and v, called the *cumulative factor*:

$$K_\Sigma = \frac{v_\Sigma}{v} = \left[2 \sin \alpha + \frac{g_P}{r_1}\left(1 + \frac{1}{u}\right)\right], \tag{3.64}$$

for which we apply also the above-mentioned sign convention.

It is noteworthy that Dowson and Higginson [20] define as rolling velocity, v_w, the *cumulative semi-velocity vector*, that is $v_w = (v_\Sigma/2)$. This velocity is thus the *average rolling velocity*, and it is a measure of the velocity at which lubricant enters the mesh. Fluid film development into the mesh zone depends on this average rolling velocity, which affects the fluid film thickness at the mesh. From this point of view, it would be appropriate to call this average rolling velocity the *entrainment velocity*.

However, we must keep in mind that the entrainment of lubricant into the contact zone between the teeth depends on the surface topology of the tooth flanks and their angular displacements about the line of action at point of contact, as well as the average rolling velocity v_w. This velocity and the sliding velocity v_g are parallel vectors only for cylindrical spur gears. In the more general case of crossed or skew gears, these vectors are no longer parallel; therefore, in this case, the angle between them, and its variability during the meshing come into play.

The *spinning velocity* is defined as the change in this angle per unit time. Since the sliding velocity vector v_g, rolling velocity vector v_w, and angle between v_g and v_w are strongly affected by the pressure angle α, spiral angle β, and shaft angle Σ, tribologists studying the rheological conditions of gearing in the presence of a lubricant introduce an *entrainment velocity* more complex than that given by the cumulative semi-velocity vector v_w. On this subject we refer the reader to specialized textbooks and the relevant scientific literature (see Dooner [18], Dooner and Seireg [19]).

Of course, since v_{t1} is generally different from v_{t2}, the two profiles slide relative to one another. We define as *sliding velocity vector* v_g between the two conjugate profiles at point P the difference between the related rolling velocities v_{t1} and v_{t2}, that is, respectively:

$$v_{g1} = (v_{t1} - v_{t2}) \quad v_{g2} = (v_{t2} - v_{t1}) = -v_{g1}. \tag{3.65}$$

In fact, everything works as if, in the relative motion between the above-mentioned equivalent cylinders, profile 2 rolled sliding on profile 1, supposed in a fixed position, and vice versa.

Taking into account Eqs. (3.59) and (3.60), from these equations, we obtain:

$$v_{g1,2} = \pm g_P v \left(\frac{1}{r_1} + \frac{1}{r_2} \right) = \pm g_P (\omega_1 + \omega_2) = \pm g_P \omega, \tag{3.66}$$

where $\omega = (\omega_1 + \omega_2)$ is the absolute value of the *relative angular velocity*, and the upper and lower signs apply for v_{g1} and v_{g2}, respectively; the sign convention for g_P is that previously indicated.

The sliding velocity is therefore equal to zero at pitch point C, where $g_P = 0$, while its absolute value increases proportionally to the distance g_P from the pitch point, reaching its maximum absolute value at points A and E (Fig. 2.10), where the contact respectively begins and ends. This sliding velocity is a quantity extremely important in the gear design. Indeed, it has a great influence on the power losses by friction and the related heating, as well as on the wear conditions and scuffing load carrying capacity.

Figure 3.7 shows the qualitative distribution curves of the values of the components v_n, v_{t1} and v_{t2} of the absolute velocities v_{P1} and v_{P2} of point of contact

P about the fixed centers O_1 and O_2, as well as the ones of the cumulative velocity v_Σ and sliding velocities v_{g1} and v_{g2}, for a given external gear pair. *Ictu oculi*, we see that v_n is a constant, v_{t1} and v_{t2} have the same direction, while v_{g1} and v_{g2} change their direction as the point of contact crosses the pitch circles. More specifically, v_{g1} is negative (i.e., it is directed from the pitch circle to the dedendum circle) during the approach path of contact, and positive (i.e., it is directed from the pitch circle to the addendum circle) during the recess path of contact. The sliding velocity v_{g2} instead behaves in the opposite way to the sliding velocity v_{g1}.

This reversal of the direction of the sliding velocities at the pitch point causes, in a contact inevitably characterized by friction, a cyclic variation of the contact forces that come into play; it is undoubtedly a source of fatigue and noise.

In the optimal condition, the maximum values of v_{g1} and v_{g2} at points A and E, where the contact begins and ends, must be equal (the reason of this is described in the next section). In this condition, a fictitious average sliding velocity, v_{gm}, is introduced [39]; it is given by the following relationship:

$$v_{gm} = \frac{v_{ga}g_{Pa} + v_{gf}g_{Pf}}{2g_\alpha} = \frac{v_{ga}^2 + v_{gf}^2}{2(v_{ga} + v_{gf})}; \tag{3.67}$$

this velocity, that is constant along the path of contact, is used for a simplified calculation of the power losses due to friction in the contact between the teeth.

Finally, as dimensionless characteristic of the toothing geometry, a sliding factor K_g is sometimes used, defined as the ratio between the absolute values of the sliding velocity v_g and tangential velocity v at pitch circle, that is:

$$K_{g1,2} = \frac{v_g}{v} = \pm g_P\left(\frac{1}{r_1} + \frac{1}{r_2}\right) = \pm \frac{g_P}{r_1}\left(1 + \frac{1}{u}\right); \tag{3.68}$$

on this factor, we can do the same considerations made for v_g, and use the same sign conventions.

The kinematic quantities concerning the internal gear pairs can be obtained with considerations similar to those described above for the external gear pairs. This work, however, is left to the reader. In this textbook, we are only going to provide the relationships relating the internal gears. Equations (3.56, 3.58 and 3.59) do not change. Equations (3.60 and 3.66) change and become, respectively:

$$v_{t2} = v\left(\sin\alpha + \frac{g_P}{r_2}\right) \tag{3.69}$$

$$v_{g1,2} = \pm g_P v\left(\frac{1}{r_1} - \frac{1}{r_2}\right) = \pm g_P(\omega_1 - \omega_2) = \pm g_P\omega, \tag{3.70}$$

where $\omega = (\omega_1 - \omega_2)$ is the absolute value of the relative angular velocity of the internal gear pair.

Fig. 3.7 Qualitative distribution curves of v_n, v_{t1}, v_{t2}, v_Σ, v_{g1}, v_{g2}, ζ_1 and ζ_2, for a given external gear pair

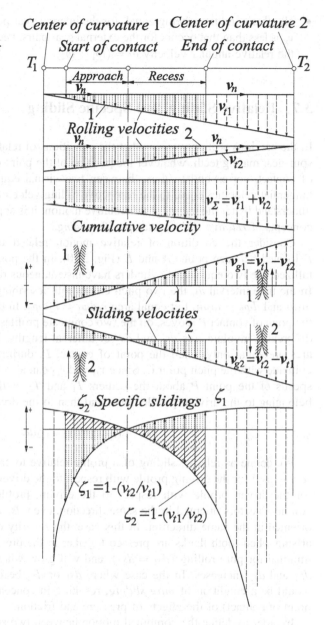

If we take into consideration Fig. 2.7 and on it we built diagrams of v_{t1}, v_{t2} and v_g corresponding to those shown in Fig. 3.7, we would notice that:

- contrary to what occurs for the external gear pairs, for the internal gear pairs both v_{t1} and v_{t2}, during the meshing, increase linearly as a function of the position of the point of contact P on the line of action, i.e., as a function of g_P;

- all other conditions being equal, for the internal gear pairs the absolute value of v_g is less than that occurs for the external gear pairs, because it is proportional to the relative angular velocity $\omega = (\omega_1 - \omega_2)$.

3.7 Relative Sliding and Specific Sliding

In the previous section, we saw that the conditions of relative motion between two spur-gear mating teeth, whatever the position of the point of contact along the path of contact, correspond to those between two virtual equivalent cylinders having tangential velocities equal to the respective rolling velocities v_{t1} and v_{t2}. In order to understand better the nature of this relative motion, it is appropriate to introduce the concepts of *relative sliding* and *specific sliding*.

Consider the condition of relative motion related to any point of contact P located between points A and E (Fig. 2.10). In the position of contact so crystallized, the two equivalent cylinders have instantaneous radii ρ_1 and ρ_2 (Fig. 3.6). In the time interval dt, the two pitch circles roll describing different angles $d\varphi_1 = \omega_1 dt$ and $d\varphi_2 = \omega_2 dt$, but equal arcs $r_1 d\varphi_1 = r_2 d\varphi_2$. In the same time interval dt, the point of contact P moves, on the two conjugate profiles, describing on them two differential involute arcs ds_1 and ds_2 of different lengths. The lengths of these two arcs are equal only when the point of contact P, during the meshing, comes to coincide with the pitch point C. Since v_{t1} and v_{t2} can also be conceived as absolute speeds of the point P about the centers T_1 and T_2, with point P thought of as belonging to the driving profile once, and then to the driven profile, we have:

$$ds_1 = v_{t1} dt \quad ds_2 = v_{t2} dt. \tag{3.71}$$

We define as relative sliding of a profile relative to each other, the differences $(ds_1 - ds_2)$ for the driving profile with respect to the driven profile, and $(ds_2 - ds_1)$ for the driven profile with respect to the driving profile. Since v_{t1} and v_{t2} are collinear vectors oriented in the same direction, even ds_1 and ds_2 are collinear and oriented in the same direction. In this case, the severity of the contact conditions arising when tooth flanks are pressed together in the presence of lubricant will be minimal for pure rolling $(ds_1 = ds_2)$, and will grow when the difference between ds_1 and ds_2 increases. In the case where ds_1 or ds_2 become equal to zero, there would be a condition of *pure sliding*, resulting in concentration at one point (the point of contact) of the effects of pressure and friction.

In order to define the combined motion between two surfaces, often the *rolling velocity ratio* is introduced, defined as the ratio between the smaller and the larger rolling velocity (see Eqs. 3.61 and 3.62). Taking algebraic sign into account, the range of possible values of this rolling velocity ratio lies between $+1$ and -1. In the case of spur, helical and bevel gears, the values always lie in the range $+1$ and 0.

Instead, to define the conditions of contact between two surfaces, the *specific sliding* and the *slide-roll ratio* are introduced. These two factors, despite the different definition, in fact coincide; for this reason, we designed them with the same symbols.

We define as specific sliding of one profile the ratio between the relative sliding of the profile itself with respect to the other and the arc along which, on the same profile, the contact is moved, and therefore, for the two profiles we have respectively:

$$\zeta_1 = \frac{(ds_1 - ds_2)}{ds_1} \quad \zeta_2 = \frac{(ds_2 - ds_1)}{ds_2}. \tag{3.72}$$

Instead we define as slide-roll ratio of a given profile the ratio between the sliding velocity of the profile itself with respect to the other at the point of contact and the rolling velocity at the same point considered belonging to the same profile, and therefore, for the two profiles we have respectively:

$$\zeta_1 = \frac{v_{g1}}{v_{t1}} = \frac{(v_{t1} - v_{t2})}{v_{t1}} \quad \zeta_2 = \frac{v_{g2}}{v_{t2}} = \frac{(v_{t2} - v_{t1})}{v_{t2}}. \tag{3.73}$$

Taking into account Eq. (3.71), the Eqs. (3.72) and (3.73) obviously coincide. Thus, recalling Eqs. (3.61 and 3.62), from Eq. (3.73) we obtain:

$$\zeta_1 = 1 - \frac{v_{t2}}{v_{t1}} = 1 - \frac{r_2 \sin \alpha - g_p}{u(r_1 \sin \alpha + g_p)} = \frac{(1 + 1/u)g_p}{r_1 \sin \alpha + g_p} \tag{3.74}$$

$$\zeta_2 = 1 - \frac{v_{t1}}{v_{t2}} = 1 - \frac{u(r_1 \sin \alpha + g_p)}{r_2 \sin \alpha - g_p} = -\frac{(1 + u)g_p}{r_2 \sin \alpha - g_p}, \tag{3.75}$$

which are respectively valid for the driving and driven gear wheels. The sign conventions to be used for the latter equations are obviously those previously described.

Equations (3.74) and (3.75) show that the specific slidings and equivalent slide-roll ratios are continuously variable along the path of contact. They are equal to zero at the pitch point and reach their absolute maximum values in the extreme points A and E of the path of contact (Fig. 2.10). These absolute maximum values are negative (negative specific slidings) and occur during the approach path of contact for the driving gear wheel, and during the recess path of contact for the driven gear wheel. In general, therefore, for the usual progressive contact that occurs in spur, helical and bevel gears, the specific slidings and slide-roll ratios are positive for all points on the addendum flanks of the teeth, and negative for all points on the dedendum flanks of the teeth.

Experimental evidence shows that the destructive effects of pressure and friction are greater on the flank surfaces having negative specific slidings and slide-roll ratios. What is more, the absolute values of these negative quantities are numerically the greater. A high negative value of these quantities, which occurs during the dedendum contact for both toothed wheels of the gear pair, means that a small arc of the dedendum profile of a tooth flank meshes with a large arc of the addendum profile of the conjugate tooth flank. Therefore, the work of the friction forces acts on the small arc of the dedendum profile of a tooth, while operates on a great arc of the addendum profile of the conjugate tooth. The resulting damages due to the wear are therefore much more pronounced on the dedendum portions of the tooth flank profiles.

If the path of contact was used until the limiting interference points T_1 and T_2 (in this case, as Fig. 2.10 shows, the radii of the addendum circles of the two gears would be respectively equal to O_2T_1 and O_1T_2), we would have $g_P = r_1 \sin \alpha$ in Eq. (3.74), and $g_P = -r_2 \sin \alpha$ in Eq. (3.75). Thus, we would have $\zeta_1 = -\infty$ at point T_1, and $\zeta_2 = -\infty$ at point T_2. Based on this result, the opportunity to exclude from the contact the portions of the involute profiles near the base circles is evident.

Figure 3.7 shows also the qualitative diagrams of the quantities ζ_1 and ζ_2, which characterize the external gear pair referred to therein. The figure shows the maximum negative values of these two quantities at point A and E, as well as their asymptotic values tending to minus infinity at both points T_1 and T_2. Scientists and industry experts attach great importance to the distribution of these quantities along the path of contact, as they have a deep effect on the gear design quality (see Giovannozzi [24], Henriot [27], Niemann and Winter [39]). The requirements considered optimum for the distributions of these quantities along the path of contact are essentially two:

- the absolute maximum values of ζ_1 and ζ_2 in the extreme points A and E of the path of contact must not be too high;
- their distribution curves should not be too different for the two members of the gear pair.

The way to achieve both of these important design goals is practically obliged: to use gears with addendum modification in terms of profile shift, which offers to the designer many opportunities of intervention.

Unlike external gears, for an internal gear pair both the rolling velocities v_{t1} and v_{t2} (given respectively by Eqs. (3.59) and (3.69)) increase as g_P increases.

During recess (g_P positive), with the pinion driving, v_{t1} is greater than v_{t2}, and the rolling velocity ratio is given by:

$$\frac{v_{t2}}{v_{t1}} = \frac{r_2 \sin \alpha + g_P}{u(r_1 \sin \alpha + g_P)}; \tag{3.76}$$

instead, during approach (g_P negative), v_{t1} is less than v_{t2}, and thus the rolling velocity ratio is:

$$\frac{v_{t1}}{v_{t2}} = \frac{u(r_1 \sin \alpha + g_P)}{r_2 \sin \alpha + g_P}. \tag{3.77}$$

The corresponding relationships to calculate the specific slidings (or the slide-roll ratios) are:

$$\zeta_1 = 1 - \frac{v_{t2}}{v_{t1}} = 1 - \frac{r_2 \sin\alpha + g_P}{u(r_1 \sin\alpha + g_P)} = \frac{(1 - 1/u)g_P}{r_1 \sin\alpha + g_P} \qquad (3.78)$$

$$\zeta_2 = 1 - \frac{v_{t1}}{v_{t2}} = 1 - \frac{u(r_1 \sin\alpha + g_P)}{r_2 \sin\alpha + g_P} = -\frac{(u - 1)g_P}{r_2 \sin\alpha + g_P}. \qquad (3.79)$$

3.8 Consideration on Wear Damage

In the previous section, we have seen that the specific sliding is one of the main causes of wear of the tooth flank surface. From the theoretical point of view, there would be no changes in the tooth profile if the abrasive wear [32] of the profile itself proceed in such a way that the ratio (s_w/r_b) was constant. This is the ratio between the tooth thickness worm away at a given point, s_w, measured along the normal to the profile, and the moment arm, r_b, of the normal to the profile at the same point with respect to the wheel axis. In this way disfigurements and changes of profile due to wear correspond to a virtual rotation of the same profile about the fixed center of the wheel [24].

However, it is to remember that the *hypothesis of Reye* [45], referring to the wear, states that the volume of material removed in a given time interval due to the wear is proportional to the work done by the friction forces in the same time interval. Consequently, the so-called *wear equation* can be written as:

$$\frac{s_w}{t} = \left(\frac{K}{H}\right)pv_g, \qquad (3.80)$$

where:

- s_w/t is the *wear velocity*, that is the quotient of the *wear depth* s_w (in mm) divided by the time t (in s);
- K is the *dimensionless wear coefficient*, which depends on the wear modes (abrasive wear, adhesive wear, fretting, etc.), material combinations (the softer of the two rubbing metals conditions the combination of the materials involved), and presence or absence of lubricant;
- H (in MPa) is the surface hardness;
- p (in MPa) is the surface interface pressure;
- v_g (in mm/s) is the sliding velocity.

For two rubbing surfaces, which are the flanks of two mating teeth, this equation implies that the rate of wear of one of the two flanks, with respect to the conjugate flank, is proportional to the wear coefficient, that of material of the tooth under consideration in contact with the material of the mating tooth. The same rate of wear is inversely proportional to the surface hardness of the tooth analyzed, and directly proportional to the rate of friction work, if we assume a constant coefficient of friction.

For a given compressive force between the flank surfaces, the volume of material worm away is independent of the area of contact. Therefore, multiplying both sides of Eq. (3.80) for the area of the contact surface, we infer that the material depth s_w removed by wear is proportional to the specific sliding ζ multiplied by the normal force $(F_t/\cos\alpha)$ acting on the tooth (F_t is the nominal transverse tangential force at reference cylinder). It should be noted that here we have used the normal force $F_t/\cos\alpha$ (Fig. 3.9a), which refers to the ideal case of frictionless. Due to the non-eliminable friction, the problem becomes more complex, as we will see in the next section.

However, for involute gearing, we have:

$$\frac{F_t}{\cos\alpha} = const \qquad r_b = r\cos\alpha = const. \qquad (3.81)$$

Therefore, if we accept the validity of the hypotheses of Reye, we cannot have a ratio $s_w/r_b = const$. In fact, if r_b is a constant, s_w is a variable, because the latter is proportional to ζ, which is continuously variable during the meshing. Thus, also the virtual rotation of the profile, corresponding to the wear depth along it, is very variable as it is proportional to ζ.

About other possible surface and sub-surface damages that can be found on the active flanks of the teeth, it is interesting to examine how the point of contact moves along the tooth profile during meshing, and what is the direction of the sliding. In this regard, we distinguish in meshing cycle the stage of approach from that of recess (see also the previous section).

During the stage of approach, the point of contact moves, on the driving profile, from the dedendum circle towards the pitch circle, and, on the driven profile, from the addendum circle to the pitch circle. Instead, the relative sliding, on the driving profile, is directed from the pitch circle to the dedendum circle, and, on the driving profile, is directed from the pitch circle to the dedendum circle. Therefore, on the driven profile, the motion of displacement of the point of contact on the profile and the motion of sliding have opposite directions, while, on the driven profile, they have equal directions. Thus, on the driving and driven profiles, during the stage of approach, we have the so-called *chamfering sliding* and respectively *stretching sliding* [39].

During the stage of recess, the point of contact moves, on the driving profile, from the pitch circle towards the addendum circle, and, on the driven profile, from the pitch circle to the dedendum circle. Instead, the relative sliding, on the driving profile, is directed from the pitch circle to the dedendum circle, and, on the driven profile, is directed from the pitch circle to the addendum circle. Therefore, on the driving profile, the motion of displacement of the point ζ contact on the profile and the motion of sliding have equal directions, while, on the driven profile, they have opposite directions. Thus, on the driving and driven profiles, during the stage of recess, we have the stretching sliding and chamfering sliding, respectively.

Fig. 3.8 shows the directions of rolling and sliding, and highlights the directions of propagation of fatigue cracks.

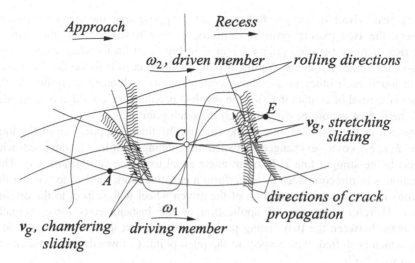

Fig. 3.8 Direction of rolling, sliding and propagation of fatigue cracks

3.9 Efficiency of Cylindrical Spur Gears

Table 1.1 shows the indicative values of efficiency of the various types of gears, used in practical applications. In this section, we want to deepen this important topic, with special reference to cylindrical spur gears. The general analytical bases to calculate the efficiency of gears were developed starting from the second half of the 19th century, with insights and generalizations gradually processed in the first half of the 20th century. In this regard, contributions specifically to be mentioned are those of [[9], 12, 34, 44, 50]. These contributions led to the definition of analytical relationships that, with little significant changes, coincide with the equations now universally used, and described below (see also Ambekar [1], Ferrari and Romiti [21], Scotto Lavina [47]).

Consider the cylindrical spur gear shown in Fig. 2.10, and provisionally assume that the transverse contact ratio is equal to unity ($\varepsilon_\alpha = 1$), whereby the path of contact, g_α, coincides with the transverse base pitch, p_{bt}. Therefore, we have a spur gear operating in the limiting condition of motion continuity, and only one mating tooth pair is in meshing along the entire path of contact. Suppose also that pinion and gear wheel are, respectively, the driving and driven wheels, and that the directions of rotation are as indicated in the same figure. We assume finally that the involute profiles of the teeth are perfectly shaped, and equally spaced along the base circle (Fig. 2.2).

Let ω_1, ω_2, T_1 and T_2, the angular velocities and nominal torques (as usual, subscript 1 and 2 refer respectively to the driving pinion, or simply driver, and the driven gear wheel or simply follower), and suppose that the point of contact between the two mating profiles is placed at any position along the path of approach, namely between points A and C in Fig. 2.10. In the ideal case of zero friction, the force $(F_t / \cos \alpha)$ that the driving profile transmits to the driven profile, which touch each other in the above current point, is directed according to the common normal of contact that, for the involute profiles here considered, coincides with the line of action, passing through the pitch point, C.

Instead, in the real case, due to the friction, the line of application of the total force $F \neq F_t / \cos \alpha$, exchanged between the mating profiles, is deviated with respect to the straight line $T_1 T_2$ of an angle equal to the *angle of friction*, φ. This deviation is in direction such as to perform a negative work (and so to counter the motion), during the relative motion of the driven wheel with respect to the driving wheel. Therefore, the line of application of the instantaneous force, actually exchanged between the two mating profiles, will intersect the center distance at a point which is shifted, with respect to the pitch point, C, towards the center of the driven wheel, O_2.

It is to be noted that, for external gears, this occurs whatever the position of the point of contact along the path of contact. In other words, the point of contact may be a point of the path of approach or a point of the path of recess. This behavior is justified by the fact that, to perform a negative work, the total force F must have, with respect to the center of instantaneous rotation of the relative motion (i.e., the pitch point, C), a torque of opposite sign with respect to that of the relative angular velocity, $\omega = (\omega_2 - \omega_1)$, of the driven wheel with respect to the driving wheel.

It is also to be noted that $\mu = \tan \varphi$ is the coefficient of sliding friction that, for dry surfaces, satisfies approximately the *Coulomb's law of friction* (see Caubet [11], Coulomb [15]), and the two *Amontons' laws* [2]. According to these laws, the kinetic friction is independent of the sliding velocity (*Coulomb's law*), and the force of friction is directly proportional to the applied load (*Amontons' first law*), and independent of the apparent area of contact (*Amontons' second law*). As we will point out in the "*Gears - Vol.3: A concise history*" of this monothematic textbook, to honor the historical reality, the two Amontons' Laws should be attributed to Leonardo da Vinci, who was the first to formulate them.

For internal gears, the torque of the total force, F, that the driving profile transmits to the driven profile must always be of opposite sign with respect to that of the relative angular velocity, $\omega = (\omega_2 - \omega_1)$ of the driven wheel with respect to the driving wheel. Therefore, this torque must have the same sign as that of $(-\omega_1)$, since $|\omega_1| > |\omega_2|$. Consequently, the line of application of the exchanged force, F, must intersect the straight line $O_1 C$, passing through the centers O_1 and O_2 of the two members of the gear under consideration, at a point located at the outside of the pitch point, C, i.e. at a point shifted towards the body of the annulus gear.

We determine now the *instantaneous efficiency*, η_i, of the external gear pair, corresponding to any position of the point of contact along the path of approach (Fig. 3.9a), assuming that the pinion is the driver. In the ideal case, if F_t is the nominal transverse tangential load, we will have $T_1 = F_t r_1$ and $T_2 = F_t r_2$, whereby the ratio between the driving torque, T_1 (i.e., the nominal torque at the pinion), and

(a)   **(b)**

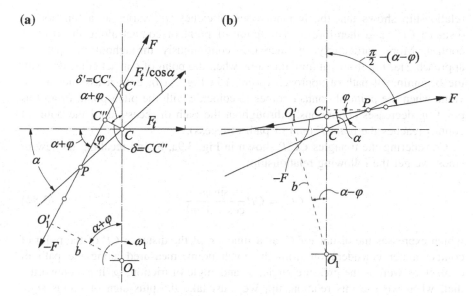

Fig. 3.9 Geometric quantities for determination of the total force exchanged between the teeth, when the point of contact coincides with: **a** A point of the path of approach; **b** A point of the path of recess

the useful resisting torque, T_2 (i.e., the nominal torque at the wheel), is equal to the ratio between the radii r_1 and r_2 of the reference pitch circles, i.e., $(T_1/T_2) = r_1/r_2$. It should be noted that T_1 and T_2 are no longer the interference points, but rather the driving and driven torques.

However, in the real case (Fig. 3.9a), we are facing an actual force $F \neq F_t/\cos\alpha$, whose line of application intersects the center distance at point C'. The component of this force applied at point C', in the direction normal to the straight line O_1O_2, is $[F\cos(\alpha+\varphi)] \neq F_t$. In addition, the position of point C' varies, depending of the position of the current point of contact P along the path of approach. Thus, the actual instantaneous driving and driven torques will be given respectively by: $T_1' = [F\cos(\alpha+\varphi)](r_1 + CC')$,   and $T_2' = [F\cos(\alpha+\varphi)]$ $(r_2 - CC')$. Thus, we get:

$$\frac{T_1'}{T_2'} = \frac{r_1 + CC'}{r_2 - CC'}. \tag{3.82}$$

Therefore, the instantaneous efficiency will be given by:

$$\eta_i = \frac{T_1}{T_1'} = \frac{(r_1/r_2)}{[(r_1 + CC')/(r_2 - CC')]} = \frac{1 - (CC'/r_2)}{1 + (CC'/r_1)}. \tag{3.83}$$

Easily, it can be shown that the relationship Eq. (3.83) is also valid when the point of contact coincides with any point of the path of recess (Fig. 3.9b). The same relationship shows that the instantaneous efficiency, η_i, varies as a function of distance CC', and therefore as a function of point of contact along the path of contact. More particularly, η_i increases continuously throughout the path of approach, starting from a minimum value when the point of contact coincides with the beginning of path of approach (point A in Fig. 2.10), until it assumes a unit value when the point of contact comes to coincide with the pitch point. From this point, η_i decreases continuously throughout the path of recess, until the point of contact reaches the point E (end of path of recess).

Considering the triangles $CC'P$ shown in Fig. 3.9a, b, and recalling the law of sines, we get the following relationship:

$$CC' = CP \frac{\sin \varphi}{\cos(\alpha \pm \varphi)}, \tag{3.84}$$

which expresses the distance CC' as a function of the distance CP of the point of contact under consideration from the pitch point, measured along the path of contact, as well as the pressure angle, α, and angle of friction, φ. It is noteworthy that, when we use this relationship, we must take the plus sign or minus sign, depending on whether the point of contact belongs to the path of approach or path of recess.

Yet we can easily prove that the relationship (3.83) expresses the instantaneous efficiency of an internal gear pair, where the pinion is the driver, as long as the radius r_2 of the pitch circle of the annulus is taken with negative sign.

Equivalently, considering the absolute value of r_2, it must replace the minus sign, which appears in the numerator of the last fraction of Eq. (3.83), with the double sign $\mp$, taking the minus sign for external gear pairs, and the plus sign for internal gear pairs.

In the case in which the gear pair under consideration had the pinion as a follower, and the gear wheel as a driver, the instantaneous efficiency would be given by the following relationship:

$$\eta_i = \frac{1 - (CC'/r_1)}{1 \pm (CC'/r_2)}, \tag{3.85}$$

where the plus sign is to be taken for external gears, and the minus sign is to be taken for internal gears.

The instantaneous efficiency is continuously variable as the point of contact moves along the path of contact. Therefore, it is necessary to calculate the *average efficiency* during a meshing cycle, which corresponds to the time interval in which the pitch circles roll on one another, without sliding, of an arc equal to the circular pitch, after which the operating conditions of the gear pair under consideration repeat itself cyclically; in this regard, we must remember that we have assumed $\varepsilon_\alpha = 1$. To this end, let us consider an external gear pair with a driving pinion; during the path of approach (Fig. 3.9a), the differential energy (or work) lost by friction, dW_μ, in the infinitesimal time interval dt is given by:

$$\frac{dW_\mu}{dt} = F|\omega_2 - \omega_1|\delta, \tag{3.86}$$

where $\delta = CC''$ is the distance of the instantaneous line of application of instantaneous force F from the pitch point. From Fig. 3.9a, we get:

$$\delta = CC' \cos(\alpha + \varphi) = \delta' \cos(\alpha + \varphi), \tag{3.87}$$

where $\delta' = CC'$.

Therefore, taking account of Eqs. (3.84) and (3.87), Eq. (3.86) becomes:

$$\frac{dW_\mu}{dt} = F|\omega_2 - \omega_1|(CP) \sin \varphi. \tag{3.88}$$

On the other hand, if F is the instantaneous force exerted by the driving profile on the driven profile, an equal and opposite force, i.e. $(-F)$, will act on the driving profile. Therefore, assuming a steady-state operating condition, which is characterized by a constant angular velocity, the torque T_1 applied by this force to the pinion will be $T_1 = Fb$, where b is the moment arm (or lever arm), that is the distance of the line of application of the force $(-F)$ from the center O_1. From Fig. 3.9a, we obtain:

$$b = O_1O_1' = (r_1 + CC') \cos(\alpha + \varphi) = r_1 \cos(\alpha + \varphi) + (CP) \sin \varphi. \tag{3.89}$$

Since $T_1 = Fb$, we get:

$$F = \frac{T_1}{b} = \frac{T_1}{r_1 \cos(\alpha + \varphi) + (CP) \sin \varphi}. \tag{3.90}$$

Therefore, Eq. (3.88) becomes:

$$\frac{dW_\mu}{dt} = |\omega_2 - \omega_1| \frac{T_1}{[r_1 \cos(\alpha + \varphi) + (CP) \sin \varphi]}(CP) \sin \varphi. \tag{3.91}$$

Since the point of contact P moves along the path of contact with a constant velocity, given by Eq. (3.58), the distance CP of the point of contact from the pitch point is expressible as:

$$CP = \varphi_1 r_1 \cos \alpha, \tag{3.92}$$

where φ_1 is the angle of rotation of the driving gear wheel (the pinion), corresponding to the displacement of point of contact from P to C.

It follows that, for contact at a current point of the path of approach, the quotient of the differential work lost by friction divided by the differential work at the same time done by the driving torque T_1 can be expressed by:

$$\frac{dW_\mu}{T_1\omega_1 dt} = \left|\frac{\omega_2}{\omega_1} - 1\right| \frac{\varphi_1 \cos\alpha \sin\varphi}{[\cos(\alpha + \varphi) + \varphi_1 \cos\alpha \sin\varphi]}$$

$$= \left|1 \pm \frac{r_1}{r_2}\right| \frac{\varphi_1 \cos\alpha \sin\varphi}{[\cos(\alpha + \varphi) + \varphi_1 \cos\alpha \sin\varphi]},$$

(3.93)

since $(\omega_2/\omega_1) = \mp(r_1/r_2)$, depending on whether the gear pair is external (minus sign) or internal (plus sign).

If contact occurs at a point of the path of recess (Fig. 3.9b), we will have:

$$b = O_1 O_1' = (r_1 + CC')\cos(\alpha - \varphi),$$

(3.94)

with

$$CC' = CP \frac{\sin\varphi}{\cos(\alpha - \varphi)}.$$

(3.95)

By repeating the previous procedure, in place of Eq. (3.93), we get:

$$\frac{dW_\mu}{T_1\omega_1 dt} = \left|1 \pm \frac{r_1}{r_2}\right| \frac{\varphi_1 \cos\alpha \sin\varphi}{[\cos(\alpha - \varphi) + \varphi_1 \cos\alpha \sin\varphi]}.$$

(3.96)

Dividing by $\cos\alpha \sin\varphi$ the numerator and denominator of the fractions to the right side of Eqs. (3.93) and (3.96), and bearing in mind that $\omega_1 dt = d\varphi_1$ and $\mu = \tan\varphi$, we can write:

$$\frac{dW_\mu}{T_1 d\varphi_1} = \left|1 \pm \frac{r_1}{r_2}\right| \frac{\mu\varphi_1}{[1 + \mu(\varphi_1 \mp \tan\alpha)]},$$

(3.97)

where, at the denominator of the last fraction, we must take the minus sign or plus sign, depending on whether the point of contact is within the path of approach or within the path of recess.

Assuming then that the driving torque T_1 is a constant, and remembering that $\varepsilon_\alpha = 1$, we infer that the work dissipated by friction, in the time corresponding to duration of contact between the two teeth, i.e. during the entire meshing cycle, is the summation of all differential works lost by friction during the paths of approach and recess. Thus, it is given by the following equation:

$$W_\mu = T_1 \left|1 \pm \frac{r_1}{r_2}\right| \mu_0 \left\{ \int_0^{e_1/r_1} \frac{(\mu/\mu_0)\varphi_1 d\varphi_1}{[1 + \mu(\varphi_1 - \tan\alpha)]} + \int_0^{e_2/r_1} \frac{(\mu/\mu_0)\varphi_1 d\varphi_1}{[1 + \mu(\varphi_1 + \tan\alpha)]} \right\}$$

$$= T_1 \left|1 \pm \frac{r_1}{r_2}\right| \mu_0 \Delta,$$

(3.98)

where e_1 and e_2 are respectively the arcs of approach and recess, given by Eqs. (3.20) and (3.21), Δ is a parameter that briefly indicates all that the bracket contains, and μ_0 is the value of coefficient of friction μ corresponding to a given point of path of contact, which can be determined conventionally.

Taking into account that, during the same meshing cycle considered for calculation of W_μ, the work done by the driving torque, W_m, is given by:

$$W_m = T_1 \frac{e_1 + e_2}{r_1}, \tag{3.99}$$

we infer that the loss of efficiency of the cylindrical spur gear pair examined here can be expressed as:

$$1 - \eta = \frac{W_\mu}{W_m} = \frac{r_1}{e_1 + e_2} \left| 1 \pm \frac{r_1}{r_2} \right| \mu_0 \Delta. \tag{3.100}$$

From Eq. (3.43), for $\varepsilon_\alpha = 1$, we get $e = e_1 + e_2 = p$. Moreover, Eq. (3.98) shows that Δ is of the order of magnitude of $(e/r_1)^2$, and thus of the order of magnitude of $(p/r_1)^2$. Therefore, if we introduce the notation:

$$\Delta^* = \Delta \left(\frac{r_1}{p} \right)^2, \tag{3.101}$$

Δ^* will be of the order of magnitude of unity. Finally, since $(r_1/r_2) = (z_1/z_2)$, Eq. (3.100), can be written in the form:

$$1 - \eta = \frac{p}{r_1} \left| 1 \pm \frac{z_1}{z_2} \right| \mu_0 \Delta^* = 2\pi \left| \frac{1}{z_1} \pm \frac{1}{z_2} \right| \mu_0 \Delta^*. \tag{3.102}$$

Therefore, the average efficiency will be given by:

$$\eta = 1 - 2\pi \left| \frac{1}{z_1} \pm \frac{1}{z_2} \right| \mu_0 \Delta^*. \tag{3.103}$$

If we assume that the coefficient of the friction, μ, varies little during the meshing cycle, for which it can be considered as a constant, the calculation of Δ^* becomes very simple. If then we neglect $[\mu(\varphi_1 \pm \tan \alpha)]$ with respect to unity (see Eq. 3.98), we obtain:

$$\Delta^* = \Delta \left(\frac{r_1}{p} \right)^2 = \frac{r_1^2}{p^2} \frac{e_1^2 + e_2^2}{2r_1^2} = \frac{1}{2} \left(\varepsilon_1^2 + \varepsilon_2^2 \right), \tag{3.104}$$

where $\varepsilon_1 = (e_1/p) = (g_{a2}/p_{bt})$ and $\varepsilon_2 = (e_2/p) = (g_{a1}/p_{bt})$ are respectively the addendum contact ratios of the wheel and pinion. Therefore, Eq. (3.103) becomes:

$$\eta = 1 - \pi\mu_0 \left| \frac{1}{z_1} \pm \frac{1}{z_2} \right| (\varepsilon_1^2 + \varepsilon_2^2). \tag{3.105}$$

This equation is known as the *formula of Poncelet* (see Ferrari and Romiti [21], Poncelet [41]). Then keeping in mind that the relationship that correlates circular pitch, p, radius of the pitch circle, r, and number of teeth z (see Eq. 2.38), it can be expressed also in the form:

$$\eta = 1 - \frac{\mu_0}{2p} \left| \frac{1}{r_1} \pm \frac{1}{r_2} \right| (e_1^2 + e_2^2). \tag{3.106}$$

Equations (3.105) and (3.106) are also valid for pinion-rack gear pairs; to do this, just put in them $z_2 = \infty$ and, respectively, $r_2 = \infty$. It should be noted that the same Eqs. (3.105) and (3.106) do not show a difference in efficiency between a spur gear pair with driving pinion and driven wheel, and the same gear pair with driving wheel and driven pinion. However, an in-depth analysis of this problem, which is left to the reader as an exercise, would highlight the fact that friction power losses in a given cylindrical spur gear pair are different depending on whether the driving member is the pinion or the gear wheel. In fact, for spur gear pairs whose members have different diameters, we would find that efficiency is greater when the smaller member is the driving wheel. This behavior justifies the fact that, for high transmission ratios, spur gear pairs are used mainly as speed reducing gears rather than as speed increasing gears.

It is to be noted that, in the conditions above described, the quantity Δ, i.e. the sum of the two integral functions appearing into the bracket of Eq. (3.98), is that shown in Fig. 3.10. On abscissa of this figure, the arc of approach, e_1, is drawn on the left of the pitch point, C, while the arc of recess, e_2, is drawn on the right. Since then Δ is proportional to $\left[(1/2)(e_1^2 + e_2^2) \right]$, proportionality factor being the constant ratio (μ/μ_0), it is obvious that from the end points, A^* and E^*, two vertical segments must to be drawn, respectively equal to e_1 and e_2. Therefore, the two straight lines CA' and CE' represent the above two integral functions. The sum of the two integrals is then proportional to the sum of the areas of two triangles CA^*A' and CE^*E', and this sum is proportional to the work lost by friction, coefficient of proportionality being the ratio (μ/μ_0). The upper boundaries of these two triangular areas are highlighted with dashed lines in Fig. 3.10.

It should be noted that the end-points A^* and E^* that delimit the arc of action measured over any of the two pitch circles correspond respectively to the end points A and E of the path of contact, shown in Figs. 1.8 and 2.10. Also, based on the correlation between quantities measured on pitch circles and quantities measured on the line of action (or in equivalent way, on the base circles), any segment on the

Fig. 3.10 Distribution curve of Δ, for $(e_1 + e_2) = p$, unless the proportionality constant (μ/μ_0)

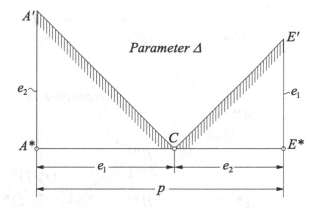

abscissa in Fig. 3.10 is equal to the quotient of the corresponding segment on the path of contact divided by $\cos\alpha$.

However, usually the arc of action $(e_1 + e_2)$ is greater than the circular pitch and, in most practical applications, it is included in the range $p < (e_1 + e_2) < 2p$. In this case, the calculation of the work dissipated by friction in the infinitesimal time interval dt (and therefore the one dissipated during an angle of rotation corresponding to the circular pitch) requires that the forces exchanged between the various mating tooth profiles are known.

The problem of determination of these forces is very complex (see Vol. 2, Sect. 1.7), because the sharing of the total load between the various tooth pairs in simultaneous meshing depends on several types of deformability, i.e.: deformability of the teeth; deformability of the gear wheel body; deformability of the shafts; deformability of the bearings; deformability of the gearbox; etc. It also depends on the local deformations due to the Hertzian pressure of contact, and the accuracy grade of the tooth profiles. In addition, it depends on the friction type, which is not a friction between dry surfaces, as hitherto assumed implicitly, but friction between lubricated surfaces.

For an approximate calculation, we assume here that the total load is distributed in equal parts between the two tooth pairs in simultaneous meshing. With this simplifying assumption, the work lost by friction will be half of what it would in the case of a single tooth pair in meshing. Figure 3.11 shows what happens in this case: on abscissa, in addition to the pitch point C, the four marked points A^*, B^*, D^*, and E^* are reported. The points B^* and D^* are defined univocally, since $A^*D^* = p$ and $B^*E^* = p$. In accordance with the afore-mentioned simplifying assumption, the points A'' and E'' are defined by the relationships $A^*A'' = A^*A'/2$ and $E^*E'' = E^*E'/2$.

In the region of single contact, between marked points B^* and D^*, only one tooth pair is in meshing, for which all the load is transmitted by means of this tooth pair. Therefore, the work lost by friction is proportional (as usual, the proportionality factor is given by the ratio μ/μ_0) to the sum of areas of the two triangles $B^*B'C$ and $D^*D'C$ (Fig. 3.11). In the regions of double contact, between the

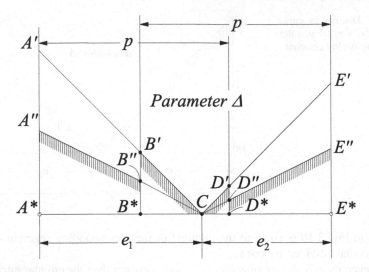

Fig. 3.11 Distribution curve of Δ, for $[p < (e_1 + e_2) < 2p]$, unless the proportionality constant (μ/μ_0)

marked points A^* and B^*, and D^* and E^*, two tooth pairs are simultaneously in meshing, whereby each of them bears half of the transmitted load. Therefore, the work lost by friction in these regions is equal to the sum of the left trapezoid area $A^*A''B^*B''$, which is half of the left trapezoid area $A^*A'B^*B'$, and right trapezoid area $D^*D''E^*E''$, which is half of the right trapezoid area $D^*D'E^*E'$.

The sum ΣA of these areas is therefore given by:

$$
\begin{aligned}
\Sigma A &= \frac{1}{4}\left[e_1^2 + e_2^2 + (p - e_1)^2 + (p - e_2)^2\right] \\
&= \frac{1}{2}\left[e_1^2 + e_2^2 + p^2 - p(e_1 + e_2)\right],
\end{aligned}
\tag{3.107}
$$

which can be expressed in the form:

$$
\Sigma A = \frac{p^2}{2}\left[\varepsilon_1^2 + \varepsilon_2^2 + 1 - (\varepsilon_1 + \varepsilon_2)\right].
\tag{3.108}
$$

Therefore, if $p < (e_1 + e_2) < 2p$, instead of the relationships Eqs. (3.106) and (3.105), which are valid for $(e_1 + e_2) = p$, we have respectively:

$$
\eta = 1 - \frac{\mu_0}{2p}\left|\frac{1}{r_1} \pm \frac{1}{r_2}\right|\left[e_1^2 + e_2^2 + p^2 - p(e_1 + e_2)\right].
\tag{3.109}
$$

$$
\eta = 1 - \pi\mu_0\left|\frac{1}{z_1} \pm \frac{1}{z_2}\right|\left[\varepsilon_1^2 + \varepsilon_2^2 + 1 - (\varepsilon_1 + \varepsilon_2)\right].
\tag{3.110}
$$

For high enough values of ε_1 and ε_2 that, in the usual practical applications, are almost next to unity, the sum $(\varepsilon_1^2 + \varepsilon_2^2)$ is not very different from the sum $(\varepsilon_1 + \varepsilon_2)$. Consequently, the value of the square bracket that appears in Eq. (3.110) is almost unitary; thus, under these conditions, Eq. (3.110) is simplified, and becomes:

$$\eta = 1 - \pi\mu_0 \left| \frac{1}{z_1} \pm \frac{1}{z_2} \right|. \tag{3.111}$$

From this last equation, we can deduce that the average efficiency of a cylindrical spur gear varies as follows: (*i*), it improves with increasing the number of teeth, z_1 and z_2; (*ii*), with the same number of teeth, it is higher for the internal gear pairs. In addition, both from Eqs. (3.105) and (3.110), and from Figs. 3.10 and 3.11, we can deduce that the average efficiency of these gears is much higher, as much as the sum $(\varepsilon_1 + \varepsilon_2)$ is low, i.e. as much as the transverse contact ratio is low; in fact, the values of ε_1 and ε_2 are generally smaller than the unity. Low values of the sum $(\varepsilon_1 + \varepsilon_2)$ mean equivalent low values of sum $(e_1 + e_2)$, i.e. smaller areas under the diagrams shown in Figs. 3.10 and 3.11.

For $(e_1 + e_2) > 2p$, three tooth pairs are simultaneously in meshing. To obtain the distribution curve of the work lost by friction along the path of contact, we can use the same procedure described for the case where $(e_1 + e_2) = p$, and $p < (e_1 + e_2) < 2p$. To this end, as Fig. 3.12 shows, first we will identify the six marked points that characterize this special case. In fact, in addition to the usual end points of path of contact, A^* and E^*, we will have the points B^* and F^*, defined by $E^*B^* = B^*F^* = p$, and the points D^* and G^*, defined by $A^*D^* = D^*G^* = p$. It is quite evident that, in this case, there is no region of single contact; instead, three

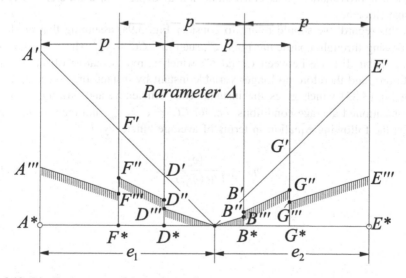

Fig. 3.12 Distribution curve of Δ, for $(e_1 + e_2) > 2p$, unless the proportionality constant (μ/μ_0)

regions of triple contact exist (those comprised between the marked points A^* and F^*, D^* and B^*, G^* and E^*), and two regions of double contact (those comprised between the marked points F^* and D^*, B^* and G^*).

Here too, for an approximate calculation, we assume that the total load is distributed in equal parts between the three or two tooth pairs in simultaneous meshing. With this simplifying assumption, in the triple and double contact regions the work lost by friction will be respectively equal to (1/3) and (1/2) of what would occur in a hypothetical region of single contact. Thus, with reference to Fig. 3.12, we will have: $A^*A''' = A^*A'/3$; $F^*F''' = F^*F'/3$; $F^*F'' = F^*F'/2$; $D^*D''' = D^*D'/3$; $D^*D'' = D^*D'/2$; $B^*B''' = B^*B'/3$; $B^*B'' = B^*B'/2$; $G^*G''' = G^*G'/3$; $G^*G'' = G^*G'/2$; and $E^*E''' = E^*E'/3$. The area between the abscissa axis and dashed segments shown in Fig. 3.12 is proportional to the work lost by friction, the proportionality factor being the ratio (μ/μ_0). This area can be expressed as a function of e_1, e_2, and p, for which the use of the previous procedure would lead to write the average efficiency in explicit form. This exercise is left to the reader.

Finally, it is to be noted that the average efficiency hitherto considered is that related only to the work lost by friction due to contact conditions between the various pairs of mating teeth that do not include the micro-slip effects. To determine the *total contact efficiency*, η_t, we would need to consider the losses due to the rolling resistance caused by the micro-slip phenomena we mentioned in Sect. 3.6.

Here, however, we will continue to address the topic in terms of macro-phenomenon, rather than micro-phenomenon, using the same procedure that allowed us to deride Eqs. (3.82) and (3.83). As we have already pointed out, this procedure reduces the micro-slip effects to resisting friction forces, which instead of passing through the centers of the gear wheels are tangent to the friction circles, and act in the opposite direction to the motion. Anyway, to define the total loss with a sufficient approximation, it is convenient first to express, in a different way, the average efficiency.

In this regard, we should return to consider Fig. 3.9a, assuming that the force F is passing through a particular point C', such that $CC' = \delta'_m$, where δ'_m indicates the particular distance between C and C', which is representative of the average conditions, and therefore no longer variable instant by instant, but constant. Thus, using Eq. (3.83) which gives the instantaneous efficiency, and writing it for the afore-mentioned average conditions, i.e. for $CC' = \delta' = \delta'_m$, and then for $\eta_i = \eta_m$, we get the following equation in terms of average efficiency, η_m:

$$\eta_m = \frac{1 - \left(\delta'_m/r_2\right)}{1 + \left(\delta'_m/r_1\right)}. \qquad (3.112)$$

From this relationship, we get:

$$\delta'_m = \frac{1 - \eta_m}{(\eta_m/r_1) + (1/r_2)}. \tag{3.113}$$

Under ideal operating conditions, the force $F_t/\cos\alpha$ passes through the pitch point, C (Fig. 3.9a), and the passive forces, i.e. the reaction forces exerted by the gearbox on the shaft pins, pass through the centers O_1 and O_2. Instead, under actual operating conditions, the force F, which represents the average conditions, passes through the particular point C', defined by $CC' = \delta' = \delta'_m$. The reaction forces exerted by the gearbox on the shaft pins are parallel to the force F, but no longer pass through the centers O_1 and O_2. They are tangent to the *circles of friction*, the radii of which are respectively ρ_1 and ρ_2 (the symbols introduced here to indicate these radii are the same used for the radii of curvature of the involute profiles, but the letters of the Greek alphabet are only twenty-four!). In addition, the points of tangency of these reaction forces with circles of friction are directed in such a way that these passive forces are opposite to the relative motions of the individual gear wheels with respect to the gearbox.

Based on these considerations, we can calculate the total contact efficiency that takes into account the losses due to contact between the mating teeth, including the rolling resistance due to the micro-slip effects. The active and passive forces described above can be considered as average forces having a constant value throughout the meshing cycle, and equivalent to the actual ones, which are variable instant by instant. Therefore, using the same procedure that allowed us to derive the relationship (3.83), we get:

$$\eta_t = \frac{[r_2 - \delta'_m - \rho_2/\cos\alpha]/r_2}{[r_1 + \delta'_m + \rho_1/\cos\alpha]/r_1} \simeq \frac{1 - (\delta'_m/r_2) - (\rho_2/r_2)}{1 + (\delta'_m/r_1) - (\rho_1/r_1)}. \tag{3.114}$$

The approximation in this relationship is due to the fact that, for the usual values of the pressure angle α, we can take $\cos\alpha \cong 1$.

Since the four ratios that appear in the above relationship are very small, we can easily demonstrate that it can be written in the form (see Scotto Lavina [47]):

$$\eta_t \cong \frac{1 - (\delta'_m/r_2)}{1 + (\delta'_m/r_1)} - \frac{\rho_1}{r_1} - \frac{\rho_2}{r_2} = \eta_m - \frac{\rho_1}{r_1} - \frac{\rho_2}{r_2}. \tag{3.115}$$

From this last relationship we deduce that the total contact efficiency, η_t, differs little from the average efficiency, as ρ_1 and ρ_2 are small amounts with respect to r_1 and r_2 respectively.

3.10 Effective Driving Torque

If the pinion is the driving member of the gear pair under consideration, in the ideal case where the friction losses are equal to zero, the nominal driving torque at the pinion, $T_1 = F_t r_1$, and the nominal driven torque (or resisting torque) at the wheel, $T_2 = F_t r_2$, must be the same, so we have:

$$T_1 = T_2 \frac{r_1}{r_2} = T_2 \frac{\omega_2}{\omega_1} = \frac{T_2}{|i|}, \qquad (3.116)$$

where $|i|$ is the absolute value of the transmission ratio, given by Eq. (1.13).

Instead, in the actual case, which is characterized by the total contact efficiency η_t, the effective driving torque T_{1e} will be given by:

$$T_{1e} = \frac{T_1}{\eta_t} = \frac{T_2}{|i|\eta_t}. \qquad (3.117)$$

This relationship allows to calculate the effective driving torque (and thus the effective driving power) to be applied to the driving shaft to win an assigned resisting torque, with a given transmission ratio, when the allowable value of total contact efficiency can be provided with sufficient precision. In other cases, where the usable driving power and the resisting torque to be won are known, the same relationship allows to obtain the required transmission ratio, by a trial procedure.

3.11 Short Notes on Cutting Methods of Cylindrical Spur Gears

We have already pointed out that appropriate design of any type of gear must take into account the cutting technologies used in its production process. Therefore, we believe that it is necessary here to describe briefly the main methods employed for the manufacturing of cylindrical spur gears. These methods can be summarized in the two following categories: form cutting methods, and generation cutting methods. Anyway, for the insights that may be need in this regard, we refer the reader to known traditional textbooks (see Björke [6], Boothroyd and Knight [7], Galassini [23], Henriot [28], Jelaska [30], Micheletti [37], Radzevich [42], Rossi [46]).

3.11.1 Form Cutting Method

This is the easiest cutting method, which is done with a milling machine, using a *form-milling cutter*, shaped exactly like the tooth space of the gear wheel to be manufactured. Therefore, the cutting edges of this form-milling cutter are shaped so that all of its sections, made with a plane passing through its axis, reproduce the tooth space between tooth and tooth of the gear wheel to be cut. In this way, it is possible to regrind the cutting edges according to a radial plane without any alteration of the shape of the same cutting edges.

To meet the above conditions, and to have a convenient *relief angle*, γ, the cutter edges of the form-milling cutter are generated by the motion of the cutter profile according to an Archimedes spiral, using special lathes. This motion ensures that the relief angle is almost exactly constant (it is to be noted that, to have the theoretical exact geometric shape, this motion should follow a logarithmic spiral). A section of the cutter's tooth with a plane perpendicular to the rake face, which delimit it in front (section $A - A$ in Fig. 3.13), has a trapezoidal shape, and thus it is characterized by a *side rake angle*, γ', which is defined by means of the following considerations. If α is the angle between the tangent at any point of the involute cutting edge of the cutter and the centerline of the form-milling cutter tooth, a differential radial shift, dr, of the trace of section $A - A$ corresponds to a decrease in half-thickness equal to $(dr \tan \alpha)$. Because of the relief angle of the cutting edge, γ, a differential shift, dr, corresponds to a differential displacement, dx, in the direction of the cutting-edge section, such that $dr = (dx \tan \gamma)$, so we will have:

$$\tan \gamma' = \frac{dr \tan \alpha}{dx} = \frac{dx \tan \gamma \tan \alpha}{dx} = \tan \gamma \tan \alpha. \qquad (3.118)$$

From this relationship, we infer that the side rake angle, γ', depends on $\tan \alpha$. Thus, at points where angle α is very small (this angle is equal to zero on the base circle of teeth with involute profiles), γ' is also very small, so over the base circle and near it unfavorable conditions to an easy and precise cutting occur. However, for teeth with involute profiles, this disadvantage has relatively modest effects because, as we have already pointed out, the portion of profile near the base circle is not normally used. It should be noted, however, that this disadvantage would be more serious for teeth with cycloid profile, for which the portion of profile near the pitch circle, where α is equal to zero, is generally the most important. However, this topic does not interest here, since in this textbook only teeth with involute profiles are considered.

With this cutting technology, the gear blank is held stationary, while the milling cutter is fed slowly in the direction of the axis of the same gear blank, as long as a tooth space is cut throughout its depth. The gear blank is then indexed to an angle exactly corresponding to one angular pitch, and a second tooth space is cut, and so on, i.e. the process is repeated until all tooth spaces are cut and the gear wheel is completed. In order to reduce the wear of the cutting edges, and thus limit the number of re-grindings, the finishing milling cutter having the appropriate geometric profile is coupled often to a roughing milling cutter (see Fig. 3.14). This roughing milling cutter has the function of cut the tooth spaces in a broadly approximate way, leaving to the finishing milling cutter the task of cutting the theoretically exact profile.

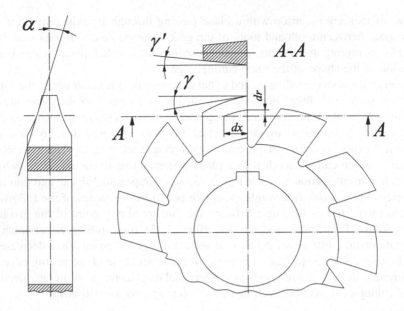

Fig. 3.13 Form-milling cutter for cutting of cylindrical spur gears

This cutting method of cylindrical spur gears has the great advantage of allowing the use of an ordinary milling machine, which is always available in small machine shops, together with the required angular indexing mechanism. However, normally, special milling machines with automatic indexing devices are used. Another advantage of this cutting method is that milling cutters are generally cheaper than the cutting tools used for other cutting methods, such as the generation cutting methods.

Faced with the above advantages, the form cutting method has many disadvantages. It should first be noted that, from the theoretical point of view, a different milling cutter is required for each combination of values of the module, m, pressure angle, α, number of teeth, z, and tooth thickness, s. In fact, it is well known that the shape of tooth space depends on all four of the above parameters. Further, this disadvantage is increased in the case of profile-shifted toothings, since the shape of the tooth space will also depend on the value of the profile shift coefficient (see Chap. 6). This disadvantage is particularly relevant if a small number of identical gear wheels are to be made, while it tends to diminish as the number of gear wheels to be cut increases, since it is then inexpensive to buy a particular cutter to perform the job.

Another disadvantage of this cutting method is when different gear wheels are to be made, using cutters that are already in stock. In order to minimize the number of cutters in stock, keeping it within acceptable limits, it is usually assumed that the cutters are used to cut gear wheels operating at the standard center distance. Therefore, the cutters are designed in such a way that the tooth thickness of each

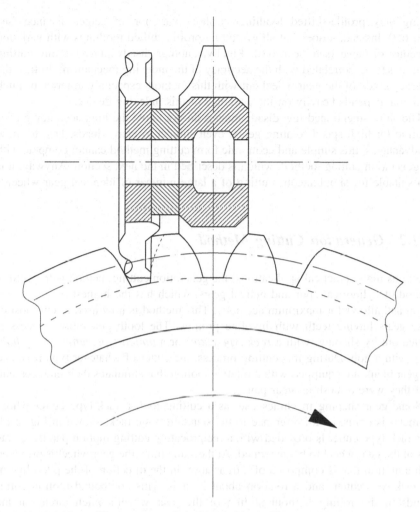

Fig. 3.14 Finishing milling cutter coupled with roughing milling cutter

gear wheel is equal to half the circular pitch, but with adequate tolerance, compatible with the necessary backlash. In addition, since the tooth space of a gear wheel with z_1 teeth is similar in shape to that of a gear wheel with z_2 teeth, provided z_1 and z_2 are sufficiently close, each cutter is usually used for a defined range of z-values. Thus, a set of eight cutters (or fifteen cutters, for very large modules) is generally considered to be adequate for each combination of values of m and α. It follows that the cutter shape is only correct for a particular z-value, while for the other z-values within the range of possible utilization, inevitable errors in the gear tooth profile are to be expected (see Galassini [23], Micheletti [37], Rossi [46]).

Other disadvantages of this cutting method of cylindrical spur gears should not go under silence. In fact, this cutting method is a very rigid method in that it allows cutting only profile-shifted toothings without variation of center distance (see Sect. 6.3). Instead, it does not allow cutting profile-shifted toothings with variation of center distance (see Sect. 6.6). Finally, another disadvantage of this cutting method is to be correlated with the accuracy of the indexing mechanism. In fact, the accuracy grade of the gear wheel cut with this method, especially in terms of pitch deviation, depends heavily on the accuracy of this indexing device.

The above-mentioned few disadvantages do not allow the high accuracy grades required for high speed rotating gears, even subject to heavy loads. Due to these disadvantages, this simple and economic form cutting method cannot compete with the generation cutting method, which is described in the next section. Anyway, it is also suitable for simultaneous cutting of a large number of identical gear wheels.

3.11.2 Generation Cutting Method

Due to its many practical applications, the generation cutting method is the cutting method of cylindrical spur and helical gears, which has the highest overall application and allows for maximum accuracy. This method is now used to cut most of these gears having teeth with involute profiles. The tooth generation process is carried out by shaping, with a *rack-type cutter* or a *pinion-type cutter*, or by *hobbing*, with a hob. During this cutting process, the cutter of whatever type it is, and the gear blank are equipped with a relative motion that simulates their engagement, as if they were a meshing gear pair.

Some gear shaping machines use, as a cutting tool, a rack-type cutter whose geometry is defined with reference to the standard basic rack, shown in Fig. 2.14. This rack-type cutter is provided with a reciprocating cutting motion parallel to the axis of the gear wheel to be generated. At the same time, the gear wheel is provided with a motion that is composed of a translation in the direction of the pitch line of the rack-type cutter, and a rotation about its axis. This compound motion corresponds to the rolling without sliding of the gear wheel's pitch circle on the rack-cutter's pitch line. Therefore, this motion simulates the relative engagement motion of the gear wheel with the rack-type cutter. In other types of gear shaping machines that use this technology, the rack-type cutter is provided not only with the reciprocating cutting motion, but also with the translation motion in the direction of its pitch line, while the gear wheel to be generated only rotates about its axis. Anyway, the relative motion between the rack-type cutter and gear wheel is that of a pure rolling between the pitch surfaces involved.

With this cutting technology, it is possible to vary the radius of the cutting pitch circle of the gear wheel to be generated. To this end, it is sufficient to vary the ratio between the translation motion in the direction of the pitch line of the rack-type cutter, regardless of whether it is given to the rack-type cutter or to the gear wheel, and the rotation of the gear wheel about its axis. This allows cutting with great easy and economy, and only with one cutter of defined sizing, gear wheels with profile-shifted toothings, both without variation of center distance, and with variation of center distance. This technology has in fact a unique limitation: that of not allowing the cutting of internal gear wheels.

However, it should be noted that this cutting method has only one disadvantage: that of not being able to get the gear wheel with a continuous generating motion, because to do this the rack-type cutter should have the same number of teeth as the gear wheel to be generated. It is not possible to meet this condition, especially when the number of teeth of the gear wheel to be generated is very large, because in this case we would have rack-type cutters of such dimensions as to be incompatible with the usual generating shaping machines. Most rack-type cutters are made with a small number of teeth (generally, six or eight teeth). Therefore, in order to be able to cut the total number of teeth of any gear wheel, the cutting process is stopped each time the gear wheel is rotated one or more angular pitches, after which the rack-type cutter is moved back the same number of angular pitches. This way of working of the gear-cutting machine not only results in an elongation of cutting time, but also results in tooth profile errors and pitch deviations.

Other gear cutting machines use, as a cutting tool, a pinion-type cutter having the shape shown in Fig. 3.15. This pinion-type cutter, also called *Fellows gear shaper*, in honor of its inventor, American entrepreneur Edwin R. Fellows. He introduced it in 1896, together with Fellows shaping machines, which gave a vital contribution to the mass production of effective and reliable gear transmissions for the nascent automotive industry. The Fellows gear shaper has a number of teeth a little less than the minimum number of teeth corresponding to the non-interference with the rack gear, so as to decrease the portion of tooth profile lost due to interference, when a gear wheel with a small number of teeth must be cut. For example, for a pressure angle of 15°, the pinion-type cutter has 24 teeth instead of 30 teeth, as the diagram shown in Fig. 3.1 indicates.

The pinion-type cutter is provided with a reciprocating cutting motion parallel to the axis of the gear wheel to be generated, and a rotation about its axis. At the same time, the gear blank is provided with a rotation about its axis. Pinion-type cutter and gear blank are driven at constant angular velocities, as if they were a meshing gear pair. In other words, the two rotations above simulate the pure rolling without sliding of the cutting pitch cylinder of pinion-type cutter over the cutting pitch cylinder of the gear wheel to be generated, and vice versa.

After the cut has been completed, the tooth shape of the gear wheel is the result of envelope of the successive positions of the pinion-type cutter. It can be easily shown that this shape is not exactly that of an involute curve, but consists of a series of arcs, whose sizes depend on the number of strokes needed to cut each tooth. However, the above-mentioned generating motion begins only when the two cutting pitch surfaces come to mutual contact. This results from the fact that the amount of material removed with each cutting stroke is limited, so it is not possible to cut the first tooth of the gear blank to its full depth. Therefore, before the generation cutting process starts, the gear blank is fed radially towards the pinion-type cutter, until the chip removal allows the cutting pitch surfaces to come in mutual tangency. At this point, the radial feed of the gear blank is stopped, and the generation cutting process starts. In some cases, a single cut is sufficient. Indeed,

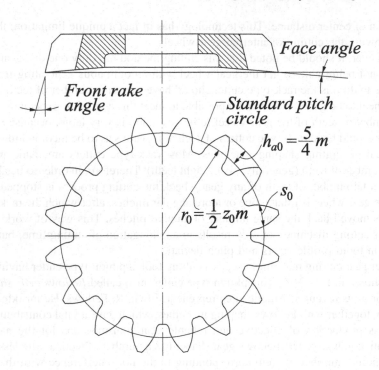

Fig. 3.15 Geometry of a pinion-type cutter

in other cases, one or more roughing cuts are necessary before the finishing cut. In other case, in addition to finishing cut, other more precise machining may be required (e.g., shaving or grinding).

This cutting method of cylindrical spur gears can be used indifferently both for external gears and for internal gears. For a pinion-type cutter having a standard sizing (addendum $h_{a0} = (5/4)m$, and dedendum $h_{f0} = m$, corresponding respectively to the dedendum, h_f, and addendum, h_a, of the gear wheel to be generated), the extension of the cutting tooth inside the base circle is done in the radial direction. With this radial shape of the cutting edge, cutting interference occurs when gear wheels with over 48 teeth have to be cut. Consequently, in this case, a portion of the tooth profile close to the tip circle is removed. This removal can be deliberately intended to obtain teeth with adequate tip relief, allowing us to achieve interesting design goals. Instead, for usual gear wheels, which have fewer teeth than 48, the cutting interferences are much smaller than those that would be, *ceteris paribus*, in the case of cutting with rack-type cutter.

A single pinion-type cutter is required for a given combination of values of the module, m, and pressure angle, α. It can be used to cut gear wheels with any number of teeth, as well as profile-shifted toothings both without variation of center distance, and with variation of center distance. Of course, the pinion-type cutter is a bit different from an ordinary pinion, not only for the above-highlighted sizing, but also because its teeth are not parallel to the cutter axis. In fact, they are relieved, in order to provide clearance behind the cutting edges. Due to this relief, the shape of the cutter teeth will change slightly after each sharpening.

Despite these variations in the cutting tooth shape, geometrically exact tooth profiles can still be cut in the gear blank. This is due to the particular configuration of the teeth of the pinion-type cutter (Fig. 3.15), which are designed in such a way that the projection of the cutting edges onto a plane perpendicular to the cutter axis is always an involute curve of the same base circle, even after each sharpening of the cutting tool. As a result, teeth on the gear blank will always be cut with the same base pitch, although after each sharpening the tooth thickness will be slightly reduced.

The most commonly used generation method for cutting cylindrical spur and helical gears is by hobbing. To this end, gear hobbing machines are used, while the cutting tool is a hob whose geometry is shown in Fig. 3.16. As this figure highlights, a hob is simply a screw with one or two threads. Each thread is cut by a number of gashes forming the cutting faces, which are similar to those of the form-milling cutter shown in Fig. 3.13. These cutting faces can either be at right angles to the threads or parallel to the hob axis. In any case, the section of a hob with a plane passing through its axis is a rack with teeth having straight profiles. We can imagine that the hob is generated by this profile when it is equipped with a screw motion, according to a mean helix having lead angle, γ_0, and belonging to the pitch cylinder (see Collins et al. [14]).

The shape of a hob can be specified by the lead angle, γ_0, or the helix angle, β_0. Usually, the lead angle is used. For a right-hand hob, the lead angle is defined as the complement of helix angle, i.e. $\gamma_0 = (90° - \beta_0)$. Generally, the lead angle is always small, especially when the hob has only one thread. For a left-hand hob, whose helix angle is negative, the lead angle is defined as $\gamma_0 = -(90° + \beta_0)$, and this in order to have a negative lead angle. However, usually, to define the geometry of a hob, the absolute value of the lead angle is given, with the addition of the specification that the hob is a right-hand or left-hand hob.

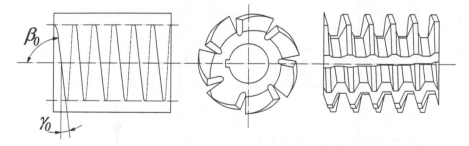

Fig. 3.16 Geometry of a hob

The cutting action of a hob is very similar to that of a rack-type cutter. The hob and the gear blank rotate about their axes with angular velocities ω_0 and ω_2 meeting the condition:

$$\frac{\omega_0}{\omega_2} = \frac{z_2}{z_0}, \qquad (3.119)$$

where z_2 is the number of teeth of the gear wheel to be generated, and z_0 is the number of threads of the hob (usually, $z_0 = 1$).

The meshing of the hob with the gear wheel being generated can be considered as a rack-gear wheel meshing, because the rotation of the hob simulates the translation of an imaginary rack-type cutter. This is evident by observing that, due to its screw shape, when the hob rotates about its axis, the threads appear to move in the direction of the same axis. In addition to rotating about its axis, the hob moves parallel to the gear wheel axis (Fig. 3.17); this is the feed motion of the hob.

However, despite the similarity between the cutting action of a hob and that of a rack-type cutter, the hobbing process has at least the following two advantages: (i), the motions of the cutting faces are continuous in both tangential and axial directions, so there is no need to move the cutting tool back after one or two teeth have been cut, as in the shaping process with a rack-type cutter; (ii), the cutting action is obtained without any reciprocating motion of the cutting tool. As Fig. 3.17 shows, the hob begins the cutting action at one end of the gear wheel, and deepens it progressively over the entire face width, seamlessly. This results in a great finishing of tooth flanks, and the practical elimination of pitch deviations.

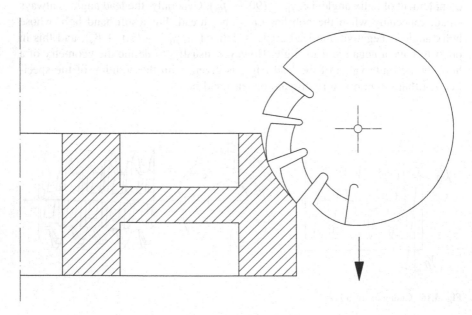

Fig. 3.17 Feed motion of the hob

The similarity between hob and rack-type cutter goes beyond the above considerations. In fact, since the cutting faces of the hob simulate the teeth of the imaginary rack-type cutter, each of them meet the gear blank at the same angle as a rack-type cutter. Therefore, to cut the teeth of a cylindrical spur gear wheel, a right-hand hob must be inclined at a small angle with respect to a plane perpendicular to the axis of the gear wheel to be generated. This angle is called *swivel angle*, and is equal to the lead angle of the mean helix, γ_0 (Fig. 3.18). In more general terms, the hob axis must be inclined with respect to the axis of the gear wheel to be generated by an angle $(90° \mp \gamma_0)$, where the minus and plus signs are respectively valid for right-hand hob and left-hand hob.

It should be noted that the hob would be completely the same as the rack-type cutter (and hence it would cut exact involute profiles) only if it had an infinite pitch radius, i.e. $r_0 = \infty$. In reality this does not happen. Therefore, the thread being not straight, but helicoidal, the profiles cut with this cutting process are approximate involutes. The approximation is bigger, the smaller the tooth depth compared to r_0 (i.e., the smaller the hob axial pitch, as this is proportional to the tooth depth, depending on the sizing adopted). Since the hob axial pitch is given by:

$$p_{a0} = 2\pi r_0 \tan \gamma_0, \tag{3.120}$$

it follows that the approximation is bigger, the smaller the lead angle, γ_0.

In practice, γ_0 does not exceed $(5 \div 6)°$, and deviations from the theoretical involute profiles are very small, so much so that this cutting process is used for gears of high accuracy grade. The gear wheel cut with this cutting method has circular pitch, p, equal to the normal pitch, p_{n0}, of the hob, that is:

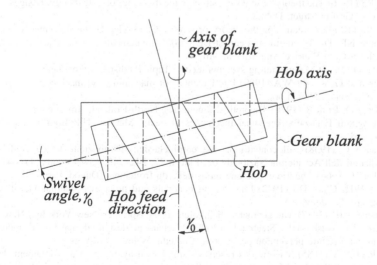

Fig. 3.18 Working position of a right-hand hob with respect to the gear blank

$$p = p_{n0} = p_{a0} \cos \gamma_0 = 2\pi r_0 \sin \gamma_0. \qquad (3.121)$$

In addition, the same gear wheel has a pressure angle, α, equal to the pressure angle, α_{n0}, in the normal section of the hob, that is:

$$\tan \alpha = \tan \alpha_{n0} = \tan \alpha_{a0} \cos \gamma_0, \qquad (3.122)$$

where α_{a0} is the pressure angle of the hob in an axial section, i.e. a section of the hob with a plane passing through its axis.

Also, for this cutting process, a single hob is required for a given combination of values of the module, m, and pressure angle, α. It can be used to cut gear wheels with any number of teeth, as well as profile-shifted toothings both without variation of center distance, and with variation of center distance.

In the chapters that follow, when we consider it necessary, further details on the three generation cutting processes described above will be given. However, for the more detailed analysis of the hobbing cutting process as well as of the two afore-mentioned shaping cutting processes, we refer the reader to more specialized textbooks (see Jain and Chitale [29], Rajput 43], Sharma [48]).

References

1. Ambekar AG (2007) Mechanism and machine theory. Prentice-Hall of India, New Delhi
2. Amontons G (1699) De la resistance causée dans les machines, tant par les frottements des parties qui les composent, que par roideur des cordes qu'on y employe, & la maniere de calculer l'un & l'autre, Mémoires de l'Académie Royale des Sciences. Histoire de l'Académie Royale des Sciences, Paris, pp 206–222
3. Ball RS (1876) The theory of screws: a study in the dynamics of a rigid body. Hodges, Foster, and Co., Grafton-street, Dublin
4. Ball RS (1900) A treatise on the theory of screws. Cambridge University Press, Cambridge
5. Belfiore NP, Di Benedetto A, Pennestrì E (2005) Fondamenti di meccanica applicata alle macchine. Casa Editrice Ambrosiana, Milano
6. Björke Ö (1995) Manufacturing systems theory. Tapir Publishers, Trondheim
7. Boothroyd G, Knight WA (1989) Fundamentals of machining and machine tools, 2nd edn. Marcell Decker, New York
8. Bottema O, Roth B (1979) Theoretical kinematics. North-Holland, Amsterdam
9. Buckingham E (1949) Analytical mechanics of gears. McGraw-Hill Book Company Inc, New York
10. Cattaneo C (1938) Sul contatto di due corpi elastici: distribuzione locale degli sforzi. Rendiconti dell'Accademia Nazionale dei Lincei, 27, 242–248, 434–436, and 474–478
11. Caubet JJ (1964) Théorie et pratique industrielle du frottement. Dunod, Paris
12. Clapp WH, Clark DS (1942) Engineering materials and processes. International Textbook Company, Scranton, PA
13. Colbourne JR (1987) The geometry of involute gears. Springer, New York Inc, New York
14. Collins JA, Busby HR, Staab GH (2009) Mechanical design of machine elements and machines: a failure prevention perspective, 2nd edn. Wiley, New York
15. Coulomb CA (1785) Théorie des machines simples en ayant égard au frottement de leurs parties. Acad Roy Sci Memo Math Phys X:16

16. Di Benedetto A, Pennestrì E (1993) Introduzione alla cinematica dei meccanismi, vol 1. Casa Editrice Ambrosiana, Milano
17. Disteli M (1914) Über des analogen der Savary schen formel und konstruktion in der kinematischen geometrie des raumes. Zeitschrift für Mathematic und Physik 62:261–309
18. Dooner DB (2012) Kinematic geometry of gearing, 2nd edn. Wiley, Chichester, UK
19. Dooner DB, Seireg AA (1995) The kinematic geometry of gearing: a concurrent engineering approach. Wiley, New York
20. Dowson D, Higginson GR (1966) Elasto-hydrodynamic lubrication. Pergamon Press, Oxford
21. Ferrari C, Romiti A (1966) Meccanica applicata alle macchine. Unione Tipografico-Editrice Torinese (UTET), Torino
22. Freudenstein F (1962) On the variety of motion generated by mechanism. ASME J Eng Ind, pp 156–160
23. Galassini A (1962) Elementi di Tecnologia Meccanica, Macchine Utensili, Principi Funzionali e Costruttivi, loro Impiego e Controllo della loro Precisione, reprint 9th edn. Editore Ulrico Hoepli, Milano
24. Giovannozzi R (1965) Costruzione di Macchine, vol II. Casa Editrice Prof. Riccardo Pàtron, Bologna
25. Gray A (1997) Modern differential geometry of curves and surfaces with mathematica, 2nd edn. CRC Press, Boca Raton, FL
26. Hain K (1967) Applied kinematics. McGraw-Hill, New York
27. Henriot G (1979) Traité théorique and pratique des engrenages, vol 1, 6th edn. Bordas, Paris
28. Henriot G (1972) Traité théorique et pratique des engrenages. Fabrication, contrôle, lubrification, traitement thermique, vol 2. Dunod, Paris
29. Jain KC, Chitale AK (2014) Textbook of production engineering, 2nd edn. PHI Learning Private Limited, New Delhi
30. Jelaska DT (2012) Gears and gear drives. Wiley, Chichester, U.K
31. Johnson KL (1985) Contact mechanics. Cambridge University Press, Cambridge
32. Juvinall RC, Marshek KM (2012) Fondamentals of machine component design, 5th edn. Wiley, New York
33. Kalker JJ (1990) Three-dimensional elastic bodies in rolling contact. Kluwer Academic Publishers, Dordrecht
34. Leutwiler OA (1917) Elements of machine design. McGraw-Hill, New York
35. Levi-Civita T, Amaldi U (1929) Lezioni di Meccanica Razionale—Vol. 1: Cinematica—Principi e Statica, 2nd edn. Reprint 1974, Zanichelli, Bologna
36. Litvin FL, Fuentes A (2004) Gear geometry and applied theory, 2nd edn. Cambridge University Press, Cambridge
37. Micheletti GF (1977) Tecnologia meccanica, 2nd edn. Unione Tipografico–Editrice Torinese (UTET), Torino
38. Mindlin RD (1949) Compliance of elastic bodies in contact. J Appl Mech 16(3):259–268
39. Niemann G, Winter H (1983) Maschinen-elemente, Band II: Getriebe allgemein, Zahnradgetriebe-Grundlagen, Stirnradgetriebe. Springer, Berlin
40. Polach O (2005) Creep forces in simulations of traction vehicles running on adhesion limit. Wear 258:992–1000
41. Poncelet JV (1874) Course de Mécanique Appliquée aux Machines. In: Posthumous (ed) vols 1, 2. Gauthier-Villars, Imprimeur-Libraire de l'École Polytechique, Paris
42. Radzevich SP (2017) Gear cutting tools: science and engineering, 2nd edn. CRC Press, Taylor & Frencis Group, Boca Raton
43. Rajput RK (2007) A textbook of manufacturing technology: manufacturing processes, 1st edn. Laxmi Publications (P) Ltd, Bangalore
44. Reuleaux F (1893) The constructor. A handbook of machine design. German edition: first published in (1861). H.H. Suplee, Philadelphia
45. Reye T (1860) Zur theorie der zapfenreibung. Der Civilingenieur 4:235–255
46. Rossi M (1965) Macchine utensili moderne. Ulrico Hoepli Editore, Milano

47. Scotto Lavina G (1990) Riassunto delle Lezioni di Meccanica Applicata alle Macchine: Cinematica Applicata, Dinamica Applicata—Resistenze Passive—Coppie Inferiori, Coppie Superiori (Ingranaggi—Flessibili—Freni). Edizioni Scientifiche SIDEREA, Roma
48. Sharma PC (1999) A textbook of production engineering. S. Chaud & Company Ltd, New Delhi
49. Veldkamp GR (1967) Canonical systems and instantaneous invariants in spatial kinematics. ASME J Mech 2:329–388
50. Weisbach J (1894) Mechanics of engineering and machinary. In: Hermann G (ed) German edition: first published in (1875). Wiley, New York

Chapter 4
Interference Between External Spur Gears

Abstract In this chapter, the problems of interference between external spur gears in both cutting and working conditions are first described and discussed. The discussion is quite general and therefore includes involute or primary interference and fillet interference. The relationships of the main geometric quantities that allow to avoid the involute interference in working conditions are obtained and design indications in this regard are provided. The reductions of the path of contact due to primary interference both in cutting and operating conditions are then quantified, and the beneficial effects of the cutting interference on the intentional modifications at the tip and root of the tooth profile are discussed. The geometry of the fillet profile generated by means of a rack-type cutter (or hob) and a pinion-type cutter is subsequently defined, and the interference effects of the rounded tip of the cutter teeth are discussed. Finally, suggestions are given to avoid the dangers of the fillet interference in operating conditions.

4.1 Introduction

In Sects. 2.4 and 3.1 we have already introduced the topic of the interference, when we defined the maximum radii of the addendum circles needed to avoid it. In this Chapter we will deepen further this interesting topic, because it is an important step of the gear design. In fact, interference results in a typical undercut tooth, which not only weakens the tooth to its root (it takes shape as a wasp-like waist), but also removes part of the useful involute profile near the base circle (see Buckingham [2], Dudley [6], Giovannozzi [11], Henriot [15], Litvin and Fuentes [20], Maitra [21], Merritt [23], Niemann and Winter [25], Pollone [26], Radzevich [29]).

It is first necessary to keep in mind that two profiles, although conjugate, cannot always be actually usable profiles of a kinematic pair. In fact, in some cases, to have the certainty that the kinematic coupling can take place, the members of a kinematic pair should be characterized by penetration of their solid parts. In other words, the members of the kinematic pair should allow that physics does not allow, that is this penetration or interference between solid bodies [8].

© Springer Nature Switzerland AG 2020
V. Vullo, *Gears*, Springer Series in Solid and Structural Mechanics 10,
https://doi.org/10.1007/978-3-030-36502-8_4

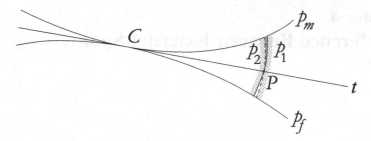

Fig. 4.1 Interference between involute profiles of two centrodes touching at point C

This happens, for example (Fig. 4.1), between the involutes p_1 and p_2 generated by the tracing point P on a straight-line t, which rolls without sliding on two *pitch curves* or *centrodes* (respectively, the fixed and moving centrodes, p_f and p_m, touching at point C, which is the instantaneous center of rotation through which the straight-line t passes. Remember, incidentally, that *centrode* in kinematics is the path traced by the instantaneous center of rotation of a rigid plane figure moving in a plane [7]. Then, in Fig. 4.1, centrodes are the loci of the pitch points for each angular position of the straight-line t when the latter rolls without sliding on the pitch curves (these are the *evolute curves*), p_f and p_m.

However, the two involutes, namely the profiles p_1 and p_2 so generated, are each (p_1) the envelope of the successive positions taken by the other (p_2) during the pure rolling motion of the centrode p_m, which is rigidly connected with profile p_2, on the other centrode. However, these involute profiles cannot be taken as conjugate profiles because, having in correspondence of the point of contact P the same center of curvature (the point C), the type of contact at point P is a contact of second order, for which the two involutes intersect each other. However, we must consider the involute profiles p_1 and p_2 not as geometric curves, but as edges that delimit real physical bodies, i.e. as boundaries of solid body sections. Figure 4.1 highlights the fact that, wherever the solid parts of the two profiles are positioned, there is always a certain region in correspondence of which the solid parts are placed at the same side; thus, the interference occurs, and the appropriate kinematic coupling between the two members is impossible.

An interference of the same type occurs all the times that two profiles have, in their tangent point, a contact of the second order, that is a point of contact where they have not only the same tangent, but also the same center of curvature and, therefore, the same curvature. However, the interference can also occur when at point of tangency between two profiles there is a simple contact of first order, because the parts of the profiles, the most distant from the point of contact, are to intersect. This condition, which occur quite often, forces to limit the extension of the profiles actually usable.

The interference, of whatever kind it may be, is never allowed during the meshing of two toothed gear wheels in their actual working conditions. In fact, if the interference would take place during operation, an unfavorable condition for transmission of motion would occur, due to jamming between the teeth pairs successively in meshing.

On the contrary, as we will specify in more detail below, during the cutting process of the teeth a certain interference between the gear wheel to be cut and the cutting tool is not only permissible, but in certain cases it is also desirable and desired. This why, by means of it, those parts of the involute profile of the teeth, which, for various reasons, may generate problems of transmission of motion, can be removed [11].

For the discussion of this important subject, we distinguish the case of external gear pairs from that of internal gear pairs. In the first case, two types of interference may occur: the *theoretical interference* (also called *involute interference* or *primary interference*), and the *fillet interference*. In the second case, five types of interference may occur (see Colbourne [3], Radzevich [28], Yu [37]):

- the *theoretical interference*;
- the *secondary interference*;
- two *fillet interferences* (the first, between the tip of the pinion and root of the annulus, and the second between the tip of the annulus and root of the pinion);
- the *trimming interference*, i.e. the interference of the radial approach of the cutter during the generation cutting process, or of the pinion during the assembly of the internal gear pair.

4.2 Theoretical Interference Between External Spur Gears

In Sect. 2.4 we have already described in detail what this type of interference consists and we have seen that, to avoid it, the outside (or tip or addendum) circles of the pinion and wheel must have radii at most equal to O_1T_2 and O_2T_1 respectively (see Fig. 2.9). In particular, we have shown that, if the tooth involute profiles are too extended in the radial direction (i.e., if the tip circles of the two members of the gear pair go beyond the point T_1 and T_2), the parts of the same profiles extended beyond these points are no longer able to transmit correctly the motion. This instead does not happen when the same profiles are in contact over the entire line of action, T_1T_2.

If the pinion teeth were higher (and therefore the radius of its addendum circle was greater than O_1T_2), by rotating the two wheels of the gear pair as if meshing properly, we would see that, beyond the point T_2, the tooth tip of the wheel 1 penetrates the tooth flank of the wheel 2. Obviously, in a case like this, when the contact reaches point T_2, the motion could no longer continue, due to the jamming of the tooth tip of the pinion 1 against the dedendum flank of the tooth of the wheel 2. In this case, therefore, the interference would occur as the impossibility of operation of the gear pair (see also Pollone [26]).

However, if we provide to remove a portion of the tooth dedendum flank of the wheel 2, the one near the tooth root and corresponding to the envelope of the tip of the tooth profile of the wheel 1 in its relative motion with respect to the wheel 2, the obstacle of the continuity of the transmission of motion is deleted, i.e. the kinematic interference is eliminated, but the tooth of the wheel 2 is weakened at its root, by the undercut.

The same problem would occur at the tooth root of the wheel 1, due to theoretical interference of the teeth of the wheel 2 when they have too high depth, i.e. when the radius of its addendum circle is greater than O_2T_1.

For both these reasons, these types of interference must be avoided, and therefore the radii of the addendum circles of the two wheels of the gear pair may be at most equal to O_1T_2 for the wheel 1, and O_2T_1 for the wheel 2.

Taking into account these factual circumstances, we consider an external spur gear pair, having number of teeth z_1 and z_2, module m, pressure angle α and center distance a. These quantities are the nominal ones, but here we leave aside the subscript 0, which indicates nominal quantities, because misunderstandings are not possible. From Eq. (2.55) we obtain:

$$r_{b1} = (mz_1/2)\cos\alpha \qquad r_{b2} = (mz_2/2)\cos\alpha, \tag{4.1}$$

while the nominal center distance a is given by Eq. (2.64).

Figure 4.2 shows the essential geometric quantities taken from Fig. 2.9. Of course, to avoid the involute interference, the radii of the addendum circles of the two wheels of the gear can have, at most, the following values:

$$O_1T_2 = \sqrt{\overline{O_1T_1^2} + \overline{T_1T_2^2}} = \frac{m}{2}\sqrt{z_1^2 + \left(z_2^2 + 2z_1z_2\right)\sin^2\alpha} \tag{4.2}$$

$$O_2T_1 = \sqrt{\overline{O_2T_2^2} + \overline{T_1T_2^2}} = \frac{m}{2}\sqrt{z_2^2 + \left(z_1^2 + 2z_1z_2\right)\sin^2\alpha}. \tag{4.3}$$

The maximum values of the addendum of the wheel and pinion must therefore satisfy respectively the inequalities:

$$h_{a2\,\max} \le \frac{m}{2}\left[\sqrt{z_2^2 + \left(z_1^2 + 2z_1z_2\right)\sin^2\alpha} - z_2\right] \tag{4.4}$$

$$h_{a1\,\max} \le \frac{m}{2}\left[\sqrt{z_1^2 + \left(z_2^2 + 2z_1z_2\right)\sin^2\alpha} - z_1\right]. \tag{4.5}$$

From these two inequalities, for given values of α, z_1 and z_2, with $z_2 > z_1$, $h_{a2\,\max}$ is less than $h_{a1\,\max}$. Therefore, if the two wheels of the gear have equal addendum, when the value of this increases, the interference of the teeth of the wheel 2 with those of the wheel 1 occurs before.

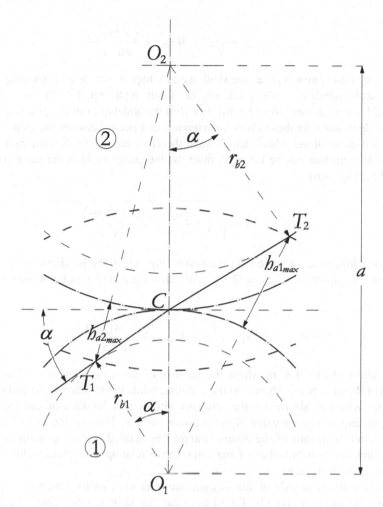

Fig. 4.2 Essential geometric quantities taken from Fig. 2.9

From the inequality (4.4), solved with respect to z_1, we get the following minimum number of teeth of the pinion, $z_{1\,min}$, that meshes, without interference, with the wheel having z_2 teeth:

$$z_{1\,min} \geq -z_2 + \sqrt{z_2^2 + 4\frac{h_{a2\,max}}{m}\frac{\left(z_2 + \frac{h_{a2\,max}}{m}\right)}{\sin^2 \alpha}}, \tag{4.6}$$

which, by introducing, in accordance with the second of Eqs. (3.9), the maximum addendum factor of the wheel 2, $k_{2\,max} = (h_{a2\,max}/m)$, can be written in the more compact form:

$$z_{1\,min} \geq -z_2 + \sqrt{z_2^2 + 4k_{2\,max}\frac{(z_2 + k_{2\,max})}{\sin^2\alpha}}. \tag{4.7}$$

In many cases, however, as we shall see in Chap. 6, we chose a toothing with normal tooth depth $(h = h_P = 2.25m)$, i.e. tooth depth equal to the one of the standard basic rack (see ISO 53 [16]), but with the addendum of the pinion greater than the dedendum. In these cases, interference can occur between the teeth of the pinion and those of the wheel, which must therefore have the following minimum number of teeth that can be obtained from the inequality (4.5) in the same way as we did for Eq. (4.6):

$$z_{2\,min} \geq -z_1 + \sqrt{z_1^2 + 4\frac{h_{a1\,max}}{m}\frac{\left(z_1 + \frac{h_{a1\,max}}{m}\right)}{\sin^2\alpha}}; \tag{4.8}$$

by introducing, in accordance with the first of Eqs. (3.9), the maximum addendum factor of the pinion 1, $k_{1\,max} = (h_{a1\,max}/m)$, this inequality can be written in the form:

$$z_{2\,min} \geq -z_1 + \sqrt{z_1^2 + 4k_{1\,max}\frac{(z_1 + k_{1\,max})}{\sin^2\alpha}}. \tag{4.9}$$

The chart of Fig. 4.3, in which the number of the teeth z_1 of one of the two wheels of the gear pair is shown on the ordinate, while the number of the teeth z_2 of the other wheel is shown on the abscissa, shows two families of curves, each corresponding to a given value of the pressure angle α. The two families of curves are separated by means of the dotted straight line defined by the equation $z_1 = z_2$, from which the two branches of the same curve relating to a given value of the pressure angle bifurcate.

Curves at the right side of this straight line refer to gears for which $z_1 < z_2$, and represent the function given by Eq. (4.6) in the case in which the center distance is the nominal one, and $h_{a2\,max}/m = 1$ $(k_{2\,max} = 1)$. Curves at the left side of this straight line instead refer to gears for which $z_1 > z_2$, and represent the function given by Eq. (4.8), also here in the case in which the center distance is the nominal one, and $h_{a1\,max}/m = 1$ $(k_{1\,max} = 1)$.

Example The following example clarifies how the chart of Fig. 4.3 must be used. Consider a toothed wheel having $z_1 = 15$ teeth and pressure angle $\alpha = 20°$. We want to know for which value of z_2 the interference occurs. The horizontal line $z_1 = 15$, parallel to the abscissa axis, intersects the two branches of the curve relating to the pressure angle $\alpha = 20°$ at the points E and F. From the chart we can deduce that: for wheels with a number of teeth z_2 which is included in the range between points D and E of this horizontal line $(z_2 < 13)$, interference occurs between the tooth tip of the addendum flank of the wheel with $z_1 = 15$ and the dedendum flank of the wheel with $z_2 < 13$ teeth. For wheels with a number of teeth

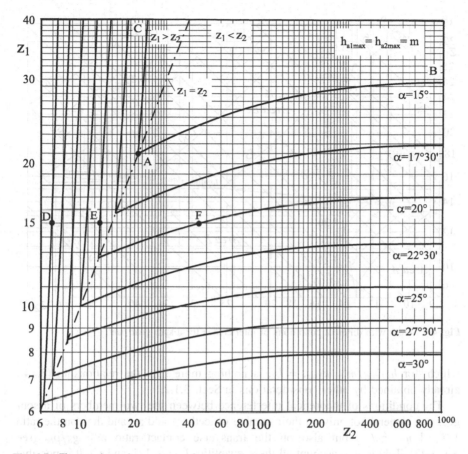

Fig. 4.3 Chart to determine $z_{1\,min}$ as a function of z_2, for $z_1 < z_2$, and $z_{2\,min}$ as a function of z_1, for $z_2 < z_1$, ($h_{a1\,max}/m = h_{a2\,max}/m = 1$, and nominal center distance)

z_2 which is included in the range between points E and F ($13 < z_2 < 45$), interference does not occur; for $z_2 > 45$ interference occurs between the tooth tip of the addendum flank of the wheel with z_2 teeth and the dedendum flank of the wheel with $z_1 = 15$ teeth.

However, the case that occurs more often is that in which a given gear ratio $u = z_2/z_1$ must be obtained by means of a gear pair having normal sizing, and with the pinion with a minimum number of teeth z_1 as possible, without the interference between the wheel and pinion occurs. In this case, processing Eq. (4.4), we obtain the following inequality:

$$z_{1\,min} \geq \frac{2h_{a2\,max}}{m} \frac{1 + \sqrt{1 + \frac{z_1}{z_2}\left(\frac{z_1}{z_2} + 2\right)\sin^2\alpha}}{\left(\frac{z_1}{z_2} + 2\right)\sin^2\alpha}, \tag{4.10}$$

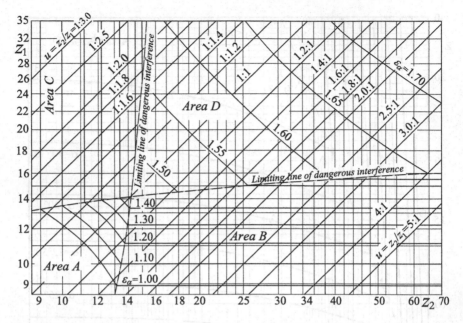

Fig. 4.4 Areas with different conditions of interference for a spur gear pair

which, for $h_{a2\,\text{max}}/m = k_{2\,\text{max}} = k_2$, is nothing more than the relationship (3.14), already obtained by other considerations in Sect. 3.1.2.

The conditions of theoretical interference between the two members of a spur gear pair depend not only on their number of teeth, z_1 and z_2, and then on the ratio $1/i = 1/u = z_1/z_2$, but also on the transverse contact ratio $\varepsilon_\alpha = g_\alpha/p_{bt}$ (see Eq. 3.43). Taking into account all these quantities $(z_1, z_2, 1/i, \text{ and } \varepsilon_\alpha)$, it is possible to identify four areas, each of which corresponds to different operating conditions. With reference to Fig. 4.4, these areas are characterized by:

- area A, with dangerous interference on both members of the gear pair;
- area B, with dangerous interference on the gear wheel 1 (the pinion);
- area C, with dangerous interference on the gear wheel 2;
- area D, without any interference on both members of the gear pair.

The interference should also be seen in comparison with the backlash between the teeth, as both influence the operating conditions of the gear. Remember that the backlash is the amount by which the space width exceeds the tooth width thickness. It is first necessary to prevent that the non-operating flanks of gear teeth come into contact. Backlash is also needed to accommodate different thermal expansions of the gear pair, tooth deflections, foreign material in the lubricant, as well as errors in the assembly, manufacture, and operation of gears in mesh. *Anti-backlash gears*, i.e. gears with zero backlash, are sometimes used in gear trains if the driving gear changes direction of rotation frequently.

Now, if considerable backlash exists between the teeth and at the same time the gear is characterized by theoretical interference, the transmission of motion is not interrupted certainly, but the contact between the teeth takes place in very bad conditions, causing variations in angular velocities, high noise, strong vibrations and very fast wear. Instead, if the backlash between the teeth does not exist or it is too small, the transmission of motion is interrupted because of the jamming.

4.3 Possibility to Realize Gear Pairs with Pinion Having Small Number of Teeth, Through the Generation Process with Rack-Type Cutter

In Sect. 3.2 we made some considerations on the theoretical interference during the cutting processes, both without generation (for example by form milling cutters) and with generation (e.g., by means of rack-type cutters, hobs, and pinion-type cutters). However, the problem of the theoretical interference in the cutting processes cannot be considered concluded with the observations made in that section. In respect of this issue, it is necessary that we make further considerations to be kept in mind in the gear design.

The cutting process without generation is performed, more frequently, by means of a single process, with milling machines using form cutters (a disk-type gear cutter, end mill) and indexing head, but also with an entire process, with a profiled cutter (an involute gear cutter), by broaching, stamping and cold drawing. Whatever the cutting process, the cutters used have profiled section as the space width between the successive teeth of the wheel to be profiled. With reference to the milling process (the same happens with the other process without generation), if the involute gear cutter has a theoretical profile that extends to the whole tooth flank depth, and the wheel has many teeth, we might have interference when it is coupled with a pinion having few teeth.

In the continuous *generation process* (also called *generation for envelope*), which is made with all purpose gear *shapers*, type Fellows, *pinion-type cutters* are used, which are cutting tools equivalent to involute toothed wheels. These gear-cutting machines, in addition to the slotting motion, also have a rolling motion relative to the wheel to be cut, with a transmission ratio defined by the ratio between the number of teeth of the *pinion-type cutter* and number of teeth of the wheel. If the number of teeth of the pinion-type cutter is not suitably chosen in relation to that of the teeth of the toothed wheel to be cut, two type of interference may occur, namely:

- interference between the addendum flank towards the tip of the pinion-type cutter and the dedendum flank towards the root of the teeth of the wheel, whereby the teeth of the latter are weakened at the root due to the *root relief*, in the case of interference not very high, or to the *undercut*, in the case of very high interference;

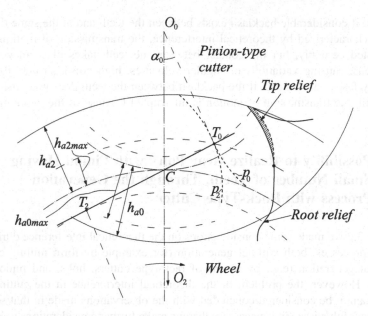

Fig. 4.5 Effects of the interference in the cutting conditions with pinion-type cutter

- interference between the addendum flank towards the tip of the wheel and the dedendum flank towards the root of the teeth of the pinion-type cutter, whereby the teeth of the wheel will have a smaller thickness towards the crest, due to the tip relief [15].

Figure 4.5 shows the effects of the interference in the cutting conditions, in the case where the cut is made with a pinion-type cutter (remember that here the subscript 0 identifies the cutting tool and, more generally, the cutting conditions). From the figure it is evident that, due to a too high addendum of the wheel to be cut ($h_{a2} > h_{a2\,max}$), which brings its addendum circle beyond the interference point T_0 of the pinion-type cutter, the interference causes a pronounced tip relief of the wheel. The same figure also shows that the pinion-type cutter has an addendum greater than the maximum value to avoid the interference ($h_{a0} > h_{a0\,max}$), so also its addendum circle goes beyond the interference point of the wheel; because of this, the interference determines a root relief of the wheel.

Example now we do an example of cutting interference, when the cutting is done with a pinion-type cutter, using for this purpose the chart shown in Fig. 4.3. Suppose to use a pinion-type cutter with $z_0 = 24$ teeth, pressure angle $\alpha_0 = 15°$, and normal sizing ($h_0 = 2.25 m_0, h_{a0} = m_0, h_{f0} = 1.25 m_0$). From Fig. 4.3 we infer that, using such a cutting tool, we can cut wheels with z_2 teeth in the range ($22 \leq z_2 \leq 44$) without the cutting interference takes place. Instead, we will have the cutting interference with the dedendum flank toward the root of the wheel for z_2 less than or equal to 21 teeth, and with the addendum flank towards the tip of the wheel for z_2 greater than 44 teeth. To cut without cutting interference a wheel with

21 teeth, the pinion-type cutter to be use, still remaining its other characteristics, must have 21 teeth instead of 24 teeth.

In the semi-continuous generation process for envelope, which is made with gear rack cutting machines, straight-sided rack-type cutters are used. With these machines, after each cutting stroke, the gear blank and rack type cutter roll slightly on their pitch circles and, when the blank and cutter have rolled a length equal to the pitch, the cutter is returned to the starting point, and the process is continued until all the teeth have been cut.

In the continuous generation process for envelope, which is made with gear hobbling machines (in these machines, the hob and blank rotate seamlessly at the proper angular velocity ratio, but the hob is also fed slowly across the face width of the blank until all the teeth have been cut), hobs are used, which are cutting tools shaped like worms. The hob teeth have straight flanks, as in the rack-type cutter. However, the hob axis must be turned through the lead angle in order to cut spur gear teeth. For this reason, the teeth generated by a hob have a slightly different shape than those generated by a rack-type cutter (this despite the fact that each longitudinal section of hob has the profile of the rack cutter). For more insights on these cutting processes, we refer the reader to Micheletti [24], Galassini [9], Rossi [31], Henriot [14], Dooner and Seireg [5], and Marinov [22].

From the point of view of the cutting interference, we can consider completely equivalent the two above-mentioned processes (the semi-continuous and the continuous one), since, in this regard, the differences are not significant. For both processes, with a normal standard sizing of the teeth, the minimum number of teeth below which we do not to go down, if we want to avoid the cutting interference, is to be deduced from the chart of Fig. 3.1 for $1/u = 0$.

We have already shown (see Sect. 3.1) that the risk of interference increases when the number of teeth z_2 of the wheel increases, and the number of teeth z_1 of the pinion decreases. The risk of the cutting interference is then maximum when we wanted to cut a pinion with a few teeth using the rack-type cutter, which can be considered as a wheel tool of infinite radius, and thus with number of teeth even infinite. Figure 4.6 shows the very high cutting interference that occurs in the case of a pinion with eight teeth, cut by means of a rack-type cutter having teeth with normal standard sizing and pressure angle $\alpha_0 = 15°$. The tooth is strongly carved at its root, with a very pronounced *undercut*, which determines the removal of material of the dedendum flank toward the root even within the fillet. This undercut occurs because the addendum line of the rack-type cutter intersects the line of action beyond the interference point T_1 of the pinion to be cut, so that $h_{a0} = h_{a0\,max}$ [15].

The undercut often constitutes a design choice, especially in cases where we want to remove the part of the involute near the base circle [11]. In fact, in correspondence of the base circle, the involute has a radius of curvature equal to zero, for which it is extremely difficult (if not impossible) to obtain accurately with the generation cutting of conventional processes, which, moreover, are the most technologically advanced. The contact at the point of regression of the involute on the base circle, which is a particular singular point, is also to be avoided due the

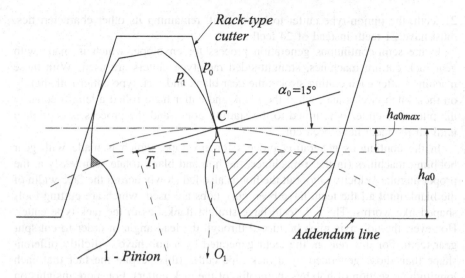

Fig. 4.6 Effect of the interference (undercut) on a pinion with eight teeth, in cutting conditions by means of a rack-type cutter

radius of curvature equal to zero. As already we have seen, this would determine specific sliding tending to infinity, and, as we shall see later, equally infinite values of the Hertz stresses (see Chap. 2 of Vol. 2). Lastly, the undercut may be desirable to accommodate post-processing, such as shaving, honing, or burnishing. The material removal in these post-processes is called *shaving stock* [29].

A safe way to get undercut is to use cutting tools with a *protuberance* (see Fig. 2.16a). The undercut determines a reduction in tooth thickness in the fillet region, for which it, based on considerations of stress concentrations and bending and fatigue strength, is not desirable. Moreover, teeth that have been undercut can determine additional discontinuities in the transverse contact ratio. However, it is equally likely that, even with cutting tools without protuberance, if interference conditions are met, one can get involute profiles characterized by cusps or slope discontinuities, with all the ensuing problems.

In conclusion, it is to be observed that, with the generation cutting of the teeth, whatever the process used (with rack-type cutters, pinion-type cutters or hobs), the tool cuts in the wheel blank space widths corresponding to their free relative motion, and therefore we will never have jamming. This is due to the relative motion between the generation cutting tool and the wheel to be cut. However, in the case in which the theoretical interference occurs, the involute profile of the tooth flank of the wheel will be altered.

With the single cutting process by milling, with form cutter, the effects of the interference during the cutting operations are not to be feared, as there is no a relative motion between the cutting tool and the wheel to be cut. However, it may possible have jamming between the two gear wheels generated with this process, when they are coupled to form a gear pair.

With the exception of the case of the generation process by means of a rack-type cutter, we must always verify that the risk of interference does not have, both during the generation of the two wheels of the gear pair, and during their mutual meshing.

Example We want to realize a gear pair having: $h_{a1\,max} = h_{a2\,max} = m$; $\alpha = 15°$; $z_1 = 22$ teeth and $z_2 = 27$ teeth. This pair can be generated without interference only by milling, with a profiled form cutter, or by generation using a pinion-type cutter having between 22 and 27 teeth. The two wheels of the gear pair cannot instead be generated without interference using a rack-type cutter.

4.4 Methods to Avoid Interference in the Cylindrical Spur Gears

In the design of spur gear pairs characterized by pinions with small number of teeth, in order to prevent the theoretical interference, we can implement different methods of solution, which emerge clearly from the foregoing considerations in the previous sections. In this regard, the most frequently used methods (they can be used not only individually, but also simultaneously, in various combinations, in order to enhance their effects), are as follows [35]:

- increase in the nominal pressure angle;
- use of a stub system (see Sect. 2.5), with teeth having depths smaller than those defined by the nominal sizing system, and given by Eq. (2.71);
- use, for a given nominal pressure angle, of teeth having nominal standard depths, but with profile shift, i.e. with a different distribution of the total depth of the tooth between the addendum and dedendum, and therefore with displacement of the addendum and dedendum circles.

With reference to the first method, from the inequalities (4.6) and (4.7) as well as from the chart shown in Fig. 4.3, it follows that the minimum number of teeth of the pinion, which meshes without interference with a given wheel, decreases when the nominal pressure angle of the toothing increases. This is referred to wheels having nominal standard sizing, and operating with the nominal center distance.

Example Wheel with $z_2 = 50$, having nominal standard sizing and operating with the nominal center distance. From inequalities (4.6) and (4.7) and chart of Fig. 4.3, we obtain: $z_{1\,min} = 25$, for $\alpha = 15°$; $z_{1\,min} = 19$, for $\alpha = 17°30'$; $z_{1\,min} = 16$, for $\alpha = 20°$; $z_{1\,min} = 13$, for $\alpha = 22°30'$.

We obtain very similar results when the ratio $1/i = z_1/z_2$ is an input data of the design; obviously in this case we use the inequality (4.10) and the chart shown in Fig. 3.1.

Example External gear pair with ratio $1/i = z_1/z_2 = 1/4$, having nominal standard sizing and operating with the nominal center distance. From inequality (4.10) and chart of Fig. 3.1, we obtain: $z_{1\,min} = 27$, for $\alpha = 15°$; $z_{1\,min} = 16$, for $\alpha = 20°$.

In front of the undoubted advantages in terms of reduction of the minimum number of teeth of the pinion to avoid interference in meshing with a given wheel (these advantages imply a consequent compactness of the gear pair, and then reductions of weight and cost), this first method of solution has quite a few drawbacks, the main of which are:

- the length of the path of contact (and then the contact ratio) is reduced when the pressure angle increases, and this reduction, however to quantify, can be too high (with consequent risk that the continuity of transmission of motion is compromised), such that a revision of the design choices is required;
- for a given power to transmit, when the pressure angle increases, the forces transmitted by the wheels of the gear pair to the shafts grow well, with repercussions not very sensitive as regards their stress states and loads on the bearings, but with effects which may not be tolerable in the case of shafts very deformable and flexible;
- with equal force exchanged between the teeth, the component that is discharged on the bearings increases, while the component that transmits the torque decreases;
- it is necessary to have, for the teeth cutting processes, numerous cutters, as many as are the possible values of the pressure angle α (in fact, this is the most consistent limit of this method, since it affects the economic aspect of the gear production).

With the second method, that is, with the use of the stub system, the tooth depth, as well as the addendum and dedendum, are lower than nominal standard. However, as the inequality (4.10) shows, the value of the minimum number of teeth, $z_{1\,min}$, of the pinion, for given values of the ratio $1/i = z_1/z_2$ and pressure angle α, is directly proportional to the addendum, $h_{a2\,max}$, of the wheel. It is therefore also evident that, using a stub sizing, the minimum number of teeth of the pinion is reduced by the ratio between the addendum selected (less than the nominal standard) and nominal standard addendum. For example, if we choose $h_{a2\,max} = 0.8\,m$, instead of $h_{a2\,max} = m$, for the same gear ratio and pressure angle, the value $z_{1\,min}$ will be equal to eighty percent of the value we would for nominal sizing.

A stub sizing of the teeth inevitably leads to a reduction of the length of the path of contact (and thus of the contact ratio). It follows the risk that the continuity of transmission of motion is compromised, for which the use of this method is not recommended for the purpose of elimination of the danger of interference, especially in the case of spur gears. Other reasons also advise against the use of stub sizing of the teeth. In fact, they have a lower resistance to surface loading than full-depth teeth (i.e., teeth with nominal standard depth), and tend to wear more quickly and to be noisy.

Their static strength, however, is about the same as that of full-depth teeth, and so they have the advantage of requiring, for the same resistance, a lesser removal of metal during cutting operations. Stub teeth may thus be suitable for gears of large module, required to operate slowly and infrequently. On the other hand, they have a greater resistance to bending of the teeth. The most negative aspect of this method is also the economic one, since the usual modular cutters cannot be used (they are standardized types of cutters corresponding to some fixed value of the *stub factor*, defined as the ratio between the stub addendum and the nominal standard addendum, equal to 0.9 or 0.8).

The third method consists of a decrease of addendum of the wheel (recall in this regard that the more restrictive condition to avoid interference is that related to the addendum circle of the wheel having a larger diameter, see Sect. 3.1), obtained approaching its addendum circle at nominal pitch circle and moving away from the latter the dedendum circle. In doing so the total depth of the wheel tooth does not change; only its sizing change, since its addendum is less than the nominal one, while its dedendum is greater than the nominal dedendum by an amount equal to the decrease of addendum. In this way, as the inequality (4.10) shows, the minimum number of teeth, $z_{1\,min}$, of the pinion decreases proportionally to $h_{a2\,max}$. Of course, the addendum of the pinion increases and its dedendum decreases to the same extent as, respectively, the dedendum of the wheel increases and addendum of the latter decreases.

With this method, the path of contact AE (Fig. 2.10) moves along the line of action, in the direction that goes from T_1 to T_2, but its length remains substantially unchanged. With this displacement of the path of contact from left to right (Fig. 2.10), the danger of interference is removed during the approach (point A moves away from T_1), but it should be checked that it is not present during the recess (point E approaches T_2). Another drawback that can occur with this solution is that of an excessive reduction of the tooth thickness of the pinion to its crest; this drawback manifests itself especially in cases in which the two wheels of the gear pair must work with nominal center distance.

In order to remedy, at least partially, to the above drawbacks, a solution a little more complex than that described above is used. It combines the aforementioned addendum modifications with a change of the center distance in the operating conditions, compared to the center distance in the cutting conditions. It follows that the pressure angle under actual operating conditions is greater than that which occurs in the cutting conditions, with all the resulting benefits. This combined method is of considerable design importance, because it enables the improvement of certain important operating quantities of the gear pair (in particular the decrease of the specific sliding), and therefore it is used widely, even in cases where the interference is not to be feared.

All three of the aforementioned methods to avoid interference require to have generation tools with different pressure angles and with various depths of the teeth. In addition, according to the type of cutting tool used, they require to verify that, during the process of generation, there is no interference between the generation tool and the gear wheel to be cut.

The processes most widely used in the cylindrical spur gear production are the generation processes, that employ rack-type cutters, hobs and pinion-type cutters as generation tools. These cutting tools are very expensive. It is therefore interesting to see how, with an usual tool characterized by certain nominal quantities (module and pressure angle), gears with the same nominal quantities of the tool can be obtained, with even small number of teeth, avoiding problems due to interference. So, we get toothing with addendum modification generated with rack-type cutters (hob, Maag cutter, Sunderland cutter), and those generated with pinion-type cutters (Fellows cutter).

4.5 Reduction of the Path of Contact Due to Cutting Interference

Let us now examine the generation cutting process, by a rack-type cutter (the same considerations are substantially valid in the case in which hobs are used for cutting), of toothed wheels having a number of teeth z less than the minimum number of teeth z_{min} that can be cut without interference. In these conditions, it is obvious that a part of the involute profile of the teeth of the gear wheel to be cut will be removed as a result of the cutting action of the addendum edge of the rack-type cutter; it fellows clearly a decrease in the path of contact.

For the analysis of the problem, we assume that (see Sect. 3.2), as regards the risk of cutting interference, things are as if the cutter addendum profile ended at the beginning of tip fillet. Consequently, we assume that this part of the addendum profile is not dangerous in respect to the interference (see Giovannozzi [11], Schiebel [33], Schiebel and Lindner [34]). In other words, we consider nonexistent the part of the rack-type cutter profile, which connects the straight-line involute from the point where it ends up to the edge on the tip line. Therefore, with reference to the effects of the cutting interference, things go as if the tooth of the rack-type cutter was truncated, sharp-edged, in that point in which the straight-line involute finishes. In these conditions, the active dedendum h_{a0} of the rack-type cutter will be equal to the difference between the dedendum $h_f = 1.25/m$ of the teeth to be cut and the dimension normal to the tip line of its curved profile at tip edge; the latter is equal to the root fillet radius, ρ_{fP}, of the basic rack. Then we have $h_{a0} = h_f - \rho_{fP}$.

In order to quantify the decrease of the path of contact due to interference, we must first identify the relative position between the cutting tool and the profile of the wheel to be cut, in the condition of incipient cutting interference. This is the position in which the virtual tip (the one defined above) with a sharp edge of the tool is for the first time in contact with the profile of the wheel, a part of which will subsequently be removed. In Fig. 4.7 this position is clearly identified by the point P, where the virtual tip line of the rack cutter intersects the involute profile of the wheel, whose point of regression on the base circle having radius r_b is the point P_0.

In the aforementioned position of incipient cutting interference, the straight-line profile of the tool is tangent at point A to the other branch of the involute (the one shown in the figure with a dashes line, which we might call the virtual involute), also with the point of regression at the point P_0 and symmetrical compared to the first branch with reference to the radial straight line OP_0. The fact that point P is precisely the point of incipient cutting interference fellows from the fact that, during

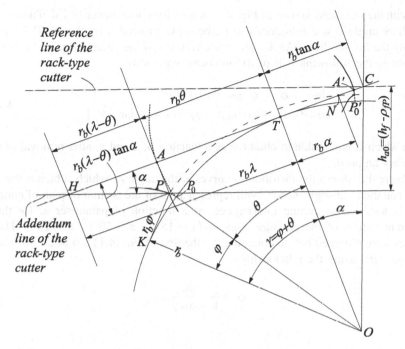

Fig. 4.7 Relative position between the rack cutter profile and profile of the wheel to be cut, in the condition of incipient cutting interference

cutting, the tool profile must be always tangent to an involute profile having the center of the curvature on the normal to the straight profile of the tool. Therefore, due to known properties to the involute, if two profiles are touching at a point A' between points T and C (i.e., within the theoretical length of the path of contact), a branch of involute exists that touches the next profile of the tool at point A. This last point is located on the line of action at a distance from the point A' equal to the base pitch p_b (see Gray [12], Lawrence [19]). The distance AA' is therefore equal to the length of the arc of base circle between the origins P_0 and P_0' of two successive involute profiles.

Figure 4.7, which shows the condition crystallized precisely in the instant in which the cutting interference begins, highlights the fact that, due to the interference, the portion of the profile corresponding to the arc PP_0 is lost. For known geometric properties of the involute, this arc corresponds to a loss of the path of contact equal to the distance PK between the points P and K, the latter being the point of tangency of the normal at point P to the involute profile with the base circle. However, the length of the segment PK is equal to that of the arc P_0K, i.e. $PK = P_0K = r_b\varphi$, where φ is the angle subtended by the arc P_0K; these lengths are also equal to the length of the segment TN, where N is the point of intersection of the line of action with the circle having fixed center O and radius OP (Fig. 4.7).

With the notations shown in Fig. 4.7, where, for convenience of calculation, the auxiliary angle θ was introduced (θ is the angle subtended by the arc P_0T), projecting the broken line $OKPA$ along the directions of the straight line CH and OT, we obtain the following pair of trigonometric equations:

$$\sin(\theta + \varphi) - \varphi \cos(\theta + \varphi) = \theta$$
$$\cos(\theta + \varphi) + \varphi \sin(\theta + \varphi) + (\lambda - \theta) \tan \alpha = 1, \quad (4.11)$$

from which it is possible to obtain φ as a function of λ and α, after removal of the auxiliary angle θ.

Figure 4.8 shows the distribution curves of the angle φ, which subtends the arc KP_0 and then the segment KP, both representative of the portion of path of contact that is lost due to cutting interference, as a function of parameter λ, for three different values of the pressure angle α ($\alpha = 15°$; $\alpha = 20°$; $\alpha = 24°$). These curves were obtained before obtaining, by the first of Eq. (4.11), φ as a function of $\gamma = (\varphi + \theta)$, using the relationship

$$\varphi = \frac{\gamma - \sin \gamma}{1 - \cos \gamma}, \quad (4.12)$$

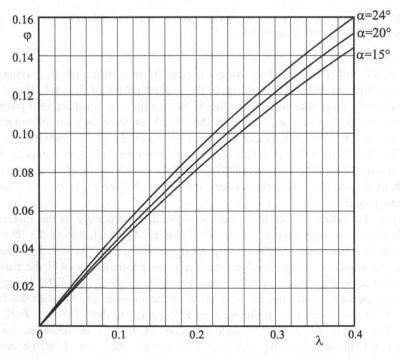

Fig. 4.8 Distribution curves of the function $\varphi = \varphi(\lambda)$, for three different values of the pressure angle α

and then the parameter λ, through the relationship

$$\lambda = (1 - \cos\gamma - \varphi\sin\gamma)\cot\alpha + (\gamma - \varphi). \tag{4.13}$$

With regard to the effects of interference, since the active addendum, h_{a0}, of the rack-type cutter is equal to the difference $(h_f - \rho_{fP})$, the length of the segment HT can be expressed as:

$$HT = \frac{h_f - \rho_{fP}}{\sin\alpha} - r_b\tan\alpha. \tag{4.14}$$

Therefore, being $r_b = (zm/2)\cos\alpha$, we obtain the following relationship that allow us to calculate the parameter λ as a function of quantities all known:

$$\lambda = \frac{HT}{r_b} = \frac{4}{z\sin 2\alpha}\frac{h_f - \rho_{fP}}{m} - \tan\alpha. \tag{4.15}$$

Once the value of the parameter λ has been so determined, we will derive φ by means of the curves shown in Fig. 4.8. For values of the pressure angle α different from those considered in Fig. 4.8, it is necessary to draw on it the corresponding curves, and then to proceed in the same way, or use directly the analytical equations described above, which are valid for any value of the angle α.

If the profiles of the teeth of both wheels of an external gear pair are cut by a rack-type cutter with a certain interference, according to what we said above, due to the removal of the corresponding profiles, two parts of the path of contact, respectively having lengths $T_1 N_1 = r_{b1}\varphi_1$ and $T_2 N_2 = r_{b2}\varphi_2$, are unusable (Fig. 4.9). The quantities φ_1 and φ_2 can be obtained using the curves of Fig. 4.7 or with the analytical relationships above mentioned.

The loss of these two portions of the theoretical path of contact affects the use of the relationship (3.44), which is used to calculate the transverse contact ratio of an external spur gear. In this respect, in fact, this relationship can be used without further considerations only in the case in which the values of the addendum factors k_1 and k_2 are of such magnitude that the addendum circles related to them intersect the line of action at points located within the segment whose end points are N_1 and N_2 (Fig. 4.9). In this case, the risk of interference between the two meshing wheels of the gear pair, under real operating conditions, is excluded. Instead, in the case in which the addendum factors k_1 and k_2 have values such that the addendum circles intersect the line of action at external points of the segment whose end points are N_1 and N_2, the values of k_1 and k_2 to be introduced in the relationship (3.44) are those that it would have if the addendum circles would pass through the points N_1 and N_2.

We must keep in mind that the cut of toothed wheels having small numbers of teeth as desired (for example, less than 8 teeth for $\alpha = 15°$, and less than 7 teeth for $\alpha = 20°$) is not possible, because below such numbers it would have pointed teeth, that is, with zero tooth thickness at top land. In gear design it is always advisable to check the tooth thickness at top land, especially in the case where a profile shift is made; we will return later on how this checking must be done.

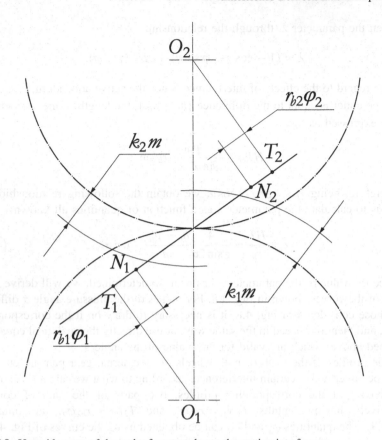

Fig. 4.9 Unusable parts of the path of contact due to the cutting interference

4.6 The General Problem of the Interference: Theoretical Interference and Fillet Interference

The problem of interference in the gears must be considered in its generality, analyzing the two basic aspects: the *cutting interference*, and the *interference in the working conditions* or *working interference*. From this last point of view, for external spur gear pairs, we must evaluate not only the effects of theoretical interference, but also those of *fillet interference*. As we mentioned in Sect. 3.2, we must keep in mind that no interference can be allowed between the two mating wheels of a gear pair in its operating conditions. On the contrary, in the generation cutting process of a gear wheel, a light interference between the cutting tool and the wheel generated by it is not only perfectly admissible, but is often desirable for the reasons already described.

The *fillet* is, by definition, that part of the tooth profile, which connects the part having an involute profile to the root (Fig. 4.10). It is shaped in such a way that this

Fig. 4.10 Fillet and fillet
circle

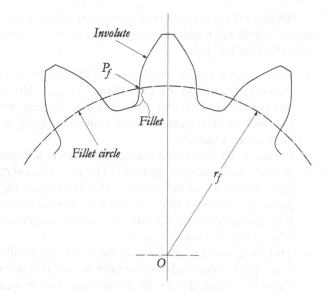

connection between the involute profile part and dedendum circle is as smooth as possible, that is made without abrupt variations of the radius of curvature.

The fillet profile, which is not an involute curve, does not take part to the contact between the two meshing wheels of the gear. If a fillet contact occurs, the tooth tips of a wheel would dig material from the root fillets of the other, and a smooth meshing between the two wheels of the gear pair would be impossible. This phenomenon is known as *fillet interference* (it is also called *trochoid interference*), and gear pairs must be designed in such a way that this type of interference does not occur.

If we want to better understand the problem, so to have the verification methods of the gears that allow us to dispel definitely the risk of fillet interference, it is necessary to deepen our understanding of the geometry of the fillet, in relation to the cutting processes. In this regard, we consider a transverse section of one member of the spur gear pair under consideration (Fig. 4.10), and call, as *fillet circle*, the circle at which the bottom point of the involute profile is connect with the top point of the fillet profile. Above this point (indicated by P_f in Fig. 4.10), where the fillet circle intersects the tooth profile, we have the involute profile, while below it we have the fillet profile, the geometry of which is closely correlated with that of tip edge of the cutting tool and the associated cutting process. The fillet circle has radius $r_f = OP_f$.

From the point of view of a greater bending strength of the tooth near its root, it would be appropriate for the fillet profile to be outside of the extension of the involute inside the fillet circle, but the cutting process is not always compatible with this design goal. Anyhow, the radius r_f depends on the cutting process and cutter type used and, generally, it is larger than the radius r_b of the base circle, because it is impossible for the involute to extend inside the base circle. In exceptional cases, r_f may be equal to r_b, but it is never smaller, that is $r_f \geq r_b$.

Whatever the generation cutting process (with rack-type cutter and, in equivalent terms, with hob, or with pinion-type cutter), we are faced with three possibilities (see also Henriot [15]):

- There is no a cutting interference, and then the bottom point of the involute profile is external to the base circle $(r_f > r_b)$. The involute part of the tooth profile near the root circle is free from a singular point in the cusp shape, the fillet and involute profile have a common tangent at point P_f, and the wheel tooth is not undercut.
- We are at the limit of the cutting interference, and then the bottom point of the involute profile reaches the base circle $(r_f = r_b)$, and P_f coincides with the point of regression, P_0, of the involute. Also, in this case, the involute part of the tooth profile near the root circle is free from a singular point in the cusp shape, the fillet and involute profile have a common tangent at point $P_f = P_0$, and the wheel tooth is not undercut.
- There is a cutting interference, for which the profile is undercut at point P_f where the involute ends and the fillet begins. This point is a singular point in the shape of a cusp, due to the discontinuity of the local tangent, which therein undergoes a sharp change; in this case it is also placed at the outside of the base circle, for which $r_f > r_b$.

Figure 4.11 shows examples of generation of spur involute teeth by a rack-type cutter. It show two different shapes of the tooth profile, respectively without (Fig. 4.11a) and with undercut (Fig. 4.11b), both obtained as the envelope of the successive positions assumed by the cutting tool during the relative motion between the cutting tool and the wheel to be cut. The involute part of the tooth profile shown in Fig. 4.11a is free from a singular point, the tooth is not undercut, and the fillet profile and involute profile have a common tangent at their connection point. Figure 4.11b instead shows a tooth with undercut, at point where the fillet profile and involute profile intersect one another, so that at this connection point the tangent undergoes an abrupt change, resulting in a singular point, a cusp.

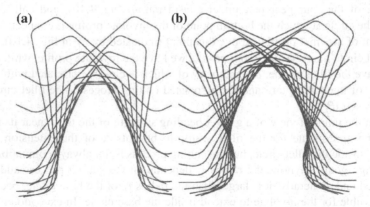

Fig. 4.11 Generation of tooth profiles by a rack-type cutter: **a** Without undercut; **b** With undercut

For the moment, we leave aside the determination of the radius r_f of the fillet circle, which refers to the cutting process, and focus our attention on the operating conditions, for which no interference (theoretical interference and fillet interference) is allowed. The gear pairs to be considered in their working conditions are those shown in Figs. 3.3a, b, which respectively show the meshing diagrams of an external gear pair and a rack-pinion pair. In both cases, the first condition to be met in order to avoid the interference is that concerning the theoretical interference.

In order to prevent the theoretical interference for an external gear pair, it is first necessary that the *limit circles* or *contact circles* intersect the line of action at point A and E (Fig. 3.3a) located inside the segment T_1T_2; in other words, the path of contact AE must be less than the line of action, T_1T_2, which represents its limit value. It is noteworthy that the limit circles coincide with the tip or addendum circles when the involute profile reaches the addendum circle; this is the hypothesis that here we do. Therefore, the necessary condition for no interference in working condition, $AT_2 < T_1T_2$, can be expressed in the following form (we assume that the pinion is the smaller wheel, whereby the position of point A with respect to point T_1 is the most restrictive):

$$\sqrt{r_{a2}^2 - r_{b2}^2} < (r_{b1} + r_{b2}) \tan \alpha. \tag{4.16}$$

Following the same procedure, we can deduce that, to prevent the theoretical interference for a rack-pinion pair (Fig. 3.3b), the first necessary condition that must be met $(AC < T_1C)$ can be expressed in the form:

$$\frac{h_{a2}}{\sin \alpha} < r_{b1} \tan \alpha. \tag{4.17}$$

If the inequalities (4.16) and (4.17), respectively valid for external gear pairs and rack-pinion pairs, are not satisfied, the theoretical interference will take place. However, even when these inequalities are satisfied, they are not always sufficient to prevent interference. In fact, the interference can still be present as fillet interference.

We defined above the fillet circle, which intersects the tooth transverse profile at point P_f, and has a radius r_f, yet to be determined. We also said that point P_f is located outside the base circle, and it is the top point of the fillet profile. We must now ensure that the top of the fillet profile does not come into contact with the teeth of the meshing wheel. To this end, the necessary and sufficient condition is that of imposing that the point where the contact on the gear tooth profile begins is outside with respect to the point P_f, namely that the radius of the limit circle, r_c, is greater than the radius of fillet circle, r_f, and therefore:

$$r_c > r_f. \tag{4.18}$$

The radius of the limit circle, $r_c = O_1A$ (see Figs. 3.3a, b), is determined on the basis of simple geometric considerations. In the case of an external gear pair (Fig. 3.3a), we have:

$$r_c^2 = \overline{O_1A^2} = \overline{O_1T_1^2} + \overline{T_1A^2} = r_{b1}^2 + \left[(r_{b1} + r_{b2})\tan\alpha - \sqrt{r_{a2}^2 - r_{b2}^2}\right]^2. \quad (4.19)$$

In the case of a rack-pinion pair (Fig. 3.3b), we have:

$$r_c^2 = \overline{O_1A^2} = \overline{O_1T_1^2} + \left(\overline{T_1C} - \overline{AC}\right)^2 = r_{b1}^2 + \left(r_{b1}\tan\alpha - \frac{h_{a2}}{\sin\alpha}\right)^2. \quad (4.20)$$

It is to be observed that, to neutralize the effects of the inevitable (albeit small) errors in the center distance, it is appropriate to design the gear pair so that the fillet circle of each wheel is smaller than the limit circle by a certain margin, usually assumed to be $0.025\,m$. Then the theoretical inequality (4.18) changes in the following design inequality:

$$r_c - 0.025m \geq r_f. \quad (4.21)$$

The inequality (4.16) and (4.21) for external gears, and (4.17) and (4.21) for rack-pinion pairs, are sufficient conditions to ensure that no interference will occur at the top of fillet profile, near the point P_f. However, there is still the possibility of interference at points of the fillet profile between the fillet circle and dedendum circle. In order to avoid this risk, which will be better detailed in the following sections, we must design the gear pair with adequate bottom clearances at each dedendum circle (or root circle).

The *bottom clearance* at the root circle of the wheel 1 is defined as the difference between the dedendum of the same wheel 1 and the addendum of the wheel 2 (and vice versa for the wheel 2). Then, for the wheel 1, we have $c_1 = \left(h_{f1} - h_{a2}\right)$, and therefore:

$$c_1 = \left(r_1 - r_{f1}\right) - \left(r_{a2} - r_2\right) = \left(r_1 + r_2\right) - \left(r_{f1} + r_{a2}\right) \quad (4.22)$$

where r_{f1} and r_{a2} are respectively the radii of the dedendum and addendum circles of the members 1 and 2 of the gear pair under consideration.

The recommended minimum value of the bottom clearance for both wheels of a gear pair is equal to $0.25\,m$, and one of the reasons for this recommendation is to help remove the risk of fillet interference. Notoriously (see Sects. 4.7 and 4.8), this type of interference is more likely to occur at the pinion fillets than at those of the wheel. Therefore, the gear designer, who must check that the above conditions of non-interference are satisfied and, simultaneously, ensure that the bottom clearances are adequate, performs both verifications for the pinion, while, for the wheel, he circumscribes his attention to verification of such bottom clearance.

4.7   Fillet Profile Generated by a Rack-Type Cutter or Hob

The *basic rack* is the *fictitious rack* that has, in its transverse cross section, the reference standard profile, and to which the dimensions of a series of toothings capable of meshing without distinction between them are referred. Then this basic rack forms part of an intermating series. The *counterpart rack*, that is the rack which can be inserted into the basic rack so that the teeth of one fit perfectly in the space widths of the other, serves as the basis of the shape of a generating cutter which will produce an intermating series of gears. It is the *rack-type cutter*, and is characterized by the so-called *axis of symmetry*, that is the *reference straight-line* or *datum line* on which the tooth thicknesses are equal to the space widths, the one and the other theoretically equal to one-half of the pitch (Fig. 2.14).

The shape of the rack-type cutter consists of two straight lines passing through the equally-spaced points along the axis of symmetry, and inclined at the pressure angle α, that generate the involute curves of the toothed wheel, an addendum straight line that generates the dedendum circle of the wheel, and the rack tip fillets that generate the wheel fillets.

In order to specify the whole profile of the rack-type cutter, additional special features are required. The addendum of the rack-type cutter must be equal to the dedendum of the wheel to be generated, and since this exceeds the addendum of the mating gear, the corners (or tips) of the tool teeth can with advantage be rounded by a radius ρ_{fP}. Incidentally, it may also be useful to ease the tips of the generated teeth by providing a corresponding thickening near the root of the rack-type cutter tooth. However, since the *tip easing* or *tip relief* does not affect the issue of interest here, we omit the discussion relating thereto.

As we mentioned in Sect. 3.2, the rounded corner of the rack-type cutter tooth, having radius of curvature ρ_{fP}, does not give rise to cutting interference. Therefore, everything works as if the addendum line of the rack-type cutter was the straight line parallel to the axis of symmetry passing through the point, L, of connection between the straight involute of the flank profile and the rounded profile of this corner, i.e. as if the rack-type cutter ended with a sharp edge at point L.

Figure 4.12 shows the tip geometry of the teeth of a rack-type cutter without undercut, in accordance with the standard basic rack of ISO 53: 1998 (E). The rounded corner of the tooth begins at point L, whose distance from the *cutter datum-line* defines the *cutter active addendum* h_{a0} of the rack-type cutter, and ends at point M, which is located on the *cutter tip line*. The center of curvature, I, of this rounded corner is located at distances ρ_{fP} and E respectively from the cutter tip line and axis of symmetry of the tooth, which bisects the thickness of the latter. The distance E is defined by (see ISO 6336-3 [17]):

$$E = \frac{\pi m_0}{4} - h_{fP} \tan \alpha_0 - (1 - \sin \alpha_0) \frac{\rho_{fP}}{\cos \alpha_0}. \tag{4.23}$$

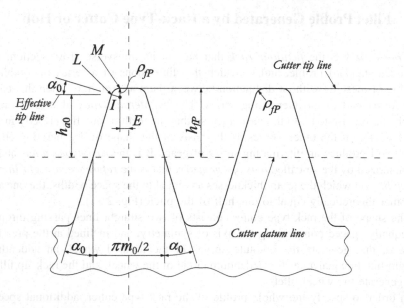

Fig. 4.12 Tip geometry of the rack-type cutter teeth

The position of point L is then uniquely identified, once that m_0, α_0, h_{fP} and ρ_{fP} are known. Consequently, also the *effective addendum line* or *effective tip line* of the rack-type cutter is identified. It is noteworthy that, for most rack-type cutters and hobs, the values of h_{fP} and ρ_{fP} are chosen so that the active addendum h_{a0} is slightly larger than m_0. However, here, we assume that h_{a0} is substantially equal to m_0.

Now we consider the relative motion between the rack-type cutter and the wheel to be generated, and assume that, in the cutting conditions, the *cutting pitch line* coincides with the cutter datum line. When this pitch line rolls without sliding on the pitch circle of the wheel to be cut, the point L, which is rigidly connected to this pitch line, but outside of it, generates the fillet profile of the wheel. This profile is a planar curve, a trochoid (so it was called by Gilles Personne de Roberval, see Auger [1], Jullien [18], Walker [36]) or cycloid.

We remember that a trochoid is the curve described by a point rigidly connected to a circle as it rolls without sliding along a planar curve, and becomes a cycloid when this planar curve is a straight line. Then the trochoid family is more general than the cycloid family. In our case, however, trochoid and cycloid coincide, so we will use the terms as synonyms.

To define the parametric equations of that particular trochoid, which is the cycloid, consider the circle of radius r and center O_1, in the initial contact at point O with a straight-line tangent to it, and that rolls without sliding on the same straight line (Fig. 4.13). Assume, as Cartesian coordinate system, the system $O(x, y)$, with origin at point O, the x-axis coinciding with the straight line on which

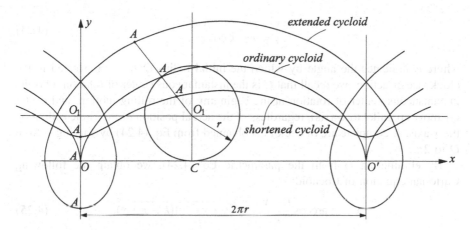

Fig. 4.13 Extended, ordinary, and shortened cycloids

the circle is rolling, and positive in the direction from left to right, and y-axis perpendicular to the x-axis, and positive in the direction that goes from the origin O to the center O_1.

Then consider a point A, located on the y-axis, in an intermediate position between points O_1 and O, or coincident with point O, or beyond point O, that is, on the negative y-semiaxis, and denote by k its distance from the center O_1 of the circle. In the chosen coordinate system, the initial coordinates of point A are thus $[O, (r - k)]$, with k taken as algebraic value.

During the rolling of the circle on the x-axis, starting from the initial position detected by the point of contact O, depending on whether point A lies inside the circle $(k < r)$, on its circumference $(k = r)$, or outside the circle $(k > r)$, the trochoid (*alias* cycloid) is described as being *shortened* (or *curtate*, or *contracted*), *ordinary* (or *conventional* or *common*), or *extended* (or *prolate*) respectively.

The trochoid, described by the tracing point A, is a periodic and transcendent planar curve, made up of an infinite number of arcs between them equal to one another, each corresponding to one complete revolution of the circle rolling on the x-axis. At the end points of each complete arc, i.e. the points of abscissa $\pm 2\pi n$, with n an integer, the ordinary cycloid has many cusps, while the extended cycloid has many loops, but the one and the other have no inflection points. Instead, the shortened cycloid has no multiple points, as the extended cycloid, but presents infinite inflection points at points of abscissa $\pm(2n + 1)\pi/2$ (Fig. 4.13). The y-axis can be considered as an axis of symmetry of the two branches of the cycloid in the right and left of it, whatever the type of cycloid (shortened, ordinary, or extended).

From the aforementioned considerations, we infer the following parametric equations of the trochoid/cycloid.

$$x = r\varphi - k \sin \varphi$$
$$y = r - k \cos \varphi, \tag{4.24}$$

where φ indicates the angle by which the initial radius $OO_1 = r$ is rotated in the clockwise direction (we recall that C is the instantaneous center of rotation, O is the instantaneous center of rotation at the beginning of the rolling motion, and center O_1 moves parallel to x-axis), regardless of the actual position of the circle rolling on the x-axis. A complete arc of cycloid is obtained from Eq. (4.24) by varying φ from O to 2π.

By eliminating φ from the parametric Eq. (4.24), we obtain the following Cartesian equation of trochoid:

$$x = r \arccos \frac{r-y}{k} - \sqrt{(k+r-y)(k-r+y)}. \tag{4.25}$$

It is to be observed that, since we are analyzing a relative motion, the shortened, ordinary, and extended trochoids of Fig. 4.13 can be generated, in an entirely equivalent manner, with the circle of center O_1 held stationary, and by rolling on it, without sliding, the x-axis to which the tracing point A, located outside of it, is rigidly connected. In relation to this way of generation and considering the case where point A is on the x-axis, some researches (see, e.g., Litvin and Fuentes [20]) call the aforementioned trochoid *shortened*, *ordinary* and *extended involute curves*, respectively.

Figure 4.14 shows the relative position between the rack-type cutter and the wheel to be generated, when the tracing point A is located on the axis of symmetry (the y-axis) of the trochoid. Any point P of this curve can be identified by a pair of Cartesian coordinates $P(x_P, y_P)$, in the Cartesian coordinate system $O(x, y)$, or by a pair of polar coordinate $P(\delta_P, r_P)$, in the polar coordinate system $O_1(\delta_P, r_P)$ having the origin at point O_1. By a shift of the axes of the Cartesian coordinate system, without rotation, leading its origin from point O to point O_1, these two coordinate systems become equivalent, and can be used interchangeably to determine, point by point, the fillet profile [15].

In the polar coordinate system, the angle δ_P between the straight line O_1P and the axis of symmetry of the trochoid is determined by observing that the point P corresponds to the position of point A after a rolling without sliding of the pitch line of the rack-type cutter on the pitch circle of the wheel, which carries the point of tangency at point T. Once a certain value of the distance $r_P = O_1P$ is chosen, and by putting $AO = AC = h_{a0} = k_0 m_0$ (h_{a0} is the active addendum of the rack-type cutter), we can write, in succession, the following relationships:

$$\delta_P = \widehat{PO_1T'} - \varphi \tag{4.26}$$

$$\tan \widehat{PO_1T'} = \frac{PT'}{O_1T'} = \frac{\sqrt{r_P^2 - (r - h_{a0})^2}}{(r - h_{a0})} \tag{4.27}$$

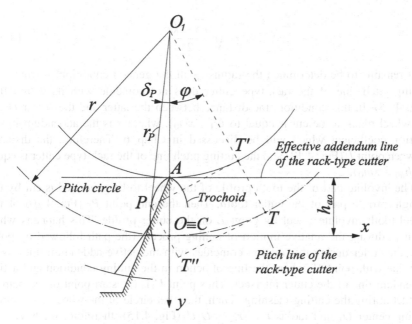

Fig. 4.14 Polar reference system, and quantities to draw, point by point, the fillet profile

$$\varphi = \frac{OT}{r} = \frac{T''T}{r} = \frac{PT'}{r} = \frac{\sqrt{r_P^2 - (r - h_{a0})^2}}{r}. \tag{4.28}$$

From these relationships, we obtain the following equation, which expresses the angle δ_P (in rad), as a function of the radius r_P:

$$\delta_P = \arctan \frac{\sqrt{r_P^2 - (r - h_{a0})^2}}{(r - h_{a0})} - \frac{\sqrt{r_P^2 - (r - h_{a0})^2}}{r}. \tag{4.29}$$

It is therefore possible to determine the trochoid, point by point, by varying r_P from its minimum value, equal to $(r - h_{a0})$, and calculating the corresponding angle δ_P by means of Eq. (4.29). Then recalling that $k_0 = h_{a0}/m_0$, and introducing the ratio $\bar{k}_0 = k_0/z$ and the dimensionless radius $\bar{r}_P = r_P/zm_0$, Eq. (4.29) takes the form:

$$\delta_P = \arctan \sqrt{\left(\frac{\bar{r}_P^2}{0.5 - \bar{k}_0}\right)^2 - 1} - 2\sqrt{\bar{r}_P^2 - (0.5 - \bar{k}_0)^2}. \tag{4.30}$$

The point of maximum interference corresponds to the maximum value of δ_P, and this quantity is maximum when the derivate $d\delta_P/d\bar{r}_P$ is zero. Therefore, carrying out the derivative of Eq. (4.30), and then imposing that it is equal to zero, we get that the maximum value of δ_P corresponds to the value of $\bar{r}_P$ given by:

$$\bar{r}_P = \sqrt{\frac{0.5 - \bar{k}_0}{2}}. \tag{4.31}$$

It remains to be determined the radius r_{fil} in the general condition in which the cutting pitch line of the rack-type cutter does not coincide with its datum line (Fig. 4.15). In this condition, the distance between the latter and the center O_1 of the wheel blank to be cut is equal to $(r_1 + xm_0)$, where x is the addendum modification coefficient which will be discussed in Chap. 6. Therefore, the distance between point L (Fig. 4.12) and the cutting pitch line of the rack-type cutter is equal to $(h_{a0} - xm_0)$.

The involute part of the tooth profile of the wheel to be generated is cut by the straight profile part of the cutter tooth. Then the end point P_f (Fig. 4.10) of the wheel tooth involute is cut by point L on the cutter profile. This happens when point L, during the relative motion of cutting process (the path followed by point L of the cutter during this motion coincides with the active addendum line), is to coincide with point L', where the line of action in the cutting condition and active addendum line of the cutter intersect. Thus point L' is the start point of the path of contact during the cutting meshing. Then, the fillet circle of the wheel is the circle having center O_1 and radius $r_f = r_{fil} = O_1 L'$ (Fig. 4.15); therefore, we have (see also Colbourne [3]):

$$r_{fil} = \sqrt{r_{b1}^2 + \left[r_{b1} \tan \alpha_0 - \frac{(h_{a0} - rm_0)}{\sin \alpha_0} \right]^2}. \tag{4.32}$$

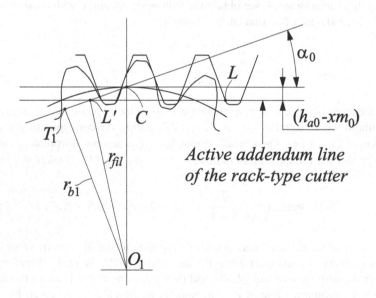

Fig. 4.15 Meshing diagram between the rack-type cutter and wheel to be cut

4.8 Fillet Profile Generated by a Pinion-Type Cutter

The generation process of gears with pinion-type cutter, first conceived by E.R. Fellows (see Galassini [9], Radzevich [30]), is a gear cutting method by shaping, whereby a reciprocating gear-like tool (this is the pinion-type cutter or shaper, which is itself a toothed wheel with hardened cutting edges), and a gear blank rotate together as if they were a gear pair. Of course, with respect to a common gear pair, the pinion-type cutter of such a gear pair is equipped with a cutting reciprocating motion in the direction of the axes, which is typical of the all-purpose gear shapers.

Suitably combining together the cutting reciprocating motion of the pinion-type cutter, and the rotation motions of the gear blank and pinion-type cutter about their axes, after a certain number of strokes for which the cutting operation is completed, we get a toothing geometry of the gear wheel so cut, which is the envelope of the successive positions of the pinion-type cutter in the said relative motion. Here we are interested in studying the tooth fillet profile of the wheel, that is, the profile at the tooth root, the one that connects the involute part of the tooth profile with the dedendum circle. Since the tooth tip of the pinion-type cutter cuts this tooth fillet, it is first necessary to define the tooth tip geometry of the cutting tool.

Figure 4.16 shows the tip geometry of a pinion-type cutter with a rounded corner at the tooth tip. The transverse section of this rounded corner is a circular arc, with radius ρ_{fP} and center I. The rounded corner of the tooth begins at point L (this point is the upper end of the involute part of the cutter tooth profile), whose distance from the *cutter reference circle* or *datum circle* defines the active addendum h_{a0} of the pinion-type cutter, and ends at point M, which is located on the cutter tip circle (or cutter addendum circle), having radius r_{a0}. The circular arc LM has a common tangent with the involute profile at point L, and with the addendum circle at point M.

Fig. 4.16 Tip geometry of the pinion-type cutter tooth

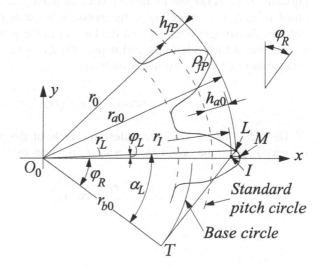

Here, it is advantageous to use a polar coordinate system (and then polar coordinates), where the position of any point is defined by the distance r from the origin O_0, and by the angle φ between r and the x-axis, coinciding with the tooth center line (Fig. 4.16). To define the tip geometry of the pinion cutter tooth, from which the fillet shape of the wheel depends, it is first necessary to determine the coordinates of points I and L. Since the circular arc LM of the cutter tip profile is tangent at point M with the addendum circle, the distance r_I from the origin O_0 to point I is equal to the difference between the radius of the addendum circle and the radius of the circular arc, i.e.:

$$r_I = r_{a0} - \rho_{fP}. \tag{4.33}$$

The polar angle φ_I between the straight line O_0I and the x-axis is given by the difference

$$\varphi_I = \alpha_I - \varphi_R, \tag{4.34}$$

where α_I is the angle $\widehat{IO_0T}$ (not shown in Fig. 4.16) and φ_R is the angle defining the polar axis position, i.e. the angle between the x-axis and the straight line O_0T. Considering the triangle IO_0T, we can write the following relationship which allows to calculate the angle α_I:

$$r_{b0} \tan \alpha_I = \sqrt{r_I^2 - r_{b0}^2}, \tag{4.35}$$

where r_{b0} is the radius of the base circle of the pinion-type cutter.

To uniquely define the polar coordinate of point I, it remains to calculate the angle φ_R. To do this, however, it is convenient to first define the polar coordinates of point L. The normal to the profile at point L passes through the center of curvature I and is tangent to the base circle at point T. The length of segment TL is equal to $r_{b0} \tan \alpha_L$, where α_L is the pressure angle at point L, that is the angle between the straight line O_0L and the tangent to the profile at point L. Therefore, considering the triangle LO_0T and noting that $LT = LI + IT$, we get the following relationship that allows us to calculate α_L:

$$r_{b0} \tan \alpha_L = \rho_{fP} + \sqrt{r_I^2 - r_{b0}^2}. \tag{4.36}$$

The radius of the effective addendum circle of the pinion-type cutter, i.e. the distance r_L from the origin O_0 to point L, is given by:

$$r_L = \frac{r_{b0}}{\cos \alpha_L}. \tag{4.37}$$

The angle φ_L between the straight line O_0L and the x-axis (Fig. 4.16), according to the considerations made in Sect. 2.5 (see also Fig. 2.12), can be written in the form:

$$\varphi_L = \frac{s_0}{2r_0} - (\text{inv}\alpha_L - \text{inv}\alpha_0), \tag{4.38}$$

where s_0 and α_0 are the nominal standard values of the tooth thickness and pressure angle at the standard pitch circle having radius r_0. The angle φ_L is then uniquely defined, when the thickness s_0 is known.

The last unknown quantities, φ_R, which appears in Eq. (4.34), is immediately defined as the difference:

$$\varphi_R = \alpha_L - \varphi_L. \tag{4.39}$$

The Cartesian coordinates of point I are given by:

$$x_I = r_L \cos \varphi_L - \rho_{fP} \sin \varphi_R \tag{4.40}$$

$$y_I = r_L \sin \varphi_L - \rho_{fP} \cos \varphi_R. \tag{4.41}$$

Therefore, instead of using Eq. (4.34), we can calculate φ_I with the following relationship:

$$\varphi_I = \arctan\left(\frac{y_I}{x_I}\right). \tag{4.42}$$

In some cases, the tooth profile of the pinion-type cutter does not have a rounded corner at the tip, but the involute profile develops up to the tip circle, so connection between the tooth profile and addendum circle configures a sharp corner. In these cases, we can use the same previous relationships, with the caveat to put in them ρ_{fP} equal to zero, and $r_I = r_L = r_{a0}$.

The pinion-type cutter shown in Fig. 4.16 is designed in such a way that, in the Cartesian coordinate system indicated therein, point I has a positive ordinate, i.e. $y_I > 0$. With this measure, the circular corner at the tip of profile smoothly connects with the tip circle. Otherwise, when $y_I < 0$, the tooth tip of the pinion-type cutter would be slightly pointed, as a consequence of a too large value of the radius ρ_{fP}.

In the design of a pinion-type cutter, sometimes we have advantages if we take the values of ρ_{fP} as high as possible, especially when the pinion-type cutter is to be used for cutting internal gears. In fact, a high value of ρ_{fP} leads to a larger radius of fillet of the wheel, and therefore to a lower stress concentration to the tooth root. The highest value of ρ_{fP} that can be used is that for which, in Eq. (4.41), $y_I = 0$ (then I is located on the tooth centerline, i.e. on the x-axis), but the calculation method of this value is not simple. Generally, we proceed by trial, choosing a value of ρ_{fP}, and then verifying that y_I is positive.

We thus uniquely defined the geometry of the tooth tip of the pinion-type cutter, including its active addendum h_{a0}. Now we consider the relative motion between this pinion-type cutter and the wheel to be generated, and assume that, in the cutting conditions, the cutting pitch circle coincides with the cutter standard datum circle. When this pitch circle rolls without sliding on the pitch circle of the wheel to be cut, the point L, which is rigidly connected to this pitch circle and is outside of it, generates the fillet profile of the wheel, which is still a trochoid, in the most general meaning of the term, and, more specifically, an *extended* (or *prolate*) *epicycloid*.

We define as epicycloid or hypocycloid the planar curves described by a tracing point rigidly connected to a mobile circle with radius r_2 as it rolls without sliding on a fixed circle having radius r_1, depending on whether the mobile circle is tangent externally or internally to the fixed circle. The epicycloids and hypocycloids are described as being *shortened* (or *curtate*), *ordinary* (or *conventional* or *common*), or *extended* (or *prolate*), depending on whether the tracing point L lies inside the mobile circle $(k < r_2)$, on its circumference $(k = r_2)$, or outside the mobile circle $(k > r_2)$. The quantities k and r_2 are respectively the distance of the tracing point from the center of the mobile circle, and the radius of the latter.

In addition, we remember that the distinction between epicycloid and hypocycloid is, in general, only formal as the Bernoulli-Euler-Goldbach double generation theorem (see Hall [13]) clearly shows that any epitrochoid (and then any epicycloid) can be expressed or generated as a hypotrochoid (and thus as a hypocycloid) and vice versa, but with some limitation, a hypocycloid can be generated as an epicycloid (in this second case, if and only if its rolling circle is larger the fixed circle).

To define the parametric equations of the epicycloid, consider a rolling circle of radius r_2 and center O_2, in contact at a point C with a fixed circle of radius r_1 and center O_1, with O_1, C and O_2 aligned along a vertical line, and on it located one over the other, in such a way that we have $O_1C = r_1$, and $CO_2 = r_2$ (Fig. 4.17). Assume, as Cartesian coordinate system, the system $O_1(x, y)$, with origin at point O_1, the y-axis coinciding with the straight line O_1O_2 and positive in the direction from point O_1 to point O_2, and the x-axis normal to the y-axis and positive in the direction from left to right. Then consider a point P_0, located on the y-axis at distance $k > r_2$ from point O_2, and rigidly connected to the rolling circle; in this chosen coordinate system, the initial coordinates of point $P = P_0$ are thus $[0, (r_1 + r_2 - k)]$.

During the rolling without sliding of the rolling circle on the fixed circle from the initial position detected by the point of contact C, the tracing point P describes an extended epicycloid (we focus only on this, because it is the one that interests us here), which consists of two symmetrical branches, with respect to the y-axis (Fig. 4.17). Generalizing the problem, in order to extend it to the case of internal gear pairs (in this case, in fact, it is interesting to consider the hypocycloid), according to the considerations mentioned above, we infer the following parametric equations of the epicycloid or hypocycloid (see Gibson [10], Rutter [32]):

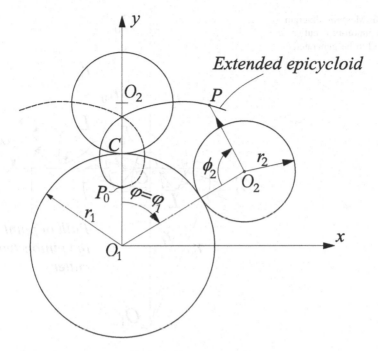

Fig. 4.17 Generation of an extended epicycloid

$$x = (r_1 \pm r_2) \sin \varphi \mp k \sin\left(\tfrac{r_1 \pm r_2}{r_2}\right) \varphi$$
$$y = (r_1 \pm r_2) \cos \varphi \mp k \cos\left(\tfrac{r_1 \pm r_2}{r_2}\right) \varphi,$$

(4.43)

where $\varphi = \varphi_1$ is the angle by which the initial segment $O_1 O_2$ is rotated in the clockwise direction (Fig. 4.17), while the upper signs apply for the epicycloids, and the lower signs for the hypocycloids. Then eliminating φ from these parametric equations, we can obtain the Cartesian equation, but this is left to the reader.

The procedure to derive the epicycloid traced from point L (Fig. 4.16) of the tooth profile of the pinion-type cutter, during the generation motion (this planar curve represents the transverse profile of the tooth fillet of the wheel obtained by the generation cutting), follows, in substance, that described in the previous section to obtain the trochoid. Leaving aside this topic, and reserving it to the reader, it remains to determine the radius r_{fil} of the fillet circle, which intersects the tooth profile of the generated wheel at the point P_f, where the involute part of the same profile ends and the fillet begins.

Figure 4.18 shows the meshing diagram between the pinion-type cutter and wheel to be generated, in the condition in which the cutting pitch circle of the pinion-type cutter coincides with its datum circle, and $r_L = r_0 + h_{a0}$ (Fig. 4.16). The involute part of the tooth profile of the wheel to be generated is cut by the

Fig. 4.18 Meshing diagram
between pinion-type cutter
and wheel to be generated

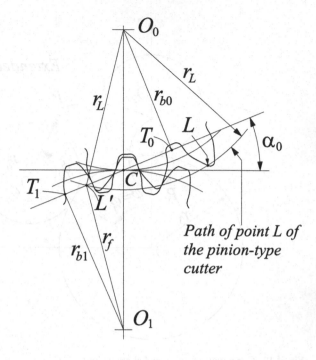

Path of point L of
the pinion-type
cutter

involute part of the cutter tooth profile. Then the end point P_f of the wheel tooth involute is cut by point L on the pinion cutter profile. This happens when point L, during the relative motion of the cutting process (the path followed by point L of the pinion-type cutter during this motion coincides with the active addendum circle), is to coincide with point L', where the line of action in the cutting condition and active addendum circle of the cutter intersect. Point L' is then the start point of the path of contact during the cutting meshing.

Therefore, the fillet circle of the wheel is the circle having center O_1 and radius $r_f = r_{fil} = O_1 L'$ (Fig. 4.18) given by:

$$r_{fil} = \sqrt{r_{b1}^2 + \left[(r_{b0} + r_{b1}) \tan \alpha_0 - \sqrt{r_L^2 - r_{b0}^2} \right]^2}. \qquad (4.44)$$

Of course, in this equation the radius r_L must be replaced with the radius r_{a0} (see Fig. 4.16) in the case for which the pinion-type cutter does not have rounding at the tooth tips.

4.9 Interference Effects of the Rounded Tip of the Cutter Teeth and Fillet Interference in the Operating Conditions

The tooth profile of most rack-type cutters, hobs and pinion-type cutters is rounded at the tip. The question is therefore legitimate to ask in relation to the effects of such rounded tips of the teeth of the cutting tools on interference in the cutting and operating conditions (see Polder and Broekhuisen [27]). In this respect, the cut with rack-type cutter (and the equivalent thereto with hob), and cutting with pinion-type cutter are to be evaluated in a different way.

When the cut is made with a rack-type cutter (the same considerations, with adequate explanations that follow, are basically valid also when the cut is made with hobs), we can show that the rounded tip of the cutter tooth does not cause a cutting interference. To prove this fact, just compare the trochoid described by the tracing point L (see Fig. 4.12), and the curve obtained as the envelope of infinite circles of radius ρ_{fp} and centers on the trochoid described by the tracing point I (Fig. 4.12), which is the center of curvature of the rounded tip of the cutter tooth. Thus, we demonstrate, unequivocally, that the trochoid described by point L and the envelope curve associated with the aforementioned infinite circles of radius ρ_{fp} and trochoid described by point I intersect the involute generated from the straight part of the rack-type cutter profile at the same point, and then determine the same fillet circle having radius r_{fil}.

Therefore, in conclusion, as regards the problem of interference during the cutting with rack-type cutter, we will consider only the point L (Fig. 4.12), where the involute straight profile ends and the rounded tip begins, and proceed as if the cutting tool were truncated with a sharp edge at point L. Of course, all other conditions being equal, the cutting interference is much higher, the lower the pressure angle. Figure 4.19, taken from Henriot [15], but reworked in the form that is of interest here, shows the tooth profiles of a pinion with standard sizing, and without profile shift, with 15 teeth, cut with four rack-type cutters, with rounded tips all characterized by the same geometry, but with pressure angles respectively equal to 15°, 20°, 25° and 30°. The four trochoids traced by point L are virtually overlapped, and then the four profiles of the root fillets practically coincide, but with $\alpha_0 = 25°$ and $\alpha_0 = 30°$ the cutting interference does not occur, while for $\alpha_0 = 15°$ and $\alpha_0 = 20°$ cutting interference and undercut occur. The same figure highlights the fact that the trochoid T_L traced by point L and the envelope curve T generated from the rounded tip of the four cutting tools intersect in the same point V the involute profile of the tooth that is generated.

Quite similar considerations can be made when hobs are used for the gear wheel cutting. As we have already mentioned in Sect. 3.1.3, few additional problem can occur in cases where, for special achievements (see, for example, the cutting by hobs for large wheels), hobs with rounded tips having radii of curvature more extensive than those corresponding to the rack-type cutter are used. In these cases,

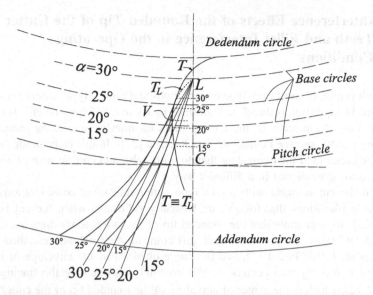

Fig. 4.19 Tooth profiles of a pinion with 15 teeth and standard sizing, generated by four rack-type cutters having different pressure angles, but rounded tips with the same geometry

effects of cutting interference of these rounded tips with the tooth profiles of the wheels to be generated cannot be excluded. In these cases, it is necessary to resort to the procedure scribed in the aforementioned Sect. 3.1.3.

We have already said that the interference in the actual operating conditions cannot be never accepted, and must be categorically excluded. The condition to be met to exclude the interference in the operating conditions is given by the inequality (4.18), or better still by the inequality (4.21). Well, when the cutting of the two toothed wheels of a gear pair is made using rack-type cutters or hobs, these inequalities are always verified.

These inequalities instead should be carefully verified in the case of a wheel cut using a pinion-type cutter, the latter with a low number of teeth, intended to mesh, under real operating conditions, with another wheel having a high number of teeth. In practice of design, the pinions generated in condition of cutting interference are very few. Then the bottom point P_f (Fig. 4.10) of the involute tooth profile is external to the base circle $(r_f > r_b)$.

However, if a pinion like this (that is, without cutting interference), under real operating conditions, meshes with a toothed gear wheel with a high number of teeth (the limit case is obviously that of a rack, and therefore a rack-pinion pair), the tooth tip of the member with high number of teeth interferes with the root fillet of the member with less number of teeth, even before the theoretical interference occurs.

Figure 4.20, taken from Henriot [15], but also reworked in the form that is of interest here, shows this state of things for a pinion with 40 teeth and standard sizing, cut using a pinion-type cutter with 20 teeth, $h_{a0} = 1.25m_0$, $h_{f0} = m_0$, and

Fig. 4.20 Fillet interference
in the operating conditions

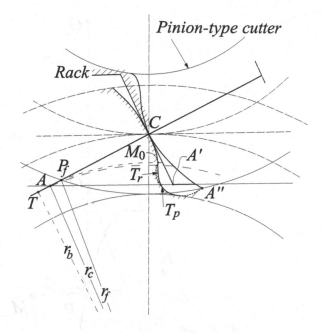

pressure angle in the cutting conditions $\alpha_0 = 20°$, and meshing, in the operating conditions, with a rack, that is, with a wheel having infinite number of teeth. The figure highlights the fact that $r_c < r_f$, albeit slightly. The inequality (4.18) is therefore not satisfied, and it follows a small interference of the tooth tip of the rack with the root fillet of the pinion, also highlighted in Fig. 4.20, where the trochoid T_r, described by the tooth tip A' of the rack, penetrates inside the trochoid T_P, described by the tooth tip A'' of the pinion-type cutter.

A fillet profile of a given wheel of center O_1, defined with geometric law which ensures non-interference with the addendum flank profile to the mating gear, having center O_2, is that proposed for the first time by Schiebel in 1912 (see also Schiebel and Lindner [34]). This fillet profile is defined as an extended epicycloid of point P (Fig. 4.21), in which the involute tooth profile ends, and the fillet profile begins, when an auxiliary circle, of radius CO_0 and center O_0, rolls without sliding on the pitch circle of the wheel to be generated. The center O_0 of this auxiliary circle (it is nothing else then the pitch circle of the pinion-type cutter to be used for cutting) is located on the line of centers O_1O_2. It is identified by the intersection between this line of centers and the axis of the segment AB, where A is the end point of the path of contact, and B is the intersection between the line of centers O_1O_2 and the dedendum circle (Fig. 4.21).

To demonstrate that this extended epicycloid does not cause interference, just to show that the path described by point P, supposed rigidly connected with the auxiliary circle, during rolling without sliding of the latter on the pitch circle of the wheel with center O_1, is at the outside of the path described by the same point P, this

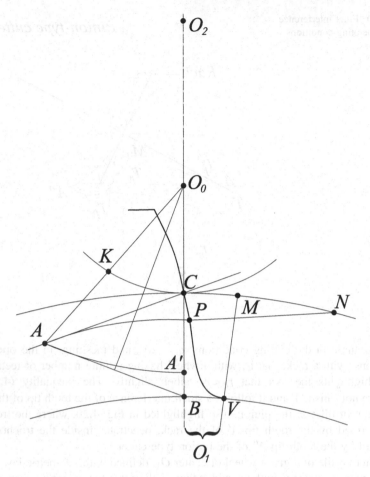

Fig. 4.21 Fillet profile in the shape of prolate epicycloid, according Schiebel

time supposed rigidly connected with the pitch circle of the wheel with center O_2, when the latter rolls without sliding on the pitch circle of the wheel with center O_1.

Since both the epicycloids are extended epicycloids described by the same point P, during pure rolling of two different circles on the same fixed circle, due to well know properties of epicycloids, it is sufficient to prove that $CO_0 < CO_2$, i.e. that the radius CO_0 of the pitch circle of the pinion-type cutter is smaller than the radius CO_2 of the pitch circle of the wheel that will mesh with the one considered here. In particular, by the Euler-Savary equation (see Dooner [4]), it is easy to verify that, at point of contact P between the two epicycloids, the condition $CO_0 < CO_2$ determines, at point P, a radius of curvature of the epicycloid related to the pitch circle of the pinion-type cutter smaller than the radius of curvature of the epicycloid related to the pitch circle of the mating wheel.

However, the inequality $CO_0 < CO_2$ is immediately verified, by noting that the distance of point B from the center O_2 is greater than the distance of point A' (this point is obtained as the intersection of the line of center O_1O_2 with the circle of radius AO_2 and center O_2, which is the addendum circle of the mating wheel) from the same center O_2. This is because a sufficient bottom clearance $A'B$ must exist between the dedendum circle of the wheel under consideration, and the addendum circle of mating wheel. Therefore, the axis of the segment AB intersects the line of centers O_1O_2 at point O_0, the distance of which from point C will be smaller than the distance, from the same point C, of point O_2 where the axis of the circular arc AA' intersects the same line of centers O_1O_2.

If we draw the tooth profile in the position in which it passes through point C (Fig. 4.21), and we want the epicycloid may link up at point P with the involute tooth profile, with a common tangent, its center of curvature should be clearly on the normal PN to the tooth profile at point P. However, also the center of instantaneous rotation, during rolling of the auxiliary circle on the pitch circle of the wheel, must be on the same normal. Consequently, the auxiliary circle must be tangent to the pitch circle at point N, when point P is located in the position shown in Fig. 4.21.

The apex of the epicycloid, i.e. its point of tangency V with the dedendum circle, will have when the auxiliary circle will have rotated, with respect to the position for which the contact was at point N, of a circular arc NM such that point P, rigidly connected with it, is located on the straight line joining point O_1 to the center of the rolling auxiliary circle. Since the segment AC is none other than the segment PN, when the point P of the latter has moved to point A, the circular arc MN we seek is equal to the circular arc KC shown in Fig. 4.21.

Therefore, reporting from the point N a circular arc $NM = KC$, we identify the point M and, then, the apex V of the epicycloid as the intersection of the straight line O_1M with the dedendum circle of the wheel. Finally, the epicycloidal profile between the end points P and V is then completed using the analytical relations given in the previous section.

References

1. Auger L (1962) Un Savant Méconnu, Gilles Personne de Roberval. Librairie Scientifique A. Blauchard, Paris
2. Buckingham E (1949) Analytical mechanics of gears. McGraw-Hill Book Company Inc, New York
3. Colbourne JR (1987) The geometry of involute gears. Springer, New York Inc, New York
4. Dooner DB (2012) Kinematic geometry of gearing, 2nd edn. Wiley, Chichester, UK
5. Dooner DB, Seireg AA (1995) The kinematic geometry of gears—a concurrent engineering approach. Wiley, New York
6. Dudley DW (1962) Gear handbook. The design, manufacture, and application of gears. McGraw-Hill Book Company, New York
7. Eckhardt HD (1998) Kinematic design of machines and mechanisms, 1st edn. McGraw-Hill, New York

8. Ferrari C, Romiti A (1966) Meccanica applicata alle macchine. Unione Tipografico-Editrice Torinese, UTET, Torino
9. Galassini A (1962) Elementi di Tecnologia Meccanica, Macchine Utensili, Principi Funzionali e Costruttivi, loro Impiego e Controllo della loro Precisione, reprint 9th edn. Editore Ulrico Hoepli, Milano
10. Gibson CG (2001) Elemntary geometry of differential, an undergraduate introduction. Cambridge University Press, Cambridge
11. Giovannozzi R (1965) Costruzione di Macchine, vol II, 4th edn. Casa Editrice Prof. Riccardo Pàtron, Bologna
12. Gray A (1997) Modern differential geometry of curves and surface with mathematica, 2nd edn. CRC Press, Taylor & Francis Group, Boca Raton
13. Hall LM (1992) Trochoids, roses, and thorns-beyond de sprirograph. Coll Math J 23(1):20–35
14. Henriot G (1972) Traité théorique et practique des engrenages—fabrication, contrôle, lubrification, traitement thermique, vol 2. Dunod, Paris
15. Henriot G (1979) Traité théorique et pratique des engrenages, vol 1, 6th edn. Bordas, Paris
16. ISO 53: 1998 (E) (1998) Cylindrical gears for general and for heavy engineering—standard basic rack tooth profile
17. ISO 6336-3:2006 (2006) Calculation of load capacity of spur and helical gears—part 3: calculation of tooth bending strength
18. Jullien V (1996) Eléments de Géométrie de Gilles Personne de Roberval. Vrin, Paris
19. Lawrence JD (1972) A catalog of special plane curves. Dover Publications, New York
20. Litvin FL, Fuentes A (2004) Gear geometry and applied theory, 2nd edn. Cambridge University Press, Cambridge
21. Maitra GM (1994) Handbook of gear design, 2nd edn. Tata McGraw-Hill Publishing Company Ltd, New Delhi
22. Marinov V (2012) Manufacturing process design, 2nd edn. Kendall/Hunt Publishing Company Inc, Dubuque, Iowa, USA
23. Merritt HE (1954) Gears, 3rd edn. Sir Isaac Pitman & Sons Ltd, London
24. Micheletti GF (1958) Tecnologie Generali: Lavorazioni dei Materiali ad Asportazione di Truciolo. Libreria Editrice Universitaria Levrotto & Bella, Torino
25. Niemann G, Winter H (1983) Maschinen-Elemente, Band II: Getriebe allgemein, Zahnradgetriebe-Grundlagen, Stirnradgetriebe. Springer, Berlin
26. Pollone G (1970) Il Veicolo. Libreria Editrice Universitaria Levrotto & Bella, Torino
27. Polder JW, Broekhuisen H (2003) Tip-Fillet Interference in Cylindrical Gears. In: ASME proceedings power transmission and gearing, paper no. DETC 2003/PTG-48060, pp 473–479
28. Radzevich SP (2013) Theory of gearing: kinematics, geometry, and synthesis. CRC Press, Taylor & Francis Group, Boca Raton
29. Radzevich SP (2016) Dudley's handbook of practical gear design and manufacture, 3rd edn. CRC Press, Taylor & Francis Group, Boca Raton
30. Radzevich SP (2017) Gear cutting tools: science and engineering, 2nd edn. CRC Press, Taylor & Francis Group, Boca Raton
31. Rossi M (1965) Macchine utensili moderne. Ulrico Hoepli Editore, Milano
32. Rutter JW (2000) Geometry of curves. Chapman Hall, CRC Mathematics, Boca Raton
33. Schiebel A (1912) Zahnräder, I Teil, Stirn-und Kegelräder mit geraden Zähnen. Springer, Berlin
34. Schiebel A, Lindner W (1954) Zahnräder, Band. I: Stirn-und Kegelräder mit geraden Zähnen. Springer, Berlin
35. Scotto Lavina G (1990) Riassunto delle Lezioni di Meccanica Applicata alle Macchine: Cinematica Applicata; Dinamica Applicata—Resistenze Passive—Coppie Inferiori; Coppie Superiori (Ingranaggi)—Flessibili—Freni. Edizioni Scientifiche Siderea, Roma
36. Walker E (1932) A study of the traité des indivisibles of Gilles Personne de Roberval. Teachers College, Columbia University, New York
37. Yu DD (1989) On the interference of internal gearing, gear technology, July/August, pp12–19; 43–44

Chapter 5
Interference Between Internal Spur Gears

Abstract In this chapter, the problems of interference between internal spur gears in both cutting and working conditions are first described and discussed. The discussion is quite general and therefore includes the five types of interference that may occur in this case, i.e.: theoretical or primary interference; secondary interference; trimming interference during the radial approach of the pinion when the internal gear pair is assembled, and of the pinion-type cutter in the conditions of the generation cutting process; fillet interference between the tip of the pinion and the root fillet of the annulus; fillet interference between the tip of the annulus and the root fillet of the pinion. For this purpose, analytical methods are used, without forgetting well-known traditional methods which, although approximate, provide reliable results of great engineering-design value. Finally, the condition for avoiding rubbing during the annulus cutting process is discussed and defined.

5.1 Introduction

To transmit the motion between two parallel shafts, which rotate in the same direction, internal gear pairs are used. An internal spur gear pair is constituted by a cylindrical pinion with external toothing and a cylindrical *internal gear* (also called *ring gear*, *annulus gear* or more simply *annulus*) with internal toothing. The geometry of an internal gear is very similar to that of an external gear. Therefore, *mutatis mutandis*, we can extend to the internal spur gears the same concepts described in the previous chapter, for external spur gears (see Buckingham [3], Dudley [7], Merritt [14], Radzevich [20], Schreier [22]). For this reason, we will focus our attention only on those aspects that differentiate the internal gears from external gears (see Tuplin [23], Polder [15–17], Litvin et al. [12], Litvin and Fuentes [11]).

Since one of the two members of an internal gear pair is an external gear wheel (the pinion), from the geometrical point of view, the first difference between an internal gear pair and an external gear pair concerns only the annulus. In fact, the teeth of this member of the internal gear pair lie outside the profiles, while those of

© Springer Nature Switzerland AG 2020   181
V. Vullo, *Gears*, Springer Series in Solid and Structural Mechanics 10,
https://doi.org/10.1007/978-3-030-36502-8_5

an external gear wheel lie inside them. In other words, the teeth of an annulus have exactly the same shape of the tooth spaces in an external gear wheel. Furthermore, the addendum circle of an annulus lies inside the reference pitch circle, while the dedendum circle lies outside it (see Pollone [19], Henriot [9]).

Consider now a ring gear with involute tooth profiles and nominal or standard sizing (i.e., with $h_{a2} = m$, and $h_{f2} = 1.25\,m$), having module m, pressure angle α, and number of teeth z_2 (as usual, the subscript 2 denotes the ring gear, and the subscript 1 denotes the pinion). The radius r_{b2} of its base circle is given by:

$$r_{b2} = \frac{mz_2}{2}\cos\alpha, \tag{5.1}$$

while its tooth profiles will have the shape of the space-widths between the teeth of the pinion, whose number of teeth is z_1.

The nominal or standard values of the diameter of the pitch circle and center distance are the ones given respectively by the first of Eqs. (2.63) and (2.65), which here are rewritten with the specific subscripts to standard symbols (see ISO 6336-1 [10]):

$$d_2 = 2r_2 = z_2 m, \tag{5.2}$$

$$a = r_2 - r_1 = \frac{z_2 - z_1}{2}m. \tag{5.3}$$

In the case of standard sizing, the diameters of the addendum circle (also called tip circle or outside circle), and dedendum circle (or root circle) are those respectively given by the second and third of Eqs. (2.63), which here are also rewritten with the specific subscripts to standard symbols:

$$d_{a2} = d_2 - 2m \qquad d_{f2} = d_2 + 2,5m. \tag{5.4}$$

However, it is to be noted that, for the internal spur gears, special sizings are often adopted. We will make mention of these special sizings, other than the standard sizing, as the opportunity presents itself.

For an internal gear pair, we have the following relationships, with the usual meaning of the symbols:

$$\omega_1 r_1 = \omega_2 r_2$$
$$\frac{r_1}{r_2} = \frac{z_1}{z_2} = \frac{\omega_2}{\omega_1} = \frac{r_{b1}}{r_{b2}}$$
$$r_1 = \frac{z_1 a}{z_2 - z_1}$$
$$r_2 = \frac{z_2 a}{z_2 - z_1}$$
$$p_{b1} = \frac{2\pi r_{b1}}{z_1} = p_{b2} = \frac{2\pi r_{b2}}{z_2}$$
$$p_1 = \frac{2\pi r_1}{z_1} = p_2 = \frac{2\pi r_2}{z_2} = \pi m$$
$$r_{b1} = r_1 \cos\alpha$$
$$r_{b2} = r_2 \cos\alpha. \tag{5.5}$$

In this chapter, however, we focus our attention on the interference problem concerning the internal spur gears. For these gears, we also face the interference problem in its dual aspects, i.e., the interference in the cutting conditions, and the interference in the operating conditions. General concepts, already described and analyzed for the external gears, with the variations of the case, retain unchanged their validity also for the internal gears. However, for internal gears, the interference problem is more complex than that of the external gears. In fact, we can be faced to five types of interference, rather than two types of interference, as in the external gears. All these five types of interference must be analyzed to prevent them from occurring.

As we mentioned in Sect. 4.1, the five types of interference that can occur in the internal gears are as follows:

- Theoretical interference (also called primary interference or involute interference).
- Secondary interference or *fouling*.
- Tertiary interference or *trimming interference*, i.e. interference of the cutter approach in the generation cutting process.
- Fillet interference between the tip of the pinion, and root fillet of the annulus.
- Fillet interference between the tip of the annulus, and the root fillet of the pinion.

In the following sections, these five types of interference are analyzed individually, and design solutions to avoid them are described.

5.2 Theoretical Interference in the Internal Spur Gears

This *theoretical interference* (also named *primary interference* or *involute interference*) in the internal spur gears corresponds to the theoretical interference in the external spur gears, already described in Sect. 4.2.

For the analysis of this type of interference, we consider the transverse section of the internal spur gear pair, shown in Fig. 5.1. This gear pair consists of a pinion 1, having fixed center O_1, radius of the base circle r_{b1}, and radius of the pitch circle r_1, and an annulus 2, having fixed center O_2, radius of the base circle r_{b2}, and radius of the pitch circle r_2.

Consider then two involute mating profiles, p_1 and p_2, related respectively to the pinion and annulus. We assume that both these profiles extend up to the respective base circles. Their origins, on these same base circles, are P_{01} and P_{02}; thus P_{01} and P_{02} are the points of regression, i.e. the points where the involute curves meet the base circles. This is a limit condition, which we use here as a hypothesis to explain how the theoretical interference occurs. We will remove this hypothesis when we will introduce the necessary restrictive conditions concerning other types of interference, such as interference between the tooth tip profiles and root fillets.

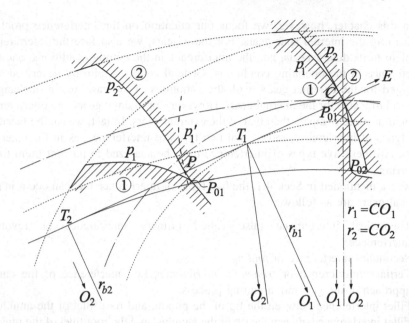

Fig. 5.1 Contact between profiles of an internal spur gear pair within the path of contact, and outside of the path of contact

The two pitch circles are tangent at the instantaneous center of rotation or pitch point C, and the line of action is the tangent to the two base circles at points T_1 and T_2. As Fig. 5.1 shows, even the two profiles p_1 and p_2 are touching at pitch point C. At this point, the center of curvature of the profile p_1 is T_1, while the center of curvature of the profile p_2 is T_2. The two centers of curvature T_1 and T_2 are located on the line of action on the same side with respect to pitch point C. The angle α between this line of action and the common tangent to the pitch circles at pitch point C is the operating pressure angle (i.e. the pressure angle at the pitch cylinders) of the internal gear pair. Obviously, the lines O_1T_1 and O_2T_2 are parallel and both perpendicular to the line of action. Therefore, each of them makes the angle α with the line of centers O_1O_2 (outside of Fig. 5.1, downward).

By extension of the concepts that we have described for the external spur gears, we can say that we will have a conjugate contact only if the point of contact between the profiles p_1 and p_2 will be between points E and T_1. Point E is the point of intersection of the line of action with the addendum circle of the pinion, located in the upper right out of the figure. Instead, there will be non-conjugate contact if the point of contact will be between points T_1 and T_2, which are the points of interference of the pinion and ring gear, respectively.

Therefore, contact cannot go beyond the point T_1, in the direction from point T_1 to point T_2. Instead, in the direction from point T_1 to point C, contact can go, at least from a theoretical point of view, up to infinity. This is why, in principle, the involute profiles of an annulus may extend to any radius, and conjugate contact is theoretically possible, however large is the radius of the addendum circle of the pinion. Actually, beyond the point C, the path of contact is limited by point E, where the addendum circle of the pinion intersects the line of action.

During the engagement, the point of contact between the conjugate profiles moves on the line of action. As long as the point of contact is located between points E and T_1, we have a conjugate contact, and the meshing has a correct kinematics. When the point of contact reaches the point T_1, it comes to coincide with the point of regression P_{01} of the involute profile p_1. Point T_1, i.e. the point of interference of the pinion, represents the limit position of the point of contact, beyond which the contact is non-conjugate, and theoretical interference occurs.

Consider now a point of contact P on the line of action, between points T_1 and T_2. In this point, which is a point of non-conjugate contact, the center of curvature of the profile p_2 is T_2. The center of curvature of the profile which is conjugate of the profile p_2 is always the point T_1. Thus, this conjugate profile has a curvature in the opposite direction compared to the one of the profile, p_2. Therefore, the profile that is conjugate of the profile p_2, is no longer the profile p_1, but the profile p_1', i.e. the fictitious branch of the profile p_1 symmetric with respect to the line O_1P_{01}, where P_{01} is the origin of the two branches p_1' and p_1 of the involute curve on the base circle of the pinion.

Therefore, the two actual profiles p_1 and p_2 intersect, and thus interference occurs, because the pinion dedendum profile near its root goes to undermine the annulus addendum profile near its tip. To avoid this type of interference, i.e. the theoretical interference, it is sufficient that the value of the radius r_{a2} of the addendum circle of the annulus is greater or equal to O_2T_1, namely $r_{a2} \geq O_2T_1$. Therefore, considering the triangle $O_2T_1T_2$ (Fig. 5.1), to avoid the primary interference, the following inequality must be satisfied:

$$r_{a2} \geq \overline{O_2T_1} = \sqrt{\overline{O_2T_2}^2 + \overline{T_1T_2}^2} = \sqrt{r_{b2}^2 + [(r_2 - r_1)\sin\alpha]^2}, \qquad (5.6)$$

as $\overline{O_2T_2} = r_{b2}$, and $\overline{T_1T_2} = \overline{CT_2} - \overline{CT_1} = (r_2 - r_1)\sin\alpha$.

To avoid this type of interference, we must limit the addendum h_{a2} of the annulus. Figure 5.2 shows the meshing diagram and two conjugate profiles in contact at the pitch point C of an internal gear pair, with module m and pressure angle α, consisting of a pinion and annulus having z_1 and z_2 teeth, respectively. The figure also shows the related radii of the base circles, r_{b1} and r_{b2}, pitch circles, r_1 and r_2, and addendum and dedendum circles, r_{a1} and $r_{f1} = r_{b1}$, r_{a2} and r_{f2}.

If we assume the pinion as driving wheel and its direction of rotation is clockwise, the line of action, which is the common tangent to the two base circles, is the one shown in Fig. 5.2. Therefore, to avoid the primary interference between the tooth tip V of the annulus and the involute profile of the pinion dedendum flank, the contact may extend, at most, up to the point T_1 where the line of action is tangent to the base circle of the pinion. Thus, the maximum value of the addendum of the annulus must satisfy the inequality:

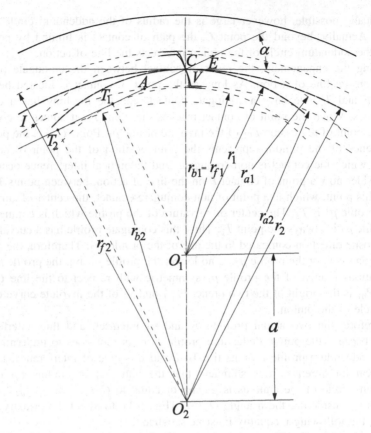

Fig. 5.2 Meshing diagram and contact between two conjugate profiles of an internal gear pair

$$h_{2a\max} \le \frac{m}{2}\left[z_2 - \sqrt{z_2^2 - \left(2z_1 z_2 - z_1^2\right)\sin^2\alpha}\right], \tag{5.7}$$

from which we obtain that, for a given addendum h_{2a} of the annulus, and for a given ratio z_1/z_2, the following relationship must be satisfied:

$$z_1 = z_{1\min} \ge 2k_2 \frac{1 + \sqrt{1 - \frac{z_1}{z_2}\left(2 - \frac{z_1}{z_2}\right)\sin^2\alpha}}{\left(2 - \frac{z_1}{z_2}\right)\sin^2\alpha}, \tag{5.8}$$

where $k_2 = h_{2a}/m$ is the addendum factor of the annulus.

Remembering that $(z_1/z_2) = (1/u)$, the inequality (5.8) coincides with the inequality (3.15). The inequality (5.8) provides the minimum number of teeth, $z_{1\min}$, of the pinion necessary to avoid the involute interference with the annulus, when the operating pressure angle α, gear ratio $u = (z_2/z_1)$, module m of the teeth, and addendum $h_{2a} = k_2 m$ of the annulus are given. In the case in which the teeth have nominal standard sizing, i.e. when $k_2 = 1$ and $h_{2a} = m$, the inequality (5.8) takes the form:

$$z_1 = z_{1min} \geq 2 \frac{u + \sqrt{u^2 - (2u - 1)\sin^2\alpha}}{(2u - 1)\sin^2\alpha}. \tag{5.9}$$

The minimum values of z_{1min} are shown in the lower part of the diagram of Fig. 3.1. From this diagram we infer that the value of z_{1min}, for given values of the operating pressure angle α and ratio $(z_1/z_2) = (1/u)$, is much greater for internal gear pairs than the one for external gear pairs.

In order to use pinions with a limited number of teeth, it is necessary to choose pressure angles α higher than those chosen for external gear pairs, or to use stub teeth since, as the inequality (5.8) shows, the minimum number of teeth z_{1min} varies proportionally to the addendum factor of the annulus $k_2 = h_{2a}/m$.

It should be noted that, generally, the solution that provides addendum modification (with consequent shifting of the addendum and dedendum circles) is not convenient because, due to the addendum modification, it is very likely that the secondary interference can manifest (see next section).

In the case where the solution would be to limit the addendum h_{2a} of the annulus, and the pressure angle α is equal to $20°$, the diagram shown in Fig. 5.3 can

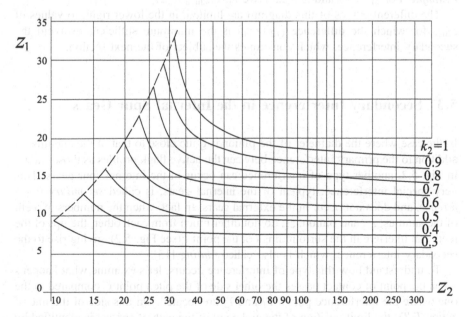

Fig. 5.3 Addendum factor k_2 of the annulus as a function of z_1 and z_2, for $\alpha = 20°$

be used (see also Henriot [9]). It allows to obtain the value of the addendum factor k_2 of the annulus as a function of the numbers of teeth z_1 and z_2, when $\alpha = 20°$.

Example With $z_2 = 60$ and $z_1 = 20$, we can choose $k_2 = 1$, i.e. a nominal standard sizing. Instead, with $z_2 = 100$ and $z_1 = 15$, the value of $k_2 = 0.80$ cannot be overcome. In this last case, it is necessary to adopt a stub gearing or an addendum modification factor $x_2 = 0.20$ (see next Chapter).

From inequality (5.7) we infer that, to avoid primary interference between an annulus having z_2 teeth and a pinion with z_1 teeth, the following relationship must be satisfied (see also Pollone [19]):

$$z_2 = z_{2min} \geq \frac{z_1^2 \sin^2\alpha - 4k_2^2}{2z_1 \sin^2\alpha - 4k_2^2}, \tag{5.10}$$

which, in the case of nominal standard sizing, i.e. for $k_2 = 1$ and $h_{2a} = m$, becomes:

$$z_2 = z_{2min} \geq \frac{z_1^2 \sin^2\alpha - 4}{2z_1 \sin^2\alpha - 4}. \tag{5.11}$$

The diagram shown in Fig. 5.4 allows to calculate, for a nominal standard sizing, the minimum number of teeth z_{2min} of the annulus as a function of the number of teeth z_1 of the pinion, for five values of the pressure angle α, in the range between $\alpha = 20°$ and $\alpha = 30°$.

Example For $z_1 = 15$ and $\alpha = 25°$, we have $z_{2min} = 27$.

The different curves of this diagram are limited, in the lower right, to values of z_{2min} for which the difference $(z_2 - z_1)$ is the minimum sufficient to avoid the secondary interference, which constitutes the subject of the next section.

5.3   Secondary Interference in the Internal Spur Gears

In the case where the diameter of the pinion is quite close to that of the annulus, in addition to the primary interference between the active flank profiles as those shown in Fig. 5.2, another type of interference can occur in the internal spur gears. This very special interference, typical of the internal gears, is called *secondary interference*, and does not occur in the external gears. In fact, when the numbers of teeth of the annulus, z_2, and pinion, z_1, do not differ much from each other, the tips of the teeth can interfere in the surroundings of the point I (see Fig. 5.2), giving rise to the secondary interference, which is also called *fouling* [13].

To understand how this type of interference occurs, let's examine what happens when the point of contact passes the other side of the pitch point C compared to the one where the interference points T_1 and T_2 are located. On this side of the line of action T_1T_2, the limit position of the end point of the path of contact is identified by

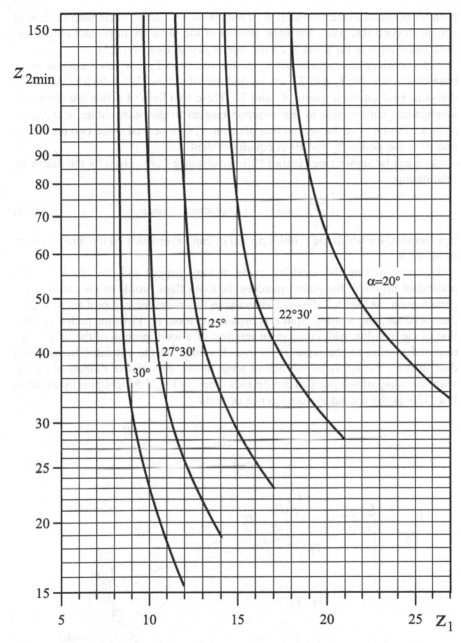

Fig. 5.4 Minimum number of teeth z_{2min} of the annulus as a function of the number of teeth z_1 of the pinion, for five values of the pressure angle α

point E, obtained as the intersection of the pinion addendum circle with the line of action. Beyond this point, the flank profiles of the teeth are in contact only along their extensions, at the point R on the line of action (see Fig. 5.5).

Let's consider now, in addition to the line of action $T_1 T_2$ which is tangent to the base circles at point T_1 and T_2, the second line of action $T_1' T_2'$, which is tangent to the same base circles at points T_1' and T_2'. These two lines of action are obviously symmetrical with respect to the common tangent to the pitch circles at the pitch point C. The fact that the second line of action $T_1' T_2'$ is normal to the two profiles p_1 and p_2 at the points P_1 and P_2 is equally evident.

As it can be shown (see Henriot [9]), the distance $P_1 P_2$, given by the following relationship:

$$\overline{P_1 P_2} = m(z_2 - z_1)\mathrm{inv}\alpha\cos\alpha = const, \tag{5.12}$$

is a constant, for which the profiles p_1 and p_2 can never touch on this line of action or above it.

However, during the meshing cycle, it may happen that the tip V_1 of the addendum profile of the pinion matches the tooth flank profile of the annulus at a certain point, as Fig. 5.5 shows. Evidently, in this point the two profiles p_1 and p_2 cannot be tangent. Starting from this instant, the tooth tip V_1 of the pinion notches that of the annulus, with consequent removal of a tooth portion of the latter. This removed portion of the annulus tooth is bounded by the extended hypocycloid described by the tip V_1 of the pinion tooth during the relative motion of rolling without sliding of two pitch circles on each other, as Fig. 5.5 highlights by a dashed line. This is the secondary interference or fouling.

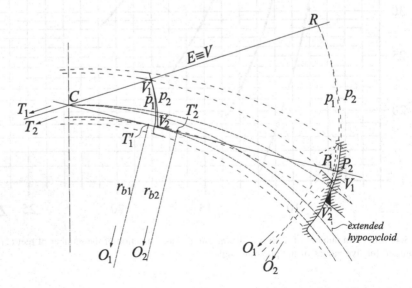

Fig. 5.5 Secondary interference or fouling: contact between the tip V_1 of the pinion tooth and the tooth flank of the annulus

Figure 5.6 shows how and to what extent this secondary interference alters the tip profile of the annulus. The portion of the annulus profile corresponding to triangles *ABC* is removed by the action exerted by the tip of the pinion tooth.

To remove the risk of secondary interference, it is necessary to verify that the extended hypocycloid described by point V_1 during this relative motion does not intersect the profile of the tooth flank of the annulus. To this end, known geometric-analytical methods may be used, based on the parametric equations of the hypocycloid given by relationships (4.43), or the Cartesian equation that is derived by them. In the next section, we will describe one of these methods, the one proposed by Colbourne [4].

Traditional methods developed by Buckingham [3], and Henriot [9] can also be used, as an alternative to such geometric-analytical methods. It is not the case to bring full the analytical developments made by these two researchers. Here we give only the results obtained by Henriot [9], as they can easily be used to verify the fact that this type of interference does not occur.

Henriot considers an internal gear pair in the limiting condition of secondary interference, which occurs when the tip V_1 of the addendum profile p_1 of the pinion touches, at a point P geometrically well-defined, the tip V_2 of the addendum profile p_2 of the annulus (with reference to Fig. 5.5, this limiting condition would come to have point P and V_1 both coinciding with point V_2). Based on geometric considerations, Henriot leads to a system of algebraic equations, where the following quantities appear:

- the numbers of teeth z_1 and z_2;
- the addendum factors k_1 and k_2;
- the angles δ_1 and δ_2 between the line of centers O_1O_2 and the lines O_1P and O_2P;
- the operating pressure angle α;
- the involutes invα_{a1} and invα_{a2}, where α_{a1} is the involute polar angle between the line O_1P and O_1P_{01}, and α_{a2} is the involute polar angle between the line O_2P and O_2P_{02}, where P_{01} and P_{02} are the origins of the involutes on the related base circles.

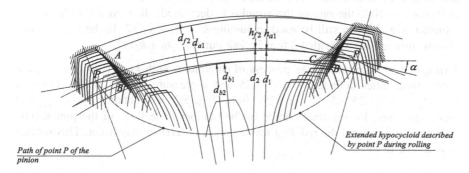

Fig. 5.6 Alteration of the annulus tip profile due to secondary interference

This system of algebraic equations allows to determine the three unknowns δ_1, δ_2 and z_1, once α, z_2, k_1 and k_2 are given.

The solution of this system of equations is far from simple, as they contain trigonometric and transcendental functions. Jean Capelle, director of the Société d'Études de l'Industrie e de l'Engrenage (SEIE) and expert in metrology and strength of gears and tooth profiles, proposed a very general graphical method (see Henriot [9]). Henriot, who succeeded Capelle in the direction of SEIE, proposed a less general method, but simpler, which however provides very satisfactory results for most of the design problems of technical interest. Henriot summarized the results obtained with this method in the diagram shown in Fig. 5.7, which allows to calculate the minimum or limiting difference $(z_2 - z_1)_{\min}$ between the numbers of teeth of the annulus and pinion, below which the secondary interference occurs, as a function of the addendum factor $k = h_{a1}/m = h_{a2}/m$ and operating pressure angle α of the toothing.

Here some calculation examples are provided. All examples show that the limiting difference $(z_2 - z_1)_{\min}$ is practicality independent of the number of teeth z_2 of the annulus.

Example 1 Input data: $z_2 = 55$; $z_1 = 45$; $\alpha = 20°$; $k = k_1 = k_2 = 1$. Result: since $(z_2 - z_1) = 10 > 8$ (see Fig. 5.7), there is no interference.

Example 2 Input data: $z_2 = 55$; $\alpha = 20°$; $k = k_1 = k_2 = 1$. Result: from Fig. 5.7, we infer $(z_2 - z_1)_{\min} = 8$; then $z_{1lim} = 47$ teeth.

Example 3 Input data: $z_2 = 90$; $\alpha = 20°$; $k = k_1 = k_2 = 1$. Result: from Fig. 5.7, we infer $(z_2 - z_1)_{\min} = 8$; then $z_{1lim} = 82$ teeth.

Example 4 Input data: $z_2 = 60$; $\alpha = 20°$; $k = k_1 = k_2 = 0.80$ (stub gearing). Result: from Fig. 5.7, we infer $(z_2 - z_1)_{\min} \cong 6$; then $z_{1lim} = 54$ teeth.

Example 5 Input data: $z_2 = 40$; $z_1 = 35$; $\alpha = 20°$. Result: from Fig. 5.7, we infer $k_{lim} = k_{1lim} = k_{2lim} = 0.625$.

It is to be noted that the conditions of secondary interference remain virtually unchanged if a profile shift is performed (for example, the addendum of the pinion is increased, while the one of the annulus is decreased). If a small profile shift is required, Fig. 5.7 can still be used, assuming $k = (k_1 + k_2)/2$. In the cutting conditions, just replace k_1 with k_0, that is, just put $k = (k_0 + k_2)/2$.

Example Consider the cutting process of an annulus with $k_2 = 1$, to be do with a pinion-type cutter defined by $k_0 = 1.25$ and a pressure angle $\alpha = 20°$. For $k = (k_0 + k_2)/2 = (1.25 + 1)/2 = 1.125$, from Fig. 5.7 we obtain $(z_2 - z_1)_{\min} = 9$ teeth. However, for the final choice of the number of teeth z_0 of the pinion-type cutter, we must dispel the risk that the tertiary interference may occur. This subject is analyzed in Sect. 5.6.

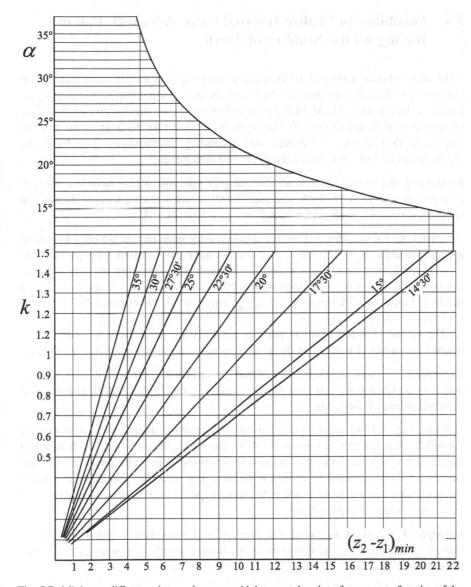

Fig. 5.7 Minimum difference $(z_2 - z_1)_{min}$ to avoid the secondary interference, as a function of the addendum factor k and operating pressure angle α

5.4   Possibility to Realize Internal Gear Pairs with Pinion Having a Low Number of Teeth

Very often, to reduce the size of the gear drives for power transmission mechanical systems, and thus their weight and their cost, the design conditions impose stringent limits on the number of teeth of the pinion. In these cases, the design choices of the internal gear pair, able to satisfy the input data, must take account of the need to dispel the double risk of primary and secondary interferences. The following examples show how to achieve the design requirements.

Example 1 We want to realize an internal gear pair with a gear ratio $u = z_2/z_1 = 2$, a pinion with $z_1 = 15$ teeth, and pinion and annulus having teeth with nominal standard sizing, i.e. with addendum factor $k = k_1 = k_2 = 1$.

From Fig. 5.4 we infer that, to avoid the primary interference, we need to choose a pressure angle $\alpha = 25°$. Since this internal gear pair consists of a pinion with $z_1 = 15$ teeth, and the annulus with $z_2 = 30$ teeth, the secondary interference does not occur. In fact, Fig. 5.7 provides, for $k = h_{a1}/m = h_{a2}/m = 1$ and $\alpha = 25°$, a limiting value of difference $(z_2 - z_1)_{min}$ equal to 5, when actually we have $(z_2 - z_1) = 15$.

The secondary interference occurs for transmission ratios close to unity, that is when the number of teeth z_1 of the pinion becomes close to that of the annulus, z_2.

Example 2 We want to realize an internal gear pair with numbers of teeth $z_1 = 55$ and $z_2 = 60$ (therefore with gear ratio $u = z_2/z_1 = 60/55 = 1.0909$), and nominal standard sizing of the teeth.

From Fig. 5.4 we infer that, to avoid the primary interference, it would be enough to choose a pressure angle $\alpha = 15°$. However, since $(z_2 - z_1) = 5$, to avoid the secondary interference, it is necessary to choose a pressure angle $\alpha = 27°30'$ (see Fig. 5.7). If we want to avoid the primary and secondary interferences during the generation cutting, to be made with pinion-type cutter, it is necessary to proceed as the following example shows.

Example 3 We want to avoid the primary and secondary interferences in the cutting conditions, with pinion-type cutter having number of teeth $z_0 = 25$, and pressure angle $\alpha = 20°$.

From Fig. 5.4 we infer that the minimum number of teeth z_{2min} of the annulus, which can be generated without primary interference, is equal to 38 teeth. From Fig. 5.7 we infer that, for $\alpha = 20°$ and $k = k_1 = k_2 = 1$, the minimum difference $(z_2 - z_1)_{min}$ is equal to 8. Therefore, since $(z_2 - z_0) = 13$, no secondary interference occurs in the cutting conditions. If the pinion-type cutter had number of teeth $z_0 = 27$, pressure angle $\alpha_0 = 20°$ and nominal standard sizing, the minimum number of teeth of the annulus z_2, which can be generated without any of the two types of interference (primary and secondary), is $z_{2min} = 35$. If, for a given pressure angle, stub teeth are used, two goals may be simultaneously achieved. The first of these goals is the reduction of the number of teeth of the pinion, z_1, suitable to avoid the primary interference. The second goal is the decrease of the minimum limiting value of the difference $(z_2 - z_1)_{min}$, below which the secondary interference manifests.

5.5   A Geometric-Analytical Method for Checking of Secondary Interference

In Sect. 5.3 we described the general concepts regarding the secondary interference; in addition, we summarized the approximate method proposed by Henriot [9] to avoid it, highlighting its great design importance. In this section, we describe the geometric-analytical method for checking of the secondary interference, proposed by Colbourne [4], also using general concept concerning the kinematic geometry of gearing (see Dooner [5], Dooner and Seireg [6]).

For this purpose, we must first define the angular positions of the pinion and annulus with respect to the centerline O_2O_1 of the internal gear pair, by means of a constraint equation that links the various quantities involved. To this end, we denote by:

- x_1 and x_2, the tooth centerlines of the pinion and annulus;
- β_1 and β_2, the angles that define the angular positions of the pinion and annulus, both measured from the centerline O_2O_1 clockwise to the x_1-axis and x_2-axis;
- s_1 and s_2, the tooth thicknesses of the pinion and annulus at the pitch circles, having radii r_1 and r_2, respectively.

Figure 5.8 shows an internal gear pair in the position where the contact point coincides with the pitch point. Obviously, in this initial position of the point of contact (the second subscript, i, refers to this initial position), $\beta_{1,i}$ is negative, while $\beta_{2,i}$ is positive. These quantities are given respectively by:

$$\beta_{1,i} = -\tfrac{s_1}{2r_1} \quad \beta_{2,i} = \tfrac{s_2}{2r_2} \tag{5.13}$$

After differential rotations $d\beta_1 = \omega_1 dt$ and $d\beta_2 = \omega_2 dt$, the instantaneous angular positions of the pinion and annulus are respectively:

$$\beta_1 = \beta_{1,i} + d\beta_1 = -\frac{s_1}{2r_1} + d\beta_1 \tag{5.14}$$

$$\beta_2 = \beta_{2,i} + d\beta_2 = \frac{s_2}{2r_2} + d\beta_2. \tag{5.15}$$

From the first of Eqs. (5.5), as $\omega_1 = d\beta_1/dt$ and $\omega_2 = d\beta_2/dt$, we get the following relationship:

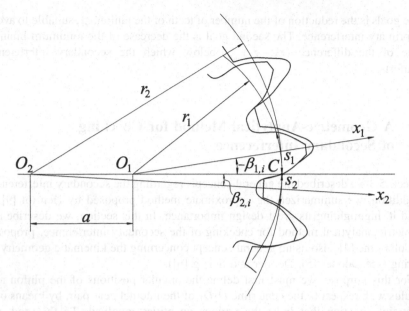

Fig. 5.8 Initial position of the two members of an internal gear pair, with contact at the pitch point

$$r_1 d\beta_1 = r_2 d\beta_2. \tag{5.16}$$

Therefore, multiplying both sides of Eqs. (5.14) and (5.15) for r_1 and r_2 respectively, with replacement, in Eqs. (5.14), of $(r_2/r_1)d\beta_2$ instead of $d\beta_1$ resulting from Eq. (5.16), and subtracting side by side the Eq. (5.15) from Eq. (5.14) so obtained, we get the following equation that correlates β_1 with β_2:

$$r_1\beta_1 - r_2\beta_2 + \frac{1}{2}(s_1 + s_2) = 0. \tag{5.17}$$

In Sect. 5.3 we have already highlighted the condition that, to avoid the secondary interference, it is necessary that the extended hypocycloid described by the tooth tip V_1 of the pinion during the meshing cycle must not intersect the profile of the tooth flank of the annulus. This hypocycloid (also called hypotrochoid), which is a convex curve, touches the annulus tooth profile at its limit circle (see Sect. 5.7) and, in a well-designed internal gear, must lie within the tooth space of the annulus, without intersecting its profile.

The path of the tooth tip V_1 of the pinion, during the meshing cycle, intersects the annulus addendum circle at the point V_1'. This position V_1' of the point V_1 should be compared with that of the tooth tip V_2 of the annulus. To avoid the secondary interference, it is necessary that the distance $V_2 V_1'$, measured along the annulus addendum circle, is greater than or at least equal to zero. In this regard, as a specific value, depending on the size and accuracy of the gears, we impose an adequate margin, i.e. that the arc length $V_2 V_1'$ is greater than or at least equal to 0.05 m.

In order to calculate this arc length, we determine first the polar coordinates of the point V_1 in the local coordinate system of the pinion, which has its origin at the center O_1 and its tooth centerline x_1 as reference axis (Fig. 5.9). The distance between V_1 and O_1 is the radius r_{a1} of the pinion addendum circle. The other polar coordinate is the angle ϑ_{a1} between the straight line $O_1 V_1$ and the x_1-axis. It can be expressed as (see Figs. 5.9 and 2.10):

$$\vartheta_{a1} = \frac{s_1}{2r_1} + \text{inv}\alpha_1 - \text{inv}\alpha_{a1}, \tag{5.18}$$

where s_1, r_1, and α_1 are the standard values of the tooth thickness, pitch circle radius and pressure angle of the pinion, and α_{a1} is the profile angle or pressure angle at point V_1, which is given by the following relationship:

$$\cos\alpha_{a1} = \frac{r_{b1}}{r_{a1}}. \tag{5.19}$$

We determine now the polar coordinates of the point V_2 in the local coordinate system of the annulus, which has its origin in the center O_2 and its tooth centerline x_2 as reference axis (Fig. 5.9). The distance between V_2 and O_2 is the radius r_{a2} of the annulus addendum circle. The other polar coordinate is the angle ϑ_{a2} between the straight line $O_2 V_2$ and the x_2-axis. To determine ϑ_{a2} as well as other interesting

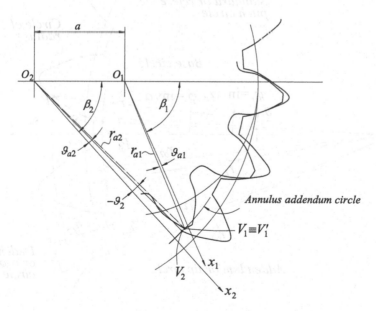

Fig. 5.9 Relative position of the pinion and annulus for checking of the secondary interference

quantities, it is convenient to generalize the problem and, with reference to Fig. 5.10, where the x_2-axis coincides with the tooth centerline of the annulus, we consider the following three points:

- point P_0 at radius r_{b2}, i.e. the origin of the involute curve, where the latter meets the base circle;
- point S at radius r_2, i.e. the point where the involute curve intersects the pitch circle;
- point P at radius r, i.e. a generic current point on the involute profile.

With the quantities shown in Fig. 5.10, all already defined, we infer that the polar coordinate ϑ_r of point P can be expressed as:

$$\vartheta_r = x_2\widehat{O_2S} - \widehat{SO_2P_0} + \widehat{PO_2P_0} = \frac{s_2}{2r_2} - \varphi + \varphi_r = \frac{s_2}{2r_2} - \mathrm{inv}\alpha + \mathrm{inv}\alpha_r, \quad (5.20)$$

where α and α_r are the profile angles at points S and P, respectively related with φ and φ_r (see also Fig. 2.1); therefore, α is the pressure angle.

The tooth thickness s_r on the circle of radius r is equal to

$$s_r = 2r\vartheta_r = r\left[\frac{s_2}{r_2} - 2(\mathrm{inv}\alpha - \mathrm{inv}\alpha_r)\right]. \quad (5.21)$$

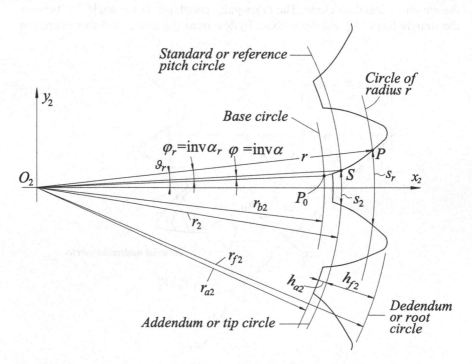

Fig. 5.10 Geometry of an internal gear, and tooth thickness s_r at radius r

Therefore, for P coincident with V_2, and bearing in mind that ϑ_{a2} is negative (Fig. 5.9), because point V_2 is located on the upper face of the tooth, and the polar angle is defined as positive when it is clockwise, we get:

$$\vartheta_{a2} = -\left[\frac{s_2}{2r_2} - \mathrm{inv}\alpha_2 + \mathrm{inv}\alpha_{a2}\right]. \tag{5.22}$$

Figure 5.9 shows the instantaneous position that is when $V_1 \equiv V_1'$, i.e. the position that is when the tip point V_1 on the pinion profile comes to coincide with the point V_1', determined as the intersection of the path of the same point V_1 with the annulus addendum circle. In this position, the distance between V_1' and O_2 is the radius r_{a2} of the addendum circle of the annulus. The angles β_1 and β_2 define the angular positions of the pinion and annulus, and ϑ_2 is the polar angle which identifies the position of the straight line $O_2V_1 \equiv O_2V_1'$ with respect to the x_2-axis. From triangle $O_2O_1V_1 \equiv O_2O_1V_1'$, we infer the following two equations:

$$\frac{\sin(\beta_1 + \vartheta_{a1})}{r_{a2}} = \frac{\sin(\beta_2 + \vartheta_2)}{r_1} \tag{5.23}$$

$$\cos(\beta_1 + \vartheta_{a1}) = -\frac{(a^2 + r_{a1}^2 - r_{a2}^2)}{2ar_{a1}}. \tag{5.24}$$

From Eq. (5.24), we obtain:

$$\beta_1 = \arccos\frac{(r_{a2}^2 - a^2 - r_{a1}^2)}{2ar_{a1}} - \vartheta_{a1}, \tag{5.25}$$

which defines the angular position of the pinion.

Therefore, the angular position of the annulus is found from Eq. (5.17), and is given by:

$$\beta_2 = \frac{1}{r_2}\left[r_1\beta_1 + \frac{1}{2}(s_1 + s_2)\right]. \tag{5.26}$$

Finally, from Eq. (5.23), we can calculate ϑ_2, which is given by the following relationship:

$$\vartheta_2 = \arcsin\left[\frac{r_{a1}}{r_{a2}}\sin(\beta_1 + \vartheta_{a1})\right] - \beta_2. \tag{5.27}$$

To avoid the secondary interference, the condition that the arc length V_2V_1' is greater than or at least equal to $0.05\,m$, leads to writing the following inequality, which must be satisfied:

$$r_{a2}(\vartheta_{a2} - \vartheta_2) \geq 0.05 \ m. \tag{5.28}$$

Generally, this inequality is satisfied for internal gear pairs for which the minimum difference $(z_2 - z_1)_{min}$ is 8 or more (it is to be noted that, in some case, this difference drops to 7 or 6). Therefore, using the inequality (5.28), we get results in full agreement with those obtained with the approximate method of Henriot [9]. Anyway, it is to keep in mind that this checking type must be carried out necessarily when the difference $(z_2 - z_1)$ is small, and that the amount of clearance depends on several quantities, such as the center distance and radii of the addendum circles.

5.6 Tertiary Interference in the Internal Spur Gears

The teeth of a ring gear is usually accomplished by generation cutting process, using shaping machines with pinion-type cutters (see Sect. 3.11.2). During this cutting process, it is also necessary to avoid, in addition to primary and secondary interferences described in previous sections, the interference that may occur during the radial penetration of the cutter into the ring gear blank. This third type of interference, called *tertiary interference* or *trimming* [14] causes the removal of the flank profile of the tooth toward the addendum circle.

It is to be noted that the two members of an internal gear pair (the pinion and annulus) can be assembled in their correct working position, by moving both axially and radially the pinion with respect to the annulus, held at rest in a fixed position. Therefore, the pinion can be brought into its meshing position with a movement, which can have the direction of its axis, or the one along a radius of the annulus.

If the internal gear pair was designed to dispel any possible type of interference (the primary, secondary and tertiary interferences, and the two types of tip interferences, which we will discuss in Sects. 5.7 and 5.8), the tooth shapes of the two members of the gear pair under consideration will allow to make the axial assembly, anyway. In some gearboxes, however, the space required to perform the axial assembly is missing. Thus, this type of assembly is impossible. *A fortiori*, the only possible method of assembling is the radial assembly, for which a checking of its feasibility is necessary in the design stage.

It is not always possible to bring the pinion into its meshing position with the mating gear ring, by moving the pinion along a radius of the annulus, held at rest. This radial approach problem is of wider interest, as it must be considered in two ways, i.e. both in the assembly condition of the two members of the internal gear pair, already cut, and in the cutting condition, in which the pinion-type cutter must cut the annulus blank. Obviously, if the tertiary interference occurs, the radial assembly of the internal gear pair is impossible. In the cutting conditions, the pinion-type cutter trims a part of the tooth profile of the annulus. Consequently, a part of the involute curves is destroyed, during the radial approach movement of the pinion-type cutter toward the ring gear blank, before the generation motion begins.

In the case of an internal gear pair, the pinion and annulus can be assembled in their meshing position by assembling the pinion in the axial direction (Fig. 5.11a). This axial assembling is not physically possible in the initial cutting conditions, when the pinion is the pinion-type cutter, and the ring gear is an internal gear blank or ring gear blank. In this case, before the motion of generation begins, the pinion-type cutter is pulled over and penetrates radially into the ring gear blank being cut, until the two cutting pitch circles are tangent at the pitch point (Fig. 5.11b). Starting from this position, the cutting motion in the axial direction is associated with the motion of generation, consisting of the pure rolling of the pitch circle of the pinion-type cutter on the pitch circle of the annulus in the cutting conditions.

The check that the tertiary interference does not take place is based on geometric considerations, in both of the following conditions:

- The centerline of a tooth of the pinion coincides with the centerline of a tooth space of the gear ring, and both these centerlines coincide with the centerline of the internal gear pair.

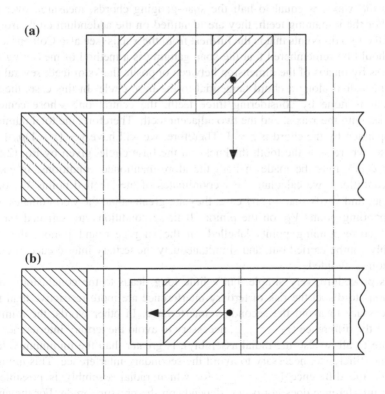

Fig. 5.11 a axial assembly, and axial motion of shaping; **b** radial assembly, and radial approach motion of the pinion-type cutter

- The centerline of a tooth space of the pinion coincides with the centerline of a tooth of the ring gear, and both these centerlines coincide with the centerline of the internal gear pair.

Obviously, for pinion we mean both the actual pinion, and the pinion- type cutter, while for internal gear pair we mean both the actual internal gear pair, and the internal gear pair in the cutting condition, depending on whether we consider the operating condition or the cutting condition.

In order to make such a check, first an orthogonal Cartesian coordinate system is chosen, with origin at the center O_2 of the ring gear, x-axis coinciding with the centerline O_2O_1 of the internal gear pair, and y-axis normal to x-axis. Then we consider the successive tip points $V_{2,i}$ (with $i = 1, 2, 3, \ldots$) of each tooth of the ring gear 2, identified on the addendum circle from the one of the two tip points that is closest to the x-axis. We consider also the successive tip points $V_{1,i}$ (with $i = 1, 2, 3, \ldots$) of each tooth of the pinion 1, which are furthest from the x-axis. For the identification of the latter it is to be observed that, for the pinion teeth closest to the x-axis, the points $V_{1,i}$ furthest from the x-axis are not the furthest tip points. They indeed correspond to the points obtained as the intersection of the tooth profiles with the tangent to the base circle, perpendicular to the x-axis; their distance y from the x-axis is equal to half the span-gauging chords, measured over three teeth. For the remaining teeth, they are identified on the addendum circle from the one of the two tip points that is the furthest from the x-axis (see also Colbourne [4]).

It should be remembered that the span gauging is a method of measuring tooth thickness by means of the dimension between opposite flanks of teeth several pitch apart, measured along a chord tangential to the base circle. In this case, the measurement is done by considering three teeth, the central one whose centerline coincides with the x-axis, and the two adjacent teeth. Therefore, the total number of teeth spanned by the chord is $z = 3$. Therefore, we will have $s = (z - 1)p_b + s_b = 2p_b + s_b$, where s_b is the tooth thickness on the base circle, given by Eq. (2.60).

The check must be made in both the afore-mentioned conditions. In each of these conditions, we calculate the y-coordinates of the labelled points $V_{2,i}$ on the ring gear, and verify that, in any case, they are greater than the y coordinates of the corresponding points $V_{1,i}$ on the pinion. If these conditions are satisfied for each pair of corresponding points labelled on the ring gear and pinion, the radial assembly can be carried out, and simultaneously the tertiary interference in cutting condition is avoided.

This procedure, very simple, but a little long, leads to the conclusion that the conditions to dispel the risk of tertiary interference are more restrictive than those are necessary to avoid the secondary interference. In other words, the minimum value of the difference $(z_2 - z_1)_{min}$, necessary to avoid the tertiary interference (and therefore to allow also the radial assembly), is greater than the minimum value of the same difference, necessary to avoid the secondary interference. This minimum value of the difference $(z_2 - z_1)_{min}$, for which radial assembly is possible and tertiary interference does not occur, depends on the pressure angle. For the cutting of a ring gear, it is necessary to use pinion-type cutters having numbers of teeth very different from those of the annulus to be cut. For this reason, the pinion-type cutters for cutting of internal gears are characterized by a very small number of teeth.

In order to avoid this long procedure, Henriot [9] suggests adopting the following simple rule: the minimum value of the difference $(z_2 - z_1)_{min}$ obtained using the diagram of Fig. 5.7 must be increased by six teeth. This approximate rule provides sufficiently reliable results.

5.7   Fillet Interference Between the Tip of the Pinion, and Root Fillet of the Annulus

The problems of interference in the internal spur gears cannot be considered completely solved, even when the conditions to avoid the primary, secondary and tertiary interferences are satisfied. Actually, the interference can still be presented as fillet interference between the tip of the pinion, and the root fillet of the annulus or as fillet interference between the tip of the annulus, and the root fillet of the pinion (see Henriot [9], Polder [17], Polder and Broekhuisen [18]). This section and the next deal with these two subjects, respectively.

Consider Fig. 5.12, which shows an internal gear pair, consisting of a driving pinion and a driven annulus. The path of contact g_α is defined as the rotation of gear members either during one meshing cycle or, in equivalent terms, as the length of the segment AE, obtained by intersecting the line of action with the addendum circles of the pinion and annulus. The length $g_\alpha = AE$ of path of contact is the sum of the approach and recess lengths, and thus can be expressed as:

$$AE = AC + CE = (T_2C - T_2A) + (T_1E - T_1C) = T_2T_1 + T_1E - T_2A. \quad (5.29)$$

Therefore, by expressing these lengths in terms of quantities shown in Fig. 5.12, we obtain:

$$g_\alpha = \left[(r_{b2} - r_{b1})\tan\alpha + \sqrt{r_{a1}^2 - r_{b1}^2} - \sqrt{r_{a2}^2 - r_{b2}^2} \right]. \quad (5.30)$$

For now, we focus our attention on what happens when the point of contact, moving on the line of action in the direction from point A to point E, arrives in the vicinity of point E. In this region of the path of contact, we must take into account the fact that the involute portion of the annulus tooth profile does not arrive until its dedendum circle, but ends at the fillet circle, having radius $r_{fil,2}$. In addition, we must consider that the end point of the active profile of the annulus tooth, which is closest to the root, is the so-called limit point, and that the circle with center O_2 and passing through this point is the limit circle of the annulus, having radius $r_{l2} = O_2E$. We remember that, for both members of an external or internal spur

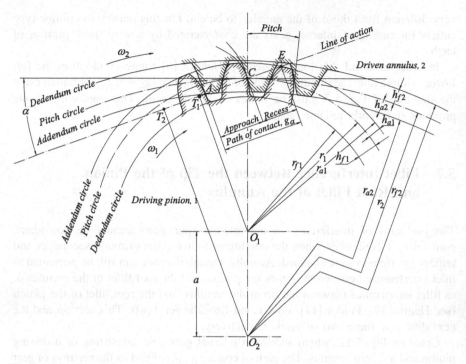

Fig. 5.12 Meshing diagram of an internal gear pair, consisting of a driving pinion and a driven annulus

gear, the active profile of a tooth is the part that actually makes contact between mating profiles.

From Fig. 5.13, which shows the meshing diagram of an internal gear pair, we infer that the radius r_{l2} of this limit circle is given by:

$$r_{l2} = \overline{O_2 E} = \sqrt{\overline{O_2 T_2}^2 + \overline{T_2 E}^2} = \sqrt{\overline{O_2 T_2}^2 + \left(\overline{T_1 E} + \overline{T_2 T_1}\right)^2}$$

$$= \left\{ r_{b2}^2 + \left[\sqrt{r_{a1}^2 - r_{b1}^2} + (r_{b2} - r_{b1})\tan\alpha \right]^2 \right\}^{1/2}. \tag{5.31}$$

To ensure that the fillet interference between the tip of the pinion, and root fillet of the annulus does not occur, since the teeth of the annulus are facing inward, the radius of its limit circle must be smaller than the radius of its fillet circle, i.e.:

$$r_{fil,2} \geq r_{l2}. \tag{5.32}$$

Here also it is to be noted that, to neutralize the effects of inevitable errors in the center distance, it is appropriate to design the internal spur gear pair so that the fillet circle of the annulus is greater than the limit circle by a certain margin, usually

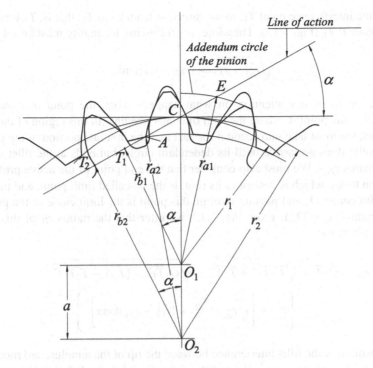

Fig. 5.13 Meshing diagram of an internal gear pair

assumed to be $0.025\,m$. Therefore, the theoretical inequality (5.32) changes, and assumes the form of the following design inequality:

$$r_{fil,2} \geq r_{l2} + 0.025\,m. \tag{5.33}$$

The radius $r_{fil,2}$ of the annulus fillet circle appearing in the inequities described above remains to be calculated. This will be done in Sect. 5.9.

5.8 Fillet Interference Between the Tip of the Annulus, and Root Fillet of the Pinion

In Sect. 5.2, we found that the contact is non-conjugate when the point of contact is located beyond the interference point T_1, in the direction from point T_1 to point T_2 (see Figs. 5.1 and 5.13). In addition, we have shown that, to avoid the theoretical or primary interference, the minimum value of the radius r_{a2} of the addendum circle of the annulus must be greater than or equal to O_2T_1, i.e. $r_{a2} \geq O_2T_1$ (Figs. 5.1 and 5.13). This amounts to saying that the end point A of the path of contact must lie beyond the interference point T_1, in the direction from T_2 to T_1, that is, T_2A must be greater than T_2T_1 (Fig. 5.13). Therefore, the following inequality must be satisfied:

$$\sqrt{r_{a2}^2 - r_{b2}^2} > (r_{b2} - r_{b1})\tan\alpha. \tag{5.34}$$

Now, we focus our attention on what happens when the point of contact is located near the point A, where the path of contact begins. In this region of the path of contact, we must take into account the fact that the involute portion of the pinion tooth profile does not arrive until its dedendum circle, but ends at the fillet circle, having radius $r_{fil,1}$. We must also consider that the end point of the active profile of the pinion tooth, which is nearest to its root, is the so-called limit point, and that the circle with center O_1 and passing through this point is the limit circle of the pinion, having radius $r_{l1} = O_1A$. From Fig. 5.13, we infer that the radius r_{l1} of this limit circle is given by:

$$r_{l1} = \overline{O_1A} = \sqrt{\overline{O_1T_1}^2 + \overline{T_1A}^2} = \sqrt{\overline{O_1T_1}^2 + \left(\overline{T_2A} - \overline{T_2T_1}\right)^2}$$
$$= \left\{ r_{b1}^2 + \left[\sqrt{r_{a2}^2 - r_{b2}^2} + (r_{b2} - r_{b1})\tan\alpha \right]^2 \right\}^{1/2}. \tag{5.35}$$

To ensure that the fillet interference between the tip of the annulus, and root fillet of the pinion does not occur, the radius of the limit circle of the pinion must be larger than the radius of its fillet circle, i.e.:

$$r_{fil,1} \le r_{l1}. \tag{5.36}$$

Here also, to neutralize the effects of inevitable errors in the center distance, it is appropriate to design the internal gear pair so that the limit circle of the pinion is greater than its fillet circle by a certain margin, usually assumed to be $0.025\,m$. Therefore, the theoretical inequality (5.36) changes, and assumes the form of the following design inequality:

$$r_{fil,1} \le r_{l1} - 0.025\,m. \tag{5.37}$$

The quantities that appear in the theoretical inequalities (5.32) and (5.36), and in the design inequalities (5.33) and (5.37) are all known, with the exception of the fillet circle radius $r_{fil,2}$ of the annulus. Instead, r_{l1} and r_{l2} are given respectively by Eqs. (5.35) and (5.31), while $r_{fil,1}$ is given by Eq. (4.44), or by Eq. (4.32), depending on whether the pinion cutting process is made with a pinion-type cutter, or with a rack-type cutter or a hob.

5.9   A Design Consideration on the Interference Between Internal Spur Gears

In conclusion, in order to avoid the risk of primary interference, inequality (5.34) must be satisfied or, in equivalent terms, the inequality (5.7), and all that follows from it. Instead, to avoid the risk of the two fillet interferences between the pinion tip and annulus root fillet, and between the annulus tip and pinion root fillet, inequalities (5.33) and (5.37) must be meet. Moreover, the two members of the internal gear pair should be designed in such a way that between the addendum circles and dedendum circles of both of these members a minimum clearance equal to $0.25\,m$ will have.

As a rule, we can say that, if a sufficient clearance exists between the dedendum circle of the annulus and the addendum circle of the pinion, the interference conditions in this region are automatically satisfied. In addition, if the interference conditions are met in the region of root fillet of the pinion, the clearance between the addendum circle of the annulus and the dedendum circle of the pinion will be more than adequate. If this general rule were invariably true, it would be sufficient to check that the interference conditions are satisfied in the region of the root fillets of the pinion, and that the clearance conditions are satisfied in the region of the root fillets of the annulus.

Unfortunately, however, these rules are not entirely invariably true, and there are cases where they are not satisfied. As a good design rule, it is always necessary to verify that all three interference conditions, given by inequalities (5.33), (5.34) and (5.37) are satisfied, and that both the above clearances are adequate. The typical case, in which the above rules are not true, concerns what happens in the region of the root fillet of the pinion. In this case, in fact, to avoid the fillet interference, it is not enough that the inequality (5.34) is satisfied (this inequality is the condition to avoid the risk that the primary interference occurs). This is because the inequality (5.34) assumes that the involute profile of the pinion reaches the base circle, and instead this never occurs.

In this regard, consider the most favorable case of a pinion cut with a rack-type cutter (in this cutting process the involute profile is more important). As Fig. 5.14 shows, the limit value of the annulus addendum to prevent the primary interference is $k_2 m$, i.e. $a_{2lim} = k_2 m$. If we consider, for example, an internal gear pair consisting of an annulus with $z_2 = 100$, and a pinion with $z_1 = 15$, from Fig. 5.3 we get $k_2 = 0.8$, and thus $a_{2lim} = 0.8m$. If we make a profile shift, without variation of center distance, with $x_2 = +0.20$ and $x_1 = +0.20$, and thus with $(x_2 - x_1) = 0$ (see Sect. 6.3.2), the fillet interference between the tip of the annulus and root fillet of the pinion will not be eliminated. Since the fillet profile of the pinion begins at fillet point P_f, we note that in any case it is necessary to trim the addendum of the annulus of an amount equal to vm, where v is the *trimming factor*, which can be obtained, for a pressure angle $\alpha = 20°$, from Fig. 5.15.

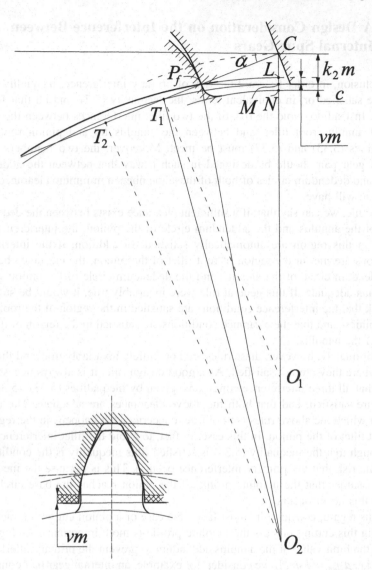

Fig. 5.14 Limit value if the addendum, a_{2lim}, of the annulus to prevent the primary interference

5.10 Annulus Fillet Profile Generated by a Pinion-Type Cutter

The shaping process with a pinion-type cutter is the cutting method almost always used to cut a ring gear. The funtamentals of this cutting method have been recalled at the beginning of Sect. 4.8, to which we refer the reader. The use of a pinion-type cutter with rounded tips is preferable since, otherwise, at the root fillet of the

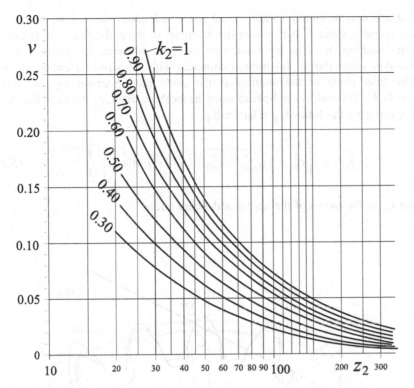

Fig. 5.15 Trimming factor of the annulus addendum as a function of z_2 and k_2, for $\alpha = 20°$

annulus, stress concentrations too high would generate, resulting from radii of curvature of the fillets too small.

The tip geometry of a pinion-type cutter with rounded corner at the tooth tip is that shown in Fig. 4.16. The end point L of the involute portion of the tooth profile has polar coordinates (r_{L0}, φ_{L0}), given by Eqs. (4.37) and (4.38), while the center I of the circular portion at the tooth tip has polar coordinates (r_{I0}, φ_{I0}), given by Eqs. (4.33) and (4.34).

To determine the radius $r_{fil,2}$ of the annulus fillet circle, which intersects the annulus tooth profile at point P_f where the involute portion of the same profile ends and the fillet begins, we consider the cutting meshing diagram between the pinion-type cutter and annulus to be generated (Fig. 5.16). The involute portion of the cutter tooth profile, which ends at point L (Fig. 4.16), cuts the involute portion of the annulus tooth profile. Thus, the end point P_f of the annulus tooth involute profile is cut by the point L on the pinion-type cutter profile. This happens when the point L, during the cutting relative motion, is to coincide with point E', where the line of action in the cutting condition and active addendum circle of the pinion-type cutter intersect. Point E' is therefore the end point of the path of contact during the cutting meshing. It is to be noted that the path followed by point L of the pinion-type cutter during this motion coincides with the active addendum circle.

The fillet circle of the annulus is the circle having center O_2 and radius $r_{fil,2} = O_2E'$. This radius can be read from the meshing diagram shown in Fig. 5.16, and is given by the following relationship:

$$r_{fil,2} = \left\{ r_{b2}^2 + \left[\sqrt{r_{a0}^2 - r_{b0}^2} + (r_{b2} - r_{b0})\tan\alpha_0 \right]^2 \right\}^{1/2}, \qquad (5.38)$$

where r_{a0} is the radius of the active addendum circle.

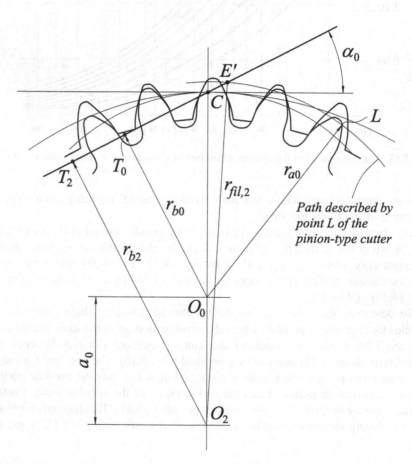

Fig. 5.16 Meshing diagram between pinion-type cutter and annulus to be generated

With reference to the meshing diagram shown in Fig. 5.16, we can define other quantities relating to the cutting condition. To this end, we denote by: z_2 and z_0, the numbers of teeth of the annulus and pinion-type cutter; α_0, the cutting pressure angle, which is the same for the pinion-type cutter and annulus to be generated; r_{b2} and r_{b0}, the radii of the base circles of the annulus and pinion-type cutter; a_0, the center distance in the cutting condition. Note that, in cases in which a possibility of misunderstanding does not exist, only the subscript 0 is used to indicate quantities that relate both to the cutting tool, and to the cutting condition. Instead, in cases in which a possibility of ambiguity exists, to avoid misunderstandings, the symbol 0 is used as the second subscript, where deemed necessary, to indicate quantities that relate to the cutting condition.

Recalling equations already shown in previous chapters, and referring to Fig. 5.16, we obtain the following relationships:

$$a_0 = \frac{z_2 - z_0}{2} m_0 \tag{5.39}$$

$$a_0 \cos\alpha_0 = (r_{b2} - r_{b0}) \tag{5.40}$$

$$r_{2,0} = \frac{z_2}{(z_2 - z_0)} a_0 \tag{5.41}$$

$$r_{0,0} = \frac{z_0}{(z_2 - z_0)} a_0 \tag{5.42}$$

$$p_0 = \frac{2\pi}{(z_2 - z_0)} a_0. \tag{5.43}$$

In these equations, m_0 is the module in the cutting condition (or *cutting module*), p_0 is the *cutting circular pitch*, and $r_{2,0}$ and $r_{0,0}$ are the radii of the cutting pitch circles of the annulus and pinion-type cutter. Of course, m_0 and p_0 are the same for the pinion-type cutter and annulus to be generated.

During the cutting process, the meshing diagram (Fig. 5.16) between the pinion-type cutter and annulus to be generated is equal to that of an internal gear pair (see Fig. 5.13), with the only feature that it is without backlash. Therefore, the tooth thickness $s_{2,0}$ of the annulus is equal to the space width of the pinion-type cutter, both measured on the cutting pitch circles; thus, we will have:

$$s_{2,0} = (p_0 - s_{0,0}), \tag{5.44}$$

where $s_{0,0}$ is the tooth thickness of the pinion-type cutter, also measured on the cutting pitch circle.

The tooth thickness of the annulus, $s_{2,0}$, at the cutting pitch circle of radius $r_{2,0}$, is related with the tooth thickness s_2 at the standard pitch circle of radius r_2. The relationship that correlates these two tooth thicknesses is obtained from Eq. (5.21), so it can be written as follows:

$$s_{2,0} = r_{2,0} \left[\frac{s_2}{r_2} - 2(\text{inv}\alpha - \text{inv}\alpha_0) \right];$$ (5.45)

this is because $\alpha_r = \alpha_0$.

Therefore, using the general Eq. (2.58), we infer that the tooth thickness of the pinion-type cutter, $s_{0,0}$, at the cutting pitch circle of radius $r_{0,0}$, is related with the tooth thickness s_0 at the reference pitch circle of radius r_0 by the following equation:

$$s_{0,0} = r_{0,0} \left[\frac{s_0}{r_0} + 2(\text{inv}\alpha - \text{inv}\alpha_0) \right].$$ (5.46)

It is noteworthy that the reference pressure angle, α, and the cutting pressure angle, α_0, of the pinion-type cutter are equal to the corresponding ones of the internal gear to be generated.

The cutting center distance a_0 and the reference cutting center distance $a_{0,s}$ are given respectively by:

$$a_0 = (r_{2,0} - r_{0,0})$$ (5.47)

$$a_{0,s} = (r_2 - r_0).$$ (5.48)

In order to obtain the tooth thickness of the annulus, s_2, at its standard pitch circle of radius r_2 as a function of $a_{0,s}$, we first substitute Eqs. (5.45) and (5.46) into Eq. (5.44); then, multiplying by the ratio $(a_{0,s}/a_0)$ both members of the equality so obtained, and bearing in mind that both the products $(a_{0,s}/a_0)(r_{2,0}/r_2)$ and $(a_{0,s}/a_0)(r_{0,0}/r_0)$ are equal to unity, we obtain:

$$s_2 = p - s_0 + 2a_{0,s}(\text{inv}\alpha - \text{inv}\alpha_0),$$ (5.49)

where p is the standard or reference circular pitch. From this equation, we obtain:

$$\text{inv}\alpha_0 = \text{inv}\alpha + \frac{1}{2a_{0,s}}(p - s_0 - s_2).$$ (5.50)

Once the value of $\text{inv}\alpha_0$ is known, the angle α_0 can be derived using, for example, the following approximate procedure, in two steps. As a first step, we set:

$$q = (\text{inv}\alpha_0)^{2/3},$$ (5.51)

and, as a second step, we obtain α_0 by using the relationship:

$$\frac{1}{\cos \alpha_0} = 1.0 + 1.04004q + 0.32451q^2 - 0.00321q^3 - 0.00894q^4 + 0.00319q^5$$
$$+ 0.00048q^6.$$

$$(5.52)$$

This procedure is affected by a maximum error equal to $0.0001°$, for values of α_0 between $0°$ and $65°$ (this range of α_0-values is sufficient for most practical applications). The coefficients in Eq. (5.52) are a simplified version of a set of coefficients developed by Polder [16].

Finally, after determining the value of α_0, from Eq. (5.40) we obtain the following expression of cutting center distance a_0:

$$a_0 = \frac{(r_{b2} - r_{b0})}{\cos \alpha_0}.$$

$$(5.53)$$

5.11 Undercut or Cutting Interference

To analyze if there is possibility of undercut or cutting interference during the annulus cutting process, we consider the meshing diagram shown in Fig. 5.16, where the point T_2 and T_0 are respectively the interference points of the annulus to be generated and pinion-type cutter. First, let's focus our attention on the point T_2, where the line of action during the cutting process touches the annulus base circle. In an internal gear, contrary to what occurs for an external gear, if the cutting point is located near the interference point, the corresponding point on the tooth profile of the gear to be cut is away from the fillet. In other words, a point near the tip of the annulus tooth profile corresponds to a cutting point near the interference point.

Now let's focus our attention on the other end point of the path of contact, the farthest from the interference point T_2. In this area, the point P_f of the annulus tooth profile, where the involute profile ends and the fillet profile begins, is cut when the cutting point coincides with the point E'. Therefore, however great may be the cutter addendum, the danger of a conventional undercut of the annulus tooth fillet does not exist.

An undercut is however possible at the tooth tips of the annulus. The theoretical involute profile of the annulus tooth is cut by means of the involute portion of the cutter tooth, and at most this involute portion can reach the base circle. Actually, however, the pinion cutter is designed with a fillet that begins slightly outside the base circle, as the radius of curvature of the involute curve is equal to zero at the base circle. Thus, the involute portion of the cutter tooth ends at the fillet circle, the radius $r_{fil,0}$ of which is therefore slightly greater than that of the base circle, or at least equal to the latter.

As Fig. 5.17 shows, the fillet circle of the pinion-type cutter intersects the line of action at point F_2 localized between T_0 and C. Since the involute portion of the cutter tooth profile begins from this fillet circle, and then develops at the outside thereof, it is clear that the cutting of the involute portion of the annulus tooth profile can take place only outside this circle. Thus, to avoid undercut, the path of contact during the cutting meshing must begin at a point of the line of action close to the point F_2, but located beyond this, i.e. between F_2 and C (or at least coincident with F_2). Therefore, the following inequality must be satisfied:

$$r_{a2}^2 \geq r_{b2}^2 + \left[\sqrt{r_{fil,0}^2 - r_{b0}^2} + (r_{b2} - r_{b0})\tan\alpha_0 \right]^2 . \qquad (5.54)$$

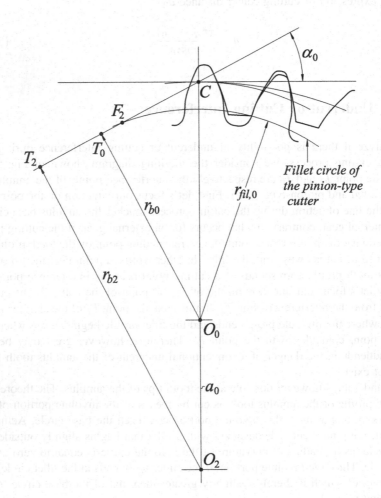

Fig. 5.17 Condition to avoid undercut by the cutter tooth fillet

If the required radius r_{a2} of the annulus addendum circle is smaller than the minimum value given by this inequality, it is necessary to use a pinion-type cutter with more teeth.

The tip interference between the pinion-type cutter and the annulus during the cutting process may cause another type of undercut of the annulus to be generated. This undercut can occur during the generating motion (in this case, the cutting center distance is equal to a_0), or during the initial phase of radial penetration of the pinion-type cutter inside the ring gear blank. During this initial radial penetration, the cutting center distance is less than a_0, and its instantaneous value a_c is variable with continuity, until reaching the maximum value equal to a_0. Actually, with the usual gear cutting machines, this center distance varies step by step. In this framework, we consider any position of the pinion-type cutter related to a given step of the initial phase of the cutting tool radial penetration, which is characterized by the defined value of the cutting center distance a_c, corresponding to the step under consideration.

So far, we have considered a pinion-type cutter with teeth rounded at the tips (Fig. 4.16). Here, to simplify the calculations related to the case of tip interference that we want to examine, we assume instead that the involute profile of the cutter tooth extends until the addendum circle, which is intersected by the involute curve at the point V_0 (Fig. 5.18). The path of this point V_0 intersects the annulus addendum circle at the point V_0'. Then we denote with V_2 the tooth tip corner of the annulus, i.e. the point in which the annulus tooth profile intersects the annulus addendum circle. To avoid this tip interference, and related undercut, it is necessary that the distance V_2V_0', measured along the annulus addendum circle, is greater than or at least equal to zero at the limiting condition; as a specified value, typically we impose that the arc length V_2V_0' is greater than or at least equal to $0.02\,m$.

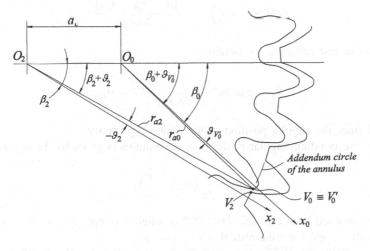

Fig. 5.18 Relative position of the pinion-type cutter and annulus to be generated, for checking of the tip interference and related undercut

In order to calculate the arc length $V_2 V_0'$, we first determine the polar coordinates of the point V_0 in the local coordinate system of the pinion-type cutter, which has its origin at the center O_0 and as reference x_0-axis its tooth centerline (Fig. 5.18). The distance between V_0' and O_0 is not more than the radius r_{a0} of the addendum circle of the pinion-type cutter. The other polar coordinate is the angle $\vartheta_{V_0'}$ between the straight line $O_0 V_0'$ and the x_0-axis; it can be expressed as (see Figs. 5.18 and 2.12):

$$\vartheta_{V_0'} = \frac{s_0}{2r_0} + \mathrm{inv}\alpha_0 - \mathrm{inv}\alpha_a, \tag{5.55}$$

where α_a is the pressure angle or profile angle at point V_0', which is given by the relationship:

$$\cos\alpha_a = \frac{r_{b0}}{r_{a0}}. \tag{5.56}$$

Figure 5.18 shows the instantaneous position that is when $V_0 \equiv V_0'$, i.e. when the tip point V_0 of the pinion-type cutter comes to coincide with the point V_0', determined as the intersection of the path of the same point V_0 with the annulus addendum circle. The distance between V_0' and O_2 is the radius r_{a2} of the addendum circle of the annulus. The angle ϑ_2 is the polar coordinate of point V_0' relative to the x_2-axis, the latter coinciding with the tooth centerline of the annulus. From triangle $O_2 O_0 V_0'$ we infer the following two equations:

$$\frac{\sin\left(\beta_0 + \vartheta_{V_0'}\right)}{r_{a2}} = \frac{\sin(\beta_2 + \vartheta_2)}{r_{a0}} \tag{5.57}$$

$$\cos\left(\beta_0 + \vartheta_{V_0'}\right) = -\frac{\left(a_c^2 + r_{a0}^2 - r_{a2}^2\right)}{2a_c r_{a0}}. \tag{5.58}$$

From this last equation, we obtain:

$$\beta_0 = \arccos\frac{\left(r_{a2}^2 - a_c^2 - r_{a0}^2\right)}{2a_c r_{a0}} - \vartheta_{V_0'}, \tag{5.59}$$

which defines the angular position of the pinion-type cutter.

The corresponding angular position of the annulus is given by the equation:

$$\beta_2 = \frac{1}{r_2}\left(r_0\beta_0 - \frac{p}{2}\right), \tag{5.60}$$

which is obtained from Eqs. (5.17) or (5.26), where we replace r_1 with r_0, β_1 with β_0, s_1 with s_0, and we remember that $s_0 + s_2 = p$.

Finally, from Eq. (5.57), we infer ϑ_2, i.e. the polar coordinate of the point $V_0 = V_0'$, which is given by the following relationship:

$$\vartheta_2 = \arcsin\left[\frac{r_{a0}}{r_{a2}}\sin\left(\beta_0 + \vartheta_{V_0'}\right)\right] - \beta_2. \tag{5.61}$$

To avoid the tip interference and related undercut, the condition that the arc length $V_2 V_0'$ is greater than or at least equal to $0.02\,m$, leads to write the following inequality, which must be satisfied:

$$r_{a2}(\vartheta_{a2} - \vartheta_2) \geq 0.02\,m, \tag{5.62}$$

where ϑ_{a2} is the polar coordinate of point V_2 on the tooth tip of the annulus, given by Eq. (5.22).

During the initial phase of radial penetration of the pinion-type cutter inside the ring gear blank, i.e. before the generating motion begins, the center O_0 of the pinion-type cutter moves along the straight line $O_2 O_0$, and the center distance a_c, initially equal to the difference $(r_{a2} - r_{a0})$, increases step by step up to the final value a_0. Therefore, to ensure that the tip interference, and related undercut, do not occur, we must carry out the checking for different values of a_c, starting from the minimum value equal to $(r_{a2} - r_{a0})$, and ending with $a_c = a_0$. If such checking shows the evidence of this tip interference for any value of a_c, between $(r_{a2} - r_{a0})$ and a_0, most likely the annulus would be not usable, since such interference would have removed a substantial portion of its involute profile.

5.12 Condition to Avoid Rubbing During the Annulus Cutting Process

Here we want to focus our attention on the phenomenon of rubbing, which is typical of the already described gear generating process by shaping. As we have already seen in Sect. 3.11.2, this is a continuous indexing gear cutting process performed by means of a reciprocating tool. During the return stroke of the pinion-type cutter, the tool must be relieved to avoid rubbing the workpiece. In fact, the rubbing would dull the cutting edges, and degrade the surface finishing of the workpiece. Here we want to define the manufacturing conditions that avoid the rubbing phenomenon, with reference to the annulus cutting process, with the warning that in the same way it will be necessary to operate for cutting of an external gear wheel by shaping processes (see Boothroyd and Knight [2], Björke [1], Radzevich [21], Gupta et al. [8]).

With reference to the annulus cutting process, regardless of whether the ring gear to be generated has zero profile-shifted toothing or profile-shifted toothing, the gear design must address the problem of rubbing that always occurs. This in order to give the manufacturer the proper indications for a satisfactory cutting process to be performed [4]. Of course, this problem must be addressed in the framework of all the other problems of cutting interference, which we have detailed in previous sections.

The annulus teeth cutting process, as we have already said elsewhere, is a shaping process, which uses a pinion-type cutter as a cutting tool (Fig. 3.15). This shaping process consists of a working stroke, during which the cutting tool must necessarily interfere with the ring gear blank in order to remove the appropriate amount of chip, and a return stroke, during which the cutting tool must be moved by some distance away from the annulus gear blank to be cut, since the relative displacement between these two elements is important. This relative displacement during the return stroke of the pinion-type cutter prevents rubbing as well as related consequences in terms of excessive cutting-edge wear, and burrs left on the tooth flanks of the annulus being cut.

The designer should check that, for the specific case of interest, which is defined by the annulus geometry to be obtained and the selected pinion-type cutter, a relative displacement direction exists for which the rubbing is eliminated. Identifying this privileged relative displacement direction is not a simple problem to solve. It is complicated by the fact that during the cutting process the tooth spaces are not yet configured in their final shape, but they are constantly becoming, starting from the ring gear blank which is gradually transformed into the desired annulus. Therefore, the pinion-type cutter can also rub against parts of the ring gear blank, which then will be cut off during the completion of the manufacturing process. In addition, during the cutting stroke, the contact points (and hence the cutting points) between pinion-type cutter and gear blank can be numerous, whereas we will have only one point of contact, and thus only one cutting point, once the annulus to be generated has reached its final shape.

To solve this problem, we assume that the ring gear blank remains fixed and that the pinion-type cutter moves away from it during the return stroke in the direction to be detected. Of course, in the opposite case, i.e. pinion-type cutter in a fixed position and gear blank moving relative to the pinion-type cutter, we would find the opposite direction on the same relative displacement line.

Let us first consider the trailing profile of one of the cutter teeth, in the position where it begins to come into contact with the gear blank at point L, i.e. the upper end of the involute part of the tooth cutter profile where it connects with the tooth rounded tip profile (Fig. 4.16). As Fig. 5.19 shows, in this position the point L, whose radial distance r_L from the center O_0 is given by Eq. (4.37), will lie on the annulus tip circle. Also in this position the tangent to the tooth cutter profile in this point will form the angle α^* with the line of centers $O_2 O_0$.

Considering the triangle $O_2 O_0 L$, we can write:

$$r_{a2}^2 = a^2 + r_L^2 - 2 a r_L \cos[\pi - (\alpha_L + \alpha^*)], \qquad (5.63)$$

from which we get the following expression

$$\alpha^* = \arccos \frac{[r_{a2}^2 - a^2 - r_L^2]}{2 a r_L} - \alpha_L, \qquad (5.64)$$

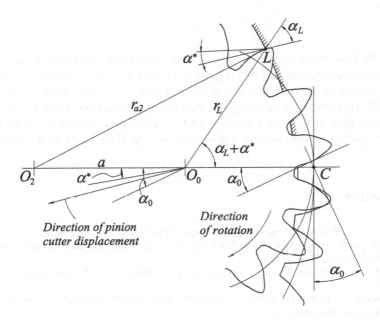

Fig. 5.19 Direction of pinion-type cutter displacement to avoid rubbing

which allows to calculate the pressure angle α^* at the upper end point L of the involute part of the tooth cutter profile.

Obviously, the angle α^* thus found represents the maximum value that the pressure angle at a point can assume, since point L is the upper end point where the involute curve of the cutting edge is truncated. As the pinion-type cutter rotates, and the cutting edge penetrates further into the ring gear blank, the pressure angles at subsequent point of contact (i.e. the cutting points) gradually decrease. Therefore, to prevent rubbing by the cutter tooth trailing profile, it is necessary that in its return stroke the pinion-type cutter is moved from the cutting zone in a direction that forms an angle greater than α^* with the line of centers.

We now consider the leading profile of the cutter tooth, which is in contact with a tooth of the ring gear blank. The angles between the tangents at points of contact and the line of centers reach their maximum values when the point of contact is the one corresponding to the lower end point of the cutting edge; in this case, this angle becomes equal to the cutting pressure angle, α_0. Therefore, to prevent rubbing by the cutter tooth leading profile, it is necessary that in its return stroke the pinion cutter is moved from the cutting zone in a direction that forms an angle less than α_0 with the line of centers.

Both of the above-mentioned requirements must be met to avoid rubbing. First, the designer should check that the pressure angle α^* at point L, given by Eq. (5.64), is less than the cutting pressure angle α_0 by at least a few degrees. Then the designer has to define the working displacement angle, α_w^*, which must be between α^* and α_0, i.e.:

$$\alpha^* < \alpha_w^* < \alpha_0. \qquad (5.65)$$

For the best possible determination of the value of this working displacement angle, it should be noted that rubbing, for a displacement angle equal to α_0 is more severe than it is for a displacement angle equal to α^*; this is apparent from Fig. 5.19. Therefore, it is more appropriate to choose a working displacement angle closer to α^* than to α_0. In case the Eq. (5.64) would give a value of α^* not less than α_0 with a sufficient margin, then a pinion cutter with fewer teeth should be used.

References

1. Björke Ö (1995) Manufacturing systems theory. Tapir Publishers, Trondheim
2. Boothroyd G, Knight WA (1989) Fundamentals of machining and machine tools, 2nd edn. Marcell Decker, New York
3. Buckingham E (1949) Analytical mechanics of gears. McGraw-Hill Book Company Inc, New York
4. Colbourne JR (1987) The geometry of involute gears. Springer-Verlag, New York Inc, New York Berlin Heidelberg
5. Dooner DB (2012) Kinematic geometry of gearing, 2nd edn. Wiley, New York
6. Dooner DB, Seireg AA (1995) The kinematic geometry of gearing: a concurrent engineering approach. Wiley, New York
7. Dudley DW (1962) Gear handbook. The design, manufacture, and application of gears, McGraw-Hill Book Company, New York
8. Gupta K, Jain NK, Laubscher R (2017) Advanced gear manufacturing and finishing: classical and modern processes. Elsevier Science Publishing Co., Inc., Academic Press Inc, San Diego, USA
9. Henriot G (1979) Traité théorique et pratique des engrenages, vol 1, 6th ed. Bordas, Paris
10. ISO 6336-1: 2006 (E): Calculation of load capacity of spur and helical gears—Part 1: Basic principles, introduction and general influence factors
11. Litvin FL, Fuentes A (2004) Gear geometry and applied theory, 2nd edn. Cambridge University Press, Cambridge
12. Litvin FL, Hsiao CL, Wang JC, Zhou X (1994) Computerized simulation of generation of internal involute gears and their assembly. ASME J Mech Des 116(3):683–689
13. Maitra GM (1994) Handbook of gear design, 2nd edn. Tata McGraw-Hill Publishing Company Ltd, New Delhi
14. Merritt HE (1954) Gears, 3th ed, Sir Isaac Pitman & Sons, Ltd., London
15. Polder JW (1969) Overcut, a new theory for tip interference in internal gears. J Mech Eng Sci 11(6):583–591
16. Polder, JW (1981) Overcut interference in internal gears. In: Proceeding international symposium on gearing and power transmissions, Tokyo
17. Polder JW (1991) Interference of internal gears. In: Townsend P (ed) Dudley's gear handbook. McGraw-Hill, New York
18. Polder JW, Broekhuisen H (2003) Tip-fillet interference in cylindrical gears, ASME Proceedings Power Transmission and Gearing, Paper No. DETC 2003/PTG-48060, pp 473–479
19. Pollone G (1970) Il Veicolo. Libreria Editrice Universitaria Levrotto & Bella, Torino
20. Radzevich SP (2016) Dudley's handbook of practical gear design and manufacture, 3rd edn. CRC Press, Taylor & Francis Group, Boca Raton, Florida

21. Radzevich SP (2017) Gear cutting tools: science and engineering, 2nd edn. CRC Press, Taylor&Francis Group, Boca Raton, Florida
22. Schreier G (1961) Stirnrad-Verzahnung: Berechnung, Werkstoffe, Fertigung. VEB Verlag Technik, Berlin
23. Tuplin WA (1967) Tip interference in internal gears. The Engineer 224(5839):827

Chapter 6
Profile Shift of Spur Gear Involute Toothing Generated by Rack-Type Cutter

Abstract In this chapter, the fundamentals of the profile shift of spur gear involute toothing generated by rack-type cutter arc first described, highlighting also the technical-applicative interest of the use of this type of profile-shifted toothing The problem is analyzed in its maximum generality, so the cases of zero profile-shifted toothing and of profile-shifted toothing without and with variation of center distance are discussed, both for external and internal spur gears In this general framework, the problems related to the minimum profile shift coefficient to avoid interference and the risk of pointed teeth arc also discussed. The problem of how to split the sum or difference of the profile shift coefficients between the pinion and mating gear wheel is then dealt with, under the double aspect of direct problem and inverse problem also with reference to the different criteria that can be used to achieve specific design goals. Finally, the problem of how to ensure the optimal value of backlash between the mating teeth of cylindrical spur gears is discussed.

6.1 Introduction

In Sect. 4.4 we have saw that, to avoid the primary interference of external spur gear pairs with pinions having small numbers of teeth, we can choose three different design solutions, namely: the increase of the standard pressure angle; the use of stub-toothed gears; the use of toothing with profile shift. We also reported that the first two solution methods have many disadvantages and drawbacks, and that the only method used is actually the third method, that is to employ involute profile-shifted toothing (see Buckingham [3], Merritt [21], Dudley [8], Giovannozzi [12], Ferrari e Romiti [11]; Pollone [25], Henriot [13], Niemann and Winter [22], Colbourne [5], Scotto Lavina [29], Maitra [20]) .

Profiles shift is closely related to the involute tooth profile. In scientific and technical literature, the term *profile shift*, exactly corresponding to the German word *Profilverschiebung*, has many synonyms, such as: *addendum modification, correction, cutter offset* and *hob offset*. None of the five synonyms can be considered without any objection.

© Springer Nature Switzerland AG 2020
V. Vullo, *Gears*, Springer Series in Solid and Structural Mechanics 10,
https://doi.org/10.1007/978-3-030-36502-8_6

Fewer objections are registered for the use of term profile shift. It is to be preferred to the term *addendum modification*, however still very widely used, but misleading, because with it we could understand a stub toothed gear, which is something quite different from a profile-shifted gear. In fact, the profile shift is a measure of the tooth thickness, rather than of the addendum length. In other word, the profile shift implies a change of the addendum length because of the tooth thickness variation, with the tooth depth that remains constant. The only change in the addendum length, without variation of the tooth thickness, like the one we have in the stub-toothed gears, does not fall within the definition of profile shift.

The term *correction* is an abbreviation of the older term *correction for undercut*. It is therefore limiting, because it would imply that the profile shift is motivated only by the need to avoid undercut. The terms *hob offset* and *cutter offset* are even more restrictive, as well as ambiguous. The reader tries to wonder what meaning the term hob offset might have when the gear is not cut by a hob. The term cutter offset is certainly an improvement over the term hob offset, but it is not suitable for a gear cut by a pinion-type cutter, since in this case the profile shift is not equal to the cutter offset (see Chap. 7).

Notoriously, the profile shift of involute toothing is the most striking stage of their geometric- kinematic design. In addition to the reduction of the minimum number of teeth to prevent the primary interference and undercut (and therefore, the size, weight and cost of the gear pair, with the same gear ratio), the profile shift allows also to achieve other desirable design goals. The main goals are as follows (see Oda et al. [23], Simon [30], Li [18], Chauhan [4]):

- shapes of teeth having greater strength, both for bending fatigue loads and surface fatigue loads;
- more uniform distribution curves of the specific sliding and slide-roll ratios for the driving and driven gears along the path of contact;
- length of path of contact as much as possible extended, for noise control;
- flanks of small pinions without undercut that, as it is well known, causes a reduction of the contact ratio, and weakens the tooth to its root; etc.

Actually, the profile shift is not only performed in cases where it is necessary to prevent the primary interference and undercut of the flanks of pinions with small number of teeth, but also in cases where it is not strictly necessary. Among these cases are, for example, those in which the design requirements impose to overcome the well-known limits related to the part of the involute curve near the base circle, where the capacity to resist surface loading is little, due to the comparatively small radius of curvature, while the sliding velocity tend to be high, and rolling velocity low. In these cases, the profile shift, although not necessary, is advantageously carried out in order to achieve better conditions of surface contact and relative motion, and a better balance of the sliding conditions between the pinion and wheel. The profile shift is also carried out in order to achieve an imposed center distance. The previously mentioned strength of profile-shifted toothing, together with its manufacturing cost, identical to that of the zero-profile-shifted toothing (or standard toothing), fully justify the fact that it is used in most of the usual technical applications, even when the problem of interference does not exist.

By definition (see ISO 1122-1: [4]) the profile shift is the distance, xm_n, measured along a common normal between the reference cylinder of the gear and the datum plane of the basis rack, when the rack and the gear are superposed so that the flanks of a tooth of one are tangent to those of the other (Fig. 6.1). By convention, which is valid for both external and internal gears, the profile shift is positive when the datum plane of the basic rack is external to the gear reference cylinder (Fig. 6.1), and negative when this datum plane cuts the gear reference cylinder. For internal gear, tooth profiles are considered those of tooth spaces. The *profile shift coefficient*, x, is the quotient of the profile shift, expressed in millimeters, divided by the normal module, m_n (in Imperial system, the profile shift coefficient is the product of the profile shift, expressed in inches, for the normal diametral pitch).

From the point of view of the history of technology, it is to be considered that the knowledge related to the profile-shifted gears were known from some time, and so, at least partially, the benefits that could be obtained with their use. However, as long as the only technology of gear cutting was the non-generation process, as the milling cutting, the profile-shifted gears could not actually find wide application, as, for the teeth cutting, it was necessary to construct two special milling cutters for each gear pair. The profile-shifted gears entered instead in the current technological practice when the gear generation cutting with rack-type cutter was introduced [9, 10, 31].

However, even when the gear generation cutting began to spread, a lot of time was needed to understand the full potential of this technology in relation to the optimal design of the gears. In fact, initially, the profile shift was used for involute gears designed to mesh at the standard center distance, and thus emulating what happened in the cycloidal gear pairs, which notoriously may mesh only at their standard center distance. In this way, the standardization, and thus the interchangeability and production of the catalogue gears, were privileged. Only in a second time the designer was able to understand the enormous benefits that could be drawn from the flexibility of use the profile shift, which, though within certain limits, allows to choose the values of the tooth thicknesses, center distance, and gear blank diameters so as to satisfy as much as possible the design requirements. The only possible restriction, consisting in the use of standard cutters for cutting profile-shifted gears, is not valid for the gear generation cutting, as one of the main advantages of the involute profile is to allow the cutting of non-standard gears by standard cutters.

Fig. 6.1 Positive profile shift, xm_n

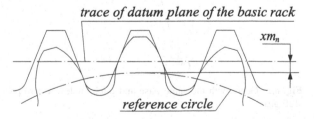

The gear cutting technology with profile shift make it possible to move the position of the cutting pitch circle with respect to the teeth. Therefore, the tooth retains its overall tooth depth, h, by varying the distance of the rack-type cutter from the axis of the gear to be generated, and leaving unaltered the speeds (translational and rotational) of the rack-type cutter and the wheel. Of course, the addendum and dedendum values change with respect to those that characterize the normal standard sizing.

A gear without profile shift will differ under many aspects compared to a gear with a positive or negative profile shift. Figure 6.2 compares two gears with the same number of teeth, z, and cut with the same rack-type cutter, and therefore with the same pitch circle (this circle coincides with the cutting pitch circle), having radius $r = d/2 = mz/2$, but the first (Fig. 6.2a) with zero-profile shift and the second (Fig. 6.2b) with positive profile shift. Since both gears have the same pitch circle and the same pressure angle, α (this does not change regardless of the straight-line parallel to the datum line of the rack-type cutter that is considered), they will have the same base circles, whereby the teeth profiles of each gear are constituted by different parts of the same involute curve. The tip and the root circles, the tooth thicknesses on the pith circles, and addendum and dedendum of the two gears are different. More particularly, the tip circle radius and tooth thickness on the pitch circle of the positive profile-shifted gear are larger compared to those of the gear with a zero-profile shift.

In addition, Fig. 6.2 shows that, in a gear with positive profile shift, the involute curves forming the opposite flanks of each tooth are more spaced than those of a gear with zero-profile shift, but have the same base and standard pitch circles. In other word, the tooth profiles of the second gear are shifted relative to those of the first gear. Then we define as profile-shifted gear, i.e. as gear cut with profile shift, any gear whose tooth thickness measured along the cutting pitch circle is not equal to $p/2 = \pi m/2$.

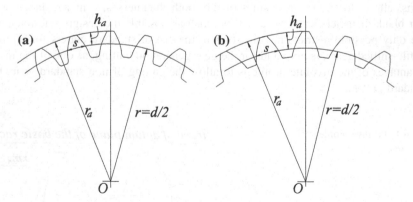

Fig. 6.2 Gears with the same base and pitch circles; **a** zero-profile shift gear; **b** positive profile shift gears

Either or both of the two members of a gear pair may have a profile-shifted toothing, and the profile shift coefficient may be positive or negative. According to this, we have two methods of applying profile shift:

(1) With no change in the center distance, which remains the same as for zero-profile-shifted gears. The profile shift coefficients of the two members of the gear pair are then equal and opposite.
(2) With a change in the center distance (usually an increase, but in some cases also a decrease). The profile shift coefficients of the two members of the gear pair are not equal and opposite, but different from each other.

Usually, the profile shift with no change in the center distance is used when the tooth-sum is large, and the center distance may be chosen to suit the tooth-sum and module. Usually, a positive profile shift coefficient is chosen for the pinion and, for the wheel, a negative profile shift coefficient, equal and opposite to that of the pinion. That is because the positive profile shift coefficient for the pinion improves its shape and strength, whilst the negative profile shift coefficient for the wheel is not so large as to impair its strength. With this choice, the resistance of the gear pair to surface loading is increased.

Instead, the profile shift with a change in their center distance is applied when the tooth-sum is comparatively small (i.e. when a negative profile shift coefficient would impair the quality of the wheel teeth). It is also used when the gear dimensions must be handled in order to agree with an imposed center distance, not strictly complied with the one related to the module and number of teeth.

These two methods together provide an infinite number of possible ways of applying profile shift, or of diving either an imposed or an arbitrary total sum of the profile shift coefficients between the pinion and mating gear wheel. With current knowledge, however, it is not possible to have precise rules that allow to achieve optimum tooth shapes, and, this is because many phenomena relating to the meshing between the teeth, especially those of contact loading, are still not completely know. For this reason, the judgment of the gear designer is still essential.

6.2 Fundamentals of Profile Shift

Figure 6.3 shows the tooth profile of generating rack-type cutter (or the axial section of a hob), consistent with the standard basis rack tooth profile. Teeth are shaped trapezoidal, with rounded tips and flanks inclined at the pressure angle α_0 to the tooth axis, which is normal to the datum line P-P; α_0 is equal to the real value of the pressure angle α_P of the standard basis rack. The *datum line* (or *reference line*) of the rack-type cutter bisects the tooth depth $h_0 = (10/4)m_0$, for which addendum and dedendum are equal between them and equal to $(5/4)m_0$.

With reference to the cutting conditions, it should be noted that the addendum of the generating rack-type cutter has to dig the tooth space width below the pitch

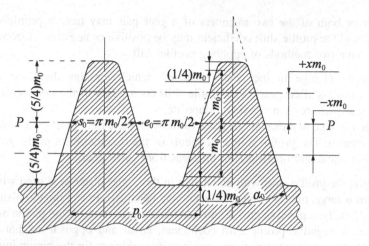

Fig. 6.3 Generating rack-type cutter

circle of the gear to be generated. Furthermore, during the tooth generation motion, the top land of the gear must not touch the bottomland of the tooth space width between two adjacent teeth of the rack-cutter. Since then the rounded tips of the rack-type cutter have a depth equal to $(1/4)m_0$, in relation to the problem of the cutting interference, the rack-type cutter shown in Fig. 6.3 will behave as if its addendum was equal to m_0.

The tooth thickness s_0 and space width e_0 on the nominal datum line are equal to half the pitch, i.e. $s_0 = e_0 = p_0/2 = \pi m_0/2$. Given the geometrical shape of the rack-cutter, the module m_0 (and hence the pitch p_0) and pressure angle α_0 do not vary whatever the straight line, parallel to the nominal datum line, which is chosen to generate a gear wheel with z teeth. Regardless of this choice, the gear will have a generation pitch circle (or cutting or nominal pitch circle) and a base circle whose diameters are given by:

$$d = 2r = zm_0 \tag{6.1}$$

$$d_b = 2r_b = zm_0 \cos \alpha. \tag{6.2}$$

Figure 6.4 shows the geometrical quantities characterizing both the generating standard rack-type cutter, and a gear with zero-profile shift, generated by it. In this case, the cutting pitch line of the rack-type cutter coincides with its datum line, and we obtain a zero profile-shifted gear wheel, i.e. a wheel with *normal standard toothing*. On the cutting pitch circle of the gear, having radius $r = (z/2)m_0$, we have: tooth thickness equal to half the pitch, i.e., $s = e = p_0/2 = \pi m_0/2$; addendum $h_a = m_0$; dedendum $h_f = 1.25\,m_0$; tooth depth $h = h_a + h_f = 2.25\,m_0$.

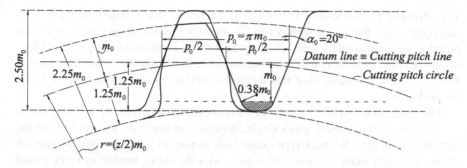

Fig. 6.4 Geometrical dimensions of a gear with zero-profile shift

To avoid misunderstanding, we remember the following definitions:

- *standard toothing* is a toothing that can be obtained by means of the generating standard rack-type cutter;
- *normal toothing*, or *zero-profile-shifted toothing*, is a standard toothing generated in the cutting conditions in which the cutting pitch line (or generation pitch line) of the rack-type cutter coincides with its datum line (Fig. 6.4);
- *profile-shifted toothing* is a non-standard toothing generated in the cutting conditions in which the cutting pitch line of the rack-type cutter does not coincides with its datum line (Fig. 6.5).

In the case of profile-shifted toothing, the position of the cutting pitch line of the rack-type cutter is defined by the *profile shift*, xm_0, relative to its datum line, where x is the *profile shift coefficient*. This coefficient is a dimensionless quantity, which we will assume positive or negative according on whether it corresponds to a displacement of the cutting pitch line towards the top land or towards the root land of the cutting tool. Figure 6.5 shows the cutting condition of a profile-shifted gear with z teeth and positive profile shift coefficient, x.

Clearly, the profile shift of the gear is equal to the distance (positive or negative) between the cutting pitch line and the datum line of the rack-type cutter.

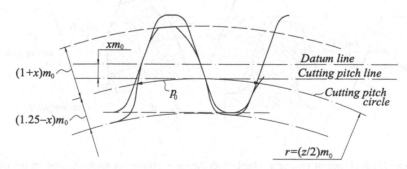

Fig. 6.5 Geometrical dimensions of a gear with positive profile shift

This distance is also called *cutter offset*. Therefore, for a gear generated by a rack-type cutter, the profile shift is equal to the cutter offset. Since then, in common practice, we assume that a gear generated by a hob is identical to that generated by a rack-type cutter (this, of course, without prejudice to the considerations described in Sect. 3.11.2), it is equally clear that the profile shift is equal to the hob offset, in the cut with a hob.

The tooth shape of the gear to be generated, characterized by a given radius $r = (z/2)m_0$ of the cutting pitch circle, depends markedly on the position of the cutting pitch line of the rack-type cutter with respect to its datum line. Figure 6.6 shows the tooth shape of two gear wheels with the same number of teeth z (and therefore with the same cutting pitch circle), both cut with the same rack-type cutter, having pressure angle α_0 and module m_0, but the one on the right with positive profile shift $(xm_0 > 0)$, and the one on the left with negative profile shift $(xm_0 < 0)$.

Figure 6.7 refers to a gear wheel with $z = 20$ teeth, cut by means of a rack-type cutter having a module m_0 and a pressure angle $\alpha_0 = 20°$. It highlights the tooth shapes due to the profile shift, with profile shift equal and opposite $(x = \pm 0.5)$, which are compared with the tooth shape related to a zero-profile shift coefficient $(x = 0)$. Figure 6.8, which refers to the same gear of Fig. 6.7, shows the change of the ratio $\rho_x/\rho_{x=0}$, as a function of the profile shift coefficient x; ρ_x and $\rho_{x=0}$ are the radii of curvature of the tooth profiles on the cutting pitch circle related respectively to any value, x, of the profile shift coefficient and to a profile coefficient equal to zero.

Both Figs. 6.7 and 6.8 show that a positive profile shift causes an increase of tooth thickness to its bottom land (but also a decrease of tooth thickness to its top land), while a negative profile shift causes a reduction in the tooth thickness to its bottom land. Therefore, with large positive profile shift we can fall into the danger of the pointed tooth, while with a large negative profile shift we can meet the danger

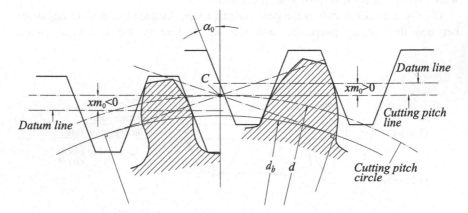

Fig. 6.6 Tooth shapes of two gear wheels with the same number of teeth, the one on the right with positive profile shift, and the one on the left with negative profile shift

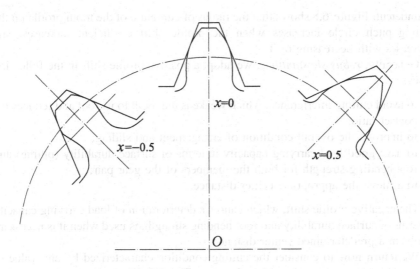

Fig. 6.7 Correlation between tooth shape and the profile shift coefficient

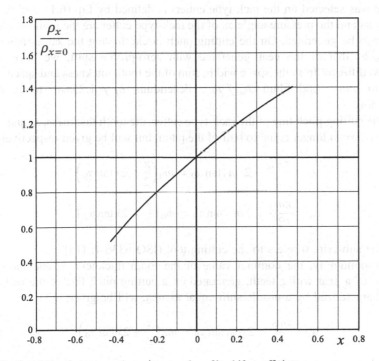

Fig. 6.8 Correlation between ratio $\rho_x/\rho_{x=0}$ and profile shift coefficient

of undercut. Figure 6.8 shows that the radius of curvature of the tooth profile on the cutting pitch circle increases when the profile shift coefficient increases, and decreases with decreasing of it.

Generally, *rebus sic stantibus*, we adopt a positive profile shift in the following cases:

- to avoid cutting interference, which weakens the tooth to its root and reduces the contact ratio;
- to improve the overall condition of engagement and sliding;
- to have good load carrying capacity in terms of surface durability (pitting) and root bending strength for both the members of the gear pair;
- to achieve the appropriate center distance.

The negative profile shift, which causes a deterioration of load carrying capacity in terms of surface durability and root bending strength, is used when it is necessary to obtain a predetermined center distance.

We return now to consider the cutting condition characterized by any value of the profile shift, xm_0, such as that shown in Fig. 6.5. We have already seen that, whatever the profile shift, xm_0, the nominal pitch circle diameter of the gear to be cut (this, during the generation motion, will roll without sliding on the cutting pith line that was selected on the rack-type cutter) is defined by Eq. (6.1), and that the module m_0 and the pressure angle α_0 of the rack-type cutter are the nominal ones of the gear to be generated. On the cutting pitch circle, having the same radius $r = (z/2)m_0$ as that of the gear generated with zero-profile shift, we have: tooth thickness different from the space width; sum of the tooth thickness and space width equal to πm_0; addendum $h_a \neq m_0$; dedendum $h_f \neq 1.25m_0$; tooth depth $h = 2.25m_0$.

On the cutting pitch line of the rack-type cutter, the tooth thickness, s_0 and space width, e_0, are no longer equal to half of the pitch, but will be given respectively by:

$$s_0 = \frac{\pi m_0}{2} - 2xm_0 \tan \alpha_0 = m_0 \left(\frac{\pi}{2} - 2x \tan \alpha_0 \right) \tag{6.3}$$

$$e_0 = \frac{\pi m_0}{2} + 2xm_0 \tan \alpha_0 = m_0 \left(\frac{\pi}{2} + 2x \tan \alpha_0 \right), \tag{6.4}$$

where the subscript 0 refers to the cutting tool (ISO 6336-1: [15]).

Correspondingly, the nominal value of the tooth thickness, s, and the space width, e, of a gear with z teeth, generated on a cutting pitch line of the rack-type cutter characterized by a profile shift equal to xm_0, will be given by:

$$s = \frac{\pi m_0}{2} + 2xm_0 \tan \alpha_0 = m_0 \left(\frac{\pi}{2} + 2x \tan \alpha_0 \right) \tag{6.5}$$

$$e = \frac{\pi m_0}{2} - 2xm_0 \tan \alpha_0 = m_0 \left(\frac{\pi}{2} - 2x \tan \alpha_0 \right). \tag{6.6}$$

The diameter of the base circle of the gear generated is given by Eq. (6.2). To calculate the tip and root circle diameters of the gear, we must keep in mind that a rack-type cutter, which is cutting a gear with a profile shift xm_0, behaves as if it had addendum h_{a0} and dedendum h_{f0} given respectively by:

$$h_{a0} = m_0(1 - x) \tag{6.7}$$

$$h_{f0} = m_0\left(\frac{5}{4} + x\right). \tag{6.8}$$

Since a clearance of $(5/4)m_0$ should remain between the tip circle of the gear and the root line of the rack-type cutter, the diameters of the tip circle and root circle of the gear will be given by:

$$d_a = 2r_a = m_0[z + 2(1 + x)] \tag{6.9}$$

$$d_f = 2r_f = m_0\left[z - 2\left(\frac{5}{4} - x\right)\right]. \tag{6.10}$$

Considering Eq. (1.18), the reference center distance, a_0 (also called standard or nominal center distance), i.e. the sum of the radii of the cutting pitch circles of the two members of an external gear pair, having respectively z_1 and z_2 teeth, is given by:

$$a_0 = m_0\left(\frac{z_1 + z_2}{2}\right). \tag{6.11}$$

When the working center distance, a', of a gear is equal to the reference center distance, a_0, the working pitch circles (or pitch circles) coincide with the cutting pitch circles. Evidently, to have the correct kinematic operating conditions when the center distance is the reference one, the sum of the tooth thicknesses of the two members of the gear pair, measured on the respective cutting pitch circles, must be equal to πm_0, that is:

$$s_{10} + s_{20} = s_1' + s_2' = \pi m_0 = p_0, \tag{6.12}$$

where s_{10} and s_{20} are the thicknesses on the cutting pitch circles, and s_1' and s_2' are the thicknesses on the pitch circles (it is to be noted that, in agreement with the ISO standards, the prime indicates quantities concerning the operating conditions; the same standards use the subscript, w, as equivalent to the prime).

Whenever $(s_1' + s_2')$ will be different from πm_0, the operating center distance, a', of the gear will differ from the reference center distance, a_0. If the two members of the gear have normal toothing, i.e. zero-profile-shifted toothing, the working center distance will be apparently equal to the reference center distance. Instead, if the two wheels have profile-shifted toothing, the two following cases may occur:

- The working center distance, a', is equal to the reference center distance, a_0; in this case, we have the profile-shifted toothing without variation of the center distance.
- The working center distance, a', is different from the reference center distance, a_0; in this case, we have the profile-shifted toothing with variation of the center distance.

6.3 Profile-Shifted Toothing Without Variation of the Center Distance: External and Internal Spur Gear Pairs

6.3.1 External Spur Gear Pairs

The easiest method to obtain profile-shifted toothings for an external spur gear pair consists of the choice of a positive profile shift coefficient, x_1, for cutting of the pinion teeth, and, for cutting of the wheel teeth, a negative profile shift coefficient equal and opposite to that of the pinion, i.e. $x_2 = -x_1$. So, with this method, we have (see Giovannozzi [12], Pollone [25], Henriot [13]):

$$(x_1 + x_2) = 0. \tag{6.13}$$

External gear pairs obtained with this method, which is also known as *long-short addendum system* [19], are nonstandard gears. With this method, the cutting pitch circles, having respectively diameters equal to $z_1 m_0$ and $z_2 m_0$, can be taken as working pitch circles, that is, as pitch circles of the two wheels in their mutual engagement. This because the pressure angle of the involute curves on the cutting pitch circles has the same value, α_0, for both the gear members, and the sum of the teeth thicknesses, $(s_{10} + s_{20})$, on these pitch circles is equal to the pitch, $p_0 = \pi m_0$. In fact, just this has to happen, so that we can have a backlash-free contact, that is a meshing without backlash and no penetration between teeth.

The tooth thickness (of course, the circular tooth thickness) of the pinion, measured along the cutting pitch circle, according to the Eq. (6.5), is given by:

$$s_{10} = \frac{\pi m_0}{2} + 2x_1 m_0 \tan \alpha_0. \tag{6.14}$$

Similarly, the tooth thickness of the wheel, measured along its cutting pitch circle, also considering the Eq. (6.13), is given by:

$$s_{20} = \frac{\pi m_0}{2} + 2x_2 m_0 \tan \alpha_0 = \frac{\pi m_0}{2} - 2x_1 m_0 \tan \alpha_0. \tag{6.15}$$

Adding member to member the Eqs. (6.14) and (6.15), and considering Eq. (6.13), we get Eq. (6.12), which therefore is satisfied. Therefore, we infer that, if the two members of an external gear pair have profile shift coefficients equal to x_1 and $x_2 = -x_1$ respectively, and tooth thicknesses respectively equal to the nominal values corresponding to the profile shifts $x_1 m_0$ and $x_2 m_0 = -x_1 m_0$, they will engage without backlash at the reference center distance, a_0, and with the pressure angle of generation, α_0.

With this method, compared with the case of zero-profile-shifted toothing, the pinion and the wheel retain unaltered the module, pitch circles and center distance. Table 6.1 summarizes, for an external spur gear pair, the characteristic quantities of the profile-shifted toothing without variation of center distance, and compares them with the corresponding quantities of the zero-profile-shifted toothing.

These characteristic quantities can also be deduced regarding the two members of the gear pair (Fig. 6.9) as if they worked on the opposite flanks of the profile of the common basic rack, and considering that all three members have a common pitch point, and common points of contact P_1 and P_2.

Table 6.1 Quantities of the profile-shifted toothing compared with the corresponding ones of the zero-profile-shifted toothing, for an external spur gear pair

Quantity	Profile-shifted toothing without variation of the center distance		Zero-profile-shifted toothing	
	Pinion	Wheel	Pinion	Wheel
Addendum	$h_{a1} = m_0(1 + x_1)$	$h_{a2} = m_0(1 - x_1)$	$h_{a1} = h_{a2} = m_0$	
Dedendum	$h_{f1} = m_0(1.25 - x_1)$	$h_{f2} = m_0(1.25 + x_1)$	$h_{f1} = h_{f2} = 1.25 m_0$	
Tooth depth	$h_1 = 2.25 m_0$	$h_2 = 2.25 m_0$	$h_1 = h_2 = 2.25 m_0$	
Profile shift coefficient	x_1	$x_2 = -x_1$	$x_1 = x_2 = 0$	
Center distance	$a' = a_0 = m_0\left(\frac{z_1 + z_2}{2}\right)$		$a_0 = m_0\left(\frac{z_1 + z_2}{2}\right)$	
Pressure angle	$\alpha' = \alpha_0$		α_0	

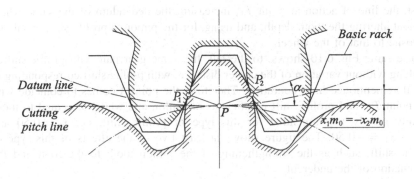

Fig. 6.9 External gear pair with profile-shifted toothings without variation of center distance

Fig. 6.10 Profile-shifted
toothing to avoid the primary
interference ($x_1 = 0.50$ and
$x_2 = -x_1 = -0.50$)

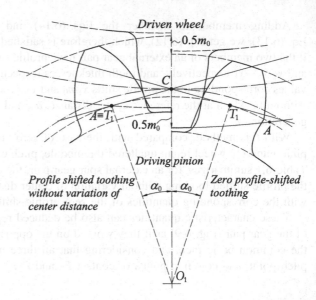

In the gears with profile-shifted toothing without variation of the center distance, the length g_α of the path of contact remains almost unchanged compared to that of the corresponding gears with zero-profile-shifted toothing. The path of contact AE (Fig. 6.10) moves on the line of action in the direction from T_1 to T_2. Now we know that the primary interference that occurs first is that related to the point T_1. This interference occurs in the approach stage of the meshing cycle when the pinion is driving, and is due to an excessive value of the wheel addendum, which brings the point A beyond the point T_1, in the direction from C to T_1 (Fig. 6.10).

Figure 6.10 shows, to the right, an external spur gear pair with zero-profile-shifted toothing. The tip circle of the wheel intersects the line of action at the point A, which is located well beyond the interference point T_1, at the outside of the segment $T_1 T_2$ (T_2 is not shown in the figure). Therefore, primary interference during the approach stage of meshing cycle occurs. To avoid this interference, it is necessary to reduce the addendum of the wheel, so that its tip circle intersects, at most, the line of action at point T_1, increasing the dedendum of the same wheel without altering the tooth depth, and using, for the pinion, a profile shift equal and opposite to that of the wheel.

The same Fig. 6.10 shows, to the left, the same gear pair with profile-shifted toothing without variation of the center distance, with profile shift corresponding to the aforementioned limit condition. Therefore, the point A comes to coincide with the point T_1, and profile shift coefficients of pinion and wheel are equal to those usually mostly recommended for this type of profile shift, i.e. $x_1 = 0.50$ and $x_2 = -x_1 = -0.50$. The figure shows at least two of the effects of this type of profile shift, such as the strengthening of the pinion tooth to the root, and the elimination of the undercut.

The profile-shifted toothing without variation of the center distance has other interesting features that we will point out every time the opportunity will present itself. We have already said that, from the point of view of the cutting technology, it is obtained with the same simple generation processes and with the same cutting tools used for the zero-profile-shifted toothing. We also said that the tooth thickness measured along the cutting pitch circles varies by an amount equal to $2xm_0 \tan \alpha_0$. This thickness variation, measured along the normal to the profile, is equal to $(2xm_0 \tan \alpha_0) \cos \alpha_0 = 2xm_0 \sin \alpha_0$, which also represents the thickness variation on the base circle.

Generally, with this type of profile shift, the continuity of the motion is not compromised, since the length of the path of contact remains substantially unchanged. It also allows, according to what we have seen above, to avoid primary interference in the stage of approach of the meshing cycle. However, since this type of profile-shift causes a displacement of the path of contact towards the interference point T_2, we may encounter the danger of interference in the stage of recess of the meshing cycle. Therefore, this type of profile shift cannot be used when an interference in the stage of recess was initially present.

Consequentially, before proceeding with the definition of the profile shift to be made, it is necessary to ensure that interference in the stage of recess is not present, verifying that the minimum number of teeth, z_{2min}, of the driven wheel is greater than that given by Eq. (4.9) for $k_{1max} = 1$. In addition, once the profile shift to do is defined, it is necessary to ensure that the interference does not appear in the recess stage. In this regard, the minimum number of teeth, z'_{2min}, of the driven wheel should be greater than that given by the same Eq. (4.9), but for $k_{1max} = (1 + x_1)$.

Another drawback that can occur with this type of profile shift is the excessive reduction of the tip thickness of the pinion tooth, measured on the addendum circle. We must therefore ensure that this thickness does not fall below the well-defined limit values, set by current standard. We will see later how to do this type of checking (see Sect. 6.5).

If we denote by z_0 the minimum number of teeth that we can cut, without interference, using a rack-type cutter having standard sizing and predetermined module, m_0, and pressure angle, α_0, for cutting without interference the teeth of the pinion and wheel characterized by profile-shifted coefficients x_1 and $x_2 = -x_1$, the latter must have minimum numbers of teeth respectively equal to $z_0(1 - x_1)$ and $z_0(1 - x_2) = z_0(1 + x_1)$. The pinion will therefore have, at minimum, $z_0(1 - x_1)$ teeth, while the wheel will have, at minimum, $z_0(1 + x_1)$ teeth. Therefore, when this method is used, the sum of the number of the teeth of the pinion and wheel cannot fall below the value $2z_0$. In practice, however, admitting a slight cutting interference according to what previously said (see Sect. 3.2, and Fig. 3.2), the sum above cannot be below of $(5/6)2z_0 = (5/3)z_0$.

Moreover, for a given value of the profile shift coefficient, $(x_1 = -x_2)$, the gear ratio $u = z_2/z_1$ cannot fall below the value given by:

$$u = \frac{z_0(1+x_1)}{z_0(1-x_1)} = \frac{1+x_1}{1-x_1}. \qquad (6.16)$$

The value of the transverse contact ratio, ε_α, is obtained from Eq. (3.44), with $k_1 = (1+x_1)$ and $k_2 = (1+x_2) = (1-x_1)$. According to Eq. (6.16), the profile shift usually most recommended, which is the one proposed by Lasche [17], characterized by $x_1 = 0.50$ and $x_2 = -x_1 = -0.50$, cannot be used for a gear ratio less than 3. In practice, this toothing is convenient for gear ratios somewhat greater than 3, and typically higher than 4.

We must also keep in mind that, with positive profile shifts, the tooth thickness at its root increases with the increase of the positive value of x, but the radius of the fillet decreases. Consequently, the tooth capacity to withstand bending loads and fatigue stress increases as long as the first effect (increase of the thickness) overrides the second effect (reduction of the fillet radius), and this occurs for small numbers of teeth. As Fig. 6.8 shows, with positive profile shifts, the radii of curvature of the profiles increases, and then the capacity to withstand surface loads grows proportionally. For negative profile shifts, the opposite effects to those described just now occur.

With appropriate choice of profile shift coefficients, we can still obtain satisfactory distribution curves of the specific sliding and sliding velocity, as well as a good load carrying capacity for the pinion and wheel. With positive profile shift, the pinion is strengthened, while the wheel with a sufficiently large numbers of teeth is only slightly weakened by the negative profile shift, and in certain circumstances even strengthened, because of the positive increase in the fillet radius. With values about the unit of the gear ratio ($u \cong 1$), gears with positive-shifted toothing without variation of the center distance are however not rational, and they are therefore to be avoided.

Finally, it is to be noted that the gears with profile-shifted toothing without variation of the center distance have not characteristics of catalogue gears; they are in fact non-standard gears. Since then the center distance is equal to that of the gear with zero-profile-shifted toothing, any fixed center distance cannot be maintained for spur gears, using standard modules.

6.3.2 Internal Spur Gear Pairs

If the number of teeth of the pinion of an internal spur gear pair is less than the minimum number of teeth to avoid interference, it is necessary to use gear with profile-shifted toothing, without or with variation of center distance. In usual technical applications, the use of internal gears with profile-shifted toothings is usually advantageous, for which the zero-profile-shifted toothing constitutes, in fact, not the norm, but rather a rarity. However, for an internal gear pairs, in contrast to what happens for an external gear pair, it is not always possible to freely share the profile shift coefficients, x_1 and x_2, and simultaneously obtain optimal values of the specific slidings and contact ratios. In this respect, in fact, the tendency of the internal gear teeth to interfere between them imposes considerable limitations (see Chap. 5, and Schreier [28], Dudley [8], Polder [24]).

For internal gears, the determination of the profile shift coefficients is not so easy. Generally, it is necessary to proceed by trial and error, case by case, since the interference problem constitutes a conditioning priority. In this regards, it should be remembered that, for external gear with standard sizing, the minimum number of teeth of the pinion to avoid interference with the mating wheel decreases when $(1/u) = z_1/z_2$ increases, and then u decreases. Instead, for internal gears, the opposite happens: the minimum number of teeth of the pinion to avoid interference with the annulus increases when $(1/u) = z_1/z_2$ increases, and then u decreases. For example, for a pressure angle $\alpha = 20°$ and standard sizing ($k_2 = 1$), from Fig. 3.1 we obtain, with $u = z_2/z_1 = 1.5$ and then $(1/u) = 0.667$, $z_{1min} = 14$, for the external gear pair, and $z_{1min} = 26$, for the internal gear pair. Instead, for $u = z_2/z_1 = 8.0$ and then $(1/u) = 0.125$, we obtain $z_{1min} = 16$, for the external gear pair, and $z_{1min} = 18$, for the internal gear pair.

In Sect. 5.3 we have shown that, to avoid the risk of secondary interference in an internal spur gear pair, the difference $(z_2 - z_1)$ between the number of teeth of the annulus and pinion must not fall below a minimum difference value $(z_2 - z_1)_{min}$, which depends on the addendum factor, k, and the operating pressure angle, α'. However, in many practical applications, it can be useful to fall below of this limit difference, and, in this case, it is necessary to resort to profile-shifted toothing. The determination of the profile shift coefficients must therefore take into account both types of interference, primary and secondary. In this section, we circumscribe the problem to the profile-shifted toothing without variation of center distance.

We have already mentioned in Sect. 6.1 that the sign convention of the profile shift of an internal gear is the same as that of an external gear. Obviously, for internal gears, the generating basic rack may be just an imaginary rack. With a profile shift equal to $x_1 m_0$, the tooth thickness of the pinion on its cutting pitch circle is given by Eq. (6.14). The tooth thickness of the wheel measured along its cutting pitch circle is instead given by:

$$s_{20} = m_0 \left(\frac{\pi}{2} - 2x_2 \tan \alpha_0 \right). \qquad (6.17)$$

By imposing the condition given by Eq. (6.12), we obtain:

$$x_2 - x_1 = 0, \qquad (6.18)$$

that is $x_2 = x_1$. The reference center distance, a_0, equal to difference $(r_{20} - r_{10})$, is given by:

$$a_0 = m_0 \left(\frac{z_2 - z_1}{2} \right). \tag{6.19}$$

Figure 6.11 shows an internal gear with profile-shifted toothing without variation of the center distance, and profile shift coefficients, x_1 and x_2, both positive and equal to 0.50. For a discussion on the choice of the profile shift coefficients, we refer to Sects. 6.8 and 6.9. Table 6.2 summarizes, for an internal spur gear pair, the characteristic quantities of the profile-shifted toothing without variation of the center distance, and compares them with the corresponding quantities of the zero-profile-shifted toothing.

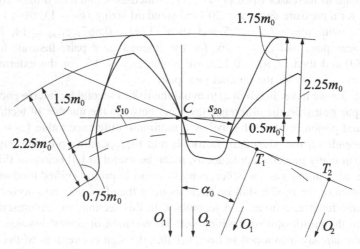

Fig. 6.11 Internal gear with profile-shiffted toothing without variation of center distance ($x_1 = x_2 = 0.50$)

Table 6.2 Quantities of the profile-shifted toothing compared with the corresponding ones of the zero-profile-shifted toothing, for an internal spur gear pair

Quantity	Profile-shifted toothing without variation of center distance		Zero-profile-shifted toothing
	Pinion	Annulus	Pinion and annulus
Addendum	$h_{a1} = m_0(1 + x_1)$	$h_{a2} = m_0(1 + x_1)$	$h_{a1} = h_{a2} = m_0$
Dedendum	$h_{f1} = m_0(1.25 - x_1)$	$h_{f2} = m_0(1.25 - x_1)$	$h_{f1} = h_{f2} = 1.25 m_0$
Tooth depth	$h_1 = 2.25 m_0$	$h_2 = 2.25 m_0$	$h_1 = h_2 = 2.25 m_0$
Profile shift coefficient	x_1	$x_2 = x_1$	$x_2 = x_1 = 0$
Center distance	$d' = a_0 = m_0 \left(\frac{z_2 - z_1}{2} \right)$		$a_0 = m_0 \left(\frac{z_2 - z_1}{2} \right)$
Pressure angle	$\alpha' = \alpha_0$		$\alpha_0.$

6.4   Minimum Profile Shift Coefficient to Avoid Interference

In Sect. 3.1 we derived the Eq. (3.10), which allows to determine the minimum number of teeth of a gear wheel, which must mesh with the rack of the corresponding series, without operating primary interference. This equation, when applied to the cutting condition of a gear with zero-profile-shifted toothing, having z teeth and cut by means of a rack-type cutter having standard sizing, becomes:

$$z_{min} \geq 2 \frac{k_0}{\sin^2 \alpha_0}, \tag{6.20}$$

where k_0 and α_0 are respectively the addendum factor and pressure angle of the rack-type cutter.

About the effects of the cutting interference and consequent undercut, a rack-type cutter, which is cutting a gear wheel with a profile shift xm_0, behaves as if it had an addendum, h_{a0}, and a dedendum, h_{f0}, given by:

$$h_{a0} = m_0(1 - x) \tag{6.21}$$

$$h_{f0} = m_0 \left(\frac{5}{4} + x \right) \tag{6.22}$$

and, therefore, as if it had $k_0 = (1 - x)$. Thus, from Eq. (6.20), we infer that the values of z_{min}, for which the cutting interference with the generating rack-type cutter is avoided, are proportional to $(1 - x)$. Therefore, the values of z_{min} can be taken from Fig. 3.1, by multiplying by $(1 - x)$ those obtained from the same figure for a ratio $(z_1/z_2) = 0$. This is because Fig. 3.1 refers to an addendum factor $k_2 - k_0 = 1$, while in this case we have an addendum factor $k_0 = h_{a0}/m_0 = (1 - x)$. Thus, we have:

$$z_{min} = z_0(1 - x), \tag{6.23}$$

where z_0 is the value derived from Fig. 3.1 for a given value of the pressure angle, α_0, and $(z_1/z_2) = 0$. Thus, we find that, compared to the values which are obtained from Fig. 3.1 for a given α_0 and $(z_1/z_2) = 0$, z_{min} decreases for positive values of the profile shift coefficient ($x > 0$), and increases for negative values of it ($x < 0$).

From Eq. (6.23) we derive that the minimum profile shift coefficient, x_{min}, and therefore the minimum profile shift, $(x_{min}m_0)$, of the cutting pitch line of the rack-type cutter with respect to its datum line, which is necessary to avoid the cutting interference when we have to cut a gear having z teeth, is given by:

$$x_{\min} = 1 - \frac{z}{z_0} = \frac{z_0 - z}{z_0}. \tag{6.24}$$

We can obtain directly this equation remembering that, generally, the cutting interference of a gear with z teeth, cut by means of a given rack-type cutter having active addendum m_0, will be avoided if the following condition is satisfied. The tip line of the rack-type cutter, i.e. the line parallel to the datum line passing through the point where the involute ends and the rounded tip connecting the involute with the top line starts, must pass through the interference point T_1 on the line of action. In this limit condition, the cutting pitch line of the rack-type cutter is positioned at a distance from the point T_1, measured along a line perpendicular to the datum line (Fig. 6.12), equal to [12]:

$$CH = T_1 C \sin \alpha_0 = r \sin^2 \alpha_0 = \frac{z m_0}{2} \sin^2 \alpha_0. \tag{6.25}$$

In this position, the profile shift, $x m_0$, of the cutting pitch line with respect to the datum line is equal to:

$$x m_0 = m_0 - \frac{z m_0}{2} \sin^2 \alpha_0. \tag{6.26}$$

Remembering the inequality (6.20), with $k_0 = 1$, in which we replace the sign of inequality with that of equality, we obtain the Eq. (6.24).

However, in practice, as we said in Sect. 3.2, a light cutting interference is not only tolerated, but also wanted. This is because, on the one hand, it causes

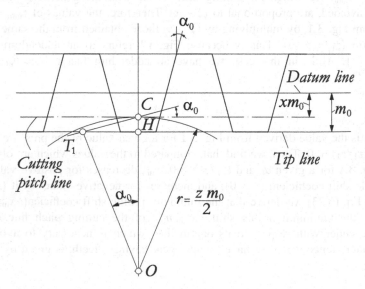

Fig. 6.12 Cutting conditions of the gear with z teeth with a rack-type cutter having active addendum m_0

negligible decreases in the contact ratio and tooth thickness at its root, and, on the other hand, it has the advantage of eliminating the involute part immediately near the base circle to which different problem are related. For this reason, instead of Eq. (6.24), the following relationship is used:

$$x_{min} = \frac{(5/6)z_0 - z}{z_0}. \tag{6.27}$$

Replacing, in Eqs. (6.24) and (6.27), the values of z_0 corresponding to the pressure angles $\alpha_0 = 15°$ and $\alpha_0 = 20°$ that are derived from Fig. 3.1 for $(z_1/z_2) = 0$, we obtain the relationships collected in Table 6.3.

Generally, the values of the profile shift coefficients fall within the following range, which defines their usual limits.

$$-0.50 \leq x \leq 1.00. \tag{6.28}$$

The lower limit, the negative one, is related to the need to avoid undercut. Actually, although gears with undercut are sometimes used, it is best to design gears which have not undercut teeth. The upper limit, the positive one, is related to the need to avoid too pointed teeth. In fact, when the profile shift coefficient increases, the corresponding increase in the addendum determines a reduction in the tooth thickness at the tip circles. This reduction can become unsustainable when it falls below the limit values fixed by the standards (in the usual practice of gear design, the minimum value of the tooth thickness at the tip circle below which we must not fall is equal to $0.25m$).

In addition to the above-mentioned requirements, which are related to the cutting conditions, other numerous and important requirements must be met, related to the assembly and operating conditions. The main ones of these requirements, which depend primarily on the tooth thickness and addendum values, are: the need to avoid operating interference, that never can be tolerated; the need to have an adequate amount of backlash; the need to ensure a suitable operating depth and contact ratio, and, simultaneously, a sufficient clearance at the root circles of each gear.

The determination of the optimum values of the profile shift coefficient based on the above requirements is of a certain complexity, but represents a striking stage of the gear design. We must however point out that, although most of the gears are designed using profile shift coefficients that fall within the limits defined by the inequality (6.26), many exceptions exist for which the aforementioned limits are not satisfied, and the number of these exceptional cases is growing in recent years.

Table 6.3 Theoretical and practical relationship to calculate x_{min}	Profile shift coefficient	$\alpha_0 = 15°$	$\alpha_0 = 20°$
	Theoretical	$x_{min} = \frac{30-z}{30}$	$x_{min} = \frac{17-z}{17}$
	Practical	$x_{min} = \frac{25-z}{30}$	$x_{min} = \frac{14-z}{17}$

6.5 Pointed Teeth and Tooth Thickness

In the previous Section, we said that the upper positive limit of the profile shift coefficient is related to the need to avoid pointed teeth. In fact, with increase of positive profile shift coefficient, the crest of pinions with small number of teeth tends to run to a point. Therefore, if the wheel to be generated has few teeth and, to avoid the cutting interference, we choose a positive profile shift coefficient too high, the tooth could may have pointed end (a cusp), that is, the tooth crest-width at tip circle may drop to zero. This pointed termination of the tooth should be avoided because it can cause breakage (especially for case-hardened gear wheels), and malfunctioning of the gear. We have therefore the need to find the relationship between x_1 and z_1 (remember that, generally, the pinion is characterized by a positive profile shift coefficient) which defines the condition of the pointed tooth.

The tooth of the gear wheel with z_1 teeth, generated with a profile shift coefficient x_1, on the cutting pitch circle has a thickness given by Eq. (6.14), to which a tooth thickness angle corresponds, given by:

$$2\vartheta_{s0} = \frac{s_{10}}{r_{10}} = \frac{\pi}{z_1} + 4\frac{x_1}{z_1}\tan\alpha_0, \tag{6.29}$$

where ϑ_{s0} is the tooth thickness half angle on the cutting pitch circle (Fig. 2.12), and $r_{10} = z_1 m_0/2$. Remembering that the angular pitch, τ_1 (in radians), of the pinion (i.e. the quotient of the angular units in a circle divided by the number of teeth of a gear) is given by

$$\tau_1 = 2\pi/z_1, \tag{6.30}$$

we obtain the following relationship which defines the space width angle:

$$2\vartheta_{e0} = \frac{\pi}{z_1} - 4\frac{x_1}{z_1}\tan\alpha_0, \tag{6.31}$$

where ϑ_{e0} is the space width half angle (note the use of double subscript: the first subscript, s or e, refers respectively to tooth thickness and space width, while the second subscript, 0, refers to cutting condition).

The tooth thickness angle, $2\vartheta_{b1}$, on the base circle of the same gear wheel is obtained (Fig. 2.12) by adding twice the angle $\varphi_0 = \mathrm{inv}\,\alpha_0$ (see Eq. 2.2) to the angle given by Eq. (6.29). Then we obtain:

$$2\vartheta_{b1} = 2\vartheta_{s0} + 2\mathrm{inv}\,\alpha_0 = \frac{\pi}{z_1} + 4\frac{x_1}{z_1}\tan\alpha_0 + 2\mathrm{inv}\,\alpha_0. \tag{6.32}$$

On the tip circle, whose radius $r_{a1} = d_{a1}/2$ is given by Eq. (6.9), the pressure angle, α_{a1}, at the tip point is given by (see Eq. 2.3):

$$\cos \alpha_{a1} = \frac{r_{b1}}{r_{a1}} = \frac{z_1}{z_1 + 2(1 + x_1)} \cos \alpha_0. \tag{6.33}$$

The condition of pointed tooth on this tip circle is equal to impose the equality to zero of the tooth thickness angle $2\vartheta_{a1}$ on the same tip circle, that is:

$$2\vartheta_{a1} = 2\vartheta_{b1} - 2\widehat{P_0 O V} = \frac{\pi}{z_1} + 4\frac{x_1}{z_1}\tan \alpha_0 + 2\text{inv}\,\alpha_0 - 2\text{inv}\,\alpha_{a1} = 0, \tag{6.34}$$

where $\widehat{P_0 O V}$ is the tooth thickness half angle at the tooth apex V (Fig. 2.12). From this equation, we get:

$$\text{inv}\,\alpha_{a1} = \text{inv}\,\alpha_0 + \frac{\pi}{2z_1} + 2\frac{x_1}{z_1}\tan \alpha_0. \tag{6.35}$$

From Eq. (6.33), we obtain:

$$z_1 = 2(1 + x_1)\frac{\cos \alpha_{a1}}{\cos \alpha_0 - \cos \alpha_{a1}}. \tag{6.36}$$

Finally, substituting this equation in Eq. (6.35), we obtain:

$$x_1 = \frac{\dfrac{\pi}{4} - \dfrac{\cos \alpha_{a1}}{\cos \alpha_0 - \cos \alpha_{a1}}(\text{inv}\,\alpha_{a1} - \text{inv}\,\alpha_0)}{\left[\dfrac{\cos \alpha_{a1}}{\cos \alpha_0 - \cos \alpha_{a1}}(\text{inv}\,\alpha_{a1} - \text{inv}\,\alpha_0) - \tan \alpha_0\right]}. \tag{6.37}$$

Using Eq. (6.36) and (6.37), it is possible to obtain the curve that gives the values of the profile shift coefficient x_1 for which a gear with z_1 teeth, generated by a rack-type cutter having pressure angle α_0, has pointed teeth. Figure 6.13 shows, left at the top, the two limit curves which give, as a function of z_1, the profile shift coefficients for which the teeth are pointed, for pressure angles $\alpha_0 = 15°$ and $\alpha_0 = 20°$. They represent the maximum theoretical values, obviously not to be exceeded. In the same figure, other four curves are shown, and precisely:

- the curve 1 and 1' that give respectively the theoretical and practical profile shift coefficients to avoid the cutting interference, when a rack-type cutter with pressure angle $\alpha_0 = 15°$ is used;
- the curve 2 and 2' that give respectively the theoretical and practical profile shift coefficients to avoid the cutting interference, when a rack-type cutter with pressure angle $\alpha_0 = 20°$ is used.

The tooth thickness s_{a1} and s_{f1} on the tip and root circles of the gear can be determined using the general Eq. (2.59), taking into account Eq. (6.56) which defines inv γ_1 (see Sect. 6.6.1). On the tip circle, the tooth thickness should not fall below the limit values set by international standards. Figure 6.14 shows, in quantitative terms, as the profile shift affects both aforesaid tooth thicknesses, s_a and s_f,

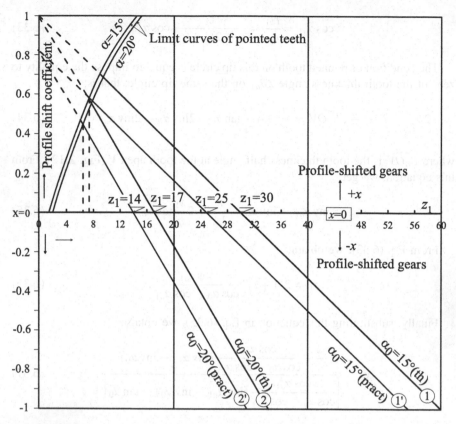

Fig. 6.13 Limit curves of pointed teeth, and values of their profile shift coefficients to avoid the cutting interference, with rack-type cutters having pressure angle $\alpha_0 = 15°$ and $\alpha_0 = 20°$

for values of the profile shift coefficients in the range of variability given by inequality (6.28).

As Eq. (6.32) shows, the maximum value of the tooth thickness angle is the one we have on the base circle, given by the same Eq. (6.32). When the root circle of the gear is inside the base circle, the parts of the tooth between the base circle and the root circle, the radius of which is then smaller than that of the first, have a decreasing thickness. Theoretically, the tooth flanks of these parts inside the base circle are made with radial profile.

In some cases, one of the requirements of the design is to get the maximum strength of the pinion teeth. In these cases, the designer must set the geometrical-kinematic design of the gear in such a way that the root circle coincides with the base circle, i.e. $d_b = d_f$. To this end, just match the diameters of these two circles, given respectively by Eqs. (6.2) and (6.10), i.e.:

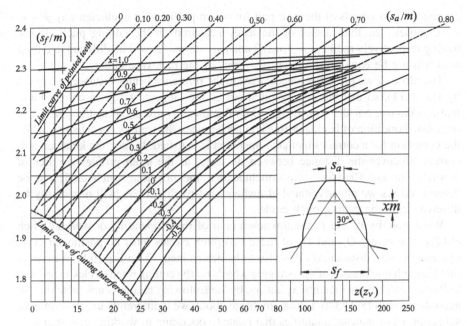

Fig. 6.14 Ratios s_a/m and s_f/m as a function of the profile shift coefficient, x, and number of teeth, z; limit curves of pointed teeth, and cutting interference

$$z_1 m_0 \cos \alpha_0 = m_0 \left[z_1 - 2 \left(\frac{5}{4} - x_1 \right) \right].$$
(6.38)

Therefore, the profile shift coefficient x_1 to be taken is given by:

$$x_1 = \frac{5}{4} - \frac{z_1}{2} (1 - \cos \alpha_0).$$
(6.39)

This type of design involves the rack railways.

6.6 Profile-Shifted Toothing with Variation of Center Distance: External and Internal Spur Gear Pairs

6.6.1 External Spur Gear Pairs

The most general method to obtain profile-shifted toothing for an external gear pair is to assume values of the profile shift coefficient, x_1 and x_2, which are not related by Eq. (6.13), but determined based on the considerations described below, and other observations that will be described in the following sections. In this case, by cutting the two wheels of the gear pair for which profile shift coefficients $x_1 \neq -x_2$ were chosen, the thicknesses of the teeth measured on the cutting pitch circles, having respectively radii $r_{10} = z_1 m_0/2$ and $r_{20} = z_2 m_0/2$, will be different and their sum will not be equal to the pitch $p_0 = \pi m_0$ of the rack-type cutter.

Therefore, if we place the wheels axes to the reference center distance, a_0, given by Eq. (6.11), so that the cutting pitch circles are tangent, we get a correct kinematic behavior, but the sum of the tooth thickness is not equal to the pitch. The operating condition of backlash-free contact between the teeth is not satisfied, while the condition for a correct kinematic operation is satisfied. In fact, the kinematics is correct whatever the distance between the axes, as the tooth profiles are arcs of involute curves. Therefore, we conclude that the operating pitch circles to be determined (as we already stated elsewhere, just pitch circles), will necessarily be different from the cutting pitch circles.

We denote by r'_1 and r'_2 the unknown radii of these operating or working pitch circles. The center O_1 and O_2 of these two pitch circles must be positioned at an operating center distance, a', this also to be determined. From Eq. (6.5), we obtain that the teeth thicknesses of the two gear wheels on the cutting pitch circles are related to the profile shift coefficients, x_1 and x_2, by the following relationships (note that, in accordance with ISO 6336-1:2006, in this textbook we indifferently use the prime, or subscript, w, to indicate quantities that relate to operating or working conditions):

$$s_{10} = m_0 \left(\frac{\pi}{2} + 2x_1 \tan \alpha_0 \right) \tag{6.40}$$

$$s_{20} = m_0 \left(\frac{\pi}{2} + 2x_2 \tan \alpha_0 \right). \tag{6.41}$$

Therefore, we can have three different cases, namely:

- The sum of the two profile shift coefficients is greater than zero $(x_1 + x_2 > 0)$, for which

$$s_{10} + s_{20} > \pi m_0; \tag{6.42}$$

in this case, the operating center distance, a', operating module, m', and operating pressure angle, α', will necessarily be greater than the corresponding reference values, that is: $a' > a_0$; $m' > m_0$; $\alpha' > \alpha_0$.

- The sum of the two profile shift coefficients is less than zero $(x_1 + x_2 < 0)$, for which:

$$s_{10} + s_{20} < \pi m_0; \tag{6.43}$$

in this case, the operating center distance, a', operating module, m', and operating pressure angle, α', will necessarily be less than the corresponding reference values, that is: $a' < a_0$; $m' < m_0$; $\alpha' < \alpha_0$.

- The sum of the two profiles shift coefficients is equal to zero $(x_1 + x_2 = 0)$. This case is the one discussed in Sect. 6.3; it therefore constitutes a special case of the general discussion that follows.

In the first two cases, those for which the sum $(x_1 + x_2)$ is different from zero, we have to determine the operating center distance $a' \neq a_0$ for which we have a meshing with backlash-free contact between the teeth, or, in equivalent terms, the operating module $m' \neq m_0$ or the operating pressure angle $\alpha' \neq \alpha_0$. Figure 6.15 shows the meshing diagram in an external spur gear pair with $(x_1 + x_2) > 0$. It highlights the two operating pitch circles that are touching at pitch point C, and all other circles and quantities that affect the solution of our problem. Note that, from now on, in accordance with the vocabulary of gear terms of the ISO 1122-1: [14], we will omit to premise the adjective *"operating"* for all the quantities of interest, but in doing so we always mean quantities related to the operating conditions; in this section, this quantities are indicated with the prime.

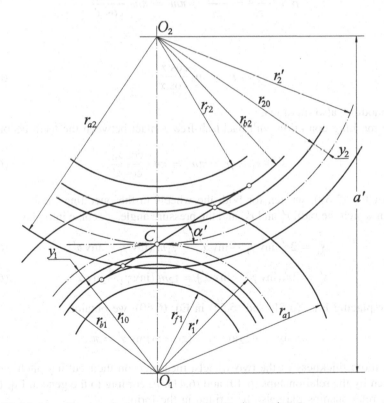

Fig. 6.15 Meshing diagram of an external gear pair with $(x_1 + x_2) > 0$

For the determination of these quantities that are interdependent between them, since they are closely related to each other (therefore, just determine one on them to obtain the other), we note first that the radii r_{b1} and r_{b2} of the base circles are given by:

$$r_{b1} = r_{10}\cos\alpha_0 = (z_1 m_0/2)\cos\alpha_0 \tag{6.44}$$

$$r_{b2} = r_{20}\cos\alpha_0 = (z_2 m_0/2)\cos\alpha_0, \tag{6.45}$$

where all known quantities appear. The radii of the pitch circles are given by:

$$r_1' = \frac{d_1'}{2} = \frac{r_{b1}}{\cos\alpha'} = r_{10}\frac{\cos\alpha_0}{\cos\alpha'} = \frac{z_1 m_0}{2}\frac{\cos\alpha_0}{\cos\alpha'} \tag{6.46}$$

$$r_2' = \frac{d_2'}{2} = \frac{r_{b2}}{\cos\alpha'} = r_{20}\frac{\cos\alpha_0}{\cos\alpha'} = \frac{z_2 m_0}{2}\frac{\cos\alpha_0}{\cos\alpha'} \tag{6.47}$$

where α' is the pressure angle, which is unknown. The circular pitch, p', also unknown, is given by:

$$p' = \frac{2\pi r_1'}{z_1} = \frac{2\pi r_2'}{z_2} = \pi m' = \pi m_0 \frac{\cos\alpha_0}{\cos\alpha'} \tag{6.48}$$

where

$$m' = m_0 \frac{\cos\alpha_0}{\cos\alpha'} \tag{6.49}$$

is the module, also unknown.

The meshing condition with backlash-free contact between the teeth becomes:

$$s_1' + s_2' = p' = \pi m' = \pi m_0 \frac{\cos\alpha_0}{\cos\alpha'}. \tag{6.50}$$

From Eq. (2.59), written for both the wheels in the operating meshing conditions, in which the radii r_1' and r_2', and the pressure angle, α', come into play, we get:

$$s_1' = 2r_1'(\mathrm{inv}\,\gamma_1 - \mathrm{inv}\,\alpha') = z_1 m'(\mathrm{inv}\,\gamma_1 - \mathrm{inv}\,\alpha') \tag{6.51}$$

$$s_2' = 2r_2'(\mathrm{inv}\,\gamma_2 - \mathrm{inv}\,\alpha') = z_2 m'(\mathrm{inv}\,\gamma_2 - \mathrm{inv}\,\alpha'). \tag{6.52}$$

By replacing Eqs. (6.51) and (6.52) in Eq. (6.50), we obtain:

$$z_1 \mathrm{inv}\,\gamma_1 + z_2 \mathrm{inv}\,\gamma_2 - (z_1 + z_2)\mathrm{inv}\,\alpha' = \pi. \tag{6.53}$$

The tooth thickness of the two wheels, measured on their cutting pitch circles, are given by the relationships (6.40) and (6.41). According to the general Eq. (2.59), these relationships may also be written in the form:

$$s_{10} = 2r_{10}(\text{inv}\,\gamma_1 - \text{inv}\,\alpha_0) = z_1 m_0(\text{inv}\gamma_1 - \text{inv}\,\alpha_0) \tag{6.54}$$

$$s_{20} = 2r_{20}(\text{inv}\,\gamma_2 - \text{inv}\,\alpha_0) = z_2 m_0(\text{inv}\,\gamma_2 - \text{inv}\,\alpha_0). \tag{6.55}$$

By equating the two relationships which express in two different but equivalent ways the thicknesses on their cutting pitch circles, i.e. equating Eq. (6.40) with Eq. (6.54), and Eq. (6.41) with Eq. (6.55), we obtain the following equations which give $\text{inv}\,\gamma_1$ and $\text{inv}\,\gamma_2$ as a function of quantities all known:

$$\text{inv}\,\gamma_1 = \text{inv}\,\alpha_0 + \frac{2}{z_1}\left(\frac{\pi}{4} + x_1 \tan\alpha_0\right) \tag{6.56}$$

$$\text{inv}\,\gamma_2 = \text{inv}\,\alpha_0 + \frac{2}{z_2}\left(\frac{\pi}{4} + x_2 \tan\alpha_0\right). \tag{6.57}$$

Finally, substituting these relationships in Eq. (6.53), we get the following equation that uniquely solves our problem.

$$\text{inv}\,\alpha' = \text{inv}\,\alpha_0 + 2\tan\alpha_0 \frac{x_1 + x_2}{z_1 + z_2}. \tag{6.58}$$

So, once calculated the pressure angle α', we have also determined the module, m', and the radii r_1' and r_2' of the pitch circles, given by Eqs. (6.49), (6.46) and (6.47), and then also the center distance, a', given by:

$$a' = (r_1' + r_2') = \frac{1}{2}(z_1 + z_2)m' = \frac{z_1 + z_2}{2}m_0 \frac{\cos\alpha_0}{\cos\alpha'} = a_0 \frac{\cos\alpha_0}{\cos\alpha'}. \tag{6.59}$$

From Eq. (6.58), we infer that:

- If $(x_1 + x_2) > 0$, we have $\text{inv}\,\alpha' > \text{inv}\,\alpha_0$, and then $\alpha' > \alpha_0$, and so from Eq. (6.59), we obtain $a' > a_0$.
- If $(x_1 + x_2) < 0$, we have $\text{inv}\,\alpha' < \text{inv}\,\alpha_0$, and then $\alpha' < \alpha_0$, and so from Eq. (6.59), we obtain $a' < a_0$.
- If $(x_1 + x_2) = 0$, we have $\text{inv}\,\alpha' = \text{inv}\,\alpha_0$, and then $\alpha' = \alpha_0$, and $a' = a_0$.

Relation (6.58) is the fundamental equation to be used for the determination of the profile shift coefficients, x_1 and x_2. It can be obtained by another procedure, which allow us to define other significant quantities of the profile-shifted gears with variation of the center distance. To this end, suppose we have generated two gear wheels with z_1 and z_2 teeth, using the same rack-type cutter, the first with profile shift $x_1 m_0$, and the second with profile shift $x_2 m_0$. The radii r_{b1} and r_{b2} of the base circles of the two gears are given by Eqs. (6.44) and (6.45). The tooth thickness angle $2\vartheta_{b1}$ of the first wheel, measured on the base circle having radius r_{b1}, is given by Eq. (6.32). The tooth thickness angle $2\vartheta_{b2}$ of the second wheel, measured on the base circle having radius r_{b2}, is given by:

$$2\vartheta_{b2} = \frac{\pi}{z_2} + 4\frac{x_2}{z_2}\tan\alpha_0 + 2\mathrm{inv}\,\alpha_0, \tag{6.60}$$

which is obtained by the same procedure used to derive Eq. (6.32).

When we place the two wheels in such a way that the flank mating profiles are in backlash-free contact, i.e. profiles are meshing without backlash and no penetration between the teeth, the radii of the operating pitch circles, r_1' and r_2', and operating center distance, a', will be related by the following relationships:

$$\frac{r_1'}{r_2'} = \frac{z_1}{z_2} \tag{6.61}$$

$$a' = r_1' + r_2'. \tag{6.62}$$

On these circles the operating pressure angle, α', is given by:

$$\cos\alpha' = \frac{r_{b1}}{r_1'} = \frac{r_{b2}}{r_2'}. \tag{6.63}$$

Since the angular pitches, τ_1 and τ_2, of the two gears are given by

$$\tau_1 = 2\pi/z_1 \tag{6.64}$$

$$\tau_2 = 2\pi/z_2, \tag{6.65}$$

the tooth thickness angles $2\vartheta_{s,1}'$ and $2\vartheta_{s,2}'$, and space width angles $2\vartheta_{e,1}'$ and $2\vartheta_{e,2}'$ of the same two gears, on their operating pitch circles, can be written in the form:

$$2\vartheta_{s,1}' = \frac{\pi}{z_1} + 4\frac{x_1}{z_1}\tan\alpha_0 + 2(\mathrm{inv}\,\alpha_0 - \mathrm{inv}\,\alpha') \tag{6.66}$$

$$2\vartheta_{s,2}' = \frac{\pi}{z_2} + 4\frac{x_2}{z_2}\tan\alpha_0 + 2(\mathrm{inv}\,\alpha_0 - \mathrm{inv}\,\alpha') \tag{6.67}$$

$$2\vartheta_{e,1}' = \tau_1 - 2\vartheta_{s,1}' = \frac{2\pi}{z_1} - 2\vartheta_{s,1}' \tag{6.68}$$

$$2\vartheta_{e,2}' = \tau_2 - 2\vartheta_{s,2}' = \frac{2\pi}{z_2} - 2\vartheta_{s,2}'. \tag{6.69}$$

The Eqs. (6.66) and (6.67) are obtained by using, for each of the two mating gears, the same procedure that allowed us to derive Eq. (6.34), while Eqs. (6.68) and (6.69) are obtained, for each gear wheel, as the difference between the angular pitches and tooth thickness angles.

To have the tooth profiles of the mating gear wheels in backlash-free contact, on the operating pitch circles the space width between the teeth of the first gear must be equal to the tooth thickness of the second gear. Therefore, it must be:

$$2r'_1 \vartheta'_{e,1} = 2r'_2 \vartheta'_{s,2} \tag{6.70}$$

from which we obtain:

$$\vartheta'_{e,1} = \frac{r'_2}{r'_1} \vartheta'_{s,2} = \frac{z_2}{z_1} \vartheta'_{s,2}. \tag{6.71}$$

From this relationship, developed taking into account Eqs. (6.66) to (6.68), we obtain the fundamental Eq. (6.58), previously deduced by another procedure.

We now analyze, for example, what happens for $(x_1 + x_2) > 0$, with x_1 and x_2 both greater than zero, when we place the two gears, cut with these profile shift coefficients, so that their center distance, a, is equal to the sum of the radii r_1 and r_2 given by (see also Henriot [13]):

$$r_1 = r_{10} + x_1 m_0 = m_0 \left(\frac{z_1}{2} + x_1 \right) \tag{6.72}$$

$$r_2 = r_{20} + x_2 m_0 = m_0 \left(\frac{z_2}{2} + x_2 \right). \tag{6.73}$$

In this example case, we obtain:

$$a = r_1 + r_2 = m_0 \frac{(z_1 + z_2)}{2} + m_0(x_1 + x_2) = a_0 + m_0(x_1 + x_2). \tag{6.74}$$

Figure 6.16 shows the two gear wheels, generated by the same rack type cutter with profile shift coefficients $x_1 > 0$ and $x_2 > 0$, and placed at center distance $O_1 O_2 = a = (r_1 + r_2)$. The figure highlights that, with this center distance, a, there is no contact between the tooth profiles of the pinion and wheel. Therefore, it is necessary to approach the two members of the gear pair up to the operating center distance, a' (hereafter, the center distance), which is the theoretical one for which the backlash becomes equal to zero. From the aforementioned considerations, we infer the following important property: *the variation of the center distance, $\Delta a' = (a' - a_0)$ i.e. the difference between the center distance, a', and the reference center distance, a_0, is always less than the sum of the profile shifts, $(x_1 + x_2)m_0$, that characterize pinion and wheel.*

Therefore, we have:

$$\Delta a' = (a' - a_0) < (x_1 + x_2)m_0. \tag{6.75}$$

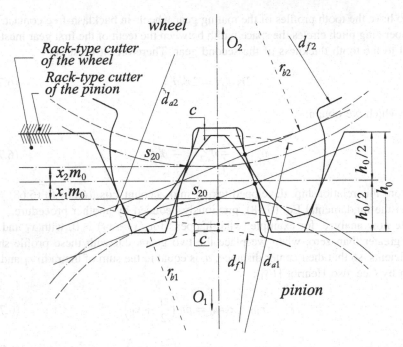

Fig. 6.16 External gear with profile shift coefficients $x_1 > 0$ and $x_2 > 0$, and gear wheels placed at center distance $a = r_1 + r_2$

If we denote by Δa the difference $(a - a_0)$, from Eq. (6.74) we obtain:

$$\Delta a = (a - a_0) = (x_1 + x_2)m_0. \tag{6.76}$$

Obtaining m_0 from Eq. (6.11) and substituting in Eq. (6.76), we get:

$$\Delta a = (a - a_0) = a_0 \frac{2(x_1 + x_2)}{(z_1 + z_2)} = a_0 X \tag{6.77}$$

where

$$X = \frac{2(x_1 + x_2)}{(z_1 + z_2)}. \tag{6.78}$$

Therefore, from Eq. (6.77), we obtain:

$$a = a_0(1 + X). \tag{6.79}$$

Taking into account the Eqs. (6.59) and (6.11), the variation of center distance $\Delta a'$ can be written in the following form:

$$\Delta a' = (a' - a_0) = a_0 \left(\frac{\cos \alpha_0}{\cos \alpha'} - 1 \right) = a_0 Y \tag{6.80}$$

where

$$Y = \left(\frac{\cos \alpha_0}{\cos \alpha'} - 1 \right). \tag{6.81}$$

Therefore, from Eq. (6.80), we obtain:

$$a' = a_0 (1 + Y). \tag{6.82}$$

Finally, subtracting Eq. (6.82) from Eq. (6.79), we get:

$$a - a' = a_0 (X - Y) = K m_0, \tag{6.83}$$

where

$$K = \frac{z_1 + z_2}{2} (X - Y) = \left[(x_1 + x_2) - \frac{z_1 + z_2}{2} \left(\frac{\cos \alpha_0}{\cos \alpha'} - 1 \right) \right]. \tag{6.84}$$

Factor K is the *truncation coefficient*, i.e. the quotient of the truncation divided by the module m_0, while $K m_0$ is the *truncation*, i.e. the reduction of the addendum, considering the addendum defined by the standard basic rack tooth profile. Truncation is also known as *topping*, and represents the addendum reduction of the two members of the gear, which is necessary if we want to have the standard clearance, c, equal to 0.25 m. Therefore, the tooth depth will be given by:

$$h = m_0 (2.25 - K), \tag{6.85}$$

while the tip circle diameters of the two wheels are respectively

$$d_{a1} = m_0 [z_1 + 2(1 + x_1 - K)] \tag{6.86}$$

$$d_{a2} = m_0 [z_2 + 2(1 + x_2 - K)], \tag{6.87}$$

where x_1 and x_2 are to be considered with their algebraic sign.

The root circle diameters of the two wheels are given by:

$$d_{f1} = m_0 \left[z_1 - 2 \left(\frac{5}{4} - x_1 \right) \right] \tag{6.88}$$

$$d_{f2} = m_0 \left[z_2 - 2 \left(\frac{5}{4} - x_2 \right) \right]. \tag{6.89}$$

Suppose, for a moment, that the truncation is not made. In this case, the clearance, c, i.e. the distance, along the line of centers, between the tip surface of the first gear and the root surface of its mating gear, can be calculated by subtracting from the operating center distance, a', given by Eq. (6.59), the tip radius of the first of the two wheels, r_{a1}, given by:

$$r_{a1} = \frac{m_0}{2}[z_1 + 2(1 + x_1)], \tag{6.90}$$

and the root radius of the other wheel, r_{f2}, given by:

$$r_{f2} = \frac{m_0}{2}\left[z_2 - 2\left(\frac{5}{4} - x_2\right)\right]. \tag{6.91}$$

Thus, the clearance, c, will be given by:

$$c = a' - r_{a1} - r_{f2} = \frac{m_0}{4} - m_0\left[(x_1 + x_2) - \frac{z_1 + z_2}{2}\left(\frac{\cos \alpha_0}{\cos \alpha'} - 1\right)\right]$$
$$= \frac{m_0}{4} - [m_0(x_1 + x_2) - (a' - a_0)] = \frac{m_0}{4} - Km_0. \tag{6.92}$$

Therefore, if $Km_0 > 0$, the clearance, c, would be less than the standard value, equal to $(m_0/4)$. In this case, the clearance is reset to the standard value with the above-described truncation. However, when the designer feels abundant this clearance standard value, at his discretion, the amount of topping can be reduced. This reduction is allowed, provided that other crest-width limitations do not condition the choice of the clearance value to be adopted. In any case, the chosen clearance value is to be checked.

The truncation Km_0 can be achieved by another method, which allow us to define other quantities of the profile-shifted gears with variation of center distance (see Giovannozzi [12]). As Fig. 6.16 shows, in the cutting conditions, the dedendum of the two members of the gear pair are respectively given by (note that the third subscript, 0, refers as usually to cutting conditions):

$$h_{f1,0} = \left(\frac{5}{4} - x_1\right)m_0 \tag{6.93}$$

$$h_{f2,0} = \left(\frac{5}{4} - x_2\right)m_0. \tag{6.94}$$

The radial distance $y_1 = \eta_1 m_0 = (r'_1 - r_{10})$, and $y_2 = \eta_2 m_0 = (r'_2 - r_{20})$ between the pitch circles and cutting pitch circles of the two gears (Fig. 6.15) are respectively given by:

$$y_1 = \eta_1 m_0 = (r_1' - r_{10}) = \frac{z_1}{2}(m' - m_0) = \frac{z_1}{2} m_0 \left(\frac{\cos \alpha_0}{\cos \alpha'} - 1\right) \qquad (6.95)$$

$$y_2 = \eta_2 m_0 = (r_2' - r_{20}) = \frac{z_2}{2}(m' - m_0) = \frac{z_2}{2} m_0 \left(\frac{\cos \alpha_0}{\cos \alpha'} - 1\right). \qquad (6.96)$$

When the pitch circles are in contact, the radial distance, y, between the cutting pitch circles is equal to:

$$y = y_1 + y_2 = (\eta_1 + \eta_2)m_0 = \frac{z_1 + z_2}{2} m_0 \left(\frac{\cos \alpha_0}{\cos \alpha'} - 1\right) = \xi m_0. \qquad (6.97)$$

The radial distances h_{f1} and h_{f2} between the pitch circles and root circles of the two members of the gear pair are:

$$h_{f1} = \left(\frac{5}{4} - x_1 + \eta_1\right) m_0 \qquad (6.98)$$

$$h_{f2} = \left(\frac{5}{4} - x_2 + \eta_2\right) m_0. \qquad (6.99)$$

Therefore, if between the tip circle of one member and the root circle of the other member of the gear pair we want to have a clearance $c = (m_0/4)$, i.e. a clearance value equal to the nominal one, the radial protrusions of the teeth beyond the pitch circles having radii r_1' and r_2' must be equal to:

$$h_{a2} = (1 - x_1 + \eta_1)m_0 \qquad (6.100)$$

$$h_{a1} = (1 - x_2 + \eta_2)m_0. \qquad (6.101)$$

Instead, according to the generation method, these radial protrusions are equal to:

$$h_{a2,0} = (1 + x_2 - \eta_2)m_0 \qquad (6.102)$$

$$h_{a1,0} = (1 + x_1 - \eta_1)m_0. \qquad (6.103)$$

Therefore, it is necessary to cut the teeth in gear blanks having reduced diameters, so that they have less depth, compared with the nominal ones, by an amount given by:

$$h_{a2,0} - h_{a2} = h_{a1,0} - h_{a1} = [(x_1 + x_2) - (\eta_1 + \eta_2)]m_0 = [(x_1 + x_2) - \xi]m_0$$

$$= \left[(x_1 + x_2) - \frac{z_1 + z_2}{2}\left(\frac{\cos \alpha_0}{\cos \alpha'} - 1\right)\right]m_0 = K m_0, \qquad (6.104)$$

as we showed before in another way.

The transverse contact ratio is calculated with Eq. (3.44), taking into account that, due to the profile shifts, x_1 and x_2, the addendum factors, k_1 and k_2, are respectively given by:

$$k_1 = (1 - x_2 + \eta_2) \frac{\cos \alpha'}{\cos \alpha_0} \tag{6.105}$$

$$k_2 = (1 - x_1 + \eta_1) \frac{\cos \alpha'}{\cos \alpha_0}. \tag{6.106}$$

Equation (6.58) correlates between them six quantities, i.e. x_1, x_2, z_1, z_2, α', and α_0. Generally, three of these quantities are known, and precisely: the pressure angle, α_0, of the rack-type cutter, and the number of teeth, z_1 and z_2. To determine the remaining three unknown quantities, x_1, x_2, and α', it is necessary to write other two equations, independent of each other and independent from Eq. (6.58). In other words, the designer must impose two conditions, translatable in as many equations that correlate the above-mentioned unknown quantities, with which he will try to find the optimal sizing, from the point of view of robustness of the teeth, the length of the path of contact, or by other points of view, described by us in the following sections.

Although we describe and utilize a method of different solution, from the historical point of view, we cannot fail to mention the two conditions set for the first time by Schiebel (see Schiebel and Lendner [26, 27]), summarized as follows:

- Contact between the mating flank surfaces, extended along the entire involute part of the tooth profile. In this regard, it is to be remembered that the generation cutting process by means of a rack-type cutter generally causes a certain interference, and thus determines the removal of the involute part near the base circle.
- Equality of the thicknesses of the teeth of the two members of the gear pair at the fillet circles, which pass through the points where the involutes end, and the fillets begin.

With these two conditions, which can take relatively simple analytical form based on calculations carried out before, together with Eq. (6.58), Schiebel formulated an algebraic system of three independent equations in the three unknowns x_1, x_2, and α', for the case $\alpha_0 = 15°$. Schiebel calculated and represented in diagrams the corresponding values of x_1 and x_2, for number of teeth of the pinion from 6 to 160, and for number of teeth of the wheel from 8 to 160. Using these diagrams, the determination of all the variables of interest here is very simple and fast.

We have thus obtained all the relationships that are used to carry out the external gear design with profile-shifted toothing and variation of center distance. It has to tackle and solve the problem of determining the sum of the profile shift coefficients and its distribution between the pinion and wheel. These subjects are covered in the next Sects. 6.7 and 6.8.

However, we must note that external gear pairs having profile-shifted toothing with variation of center distance are mainly characterized by positive profile shift coefficients (i.e. $x_1 > 0$ and $x_2 > 0$). In principle, with these types of gears, we use the most favourable properties of the profile shift for both members of the gear, the pinion and wheel. The limitations of the profile-shifted toothing without variation of center distance are eliminated and, for a given transmission ratio, the predetermined center distance can be maintained using standard modules, but choosing the corresponding sum $(x_1 + x_2)$ of profile shift coefficients. It is also possible to obtain gear wheels having characteristics of catalogue wheels. The calculations are however more complex. Compared to an external gear with zero-profile-shifted toothing, if $(x_1 + x_2) > 0$, the contact ratio and radial component of the load are higher. On the contrary, if $(x_1 + x_2) < 0$, the contact ratio and radial component of the load are lower.

6.6.2 Internal Spur Gear Pairs

In Sect. 5.3 we have shown that, to avoid the risk of secondary interference in an internal spur gear pair, the difference $(z_2 - z_1)$ between the numbers of teeth of the annulus and pinion must not fall below a minimum difference value $(z_2 - z_1)_{min}$, which depends on the addendum factor, k, and the operating pressure angle, α'. However, in many practical applications, it can be useful to fall below of this limit difference, and in this case, it is necessary to resort to profile-shifted toothing.

For an internal gear pair with profile shift coefficients $x_1 > 0$ and $x_2 > 0$, the tooth thicknesses, s_{10} and s_{20}, on the two cutting pitch circles, having radii $r_{10} = (z_1/2)m_0$ and $r_{20} = (z_2/2)m_0$, are given respectively by Eqs. (6.14) and (6.17). Considering the two circles with radii r_1 and r_2 given by Eqs. (6.72) and (6.73), and locating their axes at a center distance, a, equal to the difference $(r_2 - r_1)$, we have:

$$a = r_2 - r_1 = m_0 \frac{(z_2 - z_1)}{2} + m_0(x_2 - x_1) = a_0 + m_0(x_2 - x_1). \qquad (6.107)$$

Figure 6.17, which shows the two gear wheels placed at center distance $O_1O_2 = a = (r_2 - r_1)$, highlights that, with this center distance, a, the teeth of the pinion and wheel intersect. Therefore, it is necessary to reduce this center distance. *Mutatis mutandis*, the property already set out for external gears here becomes: *the variation of center distance $\Delta a' = (a' - a_0)$, i.e. the difference between the center distance and reference center distance, is always less than the difference of the profile shifts $(x_2 - x_1)m_0$.* Thus, we have:

$$\Delta a' = (a' - a_0) < (x_2 - x_1)m_0. \qquad (6.108)$$

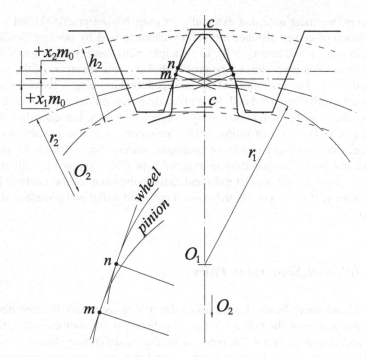

Fig. 6.17 Internal gear with profile shift coefficients $x_1 > 0$ and $x_2 > 0$, and gear wheels placed at center distance $a = (r_2 - r_1)$

Always changing what needs to be changed, by applying to the internal gear the same method that allowed us to infer the Eq. (6.58), we find that the latter is transformed into the following equation:

$$\operatorname{inv} \alpha' = \operatorname{inv} \alpha_0 + 2 \tan \alpha_0 \frac{x_2 - x_1}{z_2 - z_1}. \tag{6.109}$$

From Eq. (6.109), we infer that:

- if $(x_2 - x_1) > 0$, we have: $(s_1' + s_2') < \pi m_0$; $a' > a_0$; $\operatorname{inv} \alpha' > \operatorname{inv} \alpha_0$; $\alpha' > \alpha_0$;
- if $(x_2 - x_1) < 0$, we have: $(s_1' + s_2') > \pi m_0$; $a' < a_0$; $\operatorname{inv} \alpha' < \operatorname{inv} \alpha_0$; $\alpha' < \alpha_0$;
- if $(x_2 - x_1) = 0$, we have: $(s_1' + s_2') = \pi m_0$; $a' = a_0$; $\operatorname{inv} \alpha' = \operatorname{inv} \alpha_0$; $\alpha' = \alpha_0$.

From Eq. (6.107), taking into account the Eq. (6.19), we get:

$$\Delta a = (a - a_0) = (x_2 - x_1)m_0 = 2a_0 \frac{(x_2 - x_1)}{(z_2 - z_1)} = a_0 X \tag{6.110}$$

where

$$X = 2\frac{(x_2 - x_1)}{(z_2 - z_1)}. \tag{6.111}$$

and then

$$a = a_0(1 + X). \tag{6.112}$$

The center distance, a', is given by:

$$a' = (r_2' - r_1') = \frac{z_2 - z_1}{2}m' = \frac{z_2 - z_1}{2}m_0\frac{\cos \alpha_0}{\cos \alpha'} = a_0\frac{\cos \alpha_0}{\cos \alpha'}. \tag{6.113}$$

From this equation, taking account of Eq. (6.19), we get:

$$\Delta a' = (a' - a_0) = a_0\left(\frac{\cos \alpha_0}{\cos \alpha'} - 1\right) = a_0 Y \tag{6.114}$$

where Y is given by Eq. (6.81). From the latter equation, we obtain yet the Eq. (6.82).

Comparing the Eq. (6.112) with Eq. (6.82), it is evident that $a' < a$. Therefore, the truncation is not necessary, i.e. it is not necessary to decrease the addendum of the two members of the internal gear pair, as in the case of an external gear pair. The tooth depth does not change and, moreover, the clearance is slightly increased, and this is certainly an advantage.

6.7 The Sum (or Difference) of the Profile Shift Coefficients: Direct Problem and Inverse Problem

In Sect. 6.6 we saw that Eqs. (6.58) and (6.109), respectively valid for external and internal gear pairs, correlate three unknown quantities, α', x_1, and x_2, and that, to determine these quantities, it is necessary to write, in addition to these equations, two other equations forming, together with Eq. (6.58) or Eq. (6.109), a system of three independent algebraic equations, all in the same unknowns. However, whatever the design criterion, we must in any case use the Eq. (6.58) or Eq. (6.109). Conversely, the other two equations can be written as a function of the goals to be achieved, and the resulting design criteria related to them. In Sect. 6.6.1 we mentioned the two Schiebel's criteria which, of course, are not the only possible. Other researchers have proposed various criteria other than Schiebel's, aimed at determining the optimal shape of the teeth according to the goals to be achieved. Even the manufacturers of generating gear cutting machines have proposed their criteria.

Therefore, the gear designer has available various formulations, which allow him to determine the profile shift coefficients, x_1 and x_2, with an accuracy more or less high, but still sufficient for this type of calculation. Some of these formulations have also been translated into schedules, tables, and diagrams, which have been developed based on different criteria and points of view. In the following, we will describe some of these formulations. For the moment, however, temporarily we put aside this very complex subject, and we focus our attention on so-called direct and inverse problems. However, it is first necessary to point out that, once the profile shift coefficients have been determined, we use Eq. (6.58) or Eq. (6.109) to determine the last unknown, i.e. the operating pitch angle, α'.

In the *direct problem*, the profile shift coefficients, x_1, and x_2, are assumed known, as determined preliminarily by one of the optimization criteria that we have mentioned above. The geometry of the rack-type cutter, defined by the module, m_0, and pressure angle, α_0, is also supposed known. The nominal center distance, a_0, of the gear, given by Eq. (6.11) or Eq. (6.19) depending on whether the gear is external or internal, is therefore uniquely defined, but we need to determine the operating center distance, a'.

For an external gear pair, sequentially the following quantities are then calculated: X, by means of Eq. (6.78); α', by means of Eq. (6.58); Y, by means of Eq. (6.81); a', by means of Eq. (6.59); K, by means of Eq. (6.84); h, by means of Eq. (6.85); d_{a1} and d_{a2}, by means of Eqs. (6.86) and (6.87).

Instead, for an internal gear pair, sequentially the following quantities are calculated: X, by means of Eq. (6.111); α', by means of Eq. (6.109); Y, by means of Eq. (6.81); a', by means of Eq. (6.113); K, by means of Eq. (6.84), in which we replace $(x_1 + x_2)$ and $(z_1 + z_2)$ with $(x_2 - x_1)$ and $(z_2 - z_1)$ respectively; h, by means of Eq. (6.85); d_{a1}, by means of Eq. (6.86); $d_{a2} = m_0[z_0 - 2(1 - x_2 - K)]$, which is obtained from Eq. (6.87) written for an internal gear.

The following examples clarify how to use the above-described procedure of the sequential calculation.

Example 1 External gear pair. Input data: $z_1 = 10$; $z_2 = 15$; $\alpha_0 = 15°$; $m_0 = 10$ mm. Profile shift coefficients, $x_1 = +(1/2)$ and $x_2 = +(1/3)$, also known as previously chosen. We will have: $a_0 = 125$ mm; $X = 0.067$; $\alpha' = 23°20'$; $Y = 0.052$; $a' = 131.46$ mm; $K = 0.187$; $h = 20.63$ mm; $d_{a1} = 126.26$ mm; $d_{a2} = 172.93$ mm.

Example 2 Internal gear pair. Input data: $z_1 = 10$; $z_2 = 5\,0$; $\alpha_0 = 20°$; $m_0 = 5$ mm. Profile shift coefficients, $x_1 = +(1/2)$ and $x_2 = 0$ (i.e. wheel with zero-profile-shifted toothing), also known as previously chosen. We will have: $a_0 = 100$ mm; $X = -0.025$; $\alpha' = 14°49'$; $Y = -0.028$; $a' = 97.20$ mm; $K = 0.06$ ($Km_0 = 0.3$ mm, for which the reduction of the tooth depth can be neglected); $h = 11.25$ mm; $d_{a1} = 65$ mm; $d_{a2} = 240$ mm.

Instead, in the *inverse problem*, we start from a value imposed of the center distance, a', and a range of standard rack-type cutters, and we must first to determine the sum $(x_1 + x_2)$ or the difference $(x_2 - x_1)$ of the profile shift coefficients, depending on whether the gear is external or internal. Subsequently, we must determine the distribution of this sum or difference between pinion and wheel. The initial data are therefore the numbers of teeth, z_1 and z_2, the center distance, a', and the range of standard rack-type cutters with which the two members of the gear pair must be generated. Each of these cutting tools is defined by the module, m_0, and the pressure angle, α_0. The center distance, a', to be met, which is different from the nominal center distance, a_0, given by Eq. (6.11) or Eq. (6.19), is that given by Eq. (6.59) or Eq. (6.113), where m' and α' are the operating module and working pressure angle.

The first thing to do is to choose the rack-type cutter from those of the standard range. For this purpose, we choose initially the cutting tool with module, m_0, immediately below one of the following two modules, respectively calculated using Eq. (6.59) or Eq. (6.113), depending on whether the gear pair under consideration is an external gear or an internal gear:

$$m' = 2a'/(z_2 + z_1) \tag{6.115}$$

$$m' = 2a'/(z_2 - z_1). \tag{6.116}$$

Once this choice has been made, for which m_0 and α_0 are defined, we will calculate the nominal center distance, a_0, by means of Eq. (6.11) or Eq. (6.19), and we will verify if it meets the operating center distance, a', imposed as input data, that is $a' = a_0$. Generally, however, we will find $a' \neq a_0$ for which, depending on whether $a' \gtrless a_0$, we will have to make a profile-shifted toothing with increase or decrease in the center distance.

For an external gear pair, sequentially we will calculate the following quantities: α', by means of Eq. (6.59) or Eq. (6.113), both written in the form

$$\cos \alpha' = \left(\frac{a_0}{a'}\right) \cos \alpha_0 \tag{6.117}$$

from which we have

$$\alpha' = \arccos\left[\left(\frac{a_0}{a'}\right) \cos \alpha_0\right], \tag{6.118}$$

where a_0 is given by Eq. (6.11) or Eq. (6.19), depending on whether the gear is external or internal; Y, by means of Eq. (6.81); X, by means of Eq. (6.78), and therefore the sum of the profile shift coefficients, $(x_1 + x_2) = X(z_1 + z_2)/2$; K, by means of Eq. (6.84); h, by means of Eq. (6.85); d_{a1}, and d_{a2} by means of Eqs. (6.86) and (6.87).

Instead, for an internal gear pair, sequentially we will calculate the following quantities: α', by means of Eqs. (6.117) and (6.118), with $a_0 = m_0(z_2 - z_1)/2$; Y, by means of Eq. (6.58); X, by means of Eq. (6.83), and therefore the difference of the profile shift coefficients, $(x_2 - x_1) = X(z_2 - z_1)/2$; K, by means of Eq. (6.84), in which we replace $(x_1 + x_2)$ and $(z_1 + z_2)$ with $(x_2 - x_1)$ and $(z_2 - z_1)$ respectively; d_{a1}, by means of Eq. (6.86); $d_{a2} = m_0[z_0 - 2(1 - x_2 - K)]$, which is obtained from Eq. (6.87) written for an internal gear.

Example 3 External gear pair, and increase of the center distance. Input data: $z_1 = 20$; $z_2 = 40$; $a' = 303$ mm; standard range of rack-type cutters. We will have: $m' = 2a'/(z_1 + z_2) = 10.10$ mm; $m_0 = 10$ mm; $\alpha_0 = 20°$ (this is the initial choice of design); $a_0 = m_0(z_1 + z_2)/2 = 300$ mm. Since $a' > a_0$, we have to make a profile shift with increase of the center distance. Continuing the calculation, we have: $\alpha' = 21°30'$; $Y = 0.0100$; $X = 0.0106$; $(x_1 + x_2) = 0.32$; $K = 0.018$ (in this case, the corresponding decrease of the tooth depth, $Km_0 = 0.18$ mm, can be neglected); $h = 22.5$ mm; $d_{a1} = 10[20 + 2(1 + x_1)]$ and $d_{a2} = [40 + 2(1 + x_2)]$ to be calculated after splitting the sum $(x_1 + x_2)$ between the pinion and wheel.

Example 4 External gear pair, and decrease of the center distance. Input data: $z_1 = 20$; $z_2 = 40$; $a' = 298$ mm; standard range of rack-type cutters. We will have: $m' = 2a'/(z_1 + z_2) = 9.93$ mm; $m_0 = 10$ mm; $\alpha_0 = 20°$ (for this initial design choice, since m' has a value close to 10 mm, it is not appropriate to choose the lower standard module $m_0 = 9$ mm, because we would obtain either a pressure angle very large, or a profile shift too high); $a_0 = m_0(z_1 + z_2)/2 = 300$ mm. Since $a' < a_0$, we have to make a profile shift with decrease of the center distance. Continuing the calculation, we have: $\alpha' = 18°55'$; $Y = -0.0067$; $X = -0.0065$; $(x_1 + x_2) = -0.195$; $K = 0.006$. From now on, the observations and calculations to be made are those of the previous examples.

Example 5 Internal gear pair. Input data: $z_1 = 10$; $z_2 = 50$; $\alpha_0 = 20°$; $m_0 = 5$ mm. Profile shift coefficients $x_2 = 0$, i.e. wheel with zero-profile-shifted toothing, and $x_1 = 0.50$ (thus, $(x_2 - x_1) = -0.50$), also known as previously chosen. We will have: $a_0 = 100$ mm; $X = -0.025$; $\alpha' = 14°49'$; $Y = -0.028$; $a' = 97.20$ mm; $K = 0.06$ (in this case, the corresponding decrease of the tooth depth, $Km_0 = 0.3$ mm, is neglected, to simplify the exercise; in fact, however, a careful analysis should be carried out on the effects of this simplifying assumption on clearance); $h = 11.25$ mm; $d_{a1} = 65$ mm; $d_{a2} = 240$ mm. From now on, the observations and calculations to be made are those of the previous examples.

Example 6 Internal gear pair, and increase of the center distance. Input data: $z_1 = 10$; $z_2 = 50$; $a' = 102$ mm; standard range of pinion-type cutters. We will have: $m' = 2a'/(z_2 - z_1) = 5.10$ mm; $m_0 = 5$ mm; $\alpha_0 = 20°$ (this is the initial choice of design); $a_0 = m_0(z_2 - z_1)/2 = 100$ mm. Since $a' > a_0$, we have to make a profile shift with increase of the center distance. Continuing the calculation, we have: $\alpha' = 22°53'$; $Y = 0.020$; $X = 0.021$; $(x_2 - x_1) = 0.42$; $K = 0.02$. From now on, the observations and calculations to be made are those of the previous examples.

However, regarding the cut of the annulus of this example, we must make the following important consideration. We assume that the difference $(x_2 - x_1) = 0.42$ has been obtained by the following design choice: $x_2 = 0.82$ and $x_1 = 0.40$. To manufacture this annulus in the design condition $(x_2 = 0.82)$, we suppose to use a pinion-type cutter with $z_0 = 20$ teeth and standard sizing $(x_0 = 0)$. Therefore, we have: $a_0 = 75$ mm; $(x_2 - x_0) = 0.82$; $X = 0.055$; $Y = 0.048$; $a'_0 = 78.56$ mm (actual center distance at the end of the cutting process); $r_{a0} = 56.25$ mm (radius of the tip circle of the pinion-type cutter, whose addendum is $h_{a0} = 1.25m_0 = 6.25$ mm); $r_{f2} = a' + r_{a0} = 134.81$ mm (actual radius of the root circle of the annulus); $r_{a1} = [(z_0 + 2x_1)m_0/2] = 32$ mm (radius of the tip circle of the pinion that meshes with the annulus); $c = r_{f2} - (a' + r_{a1}) = 0.81$ mm (clearance between the root circle of the annulus and tip circle of the pinion).

Therefore, we infer that the actual clearance is decreased compared to the standard clearance; in fact, it passes from $c = 1.25$ mm to $c = 0.81$ mm. This decrease would have been even greater if the number of teeth of the pinion-type cutter used had been smaller. This result is entirely consistent with what we have said previously, namely that, in an internal gear pair, the clearance tends to increase if we carry out a profile shift with variation of center distance. Therefore, if we consider the pinion-type cutter instead of the mating pinion, we find that the cutting tool does not cut deep enough. In certain cases, the truncation of the addendum of the pinion teeth is therefore desirable.

6.8   Determination of the Profile Shift Coefficients of External Gear Pairs

It is time to address the problem of determining the profile shift coefficients, x_1 and x_2, or the problem of the distribution between pinion and wheel of their sum $(x_1 + x_2)$ or difference $(x_2 - x_1)$, depending on whether the gear is external or internal. We have already anticipated that this problem, very complex, can be approached from different points of view, to each of which a well-defined criterion corresponds. Each of these criteria has advantages and disadvantages, and none of them claims to be the best. The results that are obtained using any of these criteria are therefore optimal from the point of view at the base of the criterion chosen, but are not as optimal from the point of view of another criterion. The designer, aware of the pros and cons associated with each criterion, must be able to extricate himself from them, and from time to time choose the one best suited to the goals to be achieved.

In Sect. 6.6.1, we have made a brief reference to the two conditions proposed by Schiebel, which together make explicit the criterion that bears his name. Instead, in this section, we describe in detail other criteria, some of which are now frequently used in the gear design.

6.8.1   Criterion to Avoid the Cutting Interference

This criterion uses the Eq. (6.24) as fundamental relationship. In this equation (see Sect. 6.4), z_0 is the number of teeth that can be cut without interference, using a rack-type cutter having a pressure angle, α_0, module, m_0, and addendum factor, $k_0 = 1$, and cutting a zero-profile-shifted toothing. If we want to realize a gear having a number $z < z_0$ without cutting interference, we have to cut the gear with a positive profile shift, xm_0, and thus with a profile shift coefficient given by Eq. (6.24).

When we use this criterion, we must distinguish the following two cases:

$$(z_1 + z_2) \geq 2z_0 \tag{6.119}$$

$$(z_1 + z_2) < 2z_0. \tag{6.120}$$

In the first case we choose, for the pinion, a positive profile shift coefficient, x_1, determined by means of Eq. (6.24), and, for the wheel, a negative profile shift coefficient, $x_2 = -x_1$, that is equal in absolute value, but of opposite sign with respect to that of the pinion. By means of equations set out in Sect. 6.4, we can easily show that, if $(z_1 + z_2) > 2z_0$, the pinion is in the limit condition of cutting interference (instead, the wheel is not yet in the limit condition), while, if $(z_1 + z_2) = 2z_0$, both pinion and wheel are in the limit condition of cutting interference. In any case, since $x_2 = -x_1$, we have a profile-shifted toothing without variation of center distance.

In the second case we choose, for the pinion, a positive profile shift coefficient, x_1, determined by means of Eq. (6.24), and, for the wheel, a positive or negative profile shift coefficient, x_2, calculated using the same Eq. (6.24). In algebraic value, however, we will always have $x_2 > -x_1$, and thus $(x_1 + x_2) > 0$. Therefore, we will have profile-shifted toothing with increase of center distance.

The following three examples are related to the use of a rack-type cutter with pressure angle $\alpha_0 = 20°$. In this case, $z_0 = 17$, and thus $2z_0 = 34$.

Example 1 Input data, $z_1 = 12$ and $z_2 = 28$. Therefore, we have: $(z_1 + z_2) = 40 > 34$; $x_1 = 5/17$; $x_2 = -x_1 = -5/17$. In this case, only the pinion is in the limit condition of cutting interference.

Example 2 Input data, $z_1 = 10$ and $z_2 = 24$. Therefore, we have: $(z_1 + z_2) = 34$; $x_1 = 7/17$; $x_2 = -x_1 = -7/17$. In this case, both the pinion and wheel are in the limit condition of cutting interference.

Example 3 Input data, $z_1 = 10$ and $z_2 = 22$. Therefore, we have: $(z_1 + z_2) = 32 < 34$; $x_1 = 7/17$; $x_2 = -5/17$; $(x_1 + x_2) = 2/17 > 0$.

It is to be noted that this criterion was the basis of the old method of the German standard DIN, which prescribed choosing the profile shift coefficients with values such as to avoid the cutting interference of the pinion, in the case where $(z_1 + z_2) > 2z_0$, or of the pinion and wheel, in the case where $(z_1 + z_2) \leq 2z_0$ (see Giovannozzi [12], Niemann and Winter [22]). It is evident that, with the values of the profile shift coefficients thus calculated, we would come to have the maximum possible length of the path of contact (it would coincide with the maximum theoretical value, equal to the line of action, i.e. the length of the segment $T_1 T_2 = a \sin \alpha$, in the case where it was $(z_1 + z_2) = 2z_0$).

It is equally evident the corresponding disadvantage of having a contact between the mating profiles that would extend until the origin of involute curve, where the radius of curvature is zero, and the contact takes place in very unfavourable conditions, as we already said. For this reason, for the two cases of $\alpha_0 = 15°$ and $\alpha_0 = 20°$, the same standard DIN subsequently prescribed to calculate the profile shift coefficients using the practical relationships, rather than the theoretical ones. All these relationships are collected in Table 6.3.

6.8.2   Criterion to Equalize the Maximum Values of the Specific Sliding and Almen Factors of the Pinion and Wheel

The use of the aforementioned old method DIN, in both formulations, theoretical and practical, is no longer recommended because the gear designer has to control not only the interference and undercut problems, which are undoubtedly very important, but also many other problems certainly no less important, among which those associated with the specific sliding and surface pressure occupy an extremely significant role. In fact, the gear strength to the surface fatigue, wear and scuffing depends on them.

Henriot [13] noted that, with the values of the profile shift coefficients calculated with the old method DIN, distribution curves of the surface pressure, σ_H, specific slidings, ζ_1 and ζ_2, sliding velocities, v_{g1} and v_{g2}, and Almen factors, $\sigma_H v_{g1}$ and $\sigma_H v_{g2}$, very unbalanced along the path of contact were obtained. Starting from this observation, he proposed a criterion and related method much more articulated, but easy to apply, which not only allows avoiding the interference, but also allows equalizing the maximum values of the specific sliding and Almen factors of the pinion and wheel [1].

The balancing condition, i.e. the condition to balance the maximum values of the specific sliding on the pinion and wheel leads to write the equality:

$$\zeta_{1A} = \zeta_{2E}. \tag{6.121}$$

where ζ_{1A} and ζ_{2E} are respectively the maximum negative values of the specific sliding of the pinion and wheel, respectively at marked points A and E (see Fig. 2.10). The marked points A and E are the end points of the path of contact, where the contact begins and ends respectively when the pinion is driving. In these points, the radii of curvature of the mating profiles are: $\rho_{1A} = T_1 A$; $\rho_{2A} = T_2 A$; $\rho_{1E} = T_1 E$; $\rho_{2E} = T_2 E$. Taking into account Eqs. (3.73), (3.59), and (3.60), the equality (6.121) becomes:

$$\frac{\omega_1\rho_{1A} - \omega_2\rho_{2A}}{\omega_1\rho_{1A}} = \frac{\omega_2\rho_{2E} - \omega_1\rho_{1E}}{\omega_2\rho_{2E}}. \tag{6.122}$$

Developing this last equation, we get:

$$\omega_1^2\rho_{1A}\rho_{1E} = \omega_2^2\rho_{2A}\rho_{2E}. \tag{6.123}$$

As we shall see in due course, when we talk about the scuffing load carrying capacity of the gears, we define, as Almen factor, the product of the contact stress, σ_H (also called surface pressure or Hertzian pressure) and sliding velocity, v_g (see Vol. 2, Chap. 7). The balancing condition of the Almen factors of the pinion and wheel, calculated respectively at marked points A and E, leads to write the equality:

$$\sigma_{HA}v_{g1A} = \sigma_{HE}v_{g2E}, \tag{6.124}$$

where σ_{HA} and σ_{HE} are the contact stresses between the mating profiles at marked points A and E, while v_{g1A} and v_{g2E} are the sliding velocities of the pinion and wheel, given by Eqs. (3.65), and evaluated at the same marked points A and E. Note that the third subscript indicates these two marked points.

We will demonstrate at the time that the Hertzian pressure σ_H at a given point of contact, between the end points A and E of the path of contact, is proportional to the square root of the sum of the curvatures of the profiles at that point. At marked points A and E, the sum of these curvatures is given by:

$$\frac{1}{\rho_{1A}} + \frac{1}{\rho_{2A}} = \frac{\rho_{1A} + \rho_{2A}}{\rho_{1A}\rho_{2A}} = \frac{T_1T_2}{\rho_{1A}\rho_{2A}} = \frac{(r_1 + r_2)\sin\alpha}{\rho_{1A}\rho_{2A}} = \frac{a\sin\alpha}{\rho_{1A}\rho_{2A}} \tag{6.125}$$

$$\frac{1}{\rho_{1E}} + \frac{1}{\rho_{2E}} = \frac{\rho_{1E} + \rho_{2E}}{\rho_{1E}\rho_{2E}} = \frac{T_1T_2}{\rho_{1E}\rho_{2E}} = \frac{(r_1 + r_2)\sin\alpha}{\rho_{1E}\rho_{2E}} = \frac{a\sin\alpha}{\rho_{1E}\rho_{2E}}. \tag{6.126}$$

Specializing the Eqs. (6.65) at these marked points, and introducing them in the equality (6.124), together with the square root of Eqs. (6.125) and (6.126), we get:

$$(\omega_1\rho_{1A} - \omega_2\rho_{2A})\sqrt{\frac{a\sin\alpha}{\rho_{1A}\rho_{2A}}} = (\omega_2\rho_{2E} - \omega_1\rho_{1E})\sqrt{\frac{a\sin\alpha}{\rho_{1E}\rho_{2E}}}. \tag{6.127}$$

Developing then this last equation and simplifying, we get:

$$\frac{\omega_1}{\sqrt{\rho_{2A}\rho_{2E}}} = \frac{\omega_2}{\sqrt{\rho_{1A}\rho_{1E}}}. \tag{6.128}$$

Evidently, this relationship coincides with Eq. (6.123). Therefore, we see that the balancing of the specific slidings at marked points A and E also ensure the balancing of the Almen factors at the same points.

Starting from the relationships above [13] proposed a calculation method of the profile shift coefficients, x_1 and x_2, which ensures the equality of the maximum values of the specific slidings and Almen factors of the two members of the gear pair, which occur at marked points A and E. To make calculations easy and fast, he worked out two diagrams, one valid for $\alpha = 20°$, and the other valid for $\alpha = 15°$. Figure 6.18 shows the diagram related to $\alpha = 20°$ (for the other diagram, we refer to [13]).

The diagram is divided substantially into two regions: the first region, to the left, almost triangular, surrounded by the ordinate axis, and the two curves AB and $A'B$, and the second region, to the right, and external to the first. The inner part of the region almost triangular in shape refers to $(z_1 + z_2) < 60$. The second region, the outer to the first, refers to $(z_1 + z_2) > 60$. The two curves AB and $A'B$, the border

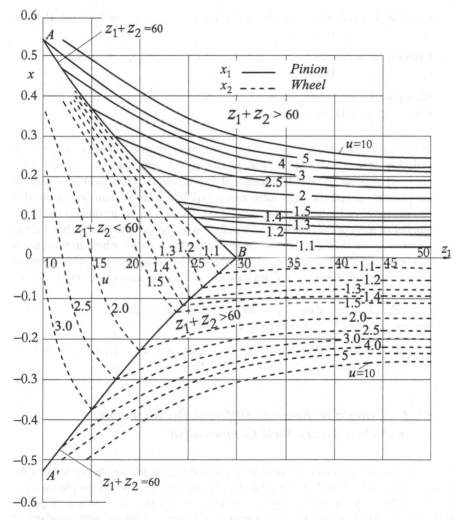

Fig. 6.18 Calculation diagram of x_1 and x_2, which ensure the equality of the maximum values of specific slidings and Almen factors for pinion and wheel, for $\alpha = 20°$ (from Henriot)

lines between the two regions, refer to $(z_1 + z_2) = 60$. Each of the curves appearing in the two regions of the diagram corresponds to a given value of the gear ratio $u = z_2/z_1$. The diagram is to be used in the following way:

(a) In the case for which $(z_1 + z_2) \geq 60$, we have profile-shifted toothing without variation of center distance, and the profile shift coefficients, x_1 and x_2, with $x_2 = -x_1$, are to be calculated as a function of the number of teeth of the pinion, z_1, and gear ratio $u = z_2/z_1$. We enter the diagram with the number of teeth z_1 of the pinion, shown on the abscissa. Therefore, with the vertical line having abscissa z_1, we intersect the curves $u = const$, which are symmetric with respect to the abscissa axis. The ordinates of the two points of intersection above are respectively x_1 and $x_2 = -x_1$.

Example 4 Input data: $z_1 = 20$; $z_2 = 60$; $u = 3$; $(z_1 + z_2) = 80 > 60$. Results: $x_1 = 0.31$; $x_2 = -0.31$.

Example 5 Input data: $z_1 = 20$; $z_2 = 40$; $u = 2$; $(z_1 + z_2) = 60$. Results: $x_1 = 0.23$; $x_2 = -0.23$.

Example 6 Input data: $z_1 = 17$; $z_2 = 90$; $u = 5.3$; $(z_1 + z_2) = 107 > 60$. Results: $x_1 = 0.41$; $x_2 = -0.41$.

(b) In the case for which $(z_1 + z_2) < 60$, we have profile-shifted toothing with variation of center distance. The profile shift coefficients, x_1 and x_2, with $(x_1 + x_2) \neq 0$, are to be still determined as a function of the number of teeth of the pinion, z_1, and the gear ratio $u = z_2/z_1$, but in a different way from that described above for the case $(z_1 + z_2) \geq 60$. We still enter the diagram with the number of teeth of the pinion, z_1. Therefore, with the vertical straight line having abscissa z_1, we intersect the two curves AB and $u = const$, which in this case is within the region almost triangular in shape. The ordinates of the two points of intersection with the curve AB and the curve $u = const$ are respectively x_1 and x_2.

Example 7 Input data: $z_1 = 20$; $z_2 = 24$; $u = 1.2$; $(z_1 + z_2) = 44 < 60$. Results: $x_1 = 0.23$; $x_2 = 0.17$; $(x_1 + x_2) = 0.46$.

Example 8 Input data: $z_1 = 20$; $z_2 = 30$; $u = 1.5$; $(z_1 + z_2) = 50 < 60$. Results: $x_1 = 0.23$; $x_2 = 0$; $(x_1 + x_2) = 0.23$.

6.8.3   Criteria to Balance Different Requirements, and Have Gears Well Compensated

One of the most important quantities in any gear design is the number of teeth of the pinion, and profile-shifted toothing has its greatest relevance when the pinion has a small number of teeth. In this regard, we have shown that the first use of profile shift was to avoid cutting interference and undercut in pinions with numbers of teeth less than those given by the inequality (6.20). It was later realized that profile-shifted toothing improves most aspects of the gear operating conditions, and today it is used for many reasons including mainly:

- avoiding cutting interference and undercut;
- avoiding narrow top lands and pointed teeth, and therefore case/core separation in case-hardened gears;
- balancing specific slidings to maximize wear strength and Hertzian fatigue strength;
- balancing flash temperature to maximize scuffing strength;
- balancing bending fatigue life to maximize bending fatigue strength;
- reducing the frictional losses to lower the contact temperature, and increase the strength to wear, macro-pitting, micro-pitting and scuffing;
- increasing or decreasing the center distance and changing tooth thickness and backlash; etc.

It is evident that not all these tasks/criteria be simultaneously fulfil, also because they are often not compatible with each other, resulting in conflicting demands. Depending on the goals to be achieved, the gear designer must make choices, favouring the balancing of the quantities that most influence the problem that he is facing, at the expense of other quantities that have more limited influence, but not nothing. In doing so, however, the problem of distribution of the sum of profile shift coefficients between the pinion and wheel would see greatly broaden his horizons, and gear design, in itself very demanding, would become even more complex.

To simplify the task of the designer, the various international and national standards have dealt with the problem, and have developed easy to apply methods, which lead to satisfactory results, although based on a reasonable compromise that ensures the balancing of a given quantity, but tries not to penalize too much other quantities, albeit a bit less influential. Here we present briefly three standardized methods for the distribution of the sum $(x_1 + x_2)$ of the profile shift coefficients between the pinion and wheel. This sum, already determined in Sect. 6.7, follows from the requirement to ensure an imposed operating center distance, and is framed within the limits recommended by standards.

Before describing these standardized methods, it is however necessary to clarify a fundamental concept, that of the operating pressure angle. In fact, a common perception is that profile shift changes the pressure angle of a gear, and this could be a concept wrong, if it was not seen in the right light. The pressure angle we specify on the gear drawing is the pressure angle of the cutting tool. The actual transverse pressure angle at a point changes along the profile, from the tooth root at tooth tip, and the operating pressure angle, i.e. the angle of the line of action, of a gear pair depends on the gear center distance and radii of the base circles only. If the sum $(x_1 + x_2)$ of the profile shift coefficients is zero, the gear pair operates at the reference center distance. If this sum is negative, the gear pair has smaller center distance, and thus it has a smaller operating pressure angle, while a positive sum requires a larger center distance, and thus an increased operating pressure angle.

Rebus sic stantibus, we can immediately calculate the operating pressure angle, α', using the Eq. (6.59), since the other three quantities, a', a_0, and α_0, that appear in this equation are all known. Once we have calculated so the value of α', by means of Eq. (6.58) we can calculate the sum $(x_1 + x_2)$ to be divided between the pinion and wheel.

6.8.3.1 Method DIN 3992

This method, standardized by DIN 3992 [6], adopts a flexible criterion that leads to toothings having load carrying capacity and sliding speed well balanced. It can also be used for predetermined center distance. The method is applicable for gears with number of teeth $z \leq 150$ (for $z > 150$, a somewhat different method is adopted, but in these cases the profile shift does not have a significant influence on the load carrying capacity of the gears). The method gives two pairs of diagrams permitting the determination of the profile shift coefficient, x_1, of the pinion by the procedure explained below. The first pair of diagrams concerns the speed reducing gears (Fig. 6.19), while the second pair regards the speed increasing gears (Fig. 6.20). The first diagram of each pair is of general validity, and common to both types of drives, that is it refers both to speed reducing gears (Fig. 6.19), and speed increasing gears (Fig. 6.20).

The first diagram of each pair does not refer, however, to the division of the sum, $\Sigma x = (x_1 + x_2)$, of the profile shift coefficients; instead it concerns the determination of this sum in case it had not been determined in advance with the method described in Sect. 6.6. For this preliminary step, we enter the first diagram of each pair with the actual sum $\Sigma z = (z_1 + z_2)$ or equivalent (or virtual) sum $\Sigma z_v = (z_{v1} + z_{v2})$ of the numbers of teeth (see Chap. 8), depending on whether the gears under consideration are parallel cylindrical spur or helical gears. We then identify a point on one of the nine characteristic lines from $P1$ to $P9$ (we could also identify a point between these lines), and we go out with a horizontal straight line up to the ordinate axis, where we read the value of the sum, $\Sigma x = (x_1 + x_2)$.

The characteristic lines $P1$ to $P9$ in Figs. (6.19) and (6.20) identify the main gear peculiarities that the designer wants to achieve, namely: region between $P1$ and $P3$, for high values of contact ratio ε_α; region between $P3$ and $P6$, for usual applications, with toothing well balanced; region between $P6$ and $P9$, for high load carrying capacity to bending and surface fatigue. This first diagram of each pair also highlights two special regions, to be avoided in the usual gear design: the lower region, which refers to the cases of small operating pressure angle, α', and large contact ratio, ε_α, and the upper region, which refers to the special cases of large operating pressure angle, α', and small contact ratio, ε_α. On the right, the diagram also shows that the radii of curvature of the profiles increase from the bottom towards the top, while the contact ratios increase from the top to the bottom.

To split out the sum Σx of the profile shift coefficients between the pinion and wheel of a speed reducing gear, we use the second diagram of Fig. 6.19. To this end, we enter the diagram with half the sum $\Sigma z/2$ or $\Sigma z_v/2$, on the abscissa, and

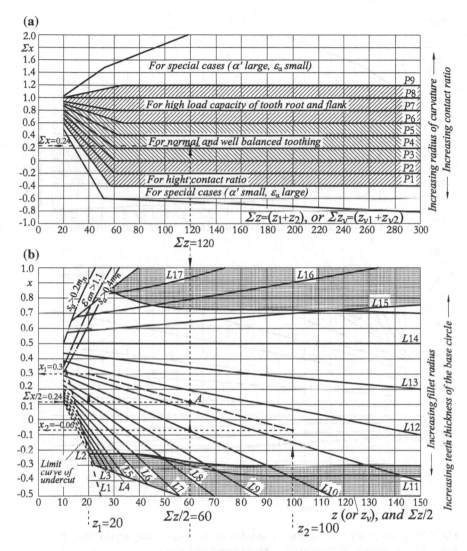

Fig. 6.19 Method DIN 3992 for external speed reducing gears: **a** diagram to choose the sum Σx; **b** diagram for partition of the sum Σx between pinion and wheel

with half the sum, $\Sigma x/2$, on the ordinate axis. So, we identify a point on this diagram, having coordinates $(\Sigma z/2, \Sigma x/2)$, which will be between the seventeen characteristic lines (lines $L1$ to $L17$, from bottom to top) that appear in the same diagram. Therefore, a graphical interpolation line is traced between two neighboring lines, and passing through the point thus identified. The ordinate of the point having abscissa, z_1, on this interpolation line gives the profile shift coefficient, x_1, of the pinion. Being determined the profile shift coefficient of the pinion, that of the wheel will be given by $x_2 = \Sigma x - x_1$.

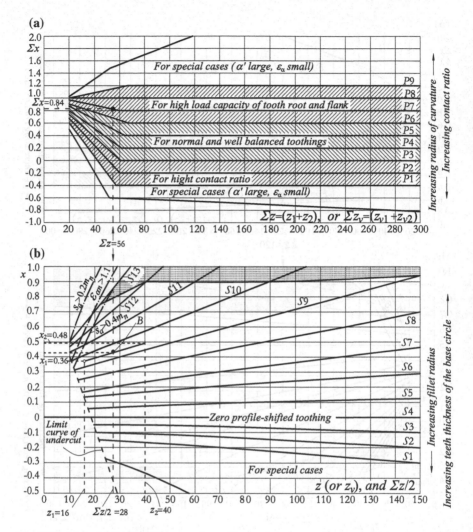

Fig. 6.20 Method DIN 3992 for external speed increasing gears: **a** diagram to choose the sum Σx; **b** diagram for partition of the sum Σx between pinion and wheel

It should be noted that the characteristic lines shown in the second diagram of Fig. 6.19 have been chosen so that the following design goals are simultaneously achieved: the bending tooth root strengths of the pinion and wheel are balanced; the sliding velocity at the tooth tip of the driving gear (usually, the pinion) is, as far as possible, a little greater than that at the tooth tip of the driven gear (this design choice is imposed by the need to reduce noise emissions); the extreme values of specific sliding are avoided. The diagram shows two regions shaded in grey, one at the top and the other at the bottom, which must be avoided. In addition, on the right, it also shows that the fillet radius increases from the top to the bottom, while the tooth thickness measured on the base circle increases from the bottom to the top. It is finally to be noted that the diagram is delimited, at the bottom left, from the limit curve of undercut and, in the upper left side, by three limit curves, corresponding respectively to: $s_a > 0.2m_n$; $\varepsilon_{an} > 1.1$; $s_a > 0.4m_n$.

To split out the sum Σx of the profile shift coefficients between the pinion and wheel of a speed increasing gear, we use the second diagram of Fig. 6.20. The use of this diagram, characterized by thirteen characteristic lines ($S1$ to $S13$, from bottom to top), a region shaded in gray at the top, and a region for special cases at the bottom, is entirely similar to the second diagram of Fig. 6.19. Finally, it is to be noted that, to facilitate the task of the designer, the abscissa of the second diagram of both Figs. 6.19 and 6.20, equal to $\Sigma z/2$ or $\Sigma z_v/2$, are coordinated with those of the first diagram of the same figures, equal to Σz or Σz_v.

Example 9 Speed reducing spur gear pair. Input data: $z_1 = 20$; $z_2 = 100$. Design goal: gears well balanced. Choice of characteristic line: between $P4$ and $P5$. From diagram shown in Fig. 6.19a, we obtain: $\Sigma x = x_1 + x_2 = 0.24$, corresponding to $\Sigma z = z_1 + z_2 = 120$. On the diagram shown in Fig. 6.19b, we identify the point A, having coordinates $(\Sigma z/2 = 60, \Sigma x/2 = 0.12)$. On the same diagram, we draw the interpolation line between $L11$ and $L12$, and passing through point A. On this interpolation line, the ordinate of point having as abscissa $z_1 = 20$ is $x_1 = 0.30$. Finally, x_2 can be obtained as the difference $x_2 = \Sigma x - x_1 = (0.24 - 0.30) = -0.06$, or as the ordinate of the point having as abscissa $z_2 = 100$ on the same interpolation line.

Example 10 Speed increasing spur gear pair. Input data: $z_1 = 16$; $z_2 = 40$. Design goal: high load carrying capacity to bending and surface fatigue. Choice of characteristic line: $P7$. From diagram shown in Fig. 6.20a, we obtain: $\Sigma x = x_1 + x_2 = 0.84$, corresponding to $\Sigma z = z_1 + z_2 = 56$. On the diagram shown in Fig. 6.20b, we identify the point B, having coordinates $(\Sigma z/2 = 28, \Sigma x/2 - 0.42)$. On the same diagram, we draw the interpolation line between $S9$ and $S10$, and passing through the point B. On this interpolation line, the ordinate of point having as abscissa $z_1 = 16$ is $x_1 = 0.36$. Finally, x_2 can be obtained as the difference $x_2 = \Sigma x - x_1 = (0.84 - 0.36) = 0.48$, or as the ordinate of the point having as abscissa $z_2 = 40$ on the same interpolation line.

6.8.3.2 Method BSI—BS PD 6457

This method, standardized by BSI—BS PD 6457 [2], uses the following general formula for the determination of the profile shift coefficient, x_1, of the pinion, in the case of an external speed reducing gear pair:

$$x_1 = C\left(1 - \frac{1}{u}\right) + \frac{\Sigma x}{1 + u}, \tag{6.129}$$

where $u = z_2/z_1$ is the gear ratio, $\Sigma x = (x_1 + x_2)$ is the sum of the profile shift coefficients, and C is a factor dependent on the criteria that are used.

For general applications, where slightly more than half of the tooth action occurs during the recess stage of the meshing cycle, the factor C is equal to 1/3, for which the Eq. (6.129) becomes:

$$x_1 = \frac{1}{3}\left(1 - \frac{1}{u}\right) + \frac{\Sigma x}{1 + u}. \tag{6.130}$$

If we want to ensure the approximate equality of the tooth bending strength factors for pinion and wheel, the factor C is equal to 1/2, for which the Eq. (6.129) becomes:

$$x_1 = \frac{1}{2}\left(1 - \frac{1}{u}\right) + \frac{\Sigma x}{1 + u}. \tag{6.131}$$

Finally, if we want to ensure the approximate balance of ratios of specific sliding or slide-roll ratio at extreme marked points A and E of path of contact, the factor C is equal to $1/\sqrt{z_v}$ for which the Eq. (6.129) becomes:

$$x_1 = \frac{1}{\sqrt{z_{v1}}}\left(1 - \frac{1}{u}\right) + \frac{\Sigma x}{1 + u}, \tag{6.132}$$

where

$$z_{v1} = \frac{z_1}{\cos^3\beta}, \tag{6.133}$$

is the equivalent or virtual number of teeth of the pinion, and β is the helix angle. Note that, for obtaining the equivalent numbers of teeth z_{v1} and z_{v2}, the helix angle of the reference helix of each member of the helical gear is considered (see Chap. 8, regarding helical gears).

6.8.3.3 Method ISO/TR 4467

This method, standardized by ISO/TR 4467 [16], employs the following formula for the determination of the profile shift coefficient of the pinion, x_1:

$$x_1 = \lambda \frac{z_2 - z_1}{z_2 + z_1} + \Sigma x \frac{z_1}{z_2 + z_1} = \lambda \frac{u - 1}{u + 1} + \frac{\Sigma x}{u + 1}, \qquad (6.134)$$

where λ is a factor to be chosen in the range $(0.50 \leq \lambda \leq 0.75)$ for speed reducing gears. If $u = z_2/z_1 > 5$, the limit calculation value is $u = 5$.

For the choice of the profile shift coefficient, x_1, we must take account of the operating conditions of the external gear pair, because things change according to whether it is a speed reducing gear or a speed increasing gear. In fact, in the first case, the smallest gear, i.e. the pinion, which is most stressed, is driving, and thus it requires a positive shift coefficient, such as to strengthen the tooth to its root, to balance the specific sliding between pinion and wheel, and to limit the contact pressure. Thus, for speed reducing gears, it is advisable to choose a factor λ in the range $(0.50 \leq \lambda \leq 0.75)$, preferably approaching to the value $\lambda = 0.75$ when the sum of the number of teeth decreases.

Example 11 (see Example 3). Input data: $z_1 = 20$; $z_2 = 40$; $\alpha_0 = 20°$; $\Sigma x = 0.32$ (this value is suitable for general mechanical applications).

Results with $\lambda = 0.75$: $x_1 = 0.36$; $x_2 = \Sigma x - x_1 = -0.04$.
Results with $\lambda = 0.50$: $x_1 = 0.28$; $x_2 = \Sigma x - x_1 = 0.04$.
Therefore, we realize that, with the choice of the λ-values coincident with the extreme of its variability range, we obtain values of x_1 and x_2 close to those obtained with the method used for the Example 3 ($x_1 = 0.32$, and $x_2 = 0$).

Example 12 Input data: $z_1 = 40$; $z_2 = 120$; $\alpha_0 = 20°$; $\Sigma x = 0.90$.

Results with $\lambda = 0.75$: $x_1 = 0.60$; $x_2 = \Sigma x - x_1 = 0.30$.
Results with $\lambda = 0.50$: $x_1 = 0.48$; $x_2 = \Sigma x - x_1 = 0.42$.
In the second case, that of speed increasing gears, the problem of determination of the profile shift coefficients, x_1 and x_2, is in general less demanding and requires no special corrective actions. However, other problems can sometimes arise, and in these cases, the design must be based on assumptions completely different. In effect, it is necessary to take into account the fact that, due to friction (so far, we have neglected the friction forces), the total force applied by the driving gear (the wheel) to the driven gear (the pinion) is not directed along the line of action. In fact, it is inclined, with respect to this, by the friction angle, φ (see Fig. 3.9), which is related to the coefficient of friction, μ, according to the relationship $\mu = \tan \varphi$.

Now it is to remember that the total resultant force, F, is the vector sum of the useful force, acting along the line of action, and friction force. The friction force is perpendicular to the useful force and has direction such as to oppose the motion. Considering even the friction force, it happens that, generally, the total resultant force, F, tends to approach the axis of the driven member (the pinion), when contact occurs in the approach path of contact AC (Fig. 6.21). In this case, it is therefore useful to have a recess path of contact wider than the approach path of contact, i.e. $CE > AC$. These favourable conditions are automatically realized with a speed reducing gear, in which the addendum of the pinion is increased, while the

Fig. 6.21 Line of application
of total resultant force, F,
which approaches the pinion
axis, due to the friction force

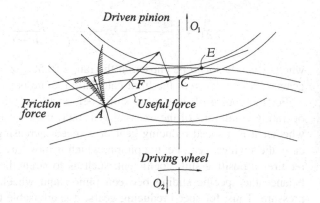

addendum of the wheel is decreased. For a speed increasing gear, however, this kind of choice of the profile shift coefficients becomes disadvantageous. In fact, if the gear ratio is large, and the number of teeth of the pinion is small, it may happen that the line of application of the total resultant force, F, approaches dangerously to the pinion axis. In this case, we could incur the risk of jamming and pour efficiency.

In these cases, the relationship (6.134) can still be used, with the caveat to choose $\lambda = 0$. For example, with $z_1 = 30$; $z_2 = 210$; $\alpha_0 = 20°$; $\Sigma x = 0$, we obtain $x_1 = 0$, and $x_2 = \Sigma x - x_1 = 0$, i.e. a standard toothing. In the case of speed increasing gear, a good solution is to use a standard toothing, associated with the choice of a number of teeth of the pinion equal to or greater than 36.

It should also be remembered that *recess action gears* could be required not only to solve the above-mentioned issues related to speed increasing gears, but also to use approximately their own interesting property. Indeed, in practice, it has been found that the performance of a gear pair is smoother during the recess path of contact rather than the approach path of contact of the meshing cycle. In Sect. 3.6 we showed that the sliding velocity changes direction and sign when the point of contact passes through the pitch point. Even the friction force changes its direction, while the total force, F, moves from side to side with respect to the line of action, as Fig. 3.9 shows. This different orientation of the forces coming into play in the recess path of contact is responsible for the smoother operation of the gear pair during the recess stage of the meshing cycle.

For a gear pair whose members do not have to change their role (in the meaning that one gear wheel is always the driving member, and the other is always the driven member), it is possible to carry out its design in order to draw advantage of the smoother *recess action*. The gear pair is designed in such a way that most or the entire path of contact is the recess path of contact. To do this, the driving member addendum must be adequately increased, and the driven member addendum must be proportionality decreased. When the recess path of contact is much broader than the approach path of contact, we say the gear pair has a *partial recess action*. Instead, when the recess path of contact constitutes the entire path of contact (that is, the length of the approach path of contact is reduced to zero), we say that the gear pair is an *only recess action gear pair*. These design goals can be achieved by choosing suitable profile shift coefficients, x_1 and x_2, using the methods described above.

6.9   Determination of the Profile Shift Coefficient of Internal Gear Pairs

In Sect. 3.2 we said that, for an internal gear pair, unlike what happens for an external gear pair, it is not always possible to freely share the profile shift coefficients, and simultaneously obtain optimal values of the specific sliding and contact ratio. In fact, the tendency of teeth of an internal gear pair to interfere with each other imposes significant limitations, so it is not easy to define a practical method to predetermine the values of profile shift coefficients. In most cases, it is necessary to proceed by trial and error, also because for internal gear pairs, contrary to external gear pairs, the minimum number of teeth of the pinion to avoid the interference is the larger, the lower is the gear ratio, $u = z_2/z_1$.

For the determination of the profile shift coefficients, the basic relationships from which we must start are those shown in Sect. 6.2. Bearing in mind that, very often, the operating center distance, a', is an input of the problem, to be observed strictly, from Eq. (6.113) we obtain the operating pressure angle, α', as follows:

$$\alpha' = \arccos\left(\frac{a_0}{a'}\cos\alpha_0\right). \tag{6.135}$$

Therefore, taking into account this relationship, by means of Eq. (6.109), we calculate the difference $(x_2 - x_1)$.

This way of proceeding is also useful even in the case where the center distance is not an input to be observed strictly. In other words, when we have to design internal gear pairs with profile-shifted toothing and variation of center distance, it is always useful to first determine the center distance, a', and then calculate the difference $(x_2 - x_1)$. For the manufacturing of the annulus, it is also important to determine (and to indicate on the gear drawing) the cutting center distance, i.e. the distance between the axes of the annulus and pinion-type cutter, in the condition that will have at the end of the cutting operation.

As regards the choice of the profile shift coefficients, it is necessary to keep in mind the following considerations:

(a) If the number of teeth, z_1, of the pinion is greater than the minimum number of teeth, z_{1min}, which depends on the ratio $(z_1/z_2) = 1/u$, and pressure angle, α (see Fig. 3.1), it is convenient to use a gear pair with profile-shifted toothing without variation of center distance $(x_2 = x_1)$. In this case, we choose a profile shift coefficient included in a narrow range, between $x_1 = 0.50$ for higher gear ratios $(8 \leq u \leq 10)$, and $x_1 = 0.60$ for lower gear ratios $(1.5 \leq u \leq 2.5)$.

The range of the favourable values of the profile shift coefficients for this type of profile-shifted toothing is in any case between $(0.50 \leq x \leq 0.65)$. With this kind of profile shift, choosing the profile shift coefficients to balance the specific sliding at the end points of the path of contact, we can obtain teeth with high load carrying capacity and favourable driving characteristics. However, in cases of predetermined center distance, this type of profile shift shows all its limitations, because it is not possible to satisfy strictly the design input data with this profile shift.

(b) If the number of teeth, z_1, of the pinion is less than the minimum number of teeth, z_{1min} (see Fig. 3.1), it is convenient to use a gear pair with profile-shifted toothing and variation of center distance, with profile shift coefficients to choose from time to time in relation to the goals to be achieved. Compared to the profile-shifted toothing without variation of center distance, those with variation of center distance do not have a significantly higher load carrying capacity, but allow the designer to have a greater freedom of choice for the achievement of predetermined design requirements. Tip thicknesses, in terms of crest-widths, space widths, and interference in the actual meshing conditions are always checked and verified. In order to ensure a predetermined center distance (this is impossible for profile-shifted toothing without variation of center distance), it is preferable to choose the profile shift coefficients in such a way that it is $(x_1 + x_2) < 0$, with x_2 conventionally considered as negative. Since the risk of interference exists, the choice above is generally necessary, even for $(z_1 + z_2) > -7$ (here also the number of teeth, z_2, of the annulus is to be considered, by convention, as negative).

From the considerations above, we infer that the profile-shifted toothing without variation of center distance do not have appreciable disadvantages, while the profile-shifted toothing with variation of center distance offers more, than the first, only the ability to freely choose the center distance. For this reason, not separated from that of considerably greater simplicity of the calculations to be performed, the designer often opts for the profile-shifted toothing without variation of center distance.

In the internal spur gears with profile-shifted toothing without variation of center distance $(x_2 = x_1)$, there is a correlation between $(r_{a2} - r_{b2})/m$ (i.e. the quotient of the radial distance between the tip circle and base circle, divided by the module) and the profile shift coefficient, x_2, which depends on the number of teeth, z_2. Figure 6.22 shows this correlation. The choice of a positive shift coefficient, x_2, for the annulus determines a tooth profile modification, which strengthens the tooth at its root, and facilitates its assembly with the pinion. However, it is to be noted that, to ensure a good condition of engagement between the pinion and annulus, a limit exists for the maximum value of the sum of the profile shift coefficients, $(x_1 + x_2)_{max}$. As Fig. 6.23 shows, this limit sum is related to the pressure angle, α, and difference $\Delta z = |z_2| - z_1$; this figure refers to an annulus to be generated with a pinion-type cutter having nominal pressure angle, $\alpha_0 = 20°$, and nominal addendum $h_0 = m_0$.

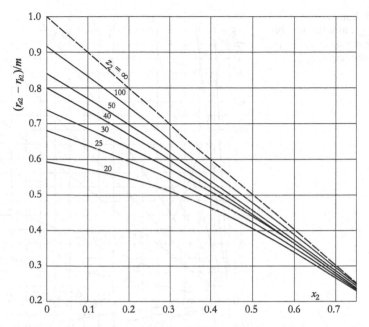

Fig. 6.22 Correlation between $(r_{a2} - r_{b2})/m$ and x_2, as a function of the number of teeth z_2

6.10 Backlash

In certain applications, such as gear trains where positioning is the design key condition, but the power to transmit is low, or lead-screws where positioning and power are both important, *backlash* (sometimes also called *lash* or *play*) is an undesirable characteristic, and should be at least minimized or even reduced to zero. In fact, the ideal theoretical condition would be to have *gears with zero backlash* or *anti-backlash gears* (see also 7).

However, in actual design practice of gears, some backlash must be necessarily allowed to prevent jamming. Reasons for the presence of backlash in a gear power transmission include allowing for manufacturing errors, thermal expansion, deflections under load, and lubrication. Here we leave aside the problems of anti-backlash design of gears, and focus our attention on those where the backlash must necessarily be allowed.

Profile shift introduced by us in the previous sections, and usually used by gear designers, is related a backlash-free contact, or zero-backlash gear set. To provide the necessary backlash, some designers reduce the profile shift, while others increase center distance. Factors affecting the amount of backlash required in a gear pair, or in a gear train, include errors in profile shape, pitch, tooth thickness, center distance, helix angle (see Chap. 8), and run-out. The greater the accuracy of manufacture and assembly, the smaller the errors mentioned above, and therefore the backlash needed.

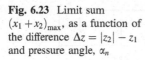

Fig. 6.23 Limit sum $(x_1 + x_2)_{max}$, as a function of the difference $\Delta z = |z_2| - z_1$ and pressure angle, α_n

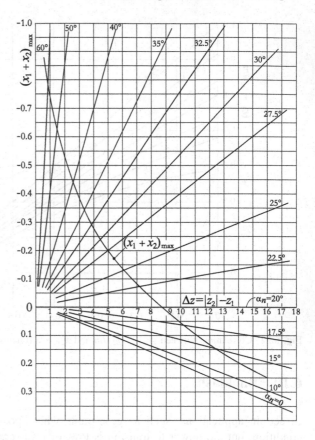

The most commonly used method to obtain the optimum value of the backlash consists in cutting the teeth deeper inside the gear blanks than the theoretical depth, corresponding to the condition of backlash-free contact. This method is the most general, and is the only one that allows satisfying a predetermined center distance required as input for the design. It is updated by changing the profile shift, that is moving the cutting tool radially to change the tooth thickness. Evidently, from this point of view, we can say that all gears have some profile shift, since the backlash must be guaranteed even in the case of zero-profile-shifted toothing.

Now, according to the sign convention for the profile shift, a positive profile shift corresponds to thicker teeth. Thus, a shift of the rack-type cutter (Fig. 6.1) towards the axis of an external gear wheel corresponds to a negative profile shift, since it makes the teeth thinner. Therefore, subtracting from the zero-backlash profile shift an additional small profile shift, corresponding to the backlash allowance, we obtain the so-called *rack shift*. When gear designer specifies only the zero-backlash profile shift (gear drawings are usually performed with respect to the zero-backlash profile shift), the gear manufacturers will thin the teeth to ensure the necessary operating backlash according to standards. Many of these standards require that the tooth thinning for backlash, called *upper allowance of size*, is a function of the module.

The values are measured on the reference cylinder, in a transverse section for spur gears, and in a normal section for helical gears (see Chap. 8). The transverse circular allowance, related to the normal allowance in the base tangent plane, is a function of allowance, center distance, and tooth accuracy.

A common convention among gear manufacturers is to reduce the normal tooth thickness of each members of the gear pair by the same amount, which may be a predetermined value, in μm, or a function of module, such as $(0.03 \div 0.05)m_n$. This maintains the same cutting depth for both gear members, and maximizes contact ratio. The direction, normal or transverse, as well as the reference circle or base tangent plane, according to which the tooth thickness reduction is to be measured, must be specified, since there is no recognized convention. Standard practice is to make allowance for half the backlash in the tooth thickness of each gear member. However, if the pinion is significantly smaller than the mating wheel, it is common practice to account for the entire backlash in the larger wheel, because this ensures the maximum strength as possible of the pinion teeth. The amount of additional material removed when manufacturing the gears depends on the pressure angle, α_0. For $\alpha_0 = 14.5°$, the cutting tool is moved towards the axis of the wheel to be cut by an amount equal to the backlash desired; this extra distance corresponds to the additional shift profile. For $\alpha_0 = 20°$, this extra distance equals 0.73 times the amount of backlash desired.

The other method consists of the increase or decrease of the center distance, depending on whether the gear is external or internal. This method is less general than the first, as it does not guarantee exactly a predetermined center distance. Furthermore, the first method comprises, as a special case, the second method, since the backlash obtained by a small change of the profile shift corresponds, in fact, to a virtual variation (increasing or decreasing respectively for external and internal gears) of the center distance.

Let us consider the example of an external gear pair, and determine the correlation between the increase of center distance (virtual increase for the first method, and actual increase for the second method) and backlash measured on the normal to the profile. To this end, it is sufficient to consider that a removal $\Delta r_1'$ of the axis of the pitch circle of the pinion along the center distance line $O_1 O_2$ (Fig. 2.10) causes a decrease in the tooth thickness equal to $2\Delta r_1' \sin \alpha'$, when it is measured along the line of action. This is because, on the pitch circle, the normal to the profile forms an angle, α', with the common tangent to the pitch circles. Similarly, a removal $\Delta r_2'$ of the axis of the pitch circle of the wheel along the center distance line $O_1 O_2$ determines a decrease of the tooth thickness equal to $2\Delta r_2' \sin \alpha'$.

Therefore, we will have a normal backlash, j_n (the backlash measured along the normal to the profile or, what is the same, along the line of action), equal to:

$$j_n = 2(\Delta r_1' + \Delta r_2') \sin \alpha' = 2\Delta a' \sin \alpha', \tag{6.136}$$

where $\Delta a' = (\Delta r'_1 + \Delta r'_2)$ is the virtual or actual increase of the center distance. The corresponding transverse backlash, j_t, i.e. the backlash measured on the pitch circle, is given by:

$$j_t = 2\Delta a' \tan \alpha'. \tag{6.137}$$

From Eq. (6.136), once the value of the normal backlash, j_n, has been fixed case by case, and assuming $\alpha' = \alpha_0$, we obtain $\Delta a'$. In reality, for a given value of j_n, varying the center distance, also α' varies, although not by much. For this reason, we should set an iterative calculation until convergence of the results is obtained.

The iterative calculations to do when the center distance, a', is predetermined are even more elaborate and complex. As in the previous case, once the value of j_n has been fixed, we calculate $\Delta a'$ by means of Eq. (6.136), in which we set $\alpha' = \alpha_0$ (actually, we should introduce a value $\alpha' > \alpha_0$ when $(x_1 + x_2) > 0$, and $\alpha' < \alpha_0$ when $(x_1 + x_2) < 0$). According to Eq. (6.59), the center distance in the condition of backlash-free contact is given by:

$$a' - \Delta a' = \frac{z_1 + z_2}{2} m_0 \frac{\cos \alpha_0}{\cos \alpha'}. \tag{6.138}$$

From this equation, in which we introduce the known values of a', z_1, z_2, m_0 and α_0, as well as the approximate value of $\Delta a'$, as determined above, we obtain α'. Therefore, by means of Eq. (6.58), we calculate the sum $(x_1 + x_2)$. The profile shift coefficients, x_1 and x_2, previously determined, must then be suitably retouched and remodulated, so that their sum coincides with that just calculated. Also α' will be different from the value introduced to compute the approximate value of $\Delta a'$. For these reasons, it will be necessary to iterate the calculations until convergence. The limit of the impossibility of ensuring a predetermined center distance remains.

References

1. Almen JO, Straub JC (1948) Aircraft gearing, analysis of test and data service. Research Laboratories Division, General Motors Corporation, Detroit, Michigan, AGMA
2. BSI-BS PD G457: Guide to the Application of Addendum Modification to Involute Spur and Helical Gears
3. Buckingham E (1949) Analytical mechanics of gears. McGraw-Hill Book Company Inc, New York
4. Chauhan V (2016) A review on effect of some important parameters on the bending strength and surface durability of gears. Int Journ Sci Res Publ 6(3):289–297
5. Colbourne JR (1987) The geometry of involute gears. Springer-Verlag, New York Inc, New York Berlin Heidelberg
6. DIN 3992: Profilverschiebung bei Stinrädern nit Aubenverzahnung (Addendum modification of external spur and helical gears)
7. Dooner DB, Seireg AA (1995) The kinematic geometry of gears—a concurrent engineering approach. Wiley, New York

8. Dudley DW (1962) Gear handbook. The design, manufacture, and application of gears. McGraw-Hill Book Company, New York
9. Dudley DW (1969) The evolution of the gear art. American Gear Manufacturers Association, Washington, D.C
10. Dudley DW, Sprengers J, Schröder D, Yamashina H (1995) Gear motor handbook. In: Bonfiglioli Riduttori SPA (Eds) Springer-Verlag, Berlin Heidelberg
11. Ferrari C, Romiti A (1966) Meccanica applicata alle macchine. Unione Tipografico-Editrice Torinese, UTET, Torino
12. Giovannozzi R (1965) Costruzione di macchine, vol II, 4th ed. Casa Editrice Prof. Riccardo Pàtron, Bologna
13. Henriot G (1979) Traité théorique et pratique des engrenages, vol 1, 6th edn. Bordas, Paris
14. ISO 1122-1 (1998) Vocabulary of gear terms
15. ISO 6336-1 Part 1: Basic principles, introduction and general influence factors
16. ISO/TR 4467 Addendum modification of the teeth of cylindrical gears for speed-addendum modification reducing and speed-increasing gear pairs
17. Lasche O (1899) Die Ausführung mit einer Profilverschiebung nach obenstchenden Augaben wird der grundlegenden, Z.d. VDI, S. 1488 auch als AEG-Verzahnung bezeichnet
18. Li S (2008) Effect of addendum on contact strength, bending strength and basic performance parameters of a pair of spur gears. Mech Mach Theory 43(12):1557–1584
19. Litvin FL, Fuentes A (2004) Gear geometry and applied theory, 2nd edn. Cambridge University Press, Cambridge
20. Maitra GM (1994) Handbook of gear design, 2nd edn. Tata McGraw-Hill Publishing Company Ltd, New Delhi
21. Merritt HE (1954) Gears, 3rd edn. Sir Isaac Pitman & Sons Ltd, London
22. Niemann G, Winter H (1983) Maschinen-elemente, band II: getriebe allgemein, zahnradgetriebe-grundlagen, stirnradgetriebe. Springer-Verlag, Berlin Heidelberg
23. Oda S, Tsubukura K, Namba C (1982) Effect of addendum modification on bending fatigue strength of spur gear with high pressure angle. Bull. Jap. Soc. Mech. Eng. 259(209):1813–1826
24. Polder JW (1991) Interference of internal gears. In: Towsend P (ed) Dudley's gear handbook. McGraw-Hill, New York
25. Pollone G (1976) Il Veicolo. Libreria Editrice Universitaria Levrotto & Bella, Torino
26. Schiebel A, Lindner W (1954) Zahnräder, Band. I: Stirn-und Kegelräder mit geraden Zähnen. Springer-Verlag, Berlin Heidelberg
27. Schiebel A, Lindner W (1957) Zahnräder, Band 2: Stirn-und Kegelräder mit schrägen Zähnen Schraubgetriebe. Springer-Verlag, Berlin Heidelberg
28. Schreier G (1961) Stirnrad-Verzahnung. VEB Verlag Technik, Berlin
29. Scotto Lavina G (1996) Riassunto delle Lezioni di Meccanica Applicata alle Macchine: Cinematica Applicata; Dinamica Applicata—Resistenze Passive—Coppie Inferiori; Coppie Superiori (Ingranaggi)—Flessibili—Freni. Edizioni Scientifiche Siderea, Roma
30. Simon V (1989) Optimal tooth modifications for spur and helical gears. J Mach Des 3(4): 611–615
31. Woodbury RS (1958) History of gear-cutting machine: a historical study in geometry and machines. MIT Press, Cambridge, Massachusetts, USA

Chapter 7
Profile Shift of External Spur Gear Involute Toothing Generated by Pinion-Type Cutter

Abstract In this chapter, the fundamentals of the profile shift of spur gears involute toothing generated by pinion-type cutter are first described, highlighting the interference and undercut problems related to this cutting process of the teeth. The limitations regarding the possibility of making external spur gears with profile-shifted toothing by means of this generation process are described, and the greatest complication of the calculations to be carried out is highlighted, with a comparison with the manufacturing process of profile-shifted toothing by rack-type cutter.

7.1 Characteristics of the Pinion-Type Cutter

The cutting tool is shaped exactly like a pinion, and is therefore called *pinion-type cutter* or *Fellows gear shaper* (see Sect. 3.11.2). However, unlike a pinion gear, the teeth of the pinion-type cutter are characterized by appropriate front rake and face angles (see Fig. 3.15), and are sharpened at their race angle (see: Micheletti [7], Galassini [2], Rossi [10]). As in a usual pinion, the teeth of the pinion-type cutter are shaped as involute curves of a base circle having diameter

$$d_{b0} = 2r_{b0} = m_0 z_0 \cos \alpha_0 \tag{7.1}$$

where z_0 is the number of teeth of this Fellows gear shaper, while m_0 and α_0 are respectively the nominal module and nominal pressure angle of its toothing; $p_0 = \pi m_0$ is its nominal pitch.

With reference to the standard (or nominal) pitch circle, having radius

$$r_0 = \frac{m_0 z_0}{2}, \tag{7.2}$$

the addendum and dedendum are both equal to $(5/4)m_0$, and the teeth have rounded tips and rounded root fillets having depth equal to $(1/4)m_0$, for which, in respect to the cutting interference, this pinion-type cutter behaves as if its addendum is only

© Springer Nature Switzerland AG 2020
V. Vullo, *Gears*, Springer Series in Solid and Structural Mechanics 10,
https://doi.org/10.1007/978-3-030-36502-8_7

m_0. The tooth thickness s_0 and space width e_0, measured on the nominal pitch circle, are both equal to $\pi m_0/2$, i.e. $s_0 = e_0 = \pi m_0/2$ (see: Buckingham [1], Merritt [6], Henriot [3], Niemann and Winter [8], Maitra [5]).

7.2   Gear Wheels Generated by a Pinion-Type Cutter

Using a pinion-type cutter, we can generate gear wheels having identical characteristics compared to those generated by a rack-type cutter. Consider a gear wheel having z_1 teeth, nominal module $m_1 = m_0$ and nominal pressure angle $\alpha_1 = \alpha_0$, generated by a rack-type cutter with a profile shift $x_1 m_0$ (then with a profile shift coefficient x_1), and suppose that it engages with another gear wheel, having z_0 teeth and zero profile-shifted toothing, i.e. $x_0 = 0$ [9].

The operating pressure angle α', in the condition of backlash-free contact between these two gear wheels, is given by Eq. (6.58). From this equation, written for the case considered here, we obtain:

$$\mathrm{inv}\alpha' = \mathrm{inv}\alpha_0 + 2\frac{x_1}{z_1 + z_0}\tan\alpha_0. \qquad (7.3)$$

Furthermore, the operating center distance, defined by Eq. (6.59) also written for the same case, is given by:

$$a' = a_0\frac{\cos\alpha_0}{\cos\alpha'} = \frac{z_1 + z_0}{2}m_0\frac{\cos\alpha_0}{\cos\alpha'}. \qquad (7.4)$$

If the gear wheel with z_0 teeth is a pinion-type cutter, and we place it to the above center distance, a', during the generation of a gear wheel having z_1 teeth, this last will have the same dimension as would have if it had been generated by the rack-type cutter, with a profile shift equal to $x_1 m_0$.

7.3   Characteristic Quantities of the Pinion-Type Cutter Referred to a Pitch Circle Shifted by xm_0 with Respect to the Nominal Pitch Circle

On a pitch circle shifted by xm_0 with respect to the nominal pitch circle, and then having radius

$$r_0' = \left(\frac{z_0}{2} + x\right)m_0, \qquad (7.5)$$

the pressure angle α' at a point corresponding to this radius and the correlated module m' are given respectively by:

$$\cos \alpha' = \frac{r_{b0}}{r_0'} = \frac{z_0}{z_0 + 2x} \cos \alpha_0 \tag{7.6}$$

$$m' = m_0 \frac{r'}{r_0} = m_0 \frac{z_0 + 2x}{z_0}, \tag{7.7}$$

where $r_0 = (z_0 m_0 / 2)$.

The tooth thickness angle $2\vartheta_{s0}$ and space width angle $2\vartheta_{e0}$ on this cutting pitch circle are given respectively by:

$$2\vartheta_{s0} = \frac{\pi}{z_0} + 2(\mathrm{inv}\alpha_0 - \mathrm{inv}\alpha') \tag{7.8}$$

$$2\vartheta_{e0} = \frac{\pi}{z_0} - 2(\mathrm{inv}\alpha_0 - \mathrm{inv}\alpha'). \tag{7.9}$$

The radii of the tip and root circles of the pinion-type cutter are given respectively by:

$$r_{a0} = \frac{z_0 m_0}{2} + 1.25 m_0 = \left(\frac{z_0}{2} + 1.25\right) m_0 \tag{7.10}$$

$$r_{f0} = \frac{z_0 m_0}{2} - 1.25 m_0 = \left(\frac{z_0}{2} - 1.25\right) m_0. \tag{7.11}$$

7.4 Characteristic Quantities of a Gear Wheel Having z_1 Teeth, and Generated with a Profile Shift Coefficient x_1

The radii of the base circle and cutting pitch circle of a gear wheel having z_1 teeth, and generated with a profile shift coefficient x_1 by means of the pinion-type cutter defined in previous section are given respectively by:

$$r_{b1} = r_{b0} \frac{z_1}{z_0} = \frac{z_1 m_0}{2} \cos \alpha_0 \tag{7.12}$$

$$r_1' = \frac{z_1 m'}{2} = \frac{z_1}{2} \frac{(z_0 + 2x_1)}{z_0} m_0 = \frac{z_1}{2} m_0 \frac{\cos \alpha_0}{\cos \alpha_1'}, \tag{7.13}$$

where the pressure angle α_1' at a point corresponding to this radius is given by:

$$\cos \alpha_1' = \frac{z_0}{z_0 + 2x_1} \cos \alpha_0. \tag{7.14}$$

During the generation, the center distance a_1' between the pinion-type cutter and the gear wheel to be generated is:

$$a_1' = r_0' + r_1' = \frac{(z_0 + z_1)(z_0 + 2x_1)}{2} \frac{m_0}{z_0} = \frac{z_0 + z_1}{2} m_0 \frac{\cos \alpha_0}{\cos \alpha_1'}, \qquad (7.15)$$

because r_0' is given by Eq. (7.5) where $x = x_1$.

On the cutting pitch circle, the tooth thickness of the gear wheel is equal to the space width between the cutter teeth, for which the following equality must be valid:

$$2\vartheta_{e0} r_0' = 2\vartheta_{s1}' r_1', \qquad (7.16)$$

where ϑ_{s1}' is the tooth thickness half angle of the gear wheel on the cutting pitch circle. From this equation, taking into account Eq. (7.5), with $x = x_1$, and Eq. (7.13), which lead to define the equality $r_0'/r_1' = z_0/z_1$, and bearing in mind the Eq. (7.9), we obtain:

$$2\vartheta_{s1}' = 2\frac{z_0}{z_1}\vartheta_{e0} = \frac{z_0}{z_1}\left[\frac{\pi}{z_0} - 2\left(\mathrm{inv}\alpha_0 - \mathrm{inv}\alpha_1'\right)\right] = \frac{\pi}{z_1} + 2\frac{z_0}{z_1}\left(\mathrm{inv}\alpha_1' - \mathrm{inv}\alpha_0\right).$$

$$(7.17)$$

On a circle having radius r_1'', the toothing pressure angle at a point is given by:

$$\cos \alpha_1'' = \frac{r_{b1}}{r_1''} = \frac{z_1 m_0}{2r_1''}\cos \alpha_0, \qquad (7.18)$$

while the tooth thickness angle $2\vartheta_{s1}''$ and space width angle $2\vartheta_{e1}''$ are given respectively by:

$$2\vartheta_{s1}'' = 2\vartheta_{s1}' + 2\left(\mathrm{inv}\alpha_1' - \mathrm{inv}\alpha_1''\right) = \frac{\pi}{z_1} + 2\left(1 + \frac{z_0}{z_1}\right)\mathrm{inv}\alpha_1' - 2\mathrm{inv}\alpha_1'' - 2\frac{z_0}{z_1}\mathrm{inv}\alpha_0$$

$$(7.19)$$

$$2\vartheta_{e1}'' = \frac{\pi}{z_1} - 2\left(1 + \frac{z_0}{z_1}\right)\mathrm{inv}\alpha_1' + 2\mathrm{inv}\alpha_1'' + 2\frac{z_0}{z_1}\mathrm{inv}\alpha_0, \qquad (7.20)$$

because

$$2\left(\vartheta_{s1}'' + \vartheta_{e1}''\right) = \frac{2\pi}{z_1}. \qquad (7.21)$$

7.5 External Gear Pair with z_1 and z_2 Teeth, Generated with Profile Shift Coefficients x_1 and x_2

From equations obtained in the two previous sections, we infer that the quantities of two external gear wheels, having respectively z_1 and z_2 teeth, and generated by a pinion-type cutter with profile shift coefficients x_1 and x_2, are given by the equations described in the following subsections. Their calculation procedure, based on the equations defined earlier as well as in the previous chapters, is not reported.

7.5.1 Pinion (First Wheel), Having z_1 Teeth, and Generated with Profile Shift Coefficient x_1

- Radius of the base circle, r_{b1}

$$r_{b1} = \frac{z_1 m_0}{2} \cos \alpha_0 \tag{7.22}$$

- Cutting pressure angle, α_1'

$$\cos \alpha_1' = \frac{z_0}{z_0 + 2x_1} \cos \alpha_0 \tag{7.23}$$

- Radius of the cutting pitch circle, r_1'

$$r_1' = \frac{r_{b1}}{\cos \alpha_1'} = \frac{z_1 m_0}{2} \frac{\cos \alpha_0}{\cos \alpha_1'} \tag{7.24}$$

- Cutting center distance, a_1'

$$a_1' = \frac{z_0 + z_1}{2} m_0 \frac{\cos \alpha_0}{\cos \alpha_1'} = m_0 \frac{(z_0 + z_1)(z_0 + 2x_1)}{2} \tag{7.25}$$

- Radius r_1'' of a circle on which the pressure angle at a point is α_1''

$$r_1'' = m_0 \frac{z_1 \cos \alpha_0}{2 \cos \alpha_1''} \tag{7.26}$$

- Tooth thickness angle $2\vartheta_{s1}''$ on the circle having radius r_1''

$$2\vartheta_{s1}'' = \frac{\pi}{z_1} + 2\left(1 + \frac{z_0}{z_1}\right) \text{inv}\alpha_1' - 2\text{inv}\alpha_1'' - 2\frac{z_0}{z_1}\text{inv}\alpha_0 \tag{7.27}$$

- Tooth space width angle $2\vartheta''_{e1}$ on the circle having radius r''_1

$$2\vartheta''_{e1} = \frac{2\pi}{z_1} - 2\vartheta''_{s1} = \frac{\pi}{z_1} - 2\left(1 + \frac{z_0}{z_1}\right)\text{inv}\alpha'_1 + 2\text{inv}\alpha''_1 + 2\frac{z_0}{z_1}\text{inv}\alpha_0 \quad (7.28)$$

- Angular pitch

$$\tau_1 = \frac{2\pi}{z_1} = 2\left(\vartheta''_{s1} + \vartheta''_{e1}\right). \quad (7.29)$$

7.5.2 Wheel (Second Wheel), Having z_2 Teeth, and Generated with Profile Shift Coefficient x_2

- Radius of the base circle, r_{b2}

$$r_{b2} = \frac{z_2 m_0}{2}\cos\alpha_0 \quad (7.30)$$

- Cutting pressure angle, α'_2

$$\cos\alpha'_2 = \frac{z_0}{z_0 + 2x_2}\cos\alpha_0 \quad (7.31)$$

- Radius of the cutting pitch circle, r'_2

$$r'_2 = \frac{r_{b2}}{\cos\alpha'_2} = \frac{z_1 m_0}{2}\frac{\cos\alpha_0}{\cos\alpha'_2} \quad (7.32)$$

- Cutting center distance, a'_2

$$a'_2 = \frac{z_0 + z_2}{2}m_0\frac{\cos\alpha_0}{\cos\alpha'_2} = m_0\frac{(z_0 + z_2)}{2}\frac{(z_0 + 2x_2)}{z_0} \quad (7.33)$$

- Radius r''_2 of a circle on which the pressure angle at a point is α''_2

$$r''_2 = m_0\frac{z_2}{2}\frac{\cos\alpha_0}{\cos\alpha''_2} \quad (7.34)$$

- Tooth thickness angle $2\vartheta''_{s2}$ on the circle having radius r''_2

$$2\vartheta''_{s2} = \frac{\pi}{z_2} + 2\left(1 + \frac{z_0}{z_2}\right)\text{inv}\alpha'_2 - 2\text{inv}\alpha''_2 - 2\frac{z_0}{z_1}\text{inv}\alpha_0 \quad (7.35)$$

- Tooth space width angle $2\vartheta''_{e2}$ on the circle having radius r''_2

$$2\vartheta''_{e2} = \frac{2\pi}{z_2} - 2\vartheta''_{s2} = \frac{\pi}{z_2} - 2\left(1 + \frac{z_0}{z_2}\right)\mathrm{inv}\alpha'_2 + 2\mathrm{inv}\alpha''_2 + 2\frac{z_0}{z_2}\mathrm{inv}\alpha_0 \qquad (7.36)$$

- Angular pitch

$$\tau_2 = \frac{2\pi}{z_2} = 2\left(\vartheta''_{s2} + \vartheta''_{e2}\right). \qquad (7.37)$$

7.5.3 Meshing Between the Two Wheels with a Backlash-Free Contact

Now we put the two wheels above so that they are meshing between them with a backlash-free contact. We denote respectively with r_{w1}, r_{w2} and α_w the radii of the two operating pitch circles and operating pressure angle (in this regard, it is noteworthy that, in accordance with the ISO 6336-1 [4], to indicate quantities related to the working, or operating, conditions, we use the subscript w, which replaces the usual prime symbol, when the latter can generate misunderstanding). The radii r_{w1} and r_{w2} are correlated by the following relationship:

$$\frac{r_{w1}}{r_{w2}} = \frac{z_1}{z_2}. \qquad (7.38)$$

On these working pitch circles, the space width between the teeth of the first wheel must be equal to the tooth thickness of the second wheel, for which

$$2r_{w1}\vartheta''_{e1} = 2r_{w2}\vartheta''_{s2} \qquad (7.39)$$

and thus

$$\vartheta''_{e1} = \frac{r_{w2}}{r_{w1}}\vartheta''_{s2} = \frac{z_2}{z_1}\vartheta''_{s2}. \qquad (7.40)$$

Substituting in this equation the expressions of ϑ''_{e1} and ϑ''_{s2} obtained from Eqs. (7.28) and (7.35), we get:

$$\mathrm{inv}\alpha_w = \mathrm{inv}\alpha''_2 = \frac{z_1 + z_0}{z_1 + z_2}\mathrm{inv}\alpha'_1 + \frac{z_2 + z_0}{z_1 + z_2}\mathrm{inv}\alpha'_2 - 2\frac{z_0}{z_1 + z_2}\mathrm{inv}\alpha_0. \qquad (7.41)$$

The operating center distance, a', in the condition of backlash-free contact, is given by:

$$a' = r_{w1} + r_{w2} = \frac{a_0}{\cos \alpha_w} = m_0 \frac{(z_1 + z_2)}{2} \frac{\cos \alpha_0}{\cos \alpha_w}. \tag{7.42}$$

Taking into account that, during generation, the cutting center distances between the pinion-type cutter and each of the two wheels are respectively given by Eqs. (7.25) and (7.33), the radii of the tip and root circles of the pinion and wheel will be given by:

$$r_{a1} = a_1' - \frac{m_0}{2}(z_0 - 2) \tag{7.43}$$

$$r_{f1} = a_1' - \frac{m_0}{2}(z_0 + 2.50) \tag{7.44}$$

$$r_{a2} = a_2' - \frac{m_0}{2}(z_0 - 2) \tag{7.45}$$

$$r_{f2} = a_2' - \frac{m_0}{2}(z_0 + 2.50). \tag{7.46}$$

Placing the two wheels to mesh with each other in the condition of backlash-free contact, we obtain that the clearance will be given by:

$$c = a' - r_{a1} - r_{f2}. \tag{7.47}$$

In the case in which such clearance would be less than $(1/4)m_0$, i.e.

$$c < 0,25m_0, \tag{7.48}$$

the tip diameters of the two wheels must be decreased by the amount

$$2(0.25m_0 - c) = 0.50m_0 - 2c. \tag{7.49}$$

7.6   Possibility of Making Profile-Shifted Toothings by Means of a Pinion-Type Cutter

First, we examine the toothing characteristics of the pinion-type cutter and wheel to be generated, on the generation, or cutting, pitch circle. We assume that the pinion-type cutter has z_0 teeth, nominal pressure angle α_0, nominal module m_0 and standard sizing. If with this pinion-tool we want to cut a wheel having z_1 teeth, with a profile shift coefficient x_1, the common module, m', and the common pressure angle, α_1', of the toothing of the pinion-type cutter and the wheel that is being generated, both measured on the cutting pitch circles, are those given by Eqs. (7.7) and (7.14), while the addendum of the pinion-type cutter and wheel will be given respectively by:

$$h_{a0} = m_0(1 - x_1) \tag{7.50}$$

$$h_{a1} = m_0(1 + x_1). \tag{7.51}$$

The profile shift coefficient x_1 must have a value such as to avoid both the interference between the tooth tip of the pinion-type cutter and the tooth profiles of the wheel to be generated, and the interference between the tooth tip of the wheel and the pinion-type cutter. The first of these two types of interference causes a weakening of the wheel to its root, because of the undercut, while the second determines a removal of part of the tooth flank profile, near the tooth tip (this is the tip relief, which sometimes is done intentionally), and consequently a reduction in the length of the path of contact.

To avoid the first type of interference, that between the tooth tip of the pinion-type cutter and the tooth root profiles of the wheel to be generated, it is necessary that the profile shift coefficient x_1 does not fall below a minimum value $x_{1,min}$. In this regard, the inequality (4.4) must be satisfied. Therefore, by writing this inequality for the case here examined and taking into account the toothing characteristics on the generation pitch circles described above, it must be:

$$2\frac{h_{a0}}{m'} \leq \sqrt{z_0^2 + (z_1^2 + 2z_1 z_0) \sin^2 \alpha_1'} - z_0; \tag{7.52}$$

from this inequality, taking into account Eqs. (7.7), (7.14) and (7.50), we obtain:

$$x_{1,min} \geq \frac{z_0}{2} \left[\frac{\sqrt{(z_0 + 2)^2 + z_1(z_1 + 2z_0) \cos^2 \alpha_0}}{(z_0 + z_1)} - 1 \right]. \tag{7.53}$$

To avoid the second type of interference, that between the tooth tip of the wheel to be generated and the pinion-type cutter, it is necessary that the profile shift coefficient x_1 does not exceed a maximum value $x_{1,max}$. In this regard, the inequality (4.5) must be satisfied. Therefore, by writing this inequality for the case here examined and taking into account the toothing characteristics on the generation pitch circles described above, it must be:

$$2\frac{h_{a1}}{m'} \leq \sqrt{z_1^2 + (z_0^2 + 2z_1 z_0) \sin^2 \alpha_1'} - z_1; \tag{7.54}$$

from this inequality, taking into account Eqs. (7.7), (7.14) and (7.51), we obtain:

$$x_{1,max} \leq \frac{z_0 \left[4(z_1 + 1) - z_0(z_0 + 2z_1) \sin^2 \alpha_0 \right]}{4(z_0 + z_1)(z_0 - 2)}. \tag{7.55}$$

Therefore, to avoid both types of interference, the values of the profile shift coefficient x_1 that can be chosen must necessarily be comprised between the above limit values, that is in the range:

$$x_{1,min} \leq x_1 \leq x_{1,max}. \tag{7.56}$$

Table 7.1 shows the values of $x_{1,min}$ and $x_{1,max}$, calculated using the relationships (7.53) and (7.55), both considered as equalities rather than inequalities, in the case of a pinion-type cutter with $z_0 = 20$ teeth, nominal pressure angle $\alpha_0 = 15°$, and standard sizing, for wheels to be generated having number of teeth z_1 in the range $14 \leq z_1 \leq 30$. The values of $x_{1,min}$ and $x_{1,max}$ shown in this table demonstrate unequivocally that, with the pinion-type cutter here considered, it is not possible to generate wheels with $z_1 \leq 19$ teeth, without the interference will not occur; instead, if $z_1 \geq 20$, it is possible to generate wheels without the interference will not occur, but the variability range of the x_1-values we can choose is very narrow.

The above limitations regarding the possibility of making with a generation process external spur gears with profile-shifted toothing, by means of a pinion-type cutter, together with the greater complication of the calculations to be performed, make reason of the fact that these gears are almost exclusively manufactured using a rack-type cutter.

Table 7.1 Values of $x_{1,min}$ and $x_{1,max}$ calculated by Eqs. (7.53) and (7.55)

z_1	$x_{1,min}$	x_1	$x_{1,max}$
14	0.143	$\leq x_1 \leq$	−0.035
15	0.114		−0.024
16	0.098		−0.013
17	0.071		−0.003
18	0.049		+0.007
19	0.038		+0.016
20	0.01	$\leq x_1 \leq$	0.025
21	−0.001		0.033
22	−0.02		0.041
23	−0.03		0.049
24	−0.04		0.056
25	−0.05		0.063
26	−0.07		0.070
27	−0.08		0.076
28	−0.09		0.082
29	−0.10		0.088
30	−0.12		0.093

7.7 Gears Having Nominal Center Distance and Profile-Shifted Toothing

We now consider a cylindrical spur gear, with pinion and wheel with z_1 and z_2 teeth, both cut by means of a pinion-type cutter, with profile shift coefficients respectively equal to x_1 and x_2. We want to define the condition that must be satisfied, in terms of correlation that links x_1 and x_2, so that these two toothing wheels are meshing with each other without variation of center distance, i.e. with their axis positioned at the nominal center distance.

Consider the first wheel, the one with z_1 teeth, generated with a profile shift coefficient x_1, by means of a pinion-type cutter having z_0 teeth, and nominal pressure angle α_0. The pressure angle α_1' on the generation pitch circle is given by Eq. (7.14). For this wheel can engage, without variation of center distance and with backlash-free contact, with the other wheel having z_2 teeth, it is necessary that the working pressure angle $\alpha_w = \alpha_2''$, given by Eq. (7.41), satisfies the condition:

$$\alpha_w = \alpha_0. \tag{7.57}$$

From Eq. (7.41) in which $\alpha_w = \alpha_0$, we obtain:

$$\mathrm{inv}\alpha_2' = \frac{1}{z_2 + z_0}\left[(z_1 + z_2 + 2z_0)\mathrm{inv}\alpha_0 - (z_1 + z_0)\mathrm{inv}\alpha_1'\right]. \tag{7.58}$$

Finally, from Eq. (7.31) we get:

$$x_2 = \frac{z_2}{2}\left(\frac{\cos \alpha_0}{\cos \alpha_2'} - 1\right). \tag{7.59}$$

Therefore, once the profile shift coefficient x_1 of the pinion has been chosen, to have a gear pair with profile-shifted toothing without variation of center distance, cut by a pinion-type cutter having given characteristics (m_0, α_0, z_0, etc.), it is necessary that $\alpha_1', \alpha_2', x_1$, and x_2 are related by the relationships (7.23), (7.58) and (7.59).

7.8 Transverse Contact Ratio

Here we examine the more general case of an external gear pair, consisting of a pinion and wheel with z_1 and z_2 teeth, having nominal module m_0 and nominal pressure angle α_0, and generated respectively with profile shift coefficient x_1 and x_2. Suppose first that is $c \geq 0,25m_0$, so it is not necessary to make a truncation, that is a reduction of addendum or, what is the same, a reduction in the diameters of the gear blanks.

To calculate the transverse contact ratio, ε_α, we use the Eq. (3.44), in which, however, the addendum factors $k_1 = h_{a1}/m_0$ and $k_2 = h_{a2}/m_0$ are to be determined. Under the working conditions corresponding to backlash-free contact, we have:

$$h_{a1} = r_{a1} - r_{w1} = a_1' - \frac{m_0}{2}(z_2 - 2) - m_0 \frac{z_1 \cos \alpha_0}{2 \cos \alpha_w} \qquad (7.60)$$

$$h_{a2} = r_{a2} - r_{w2} = a_2' - \frac{m_0}{2}(z_0 - 2) - m_0 \frac{z_2 \cos \alpha_0}{2 \cos \alpha_w}. \qquad (7.61)$$

Taking into account Eqs. (7.25) and (7.33), and substituting them inside these last relationships, we obtain:

$$k_1 = \frac{1}{2z_0}[z_0(z_1 + 2 + 2x_1) + 2z_1x_1] - \frac{z_1 \cos \alpha_0}{2 \cos \alpha_w} \qquad (7.62)$$

$$k_2 = \frac{1}{2z_0}[z_0(z_2 + 2 + 2x_2) + 2z_2x_2] - \frac{z_2 \cos \alpha_0}{2 \cos \alpha_w}. \qquad (7.63)$$

The calculation of the transverse contact ratio ε_α by means of Eq. (3.44), in which we place $\alpha = \alpha_w$, is therefore fully defined, since all the quantities that appear in it are known.

It is finally to keep in mind that, in the case where is $c < 0,25m_0$, it is necessary to reduce the tip diameters of the amount defined by relationship (7.49). In this case, in order to determine the addendum factors k_1 and k_2 to be introduced into Eq. (3.44) we can use the same procedure described above, but we must remember that h_{a1} and h_{a2}, given by Eqs. (7.60) and (7.61), must both be decreased by $(0.25m_0 - c)$, where c is given by Eq. (7.47). Consequently, also the addendum factor k_1 and k_2 will be reduced, and then the transverse contact ratio will be decreased.

References

1. Buckingham E (1949) Analytical mechanics of gears. Book Company Inc, McGraw-Hill New York
2. Galassini A (1962) Elementi di Tecnologia Meccanica: Macchine Utensili – Principi Funzionali e Costruttivi, Loro Impiego e Controllo della loro Precisione, Reprint 9th edn. Editore Ulrico Hoepli, Milano
3. Henriot G (1979) Traité théorique and pratique des engrenages, vol 1, 6th edn. Bordas, Paris
4. ISO 6336-1 (2006) Part 1—Basic principles, introduction and general influence factors
5. Maitra GM (1994) Handbook of gear design, 2nd edn. Tata McGraw-Hill Publishing Company Ltd, New Delhi
6. Merritt HE (1954) Gears, 3rd edn. Sir Isaac Pitman&Sons, London
7. Micheletti GE (1958) Tecnologie Generali: Lavorazioni dei Metalli ad Asportazione di Truciolo. Libreria Editrice Universitaria Levrotto & Bella, Torino

8. Niemann G, Winter H (1983a) Mashinen-Elemente, Band II: Getriebe allgemein, Zahnradgetriebe-Grundlagin, Stirnradgetriebe. Springer, Heidelberg
9. Pollone G (1970) Il Veicolo. Libreria Editrice Universitariaa Levrotto & Bella, Torino
10. Rossi M (1965) Macchine Utensili Moderne. Ulrico Hoepli, Milano

Chapter 8
Cylindrical Involute Helical Gears

Abstract In this chapter, the fundamentals of cylindrical involute helical gears geometry are given and the related main quantities are defined, including the total length of the line of action. The equivalent parallel cylindrical spur gears and their virtual number of teeth are also determined. The fundamentals of the cylindrical helical gears profile shift are then described, in its most general terms related to the variation of center distance. The load analysis of these gears is then performed and the thrust characteristics on shafts and bearings are defined. Short notes are also made on the parallel cylindrical double-helical gears. Subsequently, the main kinematic quantities related to the rolling and sliding motions of the mating teeth surfaces are defined and the instantaneous and average efficiencies of these gears are analytically determined. Finally, short notes on the main cutting processes of parallel cylindrical helical and double-helical gears are given.

8.1 Introduction

Cylindrical involute spur gears with parallel axes analysed in the previous chapters have some disadvantages, the main of which consists of the fact that only a part of each meshing cycle is carried out with two pairs of teeth in contact, while the remaining part is carried out with only one pair of teeth in contact. This determines sudden variations in the total length of the contact line (this varies suddenly from $2b$ to b, depending on whether two tooth pairs or one tooth pair are in contact), and abruct changes in the mesh stiffness, resulting in a very noisy operation and working irregularity. For more, a tooth pair comes into contact over its entire face width, b, and also loses contact for its entire face width, thus accentuating the discontinuity of working conditions [5, 15, 26, 27].

Parallel cylindrical helical gears, or simply cylindrical helical gears, were born and developed to overcome the disadvantages of the cylindrical spur gears. They can be considered as the evolution of the *stepped gear wheel* proposed for the first time by Hooke, and therefore called the *Hooke's wheel* [4, 10, 13, 28]. This stepped gear wheel (Fig. 8.1) is however still a cylindrical spur gear in the proper meaning

V. Vullo, *Gears*, Springer Series in Solid and Structural Mechanics 10,
https://doi.org/10.1007/978-3-030-36502-8_8

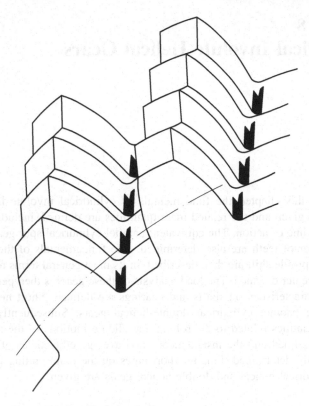

Fig. 8.1 Stepped gear wheel or Hooke's wheel

of the term, as it is constituted by a discrete set of spur gear, all the same and connected to the same shaft, but rotated by a certain angle relative to one another. By choosing this angle with value much lower than that corresponding to half of the angular pitch of the teeth, Hooke obtained a gear characterized by a significant increase in the length of the path of contact; this is because, when the first wheel of the set thus formed comes out of contact, the others are still in contact.

A cylindrical helical gear may be considered as the limit configuration of the Hooke's wheel. To this end, let us first consider the infinite number of gear wheels, having differential face widths, db, obtained by cutting a conventional cylindrical spur gear with planes perpendicular to its axis. Now, let us rotate these gear wheels, one with respect to the other, in the tangential direction, of a differential arc of pitch circle, proportional to the increment of coordinate dz along the axis of the same wheel. In this way, we get a cylindrical helical gear.

A conventional cylindrical spur gear can be considered as generated by a planar wheel, i.e. one of these infinitesimal gear wheels, when it moves, parallel to itself (that is, remaining invariably oriented) along its axis. However, this gear wheel with infinitely small thickness, while moving along its axis, also rotates with respect to it, the wheel so generated is a cylindrical helical gear. If then, in this motion of generation, the ratio between the displacement speed along the axis and the rotation speed about it remains constant, the helicoids, which constitute the tooth flanks, have a constant axial pitch. This is the case of almost all the cylindrical helical gears, which today are manufactured.

It is noteworthy that the Hooke's wheel is, in effect, a cylindrical spur gear, because it does not give rise to axial loads on the shaft, but only generates radial and tangential loads. The cylindrical helical gear instead generates, in addition to the radial and tangential loads, also an axial load, correlated to the helix angle, to be supported by suitable bearings. It is to be noted that the profiles of the teeth of the cylindrical helical gears that characterize practical applications are involute profiles. Therefore, we focus our attention on the cylindrical involute helical gears.

It is also to be noted that cylindrical involute helical gears are used both to transmit motion between parallel axes, and to transmit motion between non-parallel and non-intersecting axes. In the first case, we have *parallel helical gears*, which have the same function as cylindrical spur gears, but with much more interesting performances, as we will see in the sections below. Instead, in the second case, we have *crossed helical gears*. In this chapter, only parallel helical gears are discussed, while crossed helical gears are dealt with in Chap. 10.

8.2 Geometry of Parallel Involute Helical Gears

Let us now examine in more detail the generation of a cylindrical involute helical wheel, obtained by means of the planar wheel that, as it moves along its axis, rotates around it, with a constant ratio between the displacement speed along the axis, and the rotation speed around to it. The surface of the tooth flanks of the cylindrical involute helical wheel thus obtained, is a portion of an involute helicoid, generated by the helical motion, with lead p_z, of an involute curve of a base circle having radius r_b, contained in a transverse plane, i.e. a plane perpendicular to the axis of the wheel.

With the above generation motion, the circles characterizing the planar gear wheel generate respectively the cylinders that characterize the parallel involute helical gear wheel, i.e.: the base and pitch cylinders, and the tip and root cylinders. The intersections of the tooth flanks with these cylinders, or any other coaxial cylinder, are cylindrical helices, each with the same lead p_z. We can regard the tooth surface as formed by a family of helices, each helix lying in a cylinder coaxial with the gear wheel, and each with the same lead. All helices lying on a cylinder of given radius have the same helix angle β, defined as the angle between the tangent to the helix, at each point, and the generatrix of the cylinder passing through the same point (or, in equivalent terms, the angle between the tangent to the helix, and the axis of the wheel). The helices lying on several coaxial cylinders have the same lead, but different helix angle [23, 37].

It should be noted that the term *helix angle* used in connection with a helical gear means, conventionally, the helix angle at the pitch cylinder, and it is usually denoted by β, without any subscript. The *lead angle* is the complementary angle to the helix angle; therefore, it is given by:

$$\gamma = (\pi/2) - \beta. \tag{8.1}$$

The lead angle γ_c related to any other cylinder of diameter $d_c = 2r_c$ can be obtained by the following general relationship:

$$\tan \gamma_c = \tan \gamma \frac{d}{d_c} = \tan \gamma \frac{r}{r_c}, \tag{8.2}$$

where $d = 2r$ is the diameter of the pitch circle of the helical gear.

According to what we said above, the section of the involute helicoid with any transverse plane, i.e. any plane perpendicular to the axis of the gear wheel, is always an involute curve, which originates at a given point of the base cylinder of radius r_b. On the surface of this cylinder, the points of regression of the various involute curves describe a helix of lead p_z, called *base helix*, and characterized by the *base helix angle*, β_b, defined by the following relationship (Fig. 8.2):

$$p_z = 2 \pi r_b \cot \beta_b = 2 \pi r_b \tan \gamma_b; \tag{8.3}$$

the lead p_z represents the *axial advance* of the helix in one complete turn.

More simply, the involute helicoid can also be thought of as generated by the paths of the points of a straight line, l (generation straight line), belonging to a plane tangent to the base cylinder, and forming the base helix angle, β_b, with the axis of the wheel, when this plane rolls without sliding on the same base cylinder (Fig. 8.3). In fact, by doing so, the generating line is wrapped, on the base cylinder, according to a helix with base helix angle β_b, while its points describe, in planes

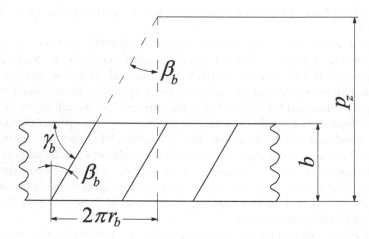

Fig. 8.2 Correlation between p_z, r_b, β_b and γ_b

Fig. 8.3 Generation of an
involute helicoid

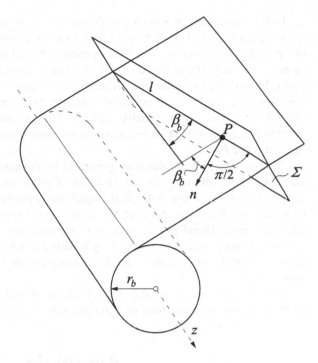

perpendicular to the wheel axis, involute curves of circles of radius r_b, having their
points of regression on this helix. This second way of generation of the involute
helicoid gives a simple explanation of its following fundamental geometry
properties:

- The involute helicoid is a ruled surface, formed by the successive positions of
  the generation straight line l. For each point P of its surface passes a straight line
  l. The involute helicoid is therefore a developable ruled surface, which can be
  also thought of as constituted by the successive tangents to the base helix. The
  tooth flanks, as developable ruled surfaces, can therefore be grinded with relative ease by grinding wheels with developable surface. An involute helicoid
  is uniquely defined by the radius r_b of the base cylinder, and the base helix angle
  β_b.
- The normal n to the helicoid surface, in one of its points P, is contained in the
  plane tangent to the base cylinder through P, and is normal to the straight line
  l through P.
- The plane tangent to the base cylinder passing through any point P of the
  involute helicoid surface intersects this last surface along a straight line; it is the
  straight line l passing through that point.
- The plane Σ tangent to the helicoid surface, in one of its point P, is normal to the
  plane tangent to the base cylinder through P; it touches the helicoid surface
  along the straight line l through P.

Two involute helicoids with parallel axes that are touching at a point P must
have the same normal. Therefore, from what we have seen above, all contacts must
occur in the plane tangent to the base cylinder, which therefore constitutes the plane
of action, because it contains the lines of action of the parallel involute helical gear
pair (Fig. 8.4). If, moreover, at a given instant of time, we have contact at a point P,
we must have contact at all points of a straight line l through P and inclined, with
respect to the axis, the base helix angle β_b. This because, in all the points of this
straight line, each of the two helicoids has the same direction of the normal that we
have at point P.

The two helicoids are therefore generated by the same straight line l (Figs. 8.3
and 8.4), during the wrapping of the plane of action on the respective base cylinders, and are then to have helix angle equal, but opposite. Therefore, if the helicoid
of a wheel is a right-hand helicoid, that of the mating gear will be left-hand helicoid,
and vice versa. Therefore, the helices on the operating pitch cylinders (or any of
coaxial cylinders between the root cylinders and tip cylinders) are of opposite
direction, but the magnitude of the helix angle (or the lead angle) is the same for
both helices.

In this regard, we recall that we call conventionally right-hand and left-hand
gears those gear wheels whose helicoid surfaces are generated by generating curves

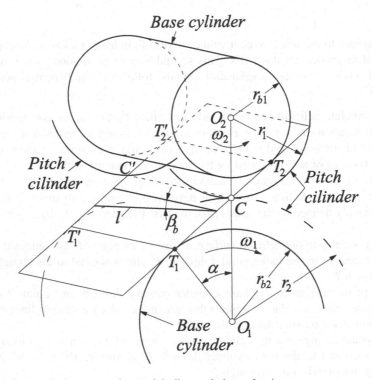

Fig. 8.4 Base cylinders, generation straight line, and plane of action

that move with right-hand and, respectively, left-hand screw motions. Figure 8.5a, b show a parallel helical gear pair, consisting of a right-hand pinion and left-hand wheel, and respectively a helical rack-pinion pair, consisting of a left-hand pinion and right-hand rack. In addition, let us remember that, according to [16] ISO 1122-1, *right-hand teeth* and *left-hand teeth* are teeth whose successive transverse profiles show clockwise displacement and respectively counterclockwise displacement with increasing distance from an observer looking along the straight-line generatrixes of the reference surface.

On the plane of action, the two helicoids are touching along a straight line, which is inclined of the angle β_b with respect to the axis, and constitutes the instantaneous line of action. During the progression of the relative motion, for well-known properties of the involute curves, this straight line moves parallel to itself on the plane of action, with a linear velocity equal to the base-line velocity (i.e. the tangential velocity at the radii of the base cylinders). Figure 8.4 shows:

- the two pitch cylinders, having radii r_1 and r_2, touching along the instantaneous axis of rotation, or pitch line CC', shown with dashes line and passing through the pitch point C;
- the two base cylinders, having radii r_{b1} and r_{b2};
- the plane of action, which is limited by the two end faces of the two gear members, whose distance measured along their axes defines the face width, b (both gear members have the same face width);
- the instantaneous line of action, in its current position on the plane of action.

The same figure shows the *theoretical rectangle of contact*, $T_1 T_1' T_2 T_2'$, which is bounded by the two aforementioned end faces, on two opposite sides, and, on the other two sides, by the two straight lines of tangency of the plane of action with the

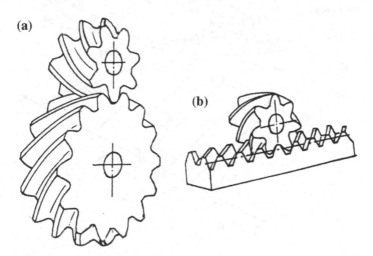

Fig. 8.5 **a**, parallel helical gear pair (right-hand pinion, and left-hand wheel); **b**, helical rack-pinion pair (left-hand pinion, and right-hand rack)

base cylinders. In an integrated view of the parallel involute gears, which includes both spur gears and helical gears, according to the aforementioned [16] ISO 1122-1, as *plane of action* we mean the plane containing the lines of action of a parallel involute gear pair; therefore, it is identified and coincides with the above-defined theoretical rectangle of contact. As well as for parallel cylinder spur gears, not the entire surface of this theoretical rectangle of contact or plane of action can be used.

The points T_1 and T_2, or better the lines $T_1 T_1'$ and $T_2 T_2'$, represent the loci of primary interference. To avoid this type of interference, the tip cylinder of the two members of the parallel helical gear pair must intersect the plane of action along lines AA' and EE'; these are on the same plane as lines $T_1 T_1'$ and $T_2 T_2'$, but are moved in the direction of the instantaneous axis of rotation CC' (Fig. 8.6). We identify so the *actual rectangle of contact*, $AA'EE'$. The two opposite sides of this rectangle AA' and EE' have length equal to the face width b $(AA' = EE' = b)$, while the other two sides AE and $A'E'$ have length equal to the length of path of contact $g_\alpha (AE = A'E' = g_\alpha)$; the latter, as shown in Fig. 8.6, is the sum of the length of path of approach, g_f, and length of path of recess, g_a, i.e. $g_\alpha = g_f + g_a$. It should be noted that all transverse sections of the actual rectangle of contact are the same, so we call as path of contact both the whole actual rectangle of contact and any of its transverse sections.

The length of the instantaneous line of action varies during the meshing cycle. In the initial stage, the length increases from zero to a maximum value, which depends

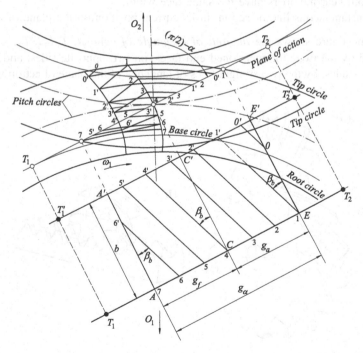

Fig. 8.6 Parallel helical gear pair: meshing diagram, rectangles of contact, and instantaneous lines of action on the tooth surface and on the actual rectangle of contact

on the face width, b, and base helix angle β_b; in the middle stage, it is maintained constant and equal to the maximum value; in the final stage, it decreases from the maximum value to zero. These lengths are shown, in real size, in Fig. 8.6 at the bottom, where the actual rectangle of contact is highlighted, after rotating the plane to which it belongs to show its contour. Figure 8.6 also shows, at the top, the lengths of the instantaneous lines of action, drawn on the tooth flank surface, and seen from the inside. The procedure for the geometrical construction of these length is quite evident. Of course, depending upon the procedure of geometrical projection, the length $00'$, $11'$, ... on the tooth flank surface are different from the actual lengths, represented on the actual rectangle of contact, shown below. Note that the segments $00'$, $11'$, ... on the tooth flank surface, if prolonged, are tangent to the base circle.

We immediately notice one of the fundamental differences between parallel spur and helical gears. In fact, the contact between two parallel spur gears in mesh takes place always along a contact line extending along the whole face width of the wheels, this line being always parallel to the axis of the gear. Instead, unlike spur gears, the instantaneous line of contact of a parallel helical gear is a diagonal across the operating flank surface of the tooth. This instantaneous line of contact (or instantaneous line of action) is the intersection of the plane of action, fixed and invariable, and the surface of the operating flank of the tooth, that instead changes its position with respect to the plane of action, instant by instant, during the meshing cycle. The inclination of these diagonal lines with respect to the axis of the wheel depends on β_b, and is much higher the greater is β_b.

We saw that the initial contact between two parallel helical gears is a point (point $A \equiv 7$, in Fig. 8.6), which gradually changes into a line as the meshing cycle proceeds. Therefore, the teeth come into meshing gradually. Similarly, they come out gradually from the meshing, because in the final stage of the meshing cycle the contact locus gradually changes from a line to a point (point E', in Fig. 8.6). With the exception of the case of wheels having face width very thin, in the parallel helical gears several pairs of teeth are simultaneously in engagement. This fact explains the many advantages we have compared to parallel spur gears. In fact, the total stiffness of the teeth (remember that this is the stiffness of all the teeth in meshing, and that its mean value is the mesh stiffness c_γ (it is better defined in Vol. 2, Sect. 1.6), fluctuates less, the greater the contact ratio. Furthermore, the rotation motion is transmitted in a uniform way, and operating noise and noise emissions are much lower.

Figure 8.7 shows one of the above instantaneous line of action on the surface of the tooth flank, in a generic position corresponding to its maximum length, l_{max}, which occurs in the middle stage of the meshing cycle. The length, l, of these diagonal lines varies from zero to the maximum value, l_{max}, and this maximum value is different depending on whether:

$$g_\alpha \cot \beta_b \gtrless b \qquad (8.4)$$

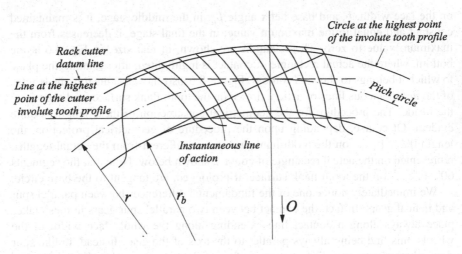

Circle at the highest point of the involute tooth profile

Rack cutter datum line

Line at the highest point of the cutter involute tooth profile

Pitch circle

Instantaneous line of action

r r_b O

Fig. 8.7 Instantaneous line of action on the tooth flank surface, having maximum length

As shown in Fig. 8.8, this maximum value is given by

$$l_{max} = \frac{b}{\cos \beta_b},$$ (8.5)

when $g_\alpha \cot \beta_b > b$, while it is given by

$$l_{max} = \frac{g_\alpha}{\sin \beta_b},$$ (8.6)

when $g_\alpha \cot \beta_b < b$.

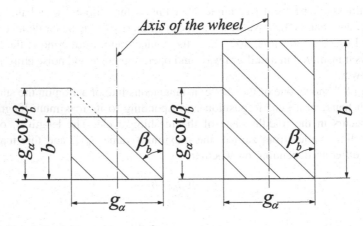

Axis of the wheel

$g_\alpha \cot \beta_b$ b β_b

$g_\alpha \cot \beta_b$ β_b b

g_α g_α

Fig. 8.8 Correlation between b, g_α and β_b

During the meshing cycle, the total length, l_t, of the line of action changes, with the position of the teeth. This aspect and its implications are discussed in detail in Sect. 8.7.

We now intersect the helicoid with two coaxial cylinders, the reference pitch cylinder and base cylinder. We get two helices, which have the same lead, p_z, but different helix angle, β and β_b. Figure 8.9 shows these two helices, having lead p_z, after having spread out in the flat the two cylinders with radii r and r_b. Since p_z is the same for the two helices, we have:

$$2\,\pi\,r \cot \beta = 2\,\pi\,r_b \cot \beta_b, \tag{8.7}$$

which coincides with Eq. (8.1), in which $\beta_c = \beta_b$, and $r_c = r_b$. Taking into account that

$$r_b = r \cos \alpha, \tag{8.8}$$

from Eq. (8.7) we obtain

$$\tan \beta_b = \tan \beta \cos \alpha; \tag{8.9}$$

this equation correlates between them the base helix angle, β_b, helix angle, β, and pressure angle, α.

8.3 Main Quantities of a Parallel Helical Gear

As for a parallel cylindrical spur gear, also the main specific quantities of a parallel cylindrical helical gear are defined with reference to the helical basic rack, which is shown in Fig. 8.10. This figure shows the profile of a tooth, both in the transverse section (section with a plane perpendicular to the wheel axis), and in the normal section (section with a plane perpendicular to the tooth axis). Of course, the tooth depth, h, is the same in both of these sections. However, the tooth appears slenderer in the normal section (its thickness, measured in the rack reference plane, is equal to

Fig. 8.9 Reference and base helices, on the cylinders with radii r and r_b, spread out in the flat

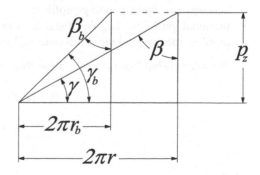

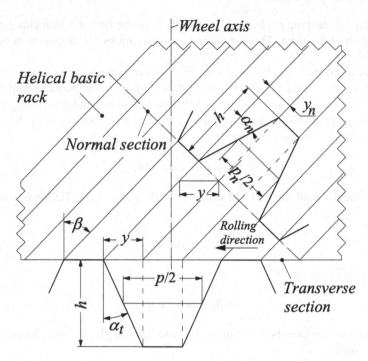

Fig. 8.10 Helical basic rack, and transverse and normal section of a tooth

half the normal pitch, p_n), and stubbier in the transverse plane (its thickness, measured always in the rack reference plane, is equal to half the transverse pitch p).

Four kinds of pitches can be considered for parallel helical gears. The first of them is the lead of the helix, p_z, which we have already defined in the previous section (see Eq. (8.3)). The other three pitches, which are correlate to the toothing, are shown schematically in Fig. 8.11. They are respectively:

- the transverse pitch, p_t, or simply pitch (of course, $p = p_t$), i.e. the length of the arc of the transverse reference circle between two consecutive corresponding profiles;
- the normal pitch, p_n, i.e. the length of the arc of a helix normal to the pitch helix, between two corresponding profiles;
- the axial pitch, p_x, i.e. the distance between two consecutive corresponding profiles, measured along a reference cylinder generatrix.

The relationship that correlate these three pitches (Fig. 8.11) are as follows:

$$p_n = p \cos \beta \qquad (8.10)$$

$$p_x = \frac{p_n}{\sin \beta} = p \cot \beta. \qquad (8.11)$$

Fig. 8.11 Transverse, normal
and axial pitches on the
reference cylinder spread out
in the flat

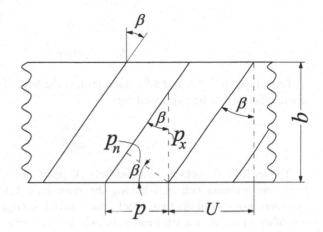

By comparison of Eqs. (8.3) and (8.11), taking into account Eq. (8.9), we obtain:

$$p_z = z p_x. \tag{8.12}$$

Two kinds of modules are introduced in the discussion of the parallel helical gears: the transverse module $m_t = m = p/\pi$, and the normal module, $m_n = p_n/\pi$. According to Eq. (8.10), these two modules are related by the relationship:

$$m_n = m \cos \beta. \tag{8.13}$$

With reference to Fig. 8.10, we can find the relationship between the transverse pressure angle, $\alpha_t = \alpha$, normal pressure angle, α_n, and helix angle, β, of the helical basic rack. We have in fact:

$$y = h \tan \alpha_t = h \tan \alpha \tag{8.14}$$

$$y_n = h \tan \alpha_n. \tag{8.15}$$

However, we have also:

$$y_n = y \cos \beta. \tag{8.16}$$

Therefore, replacing in this last relationship the Eqs. (8.14) and (8.15), we get:

$$\tan \alpha_n = \tan \alpha \cos \beta. \tag{8.17}$$

The diameters of the reference pitch cylinder and base cylinder of a wheel with z_1 teeth, which meshes with the aforementioned helical basic rack, in the condition where the reference plane of the basic rack is tangent to the reference pitch cylinder of the wheel, are respectively given by:

$$d_1 = z_1 m \tag{8.18}$$

$$d_{b1} = z_1 m \cos \alpha. \tag{8.19}$$

From Eqs. (8.3, 8.8 and 8.9), we infer that the *lead* of the helix on the reference pitch cylinder can be expressed as:

$$p_z = \frac{\pi z_1 m}{\tan \beta}. \tag{8.20}$$

The teeth flank surfaces of a parallel cylindrical helical gear pair, with z_1 and z_2 teeth, are involute helicoids having the same base helix angle, β_b, on the base cylinders. For one of the two wheels, the helicoid is a right-hand helicoid, while for the other wheel it is a left-hand helicoid, or vice versa. Therefore, the transverse pitches of the two helicoids, measured on the reference pitch cylinders, will be given respectively by:

$$p_1 = \frac{\pi z_1 m}{\tan \beta} = \frac{\pi z_1 m \cos \alpha}{\tan \beta_b} \tag{8.21}$$

$$p_2 = \frac{\pi z_2 m}{\tan \beta} = \frac{\pi z_2 m \cos \alpha}{\tan \beta_b}, \tag{8.22}$$

where m is the module, while α and β are respectively the pressure angle and angle of inclination of the teeth of the generation rack-type cutter, with respect to the axes of the two wheels. Equations (8.21) and (8.22) show that the transverse pitches of the two wheels are proportional to their number of teeth. Of course, the normal pitch, p_n, is the same for the two wheels, as it is easy to show bearing in mind Eqs. (8.21, 8.22 and 8.10), and considering that $z_1/z_2 = d_1/d_2$.

Referring to one of the two members of the parallel cylindrical helical gear pair, the transverse base pitch and normal base pitch on the base cylinder will be given respectively by:

$$p_{bt} = p_t \cos \alpha_t = p \cos \alpha \tag{8.23}$$

$$p_{bn} = p_n \cos \alpha_n. \tag{8.24}$$

Another key advantage of the parallel cylindrical helical gears against the parallel cylindrical spur gears consists of the significant increase in the contact ratio. In fact, for these gears, we have:

$$\varepsilon_\gamma = \varepsilon_\alpha + \varepsilon_\beta \tag{8.25}$$

where ε_γ, ε_α and ε_β are respectively the *total contact ratio, transverse contact ratio* and *overlap ratio*. Therefore, the total contact ratio is equal to the sum of the transverse contact ratio and overlap ratio.

Referring to Fig. 8.6, we can define the transverse contact ratio similarly compared to what we have already done for parallel cylindrical spur gears, that is, as:

$$\varepsilon_\alpha = \frac{g_\alpha}{p \cos \alpha},$$
(8.26)

where g_α is the length of the path of contact. This quantity is given respectively by the Eqs. (3.24, 3.33 and 3.40), depending on whether the parallel cylindrical helical gear pair is an external gear, a rack-pinion gear pair, or an internal gear.

To determine the overlap ratio, it is convenient to introduce the *face advance* or *tooth advance* or *offset*, U. This is defined as the distance, measured on the transverse pitch circle, through which a helical tooth moves from the initial position at which the contact begins at one end face of the wheel to the final position at the other end face where the contact ceases. This face advance, equal to $U = b \tan \beta$ (Fig. 8.11), is due to the helical orientation of the tooth course along the face width. Therefore, we define as overlap ratio, ε_β, the quotient of the face advance of a helical tooth divided by the transverse circular pitch, or, what comes to the same thing, the quotient of the face width divided by the axial pitch. Thus, we have:

$$\varepsilon_\beta = \frac{U}{p} = \frac{b \tan \beta}{p} = \frac{b}{p_x} = \frac{b \sin \beta}{\pi m_n}.$$
(8.27)

Generally, the face width b of a parallel cylindrical helical gear is made sufficiently large, so that the face advance U corresponding to a given helix angle β is greater than the transverse circular pitch, p. In this case we have $\varepsilon_\beta > 1$. Thus, from the theoretical point of view, it would be possible to ensure the continuity and uniformity of the motion even if ε_α were zero, that is, with a tooth depth $h = 0$. In fact, this happens in some special toothings, as the (W/N)-toothings (Wildhaber/Novikov toothings), which are characterized by a transversal contact ratio equal to zero ($\varepsilon_\alpha = 0$).

We will not talk of these special toothings (on this topic, we refer the reader to: Niemann and Winter [27], Litvin and Fuentes [24], Radzevich [29, 30]), and focus our attention on the parallel cylindrical involute helical gears for the usual practical applications, which are always characterized by both the transversal contact ratio, and the overlap ratio. Of course, from the kinematic point of view, the transmission of a uniform motion is ensured if $\varepsilon_\gamma \geq 1$. In general, however, it is required that ε_α and ε_β are both greater than one ($\varepsilon_\alpha > 1$, and $\varepsilon_\beta > 1$), for which $\varepsilon_\gamma \geq 2$, and this in order to obtain uniform and silent meshing condition. It is noteworthy that, to ensure continuity of motion only with the overlap ratio ε_β, the limiting value of the face advance must be $U = p$. For the sake of safety, it is customary to increase the corresponding value of the face width b by at least 15%, so that from Eq. (8.27) we obtain:

$$b \geq \frac{1.15 p}{\tan \beta}.$$
(8.28)

Finally, we must determine the two principal radii of curvature at any point P on the surface of the tooth flank, because they are used in calculation of load capacity of helical gears, and particularly in calculation of surface durability (pitting). These two principal radii of curvature are one maximum and the other minimum. Since the helicoid is a developable ruled surface, they occur, the first in the plane passing through point P and tangent to the base cylinder, and the second in the normal plane, i.e. in the plane passing through point P, perpendicular to the previous one and containing the normal n (Fig. 8.3). Of course, the first principal radius of curvature, the one in the plane passing through point P and tangent to the base cylinder, is the maximum principal radius of curvature, and is equal to infinity.

To calculate the second principal radius of curvature, the minimum one in the normal plane passing through point P, we consider the two planes passing through P, the transverse plane and normal plane, both perpendicular to the plane of action (Fig. 8.3). The angle between these two planes is β_b. According to well-known properties of differential geometry of surfaces (see do Carmo [8, 9], Gauss [12], Smirnov [34], Stoker [35]), the radii of curvature in two planes belonging to the same star of planes, one of which is a principal plane, differ from one another by an amount equal to the cosine of the angle between them. In other words, with reference to the case of interest here, if we indicate with ρ_t the radius of curvature in the transverse plane, the second, which is the minimum principal radius of curvature in the normal plane, $\rho_{n,min} = \rho_1$, will be given by:

$$\rho_1 = \rho_{n,min} = \frac{\rho_t}{\cos \beta_b}. \tag{8.29}$$

However, from Eq. (3.49), we have

$$\rho_t = r \sin \alpha + g_p, \tag{8.30}$$

for which the Eq. (8.29) becomes:

$$\rho_1 = \rho_{n,min} = \frac{r \sin \alpha + g_p}{\cos \beta_b}. \tag{8.31}$$

8.4 Generation of Parallel Cylindrical Helical Gears and Their Sizing

Usually, parallel cylindrical helical gears are generated by generation cutting process, carried out by means of gear cutting machines (gear shapers, gear hobbers, or Fellows gear shapers) that use cutting tools configured in accordance with the helical basic rack (Fig. 8.10). These cutting tools can be rack-type cutters, hobs or helical pinion-type cutters; these last, during the generation cutting, mesh with the helical gear wheel to be cut. The helical teeth of the gear wheel to be cut have helix angles equal to those of the teeth of the generating cutter. A little more detail of the cutting processes of cylindrical helical gear wheeels is made in Sect. 8.11.

Using gear cutting machine that, as cutting tools, use rack-type cutters or hobs (then, gear shapers and gear hobbers), we can vary the helix angle of the toothing without changing the cutting tool. Instead, using gear cutting machines that, as cutting tools, use helical pinion-type cutters (thus, Fellows gear shapers), to vary the helix angle of the toothing it is necessary to change the generating cutter as well as the helical guide of the Fellows gear shaper.

The generation of parallel cylindrical helical gears with the so-called method of the form cutting, using a cutting tool shaped exactly like the tooth space between two adjacent teeth of the wheel, is now rarely used. However, with this method, to obtain the normal section of the tooth space between two adjacent teeth of a wheel with z_1 teeth, reference helix angle β, and normal module m_n, we must use the same milling cutter having normal module m_n, which we use to obtain a parallel cylindrical spur gear wheel having number of teeth z_v, given by (see next section):

$$z_v = \frac{z_1}{\cos^3 \beta_b}. \tag{8.32}$$

When rack-type cutters generate parallel cylindrical helical gears, the same cutting tools used for cutting parallel cylindrical spur gears are used. The gear-tooth sizing of parallel cylindrical helical gear wheels is therefore carried out using the nominal values of the normal modules m_n (i.e. in the normal section to the axis of the teeth) standardized by the ISO 54 [17]. The gear-tooth sizing most widely used for parallel cylindrical helical gears also provides a standard addendum equal to module m_n (in mm) and a standard dedendum equal to 125 times the module m_n, that is:

$$\begin{aligned} h_a &= m_n \\ h_f &= 1,25 m_n = (5/4) m_n \\ h = h_a + h_f &= 2,25 m_n = (9/4) m_n. \end{aligned} \tag{8.33}$$

As Fig. 8.10 shows, in transverse section the teeth appear to have stub sizing, because in this section, according to Eq. (8.13), the module is equal to

$$m = \frac{m_n}{\cos \beta}, \tag{8.34}$$

and varies with the helix angle β. If we denote by α_n the pressure angle of the cutting tool tooth, the pressure angle α of the tooth profile in the transverse section is greater because, due to Eq. (8.17), we have:

$$\tan \alpha = \frac{\tan \alpha_n}{\cos \beta}. \tag{8.35}$$

Parallel cylindrical helical gears can be generated by helical pinion-type cutters, which can generate only teeth having a predetermined value of the base helix angle β_b. With this cutting process, the gear-tooth sizing is in general carried out in such a way that the pressure angle α in the transverse section of the profile has a given value, and the transverse module m assumes one of the values standardized by the aforementioned ISO 54 [17]. For example, with some cutters, (Sykes type cutters), a gear tooth stub sizing in the transverse section is used, namely:

$$h_a = 0.8m \quad h_f = 1.1m. \tag{8.36}$$

Example 1 Input data: cylindrical helical wheel to be generated by rack-type cutter, with: $z = 20$; $\alpha_n = 15°$; $\beta = 30°$; $m_n = 3$ mm. We will have: $m = m_n/\cos\beta = 3.46$ mm; $d = zm_n/\cos\beta = 69.28$ mm; $\alpha = 17°11'$; $h_a = 3.0$ mm; $h_f = 3.75$ mm.

Example 2 Input data: cylindrical helical wheel to be generated by a pinion-type cutter with: $z = 20$; $\alpha = 20°$; $\beta = 30°$; $m = 3.5$ mm. We will have: $d = zm = 70$ mm; $h_a = 0.8m = 2.80$ mm; $h_f = 1.1m = 3.85$ mm.

Now consider an external parallel helical gear constituted by two cylindrical helical wheels having z_1 and z_2 teeth, reference normal module m_n, and reference helix angle β. Taking into account the Eqs. (8.13) and (8.18), the diameters of the reference pitch circles can be expressed as:

$$d_1 = \frac{z_1 m_n}{\cos\beta} \tag{8.37}$$

$$d_2 = \frac{z_2 m_n}{\cos\beta}. \tag{8.38}$$

Therefore, the center distance, a, is given by:

$$a = \frac{d_1 + d_2}{2} = \frac{z_1 + z_2}{2}\frac{m_n}{\cos\beta}. \tag{8.39}$$

Thus, it can be varied by varying the helix angle β. The minimum value that the center distance can assume is the nominal one, namely that of the cylindrical spur gear pair, for which $\beta = 0$; this minimum value is given by:

$$a_{min} = \frac{z_1 + z_2}{2}m_n. \tag{8.40}$$

If the parallel cylindrical helical gear is an internal helical gear pair consisting of a pinion and an internal helical gear wheel, respectively having reference diameters d_1 and d_2, given by Eqs. (8.37) and (8.38), its center distance, a, is given by:

$$a = \frac{d_2 - d_1}{2} = \frac{z_2 - z_1}{2} \frac{m_n}{\cos \beta}. \tag{8.41}$$

Its minimum value, corresponding to that of the internal spur gear ($\beta = 0$), will instead be given by:

$$a_{min} = \frac{z_2 - z_1}{2} m_n. \tag{8.42}$$

By Eqs. (8.39) and (8.40), we obtain

$$\cos \beta = \frac{z_1 + z_2}{2a} m_n = \frac{a_{min}}{a}. \tag{8.43}$$

By Eqs. (8.41) and (8.42), we obtain:

$$\cos \beta = \frac{z_2 - z_1}{2a} m_n = \frac{a_{min}}{a}. \tag{8.44}$$

If, for a given value of the center distance, a, the helix angle β is too large, the axial loads on shafts and bearings will be equally large, as shown later. In order to reduce these axial loads within acceptable values, there are two possible solutions: with the first solution, we can increase z_1 and z_2, holding constant the gear ratio $u = z_2/z_1$, and so respecting an input generally imposed; with the second solution, we can impose an acceptable value of the reference helix angle β, and make use of profile-shifted toothings (see Sect. 8.6).

For the evaluation of all the remaining quantities, we proceed in a similar way to what we said for parallel cylindrical spur gears. Table 8.1 summarizes the main quantities of an external parallel cylindrical helical gear.

Table 8.1 Main quantities of an external parallel cylindrical helical gear generated by rack-type cutter

Quantity	Pinion	Wheel
Number of teeth	z_1	z_2
Reference pitch circle diameter	$d_1 = z_1 m_n / \cos \beta = z_1 m$	$d_2 = z_2 m_n / \cos \beta = z_2 m$
Tip circle diameter	$d_{a1} = d_1 + 2m_n$	$d_{a2} = d_2 + 2m_n$
Root circle diameter	$d_{f1} = d_1 - 2.50 m_n$	$d_{f2} = d_2 - 2.50 m_n$
Base circle diameter	$d_{b1} = d_1 \cos \alpha$	$d_{b2} = d_2 \cos \alpha$
Nominal tooth thickness and space width in the normal section	$s_n = e_n = p_n/2 = \pi m_n/2$	
Nominal tooth thickness and space width in the transverse section	$s = e = p/2 = \pi m/2$	
Reference center distance	$a = (d_1 + d_2)/2 = [(z_1 + z_2)/2](m_n/\cos \beta)$	

The tooth thickness s and s' in transverse plane, measured on cylinders having radii r and r', are related by the Eq. (2.58), here rewritten:

$$s' = r'\left[\frac{s}{r} + 2(\text{inv } \alpha - \text{inv } \alpha')\right], \qquad (8.45)$$

in which symbols have the same meaning as described in Sect. 2.5.

8.5 Equivalent Parallel Cylindrical Spur Gear and Virtual Number of Teeth

In accordance with the geometry of the parallel cylindrical involute helical gears described in Sect. 8.2, it is evident that the involute profile of the tooth occurs only in transverse section; in the normal section instead the tooth profile is not an involute curve. Therefore, the meshing process between conjugate profiles takes place only in the transverse section, for which, for the calculation of the center distance, as well as for the calculations to be done in connection with the use of profile-shifted toothings, the profiles to be considered are those in the transverse section. In addition, it is to be noted that the tooth flanks roll and slid only in the direction of the tooth depths.

Rebus sic stantibus, consider the meshing in the reference condition (in this case, the datum plane of the rack is tangent to the reference cylinder of the wheel), between a cylindrical helical wheel and the mating rack, and let us ask ourselves the question about the meaning of Eq. (8.17), which correlates α_n, α, and β. Both for the rack, and for the wheel, it certainly has a meaning to speak of pressure angle, α, in the transverse section. Similarly, for the rack, it also has a well-defined meaning to speak of pressure angle, α_n, in the normal section. For the wheel instead, at least from the theoretical point of view, it has no meaning to speak of pressure angle, α_n, in the normal section, since in this section the tooth profile is not an involute curve, and notoriously the pressure angle is related to the involute.

On the other hand, the Eqs. (8.10) and (8.13) have meaning for both members of the gear under consideration, i.e. the wheel and mating rack. In fact, for the wheel, the normal pitch, p_n, is the length of a portion of a helix on the reference cylinder, perpendicular to the pitch helices, between two adjacent pitch helices. This portion of helix corresponds to the length of wrapping of the normal pitch, p_n, of the rack, when the datum plane of the latter rolls without sliding on reference cylinder of the wheel.

We see now to define a parallel cylindrical spur gear that, especially from the point of view of the load carrying capacity calculations to do, but also from other points of view that will clarify in turn, is equivalent to the parallel cylindrical helical gear that we are analyzing. The introduction of an equivalent parallel cylindrical spur gear is very useful, as it is thus possible to use, albeit with the necessary modifications, the relevant spur gear formulae, applying them to parallel cylindrical helical gears, which are analogous to spur gears.

To this end, we return to consider the meshing, in the reference condition, between a cylindrical helical wheel and the mating rack (Fig. 8.12). The helix angle is β. A plane which is normal to the rack tooth (evidently, it is also normal to the wheel tooth) intersects the datum plane of the rack along a straight line, $l = MN$, which is inclined with respect to the axis of the wheel by an angle equal to $[(\pi/2) - \beta]$. The same plane intersects instead the reference pitch cylinder of the wheel according to an ellipse, having minor semi-axis equal to the radius, r, of the reference cylinder (CO, in Fig. 8.12), and major semi-axis equal to $r/\cos\beta$ (MO, in Fig. 8.12).

During the relative motion, characterized by the rolling without sliding of the datum plane of the rack on the reference pitch cylinder of the wheel, the straight line l is wrapped on this cylinder describing a helix, which is a skew curve, i.e. a curve in three-dimensional space. Therefore, from a theoretical point of view, it is not possible to speak of meshing in a normal section. Instead, we can talk of meshing in this normal section in approximate terms, if we confuse, within the small portion of contact that concerns us, the above helix with its osculating circle. As known from differential geometry of curves in two- and three-dimensional space, the radius of this osculating circle is equal to the quotient of the square of the major semi-axis of the ellipse, divided by the minor semi-axis, and therefore equal to (see, for example: Gray et al. [14], Kreyszig [22]):

$$r_v = \frac{(r/\cos\beta)^2}{r} = \frac{r}{\cos^2\beta}. \tag{8.46}$$

On this osculating circle, considered as a virtual pitch circle of radius r_v, the module is clearly the nominal module, m_n, while the number of teeth will be a virtual number of teeth, given by:

$$z_v = \frac{2r_v}{m_n} = \frac{2r}{m\cos^3\beta} = \frac{z}{\cos^3\beta}. \tag{8.47}$$

With reference to the normal section, a cylindrical helical gear wheel of z teeth can be considered approximately equivalent to a virtual cylindrical spur gear wheel having the axis perpendicular to the plane of the normal section, module m_n, diameter $z_v m_n$, and pressure angle α_n. Therefore, the study of the meshing of a parallel cylindrical helical gear pair can be made also with reference to the normal section, introducing the equivalent virtual parallel cylindrical spur gear pair, characterized by: pressure angle, α_n; module, m_n; numbers of teeth $z_{v1} = z_1/\cos^3\beta$, and $z_{v2} = z_2/\cos^3\beta$.

The approximation above is, in some respects, similar to the well-known Tredgold's approximation (see, for example: Budynas and Nisbett [3], Giovannozzi [13], Niemann and Winter [27], Collins et al. [6]) , which we will describe in the next chapter on bevel gears (see Buchanan [2]). It is related to the fact that the generation cutting of the toothing of a cylindrical helical wheel is performed by

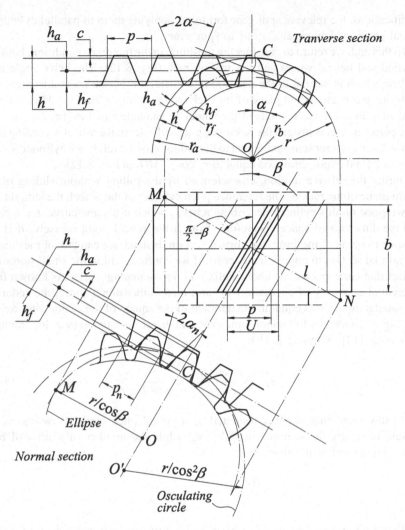

Fig. 8.12 Meshing between rack and cylindrical helical wheel: transverse and normal sections; ellipse and osculating circle in normal section

cutting tools having standard sizing in the normal section. Without the fact that calls into question the technological process of the gear generation, this approximation is not justified. It would be much easier to consider the profiles in transverse section; for these transverse profiles in fact all the formulae, considerations and gearing laws of the cylindrical spur gear are valid.

However, for calculation that do not require very high precision, this approximation allows using the formulae, tables and diagrams relating to the cylindrical spur gears without profile-shifted toothings, with the only change consisting of the substitution of α_n to α, and z_{v1} and z_{v2} to z_1 and z_2. In any case, this approximation cannot be used for calculations that require great precision, which are those concerning parallel cylindrical helical gears with profile-shifted toothing, and the quantities related to them.

The above mathematical discussion, which led us to define the virtual number of teeth, z_v, of a cylindrical helical gear wheel, in the terms expressed by Eq. (8.47), is intended to be sufficiently accurate for all practical applications. To obtain the exact value of z_v, it is necessary to consider the section passing along the course of the helix, and this results in a surface in three-dimensional space as it is a helicoid, and not in a two-dimensional section, i.e. a plane section as the normal section that we introduced above. This exact value of z_v is given by:

$$z_v = \frac{z}{\cos^2 \beta_b \cos \beta}. \tag{8.48}$$

Since, due to technological reasons described above, the nominal sizing of the toothing is the one we have in the normal section, the transverse profile of the tooth, in relation to the transverse module m, is wider and less deep than it is with reference to the nominal sizing in the normal section. In fact, the addendum factor, k, in the transverse section will be given by:

$$k = \frac{h_a}{m} = \frac{h_a}{m_n} \cos \beta = k_n \cos \beta, \tag{8.49}$$

where k_n (of course, $k_n > k$ as $\cos \beta < 1$) is the addendum factor in the normal section. In the same way, we shall have:

$$\frac{h_f}{m} = \frac{h_f}{m_n} \cos \beta \tag{8.50}$$

$$s = \frac{s_n}{\cos \beta}. \tag{8.51}$$

All this is due to the fact that, as Fig. 8.12 shows, the addendum, dedendum, and tooth depth of the cylindrical helical gear wheel under consideration do not change in the normal and transverse sections.

Also, the minimum number of teeth that can be obtained by generation cutting, using a rack-type cutter, without incurring the risk of primary interference and undercut, is reduced by a factor equal to $\cos^3 \beta$. In fact, with reference to this type of interference, the cylindrical helical wheel of z teeth, normal module m_n, and helix angle β, acts as a cylindrical spur wheel having normal module m_n and virtual number of teeth $z_v = z/\cos^3 \beta$.

As for the cylindrical spur gear (see Sect. 3.2), also in this case it is admitted that, in practice, the minimum number of teeth that can be cut by generation cutting using a rack-type cutter is equal to 5/6 of the theoretical values. Figure 8.13, which is valid for cylindrical helical wheels obtained by generation cutting using a rack-type cutter, shows the variation of the theoretical and practical (in this case a

Fig. 8.13 Theoretical and practical minimum number of teeth, z_{min}, as a function of the normal pressure angle, α_n, and helix angle, β, for cylindrical helical wheels cut with rack-type cutter

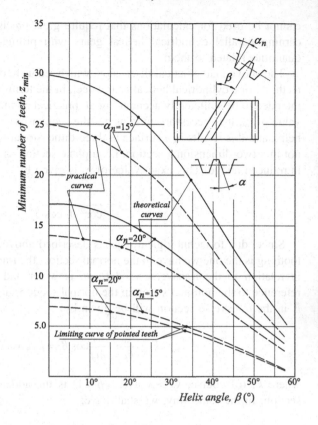

marginal amount of undercut is allowed) values of z_{min} as a function of the normal pressure angle, α_n, and helix angle, β.

The reduction by a factor equal to $\cos^3 \beta$ of the minimum number of teeth, that can be cut without primary interference using a rack-type cutter, is also evident from Eq. (3.10), in which we put $k_2 = k_n \cos \beta$, and, according to Eq. (8.17), we believe it approximately $\sin \alpha_n \cong \sin \alpha \cos \beta$.

By Eqs. (8.9) and (8.17) we obtain the following relationship:

$$\sin \beta_b = \sin \beta \cos \alpha_n \qquad (8.52)$$

which correlates β_b, β and α_n.

It should be noted finally that, when the generation cutting is performed using a helical pinion-type cutter, the conditions of primary interference must be examined by the methods described in Sect. 4.2. In this way, we assume that the pinion-type cutter and the wheel to be cut behave, during the generation motion, as two cylindrical spur wheels with parallel axes, having the features that are found in the transverse sections.

8.6 Parallel Cylindrical Helical Gears with Profile-Shifted Toothing and Variation of Center Distance

Like cylindrical spur gears, parallel cylindrical helical gears can also have profile-shifted toothings when needed. The same reasons for which we use cylindrical spur gears with profile-shifted toothings are also valid for parallel cylindrical helical gears; in this regard, we refer the reader to what we have described in the Chap. 6. Here the subject is dealt with in its most general terms, i.e. with reference to the case of profile-shifted toothings with variation of center distance, as the case of profile-shifted toothings without variation of center distance is nothing but a special case of the general subject. In addition, we will focus our attention only on special features that characterize the parallel cylindrical helical gears compared to the cylindrical spur gears.

The study of these profile-shifted toothings could be made with reference to the transverse profiles of the wheel and transverse section of the mating rack. However, in doing so, the formulae, tables and diagrams that we have introduced previously (see Chap. 6), concerning the profile-shifted toothing of cylindrical spur gears, could not be used, since they are inherent to generation cutting performed by a rack-type cutter sized with reference to the normal section. In order to use these formulae, tables and diagrams, it is therefore convenient to refer to the normal profiles of the wheel teeth, for which the mating rack and the rack-type cutter have their sizing referred to the normal section.

The reference framework is therefore that related to the aforementioned generation cutting process, with cutting tools sized in normal section. For the study of parallel cylindrical helical wheels, generated with reference to a cutting plane of the rack-type cutter shifted with respect to its datum plane, we proceed as if the to-be-cut wheel and the generation rack-type cutter were a rack-wheel spur gear pair, whose characteristic quantities correspond to those of the equivalent virtual gear shown in Fig. 8.12.

It should be noted here that the subject is discussed in a traditional way, i.e. assuming that the two members of the parallel helical gear pair are free from assembly and manufacture errors. In this theoretical case, the instantaneous line of contact between the mating tooth flank surfaces is a straight line. However, the practical applications of any gear pair are always characterized by errors and deviations from the theoretical conditions. The cylindrical parallel gears (both spur and helical gears) with involute tooth profiles are particularly sensitive to the above-mentioned errors. The parallel helical gears are sensitive to deviations from axis parallelism and helix errors, which cause discontinuous linear functions of transmission errors, resulting in vibration and noise. To reduce as far as possible the effects of these deviations and errors, other types of modifications of conventional involute helical gears can be made, such as the crowning of the pinion in the profile and longitudinal directions, which was proposed by Litvin et al. [25]. Here the topic is confined to the profile shift only.

Consider first an external parallel cylindrical helical gear, consisting of a pinion and a gear wheel, and examine what happens in transverse section. If the generation rack-type cutter has normal module m_{n0} and normal pressure angle $\alpha_{n0} = \alpha_0$, and the two members of the gear pair to be generated have reference helix angle β, in the transverse section the quantities that characterize the two members of the helical gear pair, having z_1 and z_2 teeth, are as follows:

- nominal transverse module, m_{t0}, given by Eq. (8.34), i.e. $m_{t0} = m_{n0}/\cos \beta$;
- nominal diameters d_1 and d_2 of the cutting pitch cylinders, given by Eqs. (8.37) and (8.38), i.e. $d_1 = z_1 m_{n0}/\cos \beta$ and $d_2 = z_2 m_{n0}/\cos \beta$;
- transverse pressure angle α on the cutting (or reference) pitch cylinders, given by Eq. (8.35), i.e. $\tan \alpha_{t0} = \tan \alpha_{n0}/\cos \beta$;
- radii r_{b1} and r_{b2} of the base cylinders that, taking into account of Eqs. (8.8) and (8.13), are given by:

$$r_{b1} = \frac{d_1}{2}\cos \alpha_{t0} = \frac{z_1 m_n}{2}\frac{\cos \alpha_{t0}}{\cos \beta} \tag{8.53}$$

$$r_{b2} = \frac{d_2}{2}\cos \alpha_{t0} = \frac{z_2 m_n}{2}\frac{\cos \alpha_{t0}}{\cos \beta}. \tag{8.54}$$

Note that, to avoid misunderstanding, if necessary, it is here used a double subscription, t and n to indicate transverse and normal sections, and 0 to indicate the cutting tool or cutting condition.

The base helix angle, β_b, is given by Eq. (8.9). Generating the two members of the helical gear pair with profile shifts respectively equal to $x_1 m_{t0}$ and $x_2 m_{t0}$ in the transverse section of the rack-type cutter, the operating transverse pressure angle, α'_t, in the meshing condition with backlash-free contact between the teeth, is given by Eq. (6.58). This equation, rewritten in the form of interest here, becomes:

$$\mathrm{inv}\alpha'_t = \mathrm{inv}\alpha_{t0} + 2\tan \alpha_0 \frac{x_1 + x_2}{z_1 + z_2}, \tag{8.55}$$

where α_0 is the reference pressure angle of the rack-type cutter.

However, as we said before, the sizing of the two wheels is carried out according to the normal module m_{n0} and the normal pressure angle, α_{n0}, that is according to the parameters of the rack-type cutter in the normal section. Therefore, the profile shifts that must be introduced are those related to the normal module, m_{n0}, that is $x'_1 m_{n0}$ and $x'_2 m_{n0}$. Of course, it must be necessarily $x'_1 m_{n0} = x_1 m_{t0}$ and $x'_2 m_{n0} = x_2 m_{t0}$, whereby the profile shift coefficients, x'_1 and x'_2, referred to the normal module m_n are given by:

$$x_1' = x_1 \frac{m_{t0}}{m_{n0}} = \frac{x_1}{\cos \beta} \tag{8.56}$$

$$x_2' = x_2 \frac{m_{t0}}{m_{n0}} = \frac{x_2}{\cos \beta}. \tag{8.57}$$

Obtaining x_1 and x_2 from these two relationships, substituting them in Eq. (8.55), and taking into account Eq. (8.35), we get:

$$\mathrm{inv}\alpha_t' = \mathrm{inv}\alpha_{t0} + 2 \tan \alpha_0 \frac{x_1' + x_2'}{z_1 + z_2}. \tag{8.58}$$

The operating center distance, a', of the helical gear pair with profile-shifted toothings and variation of center distance is given by:

$$a' = \frac{r_{b1} + r_{b2}}{\cos \alpha_t'} = \frac{z_1 + z_2}{2} \frac{\cos \alpha_{t0}}{\cos \alpha_t'} \frac{m_{n0}}{\cos \beta}. \tag{8.59}$$

Since the operating center distance is related to the reference center distance, a_0, given by Eq. (6.11), by means of Eq. (6.59), we can write:

$$d' = a_0 \frac{\cos \alpha_{t0}}{\cos \alpha_t'} \tag{8.60}$$

where

$$d' = a_0(1 + Y) \tag{8.61}$$

$$Y = \left(\frac{\cos \alpha_{t0}}{\cos \alpha_t'} - 1 \right). \tag{8.62}$$

To determine the addendum reduction factor, K, and truncation Km_{n0}, we use the same procedure described in Sect. 6.6.1, for which we have:

$$a = m_{t0} \left[\frac{z_1 + z_2}{2} + (x_1 + x_2) \right] = m_{n0} \left[\frac{z_1 + z_2}{2 \cos \beta} + (x_1' + x_2') \right] \tag{8.63}$$

$$d' = m_{t0} \frac{z_1 + z_2}{2} \frac{\cos \alpha_{t0}}{\cos \alpha_t'} = m_{n0} \frac{z_1 + z_2}{2 \cos \beta} \frac{\cos \alpha_{t0}}{\cos \alpha_t'} \tag{8.64}$$

$$K = \frac{a - a'}{m_{n0}} = (x_1' + x_2') - \frac{z_1 + z_2}{2 \cos \beta} \left(\frac{\cos \alpha_{t0}}{\cos \alpha_t'} - 1 \right) = \frac{z_1 + z_2}{2} \left(X' - \frac{Y'}{\cos \beta} \right) \tag{8.65}$$

where

$$X' = 2\frac{x_1' + x_2'}{z_1 + z_2} \tag{8.66}$$

$$Y' = \left(\frac{\cos\alpha_{t0}}{\cos\alpha_t'} - 1\right). \tag{8.67}$$

The radii r_1 and r_2 of the cutting pitch cylinders are given by:

$$r_1 = \frac{d_1}{2} = \frac{z_1 m_{t0}}{2} = \frac{z_1 m_{n0}}{2\cos\beta} \tag{8.68}$$

$$r_2 = \frac{d_2}{2} = \frac{z_2 m_{t0}}{2} = \frac{z_2 m_{n0}}{2\cos\beta}. \tag{8.69}$$

The radii r_1' and r_2' of the operating pitch cylinders are given by:

$$r_1' = \frac{z_1 m_{t0}}{2}\frac{\cos\alpha_{t0}}{\cos\alpha_t'} = m_{n0}\frac{z_1}{2\cos\beta}\frac{\cos\alpha_{t0}}{\cos\alpha_t'} \tag{8.70}$$

$$r_2' = \frac{z_2 m_{t0}}{2}\frac{\cos\alpha_{t0}}{\cos\alpha_t'} = m_{n0}\frac{z_2}{2\cos\beta}\frac{\cos\alpha_{t0}}{\cos\alpha_t'}. \tag{8.71}$$

Finally, the helix angle β' on the operating pitch cylinders is given by:

$$\tan\beta' = \tan\beta\frac{r_1'}{r_1} = \tan\beta\frac{r_2'}{r_2}. \tag{8.72}$$

In the case of an internal parallel cylindrical helical gear, consisting of a pinion and annulus, the discussion to be carried out is entirely analogous to that described in Sect. 6.6.2, with the modifications described above, which concern the need to consider what happens not only in the transverse section, but also in the normal section. In doing so, instead of the Eq. (8.58), we find the following fundamental relationship:

$$\mathrm{inv}\alpha_t' = \mathrm{inv}\alpha_{t0} + 2\tan\alpha_0\frac{x_2' - x_1'}{z_2 - z_1}. \tag{8.73}$$

where: α_t', is the operating transverse pressure angle; α_{t0}, is the cutting (or generating) transverse pressure angle; α_0, is the reference angle of the rack-type cutter (this cutter is an actual cutter for the pinion, and a virtual cutter for the annulus); x_1' and x_2', are the profile shift coefficients referred to the normal module m_n; z_1 and z_2, are the numbers of teeth of pinion and annulus. The development of the other equations is left to the reader.

For external and internal parallel cylindrical helical gears, the same remarks that follow are valid. Of course, for internal parallel cylindrical helical gears, some appropriate variations are necessary. With reference to external parallel cylindrical helical gears, considering a given pressure angle $\alpha_{n0} = \alpha_0$ of the rack-type cutter, it is possible to vary the helix angle and the profile shift coefficients, x_1' and x_2', so as to avoid undercut during the generation process, and therefore to determine the operating center distance a' and the operating pressure angle, α_t', by means of previous equations. In particular, when the center distance of a helical gear pair having z_1 and z_2 teeth, normal module m_n and normal pressure angle α_n are given, from Eq. (8.43) we calculate the helix angle β, required to achieve the given center distance. If this value of β is too high, in order to limit the axial loads on the shaft and bearings, a smaller helix angle β can be chosen and with Eq. (8.35) we calculate the transverse pressure angle α_t; then we calculate α_t' by means of Eq. (8.59), and finally the sum $(x_1' + x_2')$ by means of Eq. (8.58). Of course, the problem of the distribution of this sum between the two members of the helical gear pair remains to be solved.

Example 1 This example concerns an external parallel cylindrical helical gear pair. Input data: $z_1 = 8$; $z_2 = 12$; $m_{n0} = 10$ mm; $\beta = 20°$; $\alpha_{n0} = \alpha_0 = 20°$. Profile shift coefficients $x_1' = 0.40$ and $x_2' = 0.30$ were chosen earlier. We will have:

a_0 (reference center distance) $= 106.30$ mm	
$X' = 0.070$	
$\alpha_t' = 28°2'0''$	
$Y' = 0.056$	
$a' = 112.30$ mm	
$K = 0.10$	
$h = 21.50$ mm	
d_{a1} (from Eq. (6.86), with replacement of $z_1/\cos\beta$ instead of z_1) $- 110.40$ mm	
d_{a2} (from Eq. (6.87), with replacement of $z_2/\cos\beta$ instead of z_2) $= 150.96$ mm	

Example 2 This example concerns an external parallel cylindrical helical gear pair, with an increase in the center distance. Input data: $z_1 = 20$; $z_2 = 40$; $a' = 105$ mm; $\beta = 30°$; standard range of rack-type cutters. We will have:

$m_n' = 3.031$mm (this value is sufficiently accurate, but approximate, since the cosine of the operating helix angle differs very little from that of the cutting helix angle)	
$m_0 = 10$ mm, and $\alpha_0 = 20°$ (this is the initial choice of design)	
$a_0 = m_0(z_1 + z_2)/2\cos\beta = 103.92$ mm	
Since $a' > a_0$, we have to make a profile shift with increase in the center distance	
$\alpha_t' = 24°9'35''$	
$Y' = 0.0103$	
$X' = 0.0120$	
$(x_1' + x_2') = 0.36.$	

Example 3 This example concerns an internal parallel cylindrical helical gear pair, with decrease in the center distance. Input data: $z_1 = 10$; $z_2 = 50$; $a' = 105$ mm; $\beta = 20°$; standard range of rack-type cutters. We will have:

$m'_n = 4.93$ mm (this value is sufficiently accurate, but approximate, since the cosine of the operating helix angle differs very little from that of the cutting helix angle)
$m_0 = 5$ mm, and $\alpha_0 = 20°$ (this is the initial choice of design)
$a_0 = m_0(z_2 - z_1)/2\cos\beta = 106.40$ mm
Since $a' < a_0$, we have to make a profile shift with decrease in the center distance.
$Y' = [(a'/a_0) - 1] = -0.014$
$X' = -0.014$
$(x'_2 - x'_1) = -0.28.$

We can choice, for example, $x'_2 = 0$, and $x'_1 = 0.28$

$K = 0$
$d_{a1} = m_0\left[\frac{z_1}{\cos\beta} + 2(1 + x'_1)\right] = 66.00$ mm
$d_{a2} = m_0\left[\frac{z_2}{\cos\beta} - 2(1 - x'_2)\right] = 255.95$ mm
$h = 2,25m_0 = 11.25$ mm.

8.7   Total Length of the Line of Action

In Sect. 8.2 we have seen that, during the meshing cycle, the instantaneous line of action moves on the surface of the tooth flank, and that its length varies from zero to a maximum, which assumes a different value according to whether $g_\alpha \cot\beta_b \gtrless b$. We have also seen that the total length, l_t, of the line of action changes with the position of the teeth. Let us now evaluate this total length, also variable during the meshing cycle, and its implications.

In parallel cylindrical spur gears, which are typically characterized by a transverse contact ratio between 1 and 2 $(1 < \varepsilon_\alpha < 2)$, the total length of the line of contact, l_t, undergoes an abrupt change, passing from the value b, when only one pair of teeth is in meshing (single contact), to the value $2b$, when two pairs of teeth are in meshing (double contact). It is to be also noted that, in the cylindrical spur gears, the flank line and the instantaneous line of contact, on the plane of action, coincide.

Instead, in the parallel cylindrical helical gears the total length of the line of contact fluctuates between a maximum value $l_{t,max}$, and a minimum value, $l_{t,min}$. In addition, the active face width, b_v (or equivalent face width) and the transverse contact ratio, ε_α, oscillate respectively between a maximum value, $b_{v,max}$, and a minimum value $b_{v,min}$, and between a maximum value $\varepsilon_{\alpha,max}$, and a minimum value, $\varepsilon_{\alpha,min}$.

It is to be noted that, also in the parallel cylindrical helical gears, the flank line and the instantaneous line of contact, on the plane of action, coincide. While, however, in the cylindrical spur gears, these lines are parallel to the axes of the two wheels, in parallel cylindrical helical gears they are inclined, in the plane of action, by an angle equal to the base helix angle, and, in the various coaxial cylinders between the tip and root cylinders, by helix angles corresponding to them.

With reference to Fig. 8.14, which shows the actual rectangle of contact, that is the path-area of contact in which some instantaneous lines of action are represented, we can express the length of the transverse path of contact, g_α, as:

$$g_\alpha = \varepsilon_\alpha p_b = (W + X)p_{bt}, \tag{8.74}$$

where p_{bt} is the transverse base pitch, W is an integer, and X is a fractional number less than 1. Thus, we have $\varepsilon_\alpha = (W + X)$. For example, in the usual case where $1 < \varepsilon_\alpha < 2$, we have $W = 1$, and $0 < X < 1$. Similarly, we can express the face width, b, as:

$$b = \varepsilon_\beta p_x = (Y + Z)p_x \tag{8.75}$$

Fig. 8.14 Actual rectangle of contact of a parallel cylindrical gear pair, with g_α and b expressed respectively as a function of p_{bt} and p_x

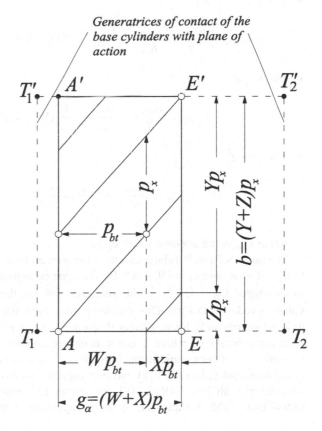

where p_x is the axial pitch, Y is an integer, and Z is a fractional number less than 1. Thus, we have $\varepsilon_\beta = (Y + Z)$. Figure 8.14 gives a clear graphic evidence of Eqs. (8.74) and (8.75), in the case in which $W = 1$ and $Y = 2$.

The actual rectangle of contact does not change during the meshing cycle, but the instantaneous lines of action, whose inclination with respect to the axes of the two wheels remains unchanged (it is equal to the base helix angle, β_b), move parallel to themselves. In relative terms, we can deal with the problem of determining the total length l_t keeping fixed the position of these instantaneous lines of action compared to the actual rectangle of contact, and by translating the latter on the plane of action, parallel to itself, in the direction perpendicular to the generatrixes of the base cylinders.

In doing so, Karas [20] obtained the following equations that express the quantities defined above: (see also [27]):

- For $Z \leq (1 - X)$

$$\varepsilon_{\alpha,min} = \frac{b_{v,min}}{b} = \frac{l_{t,min}}{l_{t,m}} = \frac{1}{\varepsilon_\beta}[(W + X)Y + WZ] \tag{8.76}$$

- For $Z > (1 - X)$

$$\varepsilon_{\alpha,min} = \frac{b_{v,min}}{b} = \frac{l_{t,min}}{l_{t,m}} = \frac{1}{\varepsilon_\beta}[(W + X)Y + WZ + (X + Z - 1)] \tag{8.77}$$

- For $X \leq Z$

$$\varepsilon_{\alpha,max} = \frac{b_{v,max}}{b} = \frac{l_{t,max}}{l_{t,m}} = \frac{1}{\varepsilon_\beta}[(W + X)Y + (X + Z)] \tag{8.78}$$

- For $X > Z$

$$\varepsilon_{\alpha,max} = \frac{b_{v,max}}{b} = \frac{l_{t,max}}{l_{t,m}} = \frac{1}{\varepsilon_\beta}[(W + X)Y + 2Z] \tag{8.79}$$

where $l_{t,m}$ is the average value of l_t.

The Eqs. (8.76–8.79) show that the minimum and maximum values of ε_α, b_v, and l_t depend on ε_β and $\varepsilon_\alpha = (W + X)$. By the same equations we infer that, for ε_β equal to an integer $(\varepsilon_\beta = 1, 2, \ldots)$, the variation between the maximum and minimum values is reduced to zero. This variation, quite pronounced for ε_β less than one, is reduced significantly for ε_β greater than one, and is the smaller, the greater is ε_β.

In the general case where ε_β is not an integer number, since the active face width b_v oscillates between a minimum value, $b_{v,min}$ and a maximum value, $b_{v,max}$, for the calculations related to the load carrying capacity to surface durability (pitting), in accordance with ISO 6336-2 [19], the *virtual face width* b_{vir} is introduced. This virtual face width is expressed by an average value between the above minimum and maximum values. The ratio b_{vir}/b then allows to calculate the contact ratio factor for pitting, Z_ε, defined by the relationship:

$$Z_\varepsilon^2 = \frac{1}{(b_{vir}/b)};\qquad(8.80)$$

we will talk about this contact ratio factor in Vol. 2, Chap. 1, dealing with surface durability (pitting) of spur and helical gears.

8.8 Load Analysis of Parallel Cylindrical Helical Gears, and Thrust Characteristics on Shaft and Bearings

Load analysis of parallel cylindrical spur and helical gears can be done with the same procedure, unique for both types of gears, bearing in mind that parallel cylindrical spur gears are a special case of parallel cylindrical helical gears. This general procedure neglects the contribution of the friction forces, but it gives results of great engineering value. Here we will discuss the more general case of parallel cylindrical helical gears, which is characterized by an additional force component (the one caused by the helix angle) with respect to the parallel cylindrical spur gears. For this more general case, we here get the related relationships, which we will eventually specialize for parallel cylindrical spur gears.

The total tooth force, or simply tooth force, F_n, acts normal to the tooth surface, and its direction (namely the normal to the instantaneous line of action in the plane of action) is the same for all points of any instantaneous line of action. This tooth force can be resolved into three components, which act at right angles to one another. In order to establish the interrelations between these components, we examine Fig. 8.15, showing the meshing diagram of a parallel cylindrical helical gear pair, as well as the actual rectangle of contact after its rotation about the common tangent to the base circles, which are not shown in the figure. In the plane of this rectangle, the force F_n can be resolved into two components: the axial component, $F_a = F_n \sin \beta_b$, acting in the direction parallel to the axes of the two wheels, and the component $F_n \cos \beta_b$, acting in the direction normal to the contact generatrix between the pitch cylinders.

This last component is inclined at an angle equal to the pressure angle, α, with respect to the common plane tangent to the two pitch cylinders, passing throught the pitch point C. This component, once translated along its own straight line of application until it meets the aforementioned generatrix at point C, can be in turn resolved into a circumferential force $F_t = F_n \cos \beta_b \cos \alpha$ (i.e. the nominal transverse tangential load at reference cylinder per mesh, as it is called by the ISO 6336-1 [18]), and a radial force $F_r = F_n \cos \beta_b \sin \alpha$. Therefore, the three components of the tooth force F_n are as follows:

Fig. 8.15 Meshing diagram of a helical gear pair, plane of action rotated in the plane of the figure, and tooth force F_n resolved into its three components F_t, F_a, and F_r

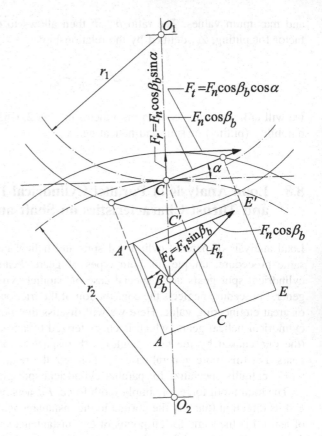

$$F_t = F_n \cos \beta_b \cos \alpha = \frac{2000T}{d} \qquad (8.81)$$

$$F_a = F_n \sin \beta_b = \frac{F_t \tan \beta_b}{\cos \alpha} = F_t \tan \beta \qquad (8.82)$$

$$F_r = F_n \cos \beta_b \sin \alpha = \frac{F_t \tan \alpha_n}{\cos \beta} = F_t \tan \alpha, \qquad (8.83)$$

where T is the driving torque (in Nm), d is the pitch circle diameter (in mm), β and β_b are the helix angles on the reference and base cylinders respectively, and α_n and $\alpha_t = \alpha$ are the pressure angles in the normal section and transverse section respectively. All the force components are expressed in newtons (N).

It is to be observed that, at equal circumferential component F_t, the total tooth force F_n has a constant intensity, but its straight line of application, from a general theoretical standpoint, changes position in dependence of the length of the instantaneous line of action, and the load sharing along it. With reference to the plane of actual rectangle of contact shown in Fig. 8.15, in the case where only one pair of teeth was in meshing, the following changes in the position of the line of application of the total tooth force F_n are evident. In fact, at the start of the meshing cycle, the force F_n would pass through point A, in the course of the meshing cycle moves parallel itself in the plane of rectangle of contact, and at the end of the meshing cycle would pass through point E'. This behaviour would cause a variation in the loads acting on the bearings, even if the intensity of the force F_n remains constant.

In practical applications, however, the parallel cylindrical helical gear pairs are characterized by a total contact ratio ε_y (sum of the transverse contact ratio, ε_α, and overlap ratio, ε_β) very high, for which several pairs of teeth are simultaneously in meshing. Therefore, with reference to the problems that interest us here, including those related to the load carrying capacity of the gear, we can assume, with good approximation, that the tooth force F_n and its three components are applied in the middle planes of the two wheels, i.e. the planes bisecting the face widths.

The axial component F_a of the tooth force F_n is the *thrust* which acts along the axis of the two members of the helical gear pair. Determining the intensity and direction of this thrust is of fundamental importance in the helical gear design. Direction of the thrust is a function of several factors, such as: the direction of helix, i.e. right-hand helix or left-hand helix; the direction of rotation of the two members of the helical gear pair; the relative positions of these two members. Direction of this thrust changes by changing any one of these three factors. It is to keep in mind that, from the design point of view, the directions of the helices of the two gear members can be fixed only when the relative position between driving and driven members has been established, and the direction of the thrust has been determined.

In accordance with Eq. (8.82), the axial force, or thrust, increases in proportion to $\tan \beta$; then the helix angle should be chosen carefully. For the usual parallel cylindrical helical gears, it is prudent to limit the helix angle β within $(20 \div 25)°$, in order to avoid excessive values of the thrust. In exceptional cases, in which the negative effects of the thrust are kept under control, the helix angle may go up to $30°$, but for normal applications it should not exceed $20°$. Of course, the intensity of the axial thrust influences the choice of the bearings that support the shafts on which the helical gears are mounted.

Finally, in accordance with what we said at the beginning of this section, it should be remembered that the Eqs. (8.81–8.83) can be used for calculating the loads acting on parallel cylindrical spur gears, as they are a special case of parallel cylindrical helical gears. To this end, just put $\beta_b = \beta = 0$ in them. So we get:

$$F_t = F_n \cos \alpha = \frac{2000T}{d} \tag{8.84}$$

$$F_a = 0 \tag{8.85}$$

$$F_r = F_n \sin \alpha = F_t \tan \alpha, \tag{8.86}$$

as $\alpha_n = \alpha_t = \alpha$. We therefore deduce that, for parallel cylindrical spur gears, at the same circumferential component F_t, the total tooth force, F_n, and the radial component, F_r, increase as the pressure angle, α, increases. For this reason, it is necessary to limit the value of the pressure angle. Furthermore, for high value of the pressure angle we can run even the danger of pointed teeth (see Fig. 3.2).

8.9 Double-Helical Gears

Double helical gears or *herringbone gears* are parallel cylindrical gears having part of the face width with right-hand teeth and part with left-hand teeth, with or without a gap between them. They are used in those cases where the axial thrust of an usual helical gear pair creates problems for the bearings or it is not sustainable for any other reason concerning the design choices. A double helical gear wheel is indeed a combination of two helical gear wheels, placed side by side, and having usually the same helix angle but of opposite hands; it should be noted, however, that the inclination of the two hands must not necessary be the same.

In accordance with the cutting method used, these herringbone gear wheels can be realized in one of the following ways:

- by cutting the double toothings (one with right-hand helix, and the other with left-hand helix, or vice versa) on the same gear blank, as Fig. 8.16a shows;
- by cutting a simple toothing (i.e. not double) on two distinct gear blanks, one with right-hand helix, and the other with left-hand helix, or vice versa, then connected between them by means of threaded connections, as Fig. 8.16b, c show;
- by connecting between them two distinct gear wheels, each characterized by a simple helical toothing, one with right-hand helix, and the other with left-hand helix, or vice versa, as Fig. 8.16d shows.

When the two double-helical gear wheels are cut using hobs, in order to relief of the cutting tool, a suitable groove is made on the gear blank between the right-hand and left-hand helical halves. Instead, when these double-helical gears are cut with gear-shaping machines that use a pair of pinion-type cutters or a pair of rack-type cutters, the toothing is continuous in the section that bisects the face width, without a groove or gap, for which teeth are configured in the shape of an arrow.

If the direction of rotation is in one direction only, it is convenient to place the arrow tips (or the apices of the teeth that converge toward the central groove) in the direction of the tangential velocity of the gear wheel. In this way, the start of the meshing and the impact of first contact related to it occur just in correspondence of these apices, where the strength of the teeth is greater, due to the convergence of the opposite teeth towards an area where the stiffness is greater. When the direction of rotation is not unique, but in both directions, the teeth do not extend up to the end faces of the gear wheel; this instead ends with two reinforcement disks at its two end faces.

Fig. 8.16 Constructive solution of double-helical gear wheels

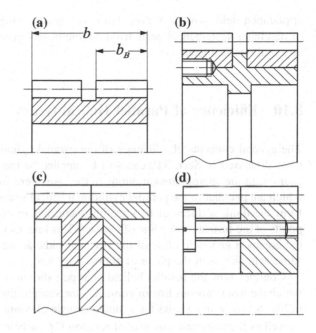

Essentially, these herringbone wheels can be considered as composite units constitutes by two simple helical gear wheels, specularly equal with respect to the plane perpendicular to the axis, which bisects the face width. It follows that, when two of these composite units, mounted on two parallel shafts, mesh with each other, the axial thrusts are counter-balanced, so that the resulting axial force is zero. For this reason, the restrictions on the maximum values of the helix angle, described in the previous section about the helical gear wheels, are no longer valid for the double-helical gear wheels; for the latter, the helix angle can reach values equal to $(30 \div 45)°$.

The practical problem of ensuring the equi-distribution of the load in the two halves of a double-helical toothing is normally solved by leaving that the pinion has the possibility of a small displacement in the axial direction. This can be achieved either with the pinion movable with respect to the shaft or by ensuring that the shaft to which the pinion is rigidly connected can move slightly along the axis, through the use a flexible coupling, which allows small axial displacements of the same shaft. In the first case, the pinion is connected to the shaft by means of a feather key or by means of a sliding mating spline, while in the second case, a corresponding small axial clearance in the bearing must be left.

Lastly, it should be recalled that in some practical applications even parallel triple-helical gears were used. They can be considered as an extension of the parallel double-helical gears, where any of the two halves is divided into two parts, which are then arranged at the opposite ends of the other half. Faced with several disadvantages related to their manufacturing processes, they do not offer any fundamental advantage over double-helical gears. For these reasons, they did not find application fields worthy of note. However, generalizing the above-mentioned way of obtaining triple-helical gears from double-helical gears, it is possible to conceive any multi-helical gear.

8.10 Efficiency of Parallel Helical Gears

The general concepts of efficiency of the parallel cylindrical spur gears, which we have discussed in Sect. 3.9, can still be applied to the parallel cylindrical helical gears, with the appropriate variations of the case. Here we dwell on these variations, which are peculiar to the parallel cylindrical helical gears. The two helicoids, which form the flank surfaces of two mating teeth, are in contact at any instant at the points of the instantaneous line of contact. This line, as we have seen in Sect. 8.2, is the common generation line of the two helicoids, as well as the intersection of the same helicoids with the plane of action (Fig. 8.6).

Consider now the parallel helical gear pair shown in Fig. 8.17, at the instant in which the instantaneous line of contact is the straight line, l. This line is highlighted in Fig. 8.17b, which shows the actual rectangle of contact $AA'EE'$ in the front view, as well as the instantaneous axis of rotation C_iC_i' where the two pitch cylinders are tangent. Let us then consider a differential segment, ds, around the generic point P of this instantaneous line of contact. Considering the friction in addition to what we said in Sect. 8.8, the instantaneous infinitesimal force exerted by the driving profile on the driven profile has two components, i.e.: a normal component, dF_n, which lies in the plane of action and is directed as Fig. 8.17b shows (see also Fig. 8.15); a friction component, dF_μ, which is directed perpendicular to the plane

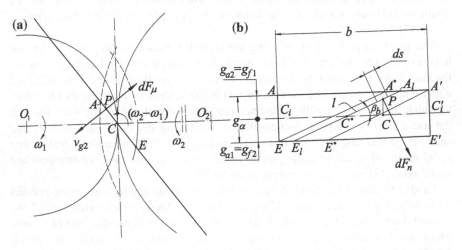

Fig. 8.17 Infinitesimal force components in a parallel cylindrical helical gear pair: **a** friction component; **b** normal component

of action and is oriented in the opposite direction to the relative rotation vector $\omega = (\omega_2 - \omega_1)$ of the driven wheel with respect to the driving wheel (Fig. 8.17a). This infinitesimal friction component is related to the infinitesimal normal component by the known relationship, $dF_\mu = \mu dF_n$, where μ is the coefficient of sliding friction.

Taking into account the relationships obtained in Sect. 3.9, in the case here examined the energy lost by friction (or lost work) on the differential segment ds in the infinitesimal time interval dt is given by:

$$\frac{dW_\mu}{dt} = \mu \frac{dF_n}{ds} ds |\omega_2 - \omega_1|(CP), \tag{8.87}$$

where CP denotes the distance of the generic point P from the instantaneous axis of rotation, measured along the direction perpendicular to the latter (it should be noted that, in order to avoid misunderstandings, in Fig. 8.17 we have indicated the instantaneous axis of rotation $C_i C_i'$, that is differently compared to Fig. 8.6 and Fig. 8.15, in which it is indicated by CC'). This distance is given by Eq. (3.92), with unchanged meaning of the symbols that appear in it. If s is the distance of point P from point C^* where the instantaneous line of contact, l, intersects the axis $C_i C_i'$ (this distance is measured along the straight line, l), from Fig. 8.17b and taking into account the Eq. (3.92), we obtain:

$$s = C^*P = CP/\sin \beta_b = \varphi_1 r_1 \frac{\cos \alpha}{\sin \beta_b}, \tag{8.88}$$

where φ_1 and r_1 are respectively the angular displacement and radius of the pitch circle of the gear wheel under consideration. Then, differentiating this last relationship with respect to φ_1, we get:

$$ds = r_1 \frac{\cos \alpha}{\sin \beta_b} d\varphi_1. \tag{8.89}$$

From Eqs. (8.87, 8.89) and (3.92), we obtain:

$$\frac{dW_\mu}{dt} = |\omega_2 - \omega_1| \frac{dF_n}{ds} \frac{\varphi_1 r_1^2 \cos^2 \alpha d\varphi_1}{\sin \beta_b} \mu. \tag{8.90}$$

However, the values of φ_1 corresponding to the points A_l and E_l (Fig. 8.17b) are equal to the approach angle, (e_1/r_1), and respectively to the recess angle, (e_2/r_1), which we have already introduced in Sect. 3.9. Therefore, assuming that (dF_n/ds) is a constant along the instantaneous line of contact, l, the instantaneous power dissipated by the friction acting on the mating tooth pair under consideration, for μ variable, will be given by:

$$\frac{dW_\mu}{dt} = |\omega_2 - \omega_1| \frac{dF_n}{ds} r_1^2 \frac{\cos^2 \alpha}{\sin \beta_b} \left[\int_0^{e_1/r_1} \mu \varphi_1 d\varphi_1 + \int_0^{e_2/r_1} \mu \varphi_1 d\varphi_1 \right], \quad (8.91)$$

while if the coefficient of friction μ is assumed as a constant, it will be given by:

$$\frac{dW_\mu}{dt} = |\omega_2 - \omega_1| \frac{dF_n}{ds} r_1^2 \frac{\cos^2 \alpha}{\sin \beta_b} \frac{\mu}{2} \left(\frac{e_1^2}{r_1^2} + \frac{e_2^2}{r_1^2} \right). \quad (8.92)$$

On the other hand, the rate at which the work carried out by the driving wheel is done (that is, the power driven by the driving wheel), corresponding to the differential segment, ds, along which the engagement between the mating tooth pair under consideration occurs, is given by:

$$\frac{dW_m}{dt} = \frac{dF_n}{ds} r_1 \omega_1 (ds \cos \beta_b \cos \alpha + \mu \varphi_1 ds \cos \alpha \mp \mu ds \sin \alpha), \quad (8.93)$$

where we must take the minus sign or the plus sign, depending on whether the differential segment, ds, belongs to the segment C^*A_l or to the segment C^*E_l (Fig. 8.17b). Consequently, taking into account the previous equations, we get:

$$\frac{dW_m}{dt} = \frac{dF_n}{ds} r_1 \omega_1 \left[\cos \beta_b \frac{\cos^2 \alpha}{\sin \beta_b} r_1 \left(\frac{e_1}{r_1} + \frac{e_2}{r_1} \right) + \mu \frac{r_1 \cos^2 \alpha}{2 \sin \beta_b} \left(\frac{e_1^2}{r_1^2} + \frac{e_2^2}{r_1^2} \right) \right.$$
$$\left. + \mu \sin \alpha \frac{r_1 \cos \alpha}{\sin \beta_b} \left(\frac{e_1}{r_1} - \frac{e_2}{r_1} \right) \right]$$
$$= \frac{dF_n}{ds} \omega_1 r_1^2 \cot \beta_b \cos^2 \alpha \left[\frac{e_1}{r_1} + \frac{e_2}{r_1} + \frac{\mu}{\cos \beta_b} \frac{e_1^2 + e_2^2}{2 r_1^2} + \frac{\mu \tan \alpha}{\cos \beta_b} \frac{e_2 - e_1}{r_1} \right].$$
$$(8.94)$$

The loss of instantaneous efficiency corresponding to the mating tooth pair at the time instant considered is therefore given by the following relationship:

$$1 - \eta_i = \frac{dW_\mu}{dW_m} = \frac{\mu}{\cos \beta_b} \left(1 + \frac{r_1}{r_2} \right) \frac{1}{2} \frac{(e_1/r_1)^2 + (e_2/r_1)^2}{\frac{e_1}{r_1} + \frac{e_2}{r_1} + \frac{\mu}{\cos \beta_b} \left(\frac{e_1^2 + e_2^2}{2 r_1^2} + \tan \alpha \frac{e_2 - e_1}{r_1} \right)}, \quad (8.95)$$

which can be expressed as follows:

$$1 - \eta_i = \frac{\pi \mu}{\cos \beta_b} \left(\frac{1}{z_1} + \frac{1}{z_2} \right) \frac{\varepsilon_1^2 + \varepsilon_2^2}{\left\{ \varepsilon_1 + \varepsilon_2 + \frac{\mu}{\cos \beta_b} \left[\pi \frac{\varepsilon_1^2 + \varepsilon_2^2}{z_1} + \tan \alpha (\varepsilon_2 - \varepsilon_1) \right] \right\}}. \quad (8.96)$$

With the approximation made to obtain the formula of Poncelet, given by Eq. (3.105), this relationship can be expressed as follows:

$$1 - \eta_i = \pi\mu_1 \left(\frac{1}{z_1} + \frac{1}{z_2} \right) \frac{\varepsilon_1^2 + \varepsilon_2^2}{\varepsilon_1 + \varepsilon_2}, \tag{8.97}$$

where $\mu_1 = \mu / \cos \beta_b$.

Since in order to obtain the formula of Poncelet we have assumed $(\varepsilon_1 + \varepsilon_2) = 1$, from the comparison of the Eqs. (3.105) and (8.97) we infer that the loss of instantaneous efficiency of the parallel cylindrical helical gear pair considered above is equal to the loss of average efficiency during the meshing cycle of the parallel cylindrical spur gear pair, multiplied by $(1 / \cos \beta_b)$. This loss of instantaneous efficiency does not vary for instantaneous lines of contact between EA^* and E^*A', i.e. for a significant portion of the meshing cycle between two mating teeth. It follows that the loss of the average efficiency of a parallel cylindrical helical gear pair can be considered almost equal to the loss of the instantaneous efficiency. Therefore, the Eq. (8.97) in which η_i is replaced by η, also expresses the loss of average efficiency.

8.11 Short Notes on Cutting Methods of Cylindrical Helical Gears

The main methods employed for the manufacture of cylindrical helical gear wheels can be also summed up in two categories: form cutting methods, and generation cutting methods (see Galassini [11], Rossi [32], Townsend [36], Davis [7], Radzevich [29, 31], Kawasaki et al. [21]).

8.11.1 Form Cutting Methods

The form cutting methods are the easiest cutting methods, which are done with milling cutting machines, using form milling cutters, such as disk-type milling cutters or end mills. Figure 8.18 shows the different arrangement of these two types of cutting tool when they are used for cutting the teeth of a cylindrical helical gear wheel.

As we already pointed out in Sect. 3.11.1, concerning the cutting of cylindrical spur gear wheels with form cutting methods, even in this case the disk-type milling cutter allows to obtain only approximate tooth flank profiles. Instead, if end mills are used, having the same profile as the tooth space in the normal section of the tooth, at least theoretically, the exact tooth flank profile can be obtained. In practice, however, major difficulties occur, because these types of cutting tools are difficult to construct and wear out quickly. Furthermore, the subsequent sharpening changes appreciably their initial profile. For these reasons, the accuracy of the gears obtained with this cutting process is extremely poor.

Fig. 8.18 Arrangements of a
disk-type milling cutter and
an end mill for cutting of a
cylindrical helical gear wheel

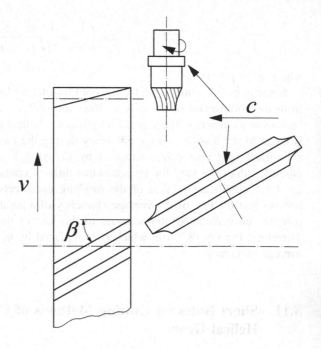

With regard to the cutting process, whatever the type of cutting tool (disk-type
milling cutter or end mill), it must have a relative helical motion, that is a screw
motion relative to the gear-blank to be cut, defined by the helix angle, β, of the
reference helix of the helical gear wheel to be manufactured. As shown in Fig. 8.18,
this screw motion is obtained by tilting the cutting tool at an angle equal to β with
respect to the gear wheel axis. It is then moved in the direction of this axis with a
velocity, c, while the gear wheel rotates with tangential velocity at the reference
circle, v. These three quantities are correlated by the following relationship:

$$c = v \cot \beta. \tag{8.98}$$

8.11.2 Generation Cutting Methods

The generation cutting methods used in the production process of cylindrical helical
gear wheels are those already described for cylindrical spur gears (see Sect. 3.11.2),
with some extra peculiarities, which are typical of these gears.

The cutting of cylindrical helical gear wheels by means of gear shaping
machines using rack-type cutters as cutting tools is quite similar to that already
described for cylindrical spur gear wheels. The only difference is that the same
rack-type cutter used to cut spur gear wheels is now tilted at an angle equal to β
with respect to the gear wheel axis (here also β is the helix angle of the reference

helix of the helical gear wheel to be cut), and its reciprocating cutting motion takes place along this direction. The relative motions between the gear wheel being cut and the cutting tool are those already described for this cutting process in Sect. 3.11.2. These relative motions simulate the pure rolling without sliding of the pitch cylinder of the gear wheel on the pitch plane of the rack-type cutter. In the normal section of the teeth so obtained, they have the sizing of the rack-type cutter.

The cutting of the cylindrical helical gear wheels by means of gear shaping machines using pinion-type cutters as cutting tools is a bit different from the one already described for cylindrical spur gear wheels (see Galassini [11], Rossi [32]). In fact, the pinion-type cutter is a helical pinion cutter, with helix angle equal to that of the gear wheel to be cut, both for external gear wheels and for internal gear wheels. In addition, the reciprocating cutting motion of the pinion cutter along the direction of the gear wheel axis does not occur parallel to this axis, but according to a helix, having the same helix angle of the gear wheel to be cut, and axis parallel to that of the latter. This reciprocating helical cutting motion is obtained by means of suitable helical guides, having lead angle equal to the reference lead angle of the gear wheel. For helix angles greater than 30°, cutting difficulties may occur due to inappropriate relief angles. Of course, for a given module and a given helix angle, two cutting tools are needed, one for right-hand helix, and the other for left-hand helix.

As we have already mentioned in Sect. 3.11.2, the most commonly used generation method for cutting cylindrical spur and helical gear wheels is by hobbing. This cutting method uses gear hobbing machines equipped with hobs whose geometry is the one shown in Fig. 3.16. However, to obtain cylindrical helical gear wheels, the hob is arranged with its axis inclined with respect to the axis of the gear wheel to be cut at an appropriate angle, ψ. This angle is determined in such a way that the reference helices of the hob and gear wheel to be cut have the same common tangent at the point where the two reference cylinders of hob and gear wheel touch each other. As Fig. 8.19 shows, this angle, ψ, is given by:

$$\psi = \gamma \pm \gamma_0, \tag{8.99}$$

where γ and γ_0 are respectively the reference lead angles of gear wheel to be cut and hob. The plus sign or the minus sign in Eq. (8.99) is to be taken, depending on whether helical gear wheel and hob have both right-hand teeth (or both left-hand teeth) or the helical gear wheel has right-hand teeth and the hob has left-hand teeth and vice versa.

To obtain a helix angle equal to $\beta = [(\pi/2) - \gamma]$, the gear wheel to be cut must have, in addition to the main rotation about its axis, an additional rotation, proportional to the advancement of the hob in the direction of the gear wheel axis (see Figs. 3.17 and 3.18). The main rotation above, which corresponds to the transmission ratio with the hob, is the only rotation of the gear wheel in the case of a cylindrical spur gear wheel. Therefore, if c is the feed rate of hob in the direction of the gear wheel axis, and v is the tangential velocity at the reference circle of the gear wheel, corresponding to the additional rotation above, the Eq. (8.98) should be satisfied. Usually, in the gear hobbing machines, the two main and additional

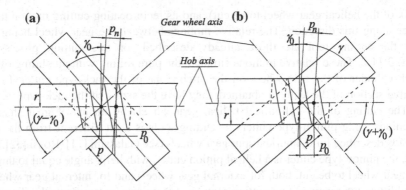

Fig. 8.19 Hob arrangement with respect to the helical gear wheel to be cut: **a** right-hand hob and left-hand gear wheel; **b** both members with right-hand teeth

rotations are added mechanically, by a very ingenious mechanism, otherwise used in other types of gear cutting machines. This mechanism transmits the rotation to the gear wheel to be cut by means of the half-shaft of a differential, whose crown wheel (or cage) is fixed, for cutting of cylindrical spur gear wheels, while it rotates at a speed proportional to the feed velocity of the hob, for cutting of cylindrical helical gear wheels.

In the normal section, the gear wheel thus obtained has a normal pitch $p_n = p_{n0}$, and normal pressure angle $\alpha_n = \alpha_{n0}$, where p_{n0} and α_{n0} are respectively the normal pitch and normal pressure angle of the hob. These two quantities are given respectively by Eqs. (3.121) and (3.122). The transverse pitch, p, and the transverse pressure angle, α, of the gear wheel are therefore defined by the Eqs. (8.10) and (8.17).

As for cylindrical spur gear wheels (see Sect. 3.11.2), even for cylindrical helical gear wheels obtained with this cutting process, the profile deviations and errors increase with the increase of the lead angle γ_0, which therefore must not exceed the value of 5°. In addition, these deviations and errors also increase with the increase of the helix angle, β, as occurs in the form cutting process using a disk-type milling cutter.

8.12 Short Notes on Cutting Methods of Double-Helical Gears

From the geometric point of view, the cutting of cylindrical double-helical gear wheel is identical to that of cylindrical helical gear wheels. Therefore, it can be performed with the same gear cutting machines as described in the previous section. In this case, we can cut separately the two halves which make up the double-helical gear wheel, and then connect them as Fig. 8.16b–d show, or we can cut their teeth on a monobloc blank, leaving a central groove in order to relief of the cutting tools, as Fig. 8.16a shows.

There are special gear cutting machines for double-helical gear wheels, which allow obtaining continuous arrow teeth, with consequent advantages in terms of their strength and loading carrying capacity. This tooth continuity can be achieved by means of a form generating process or by a generating cut process.

8.12.1 Form Cutting Methods

The form cutting methods of double-helical gear wheels are similar to those described in Sect. 8.11.1 for helical gear wheels. End mills are used as cutting tools. These are made to move in the direction of the gear wheel axis, with speed, c, which is still correlated to the tangential velocity, v, at the reference pitch cylinder of the gear wheel being cut, by the Eq. (8.98). However, when the cutting tool reaches the mid-plane of the gear wheel, this tangential velocity reverses its direction.

With this operating way of the gear cutting machines, at the tooth arrow point a tooth space is formed, whose chordal dimension, l, is less than the chordal tooth thickness, L, in the transverse mid-plane of the gear wheel. It is to remember that the chordal tooth space and the chordal tooth thickness are the distances between symmetrical points on two opposite flanks of the same tooth space, and respectively on opposite flanks of the same tooth, both measured along a chord tangent to the pitch circle. The size l of this chordal tooth space should therefore be increased to the length L. This increase can be obtained by rounding the convex edge, A, or the concave edge, B, as shown in Fig. 8.20.

In relation to the other peculiarities of these double-helical gears, which can be obtained with these form cutting processes (for example, the reinforcement disks at their end faces), we refer back to what we have already said in Sect. 8.9.

8.12.2 Generation Cutting Methods

For the cutting of cylindrical double-helical gear wheels, gear cutting machines are used, which update the same generating cutting methods already described in Sect. 8.11.2 for cutting cylindrical helical gear wheels. Of course, these machines, having to cut simultaneously the two halves of the same double-helical gear wheel, have slightly different features than those used for cutting helical gear wheels. Here we just briefly describe these peculiarities.

Two types of gear shaping machines are used: one type uses two rack-type cutters, while the other type uses two pinion-type cutters. In both types of cutting machines, the two cutting tools are mounted on the same slide, and are disposed in opposition to each other. One of the two cutting tools performs the forward stroke

Fig. 8.20 Possible rounding of the convex edge, A, or concave edge, B, to have equal chordal dimensions of the tooth space and tooth thickness, in the transverse mid-plane of the gear wheel

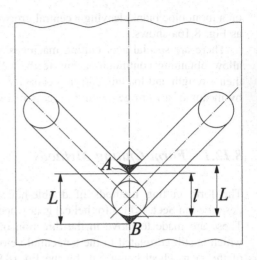

by cutting one of the two halves of toothing (e.g., half with right-hand helix), while the other cutting tool perform the return stroke, which is the idle stroke. The forward stroke of the first cutting tool is stopped exactly at the transverse mid-plane of the gear wheel being cut. At this point, the motion is reversed and the two cutting tools exchange their functions, in the sense that the cutting tool that first performed the idle stroke now makes the forward stroke and vice versa; so also the other half of the helical toothing is cut. The tooth is shaped like an arrow with a sharp edge. Usually the helix angle in the cutting conditions is assumed equal to 30°.

The gear hobbing machines also use two hobs, the way in which they operate has many similarities to that of the gear shaping machines described above. In this case, however, the two halves of the double-helical toothing are not symmetrical with respect to the mid-plane of the gear wheel, but are offset by a half circular pitch. In this way, at this mid-plane, the tooth thickness of one-half of the toothing faces the tooth space of the other half of the toothing.

This offset by a half circular pitch results in a very smooth and silent operation. The helix angle in the cutting conditions is usually equal to 23°. It should be noted, however, that during the forward strokes, the axes of the two hobs reach the mid-plane of the gear wheel being cut, so they remove a small portion of the tooth end faces of the other half toothing of the gear wheel.

Finally, among the generating cutting methods of cylindrical herringbone gear wheels, it is worth mentioning a special method, which allows to cut arrow teeth. The gear wheels obtained with this cutting method are not double-helical gear wheels in the strict meaning of the word, but gear wheels with arrow teeth.

The gear cutting machines that implement this cutting method use a special mechanism, conceived by Böttcher [1] and essentially referring to a planetary gear pair (see Chap. 13). It is made up of a rotating sun gear, whose reference pitch circle has a radius $(r_0 - e)$, and a fixed annulus having a reference pitch circle with radius r_0. The sun gear is mounted on the pin of a crank, rotating about its fixed axis, so the axis of the sun gear is mobile and describes a cylinder with radius, e. Instead, a point P rigidly connected to the sun gear, and placed to the outside of its reference pitch circle, described an extended hypocycloid. In case $r_0 = 4e$, this extended hypocycloid becomes a closed curve, namely a closed curvilinear quadrilateral like that shown in Fig. 8.21.

If now a trapezoidal cutter tool is placed in place of the point P, it will describe the tooth of a virtual generation rack, which will have in the plan view the shape of the extended closed hypocycloid defined above. If the gear wheel to be cut is equipped with suitable translation and rotation motions, corresponding to the pure rolling without sliding of its pitch cylinder on the pitch plane of the virtual generation rack, one tooth space between two adjacent teeth of the gear wheel blank is generated. In order to generate, with a continuous cutting process, all teeth of the gear wheel, the latter must have, in addition to the two afore-mentioned motions, a further motion corresponding to a rotation equal to a circular pitch per revolution of the sun gear, which has the toolholder wheel function.

In this way, however, the continuous curve described by the tooth of the virtual rack cutter is slightly asymmetrical with respect to the mid-plane of the gear wheel, as shown in Fig. 8.22. On the other hand, we have the great advantage of cutting, with a continuous process, all the teeth of the gear wheel, with all the related positive consequences in terms of the accuracy grade of the gear wheel thus obtained. In practice, three cutting tools work simultaneously: one of these gashes the tooth spaces, while the other two finish the concave and convex flanks of the teeth (see also Schiebel [33]).

Fig. 8.21 Extended closed hypocycloid described by a point P rigidly connected with the sun gear and placed at the outside of its pitch circle

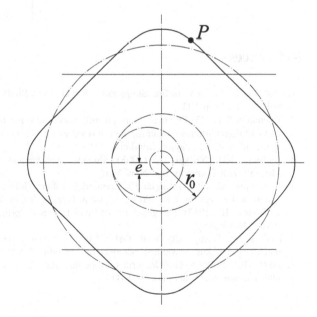

Fig. 8.22 Slightly asymmetrical paths of the cutting tool of the gear cutting machines that implement the Böttcher process

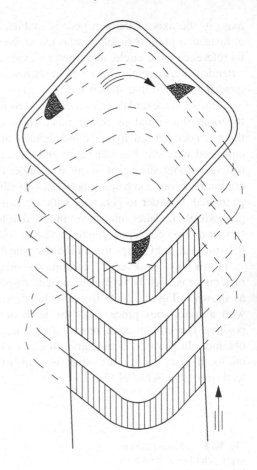

References

1. Böttcher P (1927) Vom Spiralkegelrad zur zyklischen Pfeilverzahnung, Der Maschinenbau, vol VI, Heft 3, p 103
2. Buchanan R (1823) Practical essay on mill work and other machinary, 2nd edn. With notes and additional articles containing new researches on various mechanical subjects by Thomas Tredgold, vol 1. J. Taylor, London:
3. Budynas RG, Nisbett JK (2008) Shigley's mechanical engineering design, 8th edn. McGraw-Hill Company Inc, New York
4. Champman A (2004) England's Leonardo: Robert Hooke and the Seventeenth-Century Scientific Revolution. CRC Press, Taylor & Frencis Group, Boca Raton, Florida
5. Colbourne JR (1987) The geometry of involute gears. Springer New York Inc, New York, Berlin, Heidelberg
6. Collins JA, Busby HR, Staab GH (2010) Mechanical design of machine elements and machines: a failure prevention perspective, 2nd edn. Wiley, New York
7. Davis JR (ed) (2005) Gear Materials, Properties, and Manufacture. Davis & Associates, ASM International, Materials Park, OH (USA)

8. do Carmo MP (1976) Differential geometry of curves and surfaces. Prentice-Hall, Englewood, New Jersey
9. do Carmo MP (1994) Differential Forms and Applications. Springer, Berlin Heidelberg
10. Ferrari C, Romiti A (1966) Meccanica Applicata alle Macchine. Unione Tipografica-Editrice Torinese, UTET, Torino
11. Galassini A (1962) Elementi di Tecnologia Meccanica: Macchine Utensili – Principi Funzionali e Costruttivi, Loro Impiego e Controllo della loro Precisione, reprint 9th edn. Editore Ulrico Hoepli, Milano
12. Gauss CF (1827) Disputationes generales circa superficies curvas, Commentationes Societatis Regiae Scientiarum Gottingesis Recentiores, vol VI, pp 99–146
13. Giovannozzi R (1965) Costruzione di Macchine, vol 2, 4th edn. Casa Editrice Prof. Riccardo Pàtron, Bologna
14. Gray A, Abbena E, Salomon S (2006) Modern differential geometry of curves and surfaces with mathematica, 3rd edn. Chapman & Hall/CRC, Taylor & Francis Group, Boca Raton, Florida
15. Henriot G (1979) Traité théorique et pratique des engrenages 1. Bordas, Sixième édition, Paris
16. ISO 1122-1 Vocabulary of gear terms—part 1: definitions related to geometry
17. ISO 54 (1996) Cylindrical gears for general engineering and for heavy engineering-modules
18. ISO 6336-1 (2006) Calculation of load capacity of spur and helical gears—part 1: basic principles, introduction and general influence factors
19. ISO 6336-2 (2006) Calculation of load capacity of spur and helical gears—part 2: calculation of surface durability (pitting)
20. Karas F (1949) Berechnung der Walzenpressung von Schrägzähnen an Stirnrädern. Knapp, Halle (Saale)
21. Kawasaki K, Tsuji I, Gunbara H (2015) Manufacturing method of double-helical gears using CNC machining center. Proc Inst Mech Eng Part C: J Mech Eng Sci 230(7):1989–1996
22. Kreyszig E (1991) Differential geometry. Dover Publications, New York
23. Lardner D (1840) A treatise on geometry and its applications in the arts. Longman and Taylor, London
24. Litvin FL, Fuentes A (2004) Gear geometry and applied theory, 2nd edn. Cambridge University Press, Cambridge, UK
25. Litvin FL, Fuentes A, Gonzalez-Perez I, Carnevali L, Kawasaki K, Handschuh RF (2003) Modified involute helical gears: computerized design, simulation of meshing, and stress analysis. Comput Methods Appl Mech Eng 192:3619–3655
26. Merritt HE (1954) Gears, 3rd edn. Sir Isaac Pitman & Sons Ltd, London
27. Niemann G, Winter H (1983) Maschinen-Elemente, Band II: Getriebe allgemein, Zahnradgetriebe-Grundlagen, Stirnradgetriebe, Springer, Berlin, Heidelberg
28. Pollone G (1970) Il Veicolo. Libreria Editrice Universitaria Levrotto & Bella, Torino
29. Radzevich SP (2012) Dudley's Handbook of Practical Gear Design and Manufacture, 2nd edn. CRC Press, Taylor & Francis Group, Boca Raton
30. Radzevich SP (2016) Dudley's handbook of practical gear design and manufacture, 3rd edn. CRC Press, Taylor&Francis Group, Boca Raton, Florida
31. Radzevich SP (2017) Gear cutting tools: science and engineering, 2nd edn. CRC Press, Taylor&Frencis Group, Boca Raton
32. Rossi M (1965) Macchine Utensili Moderne. Editore Ulrico Hoepli, Milano
33. Schiebel A (1934) Zahnräder-Zweiter Teil: Stirn-und Kegelräder mit schägen Zähnen. Springer, Berlin Heidelberg
34. Smirnov V (1970) Course de Mathématiques Supérieurs, Tome II, (traduit du Russe). Éditions MIR, Moscou
35. Stoker JJ (1969) Differential geometry. University of Toronto, Toronto
36. Townsend DP (1991) Dudley's Gear Handbook, The design, manufacture and application of gears, 2nd edn. McGraw-Hill, New York
37. Visconti A (1992) Introductory Differential Geometry for Physicists. World Scientific, Singapore

Chapter 9
Straight Bevel Gears

Abstract In this chapter, the fundamentals of straight bevel geometry are given and the related main quantities are defined. Geometry of spherical involute and octoid toothing is introduced and implications on the cutting processes of these gears are analyzed. Tredgold approximation is then used to define the equivalent cylindrical gears and their main characteristic quantities, including the minimum number of teeth to avoid interference. Short notes are given on the reference profile modifications. The problem of straight bevel gears with profile-shifted toothing and variation of shaft angle is therefore dealt with in detail in an analytical way, also considering the effects of transverse tooth thickness modifications. The load analysis of these gears is then performed and the thrust characteristics on shafts and bearings are defined. Subsequently, the main kinematic quantities related to the rolling and sliding motions of the mating teeth surfaces are defined and the instantaneous and average efficiencies are analytically determined. Finally, short notes on the main cutting processes of these types of gears are given and suggestions for their construction and assembly are provided.

9.1 Introduction

Straight, helical or skew, and spiral bevel gears are most commonly used for geared power transmission systems between two intersecting axes. In this chapter, we focus our attention on the *straight bevel gears*. Helical or skew, and spiral bevel gears, together with hypoid gears, will discussed in the Chap. 12. Straight bevel gears, helical and spiral bevel gears, and hypoid gears can be used for both speed-reducing and speed-increasing gear drives. Their gear ratio can be as low as 1, but it should not exceed about 10, although in machine tool design, where precision hypoid gears are required, most high values of the gear ratio, in the range between 10 and 20, have been required. However, for usual speed-increasing gear drives, it is appropriate that this gear ratio does not exceed 5 (see Buckingham [3], Merritt [23], Henriot [15], Niemann and Winter [25], Maitra [22], Radzevich [31]).

© Springer Nature Switzerland AG 2020

V. Vullo, *Gears*, Springer Series in Solid and Structural Mechanics 10,

https://doi.org/10.1007/978-3-030-36502-8_9

Basically, a straight bevel gear pair (and so also helical and spiral bevel gear pairs) is similar to a friction drive consisting of two conical rollers in which the conical surface of one roller drives that of the other roller by friction (see Pollone [27], Brar and Bansal [1], Klebanov and Groper [18]). Figure 9.1 shows two pairs of friction conical rollers, the first of which form an external friction drive (Fig. 9.1a) and the second an internal friction drive (Fig. 9.1b). In both drives, the two friction wheels of each pair are touching each other along the common generatrix passing through the apex or cone center, O, where the axes of the two cones intersect. Depending on whether the friction conical pair is external or internal, the shaft angle Σ between the axes is equal to the sum or, respectively, the difference, of the pitch cone angles or simply pitch angles, δ_1 and δ_2, of the pinion and wheel, that is:

$$\Sigma = \delta_1 + \delta_2. \tag{9.1}$$

$$\Sigma = \delta_2 - \delta_1. \tag{9.2}$$

Notoriously, when the two friction conical rollers of a pair are pressed against each other, they may transmit, without sliding, maximum tangential forces equal to the friction forces, which occur in the contact. However, the tangential friction forces that can be transmitted with these pairs of friction conical rollers are not high. Therefore, when higher tangential forces must be transmitted, the two conical rollers of the kinematic pair are equipped with teeth, the shape of which must be such the teeth of one member engage, during the rotation, with the teeth of the other member, so transmitting the motion with the same kinematics of the two aforementioned friction conical rollers. In this simple way, we pass from a friction conical pair to a bevel gear pair.

The two circular cones of the two fictitious friction wheels are the *axodes*, i.e. the *operating pitch cones* of the two members of the bevel gear pair [21]. These pitch

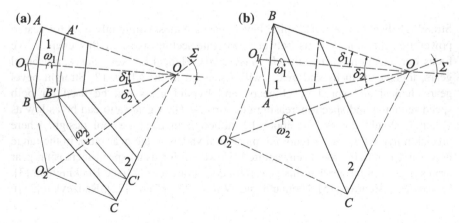

Fig. 9.1 Friction cone wheel pairs: **a** external pair; **b** internal pair

cones are in contact along a common generatrix and meet at a common point, the cone center or cone apex, O. This is also the *crossing point* O_c, i.e. the point of intersection of axes of two bevel gear members and their shafts. Therefore, crossing point and cone apex coincide ($O \equiv O_c$). The axodes are the *loci* of instantaneous axes of rotation in the movable coordinate systems $O_i(x_i, y_i, z_i)$, which are rigidly connected to rotating gear wheel i, with $i = (1,2)$. Assuming that one of the two members of the gear pair is fixed and the other movable, during their relative motion the movable cone (also called *movable axode* or *polodia axode*) rolls without sliding on the fixed cone (also called *fixed axode* or *herpolodia axode*). For further details on this subject, see Ferrari and Romiti [10].

The intersections of these pitch cones (the herpolodia axode and polodia axode) with a spherical surface having as its center the crossing point $O \equiv O_c$, are the *herpolodia* and *polodia*, i.e. the pitch circles of the two gear wheels. In other words, herpolodia and polodia are the directrices of two herpolodia axode and polodia axode, which are cones having a common apex at point $O \equiv O_c$. Since all toothing data of a bevel gear are given with reference to the heel (or large end), the pitch circles of the two members of the gear pair are those corresponding to the large end of the two cones, while AB and BC are the pitch diameters of the two gear wheels (Fig. 9.1).

In their rolling motion without sliding, the pitch cones have spherical motion, for which every point of a bevel gear remains at a constant distance from the crossing point during the motion. The two circles having diameters AB and BC, and then the two spherical segments delimited by them, roll against each other without sliding, remaining on the sphere. In order for the motion can take place in this way, the two wheels having the shape of spherical segments must be equipped with teeth profiled so that the maximum circle, normal to the profiles at point of contact, passes through the point of contact between the pitch circles. In order to shape the teeth profiles, only the spherical involutes are generally taken into consideration among all the curves that satisfy this condition. These are the curves generated by a point of a maximum circle when it rolls without sliding on a smaller circle, always keeping on the sphere [14, 27, 36].

The maximum wheel with the shape of a spherical segment is a hemispherical wheel, having as pitch circle a maximum circle. This hemispherical wheel is the so-called *crown wheel* or *crown gear*, i.e. a bevel gear with a reference cone angle of 90°. For the reasons to be described in the following sections, in which we will deepen well the generation of the spherical involute curves, it is not customary to give the spherical shape to the heel, that is the large end of a bevel gear. Instead, this heel is usually conical, with generatrices that are tangent to the theoretical sphere at the pitch circle diameter: this is the so-called *back cone*.

By doing so, the wheels having the shape of spherical segments, and face width equal to the width over the toothed part measured along a generatrix of their reference cones, are in fact replaced with the bevel gears tangent to the sphere in correspondence with the pitch circles of the same wheels. These wheels, like Fig. 9.1 shows, belong to back cones, i.e. the cones AO_1B and BO_2C, having apexes on the wheel axes and generatrices normal to those of the pitch cones.

Of course, this is an approximate way to analyze the straight bevel gears; it gives very reliable results, and is known as *Tredgold approximation*, as Tredgold conceived it (see Buchanan [2]).

The back cone corresponding to the crown wheel, and therefor tangent to the pitch circle of the maximum wheel having the shape of a spherical segment (the hemispherical wheel introduced above), degenerates into a cylinder. Since the teeth have a limited depth, on this cylinder we can replace the profile in the shape of spherical involute, with (see Giovannozzi [14], Pollone [27]):

- the straight profile of the standard basic rack, for straight bevel gears;
- the cylindrical helical profile, having with respect to the pitch plane the same inclination of the maximum osculating circle to the spherical involute, for helical bevel gears.

The back cones are developable surfaces. Therefore, they permit to bring the study of the tooth profiles of the bevel gears to the one of the teeth of equivalent cylindrical gears. Developing on a plane the cylinder in which the back cone of the crown wheel degenerated, we get a rack having straight profile flanks and pressure angle α. Then the crown wheel bears the same correlation to a bevel gear as a rack does to a cylindrical spur gear. Developing instead, always on a plane, the back cones of the two members of the bevel gear (Fig. 9.1), we get two cylindrical gear wheels having radii O_1B and O_2B, respectively.

9.2 Geometry and Characteristics of the Bevel Gears

Figure 9.2 shows two typical external bevel gears, each consisting of two bevel wheels which, in order to be able to operate with correct kinematic conditions, i.e. pure rolling without sliding between the operating pitch cones, must have the apex of the pitch cones in common. The same figure shows the most significant cones, and the main geometric entities and quantities characterizing the wheels of the bevel gear pairs and their toothings, namely:

- *Pitch cones*, i.e. the pitch surfaces of both wheels of the bevel gear pair, and *pitch angles* δ_1 and δ_2, i.e. the angles between the axes and pitch cone generatrices, which intersect at the *pitch cone apex*, coinciding with the *crossing point, O*. As in the case of cylindrical gears, here too we must be aware that, without a specific qualification, we mean the *operating* (or *working*) *pitch cones*. When necessary, to avoid misunderstanding, we add the specific qualification, so we will have the *reference pitch cones* or simply *reference cones*, and *operating pitch cones*. This notation also applies, unchanged, for many of the geometric entities and quantities defined below.
- *Back cones*, i.e. the cones at the outer end (the *heel* of the face width, whose generatrices are perpendicular to those of the reference cones, and *back cone angles*, i.e. the acute angles between the axes and generatrices of the back cones.

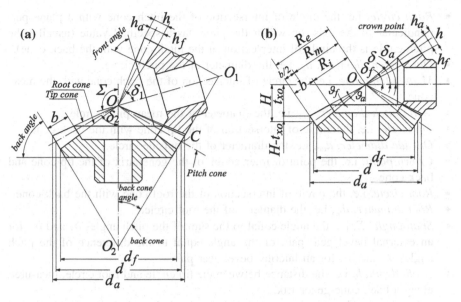

Fig. 9.2 External straight bevel gears: **a** shaft angle $\Sigma > 90°$; **b** shaft angle $\Sigma = 90°$

- *Tip cones*, i.e. the tip surfaces of both wheels of the bevel gear pair, and *face angles* or *tip angles*, δ_{a1} and δ_{a2}, i.e. the angles between the axes and tip cone generatrices.
- *Root cones*, i.e. the root surfaces of both wheels of the bevel gear pair, and *root angles* δ_{f1} and δ_{f2}, i.e. the angles between the axes and root cone generatrices.
- *Inner cones* (not shown in Fig. 9.2), i.e. the cones at the inner end (the *toe*) of the face width, whose generatrices are perpendicular to those of the pitch cones.
- *Mean cones* (not shown in Fig. 9.2), i.e. the cones at the mean point (i.e. the mean point of the face width), whose generatrices are perpendicular to those of the pitch cones.
- *Face width, b*, i.e. the width of the toothed part of a gear wheel, measured along a generatrix of its reference cone.
- *Cone distance*, i.e. the distance from the cone apex to the specified cone (then we have: mean cone distance, R_m; outer cone distance, R_e; inner cone distance, R_i; etc.), measured along a reference cone generatrix; CO_1 and CO_2 in Fig. 9.2a are the *back-cone distances*.
- *Locating face*, i.e. the plane face perpendicular to the axis of the gear wheel to be cut, by which its axial position is determined.
- *Mounting distance, H*, i.e. the axial distance from the locating face to the *pitch cone apex*.
- *Crown to crossing point, t_{x0}*, i.e. the distance along the gear axis from the *crown point* to the pitch cone apex.
- *Tip distance*, equal to the difference $(H - t_{x0})$, i.e. the distance along the gear axis from the locating face to the plane containing the tip circle.

- *Pitch circle*, i.e. the circle of intersection of the pitch cone with a plane perpendicular to the axis, on which the pitch has a specified value (usually, the pitch circle is the circle of intersection of the pitch cone with the back cone).
- *Outer pitch diameter*, d_e, i.e. the diameter of the pitch circle.
- *Mean pitch circle*, i.e. the circle of intersection of the pitch cone with the mean cone.
- *Mean pitch diameter*, d_m, i.e. the diameter of the mean pitch circle.
- *Tip circle*, i.e. the circle of intersection of the tip cone with the back cone.
- *Outside diameter*, d_{ae}, i.e. the diameter of the outside circle.
- *Crown point*, i.e. the point of intersection of the generatrix of the tip cone and back cone.
- *Root circle*, i.e. the circle of intersection of the root cone with the back cone.
- *Root diameter*, d_{fe}, i.e. the diameter of the root circle.
- *Shaft angle*, Σ, i.e. the angle equal to the sum of the pitch angles δ_1 and δ_2, for an external bevel gear pair, or the angle equal to the difference of the pitch angles δ_2 and δ_1, for an internal bevel gear pair.
- *Tooth depth*, h, i.e. the distance between the tip circle and root circle, measured along a back cone generatrix.
- *Addendum*, h_a, i.e. the distance between the tip circle and pitch circle, measured along a back cone generatrix.
- *Addendum angle*, $\vartheta_a = (\delta_a - \delta)$, i.e. the difference between the face angle and pitch angle.
- *Dedendum*, h_f, i.e. the distance between the pitch circle and root circle, measured along a back cone generatrix.
- *Dedendum angle*, $\vartheta_f = (\delta - \delta_f)$, i.e. the difference between the pitch angle and root angle.

Usually, the shaft angle Σ, which depends on the drive conditions, is equal to 90° (Fig. 9.2b), but it can assume value different from 90°. Figure 9.3 shows the range of the possible bevel gear pairs formed by a pinion, whose operating pitch angle is δ_1, and several wheels, characterized by different operating pitch angles δ_2, namely:

- Shaft angle $\Sigma_a < 90°$, with bevel gear pair formed by the pinion 1 and the wheel 2a.
- Shaft angle $\Sigma_b = 90°$, with bevel gear pair formed by the pinion 1 and the wheel 2b.
- Shaft angle $\Sigma_c > 90°$, with bevel gear pair formed by the pinion 1 and the crown wheel 2c.
- Shaft angle $\Sigma_d > 90°$, with bevel gear pair formed by the pinion 1 and the internal bevel wheel 2d.

It is first to be noted that, generally, the operating pitch angles, for the reasons that gradually we will specify better in the following sections, coincide with the reference pitch angles. It is then to be noted that the wheel 2c has an operating pitch angle of 90°: it is the crown wheel, i.e. a wheel with operating pitch angle of 90°,

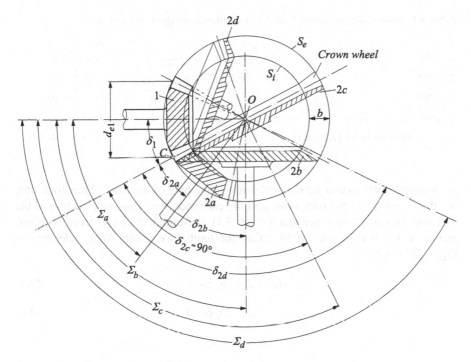

Fig. 9.3 Possible bevel gear pairs formed by a given pinion, and several wheels

and with tip and root angles respectively greater and smaller than 90°. This crown wheel can be an actual wheel, but can also be thought of as the common bevel wheel, which meshes properly with all the other wheels. In other words, it constitutes the equivalent of the standard basic rack of the parallel cylindrical spur and helical gears; it is of fundamental importance in the study of the bevel gears and their behavior. Therefore, this crown wheel, as the basic rack for cylindrical gears, is assumed as the generation wheel of all other bevel wheels.

It should finally be noted that an internal bevel gear pair, consisting of a pinion and an internal bevel wheel (2d in Fig. 9.3), although possible from the theoretical point of view, is to be considered an exceptional case that, in fact, does not have practical applications, given the difficulties that we have in making an internal bevel toothing. Therefore, in the following discussion, we consider only external bevel gear pairs. The relationships that correlate Σ, δ_1, δ_2, z_1 and z_2 (or $u = z_2/z_1$), in the different cases of straight bevel gears shown in Fig. 9.3, are summarized in Table 9.1.

In cases where tip and root cone generatrices converge in the pitch cone apex (see Sect. 9.5), the teeth of a bevel wheel taper off as they approach the apex. Thus, the tooth size decreases from the heel towards the toe, and the tooth profile also varies accordingly. For this reason, the quantities of interest, such as the addendum, dedendum, tooth depth, module, pitch circle diameter, tip circle diameter, etc., are

Table 9.1 Reference cone angles δ_1 and δ_2 for different straight bevel gear pairs

Σ	δ_2	δ_1
$\Sigma < 90°$	$\tan \delta_2 = \frac{\sin \Sigma}{(z_1/z_2) + \cos \Sigma}$	$\delta_1 = \Sigma - \delta_2$
$\Sigma = 90°$	$\tan \delta_2 = \frac{z_2}{z_1} = u$	$\delta_1 = \frac{\pi}{2} - \delta_2$
$\Sigma > 90°$	$\tan \delta_2 = \frac{\sin(\pi - \Sigma)}{(z_1/z_2) - \cos(\pi - \Sigma)}$	$\delta_1 = \Sigma - \delta_2$
$\Sigma > 90°$ (crown wheel)	$\delta_2 = 90°$	$\delta_1 = \Sigma - \frac{\pi}{2}$
$\Sigma > 90°$ (internal bevel gear pair)	$\tan \delta_2 = \frac{\sin \Sigma}{\sin \Sigma - (z_1/z_2)}$	$\delta_1 = \Sigma - \delta_2$

all conventionally measured with reference to the heel of the tooth. Therefore, they are those relating to the back cone and the corresponding transverse section of the toothing. In some cases (see again Sect. 9.5), these quantities are given with reference to the normal section bisecting the tooth face width. With reference to Fig. 9.2b, we have:

$$d_a = d + 2h_a \cos \delta \tag{9.3}$$

$$d_f = d - 2h_f \cos \delta \tag{9.4}$$

$$h = h_a + h_f \tag{9.5}$$

$$\delta_a = \delta + \vartheta_a \tag{9.6}$$

$$\delta_f = \delta - \vartheta_f \tag{9.7}$$

$$\tan \vartheta_a = \frac{2h_a \sin \delta}{d} = \frac{2h_a \sin \delta}{mz} \tag{9.8}$$

$$\tan \vartheta_f = \frac{2h_f \sin \delta}{d} = \frac{2h_f \sin \delta}{mz} \tag{9.9}$$

$$H = R_e \cos \delta = \frac{d}{2} \cot \delta. \tag{9.10}$$

Toothing of a bevel gear is sized using the same sizing of the parallel cylindrical spur gears. The nominal sizing most widely used is the following:

$$h_a = m \quad h_f = \tfrac{5}{4}m \quad h = h_a + h_f = \tfrac{9}{4}m. \tag{9.11}$$

It can also be used the standard stub sizing, given by:

$$h_a = 0.8m \quad h_f = m \quad h = 1.8m, \tag{9.12}$$

or other special sizing.

In order not to have too different profiles of the teeth, due to the taper of the tooth from heel to toe, the ratio (b/R_e) is limited within the range:

$$b/R_e \leq (0,25 \div 0,35). \tag{9.13}$$

9.3   Main Equations of Straight Bevel Gears

Consider any external bevel gear pair characterized by a given shaft angle Σ, and denote by:

- z_1 and z_2, the numbers of teeth of the pinion and wheel;
- ω_1 and ω_2 (Fig. 9.4a), the absolute values of angular velocities of the two members of the bevel gear pair, which generally are to be considered as vectors (note that vectors are indicated with bold letters);
- $\omega = (\omega_1 - \omega_2)$, the relative angular velocity, which is a vector having, as straight line of application, the common generatrix along which the operating pitch cones of the two members of the bevel gear pair touch (Fig. 9.4b);
- δ_1 and δ_2, the operating pitch cone angles that, as a rule, as we have already noted, are at the same time the reference pitch cone angles or, what is the same, the generation pitch cone angles (Fig. 9.4a);
- $d_1 = 2r_1$ and $d_2 = 2r_2$, the diameters of the two operating pitch circles, which are obtained as the intersections of the operating pitch cones with the back cones (note that the two back cones have a common generatrix passing through point C, as Fig. 9.2a shows, and that the operating pitch cone and back cone of each of the two members of a bevel gear pair have the same axis);
- R_e, R_i and R_m, the cone distances of the back, inner and mean cones from the cone apex, measured along a reference cone generatrix (Fig. 9.2b).

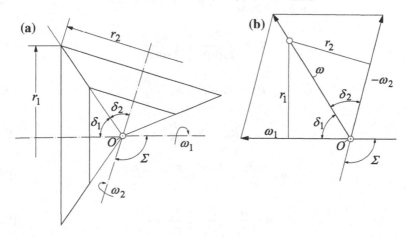

Fig. 9.4 Kinematic diagram of an external bevel gear pair, angular velocities and relative angular velocity vectors

With reference to Fig. 9.4, which shows the kinematic diagram of an external bevel gear pair, as well as the relative angular velocity vector $\omega = (\omega_1 - \omega_2)$, obtained as the difference of the vectors of the two angular velocities, ω_1 and ω_2, and to Fig. 9.2, we get the following relationships, which are valid for zero-profile-shifted bevel gears, and bevel gears with profile-shifted toothing without variation of shaft angle (see Sect. 9.8):

$$\Sigma = \delta_1 + \delta_2 \tag{9.14}$$

$$u = \frac{z_2}{z_1} = \frac{d_2}{d_1} = \frac{\omega_1}{\omega_2} = \frac{\sin \delta_2}{\sin \delta_1} \tag{9.15}$$

$$\omega = \sqrt{\omega_1^2 + \omega_2^2 + 2\omega_1\omega_2 \cos \Sigma} \tag{9.16}$$

$$\frac{\omega_1}{\sin \delta_2} = \frac{\omega_2}{\sin \delta_1} = \frac{\omega}{\sin[\pi - (\delta_1 + \delta_2)]} = \frac{\omega}{\sin \Sigma} \tag{9.17}$$

$$\sin \delta_1 = \frac{\omega_2}{\omega} \sin \Sigma = \frac{\sin \Sigma}{\sqrt{1 + u^2 + 2u \cos \Sigma}} = \frac{z_1 \sin \Sigma}{\sqrt{z_1^2 + z_2^2 + 2z_1 z_2 \cos \Sigma}} \tag{9.18}$$

$$\sin \delta_2 = \frac{\omega_1}{\omega} \sin \Sigma = \frac{u \sin \Sigma}{\sqrt{1 + u^2 + 2u \cos \Sigma}} = \frac{z_2 \sin \Sigma}{\sqrt{z_1^2 + z_2^2 + 2z_1 z_2 \cos \Sigma}} \tag{9.19}$$

$$\tan \delta_1 = \frac{\sin \Sigma}{u + \cos \Sigma} \quad (\text{for } \Sigma = 90°, \quad \tan \delta_1 = \frac{1}{u}) \tag{9.20}$$

$$\tan \delta_2 = \frac{u \sin \Sigma}{1 + u \cos \Sigma} \quad (\text{for } \Sigma = 90°, \quad \tan \delta_2 = u) \tag{9.21}$$

$$\frac{1}{\cos \delta_2} = \frac{1}{\sin \delta_1} = \sqrt{u^2 + 1} \quad \text{for } \Sigma = 90° \tag{9.22}$$

$$R_e = \frac{d_1}{2 \sin \delta_1} = \frac{d_2}{2 \sin \delta_2} \tag{9.23}$$

$$R_i = R_e - b = R_m - \frac{b}{2} \tag{9.24}$$

$$R_m = R_e - \frac{b}{2} \tag{9.25}$$

$$p = \frac{\pi d_1}{z_1} = \frac{2\pi r_1}{z_1} = \frac{\pi d_2}{z_2} = \frac{2\pi r_2}{z_2} \tag{9.26}$$

$$m = \frac{d_1}{z_1} = \frac{2r_1}{z_1} = \frac{d_2}{z_2} = \frac{2r_2}{z_2} \tag{9.27}$$

$$d_{m1} = d_1 - b \sin \delta_1 \qquad (9.28)$$

$$d_{m2} = d_2 - b \sin \delta_2 \qquad (9.29)$$

$$d_{a1} = d_1 + 2h_{a1} \cos \delta_1 = m \left(z_1 + 2\frac{h_{a1}}{m} \cos \delta_1 \right) \qquad (9.30)$$

$$d_{a2} = d_2 + 2h_{a2} \cos \delta_2 = m \left(z_2 + 2\frac{h_{a2}}{m} \cos \delta_2 \right) \qquad (9.31)$$

$$d_{f1} = d_1 - 2h_{f1} \cos \delta_1 = m \left(z_1 - 2\frac{h_{f1}}{m} \cos \delta_1 \right) \qquad (9.32)$$

$$d_{f2} = d_2 - 2h_{f2} \cos \delta_2 = m \left(z_2 - 2\frac{h_{f2}}{m} \cos \delta_2 \right). \qquad (9.33)$$

In the above relationships, the meaning of the symbols already defined while remaining unchanged, d_{m1} and d_{m2} are the diameters of the reference mean pitch circles of the pinion and the wheel, which are used for strength calculations.

We have also the following relationships:

$$\frac{z_1}{\sin \delta_1} = \frac{z_2}{\sin \delta_2} \qquad (9.34)$$

$$\tan \vartheta_{a1} = 2\frac{h_{a1} \sin \delta_1}{mz_1} \qquad (9.35)$$

$$\tan \vartheta_{f1} = 2\frac{h_{f1} \sin \delta_1}{mz_1} \qquad (9.36)$$

$$\tan \vartheta_{a2} = 2\frac{h_{a2} \sin \delta_2}{mz_2} \quad (\vartheta_{a2} = \vartheta_{f1}) \qquad (9.37)$$

$$\tan \vartheta_{f2} = 2\frac{h_{f2} \sin \delta_2}{mz_2} \quad (\vartheta_{f2} = \vartheta_{a1}). \qquad (9.38)$$

9.4 Spherical Involute Toothing and Octoidal Toothing, and Their Implications on the Cutting Process

In the introduction (see Sect. 9.1) we have already seen that the motion of the two pitch cones, which roll on each other without sliding, is a spherical motion, for which, during this motion, every point of a bevel gear remains at a constant distance from the apex. Therefore, any point of the tooth flank of a bevel gear wheel moves on a sphere, whose center is the crossing point, $O = O_c$, where the axes meet, and whose radius is the distance of this point from the cone apex. Of course, this point describes, on the sphere surface, a curve that is not a planar curve, but a three-dimensional curve. The profile of the tooth is obtained as the intersection of the tooth flank of the bevel gear wheel with the surface of this sphere (see Giovannozzi [14], Ferrari and Romiti [10], Pollone [27], Radzevich [30]).

We first examine bevel gears having spherical involute toothing. In this regard, without less of generality, we consider a sphere of unit radius and center O, and two cones (the base cones) that do not touch each other, with base cone angles δ_{b1} and δ_{b2} and having their apices coinciding with point O (Fig. 9.5). These two cones intersect the surface of the sphere according two circles, indicated with c_1 and c_2 in Fig. 9.5. Consider then a plane passing through the point O and tangent to both base cones; it intersects the sphere according to the maximum circle c_{max}, that is having maximum diameter.

Let us now roll this maximum circle c_{max} with a uniform angular velocity ω about an axis passing through the point O and perpendicular to the plane that contains it. We also suppose that, during this motion, the circle c_{max} drags by friction, without sliding, the two not-maximum circles c_1 and c_2, as well as the base cones having base cone angles δ_{b1} and δ_{b2}, with angular velocities ω_1 and ω_2 about their axes. Under the hypothesis that this drag motion between the three circles

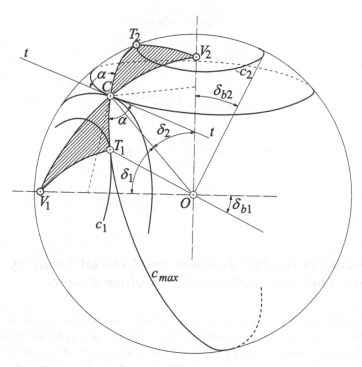

Fig. 9.5 Generation of spherical involutes and tooth profiles

c_{max}, c_1 and c_2 is a pure rolling motion, without sliding, their tangential velocities must be equal; therefore, since the sphere has a unit radius, we can write the following equation:

$$\omega = \omega_1 \sin \delta_{b1} = \omega_2 \sin \delta_{b2}. \tag{9.39}$$

However, the following relationship can be written for the operating pitch cones to be find, which in Fig. 9.5 are tangent along the straight line OC:

$$\omega_1 \sin \delta_1 = \omega_2 \sin \delta_2. \tag{9.40}$$

Therefore, from a comparison of Eqs. (9.39) and (9.40), we obtain:

$$\frac{\omega_1}{\omega_2} = \frac{\sin \delta_2}{\sin \delta_1} = \frac{\sin \delta_{b2}}{\sin \delta_{b1}}. \tag{9.41}$$

Consider now the plane containing the axes of the two cones. It intersects the sphere according to another circle having a maximum diameter, which passes through the intersection points, V_1 and V_2, of the axes of the cones with the sphere, and intersects at point C the other circle of maximum diameter, c_{max}. We denote by $[(\pi/2) - \alpha]$ the angle at point C formed, on the sphere, between the two circles of maximum diameters, and between the planes that contain them. Then we denote by α the pressure angle, i.e. the angle between the circle c_{max} and the common tangent $t - t$ at point C to the two pitch circles, which is perpendicular to the plane passing through points V_1, C, and V_2. Finally, we denote by T_1 and T_2 the points were the circle c_{max} touches respectively the circles c_1 and c_2. Applying the law of sines of spherical trigonometry to spherical triangles CT_1V_1 and CT_2V_2, which are respectively rectangles at points T_1 and T_2, we obtain the relationships:

$$\sin \delta_1 = \frac{\sin \delta_{b1}}{\cos \alpha} \quad \sin \delta_2 = \frac{\sin \delta_{b2}}{\cos \alpha}. \tag{9.42}$$

Let us now roll without sliding the plane containing the circle c_{max} on the base cones. During this rolling motion, a point of the circle c_{max} between T_1 and T_2 describes, on the sphere, two spherical involutes, tangent to each other, and therefore conjugate, which are the profiles of the teeth, i.e. the intersections of the tooth flanks with the sphere. Such spherical involutes touch at a point of the circle c_{max}, which therefore is, on the sphere, the curve of action, while the arc T_1T_2 is the arc of action.

If we join any point of contact on the circle c_{max} with the center O of the sphere, we get the instantaneous line of contact, which, like the circle c_{max}, rotates, about the axis of the latter with angular velocity ω. Therefore, the plane containing the circle c_{max}, which is tangent to the base cones, is the *locus* of points of contact, i.e. the *surface of contact*. During the rolling without sliding of the plane of contact on the base cones of the two gear wheels, the instantaneous line of contact describes and generates the flanks of the mating teeth.

The intersections of these flanks with the countless spheres between the back cone and inner cone, which are respectively tangent to the heel and toe of the tooth (these spheres are indicated with S_e and S_i in Fig. 9.3), are spherical involute curves, the geometry of which varies as a function of the radius of the sphere under consideration. The tooth profiles obtained as intersections of the tooth flank with each of these numberless spheres are conjugate profiles, and therefore ensure, from a theoretical point of view, an appropriate kinematics. Their geometry, however, varies from one to another sphere, and this has important implications, which call into question the choice of the virtual generating bevel wheel.

In relation to the cutting process, it is first to remember that even the toothing of the bevel gears can be broadly classified into two main groups: the one related to catalogue gears, and the one for gears that are not catalogue gears, with toothing for single gears (see Galassini [12], Rossi [33], Niemann and Winter [25], Stadtfeld [39]).

In the first case, both members of the gear pair have in common the reference mating crown wheel, i.e. the generation crown wheel, which is a virtual wheel that simulates the motion of the cutting tool. In this regard, Fig. 9.6 shows the materialization of the profiling motion of two planer tools with straight flanks on the corresponding crown wheel, having also straight flanks, for cutting of two straight bevel wheels. These wheels, being generated by the same wheel, i.e. the virtual generation wheel or simply the generation wheel, are able to mesh between them with a proper kinematics. This applies to all the possible bevel wheels mating with the generation crown wheel. Any two of the bevel wheels thus obtained are

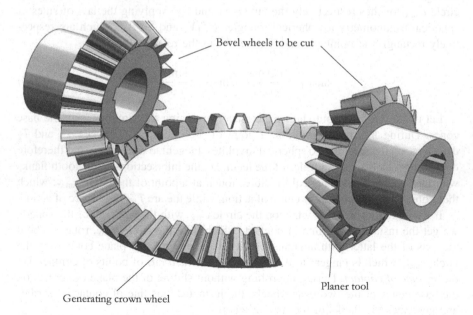

Fig. 9.6 Cutting of the straight bevel wheels with gear planer

conjugate between them, and are therefore able to ensure an appropriate kinematics, with a linear contact between the tooth flanks. In this way, we obtain bevel wheels that have the characteristics of catalogue gears.

The generation crown wheel of a batch of bevel wheels produced with these characteristics is uniquely defined (Fig. 9.3) by the following data:

- external cone distance, $R_e = OC = d_e/2 \sin \delta$;
- face width, b;
- shape of the tooth profile, geometrically defined by the reference crown wheel tooth profile;
- geometric shape and direction of the *tooth trace*, that is the line of intersection of the tooth flank with the reference surface (see Sect. 12.1);
- tip and root cone surfaces.

The various gear wheels thus obtained are characterized by different operating pitch cone angles δ, for which it is possible to derive gears for transmission drives between axes with different shaft angles, Σ. The bevel gear pair is so univocally defined once δ, Σ and the reference crown wheel are known.

Instead, in the second case, the one of gears that are not catalogue gears, the toothing is not determined with reference to a common generation crown wheel. The straight bevel gear pair consists of a generated pinion and a non-generated wheel, and the related cutting process with which the gear pair is obtained is known as *formate-cut process*. The wheel of the gear pair is produced without generation motion, with a form cutting process, by means of inserted-blade milling cutters that are different depending on the cutting process involved, which generally use teeth with straight profile flanks. The pinion is instead obtained with a generation motion, by means of a profiling tool able to simulate the pure rolling of the operating pitch cone of the same pinion. In this way, as Fig. 9.7 shows, the straight flanks of the

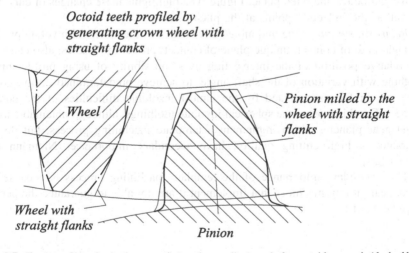

Octoid teeth profiled by
generating crown wheel with
straight flanks

Wheel

Pinion milled by the
wheel with straight
flanks

Wheel with
straight flanks

Pinion

Fig. 9.7 Teeth profiles of a single gear pair (continuous line), and of an octoid gear pair (dashed line)

teeth of the generation wheel generate the curved flanks of the pinion teeth. Thus, we have a *single gear pair*, in the sense that only that pinion and only that wheel are able to ensure an appropriate kinematics, if the assembly and actual operating conditions strictly fulfill the cutting conditions (see Stadtfeld [39–41].

We return now to consider the catalogue bevel gears produced by a generation process. In this process, the tooth space of the wheel to be cut is the envelope of the successive positions assumed by the conjugate tooth of the generation wheel, during the relative motion of rolling without sliding of the cutting pitch cone of the generation wheel on the cutting (or operating) pitch cone of the to-be-generated gear wheel. In this way, all the wheels derived from the same generation wheel are conjugate between them. However, using such a technological process, it is not possible to choose a generation wheel having any tooth profile, and any cutting cone angle, δ. This would be possible only in the case of gears obtained by a hot-rolling process, e.g. using the Anderson gear rolling machine or similar machines (see Galassini [12]).

It is necessary to consider that the tooth of the generation wheel tapers off as it approaches the apex, and that its profile has variable size and shape along its entire face width. Therefore, it is not possible to simulate this tooth, having variable profile, replacing it with a planer tool with constant profile, equipped with cutting motion that will make it run from the heel to toe along a guide inclined of an angle δ. Significant cutting errors and profile deviations would be. To solve this problem, it is necessary to choose, as generation wheel, a particular bevel wheel whose teeth do not change the shape of their profile from heel to toe. This wheel is the crown wheel, which has a cone angle $\delta = 90°$, and reference pitch cone that degenerates into a plane (the pitch plane). However, it must have straight tooth profiles. On the back cone of the crown wheel, as well as on the numberless cones included between the back and inner cones of the same crown wheel, all degenerated into cylinders, the spherical involute profiles vary from heel to toe, also changing their curvature above and below the pitch plane. Figure 9.8a highlights these changes of curvature as well as the inflection points at the pitch plane.

Rebus sic stantibus, the undoubted advantages of the spherical involute profiles (straight path of contact; unique plane of contact; proper kinematics, also chancing the relative position of the intersecting axes; possibility of using profile-shifted toothing with variation of the shaft angle, to achieve predetermined design goals; etc.) are, in fact, annihilated by the almost insoluble difficulties of the cutting process. For this reason, the spherical involute toothings, although obtainable using bevel gear planers, which implement a template machining process, but do not guarantee a high cutting quality, have secondary importance (Niemann and Winter [25]).

This secondary importance of the template machining process is because the bevel gear cutting machines that implement it are not able to guarantee the perfect alignment -of-:

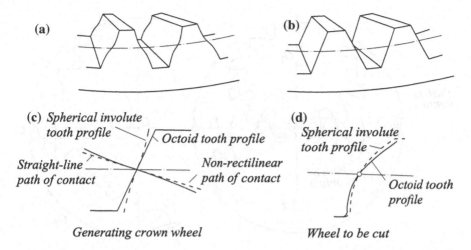

Fig. 9.8 Generation crown wheels with spherical involute tooth profiles (**a**), and with octoid tooth profiles (**b**); tooth profile shapes of the two generation crown wheels (**c**), and the generated wheels (**d**); paths of contact on the sphere (**c**)

- the feeler pin or tracer, whose tracing point follows the contour of the three-dimensional master model or template, reproducing the spherical involute;
- the shaper cutting edge, following the path taken by the tracer to machine the spherical involute shape;
- the center of the spherical motion, which coincides with point where the axes of the bevel wheel being cut and virtual crown wheel that is simulated by the shaping cutter intersect.

However, the bevel gears obtained with this template machining process, which was used sporadically as long as the CAD/CAM systems become widespread for machining complex three-dimensional shapes [26], such as spherical involutes, are of poor accuracy grade because, due to the shaping cutter wear, the above alignment could not be guaranteed. In practical reality, shaper wear determines significant deviations of the profiles relative to the theoretical geometry of the spherical involute. In addition, the inevitable pitch errors, due to the use of the indexing head for the single cut of each tooth, are added to these deviations. It should be noted that this template cutting process, even upgraded with CAD/CAM systems (these reduce and even eliminate indexing errors), is always of secondary importance for the manufacture of straight bevel gears, as it is not suitable for large series productions.

Anyway, the toothings that are actually produced are *octoid toothings*. They are related to a generation crown wheel with trapezoidal teeth (Fig. 9.8b). In this case, the tooth profile is a straight line, and corresponds to that of the planar tool (Fig. 9.6). The latter, in its cutting motion, will be engaged to a depth decreasing from the heel to the toe, in the case of tooth that tapers off as it approaches the apex, and to a constant depth, in the case of tooth with constant depth (see next section). According to Tredgold approximation, also briefly recalled in the next section, on

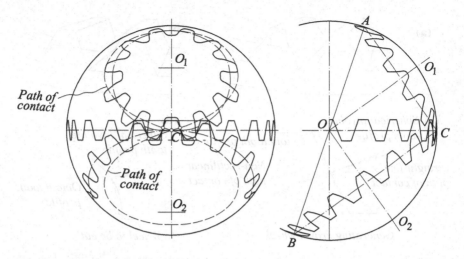

Fig. 9.9 Reference octoid crown wheel and bevel gear pair with octoid toothing; path of contact in the shape of eight on the sphere (the octoid)

the cylinder in which the back cone of the octoid generation crown wheel is degenerated, we have a basic rack with straight-line tooth flanks. Then the generation of the octoid toothing corresponds, essentially, to that of the involute tooth profiles of the cylindrical spur gear wheels. Substantially, in this generation cutting method, a straight sided cutting tool, simulating the generation crown wheel, and the cutting cone of the blank of the bevel wheel to be cut roll on each other producing the desired bevel wheel.

From a conceptual point of view, the two generation cutting processes, the hypothetical one that would use a crown wheel with spherical involute tooth profiles, and the one actually used, which employs a crown wheel with octoid tooth profiles, are quite similar. The difference consists in the fact that, considering the rolling on the sphere, in the first case the path of contact is a straight line, while in the second case the path of contact deviates from the straightness, albeit slightly (Fig. 9.8c). If we consider the two wheels of a bevel gear and the common generation octoid crown wheel, this path of contact develops, on the sphere, according to a curve in the shape of eight (Fig. 9.9). The geometry of this curve is determinable using the known methods of the mating profiles (see Levi-Civita and Amaldi [20], Ferrari and Romiti [10], Scotto Lavina [36]).

Despite the deviation of the path of contact from the straight line, the kinematics of operation of the bevel gears with octoid toothings is correct for zero profile-shifted toothing and profile-shifted toothing without variation of shaft angle. The cutting process of bevel gears with octoid toothing is very precise, and leads to high quality gears, unimaginable with a template shaping process, with which we would get wheels with spherical involute toothing.

For brief information on the cutting methods of straight bevel gears and gear cutting machines that implement them, we refer the reader to Sect. 9.12.

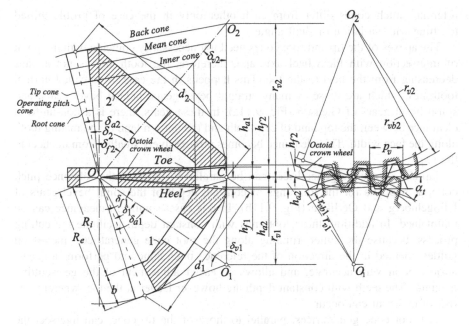

Fig. 9.10 Bevel gear with zero profile-shifted toothing, and equivalent cylindrical gear at the back cone

9.5 Main Conical Surfaces and Equivalent Cylindrical Gear

As Fig. 9.10 shows, various conical surfaces are identifiable in a bevel gear pair, the main of which are:

- the operating pitch cones, among them tangent along the common generatrix $OC = R_e$, and defined by the cone angles δ_1 and δ_2;
- the reference pitch cones, which coincide with the cutting pitch cones (the operating pitch cones in cutting conditions), and are therefore the cones that, in cutting conditions, roll without sliding on the pitch plane of the common generation crown wheel;
- the tip and root cones, defined by the tip and root angles $\delta_{a1,2}$ and $\delta_{f1,2}$;
- the back, mean and inner cones, defined by the back-cone angles $[(\pi/2) - \delta_1]$ and $[(\pi/2) - \delta_2]$.

It is to be noted that the back and mean cones are used respectively to define the equivalent cylindrical gear and to set the bending strength calculations of the teeth. It should also be noted that the operating pitch cones and reference pitch cones of a bevel gear pair coincide in the two cases of zero profile-shifted toothing, and profile-shifted toothing without variation of shaft angle. Of course, in these two cases, also the corresponding cone angles coincide. The operating pitch cones and reference pitch cones differ from each other only in the case of profile-shifted toothing with variation of shaft angle.

The apexes of the tip and root cones need not necessarily coincide with the point of intersection with the wheel axis and, therefore, the tooth size is not always decreasing from the heel to the toe. This happens in the case of so-called normal toothings, which are those of many straight bevel gears, helical bevel gears, and spiral bevel gears of Gleason (Fig. 9.11a). In these gears, there is a non-uniform clearance between the top land of one tooth and the bottom land of the mating tooth along the face width. This clearance becomes progressively smaller from the heel to the toe.

Tip and root cone generatrices can be parallel to those of the reference pitch cone, as happens for the toothings of constant depth of the spiral bevel gears of Klingelnberg and Oerlikon (Fig. 9.11b). In this case a uniform clearance can be maintained. In addition, these toothings with constant depth facilitate the cutting process, because the cutter, running along the root cone generatrices, moves on guides oriented in the direction of the reference pitch cone and performs a generation motion very accurate, and allows an easier adjustment of the gear-cutting machine. The teeth with constant depth are however leaner at the toe, whereby the risk of undercut can occur.

The root cone generatrices, parallel to those of the tip cone, can intersect the generatrices of the reference pitch cone, as happens for the toothings with constant depth of the spiral bevel gears of Klingelnberg-Palloid (Fig. 9.11c). Finally, the tip and root cone generatrices can degenerate into the tip and root cylinder generatrices, parallel to the axis to the wheel, which intersect the reference pitch cone generatrices at the two end faces of the wheel, as occurs in the cylindrical-bevel gears (Fig. 9.11d).

Figure 9.10 shows, in addition to the back-cone distance, R_e, also the inner cone distance, R_i. It also highlights the back cone and mean cone, i.e. two of the numberless cones between the back cone and inner cone. Usually, the study of tooth profiles of a bevel gear pair is led back to the study of the tooth profiles of an equivalent cylindrical gear pair, which is obtained by developing on a plane the back cones. This is an approximate method, proposed for the first time by Tredgold,

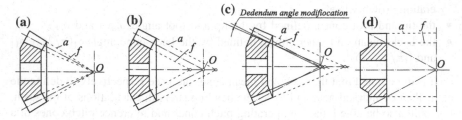

Fig. 9.11 Tip and root cone generatrices: **a** converging in the operating pitch cone apex; **b** parallel to pitch cone generatrices; **c** root cone generatrices intersecting those of the pitch cone and parallel to those of the tip cone; **d** tip and root cone generatrices parallel to the wheel axis

in 1823, and then known as *Tredgold approximation* or *Tredgold method* (see Buchanan [2]). Results obtained using this method, albeit approximate, are more than satisfactory for calculations inherent in this type of gears.

The approximate Tredgold method is based on replacing the small portion of the sphere (corresponding to the toot depth measured along a back cone generatrix), on which the spherical involutes previously described develop, with the back cones of the two wheels. Figure 9.12 shows a bevel gear pair with shaft angle $\Sigma = 90°$, and the virtual (or equivalent) cylindrical gear pair obtained by the Tredgold approximation.

Whatever is the shaft angle Σ (it is therefore not limitative consider bevel gear pairs with shaft angle $\Sigma = 90°$, such as those shown in Figs. 9.10 and 9.12), in the development on a plane of the back cones of the two wheels we get an equivalent parallel cylindrical spur gear pair having:

- the same module, $m = d_1/z_1 = d_2/z_2$ and, therefore, the same transverse pitch, p_t (in the aforementioned development, it does not change);
- the same transverse pressure angle, $\alpha_t = \alpha$;
- the same transverse tooth thickness, $s_t = s$;
- the same tooth depth, $h = h_a + h_f$, of the two bevel wheels.

The radii of the pitch circles of the equivalent cylindrical wheels, equal to the cone distances of the two back cones, are then given respectively by:

Fig. 9.12 Bevel gear pair with shaft angle $\Sigma = 90°$, and virtual cylindrical gear pair obtained by the Tredgold approximation

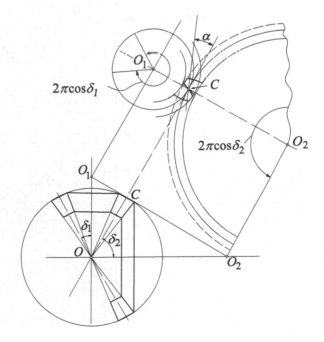

$$r_{v1} = \frac{d_1}{2\cos\delta_1} \quad r_{v2} = \frac{d_2}{2\cos\delta_2};\tag{9.43}$$

therefore, the virtual (or equivalent) numbers of teeth of the two virtual cylindrical wheels are given by:

$$z_{v1} = \frac{z_1}{\cos\delta_1} \quad z_{v2} = \frac{z_2}{\cos\delta_2}.\tag{9.44}$$

The diagram in Fig. 9.13 allows to calculate immediately the virtual number of teeth z_v as a function of the actually number of teeth, z, and the reference cone angle, δ.

Therefore, the gear ratio of the equivalent cylindrical gear pair is given by the relationship:

$$u_v = \frac{z_{v2}}{z_{v1}} = \frac{z_2\cos\delta_1}{z_1\cos\delta_2} = u\frac{\cos\delta_1}{\cos\delta_2}.\tag{9.45}$$

Fig. 9.13 Virtual number of teeth z_v as a function of z and δ

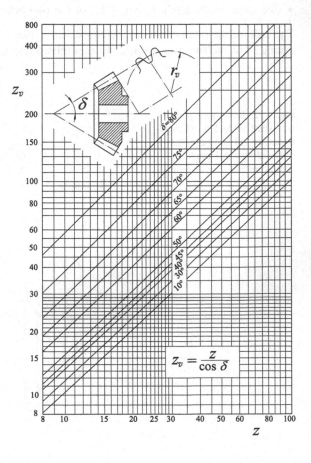

For $\Sigma = (\delta_1 + \delta_2) = 90°$, Eqs. (9.44) and (9.45) become respectively:

$$z_{v1} = z_1 \frac{\sqrt{u^2+1}}{u} \qquad z_{v2} = z_2 \sqrt{u^2+1} \qquad (9.46)$$

$$u_v = \left(\frac{z_2}{z_1}\right)^2 = u^2. \qquad (9.47)$$

It is to be noted that, for the calculations of load carrying capacity of bevel gears, the virtual parallel cylindrical gear pair to be considered, according to the Tredgold approximation, is that corresponding to the mean cones, i.e. the cones whose generatrices bisect the face width, and are therefore equidistant and parallel to the generatrices of the back and inner cones. With reference to Fig. 9.14, we have the following quantities related to the mean cone (subscripts m and e indicate quantities respectively related to the mean and back cones):

- Reference diameters, d_{vm}:

$$d_{vm1} = \frac{d_{m1}}{\cos \delta_1} = \frac{R_m d_{e1}}{R_e \cos \delta_1} \qquad d_{vm2} = \frac{d_{m2}}{\cos \delta_2} = \frac{R_m d_{e2}}{R_e \cos \delta_2} \qquad (9.48)$$

and, for $\Sigma = 90°$:

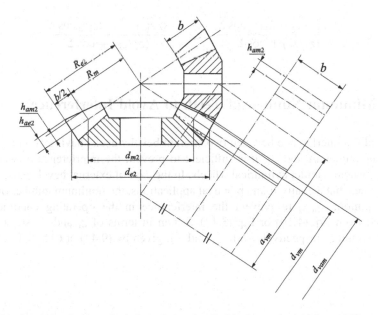

Fig. 9.14 Quantities for the calculation of virtual cylindrical gears at mid-face width

$$d_{vm1} = d_{m1}\frac{\sqrt{u^2+1}}{u} \qquad d_{vm2} = u^2 d_{vm1}. \tag{9.49}$$

- Center distance, a_{vm}:

- $$a_{vm} = \frac{1}{2}(d_{vm1} + d_{vm2}). \tag{9.50}$$

- Tip diameters, d_{vam}:

$$d_{vam1} = d_{vm1} + 2h_{am1} \qquad d_{vam2} = d_{vm2} + 2h_{am2}. \tag{9.51}$$

- Base diameters, d_{vbm}:

$$d_{vbm1} = d_{vm1}\cos\alpha_{vt} \qquad d_{vbm2} = d_{vm2}\cos\alpha_{vt}, \tag{9.52}$$

where $\alpha_{vt} = \alpha_t = \alpha$ is the pressure angle in the transverse section.

According Eqs. (9.18) to (9.21), the virtual numbers of teeth of the equivalent cylindrical wheels, given by Eq. (9.44), can also be expressed as function of z_1, z_2 and Σ, for which we have:

$$z_{v1} = \frac{\sqrt{z_1^2 + z_2^2 + 2z_1 z_2 \cos\Sigma}}{(z_2/z_1) + \cos\Sigma} \qquad z_{v2} = \frac{\sqrt{z_1^2 + z_2^2 + 2z_1 z_2 \cos\Sigma}}{(z_1/z_2) + \cos\Sigma}. \tag{9.53}$$

9.6 Minimum Number of Teeth to Avoid Interference

To avoid the interference between two straight bevel wheels, having z_1 and z_2 teeth and cone angles δ_1 and δ_2, it is sufficient to prevent the interference between the two equivalent parallel cylindrical wheels. In the case of external bevel gears, which are the ones that actually found practical applications, the minimum number of teeth of the pinion, z_{1min}, to prevent the interference in the operating conditions is obtained from Eq. (4.10) or Eq. (3.14) written in terms of z_1 and z_2, substituting therein z_1 and z_2 respectively with z_{v1} and z_{v2}, given by (9.44) or (9.53). Therefore, we have:

$$z_{1min} \geq 2k_2 \cos \delta_1 \frac{1 + \sqrt{1 + \frac{z_1 \cos \delta_2}{z_2 \cos \delta_1}\left(2 + \frac{z_1 \cos \delta_2}{z_2 \cos \delta_1}\right) \sin^2 \alpha_t}}{\left(2 + \frac{z_1 \cos \delta_2}{z_2 \cos \delta_1}\right)\sin^2 \alpha_t}, \quad (9.54)$$

where k_2 is the addendum factor of the largest wheel. Thus, the minimum number of teeth of the pinion, z_{1min}, varies linearly with k_2, and depends on the ratio $(z_1/z_2) = (1/u)$, transverse pressure angle α_t and shaft angle $\Sigma = (\delta_1 + \delta_2)$. Recalling Eqs. (9.45), (9.54) can be written also in the following more compact form:

$$z_{1min} \geq \frac{2k_2 \cos \delta_1}{\left[\sqrt{u_v^2 + (1 + 2u_v) \sin^2 \alpha_t} - 1\right]} = 2k_2 \cos \delta_1 \frac{u_v + \sqrt{u_v^2 + (1 + 2u_v) \sin^2 \alpha_t}}{(1 + 2u_v) \sin^2 \alpha_t}.$$

$$(9.55)$$

It is then necessary to check that the cutting interference does not occur during the generation process, when cutting of the bevel wheel is done by means of a gear cutting machine that simulates the generation crown wheel. Since the cone angle δ_2 of the generation crown wheel is equal to $\pi/2$, from Eq. (9.54) we infer that, to avoid the cutting interference it must be:

$$z_{1min} \geq 2k_{20} \frac{\cos \delta_1}{\sin^2 \alpha_t}, \quad (9.56)$$

where k_{20} is the addendum factor of the cutting tooth. Therefore, in this condition, z_{1min} varies linearly with k_{20} and depends on δ_1, and thus on $\Sigma = (\delta_1 + \delta_2)$, and transverse pressure angle α_t.

Diagrams in Figs. 9.15 and 9.16, and in Fig. 9.17, respectively show the curves related to Eq. (9.54) or the equivalent Eqs. (9.55), and (9.56). The diagrams of Figs. 9.15 and 9.16, respectively valid for $\alpha_t = 15°$ and $\alpha_t = 20°$, highlight, for $k_2 = 1$, the variation of z_{1min} in the operating conditions, as a function of the shaft angle Σ, shown on the abscissa, and the ratio $(z_1/z_2) = (1/u)$, which is the parameter related to each curve. The diagram of Fig. 9.17 shows in a single synoptic, for $k_{20} = 1$, the double family of curves (top, one inherent to $\alpha_t = 15°$, and below that relating to $\alpha_t = 20°$) that give, in the cutting conditions performed with generation crown wheel, the variation of z_{1min} as a function of the shaft angle $\Sigma = (\delta_1 + \delta_2)$ and ratio $(1/u) = (z_1/z_2)$.

The number of teeth, z_{20}, of the virtual generation crown wheel in the cutting conditions, bearing in mind the Eq. (9.15), and remembering that $\delta_{20} = \delta_0 = \pi/2$, is given by:

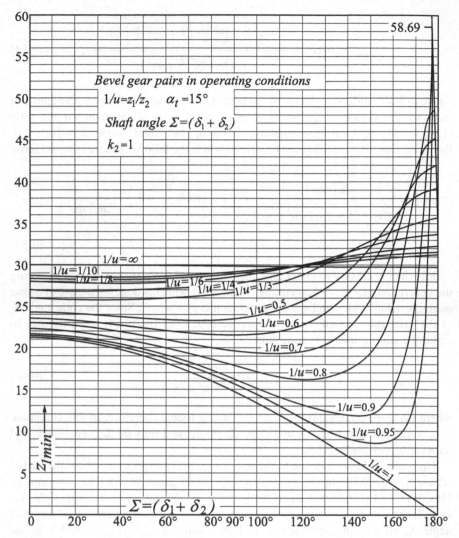

Fig. 9.15 Distribution curves of z_{1min} given by Eq. (9.54), as a function of $\Sigma = (\delta_1 + \delta_2)$ and $(1/u) = (z_1/z_2)$, for $k_2 = 1$ and $\alpha_t = 15°$

$$z_{20} = \frac{z_1}{\sin \delta_1} = z_0. \tag{9.57}$$

Of course, all this is true for bevel gears with zero profile-shifted toothing. Comparing the results obtained using the relationships (9.54) and (9.56), or the diagrams shown in Figs. 9.15, 9.16 and 9.17, we can deduce that the more restrictive conditions are those inherent in the cutting conditions. For bevel gear pairs with perpendicular axes ($\Sigma = 90°$), and with $k_1 = k_2 = 1$, the values of z_{1min}

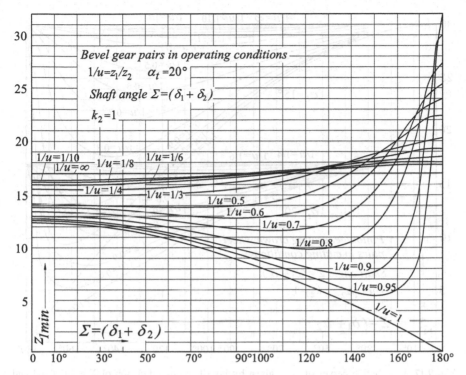

Fig. 9.16 Distribution curves of z_{1min} given by eq. (9.54), as a function of $\Sigma = (\delta_1 + \delta_2)$ and $(1/u) = (z_1/z_2)$, for $k_2 = 1$ and $\alpha_t = 20°$

that are suitable to avoid both the operating interference with the wheel of z_2 teeth, and the cutting interference with the virtual generation crown wheel, are those collected in Table 9.2.

It is to remember that the values of z_{1min} taken from diagrams shown in Figs. 9.15, 9.16 and 9.17 and those collected in Table 9.2 refer to teeth with addendum factor $k_2 = k_{20} = 1$. The relationships (9.54) and (9.56) show, however, that z_{1min} is proportional to the addendum factor. Therefore, if different sizing were adopted for the teeth, the values of z_{1min} to be choose are those derived from diagrams and table above, multiplied by the actual addendum factor.

As for the cylindrical spur gear (see Sect. 3.2) and cylindrical helical gears (see Sect. 8.5), also in this case it is admitted that, in practice, the minimum number of teeth z_{1min} that can be cut by generation cutting using a virtual generation crown wheel is equal to $5/6$ of the theoretical values. The diagram shown in Fig. 9.18 allows to calculate the minimum number of teeth, z_{1min}, that can be cut without interference, as a function of the reference pitch cone angle δ_1, shown on the abscissa, and transverse pressure angle α_t, shown on each curve. The broken line curves represent the theoretical values of z_{1min}, while the solid line curves provide the practical values z'_{1min} of z_{1min}. With z'_{1min} a marginal amount of undercut is allowed.

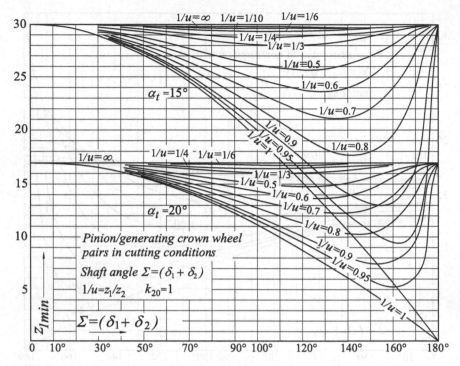

Fig. 9.17 Distribution curves of z_{1min} given by Eq. (9.56), as a function of $\Sigma = (\delta_1 + \delta_2)$ and $(1/u) = (z_1/z_2)$, for $k_{20} = 1$ and $\alpha_t = 15°$, and $\alpha_t = 20°$

Table 9.2 Values of z_{1min} to avoid both operating and cutting interference, for bevel gear pair with $\Sigma = 90°$, and $k_1 = k_2 = k_{20} = 1$

z_1/z_2	z_{1min}		z_1/z_2	z_{1min}	
	$\alpha_t = 15°$	$\alpha_t = 20°$		$\alpha_t = 15°$	$\alpha_t = 20°$
1/1,00	21	12	1/3	28	16
1/1,05	22	12	1/4	29	17
1/1,10	22	13	1/5	29	17
1/1,25	23	13	1/6	29	17
1/1,43	24	14	1/8	30	17
1/1,67	26	15	1/10	30	17
1/2,00	27	16			

Figure 9.18 also shows the advantage of the straight bevel gears with respect to the parallel cylindrical spur gears, in terms of minimum number of teeth that can be cut without interference. For example, for $\alpha_t = 20°$ and $\delta_1 = 30°$, z'_{1min} drops from 14 teeth to 12 teeth. Here, too, it is obvious that, when we wanted to use pinions characterized by a number of teeth even lower, and together dispel the risk of the

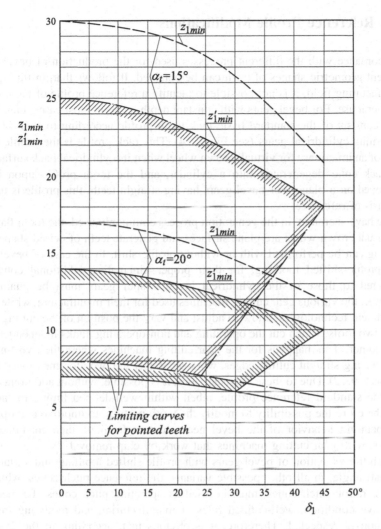

Fig. 9.18 Theoretical (z_{1min}) and practical (z'_{1min}) minimum number of teeth as a function of the reference pitch cone angle, δ_1, and transverse pressure angle, α_t, for bevel gears cut with virtual generation crown wheel

operating interference (this, in contrast with the cutting interference, is absolutely to ward off), the only way is to use bevel gears with profile-shifted toothing.

Example Input data: straight bevel gear pair, with: $\Sigma = 90°$; $(1/u) = (z_1/z_2) = 1/2$; $\alpha_t = 15°$; $k_1 = k_2 = k_{20} = 1$. Results: (*i*), from the diagram of Fig. 9.15, we infer that, in order to avoid the operating interference, it must be $z'_{1min} \geq 24$; (*ii*), from the diagram of Fig. 9.17, we infer that, in order to avoid the cutting interference, i.e. the undercut, it must be $z'_{1min} \geq 28$.

9.7 Reference Profile Modifications

In accordance with the different processes used for the production of bevel gears, different geometric shapes of teeth can be obtained. It follows that, in this specific manufacturing field, it is not possible to identify a reference profile of the toothing of general use. For bevel gears with constant depth teeth, in most cases, the starting point consists of the standard basic rack tooth profile according to ISO 53: [17], concerning cylindrical gears (see Fig. 2.14). This rack profile is the profile of the tooth of an imaginary (or virtual) crown wheel when its cylindrical back surface, i.e. the back cone degenerated into a cylinder, and the tooth profile upon it, are developed on a plane. For bevel gears having straight teeth, this profile is used as the basis of reference.

We have seen that, in the generation process commonly used, the tooth flanks of the virtual crown wheel are plane surfaces and generate teeth of octoid shape. This toothing can be performed with or without profile shift. In the case of bevel gears with profile-shifted toothing, for their proper kinematics, additional conditions compared to those of the cylindrical involute spur gears must be guaranteed. However, the various gear cutting machines used for their manufacture, while based on different technologies, allow to adjust and vary the position of the cutting edges of the two tools, which cut the operating and non-operating flanks independently of one another. This happens for the gear cutting machines where the two finishing tools, having straight cutting edges, work separately the two aforementioned flanks (see Sect. 9.12). Due to these characteristics, it is possible, without additional costs, vary the standard reference profile, albeit within well-defined limits. In fact, we have therefore the possibility to modify the tooth profiles, to improve and optimize the operating behavior of the bevel gears, and to facilitate their manufacturing process, with gear cutting machines that work by chip removal.

With the exception of bevel gears with profile-shifted toothing and variation of the shaft angle, in all other possible variants, the reference pitch cones, which are the generation pitch cones, must be used as operating pitch cones. To meet this restrictive condition, well-defined rules of manufacturing and mounting must be scrupulously respected. Therefore, it is obvious that, according to the Tredgold approximation, the imaginary rack-type cutter, with which the wheel is cut, will be the mirror image, compared to the reference pitch line, of the equally imaginary rack-type cutter with which the pinion is cut. The imaginary rack-type cutter is achieved by developing on a plane the back cone of the generation octoid crown wheel, which has degenerated into a cylinder.

For the above reasons, the range of families of bevel gears are much broader than that of cylindrical gears. Indeed, we can have the following types of bevel gears (see Fig. 9.19):

(A) Bevel gears with zero profile-shifted toothing, that is constituted by two bevel wheels, both without profile shift.

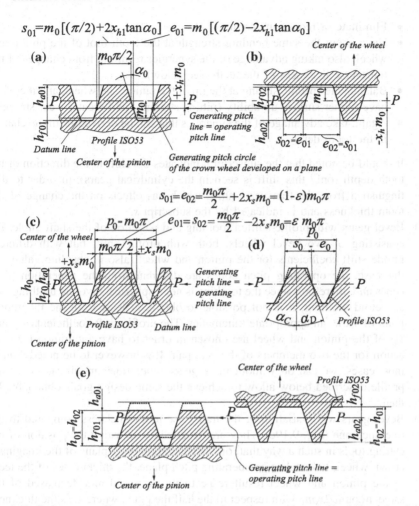

$$s_{01}=m_0[(\pi/2)+2x_{h1}\tan\alpha_0] \quad e_{01}=m_0[(\pi/2)-2x_{h1}\tan\alpha_0]$$

$$s_{01}=e_{02}=\frac{m_0\pi}{2}+2x_s m_0=(1-\varepsilon)m_0\pi$$

$$e_{01}=s_{02}=\frac{m_0\pi}{2}-2x_s m_0=\varepsilon\, m_0\pi$$

Fig. 9.19 Reference profile modifications of bevel gears: **a** positive profile shift of the pinion; **b** negative profile shift of the wheel, in absolute value equal to that of the pinion; **c** increase of the transverse thickness of the pinion tooth; **d** variation of the tooth depth of pinion and wheel; **e** variation of the nominal pressure angle of the operating and non-operating flanks (typically, $\alpha_D \geq 9°$ for the operating flank, and $\alpha_C \leq 31°$ for the non-operating flank)

(B) Bevel gears with profile-shifted toothing without variation of the shaft angle, i.e. consisting of two bevel wheels, both with profile shift, positive for the pinion, and negative for the wheel, but equal in absolute value, as Figs. 9.19a and 9.19b show. As for the cylindrical gears, the profile shift coefficient x_h is chosen so as to achieve different design goals, such as:

- Eliminate or reduce undercut.
- Have about the same bending strength at the tooth root of the pinion and wheel, also taking advantage of the synergies resulting from changes of the transverse thickness of the teeth (see below).
- Balance the specific sliding at the tip of the pinion and wheel, and therefore have higher surface durability (pitting strength) and greater wear strength.
- Again, take advantage of the synergies that can be derived from the change in the tooth depth.

It should be noted that the subscript h indicates the shift in the direction of the tooth depth (only this shift is seen in the cylindrical gears), in order to distinguish it from the transverse shift, which has effects on the change of the tooth thickness and is indicated by the subscript s.

(C) Bevel gears with profile-shifted toothing and variation of the shaft angle, i.e. consisting of two bevel wheels, both with profile shift, but with different profile shift coefficients for the pinion and wheel (also in absolute value). In this case, the operating pitch cones are different from the generation pitch cones and, therefore, since the toothing is not a spherical involute toothing, but an octoid toothing, it is not possible to obtain, at least from the theoretical point of view, appropriate kinematics. The profile shift coefficients x_{h1} and x_{h2} of the pinion and wheel are chosen in order to have a common plane of action for the two members of the gear pair. It is however to be noted that, in most cases, we can do without these gears, since other modifications of the profile described below allow to achieve the same design goals obtainable by them.

(D) Bevel gears with variation of the transverse tooth thickness compared to the reference one (Fig. 9.19c). They are obtained by adjusting the position of the cutting tools in such a way that, on the generation pitch plane of the imaginary crown wheel, to be taken as operating pitch plane, the thicknesses of the teeth of the pinion and wheel result respectively increased and decreased of the same amount $2x_s m_P$ with respect to the half the pitch, where x_s is the thickness modification coefficient. It is noteworthy that the subscript P refers to the standard basic rack tooth profile.

(E) Bevel gears with variation of the depth of the reference profile and, therefore, with variation of the tooth depth (Fig. 9.19d). In this case, the tooth thickness on the generation pitch plane of the imaginary crown wheel remaining unchanged, the addendum h_{aP} and dedendum h_{fP} are changed, independently of one another, increasing them or reducing them with respect to the standard reference values. Stub teeth or raised teeth are thus obtained.

(F) Bevel gears with variation of the nominal pressure angle of the tooth profile (Fig. 9.19e). They are obtained by adjusting the position of the cutting tools in such a way that, the tooth thickness on the generation pitch plane of the imaginary crown wheel remaining unchanged, the operating flank and the non-operating flank have pressure angles respectively less and greater than the nominal pressure angle. In this way, we get quieter gears, since they are characterized by a greater transverse contact ratio. Often, this type of profile is used for hypoid gears, so also because the cutting process is simpler. The ISO standard 23509: 2016 (E) [16] calls the operating and non-operating flanks *drive side* and *coast side* respectively, and indicates the pressure angles related to them respectively with the subscripts D and C.

To achieve important design goals, it is possible to use a combination of two or more of the solutions described at points (A) to (F), with the exception of solution (C) that, in fact, is rarely practiced. By doing so, the synergistic effects of the various modifications of the toothing profile above envisaged are enhanced, and it is therefore possible to obtain optimal configurations of the toothing in relation to various design requirements.

9.8   Straight Bevel Gears with Profile-Shifted Toothing and Variation of Shaft Angle

As we have said elsewhere, the use of gears with profile-shifted toothing is related to the involute profile. In fact, two conjugate involutes of two base circles continue to have an appropriate kinematics in whatever way we change the relative position of the axes of the two wheels, provided that the teeth touch each other: only the pressure angle changes. We have also seen that, in the case of bevel wheels, the spherical involute profiles effectively do not exist, since the bevel gears that have a real practical interest are those with toothing having octoid profile.

The octoid profile is different, albeit not by much, from that of the spherical involute profile, whereby the wheels with octoid toothing has a kinematic behavior that only approximately emulates that of the wheels with toothing having spherical involute profile. Their kinematic behavior is correct only when the generating condition is respected, that is when operating condition and cutting condition, i.e. the generating condition, coincide. Even modest differences between these two conditions determine an incorrect kinematic behavior.

From a strictly theoretical point of view, it is therefore improper to speak of profile shift for bevel gears. De facto, we speak of profile-shifted toothing of the bevel gears, believing that the octoid profiles approximate well enough the spherical involute profiles, so much so that the octoid toothing is confused with the spherical involute toothing. In this framework, in approximate terms and with all the limitations resulting from it, we continue to speak of profile shift of the bevel gears by extending to them, *mutatis mutandis*, the concepts that we have already described for cylindrical involute gears.

As for cylindrical involute gears, to achieve specific design goals, it is necessary to resort to the profile shift, changing the position of the pitch cone surfaces with respect to the tooth depth, and therefore varying the operating pitch cone angle. Also, for straight bevel gears, the design goals to be achieved with profile shift are:

- the reduction of the minimum number of teeth below the one defined in Sect. 9.6, to avoid undercut and interference in operating conditions;
- greater bending strength of the teeth;
- higher contact ratio, and therefore less noisy gears;
- more balanced specific sliding between pinion and wheel;
- higher strength to surface fatigue and wear; etc.

To further study, we refer the reader to specialized textbooks, and the instructions and suggestions of manufacturers of bevel gear cutting machines. We here set the problem in its essential lines, but in entirely general terms, with the purpose of use results obtained with cylindrical involute gears, in the framework of the Tredgold approximation. As we have said, in almost all practical applications, bevel gear pairs with profile-shifted toothing without variation of shaft angle are used, i.e. bevel gear pairs with profile shift coefficients for pinion and wheel equal, but opposite; this is the profile shift type Lasche [19], which therefore satisfies the relationship $(x_{h1} + x_{h2}) = 0$. Bevel gear pairs with profile-shifted toothing and variation of shaft angle are rarely used. This is because, to achieve predetermined design goals, unlike the cylindrical gears, we can use other modifications in the teeth profiles, such as those described in the previous section.

Therefore, the general discussion that follows has no condition as regards the sum of profile shift coefficients of the pinion and wheel, which can assume any value, i.e. $(x_{h1} + x_{h2}) \neq 0$. This is because, using gear-cutting machines based on the generation cutting process, we can choose, for the pinion and wheel, the profile shift considered more advantageous with regard to different adoptable design criteria. In this framework, we suppose to use as virtual cutting tool, a generation crown wheel having pitch circle diameter $d_0 = 2r_0$, virtual number of teeth z_0, and therefore module given by:

$$m_0 = \frac{d_0}{z_0} = \frac{2r_0}{z_0}. \tag{9.58}$$

It is to be noted that, as for the other gears, the subscript 0 indicates quantities relating to the cutting tool. In addition, it is to be remembered that the reference pitch circle, which is also the operating pitch circle during the cutting without profile shift, is the one that we get as the intersection of the reference pitch plane of the generation virtual crown wheel with the back cylinder. The pitch plane of the generation virtual crown wheel is the plane in which the reference pitch cone is degenerated, and the back cylinder is the cylinder in which the back cone is degenerated.

To have bevel gears with zero profile-shifted toothing, the operating pitch cone of the virtual crown wheel during the cutting process has cone angle $\delta_0 = 90°$. This operating pitch cone is the generation pitch cone that, in this case, coincides with the reference pitch plane of the same crown wheel. Instead, to have bevel gears with profile-shifted toothing, the generation pitch cone angles of the virtual crown wheel will be different from $(\pi/2)$ and, for the pinion and wheel, will be equal to:

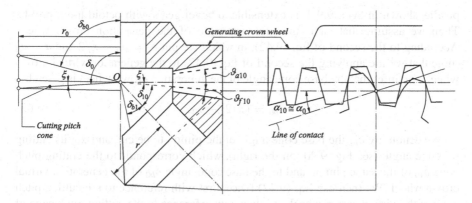

Fig. 9.20 Cutting condition of a straight bevel gear wheel with profile-shifted toothing, with respect to the generation virtual crown wheel

$$\delta_{01} = \frac{\pi}{2} + \xi_1 \quad \delta_{02} = \frac{\pi}{2} + \xi_2, \tag{9.59}$$

where ξ_1 and ξ_2 are, respectively, the deviations of the generation pitch cone angles of pinion and wheel from 90°.

However, in correspondence of the cylinder of diameter d_0, corresponding to the back cone of the crown wheel, degenerated in a cylinder, the module is everywhere m_0. Therefore (Fig. 9.20), the diameters of the pitch circles of the pinion and wheel in the cutting conditions will be given by the relationships:

$$d_1 = z_1 m_0 \quad d_2 = z_2 m_0. \tag{9.60}$$

On the other hand, the operating pitch cone angles δ_1 and δ_2 of the pinion and wheel, meshing with backlash-free contact, determined by the Eqs. (9.20) and (9.21) as a function of u and Σ, which are design data, differ from the generation pitch cone angles δ_{01} and δ_{02} of the amounts $\Delta\delta_1$ and $\Delta\delta_2$, for which we shall have:

$$\delta_1 = \delta_{01} + \Delta\delta_1 \quad \delta_2 = \delta_{02} + \Delta\delta_2. \tag{9.61}$$

Therefore, the shaft angle Σ of the bevel gear pair meshing with backlash-free contact will be given by:

$$\Sigma = \delta_{01} + \delta_{02} + \Delta\delta_1 + \Delta\delta_2. \tag{9.62}$$

It is obvious that, to have the backlash necessary to ensure good operating conditions of the bevel gear pair, this shaft angle must be increased by $\Delta\Sigma$, according to the considerations mentioned below.

Now we focus our attention on the pinion (of course, the same reasoning is to do for the wheel), and assume that, according to the approximations described above, the validity of equations related to the bevel wheels with spherical involute tooth profile, shown in Sect. 9.9.4, is extensible to bevel gears with octoid tooth profile. Then we assume that α_0 is the pressure angle of the generation crown wheel. According to the second of Eq. (9.42), in which we put $\delta_2 = \delta_0 = \pi/2$ and $\alpha = \alpha_0$ (note that we are applying the second of Eq. (9.42) to the generation virtual crown wheel), we infer that the base cone angle of the generation crown wheel is given by:

$$\delta_{b0} = (\pi/2) - \alpha_0. \tag{9.63}$$

We denote by δ_{b1} the base cone angle of the pinion to be cut, and α_{10} its cutting pressure angle (see Fig. 9.20, on the right), which corresponds to the cutting pitch cone δ_{10} of the same pinion, and to the base cone angle δ_{b0} of the generation virtual crown wheel. We then use Eq. (9.42) twice, first with reference to the cutting pitch cone of the virtual crown wheel and then with reference to the cutting pitch cone of the pinion to be cut. Therefore, we get the following relationships:

$$\sin \delta_{00} = \frac{\sin \delta_{b0}}{\cos \alpha_{10}} \quad \sin \delta_{10} = \frac{\sin \delta_{b1}}{\cos \alpha_{10}}, \tag{9.64}$$

where the first of the two subscripts refers to the wheel under consideration (0, for the virtual crown wheel; 1, for the pinion; 2, for the wheel), while the second subscript, 0, identifies the cutting condition. Obtaining $\cos \alpha_{10}$ from the first of these equations and replacing what was found in the second, and recalling Eq. (9.63) and the first of Eq. (9.59), we obtain:

$$\sin \delta_{b1} = \sin \delta_{10} \frac{\cos \alpha_0}{\sin \delta_{00}} = \sin \delta_{10} \frac{\cos \alpha_0}{\cos \xi_1}. \tag{9.65}$$

With similar procedure applied to the wheel to be cut, we get:

$$\sin \delta_{b2} = \sin \delta_{20} \frac{\cos \alpha_0}{\cos \xi_2}. \tag{9.66}$$

The cutting pressure angles α_{10} and α_{20}, respectively used for the pinion and wheel, are defined by the relationships:

$$\cos \alpha_{10} = \frac{\sin \delta_{b0}}{\cos \xi_1} \quad \cos \alpha_{20} = \frac{\sin \delta_{b0}}{\cos \xi_2}. \tag{9.67}$$

From a strictly theoretical point of view, to be unrelated to the actual cutting conditions of wheels with octoid toothing, we can think that, being the angles ξ_1 and ξ_2 very small, for which $\cos \xi_1 \cong \cos \xi_2 \cong 1$, everything works as if it was $\alpha_{10} = \alpha_{20} = \alpha_0$, that is, as if the teeth of the generation crown wheel had pyramidal shape. In fact, the cutting tool that materializes the generation crown wheel has a cutting edge with a given pressure angle α_0, and the gear wheels generated by it are wheels with octoid toothing, corresponding to the pyramidal profile of the tooth of the generation virtual crown wheel.

Fig. 9.21 Pinion and wheel
in their relative position of
meshing with backlash-free
contact, cutting pitch cones
having cone angles δ_{10} and
δ_{20}, and their back cones
having cone angles
$[(\pi/2) - (\delta_{10} + \xi_1)]$ and
$[(\pi/2) - (\delta_{20} + \xi_2)]$

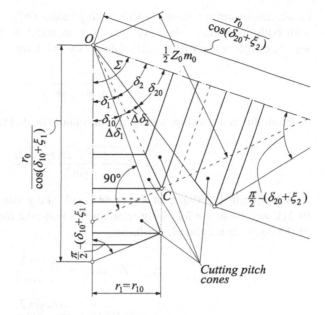

In relation to the profile shift, being known δ_1 and δ_2, which are determined as a
function of the design data, u and Σ, using Eqs. (9.20) and (9.21), it is necessary to
determine the other unknown quantities. To this end, we consider the pinion and
wheel in their relative position corresponding to the meshing condition with zero
backlash contact (Fig. 9.21). In this position, the operating pitch cones (i.e. the
cones with cone angles δ_1 and δ_2, which are different from the cutting pitch cones
having cone angles δ_{10} and δ_{20}) of the bevel gear pair are tangent to each other. The
same figure shows the back cones of the two wheels, i.e. the cones whose common
generatrices form, with the axes of the same wheels, the cone angles given
respectively by $[(\pi/2) - (\delta_{10} + \xi_1)]$ and $[(\pi/2) - (\delta_{20} + \xi_2)]$.

However, the intersections between the above-mentioned back cones, and the
cylinder of radius r_0, which delimits the generation virtual crown wheel and represents its back cone degenerated in this cylinder, are the pitch circles of the two
wheels in the cutting condition; they both have radius r_0. However, it is to keep in
mind that, according to the Tredgold approximation, for the study of meshing in the
cutting conditions, as well as for the analysis of the forces exchanged between the
teeth, it is convenient to refer precisely to these back cones. This because, in their
development on a plane, the toothing of the generation virtual crown wheel is
developed in the standard rack-type cutter with pressure angle α_0.

For ease of calculation, it is then convenient to consider, as operating pitch
circles, those having the same radii $r_1 = r_{10}$ and $r_2 = r_{20}$ of the cutting pitch circles,
for which the module is m_0. It is also convenient to introduce the number of teeth Z_0

of a fictitious cutting crown wheel, having radius $OC = (Z_0 m_0)/2$, which meshes with both members of the bevel gear pair arranged in the meshing position with zero backlash contact (Fig. 9.21). Obviously, we have:

$$Z_0 = \frac{z_1}{\sin \delta_1} = \frac{z_2}{\sin \delta_2}. \tag{9.68}$$

From this relationship, taking into account Eqs. (9.18) and (9.19), we obtain:

$$Z_0 = \frac{\sqrt{z_1^2 + z_2^2 + 2z_1 z_2 \cos \Sigma}}{\sin \Sigma}. \tag{9.69}$$

Substituting Eq. (9.68) in Eq. (9.44), and taking into account Eqs. (9.20) and (9.21), we obtain the following relationships that give the virtual numbers of teeth of the equivalent cylindrical wheels:

$$z_{v1} = Z_0 \tan \delta_1 = \frac{Z_0 \sin \Sigma}{u + \cos \Sigma} \tag{9.70}$$

$$z_{v2} = Z_0 \tan \delta_2 = \frac{u Z_0 \sin \Sigma}{1 + u \cos \Sigma}. \tag{9.71}$$

Following the same procedure described for cylindrical gears, and using the above virtual numbers of teeth, as well as the relationships or diagrams and charts available, we obtain the optimal values of the profile shift coefficients x_{h1} and x_{h2}, and then the pressure angle, α', using the relationship:

$$\mathrm{inv}\alpha' = \mathrm{inv}\alpha_0 + 2 \tan \alpha_0 \frac{x_{h1} + x_{h2}}{z_{v1} + z_{v2}}, \tag{9.72}$$

which is similar to Eq. (6.58), deduced and used for cylindrical gears. To this purpose, we can use, for example: the diagrams of Schiebel (see Schiebel and Lindner [34, 35], the relationships and diagrams of Henriot [15]; the diagrams of Niemann and Winter [24]; etc.

Let us consider Fig. (9.20) to the left, in which however we take into account, in place of the generation virtual crown wheel that is bounded by the cylinder of radius r_0 (the back cone degenerated in this cylinder), the fictitious crown wheel bounded by the cylinder of radius $OC = (Z_0 m_0)/2$. From this modified figure, we infer that the cone angle variations ξ_1 and ξ_2 of the cutting pitch cones of the fictitious crown wheel compared with the right angle, which we would have in the absence of profile shift, are given by the relationships:

$$\tan \xi_1 = \frac{x_{h1} m_0}{OC} = \frac{2x_{h1}}{Z_0} \qquad \tan \xi_2 = \frac{x_{h2} m_0}{OC} = \frac{2x_{h2}}{Z_0}, \tag{9.73}$$

which are respectively valid for the pinion and wheel.

Since δ_1 and δ_2 are known from the design data, once the pressure angle α' in the operating conditions is determined by the Eq. (9.72), the base cone angles and cutting pitch cone angles of the two wheels, and the differences $\Delta\delta_1$ and $\Delta\delta_2$ which appear in Eq. (9.61) remain to be calculated.

From Eq. (9.42), we get:

$$\sin \delta_{b1} = \sin \delta_1 \cos \alpha' \quad \sin \delta_{b2} = \sin \delta_2 \cos \alpha'; \tag{9.74}$$

these relationships, taking into account Eq. (9.68), can be written also in the following form:

$$\sin \delta_{b1} = \frac{z_1}{Z_0} \cos \alpha' \quad \sin \delta_{b2} = \frac{z_2}{Z_0} \cos \alpha'. \tag{9.75}$$

Comparing Eq. (9.75) with Eqs. (9.65) and (9.66), we get the following relationships that allow us to calculate the cutting pitch cone angles of the two wheels:

$$\sin \delta_{10} = \frac{z_1}{Z_0} \frac{\cos \alpha'}{\cos \alpha_0} \cos \xi_1 \quad \sin \delta_{20} = \frac{z_2}{Z_0} \frac{\cos \alpha'}{\cos \alpha_0} \cos \xi_2. \tag{9.76}$$

From Eq. (9.61), we infer the relationships:

$$\Delta\delta_1 = \delta_1 - \delta_{10} \quad \Delta\delta_2 = \delta_2 - \delta_{20}. \tag{9.77}$$

From Fig. (9.20), we then obtain the following relationship that relates the radius r_0 with the radius $r_1 = d_1/2$ of the cutting pitch circle of the pinion:

$$r_{10} = \frac{r_0}{\cos \xi_1} \sin \delta_{10}; \tag{9.78}$$

therefore, we have:

$$z_1 = \frac{Z_0}{\cos \xi_1} \sin \delta_{10}. \tag{9.79}$$

Substituting in this last equation the ratio $(\sin \delta_{10} / \cos \xi_1)$ that we derive from the first of Eq. (9.76), we obtain the relationship:

$$z_0 = Z_0 \frac{\cos \alpha_0}{\cos \alpha'}, \tag{9.80}$$

which correlates the number of teeth, z_0, of the generation virtual crown wheel with the number of teeth, Z_0, of the fictitious crown wheel, which we introduced for convenience of calculation.

It is now necessary to reconsider the hypothesis, previously put forward, that the two members of the bevel gear pair under consideration mesh with zero backlash contact. In fact, as we have said for cylindrical spur gears, to ensure optimal operating conditions of the bevel spur gears, it is also necessary to leave a suitable backlash between the flanks of the teeth in mesh. This suitable backlash both ensures good lubrication conditions, and prevents possible interlocking of the teeth of a wheel in space widths of the mating wheel, due to different thermal expansions between pinion and wheel. This backlash is obtained by increasing the shaft angle, Σ, of a suitable amount, $\Delta\Sigma$, to be determined.

To give an idea of the calculation procedure to be made, consider the case of a straight bevel gear characterized by tapered toothing from the heel toward the toe, and assume that the clearance c_P is decreasing linearly in the same direction above, namely that the angular clearance is constant.

According to the generation cutting process described above, between the operating pitch cones and root cones of the pinion and wheel we have the following dedendum angles:

$$\vartheta_{f1} = \vartheta_{f10} + \Delta\delta_1 \quad \vartheta_{f2} = \vartheta_{f20} + \Delta\delta_2, \tag{9.81}$$

where ϑ_{f10} and ϑ_{f20} are the dedendum angles in the cutting condition (see Fig. 9.20). With a nominal sizing, i.e. with addendum and dedendum respectively equal to $1.00m_0$ and $1.25m_0$, we will have:

$$\tan\left(\vartheta_{f10} + \xi_1\right) = \tan\left(\vartheta_{f20} + \xi_2\right) = 1.25\frac{m_0}{r_0} = \frac{2.50}{z_0}. \tag{9.82}$$

Since between the tip cone of the pinion and root cone of the mating wheel, and vice versa, we want to have an angular clearance equal to $(m_0/4r_0) = (1/2z_0)$, it is necessary that the angular addendum between the operating pitch cones and tip cones of the pinion and wheel are respectively equal to:

$$\vartheta_{a1} = \vartheta_{f20} + \Delta\delta_2 - \frac{1}{2z_0} \quad \vartheta_{a2} = \vartheta_{f10} + \Delta\delta_1 - \frac{1}{2z_0}. \tag{9.83}$$

Instead, according to this generation cutting process, the two actual angular addendums are respectively equal to:

$$\vartheta_{a1,e} = \vartheta_{a10} - \Delta\delta_1 \quad \vartheta_{a2,e} = \vartheta_{a20} - \Delta\delta_2, \tag{9.84}$$

where ϑ_{a10} and ϑ_{a20} are the addendum angles in the cutting condition (see Fig. 9.20). With a nominal standard sizing, we will have:

$$\tan(\vartheta_{a10} - \xi_1) = \tan(\vartheta_{a20} - \xi_2) = \frac{m_0}{r_0} = \frac{2.00}{z_0}. \tag{9.85}$$

To ensure the proper clearance, it is therefore necessary to cut the toothing in blanks with a reduced tip cone angle, so that the angular addendum of the pinion and wheel are smaller than nominal ones, of amounts $\Delta\vartheta_{a1}$ and $\Delta\vartheta_{a2}$ given respectively by:

$$\Delta\vartheta_{a1} = \vartheta_{a10} - \vartheta_{f20} - (\Delta\delta_1 - \Delta\delta_2) + \frac{2.50}{z_0}$$
$$\Delta\vartheta_{a2} = \vartheta_{a20} - \vartheta_{f10} - (\Delta\delta_1 - \Delta\delta_2) + \frac{2.50}{z_0}. \tag{9.86}$$

To obtain between the meshing teeth a suitable normal backlash (this is the backlash measured along the direction of the normal to the contact surface), under the hypothesis that the toothing modification is only that described above, i.e. the profile shift, it is necessary to increase the shaft angle Σ by an amount $\Delta\Sigma$ in such a way to have:

$$j_n = 2\Delta\Sigma\frac{Z_0 m_0}{2}\sin\alpha', \tag{9.87}$$

from which follows:

$$\Delta\Sigma = \frac{j_n}{Z_0 m_0 \sin\alpha'} \cong \frac{j_n}{Z_0 m_0 \sin\alpha_0}. \tag{9.88}$$

With the procedure described above, all the quantities regarding gear wheels with profile-shifted toothing are completely defined. Synthetically, the various steps of the calculation procedure of interest here are the following:

- First, we fix the nominal backlash j_n as a function of the module m (this can be the external transverse module, m_{et}, or the mean normal module, m_{mn}, depending on the case under consideration), using data or diagrams found in the scientific literature (e.g. Niemann and Winter [25]).
- Then we calculate $\Delta\Sigma$ by Eq. (9.88), assuming $\alpha' \cong \alpha_0$.
- Subsequently we calculate Z_0 by means of Eq. (9.69), in which we replace the actual shaft angle Σ with a calculation shaft angle, equal to $(\Sigma - \Delta\Sigma)$.
- We then determine z_{v1} and z_{v2} by means of Eqs. (9.70) and (9.71).
- Depending on the optimization method chosen, we determine the optimal values of the profile shift coefficients, x_{h1} and x_{h2}.
- We then calculate ξ_1 and ξ_2 by means of Eq. (9.73).
- We determine the operating pressure angle, α', by means of Eq. (9.72).
- Finally, we calculate the operating pitch cone angles and tip cone angles of the two wheels of the straight bevel gear pair, by means of Eqs. (9.80), (9.82), (9.85) and (9.86).

For better orientation within the profile shift of straight bevel gears, in addition to what we said at the beginning of this section, it should be noted that, for gear ratios next to the unit, the profile shift is not necessary. For gear ratios very different from the unit, the meshing and operating conditions of a straight bevel gear are instead very unfavorable, whereby the profile shift becomes necessary. We have already said that the bevel gear pairs mostly used are those with profile-shifted toothing without variation of shaft angle. This corresponds to profile shift coefficients, which

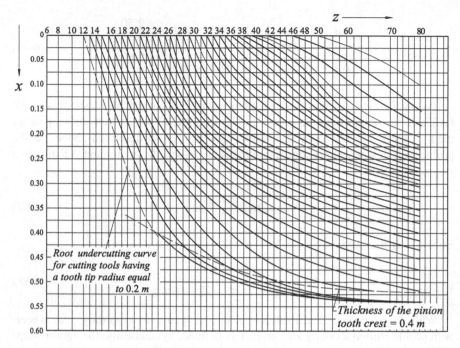

Fig. 9.22 Diagram to determine the profile shift coefficients of straight bevel gear pairs with shaft angle $\Sigma = 90°$, pressure angle $\alpha_0 = 20°$, and profile-shifted toothing without variation of the shaft angle

satisfy the relationship $(x_{h1} + x_{h2}) = 0$, i.e. profile shift coefficients equal in absolute value, but positive for the pinion and negative for the wheel.

For the choice of the profile shift coefficients of straight bevel gear pairs with shaft angle $\Sigma = 90°$, pressure angle $\alpha_0 = 20°$ and profile-shifted toothing without variation of shaft angle, we can use the chart shown in Fig. 9.22. In this chart we enter with the number of teeth of the wheel, reported on the abscissa, and with a vertical line downward cross the curve of the number of teeth of the pinion. The horizontal line through this crossing point identifies, on the ordinate on the left, the profile shift coefficient to be choose. Note that the diagram is delimited, towards the bottom, by the limiting curve of the crest thickness of the pinion tooth and, on the left, by the root-undercut curve for cutting tools having a tooth tip radius equal to $0.2m$.

We also reported that the straight bevel gear pairs with profile-shifted toothing and variation of shaft angle are rarely used. However, in the case we want to design bevel gears of this type, that is with $(x_{h1} + x_{h2}) \neq 0$,, the optimum values of the profile shift coefficients can be determined by reducing the bevel gear pair to an equivalent virtual cylindrical gear pair corresponding, in the Tredgold approximation, to the mean cones. In this case, it is necessary first to determine the numbers of teeth z_{v1} and z_{v2} using the relationships (9.70) and (9.71), and then use the criteria and diagrams available for the cylindrical gears. To this purpose, for example, the diagrams of Schiebel and Lindner [15, 35], Henriot [15], and Niemann and Winter [24], mentioned above can be used.

Finally, it should be noted that the geometry of straight bevel gears can be defined using the criteria described in Chap. 12, concerning spiral bevel and hypoid gears. In fact, straight bevel gears are a special case of the spiral bevel gear family, and of the more general hypoid gear family. Furthermore, computerized methods can be used to optimize tooth profiles (see, for example, Fuentes et al. [11]).

9.9 Straight Bevel Gears with Profile-Shifted Toothing and Variation of Shaft Angle, as Result of a Transverse Tooth Thickness Modification

In Sect. 6.1 we have already indicated that the profile shift is a measure of tooth thickness, highlighting the fact that the addendum modification without tooth depth variation is the result of a tooth thickness variation. In this section, we examine the same subject from a different point of view than discussed in the previous section.

This different point of view is related to a unique ability offered by the bevel gear cutting machines. This ability consists in the fact that, with the tooth depth remaining constant, we can change not only the addendum and dedendum of the generation crown wheel (this is the profile shift, previously discussed), but we can also change, simultaneously and independently from the change above, the tooth space between the teeth and their thicknesses, the pitch remaining unchanged. This is the *transverse tooth thickness modification*. In other words, the toothing modifications of which we are speaking are given by the combination of those shown in Figs. 9.19a and 9.19b, however with $(x_{h1} + x_{h2}) \neq 0$, and in Fig. 9.19c. Since all these modifications in any case determine a tooth thickness change, without tooth depth modification, we address here the subject in terms of transverse tooth thickness modification.

Figure 6.3 best represents the actual geometry of the generation rack-type cutter, obtained developing on a plane the cylinder of radius r_0, which delimits the generation crown wheel (Fig. 9.20). Also taking into account this Fig. 6.3, from Fig. 9.19c we infer the following relationships, which provide respectively the space width e_0 and tooth thickness s_0, both measured along the generation pitch line of the basic rack:

$$e_0 = \frac{\pi m_0}{2} - 2x_s m_0 = \varepsilon \pi m_0 \quad s_0 = \frac{\pi m_0}{2} + 2x_s m_0 = (1 - \varepsilon)\pi m_0, \tag{9.89}$$

where

$$\varepsilon = \frac{1}{2} - \frac{2x_s}{\pi} \tag{9.90}$$

is the so-called *thickness factor*, of course less than 1.

We now face the problem of generating a straight bevel gear pair with gear ratio $u = z_2/z_1$, tooth depth $h = (9/4)m$, and pressure angle α, by means of a generation crown wheel corresponding to the rack-type cutter shown in Fig. 6.3. If we choose as the cutting pitch line of the generation rack-type cutter its datum line (as Fig. 6.3 shows, this datum line divides the tooth depth equally), to avoid cutting interference with the generation crown wheel having an addendum equal to the module, m_0, we should have a pinion with a minimum number of teeth, z_{1min}, which is deduced from curves shown in Fig. 9.17.

However, the minimum number of teeth suitable the avoid the cutting interference with the generation crown wheel is proportional to the addendum factor, $k_0 = (h_{a0}/m)$ of the same virtual generation crown wheel. Therefore, with a profile shift equal to $x_1 m_0$, the addendum of the generation rack-type cutter will be $h_{a0} = (1 - x_1)m_0$, so we can choose a pinion having a minimum number of teeth $z_{1min}^* = z_{1min}(1 - x_1)$, indeed of z_{1min}. From this last relationship, which relates z_{1min}^* with z_{1min}, we infer that the profile shift coefficient, x_1, which is necessary to avoid the cutting interference between the pinion of a straight bevel gear pair having a gear ratio u, and the generation crown wheel, is given by the equation:

$$x_1 = \frac{z_{1min} - z_{1min}^*}{z_{1min}}, \tag{9.91}$$

where z_{1min} is the value obtained from Fig. 9.17 for the given values of $1/u = z_1/z_2$, α_t and Σ. Of course, the pinion generated with a profile shift coefficient, x_1, will have addendum and dedendum given respectively by $h_{a1} = (1 + x_1)m_0$ and $h_{f1} = (1.25 - x_1)m_0$.

Example Input data: straight bevel gear pair, with: $\Sigma = 80°$; $1/u = z_1/z_2 = 1/2$; $\alpha_t = 15°$; $z_1 = z_{1min} = 20$. As a design goal, it is necessary to avoid the cutting interference with the generation crown wheel. Results: (*i*), from curves shown in Fig. 9.17, for $k_0 = (h_{a0}/m) = 1$, we obtain $z_1 = z_{1min} = 27$; (*ii*), from Eq. (9.91), we infer that, in order to avoid the cutting interference, the required profile shift coefficient is equal to $x_1 = [(27 - 20)/20] = 0.259$.

After this necessary premise on cutting interference, we return to the topic outlined at the beginning of this section. Thanks to the operating flexibility of the bevel gear cutting machines, we can vary, as best we believe from the point of view of the design optimization, the tooth thickness s_0 and space width e_0 on the generation pitch line of the rack-type cutter, so as to prevent the teeth of the two members of the gear pair are too weak. This rack-type cutter is obtained by developing the generation crown wheel on a plane.

Now, let us consider a cutting process related to an equivalent virtual crown wheel, having nominal module m_0 and nominal pressure angle α_0. Using as cutting plane the reference pitch plane, we want to generate the two members of a straight bevel gear pair having predetermined numbers of teeth, z_1 and z_2, operating cone angles δ_1 and δ_2, and profile shift coefficients x_{h1} and x_{h2} in the case in which the teeth of the virtual crown wheel have thickness factors ε_1 and ε_2 respectively for the pinion and wheel. Note that the tooth thicknesses of the pinion and wheel correspond to the space widths of the related virtual crown wheels. For this reason, we speak of thickness factor with reference to the space width of the crown wheel, because we are interested in what is produced by the virtual crown wheel, that is, the actual bevel gear wheels.

As usual, we reduce the problem to the equivalent cylindrical gear pair, which we obtain developing on a plane the cylinder of radius r_0 (Fig. 9.20) that delimits the generation crown wheel. This equivalent cylindrical spur gear pair has virtual numbers of teeth z_{v1} and z_{v2}, given by Eq. (9.44), while the diameters of their pitch circles and the thicknesses of their teeth, for the pinion and wheel, will be respectively given by:

$$d_{v1} = \frac{z_1 m_0}{\cos \delta_1} \quad d_{v2} = \frac{z_2 m_0}{\cos \delta_2} \tag{9.92}$$

$$s_1 = \varepsilon_1 \pi m_0 \quad s_2 = \varepsilon_2 \pi m_0. \tag{9.93}$$

Bearing in mind the relationships (6.66) and (6.67), in which we put $x_1 = x_2 = 0$, and taking into account Eqs. (9.92) and (9.93), since the first terms of the second members of Eq. (6.66) and (6.67) are respectively equal to the ratios $2s_1/d_{v1}$ and $2s_2/d_{v2}$, we obtain the following equations that give the tooth thickness angles corresponding to the pitch circles on which the pressure angle is α':

$$2\vartheta'_{s,1} = \frac{2\pi}{z_1} \varepsilon_1 \cos \delta_1 + 2(\mathrm{inv}\alpha_0 - \mathrm{inv}\alpha') \tag{9.94}$$

$$2\vartheta'_{s,2} = \frac{2\pi}{z_2} \varepsilon_2 \cos \delta_2 + 2(\mathrm{inv}\alpha_0 - \mathrm{inv}\alpha'). \tag{9.95}$$

Since the angular pitches, τ_{v1} and τ_{v2}, of the two equivalent cylindrical wheels are given by:

$$\tau_{v1} = 2\pi/z_{v1} \quad \tau_{v2} = 2\pi/z_{v2}, \tag{9.96}$$

the space width angles $2\vartheta'_{e,1}$ and $2\vartheta'_{e,2}$ of the same two wheels on the pitch circles above will be given by:

$$2\vartheta'_{e,1} = \tau_{v1} - 2\vartheta'_{s,1} = \frac{2\pi}{z_{v1}} - 2\vartheta'_{s,1} \tag{9.97}$$

$$2\vartheta'_{e,2} = \tau_{v2} - 2\vartheta'_{s,2} = \frac{2\pi}{z_{v2}} - 2\vartheta'_{s,2}. \tag{9.98}$$

To have the profiles of the teeth of the two mating wheels in backlash-free contact, on the above-mentioned pitch circles, the space width between the teeth of the first wheel must be equal to the tooth thickness of the second wheel, that is the Eq. (6.71) must be valid, in which we replace z_{v2}/z_{v1} instead of z_2/z_1.

Developing the Eq. (6.71) so modified, taking into account Eqs. (9.97), (9.94) and (9.95), we obtain the following fundamental equation, that allows to calculate α' as a function of quantities all known:

$$\mathrm{inv}\alpha' = \mathrm{inv}\alpha_0 + \frac{\pi(\varepsilon_1 + \varepsilon_2 - 1)}{\frac{z_1}{\cos\delta_1} + \frac{z_2}{\cos\delta_2}}. \tag{9.99}$$

Once α' has been determined this way, the diameters of the pitch circles in the condition of backlash-free contact are those given by the Eq. (6.46) and (6.47), which here become:

$$d'_{v1} = \frac{z_1 m_0 \cos\alpha_0}{\cos\delta_1 \cos\alpha'} \quad d'_{v2} = \frac{z_2 m_0 \cos\alpha_0}{\cos\delta_2 \cos\alpha'}. \tag{9.100}$$

If, for the generation of the two members of the bevel gear pair, with profile-shifted toothing and variation of the shaft angle, we take profile shift coefficients respectively equal to x_{h1} and x_{h2}, the maximum nominal addendum, compared to the nominal pitch circles whose diameters are given by Eq. (9.92), will be given by:

$$h_{a1} = m_0(1 + x_{h1}) \quad h_{a2} = m_0(1 + x_{h2}), \tag{9.101}$$

while the nominal dedendum will be given by:

$$h_{f1} = m_0\left(\frac{5}{4} - x_{h1}\right) \quad h_{f2} = m_0\left(\frac{5}{4} - x_{h2}\right). \tag{9.102}$$

However, since we wish to have, in the meshing condition of backlash-free contact, a clearance at least equal to $(1/4)m_0$, in the operating conditions the actual operating values h'_{a1} of h_{a1}, and h'_{a2} of h_{a2} cannot exceed, for the pinion and wheel, respectively the following maximum values:

$$h'_{a1\max} = m_0\left[\frac{1}{2}\left(\frac{z_1}{\cos\delta_1} + \frac{z_2}{\cos\delta_2}\right)\left(\frac{\cos\alpha_0}{\cos\alpha'} - 1\right) + (1 + x_2)\right] \tag{9.103}$$

Table 9.3 Main data concerning a straight bevel gear pair with shaft angle Σ, z_1 and z_2 teeth, profile shift coefficients x_{h1} and x_{h2}, and thickness factors ε_1 and ε_2, generated by a virtual crown wheel having module m_0 and pressure angle α_0

Quantity	Pinion	Wheel
Number of teeth	z_1	z_2
Profile shift coefficient	x_{h1}	x_{h2}
Thickness factor	ε_1	ε_2
Pitch cone angle	$\tan\delta_1 = \frac{(z_1/z_2)\sin\Sigma}{1+(z_1/z_2)\cos\Sigma}$	$\tan\delta_2 = \frac{(z_2/z_1)\sin\Sigma}{1+(z_2/z_1)\cos\Sigma}$
Operating pressure angle α' in the condition of backlash-free contact	$\mathrm{inv}\alpha' = \mathrm{inv}\alpha_0 + \frac{\pi(\varepsilon_1+\varepsilon_2-1)}{(z_1/\cos\delta_1)+(z_2/\cos\delta_2)}$	
Diameter of the pitch circle in the condition of backlash-free contact	$d_1 = z_1 m_0(\cos\alpha_0/\cos\alpha')$	$d_2 = z_2 m_0(\cos\alpha_0/\cos\alpha')$
Diameter of the tip circle	$d_{a1} = z_1 m_0 + 2h_{a1}\cos\delta_1$	$d_{a2} = z_2 m_0 + 2h_{a2}\cos\delta_2$
Diameter of the root circle	$d_{f1} = z_1 m_0 - 2h_{f1}\cos\delta_1$	$d_{f2} = z_2 m_0 - 2h_{f2}\cos\delta_2$
Tip cone angle	$\delta_{a1} = \delta_1 + \frac{2h_{a1}}{m_0}\frac{\sin\delta_1}{z_1}$	$\delta_{a2} = \delta_2 + \frac{2h_{a2}}{m_0}\frac{\sin\delta_2}{z_2}$

Note h_{a1} has the lowest value among those given by the first of Eqs. (9.101) and (9.103); h_{a2} has the lowest value among those given by the second of Eqs. (9.101) and (9.104)

$$h'_{a2max} = m_0\left[\frac{1}{2}\left(\frac{z_1}{\cos\delta_1} + \frac{z_2}{\cos\delta_2}\right)\left(\frac{\cos\alpha_0}{\cos\alpha'} - 1\right) + (1+x_1)\right]. \qquad (9.104)$$

The actual dedendum of the two wheels coincide with the nominal values given by Eq. (9.102).

Table 9.3 shows the main data concerning a straight bevel gear pair with z_1 and z_2 teeth, shaft angle Σ, profile shift coefficients x_{h1} and x_{h2}, and thickness factors ε_1 and ε_2, generated by means of a virtual crown wheel having module m_0 and pressure angle α_0.

Of course, when the sum of the thickness factors is equal to unit, that is, when:

$$(\varepsilon_1 + \varepsilon_2) = 1, \qquad (9.105)$$

we infer that (see Eqs. 9.99 and 9.100), in the operating conditions of backlash-free contact, the values of the pressure angle and the diameters of the pitch circles are the nominal ones.

Some manufacturers of straight bevel gears use thickness factors related to the profile shift coefficients by the same relationships valid for parallel cylindrical spur gears, and therefore given by:

$$\varepsilon = \frac{1}{2} + \frac{2}{\pi}x\tan\alpha. \qquad (9.106)$$

In this case, suppose we want to make a bevel gear pair with shaft angle Σ, and characterized by a ratio $(1/u) = (z_1/z_2)$ and nominal values of the pressure angle and diameters of the pitch circles. From the diagrams of Fig. 9.17, we derive, for

$k_{20} = 1$, the minimum number of teeth, z_{10}, of the pinion to avoid cutting interference with the generation crown wheel, whereby if

$$z_{10} + z_{20} = z_{10}(1 + u) \geq 2z_1, \tag{9.107}$$

we can adopt, for the pinion, a profile shift coefficient equal to:

$$x_1 = \frac{z_1 - z_{10}}{z_1} \tag{9.108}$$

and, for the wheel, a profile shift coefficient equal to:

$$x_2 = -x_1. \tag{9.109}$$

Since, in accordance with Eq. (9.106), we have

$$\varepsilon_1 = \frac{1}{2} + \frac{2}{\pi} x_1 \tan \alpha_0 \quad \varepsilon_2 = \frac{1}{2} + \frac{2}{\pi} x_2 \tan \alpha_0, \tag{9.110}$$

we see that Eq. (9.105) is satisfied, for which the values of the pressure angle and diameters of the pitch circles will be equal to the nominal ones.

Example Input data: straight bevel gear pair with $\Sigma = 90°$, ratio $(1/u) = (z_1/z_2) = 1/3$, and nominal pressure angle $\alpha_0 = 20°$. As a design choice, we want to use a pinion with $z_1 = 12$ teeth, for which $z_2 = 36$ teeth. Thus, we have $(z_{10} + z_{20}) \geq 2z_1$. Therefore, we will adopt, for the pinion, a profile shift coefficient equal to $x_1 = (15 - 12)/15 = 0.2$ and, for the wheel, a profile shift coefficient $x_2 = -x_1 = -0.2$. The related thickness factors are those given by Eq. (9.110).

9.10 Load Analysis for Straight Bevel Gears and Thrust Characteristics on Shaft and Bearings

Load analysis for straight bevel gears can be made with the same procedure used for parallel cylindrical spur and helical gears, which neglects the contribution of the friction forces. The total tooth force, $F = F_n$, exchanged between the meshing teeth acts along the direction of the normal to the tooth surface. As we said in the Sect. 9.4 on the geometry of the spherical involute, we assume that the kinematics of bevel gears with octoid toothing is comparable, with sufficient approximation, to that of the bevel gears with spherical involute toothing. Under this assumption, for a given position, the teeth are touching along the straight line of the plane of contact, which passes through the apex of the pitch cones (Fig. 9.5). Thus, the normal to the tooth surface in any point of this line is directed along the normal to this line, on the plane of contact.

Fig. 9.23 Plane of contact
rotated with respect to contact
generatrix, total tooth force
$F = F_n$, and its components

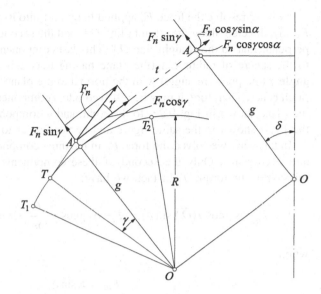

Therefore, the total tooth force has, as its line of application, the normal to the instantaneous line of contact through the middle of the face width, and thus coincident with a mean cone generatrix, on the plane of contact. This force, however, actually occurs somewhere between the mean and back cones, but the error due to the above assumption is marginal.

We denote by R the distance, which does not vary and remain constant, between the line of application of the force F_n and the apex O of the pitch cone. In the plane of contact, this force, constant in magnitude, changes its direction during the meshing cycle, remaining however tangent to the circle of radius R. Referring to Fig. 9.23, we denote with g the contact generatrix of the operating pitch cones, and with δ the cone angle of one of these cones. In the same figure, we see, to the left, the plane of contact (it is tangent to the base cones along the straight lines OT_1 and OT_2, see also Fig. 9.5) rotated with respect to line g. In this plane, the instantaneous position of the force F_n is defined by the angle γ between the normal to the force F_n (this normal is the instantaneous line of contact, OT in the figure) and the common generatrix of contact g.

The total tooth force F_n can be resolved into three components, which act at right angles to one another. These components are, respectively, the tangential force or transmitted load, F_t, the axial force, F_a, and the radial force, F_r. In order to establish the interrelationships between these components, consider the contact corresponding to the instantaneous line of contact OT (Fig. 9.23). The force F_n is applied at point T, and in this point it is tangent to the circle of radius R, and thus, it is normal to the instantaneous line of contact. We now translate the force F_n parallel to itself, to apply it at point A, whose distance from the apex O is given by $OA = R/\cos\gamma$ and is located on the common contact generatrix of the operating pitch cones.

Now we resolve the force F_n applied to point A into its two components, the first, $F_n \sin \gamma$, directed along the straight line OA, and the second, $F_n \cos \gamma$, in the direction perpendicular to the straight line OA. This last component lies in the plane tangent to the sphere of radius OA (the plane having trace t in Fig. 9.23) and forms the angle α (the pressure angle) with the normal to the plane formed by the axes of the pitch cones. Therefore, it may be resolved into a component, $F_n \cos \gamma \cos \alpha$, normal to the plane shown in Fig. 9.23, to the right, and a component, $F_n \cos \gamma \sin \alpha$, lying in the plane shown in the same figure and perpendicular to the straight line OA.

In this way, we solved the force F_n in the three components $F_n \sin \gamma$, $F_n \cos \gamma \cos \alpha$ and $F_n \cos \gamma \sin \alpha$. Only the second of these components contributes to the transmission of the torque T. In fact, we have:

$$T = F_n \cos \gamma \cos \alpha \left(\overline{OA} \sin \delta\right) = F_n \cos \gamma \cos \alpha \frac{R}{\cos \gamma} \sin \delta = F_n R_m \cos \alpha, \quad (9.111)$$

where

$$R_m = R \sin \delta \tag{9.112}$$

is the radius corresponding, on the operating pitch cone generatrix, to the distance R from the apex. The *tangential component* or *tangential force*

$$F_t = F_n \cos \gamma \cos \alpha \tag{9.113}$$

of the total force F_n, being the only one that transmits the torque T, is also called *transmitted load*.

It is first to be noted that the force F_n is constant in absolute value only if the torque applied to the wheel is constant. The axial and radial forces, respectively directed parallel and radially to the axis of the wheel, are instead not constant. As we can deduce from Fig. 9.23, these forces are given by the following relationships:

$$F_a = F_n(\cos \gamma \sin \alpha \cos \delta - \sin \gamma \sin \delta) \tag{9.114}$$

$$F_r = F_n(\cos \gamma \sin \alpha \sin \delta + \sin \gamma \cos \delta), \tag{9.115}$$

and therefore, vary as a function of δ during the meshing cycle, even if F_n and therefore the torque T remain constant.

Therefore, unlike what happens in the parallel cylindrical spur gears with involute tooth profiles, in which the force exchanged between the meshing teeth is constant in the absolute value and in direction along the path of contact, in this case the force F_n is constant in absolute value, being equal to

$$F_n = \frac{T}{R_m \cos \alpha}, \tag{9.116}$$

but it continuously varies its direction, moving with uniform angular velocity in the plane of action, and keeping in this plane constantly tangent to the circle of radius $R = R_m/\sin\delta$. The increased noise of the straight bevel gears with respect to the parallel cylindrical gears depends, at least in a partial amount, by this variable direction of the line of application of the force F_n.

For strength calculation of straight bevel gears, the instantaneous line of contact characterized by $\gamma = 0$ is usually considered (thus it coincides with the contact generatrix of the operating pitch cones). In this position, the Eqs. (9.114) and (9.115) are simplified and take the form:

$$F_a = F_n \sin\alpha\cos\delta \qquad\qquad (9.117)$$

$$F_r = F_n \sin\alpha\sin\delta. \qquad\qquad (9.118)$$

Bearing in mind the Eq. (9.17) and introducing the nominal transverse tangential load F_{nt} referred to the radius R_m, i.e. the tangential force on the reference pitch cone at the middle of face width, given by

$$F_{nt} = \frac{T}{R_m}, \qquad\qquad (9.119)$$

the Eqs. (9.117) and (9.118) can be written in the following form, which is the one usually used in practical applications:

$$F_a = F_{nt} \tan\alpha\cos\delta \qquad\qquad (9.120)$$

$$F_r = F_{nt} \tan\alpha\sin\delta. \qquad\qquad (9.121)$$

The introduction of the force F_{nt} is justified by the fact that a more precise determination of the load distribution along the face width, although possible from the theoretical point of view, is not convenient because of the unreliability of the results arising from the inevitability to take into account the cutting and assembly errors.

9.11 Efficiency of Straight Bevel Gears

The general concepts of efficiency of the parallel cylindrical spur and helical gears, discussed respectively in Sects. 3.9 and 8.10, can still be applied to the straight bevel gears, with the appropriate variations of the case. Here we dwell on these variations, which are peculiar to the straight bevel gears.

To address the problem, we assume that only one tooth pair is in engagement, and that the mating surfaces are generated according to the procedures described in previous sections, so the geometric properties of toothing are perfectly known. In particular, the following two geometric entities are known:

- The plane of action, that is the portion of plane having the shape of a circular ring sector, delimited by the two circles with radii R_e and R_i, and the interference lines OT_1T_1' and OT_2T_2'. Therefore, the plane of action defines the theoretical circular ring sector of contact, analogously to the theoretical rectangle of contact we introduced in Sect. 8.2 for parallel cylindrical helical gears.
- The path of contact, that is the portion of plane having also the shape of a circular ring sector, delimited by the same two circles with radii R_e and R_i, and the straight lines OAA' and OEE'. These straight lines are the intersections of the plane of action with the tip cones of the two members of the straight bevel gear pair under consideration. Therefore, the path of contact defines the actual circular ring sector of contact.

These two geometric entities are shown in Fig. 9.24, where they are respectively indicated with $T_1T_1'T_2T_2'$ and $AA'EE'$. Figure 9.24 also shows the instantaneous line of contact $l = P_iP_e$, defined by the angle φ that it forms with the instantaneous axis of rotation OC_iC_i', and the differential normal component, dF_n, that the pinion transmits to the wheel along a differential segment, ds, of the instantaneous line of contact. It acts on the plane of action, and is perpendicular to the instantaneous line of contact. However, in addition to this differential normal component, dF_n, we will have a differential passive loss component, $dF_\mu = \mu dF_n$, due to friction, acting on the same differential segment, ds. This component, also called differential friction

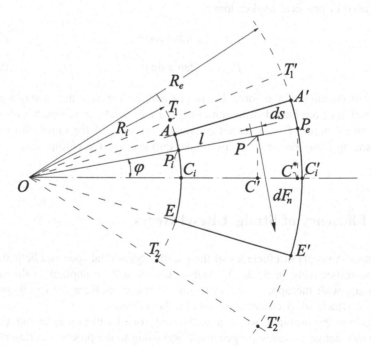

Fig. 9.24 Differential normal component dF_n acting on a differential segment ds of the instantaneous line of contact

force, is directed according to the relative velocity vector v_r of the driven wheel with respect to the driving pinion at point P, so it is directed perpendicularly to the plane of action, and oriented in such a way as to obstruct the relative motion.

The relative velocity vector, v_r, is given by the relationship:

$$v_r = |\omega_2 - \omega_1|(PC'), \tag{9.122}$$

where ω_2 and ω_1 are the angular velocity vectors of the two gear wheels, and PC' is the distance of point P from the contact generatrix of the two operating pitch cones, that is the instantaneous axis of rotation. Similarly to what we have already done in Sects. 3.9 and 8.10, in the case here examined the energy lost by friction, or lost work, on the differential segment ds in the infinitesimal time internal dt is given by:

$$\frac{dW_\mu}{dt} = \mu \frac{dF_n}{ds} ds |\omega_2 - \omega_1| PC', \tag{9.123}$$

If we measure the abscissa s of point P along the instantaneous line of contact, starting from the apex O, and denote with $P_e C^*$ the distance of point P_e from the instantaneous axis of rotation (Fig. 9.24), we can write:

$$PC' = \frac{P_e C^*}{R_e} s. \tag{9.124}$$

Substituting this last relationship into the Eq. (9.123), and assuming that (dF_n/ds) is a constant along the instantaneous line of contact, we get:

$$\frac{dW_\mu}{dt} = \mu \frac{dF_n}{ds} \frac{P_e C^*}{R_e} |\omega_2 - \omega_1| \frac{R_e^2 - R_i^2}{2} = \mu F_n \frac{P_e C^*}{R_e} |\omega_2 - \omega_1| \frac{R_e + R_i}{2}. \tag{9.125}$$

From Fig. 9.24 we also obtain:

$$\frac{P_e C^*}{R_e} = \sin \varphi. \tag{9.126}$$

On the other hand, while the two members of the straight bevel gear pair rotate about their axes at angular velocities ω_1 and ω_2, the instantaneous line of contact rotates about the normal to plane of action passing through point O, with angular velocity $\omega_1 \cos \alpha \sin \delta_1 = \omega_2 \cos \alpha \sin \delta_2$. Obviously, the instantaneous line of contact is the common straight line along which the conical surfaces of the two mating tooth flanks intersect the plane of action. Therefore, if φ_1 is the rotation angle of the pinion about its axis (it is obtained by integrating the first Eq. (3.55) for $\omega_1 = const$), the corresponding angle on the plane of action described by the contact generatrix of conical surfaces of the mating teeth when it moves from the position $OP_i P_e$ to position $OC_i C_i'$ (Fig. 9.24) is given by:

$$\varphi = \varphi_1 \cos \alpha \sin \delta_1. \tag{9.127}$$

Developing Eq. (9.126) in Maclaurin series (see Tricomi [43]; Buzano [4]; Smirnov [38], considering only the first term of this series development, and taking into account the Eqs. (9.127), (9.125) becomes:

$$\frac{dW_\mu}{dt} = \mu F_n \varphi_1 \cos \alpha \sin \delta_1 |\omega_2 - \omega_1| \frac{R_e + R_i}{2}. \tag{9.128}$$

Now let us calculate the work done by the driving wheel. The torque resulting from differential normal force component, dF_n, applied to the differential segment, ds, of the instantaneous line of contact, calculated with respect to the crossing point, O, is a vector perpendicular to the plane of action, having an absolute value equal to $F_n(R_e + R_i)/2$. The component of this vector in the direction of the pinion axis is given by $[-F_n \sin \delta_1 (R_e + R_i)/2]$, since the angle between the normal to the plane of action and the pinion axis is equal to $(90° - \delta_{b1})$.

On the other hand, the torque resulting from the differential friction force components, also calculated with respect to the crossing point, O, is a vector perpendicular to the instantaneous line of contact, OP_iP_e, having an absolute value equal to $[\mu F_n(R_e + R_i)/2]$. As shown in Fig. 9.25, this torque can be resolved in two components: the first is directed according to normal to the straight line OC_iC_i', and is equal to $[\mu F_n(R_e + R_i)\cos \varphi/2]$; the second is directed according to the straight line OC_iC_i', and is equal to $[\mu F_n(R_e + R_i)\sin \varphi/2]$. Considering the components of these two torque components according to the pinion axis, and taking into account the Eq. (9.127), we get the following relationship, which expresses the resulting torque $T_{\mu 1}$ in the direction of the pinion axis, due to differential friction force components:

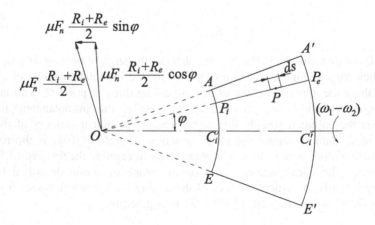

Fig. 9.25 Torque resulting from the differential friction component dF_μ

$$T_{\mu 1} = \mu F_n \frac{R_e + R_i}{2} [\pm \mu \sin \delta_1 \sin \alpha \cos(\varphi_1 \cos \alpha \sin \delta_1) - \cos \delta_1 \sin(\varphi_1 \cos \alpha \sin \delta_1)],$$

$$(9.129)$$

where the plus and minus signs must be taken, depending on whether the contact occurs during the path of approach or the path of recess.

To obtain the above torque components, we have assumed that the angular velocities ω_1 and ω_2 are constant, that is steady state operating conditions. Now, taking the torque components of all the forces applied to the pinion in the direction of its axis, we get the following relationship that expresses the work done by the driving wheel, i.e. the driving torque applied to the pinion:

$$T_1 = F_n \frac{R_e + R_i}{2} [\sin \delta_1 \cos \alpha \mp \mu \sin \delta_1 \sin \alpha \cos(\varphi_1 \cos \alpha \sin \delta_1)$$
$$+ \mu \cos \delta_1 \sin(\varphi_1 \cos \alpha \sin \delta_1)]. \tag{9.130}$$

Depending on the geometry of the straight bevel gear pair defined in Sects. 9.2 and 9.3, we can write:

$$\frac{R_e + R_i}{2} \sin \delta_1 = \frac{r_1 + r_{1,i}}{2} \quad \frac{R_e - R_i}{2} \sin \delta_1 = \frac{r_1 - r_{1,i}}{2}, \tag{9.131}$$

where $r_1 = r_{1,e}$ and $r_{1,i}$ are respectively the radii of circles obtained by intersecting the pitch cone of the pinion with the spheres having both centers O, and radii R_e and R_i (r_1 is the already defined radius of the reference pitch circle of the pinion).

Therefore, combining the two Eqs. (9.128) and (9.130), and taking into account the first Eq. (9.131), we get:

$$\frac{dW_\mu}{T_1 dt} - \frac{\mu \cos \alpha |\omega_2 - \omega_1| \varphi_1}{\cos \alpha + \mu [\mp \sin \alpha \cos(\varphi_1 \cos \alpha \sin \delta_1) + \cot \delta_1 \sin(\varphi_1 \cos \alpha \sin \delta_1)]}.$$

$$(9.132)$$

From Fig. 9.4 as well as from Eq. (9.16), we obtain:

$$|\omega_2 - \omega_1| = \omega_1 \left(1 + \frac{\omega_2^2}{\omega_1^2} + 2\frac{\omega_2}{\omega_1} \cos \Sigma\right)^{1/2} = \omega_1 \left(1 + \frac{z_1^2}{z_2^2} + 2\frac{z_1}{z_2} \cos \Sigma\right)^{1/2}.$$

$$(9.133)$$

Assuming that the driving torque T_1 is a constant, we infer that, under the conditions considered, the work dissipated by friction, in the time corresponding to duration of contact between the two teeth, i.e. during the complete meshing cycle, is given by the following equation:

$$W_\mu = \mu_0 T_1 \left(1 + \frac{z_1^2}{z_2^2} + 2\frac{z_1}{z_2}\cos \Sigma\right)^{1/2} \left\{ \int_0^{e_1/r_1} \frac{(\mu/\mu_0)\varphi_1 d\varphi_1}{\left(1 + \mu\left[\frac{\cot \delta_1}{\cos \alpha}\sin(\varphi_1 \cos \alpha \sin \delta_1) - \tan \alpha \cos(\varphi_1 \cos \alpha \sin \delta_1)\right]\right)} \right.$$

$$\left. + \int_0^{e_2/r_1} \frac{(\mu/\mu_0)\varphi_1 d\varphi_1}{\left(1 + \mu\left[\frac{\cot \delta_1}{\cos \alpha}\sin(\varphi_1 \cos \alpha \sin \delta_1) + \tan \alpha \cos(\varphi_1 \cos \alpha \sin \delta_1)\right]\right)} \right\}$$

$$= \mu_0 T_1 \Delta \left(1 + \frac{z_1^2}{z_2^2} + 2\frac{z_1}{z_2}\cos \Sigma\right)^{1/2},$$

$$(9.134)$$

where e_1 and e_2 are respectively the lengths of arcs of approach and recess on the sphere with radius R_e, Δ briefly indicates all that the bracket contains, and μ_0 is the value of coefficient of friction μ corresponding to a given point of path of contact, which can be determined conventionally. This last equation is similar to Eq. (3.98), and reduces to it for $\delta_1 = 0$.

The loss of efficiency of the straight bevel gear pair examined here can be expressed as:

$$1 - \eta = \frac{W_\mu}{W_m} = \frac{W_\mu}{T_1\left(\frac{e_1 + e_2}{r_1}\right)} = \frac{p\mu_0\Delta^*}{r_1}\left(1 + \frac{z_1^2}{z_2^2} + 2\frac{z_1}{z_2}\cos \Sigma\right)^{1/2}, \qquad (9.135)$$

where W_m, given by Eq. (3.99), is the work done by the driving torque during the same meshing cycle considered for calculation of W_μ, while Δ^* is still given by Eq. (3.101).

Equation (9.135) can be written in the form:

$$1 - \eta = 2\pi\mu_0\Delta^*\left(1 + \frac{z_1^2}{z_2^2} + 2\frac{z_1}{z_2}\cos \Sigma\right)^{1/2}. \qquad (9.136)$$

If we assume that the coefficient of friction is a constant during the meshing cycle, and if in the bracket of Eq. (9.134), which defines Δ, we neglect the terms containing μ compared to unit, we obtain:

$$\Delta^* = \Delta\frac{r_1^2}{p^2} = \frac{r_1^2}{p^2}\frac{(e_1^2 + e_2^2)}{2r_1^2} = \frac{1}{2}\left(\varepsilon_1^2 + \varepsilon_2^2\right), \qquad (9.137)$$

which coincides with Eq. (3.104) obtained for parallel cylindrical spur gear pairs.

Therefore, we can conclude that, within the approximation in which Eq. (9.137) is valid, the efficiency of a straight bevel gear pair is given by the same relationship found for a parallel cylindrical spur gear pair, replacing $\left(\frac{1}{z_1} \pm \frac{1}{z_2}\right)$ with $\left(\frac{1}{z_1^2} + \frac{1}{z_2^2} + \frac{2}{z_1 z_2}\cos \Sigma\right)^{1/2}$.

It should be noted that the above-described procedure has general validity. Therefore, it is extendable to spiral bevel gears, with very negligible variations, on which we do not think to dwell on Chap. 12, regarding these types of gears.

9.12 Short Notes on Cutting Method of Straight Bevel Gears

The main methods used for the manufacture of straight bevel gears can be also summarized up in two categories: non-generation methods or form cutting methods, and generation methods. However, it should be noted that the cutting processes of the straight bevel gears, and more generally the ones of the curved-toothed bevel gears, as far as they are similar to those already described for the parallel cylindrical spur and helical gears, have peculiar features, which deserve to be briefly highlighted. These peculiarities concern both gear-cutting machines and cutting tools.

9.12.1 Form Cutting Methods

In Sect. 9.4 we have already introduced the concepts underlying the template machining process for manufacturing straight bevel gears. This process is implemented by means of bevel gear planers, variously named: template-shaping machines, tracer shaping machines, profiling shaping machines, etc. In the same Sect. 9.4 we have also described some limitations of this non-generation cutting method, which is still used for low production of large straight bevel gears.

Two templates are used, one for each of the two opposite flanks of two successive teeth. From a theoretical point of view, two templates would be needed for each gear ratio. However, for obvious reasons of economy, one template pair covers a small range of gear ratios, so we have the same disadvantages of profile accuracy, as we have described in Sect. 3.11.1 regarding the cutting of cylindrical spur gear wheels by means of disk-type milling cutters. The cut is done with two simple single-point cutting tools, each driven by the related template follower resting on a straight guide (see [7, 12, 14]).

The tooth spaces are machined one at time. Thus, once a tooth space is finished, the next tooth space is indexed with the indexing device and machined, and so on until the entire tooth spaces are cut, and the teeth are formed. A dual machining cycle is needed: in fact, tooth spaces are first roughed, using slotting tools or other types of roughing tools, and then machined by finishing. With a proper set up, today's template shaping machines, which are almost all equipped with CAD/CAM systems, allow to minimize the disadvantages described in Sect. 3.4. These systems also allow to obtain teeth with lengthwise crowning, which is achieved by a slight motion of the tool arm as the cutting tool moves along the face width. This crowning allows optimal location of the tooth bearing contact area.

Another form cutting method of straight bevel gears is milling. However, it is not widely used because of the remarkable limits of accuracy that characterize it. In fact, other specific limitations of this method are added to the above-mentioned limits of the indexing devices. The proper technological limits are mainly because the form-milling cutter (the disk-type cutter or end mill) has a constant profile, while the tooth profile varies from the inner cone to the back cone. Obviously, the milling cutter cannot have greater thickness than the tooth space at the inner cone and, given the variability in the width of the tooth space in the direction of the face width, it can cut only on one side.

The operation of the gear-cutting machine that implements this cutting process is time consuming, and the cut is very approximate, so this non-generation cutting method is sometimes used to rough straight bevel gears, which are then finished by other methods and cutting machines. Often, to limit the aforementioned drawbacks and have a little better-quality cut, several cutting tools are used. Nevertheless, this method, when used to obtain straight bevel gears ready for use, is to be regarded as a workshop gadget rather than a real cutting method.

9.12.2 Generation Cutting Methods

For a more in-depth knowledge of the generation cutting methods of the straight bevel gears, we refer the reader to specialized textbooks (see Galassini [12], Rossi [33], Dudley [9], Townsend [42], Davis [7], Radzevich [28, 29, 32]). However, to give an idea of these methods, we think it is useful to provide here some basic concepts on the main generation cutting processes used in this field.

One of the traditional generation processes is the one known as *planning generating method* or *planning generator*. This method is very versatile, as it allows to cut both straight-teeth and curved-teeth bevel gears. Moreover, it allows the cutting of hypoid gears, with the addition of special heads to standard gear planers. These gear planers use two planer cutters with straight cutting edges (Fig. 9.6), which move in a reciprocating motion (a back and forth motion, with each of the two cutters working during the return stroke of the other, and cutting off the flanks of the same tooth). The cutters are mounted on two adjustable guides that are carried on a cradle, which materializes the generation crown wheel and rotates about its axis, driven by a mechanism made up of a crank and a connecting rod. Even the bevel gear wheel being cut rotates about its stationary axis, with a rotational motion corresponding to the rolling without sliding of its pitch cone on the pitch plane of the virtual generation crown wheel. This motion is transmitted from the cradle to the bevel gear wheel via a gear drive system.

In order to realize the generation cutting process, the bevel gear being cut rotates with a constant angular velocity, synchronized with the reciprocating motion of the cutter, and thus with the motion of the cradle, so that the cutter go right to finish in the next tooth space. Therefore, at any instant, all the teeth are in the same generation stage. For straight bevel gears, the two tool-holder guides are oriented in the radial direction. Moreover, they are adjustable, and are mounted on the cradle to allow small oscillations of the working direction of the two cutters, for which it is possible to obtain crowned teeth, with all the advantages already described in terms of proper location of the tooth bearing contact area.

With this generation cutting method, large gear wheels, to be produced in small and medium series, can be obtained. Depending on the tooth depth and shape, several steps are required to complete the two members of a bevel gear. Generally, the gear wheel flat blank is first roughed without generation cutting process, using suitable roughing tools. Subsequently, the rough gear wheel is finished with at least two finishing operations on each tooth flank, both made with a generation cutting process. Similar machining processes are performed for cutting of the pinion, but both roughing and finishing operations are carried out with generation cutting processes. Of course, the generation mechanism, which synchronizes the cutters and work-piece motions according to a definite timing, is deactivated for those operations that are done without it.

Another traditional generation process is the one known as *interlocking process*, which is realized with gear milling machines using two flat interlocking disk-type cutters (Fig. 9.26). Therefore, due to this flat shape of the cutters, the corresponding virtual generation crown wheel has trapeze shaped teeth, such as that shown in Fig. 9.8b. Consequently, this process can only be used to cut straight bevel gears, so it is much less versatile than the one described above. It is also known as *completing cutting method*, as it is able to generate the teeth of the two members of a straight bevel gear from the corresponding solid blanks in one operation.

The two milling cutters have interlocked teeth, that is their shape and size are such that the teeth of a cutter fit the tooth space of the other. As shown in Fig. 9.26, the cutting edges of the two cutters reproduce the tooth shape of the generation

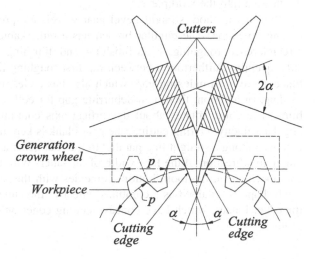

Fig. 9.26 Interlocking cutting process of a straight bevel gear wheel

crown wheel. They are on two planes, each of which is perpendicular to the axis of its cutter, and both cut in the same tooth space. In order to obtain the lenghtwise crowning, and thus to have an optimal location of the tooth bearing contact area, the cutting edges have a concave cutting surface that removes more metal at the ends of the tooth. The two interlocking cutters are mounted on a cradle, with their rotating axes inclined to the platform of the same cradle. The work-piece is mounted on a work spindle rotating about its axis at a synchronized rotational speed at an appropriate timing rate relative to the cradle speed, to simulate the relative rolling motion without sliding of the pitch cone of the bevel gear wheel being cut on the cutting pitch plane of the generation crown wheel.

For reasons of brevity, we cannot describe here neither the operating modes of the gear cutting machines using this interlocking cutting process, or the ingenious mechanisms that implement and control the relative motion between the tool-holder cradle and work-piece spindle. In this regard, we refer the reader to specialized textbooks, such as those mentioned above. However, it should be noted that the latest gear milling machines using this cutting method have CAD/CAM systems, which allow to generate tooth flank surfaces of straight bevel gears with a combination of profile and lengthwise crowning. For further information on this interesting topic, we refer the reader to specialized textbooks and scientific literature (see, e.g.: Stadtfeld [40], Radzevich [32], Shih and Hsieh [37]).

Finally, another generation cutting method deserves to be mentioned here, as it is used for mass production of straight bevel gears even of considerable size and good commercial quality at the lowest cost and in the fastest way. This method is known as *revacycle generation process*. In essence, it is a generating milling method, which uses a special cutter, which we can define as circular broach. This is because the cutter blades, which extend radially outward from the cutter head, are distributed along the cutter circumference so that each subsequent blade is more protruding than the preceding one, as in a traditional broach. Since each cutter blade is progressively longer than the one before it, there is no need to feed the cutter depth-wise into the workpiece.

With this method, straight bevel gear wheels are produced with a single completing operation. The circular broach has a remarkable diameter, since along its circumference roughing, semi-finishing and finishing blades must be arranged subsequently. Furthermore, between the first roughing blade and the last finishing blade, there is an indexing gap, which also has a clearance function for automatic loading. In addition, there is a deburring gap for collecting and expelling the tool burrs. The cutter rotates about its vertical axis continuously and with a uniform angular velocity. During cutting, the gear blank is kept motionless, while the cutter is moved along a straight line parallel to the root generatrix of the gear wheel being cut, in the direction of the face width of the latter. With this cutter motion, due to a cam mechanism, the tooth bottom coincides with the root cone, while the desired tooth shape is obtained as a combined effect of the cutter motion and the shape of the cutter blades, which have concave cutting edges able to produce convex tooth profiles.

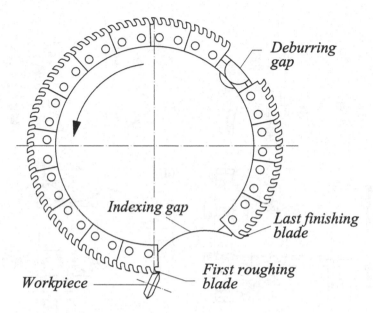

Deburring gap

Last finishing blade

First roughing blade

Indexing gap

Workpiece

Fig. 9.27 Revacycle generation process scheme

Most bevel gear wheels are completed in one operation. One revolution of the cutter completes each tooth space, and the next tooth space is indexed in the gap between the last finishing blade and the first roughing blade (Fig. 9.27). In the few cases where the tooth depth is too large, so the teeth could not be completed in one cut, separated roughing and finishing operations are used. Each of these operations is similar to the one described above, which uses a completing cutter, but requires cutters and setups of the gear cutting machine. The first operation is roughing, and is made with a cutter provided only with roughing blades. The second operation is finishing, and is made with a cutter provided only with semi-finishing and finishing blades.

Of course, the various motions described above simulate the relative rolling motion without sliding of the pitch cone of the straight bevel gear wheel to be cut on the pitch plane of the generation crown wheel. For reasons of brevity, here also we cannot describe the mechanisms that implement and control this relative rolling motion. For any further details, we refer the reader to specialized textbooks and scientific literature (see Stadtfeld [41], Radzevich [32]).

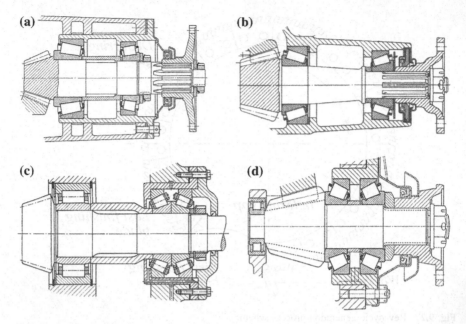

Fig. 9.28 Pinion cantilever mounted: **a** two preloaded tapered roller bearings, and adjustment bush; **b** two preloaded tapered roller bearings, without adjustment bush; **c** simple supported pinion with a cylindrical roller bearing on the left, and rigid pair consisting of two tapered roller bearings and related adjustment device on the right; **d**, straddle supported pinion with a cylindrical roller bearing on the left, and rigid pair consisting of two tapered roller bearings and related adjustment device on the right

9.13 Construction and Assembly Solutions for Bevel Gears

In the design of the straight (and curved-toothed) bevel gears, we must keep in mind that the sizing, manufacturing, control, assembly and mounting are more difficult than is the case of parallel cylindrical gears. Usually, the pinion is cantilever mounted, so it is necessary to adjust its position with respect to that of the mating gear wheel. In this regard, several construction and assembly solutions can be used, four of which are shown in Fig. 9.28. For details of these solutions and their other peculiarities, we refer the reader to traditional textbooks on this subject (see Giovannozzi [13], Conti [6], Deutschman et al. [8], Chévalier [5], Niemann and Winter [25]).

From these construction and assembly solutions, further possibilities of errors than those associated with the parallel cylindrical gears result. They are primarily due to the error of axial position of the pinion and wheel, as well as to the deviation of the actual value of the shaft angle from the nominal value, especially because of the above-mentioned cantilever mounted pinion. These errors, if not adequately contained and controlled, can determine uneven distribution of the load along the face width, with the risk of load carrying capacity only on edges, irregular running conditions, and, in extreme case, even interlocking of the bevel gear pair, due to the total disappearance of the backlash.

To avoid or to mitigate these risks, we must first limit the face width, by imposing the conditions $b \leq 0, 30R_e$ and $b \leq 10m_{et}$, and possibly use crowned teeth, with recommended crowning depth in the range from $(b/300)$ to $(b/500)$, for case-hardened and grinded toothings, or equal to about $(b/1000)$, for machining finished toothings. The lengthwise crowning, even if does not allow to completely remedy the misalignment error of the axes, which no longer converge at the apex, allows to bring the contact at the middle of the face width, thus avoiding contact on the edges. It is also appropriate that face widths of the two members of the bevel gear pair are equal, in order to prevent that sharp edges are formed during running-in, or due to the smoothing and polishing exerted by a case-hardened or nitride pinion $(HRC > 50)$ on the tooth flanks of the through hardened mating wheel.

It is then necessary to adjust the axial positions of the pinion and wheel, so as to get as close as possible to that of correct kinematics, which from the theoretical point of view is only one, to work together the two members of the bevel gear, and to mount and replace them in pairs. A complete interchange ability is only possible with mass-produced bevel gears, and with adequate experience in the production process and choice of materials and heat treatments. When adjusting the axial position of the two members of the bevel gear pair, it is necessary to check that clearance remains constant. For mounting, the shaft angle must be the same as the design shaft angle, and the axes of the two bevel gear members must intersect. Though small deviations from the theoretical conditions cannot be avoided in practical applications, care must be taken to reduce these deviations to as small a value as possible, as otherwise serious running difficulties are encountered.

To minimize the detrimental effect of the bending deflections, bearings must be properly spaced. The spacing depends on the stiffness of the shafts carrying the gear wheels, and the mounting type, i.e. whether the wheels are straddle mounted or overhung mounted. It is also necessary to preload the bearings, reducing their clearances, and so have smaller lateral forces and greater stiffness of the assembly, to which an appropriate support sizing, perhaps with suitable ribs, makes a significant contribution. Always to have the support maximum stiffness, and to avoid or to limit the lateral forces, it is appropriate that the distances l_1 and l_2 between the middle-planes of the bearings that support the shafts of the pinion and wheel have suitable values, and precisely: $l_1 = (2.0 \div 2.5)d_{m1}$ for $u = (3 \div 6)$, and $l_1 = (1.2 \div 2.0)d_{m1}$ for $u = (1 \div 6)$, for the pinion; $l_2 > 0.7d_{m2}$, for the wheel.

Usually, pinion and its shaft make up a single piece. For pinion keyed on its shaft, to avoid dangerous strength decreases to the tooth roots, it must be ensured that the minimum rim thickness, s_R, is at least equal to $2m_n$. In this way, the maximum diameter of the shaft, on which the pinion is keyed, is determined. When possible, to avoid the afore-mentioned errors, it is good to bear the shaft of the pinion from both sides of the toothed part. Generally, for bevel gears with converging axes, this is not possible. In these cases, in addition to reducing the face width, it is necessary that one of the two bearings that support the shaft of the pinion cantilever mounted is as near as possible to the heel of the pinion, to limit the bending deflection and, therefore, an uneven load distribution on the teeth (see Giovannozzi [13].

In some cases, the pinion is supported from both sides thereof itself; this is often possible with hypoid gear pair. In these cases, it is necessary to size the shaft ends of the pinion in such a way that, during the generation cutting of the teeth, the cutting tool does not interfere with the keying surface of the bearings (see Niemann and Winter [25].

Wheels of the bevel gears are usually keyed to the shaft, and the shaft is supported by bearings located on both sides of the wheel itself. Wheels of a relatively small diameter are made in one piece. Beyond certain diameter (800 mm), the wheels are made of two pieces, being made up of a toothed crown ring in case-hardened steel, having rim thickness sufficient to provide full support for the tooth roots, and a hub in through hardened steel, joined by threaded fasteners.

Mountings must be rigid, so that the relative displacements of the two members of the bevel gear pair under operating conditions are kept within allowable limits. Detrimental effects of misalignment must be minimized, ensuring a proper alignment of the two wheels. Besides, bearing housing and mountings must be accurately machined, couplings accurately mounted, and keys must be properly fitted.

References

1. Brar JS, Bansal RK (2004) A textbook of theory of machines. Laxmi Publications (P), Ltd., New Delhi
2. Buchanan R (1823) Practical essays on mill work and other machinary, with notes and additional articles, containing new researches on various mechanical subjects by Thomas Tredgold. J Taylor, London
3. Buckingham E (1949) Analytical mechanics of gears. McGraw-Hill Book Company Inc, New York
4. Buzano P (1961) Lezioni di analisi matematica, 5th edn. Libreria Editrice Universitaria Levrotto&Bella, Torino
5. Chévalier A (1983) Manuale del Disegno Tecnico. In: Chirone E, and Vullo V (Eds) Revised and expanded Italian. Società Editrice Internazionale, Torino
6. Conti G (1969) I Cuscinetti a Rotolamento, vol. I and vol. II, 4th. Editore Ulrico Hoepli, Milano
7. Davis JR (Ed) (2005) Gear materials, properties, and manufacture. J.R. Davis, Davis & Associates, ASM International, Materials Park, OH, USA
8. Deutschman AD, Michels WJ, Wilson C (1975) Machine design: theory and practice. Macmillan Publishing Co., Inc, New York
9. Dudley DW (1984) Handbook of practical gear design. McGraw-Hill Book Company, New York
10. Ferrari C, Romiti A (1966) Meccanica applicata alle macchine. Unione Tipografica-Editrice Torinese (UTET), Torino
11. Fuentes A, Gonzalez I, Pasapula HK (2017) Computerized design of straight bevel gears with optimized profiles for forging molding, or 3D printing. Thermal processing magazine, mar

12. Galassini A (1962) Elementi di Tecnologia Meccanica, Macchine Utensili, Principi Funzionali e Costruttivi, loro Impiego e Controllo della loro Precisione, reprint 9th ed., Editore Ulrico Hoepli, Milano
13. Giovannozzi R (1965) Costruzione di Macchine, vol I, 2th edn. Casa Editrice Prof. Riccardo Pàtron, Bologna
14. Giovannozzi R (1965) Costruzione di Macchine, vol II, 4th edn. Casa Editrice Prof. Riccardo Pàtron, Bologna
15. Henriot G (1979) Traité Théorique et Pratique des engrenages, vol 1, 6th ed, Bordas, Paris
16. ISO 23509: 2016 (E) Bevel and hypoid gear geometry
17. ISO 53 (1998) Cylindrical gears for general and for heavy engineering—standard basic rack tooth profile
18. Klebanov BM, Groper M (2016) Power mechanisms of rotational and cyclic motions. CRC Press, Taylor & Frencs Group, Boca Raton, Florida
19. Lasche O (1899) Die Ausführung mit einer Profilverschiebung nach obenstchenden Augaben wird der grundlegenden, Z.d. VDI S. 1488 auch als AEG-Verzahnung bezeichnet
20. Levi-Civita T, Amaldi U (1929) Lezioni di Meccanica Razionale—vol 1: Cinematica—Principi e Statica, 2nd ed., reprint 1974, Zanichelli, Bologna
21. Litvin FL, Fuentes A (2004) Gear geometry and applied theory 2nd ed. Cambridge University Press, Cambridge
22. Maitra GM (1994) Handbook of gear design, 2nd edn. Tata McGraw-Hill Publishing Company Ltd, New Delhi
23. Merritt HE (1954) Gears 3th ed. Sir Isaac Pitman&Soins Ltd., London
24. Niemann G, Winter H (1983) Maschinen-elemente, band II: getriebe allgemein, Zahnradgetriebe-Grundlagen, Stirnradgetriebe. Springer-Verlag, Berlin Heidelberg
25. Niemann G, Winter H (1983) Maschinen-elemente, band III: Schraubrad-, Kegelrad-, Schnecken-, Ketten-, Rienem-, Reibradgetriebe, Kupplungen, Bremsen, Freiläufe. Springer-Verlag, Berlin Heidelberg
26. Nof SY (ed) (2009) Springer handbook of automation. Technology & Engineering, Springer-Verlag, Berlin Heidelberg
27. Pollone G (1970) Il veicolo. Libreria Editrice Universitaria Levrotto & Bella, Torino
28. Radzevich SP (2010) Gear cutting tools: fundamentals of design and computation. CRC Press, Taylor & Francis Group, Boca Raton, Florida
29. Radzevich SP (2010) A new angle on cutting bevel gears. Gear Solutions
30. Radzevich SP (2013) Theory of gearing: kincmatics, geometry, and synthesis. CRC Press, Taylor & Francis Group, Boca Raton, Florida
31. Radzevich SP (2016) Dudley's handbook of practical gear design and manufacture, 3rd edn. CRC Press, Taylor & Francis Group, Boca Raton, Florida
32. Radzevich SP (2017) Gear cutting tools: science and engineering, 2nd edn. CRC Press, Taylor & Francis Group, Boca Raton, Florida
33. Rossi M (1965) Macchine Utensili Moderne. Editore Ulrico Hoepli, Milano
34. Schiebel A, Lindner W (1954) Zahnräder, Band 1: stirn-und Kegelräder mit geraden Zähnen. Springer-Verlag, Berlin Heidelberg
35. Schiebel A, Lindner W (1957) Zahnräder, Band 2: Stirn-und Kegelräder mit schrägen Zähnen Schraubgetriebe. Springer-Verlag, Berlin Heidelberg
36. Scotto Lavina G (1990) Riassunto delle Lezioni di Meccanica Applicata alle Macchine: Cinematica Applicata, Dinamica Applicata—Resistenze Passive—Coppie Inferiori, Coppie Superiori (Ingranaggi—Flessibili—Freni). Edizioni Scientifiche SIDEREA, Roma
37. Shih YP, Hsieh HY (2016) Straight bevel gear generation using the dual interlocking circular cutting method on a computer numerical control bevel gear-cutting machine. ASME, J Manuf Sci Eng 138 (2)
38. Smirnov V (1969) Course de Mathématiques Supérieurs Tome I. Édition MIR, Moscou (traduit du Russe)
39. Stadtfeld HJ (1993) Handbook of bevel and hypoid gears: calculation, manufacturing and optimization. Rochester Institute of Technology, Rochester, New York

40. Stadtfeld HJ (2007) Straight bevel gears on phoenics R-machines using coniflex R-tools. The Gleason Works, Rochester, New York
41. Stadtfeld HJ (2010) Coniflex plus straight bevel gears manufacturing. Gear Solutions, pp 44–55
42. Townsend DP (1991) Dudley's gear handbook: the design, manufacture and application of gears. McGraw-Hill, New York
43. Tricomi FG (1956) Lezioni di Analisi Matematica, Parte Seconda, CEDAM— Padova: Casa Editrice Dott. Antonio Milani

Chapter 10
Crossed Helical Gears

Abstract In this chapter, the fundamentals of general rigid kinematic pairs are first described and the hyperboloid pitch surfaces are defined. These concepts are then applied to the generation of cylindrical crossed helical gear pairs. The main geometrical quantities of these types of gears are then defined and their kinematic quantities are determined, particularly those correlated with the lengthwise sliding and sliding velocity. Subsequently, the load analysis of this gears is performed and the thrust characteristic on shafts and bearings are defined. Other important kinematic quantities related to the rolling and sliding motions of the mating teeth surfaces are then defined and the instantaneous and average efficiencies of these gears are analytically determined. Finally, short notes are given regarding the fundamentals of profile-shifted toothing for crossed helical gears.

10.1 Fundamentals of General Rigid Kinematic Pairs

By definition, *crossed helical gears* are cylindrical gear pairs consisting of mating helical members with crossed axes. However, the tooth flank surfaces of the parallel cylindrical involute helical gears are in line contact, while those of the crossed involute helical gears are in point contact. In this chapter, we will discuss only cylindrical involute helical gears with crossed axes, without considering worm gears, which are a special case of crossed helical gears. Worm gears will be discussed in the next chapter.

The discussion is here carried out in accordance with the classical one, available in more or less extensive form in traditional textbooks (see Buckingham [2], Colbourne [4], Dobrovolski et al. [5], Dudley [6], Giovannozzi [9], Henriot [10], Litvin and Fuentes [13], Merritt [14], Niemann and Winter [17], Pollone [18], Radzevich [20], Townsend [27]). However, to better understand the kinematics of these gear pairs, we think it is appropriate to first introduce two introductory sections to the topic that interest us. Therefore, in this perspective, this and the next section deal with the general rigid kinematic pairs and hyperboloid gear pairs, respectively.

© Springer Nature Switzerland AG 2020
V. Vullo, *Gears*, Springer Series in Solid and Structural Mechanics 10,
https://doi.org/10.1007/978-3-030-36502-8_10

Notoriously, a cylindrical *crossed helical gear pair* can be imagined as originated by a *hyperboloid gear pair*, which represents a more general case of mating gears with crossed axes. It is also known that *hyperboloid gears* are a special case of the most *general rigid kinematic pair* for transmission of motion between two bodies, 1 and 2, with crossed axes (note that 1 and 2 indifferently indicate the two bodies or their axes of rotation). In this general rigid kinematic pair, the *axodes*, i.e. the pitch surfaces, that define the relative motion are two ruled surfaces, at all-time tangent along a straight line. This straight line is the *instantaneous axis of rotation*, also called with synonyms *axis of the helical relative motion*, or *axis of screw motion*, or *Mozzi's axis* (see Ball [1], Ceccarelli [3], Euler [7], Levi-Civita and Amaldi [12], Minguzzi [15], Mozzi del Garbo [16], Scotto Lavina 24]).

While bodies 1 and 2 are rotated about their axes, the instantaneous axis of the screw motion generates, in two Cartesian coordinate systems $O_i = (x_i, y_i, z_i)$ rigidly connected to the same rotating bodies i (with $i = 1, 2$), two surfaces which are two *hyperboloids of revolution*. These hyperboloids are the axodes for transformation of rotation between crossed axes. Assuming, for convenience, that one of the two bodies is mobile and the other motionless, the motion of the mobile axode with respect to the motionless one is an *instantaneous helical motion*, or *instantaneous screw motion*, with relative angular velocity $\omega_{r1,2} = (\omega_1 - \omega_2)$ about the aforementioned axis, and a translation along the same axis, with velocity v_r. This is notoriously the more general motion of any rigid system (see Ferrari and Romiti [8], Poritsky and Dudley [19]).

To define the two axodes, which are ruled surfaces, the position of the Mozzi's axis at each instant must be first determined. To this end, we consider the case of an external gear pair, with shaft angle $\Sigma < \pi/2$. It is to be noted that, de facto, external gear pairs are the only of interest in practical applications. Suppose that the following quantities characterizing this external gear pair are known: the angular velocity vectors, ω_1 and ω_2, about the axes 1 and 2; the shortest distance or center distance, $a = O_1O_2$, between the two crossed axes; the shaft angle Σ between the direction of vectors ω_1 and $-\omega_2$.

We now look at what happens in two orthogonal projections (Fig. 10.1), the first on the plane normal to the shortest distance straight line, and the second on a plane normal to that plane. The quantities related to the first and second projection are indicated respectively with one and with two indices. In the first projection (Fig. 10.1b), we will get the straight lines $1'$ and $2'$, which converge at the point $O_1' \equiv O_2' \equiv O'$, and have the directions of the vectors ω_1 and ω_2 respectively, and therefore form the angle Σ. In the second projection (Fig. 10.1a), we will have the parallel lines $1''$ and $2''$ respectively through the outermost points O_1'' and O_2'' of the center distance $a = O_1''O_2'' = O_1O_2$.

The Mozzi's axis, m, will lie on a plane parallel to axes 1 and 2, will meet the center distance $a = O_1O_2$ at a point O, located at a distance $r_1^* = OO_1$ from the axis 1, and a distance $r_2^* = OO_2$ from the axis 2, and will form the angles β_1^* and β_2^* with the projections $1'$ and $2'$ in the first plane of projection (Fig. 10.1b). To uniquely define the position of the Mozzi's axis, m, it is therefore necessary to determine the distances r_1^* and r_2^*, and the angles β_1^* and β_2^*. We denote by m' and m'' the

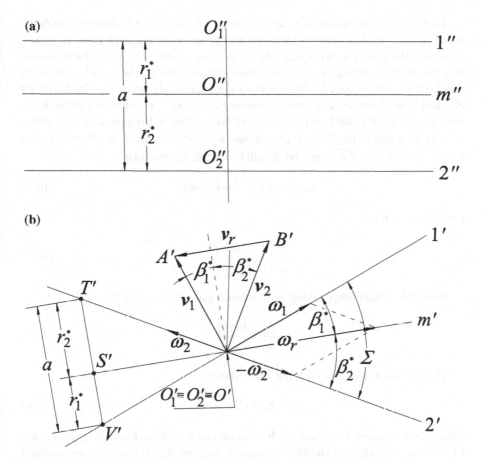

Fig. 10.1 Rigid kinematic pair, and Mozzi's axis, in two orthogonal projections

projections of m-axis respectively in the first and second projection planes. Since the vector of the relative angular velocity $\omega_{r1,2} = (\omega_1 - \omega_2)$ is directed along the Mozzi's axis, if we compose, in the first projection plane (Fig. 10.1b), the vector ω_1 with the vector $-\omega_2$, we get the direction of m', and then the angles β_1^* and β_2^*. Therefore, we have:

$$\frac{\omega_1}{\omega_2} = \frac{\sin \beta_2^*}{\sin \beta_1^*}. \tag{10.1}$$

This equation, together with the condition

$$\beta_1^* + \beta_2^* = \Sigma, \tag{10.2}$$

allows us to determine the angles β_1^* and β_2^*, as we have already seen in the previous chapter, regarding the straight bevel gears, which are gears with intersecting axes.

To determine the distances r_1^* and r_2^*, let us consider point O through which the m-axis passes, as well as the velocity vectors v_1 and v_2 of the same point O, whose absolute values are given respectively by $v_1 = \omega_1 r_1^*$ and $v_2 = \omega_2 r_2^*$. Then assume that this point O belongs once to the ruled surface having axis 1, and once to the ruled surface having axis 2. These vectors are respectively normal to the directions of 1 and 2, while the relative velocity vector $v_{r1,2} = (v_1 - v_2)$ must be parallel to m-axis. Therefore, if in the first projection we bring, starting from point O', the vector $v_1 = \overline{O'A'}$, normal to $1'$, and the vector $v_2 = \overline{O'B'}$, normal to $2'$, the vector $v_{r1,2} = (v_1 - v_2) = \overline{B'A'}$, must be parallel to m'. It follows that

$$\omega_1 r_1^* \cos \beta_1^* = \omega_2 r_2^* \cos \beta_2^*, \tag{10.3}$$

for which we have:

$$\frac{\omega_1}{\omega_2} = \frac{r_2^* \cos \beta_2^*}{r_1^* \cos \beta_1^*}. \tag{10.4}$$

From this relationship, taking into account Eq. (10.1), we get:

$$\frac{r_1^*}{r_2^*} = \frac{\tan \beta_1^*}{\tan \beta_2^*}. \tag{10.5}$$

This equation, together with the relationship

$$r_1^* + r_2^* = a, \tag{10.6}$$

allows us to determine r_1^* and r_2^*. It is to be noted that, as Eq. (10.5) shows, the Mozzi's axis divides the shortest distance, a, in parts that are directly proportional to the trigonometric tangents of angles β_1^* and β_2^*, which the ω_r-vector forms respectively with the rotational axes of the two bodies under consideration.

The relative velocity vector $v_{r1,2} = (v_1 - v_2) = \overline{B'A'}$, which is notoriously the sliding velocity of the points lying on Mozzi's axis, has an absolute value equal to:

$$v_r = \omega_1 r_1^* \sin \beta_1^* + \omega_2 r_2^* \sin \beta_2^*, \tag{10.7}$$

which, taking into account Eq. (10.1), can be written as:

$$v_r = a\omega_1 \sin \beta_1^* = a\omega_2 \sin \beta_2^*. \tag{10.8}$$

Therefore, the ruled pitch surfaces, i.e. the axodes, and the characteristics of the relative motion, ω_r and v_r, are thus completely defined. It should be noted that there are no points in which the relative velocity is zero. The minimum relative velocity possible is the one calculated above, and refers to points of Mozzi's axis. The points located outside of this axis have a greater relative velocity, which increases with their distance from this axis.

10.2   Hyperboloid Pitch Surfaces

Let us now consider a hyperboloid gear pair. If the transmission ratio $i = \omega_1/\omega_2$ is a constant, from Eqs. (10.1) and (10.4) we infer that the values of the angles β_1^* and β_2^*, and distances r_1^* and r_2^* are also constant. Therefore, the *ruled pitch surfaces* become two *double ruled hyperboloids of one sheet*, also called *one-sheeted double ruled hyperboloids*, of which r_1^* and r_2^* are the radii of their *symmetry cross sections* or *throat cross sections* (Fig. 10.2). The ruled pitch surface, which constitutes the axode of gear wheel 1, is the locus of the instantaneous axis of screw motion, m, which forms, with the direction of the axis 1, an angle β_1^* constant, and leans on a circumference of radius r_1^* lying in the plane normal to the axis 1, and passing through the shortest distance, a. The same definition applies, *mutatis mutandis*, for the ruled pitch surface, which constitutes the axode of the gear wheel 2.

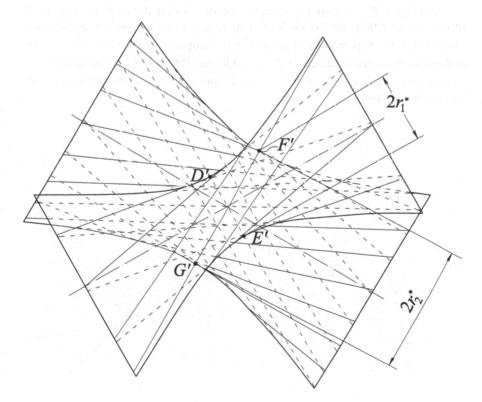

Fig. 10.2 Pair of double ruled hyperboloids of one sheet

By analogy with the other types of gear pairs, and assuming that the gear wheel 1 is the driving one, the absolute value of the transmission ratio $i = \omega_1/\omega_2$, taking into account Eq. (10.1), will be given by:

$$i = \frac{\omega_1}{\omega_2} = \frac{\sin \beta_2^*}{\sin \beta_1^*}. \tag{10.9}$$

For the determination of the characteristic quantities of the two hyperboloids, we consider the same two projections used in the previous section. The radii r_1^* and r_2^* of the two symmetry cross sections are obtained, in the first projection (Fig. 10.1b, left), drawing, in the direction normal to the m'-axis, the segment $a = T'V'$, whose end points T' and V' lie on the projected axes $1'$ and $2'$. Taking into account Eqs. (10.5) and (10.6), we have: $V'S' = r_1^*$ and $T'S' = r_2^*$. The throat cross sections of the two hyperboloids (Fig. 10.3) are the two circles whose diameters are projected, in this projection, in segments $D'E' = 2r_1^*$ and $F'G' = 2r_2^*$.

Consider now the cross sections of two hyperboloids made with planes perpendicular to their axes, and passing through the same point P of the m-axis, placed at a distance $OP = l$ from the straight line of shortest distance. As Fig. 10.3b shows, these sections are the circles having radii r_{1P}^* and r_{2P}^*, which can be easily calculated as the hypotenuses of the rectangle triangles $O_3'P'R$ and $O_4'P'S$. The first of these triangles has as cathets $O_3'P' = l \sin \beta_1^*$ and $P'R$ equal to the throat radius r_1^*, and the second has as cathets $O_4'P' = l \sin \beta_2^*$ and $P'S$ equal to the throat radius r_2^*, for which we have:

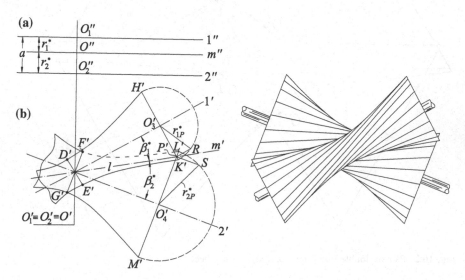

Fig. 10.3 Hyperboloid gear pair, and Mozzi's axis, in two orthogonal projections

$$r_{1P}^* = \sqrt{\left(r_1^*\right)^2 + l^2 \sin^2 \beta_1^*} \qquad r_{2P}^* = \sqrt{\left(r_2^*\right)^2 + l^2 \sin^2 \beta_2^*}. \tag{10.10}$$

In the first projection (Fig. 10.3b), the projection of the two circles having radii r_{1P}^* and r_{2P}^* are represented by the two segments $H'K' = 2r_{1P}^*$ and $L'M' = 2r_{2P}^*$, respectively drawn from point P' normally to the projected axes $1'$ and $2'$. Also in the first projection, the generatrices of the two hyperboloids are projected in the hyperbolas $D'H'$, $E'K'$ and $F'L'$, $G'M'$, which can be drawn point by point, performing the intersections of the two hyperboloids with the planes, normal to the axes, conducted through other points of the m-axis, as we did for point P. The first hyperbola has, as asymptotes, the straight line m' and its symmetrical line with respect to $1'$, while the second hyperbola has, as asymptotes, the straight line m' and its symmetrical line with respect to $2'$.

If we refer the hyperbola $D'H'$, $E'K'$ to a Cartesian coordinate system having its origin at the point O', the abscissa axis $x \equiv O'D'$, and ordinate axis $y \equiv 1'$, we infer that the coordinates of one of its current point H', corresponding to the section at a distance $OP = l$ from the straight line of shortest distance, are [29]:

$$x = \sqrt{\left(r_1^*\right)^2 + l^2 \sin^2 \beta_1^*} \qquad y = l \cos \beta_1^*. \tag{10.11}$$

for which the Cartesian equation of the hyperbola is:

$$x^2 - y^2 \tan^2 \beta_1^* = \left(r_1^*\right)^2. \tag{10.12}$$

The equation of the other hyperbola $F'L'$, $G'M'$ is obtained in a similar manner.

If the axes of rotation 1 and 2 are orthogonal (this is a very frequent case in practical applications), we have $\Sigma = \pi/2$, and $\beta_2^* = \left[(\pi/2) - \beta_1^*\right]$, for which the Eqs. (10.9) and (10.5) become respectively:

$$i = \frac{\omega_1}{\omega_2} = \cot \beta_1^* \tag{10.13}$$

$$\frac{r_1^*}{r_2^*} = \tan^2 \beta_1^* = \frac{1}{i^2}. \tag{10.14}$$

This last equation shows that, in the case where $\Sigma = \pi/2$, the transmission ratio $i = \omega_1/\omega_2$ is equal to the square root of the ratio r_2^*/r_1^*. The same equation shows that, with a small transmission ratio, the radius of the throat cross section of the greater gear wheel may be much larger than that of the other gear wheel. For example, with $i = 4$, we would have $r_2^* = 16r_1^*$. Moreover, always in the case where $\Sigma = \pi/2$, taking into account Eqs. (10.13) and (10.14), one of the following relationships can express the absolute value of the relative velocity along the Mozzi's axis, given by Eq. (10.7):

$$v_r = \omega_1 r_1^* \sin \beta_1^* (1 + \cot^2 \beta_1^*) = \frac{\omega_1 r_1^*}{\sin \beta_1^*} = \omega_2 r_2^* \sin \beta_2^* (1 + \cot^2 \beta_2^*) = \frac{\omega_2 r_2^*}{\sin \beta_2^*}.$$

$$(10.15)$$

In the general case where $\Sigma \neq \pi/2$, if appropriate parts of the two pitch hyperboloids are provided with teeth, we get a hyperboloid gear pair. These toothed parts can be delimited by sections normal to the axis of rotation, located on opposite sides or the same side with respect to the throat cross section. In the first case, we have the so-called *throat hyperboloid wheels* (Fig. 10.4a) which, for limited values of the length of the toothed part, are approximately cylindrical, and therefore resemble cylindrical helical wheels. In the second case, we have the *pseudo-bevel hyperboloid wheels* (Fig. 10.4b), so named for their resemblance to the helical and spiral bevel wheels, the characteristics of which tend to approach the more, the greater the distance of the sections which delimit the toothed part from the throat cross section.

The throat hyperboloid wheels are defined by the face width (the width over the toothed part of the wheel, measured along a hyperboloid reference generatrix), and the values of the radii r_1^* and r_2^*, and angles β_1^* and β_2^*, which are calculated using the relationships above. These relationships cannot however be applied for the pseudo-bevel hyperboloid wheels, since in these wheels it is not detectable any pair of corresponding circular cross sections whose radii are given by the above equations. However, it is possible to show that the ratio between the two radii tends to the transmission ratio (as in the cylindrical gears, and in the bevel gears), when the distance of the corresponding cross sections from the throat cross sections of the two hyperboloids tends to infinity.

The determination of the surfaces delimiting the toothing of a hyperboloid wheel is carried out using the known general rules of Applied Mechanics, concerning the analytical definition of the mating surface of a general rigid kinematic pair.

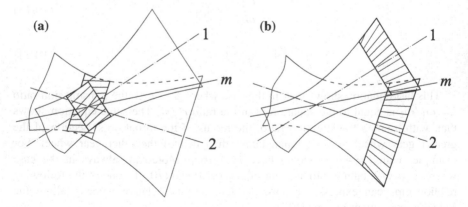

Fig. 10.4 Hyperboloid gear pair: **a** throat hyperboloid gear pair; **b** pseudo-bevel hyperboloid gear pair

The practical solution of this problem, however, gives rise to considerable technological difficulties, due mainly to the impossibility to identify a particular hyperboloid wheel that, for the geometric simplicity of the active surfaces of its teeth, can be assumed as generation wheel capable of generating, by a generation cutting process, the hyperboloid mating wheel. This fact, entirely new, does not occur for the cylindrical and bevel wheels for which, as we have seen in previous chapters, it was possible to identify respectively the reference basic rack, and the reference crown wheel.

It is to be noted that the hyperboloid wheels have theoretically two main advantages:

- the contacts between the conjugate surfaces of the meshing teeth are line contacts;
- in such contacts, when they occur on the hyperboloid pitch generatrices and, by extension, in their immediate vicinity, the minimum sliding velocities occur.

However, because of the above-mentioned technological difficulties, which arise for a correct cutting of the teeth, their use is very limited. In practice, it is preferred to forego the aforementioned advantages, in favor of a simpler and more economical cutting process.

We can realize the above-mentioned impossibility of identifying a particular hyperboloid wheel to be used as a generation gear wheel, considering the special case of hyperboloid gear pair in which the shaft angle is greater than $\pi/2$ $(\Sigma > \pi/2)$, and $\beta_2^* = \pi/2$. This special case is shown in Fig. 10.5, which highlights these two angles in the side view (Fig. 10.5a). For $\beta_2^* = \pi/2$, from Eq. (10.9) we infer $\sin \beta_1^* = \omega_2/\omega_1$, so from Eqs. (10.5) and (10.6) we get $r_1^* = 0$, and $r_2^* = a$. Therefore, the Mozzi's axis, m, lies on the plane perpendicular to the shortest distance straight line containing the 1-axis, intersects the latter at point P $(O_1P = 0$, and $O_2P = a)$, and is normal to the projection on this plane of the 2-axis, so m' is normal to $2'$.

Rebus sic stantibus, it follows that the ruled pitch surface of the wheel 1 degenerates into the cone with apex P and cone angle $\beta_1^* = (\Sigma - \pi/2)$, while the pitch cone surface of the wheel 2 degenerates into a ruled plane. This plane is formed by the straight lines that are tangent to the circle having as its center P (these tangents are nothing more than the successive positions of the Mozzi's axis, m), and radius equal to the shortest distance, a. The second wheel is the hyperboloid crown wheel. Figure 10.5b shows what we have just said, in the projection made in the direction of the 2-axis. The transmission ratio of this special hyperboloid gear pair is $i = (\omega_1/\omega_2) = (1/\sin \beta_1^*)$, while in accordance with Eq. (10.7), the absolute value of the sliding velocity is given by $v_r = \omega_2 a$.

Assume now the ruled pitch surfaces defined above are provided with toothing. We have so a power transmission between crossed axes with a gear pair constituted by a straight bevel wheel and a crown wheel with curved teeth, whose intersections with the reference plane of the same crown wheel are straight lines that are tangent to the circle defined above, as Fig. 10.5b shows. However, it should be noted that it

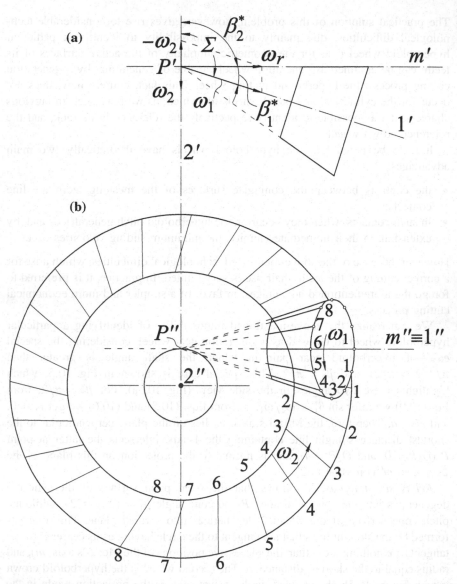

Fig. 10.5 Hyperboloid gear pair with $\Sigma > \pi/2$ and $\beta_2^* = \pi/2$: **a** projection in the side view; **b** projection in the direction of the 2-axis

is impossible to use this gear pair for transmitting motion between orthogonal crossed axes. This is because the shaft angle, Σ, between the crossed axes 1 and 2 cannot be equal to $\pi/2$, because in that case we should have $\Sigma = \pi/2$ and $\beta_1^* = 0$. It follows the impossibility of identifying a particular hyperboloid crown wheel to be used as a generation gear wheel for hyperboloid gear pairs.

10.3   Generation of a Crossed Helical Gear Pair

If we want to avoid the above technological difficulties that characterize the hyperboloid wheels, to transmit motion between crossed axes, we can use cylindrical helical gear pairs of the same type as those analyzed in Chap. 8. Using these gear pairs, we give up the advantage of having mating surfaces with line contacts and minimum sliding velocities, but in return their cutting method and production process become extremely simple, as is the case for cylindrical helical gears. Both members of the gear pair are involute helicoids cut on cylindrical blanks, but connecting skew shafts, for transmission of rotation between crossed axes.

The basic relationship that correlates the shaft angle Σ and the helix angles β_1 and β_2 of the two members of the gear pair is given by:

$$\Sigma = \beta_1 \pm \beta_2, \tag{10.16}$$

where the plus and minus signs are respectively for helices that have the same direction (i.e., both right-hand helices or both left-hand helices), and for helices that have opposite directions (i.e., one right-hand helix and the other left-hand helix or vice versa). Four different combinations are therefore possible, and precisely:

- Right-hand driving wheel meshing with right-hand driven wheel.
- Left-hand driving wheel meshing with left-hand driven wheel.
- Right-hand driving wheel meshing with left-hand driven wheel.
- Left-hand driving wheel meshing with right-hand driven wheel.

In most cases, the helices have the same direction, and the driving wheel is the one with greater helix angle. Only for small values of the shaft angle, the helices have opposite directions. The direction of rotation of the driven wheel depends on the helix direction, and the direction of rotation of the driving wheel.

In the usual practical applications, the crossed axes are orthogonal ($\Sigma = \pi/2$). Contrary to what happens in parallel cylindrical helical gears, in which the helix angles are equal, but of opposite directions, in the crossed helical gears the helix angles are generally different from each other, and the wheel with greater helix angle is the driving wheel. From a theoretical point of view, the teeth of a crossed helical gear have a point contact between the mating tooth flank surfaces.

Actually, due to the Hertzian contact deformations and wear after some time of operation, the theoretical point of contact develops into an ellipse of contact, which extends in length, constituting almost a line contact. However, despite this positive extension of the area of contact, these gears are used only for transmission of small loads. They therefore have a limited use, and are mainly used, for example, as secondary drives for textile machines, speedometer drives, pump drives, instrumentation, distributor drives of internal combustion engines, and other similar applications.

When the load to be transmitted is high, due to sliding, these gears will wear out quickly, even if the rotational speed is not high. If then also the rotational speed is high, the scuffing becomes the main cause of deterioration of their teeth.

If the shaft angle increases, the sliding velocity along the tooth trace increases as well, because we have here a screw sliding. Compared to a parallel cylindrical helical gear, where sliding is only the one in the direction of the tooth depth, we have here the worst sliding conditions, as the *lengthwise* or *longitudinal sliding* in the direction of the tooth axis is superimposed to the *profile sliding* in the direction of the tooth depth. With small values of the shaft angle ($\Sigma < 25°$), due to the increasing wear, the ellipse of contact becomes increasingly extensive, thereby enhancing the load carrying capacity appreciably.

As for a general rigid kinematic pair or a hyperboloid gear pair, for the generation of a crossed helical gear pair we can consider the same two orthogonal projections used in Sects. 10.1 and 10.2 (Fig. 10.6). Also, in this case we indicate with: 1 and 2, the skew shaft axes; Σ, the shaft angle; $a = O_1 O_2$, the *center distance*

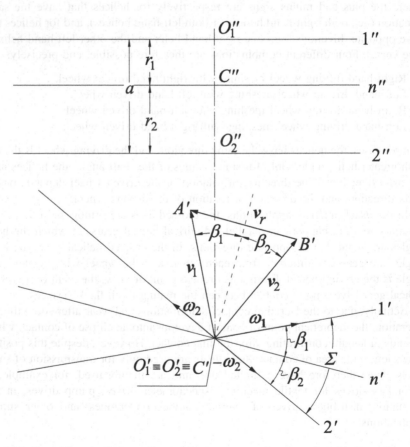

Fig. 10.6 Crossed external gear pair, and instantaneous axis of rotation, in two orthogonal projections

or *shortest distance* or *minimum distance* between the two axes; ω_1 and ω_2 the angular velocity vectors. If we choose an arbitrary point C of the center distance $a = O_1O_2$ (this point is generally distinct from the point O through which the Mozzi's axis passes), and denote with $r_1 = CO_1$ and $r_2 = CO_2$ the distances of the point C from the two axes, we can write:

$$a = r_1 + r_2. \tag{10.17}$$

Assume the cylinders having axes 1 and 2, and radii r_1 and r_2, as *pseudo-pitch cylinders* of the two wheels. Strictly, they are not pitch surfaces. In this regard, we remember in fact that, from a theoretical point of view, the pitch surfaces, i.e. the axodes, are the loci of the instantaneous axes of rotation, and their cross sections are the centrodes, i.e. the loci of the instantaneous centers of rotation. For this reason, the cylinders above are called pseudo-pitch surfaces. These pseudo-pitch surfaces are used here as pitch surfaces of the corresponding toothing, thus fulfilling the function that the cylindrical pitch surfaces had in parallel cylindrical helical gears. We will clarify this concept soon. In any case, hereafter, when we talk of pitch surfaces of the crossed helical gears, we refer to these pseudo-pitch surfaces.

Let us now consider the common Π-plane (Fig. 10.7), which is tangent to the aforementioned cylinders (and therefore passing through the point C and normal to the center distance) and, on it, the straight line n, having the direction of the relative velocity vector of point C, $v_r = (v_1 - v_2)$. If $v_1 = \omega_1 r_1$ and $v_2 = \omega_2 r_2$ are the absolute values of the velocities of point C, considered as belonging respectively to the first and second cylinder, the vector $v_1 = \overline{C'A'}$ (see Fig. 10.6) will be normal to the axis $1'$, and the vector $v_2 = \overline{C'B'}$ will be normal to the axis $2'$, for which the vector of the aforementioned relative velocity will be $v_r = (v_1 - v_2) = \overline{B'A'}$.

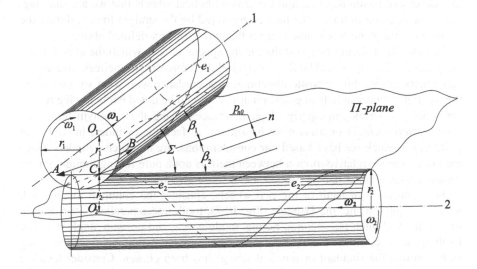

Fig. 10.7 Pseudo-pitch cylinders, and generation straight line of the pitch helices

Therefore, the first projection n' of n will pass through the point C' and shall be parallel to $\overline{B'A'}$, for which the angles β_1 and β_2 between n' and $1'$ and, respectively, between n' and $2'$ will have to satisfy the following condition:

$$\omega_1 r_1 \cos \beta_1 = \omega_2 r_2 \cos \beta_2; \tag{10.18}$$

thus, the transmission ratio, i, is given by:

$$i = \frac{\omega_1}{\omega_2} = \frac{r_2 \cos \beta_2}{r_1 \cos \beta_1} = \frac{r_2 \sin \gamma_2}{r_1 \sin \gamma_1}. \tag{10.19}$$

The one or the other of the two equations above, together with Eq. (10.16), allow determining the position of the straight line, n, which satisfies the condition imposed. Wrapping the Π-plane containing the straight line, n (Fig. 10.7), respectively on the first and second of the two aforementioned cylinders, we will obtain, on the first cylinder, the helix e_1 with helix angle, β_1, and, on the second cylinder, the helix e_2 with helix angle, β_2. The helices e_1 and e_2 thus obtained, both corresponding to the straight line, n, are taken as tooth traces respectively of the wheel with axis 1 and of the one with axis 2. In the case shown in Fig. 10.7, they are both left-hand helices (of course, they could be both right-hand helices), for which Eq. (10.16) must be considered with the plus sign.

It should be noted that the difference between the crossed helical gears and hyperboloid gears lies in the aforementioned generation method of tooth traces. When the point C arbitrary chosen coincides with the point O considered in Sects. 10.1 and 10.2 (in this case we have $r_1^* = r_1$ and $r_2^* = r_2$), the straight line, n, coincides with the Mozzi's axis, m. Also, in this particular case, a substantial difference exists between the hyperboloid wheels, whose tooth traces coincide with the successive positions of m, and the crossed helical wheels that we are studying, which have, as tooth traces, the helices generated by the straight line, n, during the rotation of the plane that contains it on the two cylinders defined above.

In addition, it should be noted that, in the special case in which the axes 1 and 2 are parallel ($\Sigma = 0$), would be $\beta_1 = -\beta_2$, and therefore the two helices would have equal helix angles, but opposite directions, in accordance with what we said about the parallel cylindrical helical gears. Finally, it should be noted explicitly the reason for which we speak improperly of pitch surfaces, using the synonym terms *improper pitch surfaces* or *pseudo-pitch surfaces*. In fact, the cylinders in contact at point C, on which we have based our considerations, are not pitch cylinders, since the instantaneous relative motion between them is not a pure rolling motion, but an instantaneous relative helical motion, i.e. a screw motion.

For generation of teeth of the two wheels, consider the common mating rack having, as pitch plane, the Π-plane (Fig. 10.7), tooth traces parallel to the straight line, n, normal pitch $p_{n0} = \pi m_{n0}$, normal module m_{n0}, and pressure angle α_{n0}. The toothing of the common mating rack will be so completely defined, once the sizing (for example, the standard or nominal sizing) has been chosen. Consider then the first of the two wheels of the crossed helical gear pair, i.e. the pinion 1, having the cylinder with axis 1 and radius r_1 as pitch surface, which meshes with the mating rack. This wheel will show, on its pitch surface, a succession of tooth traces consisting of helices e_1 having normal pitch $p_{n1} = p_{n0}$, and sizing based on the normal module m_{n0}, while its transverse pitch is given by:

$$p_{t1} = \frac{p_{n1}}{\cos \beta_1};$$ (10.20)

thus, if z_1 is its number of teeth, we will have:

$$p_{t1}z_1 = 2\pi r_1.$$ (10.21)

It is to be noted that, when we consider the gear pair constituted by the pinion to be cut and the rack-type cutter, the cylinder of radius r_1, in effect, is the pitch surface, as it meets the theoretical definition by us previously callback. The same thing we can say when we consider the gear pair constituted by the wheel to be cut and the rack-type cutter. This second wheel of the crossed helical gear pair, i.e. the wheel 2, has as pitch surface the cylinder with axis 2 and radius r_2. This wheel meshes with the mating rack, and shows on its pitch surface a succession of tooth traces consisting of helices e_2 having normal pitch $p_{n2} = p_{n0}$, sizing based on the normal module m_{n0}, and transverse pitch given by:

$$p_{t2} = \frac{p_{n2}}{\cos \beta_2};$$ (10.22)

if z_2 is its number of teeth, we will have:

$$p_{t2}z_2 = 2\pi r_2.$$ (10.23)

Since the pinion and the wheel so obtained are conjugate with the common mating rack, they will be conjugate between them. Figure 10.8 shows a crossed helical gear pair with $\Sigma < \pi/2$, in the projection on a plane normal to the shortest distance straight line. The ranks of tooth traces, consisting of right-hand helices e_1 and e_2, which appear in this figure, are those that are found on the front halves of the cylinders having radii r_1 and r_2, for which the point C of contact between the two cylinders and the straight line, n, that generates these helices, are hidden to the observer.

10.4 Fundamental Kinematic Properties

Let us denote, as usual, with r_{b1} and r_{b2} the radii of the base cylinders, and with β_{b1} and β_{b2} the base helix angles, and show first the following properties, quite unique, of the involute helical gears with crossed axes. One of these properties consists in the fact that, by varying in any way the relative position between the transmission

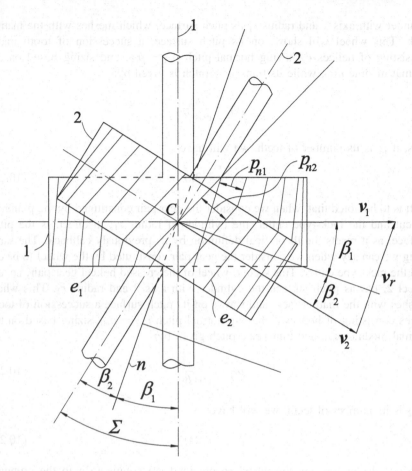

Fig. 10.8 Crossed helical gear pair with $\Sigma < \pi/2$

axes 1 and 2, and thus causing them to translate or rotate relative to one another, the transmission ratio, i, remains constant and is equal to (see Giovannozzi [9], Herrmann [11], Schiebel [21]):

$$i = \frac{\omega_1}{\omega_2} = \frac{r_{b2} \cos \beta_{b2}}{r_{b1} \cos \beta_{b1}} = \frac{r_{b2} \sin \gamma_{b2}}{r_{b1} \sin \gamma_{b1}}. \tag{10.24}$$

Obviously, this property is valid within the limits allowed by the tooth depth, so that the mating teeth can touch each other. In this framework, we consider any point of contact, P, between the active tooth flank surfaces of the two wheels, and the two planes tangent to the base cylinders passing through point P (to be exact, one of the two possible planes tangent to each base cylinder, the one that, by wrapping on the base cylinder, generated the tooth surface). As we saw in Chap. 8, both of these planes contain the common normal to the tooth surfaces through point P, and therefore this normal cannot be anything other than the intersection of the two aforementioned planes.

The transmission ratio is defined by the equality of the components v_{1Pn} and v_{2Pn} of the velocity vectors $\boldsymbol{v}_{1P}$ and $\boldsymbol{v}_{2P}$ at point P in the direction of the common normal. It is to be noted that the third subscript, n, indicates the direction of this normal. In the plane of action passing through the point P and tangent to the base cylinder having radius r_{b1} (see Fig. 10.9, where this plane is revolved in the relevant view), the component of the velocity at the point P, supposed belonging to the corresponding wheel, is equal to $v_{1P} = r_{b1}\omega_1 = v_{2P} = r_{b2}\omega_2$ and it is normal to the generatrix of contact of the plane considered with the base cylinder.

Recalling that the normal, n, to the helicoid lies in the same plane and is normal to the generation straight line, which is inclined by the helix angle, β_b, with respect to the axis of the wheel under consideration, for the first and second wheel we can write, respectively, the following relationships:

$$v_{1Pn} = r_{b1}\omega_1 \cos \beta_{b1} \tag{10.25}$$

$$v_{2Pn} = r_{b2}\omega_2 \cos \beta_{b2}. \tag{10.26}$$

From the condition $v_{1Pn} = v_{2Pn}$ follows Eq. (10.24), which is therefore demonstrated, whatever the relative position of the axes of the two wheels, since the reasoning applies whatever the point P, and a point of contact is certainly possible to find, however the axes of the two wheels are arranged. Of course, in relation to the radii of the base cylinders and tooth depth, there are limits to the displacements of the axes, on which it is not the case here to dwell upon.

For a given position of the axes of the two wheels, the locus of the points of contact is a straight line (the *line of action*), and precisely the common normal to the tooth surfaces, in their point of contact. If the wheels have constant angular velocities, the instantaneous point of contact moves, on this line of action, with a constant velocity. This line of action is tangent to the base cylinders as well as to the base helices of the two wheels. Of course, we can have two different lines of action, each of which corresponds to the meshing of the respective sides of the tooth surfaces. In the purely theoretical case of an involute crossed helical gear pair without errors, and with infinitely high tooth stiffness, these two lines of action intersect each other at a point belonging to the shortest center distance, a. The design of an involute helical gear pair with crossed axes that satisfies these theoretical conditions is called *canonical design*, and the gear is called *canonical crossed helical gear*.

This theoretical behavior is immediately evident in that, for the common mating rack of the two wheels, the planes of contact, or *planes of action*, related to the two rack-wheel pairs are the planes passing through the point of contact P and tangent to the base cylinders. Moreover, the intersection of these planes is the common

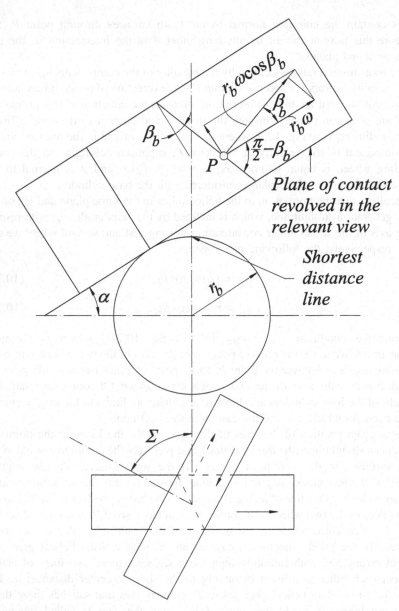

Fig. 10.9 Plane of contact revolved in the relevant view, and components of the velocity of the point P

normal to the two helicoids at point P. If we disregard friction, for a given position of the axes of the two wheels, the line of application of the force that the teeth transmit remains therefore in a fixed position in three-dimensional space, and it is the same line of action.

Given the property expressed by Eq. (10.24), and contrary to what is sometimes affirmed, the kinematics of the crossed helical gear pairs is always theoretically correct, even if the relative displacements between the axes occur. The mounting tolerances, from this point of view, can be very large. A translation without rotation of the axes of the wheels does move in three-dimensional space, parallel to itself, the line of action. A rotation without translation of the same axes instead does generally translate and rotate the line of action.

Since the contact is a point contact, the crossed helical gears are more insensitive to small errors in the helix angles than the parallel cylindrical helical gears. Instead, they are sensitive to errors in center distance as, in this case, the sum of the helix angles on the pitch cylinders is not equal to the shaft angle. In addition, the wheels of a crossed helical gear can be moved in the axial direction without affecting the tooth meshing, provided the face width is sufficient. Finally, it is to be noted that the crossed helical gears, such as the parallel cylindrical helical gears, have a very good and silent operation, and assure a remarkable regularity of the transmission of motion. In addition, because they allow a wide possibility of choice of the radii of their pitch cylinders for the same transmission ratio, it is possible to realize wheels having limited overall dimensions.

Unfortunately, these theoretical kinematic conditions are never met in practical applications. It is therefore necessary to deal with the misalignment problems, always present. When the rules of canonical design are not satisfied, the two above-mentioned lines of action do not intersect each other, but they are crossed lines, even if each of them is still a tangent to both base cylinders and base helices. The crossing of the two possible lines of action is the result of an error $\Delta \Sigma$ of the nominal value of the shaft angle, Σ, or the result of an error Δa of center distance, a. The combined effects of these two errors are included in the term *misalignment*.

Errors of alignment, i.e. the changes of shaft angle and center distance, cause the shift of the line of action. Depending on the amount of this shift, various effects may occur, the main of which are:

- Shift of the bearing contact area far from the mid-face width of the two members of the crossed helical gear pair.
- Risk of edge contact, when this shift exceeds a threshold value; if sufficient face width of gears is not provided, edge contact is inevitable.
- High levels of vibrations and noise, increasing with the increase of this shift.

Despite the studies and research on this subject, controlling the misalignment of involute helical gear pairs with crossed axes still constitutes a concern of designers and manufacturers. To correct the defects arising from misalignment, some modifications of the gear geometry have been proposed, such as:

- Modification of the helix angle of the pinion, which, however, weighs down the manufacturing costs, since it involves the regrinding.
- Teeth of the two members of the gear pair characterized by tip relief and end relief, determined according to the experience of manufacturers.
- Use of nonstandard crossed helical gears, i.e. with profile-shifted toothing.

More recently, Litvin and Fuentes [13] have proposed the use of gear pairs, constituted of standard gear wheels to be coupled to pinions characterized by profile-crowning tooth surfaces. Subsequently, the same authors proposed the use of pinions having tooth surfaces characterized not only by profile crowning, but also by lengthwise crowning.

Here we do not dwell on this subject, even if it is very interesting. In this regard, we refer the reader to more specialized textbooks. The sections that follow therefore concern only the canonical design of involute helical gears with crossed axes. In any case, it should be noted the fact that the effects caused by errors $\Delta\Sigma$ and Δa, in terms of shift of the line of action, increase almost exponentially with decreasing shaft angle. From this point of view, the involute cylindrical helical gears with parallel axes are obviously those more sensitive to the above-mentioned errors.

10.5   Determination of Other Characteristic Quantities of the Crossed Helical Gear Pairs

We assume the canonical design conditions, for which the line of action intersects the nominal center distance $O_1 O_2$ between the axes of the two wheels, as shown in Fig. 10.10. This assumption would not be necessary, and in any case it does not affect the equations already written and those we will write below. In practice, however, the canonical determination of the characteristic quantities of the crossed helical gear pairs is performed under the assumption that, on the nominal center distance $O_1 O_2$, a point of contact C exists. In this case, the helices belonging to the tooth flank surfaces and passing through point C are the pitch helices, and they are certainly tangent to each other at point C.

Let us consider two external cylindrical helical wheels with z_1 and z_2 teeth, having toothing with the same normal module m_n, equal nominal pressure angle α_n of the normal profiles, pitch helix angles β_1 and β_2, and pitch helices of the same direction. These wheels can work properly when they mesh mounted on crossed axes, with shaft angle Σ equal to the sum $(\beta_1 + \beta_2)$ of the pitch helix angles, if the conditions described in Sects. 10.3 and 10.4 are met. The relationship that correlates the pitch angles β_1 and β_2, and the shaft angle Σ is Eq. (10.16), taken with the plus sign.

In the normal section passing through point C, which is perpendicular to the pitch helices, the normal pitch must be, with reference to the common mating rack, the same for the two wheels, for which we can write:

$$p_n = p_{n1} = p_{n2} = p_{n0} = p_{t1} \cos \beta_1 = p_{t2} \cos \beta_2. \tag{10.27}$$

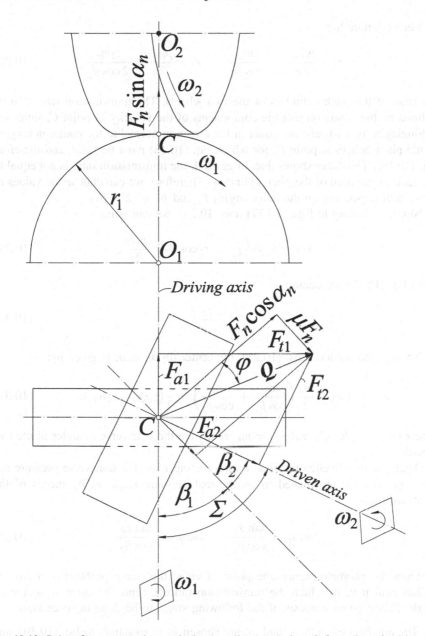

Fig. 10.10 Point of contact on the nominal center distance, and tooth force F resolved into its three components, F_t, F_a, and F_r

Let us denote by:

$$r_1 = \frac{p_{t1}z_1}{2\pi} = \frac{z_1 m_n}{2\cos\beta_1} \qquad r_2 = \frac{p_{t2}z_2}{2\pi} = \frac{z_2 m_n}{2\cos\beta_2} \qquad (10.28)$$

the radii of the pitch cylinders of the two wheels. The transmission ratio will be defined by the condition that the components of the velocity of point C, supposed belonging to two wheels, are equal in the direction normal to the common tangent to the pitch helices at point C, for which Eq. (10.18) must be valid, and therefore Eq. (10.19). This latter shows that, in general, the transmission ratio is not equal to the ratio of the radii of the pitch cylinders. Therefore, we can choose the values of these radii depending on the helix angles β_1 and $\beta_2 = (\Sigma - \beta_1)$.

Since, according to Eqs. (10.27) and (10.28), we can write

$$r_1\cos\beta_1 = \frac{p_n z_1}{2\pi} \qquad r_2\cos\beta_2 = \frac{p_n z_2}{2\pi}, \qquad (10.29)$$

from Eq. (10.19) we obtain:

$$i = \frac{z_2}{z_1}. \qquad (10.30)$$

Taking into account Eqs. (10.28), the center distance, a, is given by:

$$a = r_1 + r_2 = \frac{m_n}{2}\left(\frac{z_1}{\cos\beta_1} + \frac{z_2}{\cos\beta_2}\right) = \frac{1}{2}(z_1 m_{t1} + z_2 m_{t2}), \qquad (10.31)$$

where $m_{t1} = m_n/\cos\beta_1$ and $m_{t2} = m_n/\cos\beta_2$ are the transverse modules of the two wheels.

Again, with reference to the common mating rack, the transverse pressure angles, α_{t1} and α_{t2}, are related to the normal pressure angle α_n by means of the relationships:

$$\tan\alpha_{t1} = \frac{\tan\alpha_n}{\cos\beta_1} \qquad \tan\alpha_{t2} = \frac{\tan\alpha_n}{\cos\beta_2}. \qquad (10.32)$$

From the geometric-kinematic point of view, the sizing problem of a crossed helical gear pair, of which the transmission ratio, i, center distance, a, and shaft angle, Σ, are given, consists of the following steps, to be done in succession:

- The numbers of teeth, z_1 and z_2, are chosen so as to satisfy to Eq. (10.30), and the radii, r_1 and r_2, of the two pitch cylinders are chosen so as to meet Eq. (10.31).
- According to the system of two algebraic equations in two unknowns, constituted by Eqs. (10.19) and (10.16), we calculate β_1 and β_2.
- From Eqs. (10.28), we deduce p_{t1} and p_{t2}.
- From Eqs. (10.27), we obtain $p_{n1} = p_{n2}$.

- Finally, once we have chosen α_n, from Eq. (10.32) we get α_{t1} and α_{t2}.

Of course, this geometric-kinematic calculation must be followed and supplemented by the strength calculations, and load carrying capacity verifications. These calculations and verifications may induce the designer to vary the module, m_n, and then the center distance, a. Obviously, during the course of design calculations, seen in their general framework together, it is possible that the designer encounters other problems that can be solved by using appropriately the afore-mentioned relationships.

Finally, the following correlations allow us to calculate the quantities relating to the base cylinders, once those relating to the pitch cylinders are known:

$$r_{b1} = r_1 \cos \alpha_{t1} \qquad r_{b2} = r_2 \cos \alpha_{t2} \tag{10.33}$$

$$\tan \beta_{b1} = \tan \beta_1 \cos \alpha_{t1} \qquad \tan \beta_{b2} = \tan \beta_2 \cos \alpha_{t2}. \tag{10.34}$$

Now we focus our attention on Eqs. (10.19) and (10.30), as well as on the following relationship:

$$\frac{p_{t1}}{p_{t2}} = \frac{r_1 z_2}{r_2 z_1} = \frac{\cos \beta_2}{\cos \beta_1}, \tag{10.35}$$

which is obtained dividing member to member the two Eqs. (10.28).

Equation (10.19) shows that, for the crossed helical gears, the transmission ratio, which is an input datum, can be realized in infinite ways, by using two different procedures, namely:

- Arbitrarily choosing the values of r_1 and r_2, so as Eq. (10.17) is satisfied, and then calculating by Eqs. (10.19) and (10.16) the corresponding values of β_1 and β_2. With this procedure, the point of contact between the pseudo-pitch surfaces of the two wheels can be any of the infinite points of the center distance, $O_1 O_2$
- Arbitrarily choosing the values of the helix angles β_1 and β_2 that define the directions of the tooth traces, so as to satisfy Eq. (10.16), and then calculating by Eqs. (10.17) and (10.19) the corresponding values of r_1 and r_2. With this procedure, the axis of the tooth can be directed according to any one of the infinite straight lines outgoing from a point of the center distance, $O_1 O_2$.

This wide possibility of choice of the radii r_1 and r_2 of the wheels and of the helix angles β_1 and β_2 of the tooth traces is a characteristic property of the crossed helical gear pairs. It constitutes a considerable advantage in cases where, for small values of the transmission ratio, the solution realized with hyperboloid gear pairs would give rise to strong differences between the radii r_1^* and r_2^* of the throat cross sections of the two wheels. This drawback of the hyperboloid gear pairs, as also we have already shown in Sect. 10.2, can present so serious in the case where the crossed axes 1 and 2 have orthogonal directions ($\Sigma = \pi/2$). This is because, in this case, the ratio between the radii of throat cross sections is equal to i^2 (see Eq. (10.14)).

If, for transmitting motion between orthogonal crossed axes, we make use of crossed helical gear pairs, since $\beta_2 = [(\pi/2) - \beta_1]$, from Eq. (10.19) we obtain:

$$i = \frac{r_2}{r_1} \tan \beta_1;$$ (10.36)

so it is possible to make gear wheels having pseudo-pitch cylinders with radii r_1 and r_2 arbitrarily chosen (possibly the same), provided that the helix angles β_1 and $\beta_2 = [(\pi/2) - \beta_1]$ of the tooth traces meet Eq. (10.36).

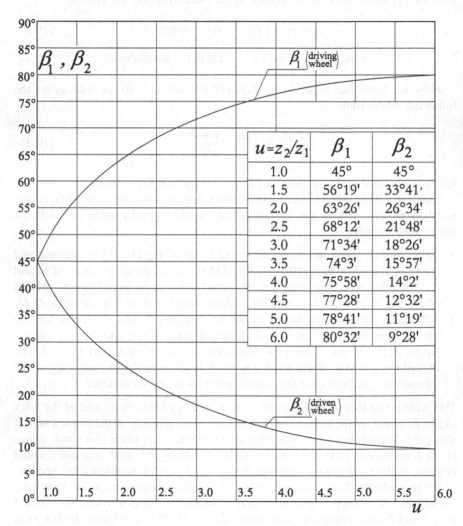

$u=z_2/z_1$	β_1	β_2
1.0	45°	45°
1.5	56°19'	33°41'
2.0	63°26'	26°34'
2.5	68°12'	21°48'
3.0	71°34'	18°26'
3.5	74°3'	15°57'
4.0	75°58'	14°2'
4.5	77°28'	12°32'
5.0	78°41'	11°19'
6.0	80°32'	9°28'

Fig. 10.11 Pitch helix angles β_1 and β_2 of crossed helical gears with shaft angle $\Sigma = \pi/2$, and $r_1 = r_2$, as a function of the gear ratio $u = z_2/z_1 = i$.

Figure 10.11 shows the change of the pitch helix angles of the driving and driven wheels (respectively, β_1 and β_2), for crossed helical gear pairs with $r_1 = r_2$ and shaft angle $\Sigma = \pi/2$, as a function of the gear ratio $u = z_2/z_1 = i$. It highlights the fact that, in the special case in which $u = i = 1$, the helix angles are equal between them and equal to $\pi/4$ ($\beta_1 = \beta_2 = \pi/4$). In this case, Eq. (10.36) becomes:

$$i = \frac{r_2}{r_1}. \tag{10.37}$$

This equation, which coincides with the fundamental relationship that expresses the transmission ratio of the parallel cylindrical gears and bevel gears, shows that only if $\beta_1 = \beta_2 = \pi/4$ the transmission ratio of the crossed helical gear pairs is equal to the ratio between the radii of the pseudo-pitch cylinders.

Example : The input data of this example, that represents a case arising frequently in practical applications, are the following: shaft angle $\Sigma = \pi/2$; transmission ratio $i = u = 2, 0$; $r_1 = r_2$. Results: from Eq. (10.36), we obtain $\cot \beta_1 = 0.5$; therefore, we have $\beta_1 = 63°26'$ and $\beta_2 = [(\pi/2) - \beta_1] = 26°34'$.

10.6   Path of Contact, Face Width, and Contact Ratio

Let us examine the special case of a crossed helical gear pair with shaft angle $\Sigma = 90°$, assuming that toothing is not characterized by any undercut, including tip rounding or tip and root relief. The following discussion, however, is quite general, but here we consider the special case of $\Sigma = 90°$, because, as we have said elsewhere, it is that we meet more frequently in practical applications. It also lends itself better to the orthogonal projection representations, which show the genesis of generally applicable relationships that we will write.

The tip cylinders of the two wheels intersect the line of action, defined in Sect. 10.4, in the usual marked points A and E. Thus, the length $g = AE$ of *path of contact* is uniquely defined. For the determination of the quantities of interest here, it is convenient to consider what happens in three orthogonal projections: the first two, on planes perpendicular to the wheel axes, and the third on the common plane tangent to the two pitch cylinders at the point of contact C. This last plane is a plane perpendicular to the shortest distance straight line (Fig. 10.12).

The path of contact, AE, is projected, on the two planes perpendicular to the wheel axes (these are two axial views), according to the straight lines $A'E'$ and $A''E''$, which are inclined by α_{t1} and α_{t2} with respect to the plane tangent at point C to the pitch cylinders of radii r_1 and r_2. The straight lines $A'E'$ and $A''E''$ are the intersections of the planes of action between each of the two wheels and the common mating rack. The same path of contact is projected, on the plane tangent at

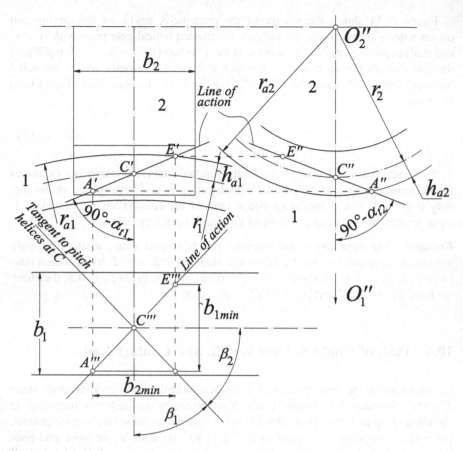

Fig. 10.12 Path of contact of a crossed helical gear with $\Sigma = 90°$, projected on three orthogonal planes

point C to the pitch cylinders, according to the straight line $A'''E'''$, which is normal to the common tangent to the pitch helices at point C.

In the axial view of the first wheel (the wheel 1), the point E' is identified, while in the axial view of the second wheel (the wheel 2), the point A'' is identified. The other two projections E'' and E''' of the point E are obtained by intersecting the other two projections of the line of action with straight lines parallel to the axes of the two wheels, and passing through point E', such as Fig. 10.12 shows. The other two projections A' and A''' of point A are obtained in a similar manner, once the point A'' is identified. The projection $A'''E'''$ of the path of contact on the plane tangent to the pitch cylinders defines the minimum face widths b_{1min} and b_{2min}, i.e. the usable face widths, of the two wheels (Fig. 10.12).

Generalizing the problem to the case of shaft angle $\Sigma \neq 90°$, and also considering the two planes of action (one of these planes is shown, revolved in the relevant view, in Fig. 10.9) and the two base cylinders, still remaining the meaning of the symbols already introduced in the previous chapters (see Figs. 1.8 and 2.10), we can write the following relationships:

$$g = AE = g_{an1} + g_{an2} = g_{fn2} + g_{fn1} \tag{10.38}$$

$$g_{fn2} = g_{an1} = \frac{g_{at1}}{\cos \beta_{b1}} = \frac{\left[\sqrt{d_{a1}^2 - d_{b1}^2} - \sqrt{d_1^2 - d_{b1}^2}\right]}{2 \cos \beta_{b1}} = CE \tag{10.39}$$

$$g_{fn1} = g_{an2} = \frac{g_{at2}}{\cos \beta_{b2}} = \frac{\left[\sqrt{d_{a2}^2 - d_{b2}^2} - \sqrt{d_2^2 - d_{b2}^2}\right]}{2 \cos \beta_{b2}} = AC, \tag{10.40}$$

where g_{an1} and g_{an2} are respectively the lengths of the paths of recess and approach in the normal section, while $g_{at1} = C'E'$ and $g_{at2} = C''A''$ are respectively the lengths of the paths of recess and approach in the transverse sections.

For determining the minimum face widths of the two wheels, it is more convenient to project the path of contact in the plane of action of the first wheel, on the generatrix of the corresponding base cylinder, and the same path of contact in the plane of action of the second wheel, on the generatrix of the corresponding base cylinder. In doing so, we obtain the following relationships (see Fig. 10.12 below):

$$b_{1min} = g \sin \beta_{b1} \qquad b_{2min} = g \sin \beta_{b2}. \tag{10.41}$$

It should be noted that the parts of toothing that, in the axial direction, lie outside of the range of minimum face width thus defined do not participate to meshing, and they are therefore unnecessary. If we then consider the reference mating rack, which is characterized by an infinite number of teeth, we obtain:

$$b_{1min} = \frac{(h_{a1} + h_{a2})}{\tan \alpha_n} \cos \beta_{b1} \qquad b_{2min} = \frac{(h_{a1} + h_{a2})}{\tan \alpha_n} \cos \beta_{b2}. \tag{10.42}$$

For $(h_{a1} + h_{a2}) = 2m_n$ and $\alpha_n = 20°$, i.e. with teeth having nominal depth, and zero shifted toothing profile or profile-shifted toothing without variation of center distance, we have:

$$b_{1min} \cong 5.5 m_n \sin \beta_{b1} \qquad b_{2min} \cong 5.5 m_n \sin \beta_{b2}. \tag{10.43}$$

Since these values are already valid for wheels having an infinite number of teeth, to take account of assembly tolerances it is sufficient to increase them, by adding $(1 \div 2)m_n$. In order to have teeth more resistant, as well as to avoid the risk of edge contact, the inequality $b \geq 6m_n$ must however always be satisfied.

Finally, the contact ratio ε_n in the normal section is equal to the quotient of the length of path of contact, AE, divided by the normal base pitch, p_{bn}, for which we have:

$$\varepsilon_n = \frac{AE}{p_{bn}} = \frac{AE}{\pi m_n \cos \alpha_n} = \varepsilon_{n1} + \varepsilon_{n2}, \tag{10.44}$$

where ε_{n1} and ε_{n2} are respectively the addendum contact ratios of the pinion and wheel, in the normal section, given by the relationships:

$$\varepsilon_{n1} = \frac{CE}{p_{bn}} = \frac{g_{an1}}{p_{bn}} \qquad \varepsilon_{n2} = \frac{AC}{p_{bn}} = \frac{g_{an2}}{p_{bn}}. \tag{10.45}$$

10.7   Longitudinal Sliding and Sliding Velocity

The contact between the teeth of the two mating members of a canonical crossed helical gear, as we have already said, is a point contact, and, during the transmission of motion, a considerable *helical sliding* occurs between the teeth, which overlaps with the *profile sliding* along the tooth depth. This helical sliding, also called *lengthwise cross-sliding*, or *longitudinal* or *lengthwise sliding*, or *axial sliding*, is evident if we consider the mode of transmission of motion from the driving wheel 1 to the driven wheel 2 of the crossed helical gear shown in Fig. 10.7.

The driving wheel 1, rotating about its own axis, causes a displacement of the mating rack (this is materialized by its pitch plane, the Π-plane) in the direction normal to the axis 1. Suppose that this displacement, referred to the unit time, is represented by vector $\overline{CA}$. This vector can be resolved into two component vectors: the first, $\overline{AB}$, is the displacement of the pitch plane of the mating rack in the direction of the straight line, n; the second, $\overline{CB}$, is the displacement of the same pitch plane in the direction normal to the axis 2. This second displacement causes the rotation of the driven wheel 2 about its own axis. It thus becomes clear that the transmission of motion from the driving wheel 1 to the driven wheel 2 is accompanied by a sliding in the direction of the tooth traces, which are helical traces. Therefore, this sliding is known as *helical sliding*.

This type of transmission of motion shows a phenomenon similar to the one described in Sect. 10.2, with reference to the mating hyperboloid gear pairs. However, we are facing a substantial difference: in the present case, in fact, the sliding no longer takes place along the Mozzi's axis, but along the straight line, n, chosen according to the criterion shown in the Sect. 10.3, and in any case distinct from the Mozzi's axis.

With reference to the sliding velocity, let us consider the points of contact C that fall on the center distance $O_1 O_2$ (Fig. 10.7), i.e. the points of contact between the pseudo-pitch surfaces. Once the choice of the straight line, n, is made, the *relative velocity vector* $v_r = (v_1 - v_2)$ has the direction of this straight line, while its absolute value is given by (see Fig. 10.6):

$$v_r = \omega_1 r_1 \sin \beta_1 + \omega_2 r_2 \sin \beta_2 = \omega_1 r_1 (\sin \beta_1 + \cos \beta_1 \tan \beta_2). \tag{10.46}$$

The second equality of this equation follows from the fact that $v_1 = \omega_1 r_1$, $v_2 = \omega_2 r_2$, and $\omega_2 r_2 = \omega_1 r_1 (\cos \beta_1 / \cos \beta_2)$ according to Eq. (10.19). For $\Sigma = 90°$, which is the most frequent case of practical application, this equation becomes:

$$v_r = \frac{\omega_1 r_1}{\sin \beta_1}. \tag{10.47}$$

Equation (10.46) gives the *sliding velocity* of any point C between O_1 and O_2. It is certainly greater than that of points belonging to the Mozzi's axis, which characterizes the relative motion of a wheel with respect to the other. This factual situation is fully justified by the following considerations. The *relative velocity*, v_r, is a function of r_1 and r_2, and therefore also of β_1 and β_2, as Eqs. (10.19) and (10.46) show. The condition for this relative velocity to be a minimum is obtained by equating to zero the differential of Eq. (10.46), that is by imposing:

$$dv_r = \omega_1 \sin \beta_1 dr_1 + \omega_1 r_1 \cos \beta_1 d\beta_1 + \omega_2 \sin \beta_2 dr_2 + \omega_2 r_2 \cos \beta_2 d\beta_2 = 0. \tag{10.48}$$

Since the sums $(r_1 + r_2) = a$ and $(\beta_1 + \beta_2) = \Sigma$ are constant, we have:

$$dr_1 = -dr_2 \qquad d\beta_1 = -d\beta_2, \tag{10.49}$$

for which Eq. (10.48) becomes:

$$(\omega_1 \sin \beta_1 - \omega_2 \sin \beta_2) dr_1 + (\omega_1 r_1 \cos \beta_1 - \omega_2 r_2 \cos \beta_2) d\beta_2 = 0. \tag{10.50}$$

However, the helix angles β_1 and β_2 have been chosen by imposing the condition given by Eq. (10.18), for which the factor within the second-round brackets in Eq. (10.50) is equal to zero. Therefore, we get:

$$\omega_1 \sin \beta_1 = \omega_2 \sin \beta_2. \tag{10.51}$$

Dividing this last equation, member to member, by Eq. (10.18), we obtain finally:

$$\frac{\tan \beta_1}{r_1} = \frac{\tan \beta_2}{r_2} \tag{10.52}$$

and thus

$$\frac{r_1}{r_2} = \frac{\tan \beta_1}{\tan \beta_2}. \tag{10.53}$$

Therefore, the relative velocity reaches its minimum value when the distances r_1 and r_2 are such that their ratio equals the one between the trigonometric tangents of the angles β_1 and β_2, that is, when they coincide with the distances r_1^* and r_2^* which define the position of the Mozzi's axis.

If we look at the same phenomenon above analyzed from another point of view, we have the opportunity to emphasize even more the substantial difference between the hyperboloid gears and crossed helical gears (see Pollone [18], Schiebel and Lindner [22, 23]). To this end, let us examine Fig. 10.13. It shows that, for a hyperboloid gear pair with axes 1 and 2, shaft angle $\Sigma = (\beta_1^* + \beta_2^*)$, center distance $a = O_1 O_2$, and angular velocities ω_1 and ω_2, the Mozzi's axis, i.e. the axis of the instantaneous relative helical motion passes through a point O lying between the end points O_1 and O_2 of the center distance $O_1 O_2$. Since $(\omega_1/\omega_2) = (z_2/z_1)$, where z_1 and z_2 are the numbers of teeth of the two wheels, according to Eq. (10.1), the angle $\beta_1^* = (\Sigma - \beta_2^*)$ between the instantaneous axis of rotation and the axis 1 of the first wheel, can be expressed with the following relationship:

$$\tan \beta_1^* = \frac{(z_1/z_2) \sin \Sigma}{1 + (z_1/z_2) \cos \Sigma} = \frac{\sin \Sigma}{u + \cos \Sigma}. \qquad (10.54)$$

Mutatis mutandis, this relationship corresponds to Eq. (9.20) of straight bevel gears.

The ratio between the distances of the axis of the relative helical motion from the axes of the two wheels, measured along the center distance, is given by Eq. (10.5), which is written here in the form:

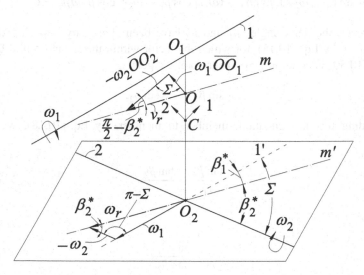

Fig. 10.13 Instantaneous relative helical motion of a hyperboloid gear with shaft angle Σ, and characteristic velocities: relative angular velocity vector, ω_r, and relative velocity vector, v_r

$$\frac{r_1^*}{r_2^*} = \frac{OO_1}{OO_2} = \frac{\tan\beta_1^*}{\tan\left(\Sigma - \beta_1^*\right)}. \tag{10.55}$$

Resolving the system formed by the two algebraic Eqs. (10.55) and (10.6), we obtain:

$$OO_1 = a\frac{\tan\beta_1^*}{\tan\beta_1^* + \tan\left(\Sigma - \beta_1^*\right)} \qquad OO_2 = a\frac{\tan\left(\Sigma - \beta_1^*\right)}{\tan\beta_1^* + \tan\left(\Sigma - \beta_1^*\right)}. \tag{10.56}$$

The instantaneous relative helical motion is defined by the relative angular velocity vector $\omega_{r1,2} = (\omega_1 - \omega_2)$ and relative velocity vector, $v_{r1,2} = (v_1 - v_2)$, along the axis of the helical motion. The absolute values of these vectors are given respectively by the relationships (see also Fig. 10.13):

$$\omega_r = \sqrt{\omega_1^2 + \omega_2^2 + 2\omega_1\omega_2\cos\Sigma} = \omega_1\sqrt{1 + \left(\frac{z_1}{z_2}\right)^2 + 2\frac{z_1}{z_2}\cos\Sigma} = \omega_1\frac{\sin\Sigma}{\sin\left(\Sigma - \beta_1^*\right)} \tag{10.57}$$

$$v_r = \sqrt{\omega_1^2\overline{OO_1}^2 + \omega_2^2\overline{OO_2}^2 - 2\omega_1\omega_2\overline{OO_1}\,\overline{OO_2}\cos\Sigma} = \omega_1\overline{OO_1}\frac{\sin\Sigma}{\cos\left(\Sigma - \beta_1^*\right)}$$

$$= \omega_1 a\frac{\tan\beta_1^*}{\tan\beta_1^* + \tan\left(\Sigma - \beta_1^*\right)}\frac{\sin\Sigma}{\cos\left(\Sigma - \beta_1^*\right)}. \tag{10.58}$$

Now, again with reference to Fig. 10.13, we replace the hyperboloid gear pair with a crossed helical gear pair, having the same number of teeth, z_1 and z_2, normal module, m_n, and helix angles, β_1 and β_2. The radii of the pseudo-pitch cylinders of the two members of this gear pair, which are tangent at point C, will be given by the following relationship:

$$r_1 = CO_1 = \frac{z_1 m_n}{2\cos\beta_1} \qquad r_2 = CO_2 = \frac{z_2 m_n}{2\cos\beta_2}. \tag{10.59}$$

The absolute value of the relative velocity (or sliding velocity) v_r' can be written as:

$$v_r' = \sqrt{\omega_r^2\left(\overline{CO_1} - \overline{O_1O_2}\right)^2 + v_r^2}. \tag{10.60}$$

From this relationship, we infer that the minimum value of the sliding velocity is reached when

$$CO_1 = \frac{z_1 m_n}{2 \cos \beta_1} = OO_1, \qquad (10.61)$$

that is when $\beta_1 = \beta_1^*$, i.e. when the radii of the pitch cylinders are respectively equal to OO_1 and OO_2. This solution, corresponding to the minimum sliding, is feasible only when the transmission ratio is slightly different from the unit. For greater transmission ratios, the diameter of the pinion would be too small, as well as its helix angle, as we can see from Eqs. (10.1) and (10.54).

It is to be remembered that the lengthwise sliding overlaps to the profile sliding, already seen in parallel cylindrical gears (see Sect. 3.6). Generally, however, especially when the shaft angle is large, the profile sliding is relatively low compared to the lengthwise sliding. The *total sliding velocity vector*, $v_{g\gamma}$, is the vector sum of two component vectors: the *profile sliding velocity vector*, $v_{g\alpha}$, in the direction of the tooth depth, and the *helical sliding velocity vector*, $v_{g\beta}$, in the direction of the tooth traces. Therefore, we have: $v_{g\gamma} = \left(v_{g\alpha} + v_{g\beta} \right)$.

This total sliding velocity is calculated at the tip cylinder of the wheel 1, because here it has a maximum value. Of course, the calculation of the total sliding velocity of the wheel 2 is still performed with the same procedure. Since the vectors $v_{g\alpha}$ and $v_{g\beta}$ are orthogonal vectors, the absolute value of the vector $v_{g\gamma 1}$ of the wheel 1 will be given by the relationship

$$v_{g\gamma 1} = \sqrt{v_{g\alpha 1}^2 + v_{g\beta 1}^2}, \qquad (10.62)$$

and must be calculated with reference to its tip cylinder.

In this regard, it is to be observed that the profile sliding velocity vector, $v_{g\alpha}$, along the tooth depth, during the meshing cycle varies continuously, as Eq. (3.66) shows, changing its sign (see Fig. 3.7) and increasing, in absolute value, linearly with the distance g_p of the point P under consideration from the pitch surface. Therefore, it assumes its maximum positive value at the marked point E, localized on the tip surface.

Instead, the helical sliding velocity vector, $v_{g\beta}$, whose absolute value is expressed by the relative velocity v_r, given by Eq. (10.46), is a constant during the meshing cycle. Consequently, the total sliding velocity vector, $v_{g\gamma}$, varies continuously during the meshing cycle, both in direction and in absolute value (note that one of the shaving process of the gears, the so-called *cross-axial shaving*, is precisely based on the above-mentioned change of direction of the total sliding velocity). Figure 10.14 shows the variation in intensity and direction of the total sliding velocity vector during the meshing cycle, that is with the change in position of the point of contact along the path of contact on the active tooth flank surface of one of the two members of the crossed helical gear pair under consideration.

Finally, it is to be emphasized that the absolute value of total sliding velocity vector $v_{g\gamma}$ constitutes a very significant quantity for security calculations against scuffing. For calculations of the power losses and efficiency, it is preferable to

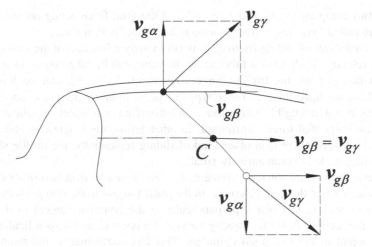

Fig. 10.14 Variation in intensity and direction of the total sliding velocity vector during the meshing cycle

introduce an average sliding velocity; its mean value can be determined, in a manner sufficiently approximated, by calculating separately the two average values of the two components, $v_{g\alpha}$ and $v_{g\beta}$, of the total sliding velocity, and then adding them together.

10.8   Load Analysis for Crossed Helical Gears and Thrust Characteristics on Shafts and Bearings

Here the load analysis for crossed helical gears is performed considering only the power losses due to lengthwise sliding, and neglecting, as we did for parallel cylindrical spur and helical gears and straight bevel gears, the power losses due to profile sliding, in the direction of the tooth depth.

In order to perform this analysis, for ease of calculation, which does not affect the reliability of the results obtained, we assume that the total force that the two members of the crossed helical gear pair transmits to each other passes through point C, where the two pitch cylinders touch one another. This force (Fig. 10.10) in the section normal to the helicoids that are touching at point C, is characterized by two components:

- A component, F_n, normal to the two surfaces in contact at point C, and inclined, with respect to the common plane tangent at point C to the two pitch cylinders, of the normal pressure angle, α_n.
- A friction component, $F_\mu = \mu F_n$, having the direction of the tangent to the pitch helices at point C, and oriented so as to counteract motion.

These two components, F_n and $F_\mu = \mu F_n$, of the total force acting on the mating teeth are called here, respectively, normal force and friction force.

The coefficient of sliding friction, μ, is notoriously a function of the material and sliding velocity. With a good lubrication, its value can be taken equal to 0.1. In a crossed helical gear, the friction force component, $F_\mu = \mu F_n$, can no longer be ignored, as we have done in other types of gears hitherto analyzed, since it has generally a not negligible magnitude, and therefore can affect significantly the transverse tangential force transmitted. In other terms, the lengthwise sliding has more pronounced effect than other kind of sliding (essentially, the profile sliding), whose magnitude is comparatively small.

In turn, the normal force component, F_n, can be resolved in two other components (Fig. 10.10): the first, $F_n \cos \alpha_n$, in the plane tangent to the two pitch cylinders at point C, and, in this plane, perpendicular to the common tangent to the pitch helices; the second, $F_n \sin \alpha_n$, passing through the point C, and perpendicular to the plane tangent to the two pitch cylinders. This last component is the radial component, F_r, of the total force, acting on both the wheels. In the same plane tangent to the two pitch cylinders, vectorially adding the other component $F_n \cos \alpha_n$ of the normal force, and the friction force, $F_\mu = \mu F_n$, we obtain the vector Q applied at point C (Fig. 10.10), and inclined, with respect to the component $F_n \cos \alpha_n$, of the angle φ, which is the angle of friction related to μ by the relationship $\mu = \tan \varphi$.

Finally, we resolve the vector Q according two different procedures:

- the first time, in the direction of the tangent to the pitch cylinder, in the middle cross section of the first wheel, passing through the point C, and in the direction of its axis;
- the second time, in the direction of the tangent of the pitch cylinder, in the middle cross section of the second wheel, passing through the point C, and in the direction of its axis.

By doing so, we obtain the other components of the total force, i.e. the transverse tangential forces F_{t1} and F_{t2}, and the axial forces F_{a1} and F_{a2} acting on the two wheels. All these components of the total force are given by:

$$F_{r1} = F_{r2} = F_n \sin \alpha_n \tag{10.63}$$

$$F_{t1} = F_n(\cos \alpha_n \cos \beta_1 + \mu \sin \beta_1) \tag{10.64}$$

$$F_{t2} = F_n(\cos \alpha_n \cos \beta_2 - \mu \sin \beta_2) \tag{10.65}$$

$$F_{a1} = F_n(\cos \alpha_n \sin \beta_1 - \mu \cos \beta_1) \tag{10.66}$$

$$F_{a2} = F_n(\cos \alpha_n \sin \beta_2 + \mu \cos \beta_2). \tag{10.67}$$

The transverse tangential force, F_{t2}, acting on the driven wheel (we assume that the wheel 1 is the driving wheel, and the wheel 2 is the driven wheel), is known. It is equal to the ratio between the *input torque*, T_2, to the driven machine (note that the load capacity rating is effectively based on this torque), and the radius r_2 of the corresponding pitch cylinder $(F_{t2} = T_2/r_2)$. For this reason, it is best to express the aforementioned force components as a function of F_{t2}. Therefore, we have:

$$F_{r1} = F_{r2} = F_{t2} \frac{\sin\alpha_n}{\cos\alpha_n \cos\beta_2 - \mu\sin\beta_2} \cong F_{t2} \frac{\sin\alpha_n}{\cos(\beta_2 + \varphi)} \tag{10.68}$$

$$F_{t1} = F_{t2} \frac{\cos\alpha_n \cos\beta_1 + \mu\sin\beta_1}{\cos\alpha_n \cos\beta_2 - \mu\sin\beta_2} \cong F_{t2} \frac{\cos(\beta_1 - \varphi)}{\cos(\beta_2 + \varphi)} \tag{10.69}$$

$$F_{a1} = F_{t2} \frac{\cos\alpha_n \sin\beta_1 - \mu\cos\beta_1}{\cos\alpha_n \cos\beta_2 - \mu\sin\beta_2} \cong F_{t2} \frac{\sin(\beta_1 - \varphi)}{\cos(\beta_2 + \varphi)} \cong F_{t1}\tan(\beta_1 - \varphi) \tag{10.70}$$

$$F_{a2} = F_{t2} \frac{\cos\alpha_n \sin\beta_2 + \mu\cos\beta_2}{\cos\alpha_n \cos\beta_2 - \mu\sin\beta_2} \cong F_{t2} \frac{\sin(\beta_2 + \varphi)}{\cos(\beta_2 + \varphi)} \cong F_{t2}\tan(\beta_2 + \varphi). \tag{10.71}$$

It should be noted that the expression preceded by the symbol $\cong$ in the above formulas are obtained by taking $\cos\alpha_n \cong 1$. They are approximate relationships.

10.9 Efficiency of Crossed Helical Gears

The concepts of efficiency of the parallel cylindrical spur and helical gears, which we have discussed in Sects. 3.9 and 8.10, can be applied also to the crossed helical gears, with the necessary addition, as these gears are an even more general case. Here we dwell on these additions, which are peculiar of the crossed involute helical gears. The two helicoids, which form the flank surfaces of the mating teeth, are in contact, at a given instant of the meshing cycle, at a single point of the path of contact. In accordance with what we did in the previous section for calculation of the forces acting on the teeth, also to calculate the friction losses due to the contact, we limit our analysis to the case where this point of contact is the same point C, first considered. According to the assumption made in the previous section, this point belongs to the shortest distance straight line, O_1O_2.

As Cartesian coordinate systems for the two wheels that make up the crossed helical gear pair under consideration, we take the right-hand coordinate systems $O_1(x_1, y_1, z_1)$ and $O_2(x_2, y_2, z_2)$, shown in Fig. 10.15. The origins of these systems are the end points O_1 and O_2 of the center distance, a. Their z-axes coincide with the axes of the two wheels; their x-axes coincide with the center distance straight line, but are oriented from origin O_1 towards point C for the first wheel, and from the origin O_2 towards point C for the second wheel; their y-axes are arranged to complete the two right-hand coordinate systems. The distances of point C from the end points of the center distance are the radii of the two pitch cylinders, that is $CO_1 = r_1$ and $CO_2 = r_2$. As we have already shown, these radii are related to the radii of the base cylinders, r_{b1} and r_{b2}, and the transverse pressure angles, $\alpha_1 = \alpha_{t1}$

and $\alpha_2 = \alpha_{t2}$, by the relationships (10.33). Figure 10.15 also shows the tangential velocity vectors v_1 and v_2 of the pitch point C, supposed to be rigidly connected first to the wheel 1, and then to the wheel 2, and the projections of the common tangent, t, to the mating tooth surfaces, which are touching at point C.

Here we assume that the shaft angle, Σ, is less than the sum of the absolute values of the base lead angles, $\gamma_{b1} = [(\pi/2) - \beta_{b1}]$ and $\gamma_{b2} = [(\pi/2) - \beta_{b2}]$, of the two gear members under consideration (β_{b1} and β_{b2} are the base helix angles), for which the following inequality is valid:

$$\Sigma < |\gamma_{b1}| + |\gamma_{b2}|. \tag{10.72}$$

Under these conditions, the cones of normals have two generatrices in common (see [8]). However, the condition for which a point P is a point of contact involves the fact that two mating tooth surfaces must have at point P the same normal. It follows that the cones of normals must have a common generatrix. However, we have two common generatrices among the cones of normals. Therefore, there will also be two possible directions of the common normal of the mating tooth surfaces: one of them will be the active normal associated with one of the two possible directions of rotation of members of the gear; the other will be the active normal when the direction of rotation is reversed.

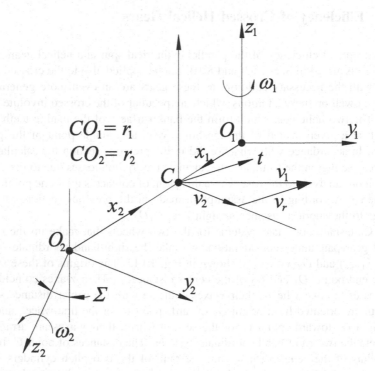

Fig. 10.15 Coordinate systems $O_1(x_1, y_1, z_1)$ and $O_2(x_2, y_2, z_2)$, pitch point C on the center distance, and velocity vectors, v_1 and v_2

Let us consider the Π-plane of Fig. 10.7, which is the plane perpendicular to the shortest distance straight line passing through point C as well as the common tangent plane to both pitch cylinders. As we have already seen, on this Π-plane, the total force exerted by the wheel 1 on the wheel 2 has two components: the component of the normal force, $F_n \cos \alpha_n$, and the friction force, $F_\mu = \mu F_n$, already defined in the previous section. These two components are shown in Fig. 10.16. The same figure shows the tangential velocity vectors, v_1 and v_2, the relative velocity vector, $v_r = (v_1 - v_2)$ of the wheel 1 with respect to the wheel 2, and the projections of the common tangent, t, and common normal, n, and projections of the z-axes and y-axes of the aforementioned coordinate systems $O_1(x_1, y_1, z_1)$ and $O_2(x_2, y_2, z_2)$, on Π-plane. Vectors v_1 and v_2 are inclined to the helix angles, β_1 and β_2, with respect to the application line of the component $F_n \cos \alpha_n$, while vector v_r has the same application line of the friction force, that is the aforementioned tangent, t, but acts in the opposite direction to counteract motion.

With reference to the coordinate systems above, the three vectors v_1, v_2, and v_r are given respectively by the following relationships (see Ferrari and Romiti [8], Spiegel et al. [25]):

$$v_1 = \omega_1 \times (C - O_1) = \omega_1 r_1 j_1 \tag{10.73}$$

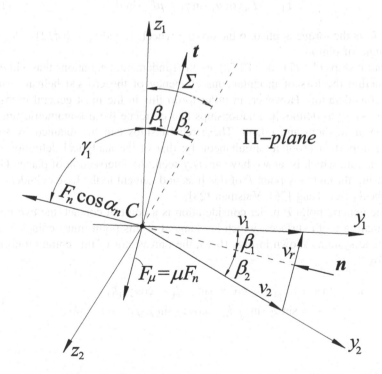

Fig. 10.16 Components on the Π-plane of the total force exerted by the wheel 1 on the wheel 2

$$v_2 = \omega_2 \times (C - O_2) = \omega_2 r_2 j_2 \tag{10.74}$$

$$v_r = v_1 - v_2 = \omega_1 \times (C - O_1) - \omega_2 \times (C - O_2) = \omega_1 r j_1 - \omega_2 r_2 j_2, \tag{10.75}$$

where j_1 and j_2 are the unit vectors of y_1-axis and y_2-axis.

Taking into account the relationships obtained in Sects. (3.9) and (8.10), in the case here examined, the energy lost by friction (or lost work) in the instant at which the contact is at point C, is given by:

$$\frac{dW_\mu}{dt} = \mu F_n v_r, \tag{10.76}$$

which, taking into account Eq. (10.46), can be expressed in the form:

$$\frac{dW_\mu}{dt} = \mu F_n \omega_1 r_1 \left(\sin \beta_1 + \frac{\cos \beta_1}{\cos \beta_2} \sin \beta_2 \right). \tag{10.77}$$

On the other hand, under steady-state conditions, for which ω_1 and ω_2 are constant, by means of an equilibrium equation of moments about the z_1-axis, we get the following relationship:

$$T_1 = (F_n \cos \alpha_1 \sin \gamma_1 + \mu F_n \sin \beta_1) r_1, \tag{10.78}$$

where T_1 is the torque applied to the driving wheel, 1, and $\gamma_1 = [(\pi/2) - \beta_1]$ is the lead angle of pinion.

Relationships (10.77) and (10.78) are the fundamental equations that will enable us to define the loss of instantaneous efficiency of the crossed helical gear pair under consideration. However, in order to do this in the most general terms, it is first necessary to define the relationships that correlate the trigonometric functions that appear in them and others. These relationships can be obtained by solving another important problem simultaneously: that of the analytical definition of the line of action, which is, as we have already seen, the intersection of planes Π_1 and Π_2 passing through any point P of this line, and tangent to the base cylinders of the two wheels (see Tang [26], Vaisman [28]).

If the generic point P under consideration is a point of contact, the two normals at the mating tooth surfaces, which are tangent at this point, must coincide. In the coordinate systems shown in Fig. 10.15, the unit vector of this common normal is given by:

$$\begin{aligned}
n &= \sin \alpha_1 \sin \gamma_{b1} i_1 - \cos \alpha_1 \sin \gamma_{b1} j_1 + \cos \gamma_{b1} k_1 \\
&= \sin \alpha_2 \sin \gamma_{b2} i_2 - \cos \alpha_2 \sin \gamma_{b2} j_2 - \cos \gamma_{b2} k_2,
\end{aligned} \tag{10.79}$$

where:

- i_1, j_1, k_1, and i_2, j_2, k_2 are the unit vectors of axes of the two coordinate systems;
- α_1 (it is measured in the direction of the positive rotations about the z_1-axis) is the angle between the plane (x_1, z_1) and the plane detected by the z_1-axis and generatrix of contact of plane Π_1 with the base cylinder of the wheel 1;
- α_2 (it is measured in the direction of the positive rotations about the z_2-axis) is the angle between the plane (x_2, z_2) and the plane detected by the z_2-axis and generatrix of contact of the plane Π_2 with the base cylinder of the wheel 2;
- $\gamma_{b1} = [(\pi/2) - \beta_{b1}]$ and $\gamma_{b2} = [(\pi/2) - \beta_{b2}]$.

Among the unit vectors of the two coordinate systems above, the following correlations exist:

$$i_1 = -i_2 \qquad j_1 = \cos \Sigma j_2 + \sin \Sigma k_2 \qquad k_1 = \sin \Sigma j_2 - \cos \Sigma k_2, \qquad (10.80)$$

where Σ is the acute angle between the z_1-axis and z_2-axis, i.e. the shaft angle.

Introducing these last relationships in Eq. (10.79), we get:

$$
\begin{aligned}
&- \sin \alpha_1 \sin \gamma_{b1} i_2 + (\cos \gamma_{b1} \sin \Sigma - \cos \alpha_1 \sin \gamma_{b1} \cos \Sigma) j_2 \\
&- (\cos \alpha_1 \sin \gamma_{b1} \sin \Sigma + cos\gamma_{b1} \cos \Sigma) k_2 \\
&= \sin \alpha_2 \sin \gamma_{b2} i_2 - \cos \alpha_2 \sin \gamma_{b2} j_2 - \cos \gamma_{b2} k_2.
\end{aligned}
\qquad (10.81)
$$

By equating the homogeneous terms in this equality, we get the following three equations:

$$
\begin{aligned}
- \sin \alpha_1 \sin \gamma_{b1} &= \sin \alpha_2 \sin \gamma_{b2} \\
\cos \gamma_{b1} \sin \Sigma - \cos \alpha_1 \sin \gamma_{b1} \cos \Sigma &= - \cos \alpha_2 \sin \gamma_{b2} \\
\cos \alpha_1 \sin \gamma_{b1} \sin \Sigma + \cos \gamma_{b1} \cos \Sigma &= \cos \gamma_{b2},
\end{aligned}
\qquad (10.82)
$$

of which only two are independent of each other.

From the last of the Eqs. (10.82), we get α_1, so we can write, in the coordinate system $O_1(x_1, y_1, z_1)$, the following equation of plane Π_1, which is tangent to the base cylinder of the wheel 1, to which the line of action to be defined belongs:

$$x_1 \cos \alpha_1 + y_1 \sin \alpha_1 = r_{b1}. \qquad (10.83)$$

Once α_1 is known, from the second of the Eqs. (10.82), we get α_2, so we can write, in the coordinate system $O_2(x_2, y_2, z_2)$, the following equation of the plane Π_2, which is tangent to the base cylinder of the wheel 2, to which the line of action to be defined also belongs:

$$x_2 \cos \alpha_2 + y_2 \sin \alpha_2 = r_{b2}. \qquad (10.84)$$

Since

$$x_2 = a - x_1 \qquad y_2 = y_1 \cos \Sigma + z_1 \sin \Sigma \qquad z_2 = y_1 \sin \Sigma - z_1 \cos \Sigma, \quad (10.85)$$

the equation of plane Π_2, written in the coordinate system $O_1(x_1, y_1, z_1)$, becomes:

$$-x_1 \cos \alpha_2 + y_1 \sin \alpha_2 \cos \Sigma + z_1 \sin \alpha_2 \sin \Sigma = r_{b2} - a + \alpha_2. \quad (10.86)$$

Equations (10.83) and (10.86) are the parametric equations of the intersection straight line between the two planes, Π_1 and Π_2; this line is the common normal between the mating tooth flank surfaces at points of contact. We have thus uniquely defined the line of action of the crossed helical gear under consideration. From Eq. (10.82), we infer the following relationships which define the angles α_1 and α_2 as a function of angles γ_{b1}, γ_{b2}, and Σ:

$$\cos \alpha_1 = \frac{\cos \gamma_{b2} - \cos \gamma_{b1} \cos \Sigma}{\sin \gamma_{b1} \sin \Sigma} \qquad \cos \alpha_2 = \frac{\cos \gamma_{b2} \cos \Sigma - \cos \gamma_{b1}}{\sin \gamma_{b2} \sin \Sigma}. \quad (10.87)$$

Now let us consider the case where the path of contact intersects the shortest distance straight line, that is the x_1-axis, at the pitch point C. At this point, we have $y_1 = y_2 = z_1 = z_2 = 0$, so from Eqs. (10.83) and (10.84), taking into account the first of the Eqs. (10.85), we get:

$$x_1 = \frac{r_{b1}}{\cos \alpha_1} = r_1 \qquad x_2 = \frac{r_{b2}}{\cos \alpha_2} = r_2 = a - r_1, \quad (10.88)$$

which coincide with Eqs. (10.33). So, the angles $\alpha_1 = \alpha_{t1}$ and $\alpha_2 = \alpha_{t2}$ are the transverse pressure angles of the two members of the crossed helical gear. Of course, in order to transmit the motion between these two members, it is necessary that Eq. (10.31) is satisfied.

Let us return to consider the Π-plane, shown in Fig. 10.16. In this figure, in addition to the quantities already defined above, also the projections of the common tangent to pitch helices, t, and the common normal, n, to the mating tooth surfaces, which are touching at point C, are highlighted. Of course, t and n are perpendicular to each other. From Eq. (10.79) in terms of coordinate system $O_1(x_1, y_1, z_1)$, it turns out that the normal, n, oriented as in the figure, has direction parameters relative to y_1- and z_1-axes, respectively equal to $-\cos \alpha_1 \sin \gamma_{b1}$ and $\cos \gamma_{b1}$, and therefore direction cosines given by:

$$\cos(n, y_1) = -\frac{\cos \alpha_1 \sin \gamma_{b1}}{\sqrt{\cos^2 \gamma_{b1} + \cos^2 \alpha_1 \sin^2 \gamma_{b1}}} = -\frac{\cos \alpha_1 \sin \gamma_{b1}}{\sqrt{1 - \sin^2 \alpha_1 \sin^2 \gamma_{b1}}}$$

$$\cos(n, z_1) = \frac{\cos \gamma_{b1}}{\sqrt{1 - \sin^2 \alpha_1 \sin^2 \gamma_{b1}}}. \quad (10.89)$$

Since, on the Π-plane (Fig. 10.16), β_1 and β_2 are the angles that the tangent, t, forms respectively with the z_1-axis and the $(-z_2)$-axis, we will have: $(n, y_1) = (\pi - \beta_1)$, $\quad (n, y_2) = (\pi - \beta_2)$, $\quad (n, z_1) = [(\pi/2) - \beta_1]$ $\quad$ and $(n, z_2) = [(\pi/2) - \beta_2]$. From the second and first Eqs. (10.89), we get:

$$\sin \beta_1 = \frac{\cos \gamma_{b1}}{\sqrt{1 - \sin^2 \alpha_1 \sin^2 \gamma_{b1}}} \qquad \cos \beta_1 = \frac{\cos \alpha_1 \sin \gamma_{b1}}{\sqrt{1 - \sin^2 \alpha_1 \sin^2 \gamma_{b1}}}. \tag{10.90}$$

Of course, we can write the two similar following relationships for the wheel 2:

$$\sin \beta_2 = \frac{\cos \gamma_{b2}}{\sqrt{1 - \sin^2 \alpha_2 \sin^2 \gamma_{b2}}} = \frac{\cos \gamma_{b2}}{\sqrt{1 - \sin^2 \alpha_1 \sin^2 \gamma_{b1}}}$$

$$\cos \beta_2 = \frac{\cos \alpha_2 \sin \gamma_{b2}}{\sqrt{1 - \sin^2 \alpha_2 \sin^2 \gamma_{b2}}} = \frac{\cos \alpha_2 \sin \gamma_{b2}}{\sqrt{1 - \sin^2 \alpha_1 \sin^2 \gamma_{b1}}}, \tag{10.91}$$

because, for the first of the Eqs. (10.82), we have $\sin^2 \alpha_2 \sin^2 \gamma_{b2} = \sin^2 \alpha_1 \sin^2 \gamma_{b1}$. From Eq. (10.78), taking into account the first of the Eqs. (10.90), we get:

$$F_n = \frac{T_1}{r_1 \left[\cos \alpha_1 \sin \gamma_1 + \mu \dfrac{\cos \gamma_1}{\sqrt{1 - \sin^2 \alpha_1 \sin^2 \gamma_1}} \right]}. \tag{10.92}$$

From Eq. (10.77), taking into account the Eqs. (10.91) and (10.92), we obtain:

$$\frac{dW_\mu}{dt} = \mu \omega_1 r_1 T_1 \frac{\dfrac{\cos \gamma_{b1}}{\sqrt{1 - \sin^2 \alpha_1 \sin^2 \gamma_{b1}}} + \dfrac{\cos \alpha_1 \sin \gamma_{b1}}{\cos \alpha_2 \sin \gamma_{b2}} \dfrac{\cos \gamma_{b2}}{\sqrt{1 - \sin^2 \alpha_1 sin^2 \gamma_{b1}}}}{\cos \alpha_1 \sin \gamma_{b1} + \mu \dfrac{\cos \gamma_{b1}}{\sqrt{1 - \sin^2 \alpha_1 \sin^2 \gamma_{b1}}}}. \tag{10.93}$$

Introducing to simplifying the notation

$$\mu_1 = \frac{\mu}{\sqrt{1 - \sin^2 \alpha_1 \sin^2 \gamma_1}}, \tag{10.94}$$

we can express the losses of instantaneous efficiency of the crossed helical gear pair here examined (they are obviously equal to the quotient of the instantaneous power dissipated, dW_μ/dt, divided by the instantaneous power supplied by the driving wheel, dW_m/dt) as follows:

$$1 - 1 - \eta_i = \frac{dW_\mu/dt}{dW_m/dt} = \frac{dW_\mu/dt}{T_1\omega_1} = \mu_1 \frac{\cos\gamma_{b1} + \dfrac{\cos\alpha_1 \sin\gamma_{b1}}{\cos\alpha_2 \sin\gamma_{b2}}\cos\gamma_{b2}}{\cos\alpha_1 \sin\gamma_{b1} + \mu_1 \cos\gamma_{b1}}. \qquad (10.95)$$

From this last relationship, with further notations

$$\mu_1^* = \frac{\mu_1}{\cos\alpha_1} \qquad \mu_2^* = \frac{\mu_1}{\cos\alpha_2}, \qquad\qquad (10.96)$$

we obtain the following equation that expresses the instantaneous efficiency:

$$\eta_i = \frac{\cos\alpha_1 \sin\gamma_{b1} - \mu_1 \dfrac{\cos\alpha_1 \sin\gamma_{b1}}{\cos\alpha_2 \sin\gamma_{b2}}\cos\gamma_{b2}}{\cos\alpha_1 \sin\gamma_{b1} + \mu_1 \cos\gamma_{b1}} = \frac{1 - \mu_2^* \cot\gamma_{b2}}{1 + \mu_1^* \cot\gamma_{b1}}. \qquad (10.97)$$

Relationships (10.96) clearly show that ratios $(\mu_1^*/\mu_1) = (1/\cos\alpha_1)$ and $(\mu_2^*/\mu_1) = (1/\cos\alpha_2)$ are to be considered as friction amplification factors due to geometry of the crossed helical gear pair.

From Eq. (10.87) we infer that, in the case of gear drives with orthogonal axes $(\Sigma = 90°)$, we have:

$$\cos\gamma_{b2} = \cos\alpha_1 \sin\gamma_{b1} \qquad \cos\gamma_{b1} = \cos\alpha_2 \sin\gamma_{b2}. \qquad (10.98)$$

From Eq. (10.24), taking into account these two last relationships as well as Eq. (8.8), applied to both gear members, we get:

$$i = \frac{r_2 \cos\alpha_2 \sin\gamma_{b2}}{r_1 \cos\alpha_1 \sin\gamma_{b1}} = \frac{r_2 \cos\gamma_{b1}}{r_1 \cos\gamma_{b2}} = \frac{\cos\gamma_{b1}}{\rho \cos\gamma_{b2}}, \qquad (10.99)$$

where $\rho = (r_1/r_2)$. Then, from Eqs. (10.99) and (10.98), we obtain:

$$\cot\gamma_{b2} = \frac{\cos\alpha_2}{i\rho} \qquad \cot\gamma_{b1} = i\rho \cos\alpha_1. \qquad (10.100)$$

Finally, introducing these two relationships into Eq. (10.97), and taking into account Eq. (10.96), the latter becomes:

$$\eta = \frac{1 - \mu_2^* \dfrac{\cos\alpha_2}{i\rho}}{1 + \mu_1^* i\rho \cos\alpha_1} = \frac{i\rho - \mu_1}{i\rho(1 + \mu_1 i\rho)}. \qquad (10.101)$$

This is the final relationship which expresses the efficiency of a crossed helical gear pair with shaft angle $\Sigma = 90°$. From this equation we first infer that, if $\rho \le (\mu_1/i)$, the transmission of motion under steady state conditions is not possible, as the efficiency becomes negative, so that $(dW_\mu/dt) > T_1\omega_1$, that is the power dissipated by friction is greater than the power available. For $\rho \to \infty$, also $\eta \to \infty$, while in the range $(\mu_1/i) \leq \rho \leq \infty$, the efficiency is always positive; thus, we infer that, for ρ-values within this range, the efficiency must have a maximum. Therefore, from condition $(d\eta/d\rho) = 0$, we get the following second degree algebraic equation:

$$i^2 \rho^2 - 2\mu_1 i\rho - 1 = 0; \qquad (10.102)$$

solving this last one, we obtain:

$$\rho = \frac{1}{i}\left(\mu_1 + \sqrt{1 + \mu_1^2}\right) \cong \frac{1}{i}(1 + \mu_1). \qquad (10.103)$$

If μ_1 is very small compared to the unit, for $\eta = \eta_{max}$, we have $\rho \cong (1/i)$. Under these conditions, from Eqs. (10.99) and (10.98) we obtain:

$$\rho i = \frac{\cos \gamma_{b1}}{\cos \gamma_{b2}} = \frac{\cos \alpha_2 \sin \gamma_{b2}}{\cos \alpha_1 \sin \gamma_{b1}} = 1; \qquad (10.104)$$

therefore, we have: $\gamma_{b1} = \gamma_{b2}$; $\alpha_1 = \alpha_2$; $\cot \gamma_{b1} = \cos \alpha_1$.

Bearing in mind Eq. (8.9), which correlates the base helix angle, β_b, the pressure angle, α, and the helix angle, β, we can write:

$$\cot \beta_1 = \frac{\cos \alpha_1}{\cos \gamma_{b1}} = 1. \qquad (10.105)$$

Therefore, we conclude that, for crossed helical gears with shaft angle $\Sigma = 90°$, and contact occurring on the shortest distance straight line, the frictional power losses are minimal when the ratio of the pitch radii of the two gear members, $\rho = (r_1/r_2)$ is about equal to $(1/i) = (\omega_2/\omega_1)$, so the helix angles are about equal, and equal to $\pi/4$.

In global terms, it is possible to express the efficiency of toothing of a crossed helical gear pair much more simply, as the ratio between the *input power* to the driven machine, or *usable power*, and the *output power* of the driving machine, or *engine power*. Therefore, taking into account Eq. (10.18) and the appropriate relationship (10.69), we can write the following equation, which is valid whatever the shaft angle:

$$\eta = \frac{F_{t2}r_2\omega_2}{F_{t1}r_1\omega_1} = \frac{\cos \beta_1 \cos(\beta_2 + \varphi)}{\cos \beta_2 \cos(\beta_1 - \varphi)} = \frac{1 - \mu \tan \beta_2}{1 + \mu \tan \beta_1}. \qquad (10.106)$$

If we make the derivative of this equation with respect to β_1, taking into account that $\beta_2 = (\Sigma - \beta_1)$, and equating to zero the derivative thus obtained, we find that, for a given value of the shaft angle Σ, the efficiency reaches a maximum for $(\beta_1 - \beta_2) = \varphi$, i.e. for:

$$\beta_1 = \frac{1}{2}(\Sigma + \varphi), \quad \beta_2 = \frac{1}{2}(\Sigma - \varphi). \tag{10.107}$$

Since the value of the angle of friction φ is very small, we come to the conclusion that the efficiency has its maximum value when

$$\beta_1 \cong \beta_2 \cong \frac{\Sigma}{2}. \tag{10.108}$$

In the special case in which the shaft angle $\Sigma = (\pi/2)$, the efficiency is maximum when $\beta_1 = \beta_2 = \pi/4$.

For a crossed helical gear pair with shaft angle Σ, z_1 and z_2 teeth, normal module m_n, and helix angle $\beta_1 = \beta_2 = \Sigma/2$, and then in the condition of maximum efficiency, the center distance, a, is given by:

$$a = \frac{1}{2}(z_1 + z_2)\frac{m_n}{\cos(\Sigma/2)}. \tag{10.109}$$

Since the pitch helix angles are equal, the pitch diameters are proportional to the numbers of teeth, i.e.:

$$d_1 = 2r_1 = \frac{z_1 m_n}{\cos(\Sigma/2)} \quad d_2 = 2r_2 = \frac{z_2 m_n}{\cos(\Sigma/2)}. \tag{10.110}$$

When the center distance, a, the shaft angle, Σ, the normal module, m_n, and the transmission ratio, i, are given, the numbers of teeth of the two wheels are given by:

$$z_1 = \frac{2a\cos(\Sigma/2)}{m_n(1+i)} \quad z_2 = \frac{2a\cos(\Sigma/2)}{m_n i(1+i)}. \tag{10.111}$$

It is to be noted that $\beta_1 > \beta_2$, and that when β_2 exceeds a certain limit value, the transmission of motion from the driving shaft 1 to the driven shaft 2 cannot take place. This limiting value can be obtained by putting $\eta = 0$ in Eq. (10.106), whence we get $\tan\beta_2 = (1/\mu) = (1/\tan\varphi) = \cot\varphi = \tan(90° - \varphi)$. Therefore, the transmission of motion is theoretically possible only when $\beta_2 < (90° - \varphi)$. The practical limiting value of β_2 is still less. When the shaft angle is $\Sigma = 90°$, the efficiency, η, is given by:

$$\eta = \frac{\tan(\beta_1 - \varphi)}{\tan\beta_1}. \tag{10.112}$$

Finally, it is to be noted that the power losses due to the lengthwise sliding previously considered do not constitute the only power losses of a crossed helical gear pair. Beside these losses, other losses occur, such as the losses due to toothing action (these losses depend on the transmission ratio, the numbers of teeth, and the normal module), the bearing losses, and other loss phenomena.

10.10   Profile-Shifted Toothing for Crossed Helical Gears

Like parallel cylindrical helical gears, crossed cylindrical helical gears can also have profile-shifted toothing when needed. The same reasons for which we use parallel cylindrical helical gears with profile-shifted toothing are also valid for crossed cylindrical helical gears (in this regard, see Chap. 8). Here the subject is discussed in its most general terms, i.e. with reference to the case of profile-shifted toothing with variation of center distance, leaving the reader the task of applying the general concepts to the special case of profile-shifted toothing without variation of center distance.

We assume to cut the two members of the crossed helical gear under consideration by means of a rack-type cutter having, in its normal section, normal module m_n, and nominal standard sizing. We continue to give to the quantities and symbols already introduced, i.e. $\alpha_n, \alpha_{t1}, \alpha_{t2}, \beta_1, \beta_2, r_1, r_2, m_n, m_{t1}$, and m_{t2}, the same meaning given to them that we have so far, and we denote with the subscript, 0, the corresponding quantities referred to the cutting pitch cylinders. The relationships (10.27), (10.28), and (10.32) are valid, but also the equations that follow apply:

$$r_{10} = \frac{z_1 m_{t0}}{2} = \frac{z_1 m_{n0}}{2 \cos \beta_{10}} \tag{10.113}$$

$$r_1 = r_{10} \frac{\cot \beta_{10}}{\cot \beta_1}; \tag{10.114}$$

this is because $m_{n0} = m_{t0} \cos \beta_{10}$, and $p_z = 2\pi r_{10} \cot \beta_{10} = 2\pi r_1 \cot \beta_1$. In addition, the following equations apply:

$$\tan \alpha_{n0} = \tan \alpha_{t0} \cos \beta_{10}, \tag{10.115}$$

$$\sin \beta_{b1} = \sin \beta_{10} \cos \alpha_{n0} = \sin \beta_1 \cos \alpha_n; \tag{10.116}$$

these equations are, respectively, in accordance with Eqs. (8.17) and (8.35).

On the cutting pitch cylinder of the first wheel, the normal tooth thickness s_{n10} and transverse tooth thickness s_{t10} are given by:

$$s_{n10} = m_{n0}\left(\frac{\pi}{2} + 2x_1 \tan \alpha_{n0}\right) \tag{10.117}$$

$$s_{t10} = \frac{s_{n10}}{\cos \beta_{10}} = m_{n0}\left(\frac{\pi}{2\cos \beta_{10}} + 2x_1 \tan \alpha_{t0}\right) = 2r_{10}(\mathrm{inv}\gamma_1 - \mathrm{inv}\alpha_{t10}). \tag{10.118}$$

From this last equation, we get:

$$\text{inv}\gamma_1 = \text{inv}\alpha_{t0} + \frac{\pi}{2z_1} + \frac{2x_1}{z_1}\tan\alpha_{n0}. \tag{10.119}$$

On the operating pitch cylinder, having radius r_1, the normal tooth thickness s_{n1} is equal to:

$$
\begin{aligned}
s_{n1} &= s_{t1}\cos\beta_1 = 2r_1\cos\beta_1(\text{inv}\gamma_1 - \text{inv}\alpha_{t1}) \\
&= m_n\left[\frac{\pi}{2} + 2x_1\tan\alpha_{n0} + z_1(\text{inv}\alpha_{t10} - \text{inv}\alpha_{t1})\right]. \tag{10.120}
\end{aligned}
$$

For the second wheel, the driven wheel, analogous relationships apply. In particular, on its operating pitch cylinder, having radius r_2, the normal tooth thickness s_{n2} is equal to:

$$s_{n2} = m_n\left[\frac{\pi}{2} + 2x_2\tan\alpha_{n0} + z_2(\text{inv}\alpha_{t20} - \text{inv}\alpha_{t2})\right]. \tag{10.121}$$

To have good operating and lubrication conditions, it is necessary that, among the mating tooth profiles, a suitable normal backlash occurs, which we may assume equal to a fraction of the normal module, i.e. $j_n = cm_n$ (usually, $c \cong 0.05$). Therefore, in the normal section, the following condition must be satisfied:

$$s_{n1} + s_{n2} + j_n = s_{n1} + s_{n2} + cm_n = \pi m_n, \tag{10.122}$$

which, given the equations written above, takes the following form:

$$z_1(\text{inv}\alpha_{t1} - \text{inv}\alpha_{t10}) + z_2(\text{inv}\alpha_{t2} - \text{inv}\alpha_{t20}) = 2(x_1 + x_2)\tan\alpha_{n0} + c. \tag{10.123}$$

Once the values of $\beta_1, \beta_2, \alpha_{n0}$ and c are chosen, and values of shift coefficients x_1 and x_2 have been determined based on the virtual numbers of teeth $z_{v1} = (z_1/\cos^3\beta_1)$ and $z_{v2} = (z_2/\cos^3\beta_2)$, using the procedures described in this regard in Chaps. 6 and 8, the value of the second member of Eq. (10.123) is uniquely defined. Taking into account the relationships (10.32), (10.115) and (10.116), the first member of the same Eq. (10.123), once the values of β_1 and β_2 are chosen, is a function only of α_n, which can be determined by a trial and error procedure. Of course, this equation can be easily solved numerically.

Always with reference to the normal section, the determination of α_n can be carried out also, in a more rapid but approximated way, using the following relationship:

$$\text{inv}\alpha_n = \text{inv}\alpha_{n0} + 2\tan\alpha_{n0}\frac{x_1 + x_2}{z_{v1} + z_{v2}}. \tag{10.124}$$

Once the value of α_n has been thus calculated, using the formulas above, we determine in succession the quantities $\beta_{10}, \beta_{20}, \alpha_{t10}, \alpha_{t1}, \alpha_{t20}$ and α_{t2}; then we modify x_1 and x_2 so that Eq. (10.123) is satisfied.

Further insights on this subject are not necessary, since the calculation procedure does not differ from that described in the already above-mentioned Chaps. 6 and 8.

References

1. Ball RS (1876) The theory of screw: a study in the dynamics of a rigid body. Hodges, Foster&Co, Dublin
2. Buckingham E (1949) Analytical mechanics of gears. McGraw-Hill Book Company Inc, New York
3. Ceccarelli M (2000) Screw axis defined by Giulio Mozzi in 1763 and early studies on helicoidal motion. Mech Mach Theory 35:761–770
4. Colbourne JR (1987) The geometry of involute gears. Springer, New York Inc, New York Berlin Heidelberg
5. Dobrovolski V, Zablonski K, Mak S, Radtchik A, Erlikh L (1971) Éléments de Machines. Édition MIR, Moscou
6. Dudley DW (1962) Gear handbook. The design, manufacture, and application of gears, McGraw-Hill, New York
7. Euler L (1765) Theoria Motus Corporum Solidorum seu Rigidorum ex primis nostrae cognitionis principiis stabilita et ad omnes motus, qui in huius modi corpora cadere possunt, accomodata, Rostochii et Gryphiswaldiae Litteris et Impensis A. F. Röse
8. Ferrari C, Romiti A (1966) Meccanica Applicata alle Macchine, Unione Tipografica – Editrice Torinese (UTET), Torino
9. Giovannozzi R (1965) Costruzione di Macchine, vol II, 4th edn. Casa Editrice Prof. Riccardo Pàtron, Bologna
10. Henriot G (1979) Traité théorique et pratique des engrenages, vol 1. 6th edn. Borda, Paris
11. Herrmann R (1928) Evolventen-Stirnrädgetriebe. Julius Springer, Berlin
12. Levi-Civita T, Amaldi U (1929) Lezioni di Meccanica Razionale—Vol. 1: Cinematica—Principi e Statica, 2nd edn. Reprint 1974, Zanichelli, Bologna
13. Litvin FL, Fuentes A (2004) Gear geometry and applied theory, 2nd edn. Cambridge University Press, Cambridge
14. Merritt HE (1954) Gears, 3rd edn. Sir Isaac Pitman & Soins Ltd, London
15. Minguzzi E (2013) A geometrical introduction to screw theory. Eur J Phys 34:613–632
16. Mozzi del Garbo GG (1763) Discorso matematico sopra il rotolamento momentaneo dei corpi, Stamperia Donato Campo, Napoli
17. Niemann G, Winter H (1983) Maschinen-Element Band III:Schraubrad-, Kegelrad-, Schneckn-, Ketten-, Riemen-, Reibradgetriebe, Kupplungen, Bremsen, Freiläufe, Springer, Berlin, Heidelberg
18. Pollone G (1970) Il Veicolo. Libreria Editrice Universitaria Levrotto & Bella, Torino
19. Poritsky H, Dudley DW (1948) Conjugate action of involute helical gears with parallel or inclined axes. Q Appl Math 6(3):193–214
20. Radzevich SP (2016) Dudley's handbook of practical gear design and manufacture, 3rd edn. CRC Press, Taylor & Francis Group, Boca Raton, Florida
21. Schiebel A (1913) Zahnräder, II Teil, Räder mit schrägen Zähnen. Verlag con Julius Springer, Berlin
22. Schiebel A, Lindner W (1954) Zahnräder, Band 1: Stirn-und Kegelräder mit geraden Zähnen, Springer, Berlin, Heidelberg

23. Schiebel A, Lindner W (1957) Zahnräder, Band 2: Stirn-und Kegelräder mit schrägen Zähnen Schraubgetriebe, Springer, Berlin, Heidelberg
24. Scotto Lavina G (1990) Riassunto delle Lezioni di Meccanica Applicata alle Macchine: Cinematica Applicata, Dinamica Applicata - Resistenze Passive - Coppie Inferiori, Coppie Superiori (Ingranaggi – Flessibili – Freni). Edizioni Scientifiche SIDEREA, Roma
25. Spiegel MR, Lipschutz S, Spellman D (2009) Vector Analysis, Schaum's Outlines, 2nd edn. McGraw-Hill, New York
26. Tang KT (2006) Mathematical methods for engineers and scientists. Springer, Berlin, Heidelberg
27. Townsend DP (1991) Dudley's gear handbook, 2nd edn. McGraw-Hill, New York
28. Vaisman I (1997) Analytical geometry. World Scientific Publishing Co, Singapore
29. Zwirner G (1961) Istituzioni di Matematiche, Parte Prima, 5th edn. CEDAM-Casa Editrice Dott. Antonio Milani, Padova

Chapter 11
Worm Gears

Abstract In this chapter, the geometry of the various types of worms and corresponding worm wheels is first described and short notes about their cutting processes are provided. The main geometric quantities of worms and worm wheels are then defined. The parametric equations of helicoid surfaces in terms of differential geometry are subsequently determined and the main kinematic quantities of these types of gears are obtained. The meshing between worm and worm wheel is then analyzed using the aforementioned concepts of differential geometry, and the instantaneous lines of contact as well as surface of contact are analytically determined. Subsequently, the two well-known graphic-analytical methods of determining the surface of contact are described, i.e. the Schiebel's method for Archimedean spiral worms and Ingrisch's method for involute worms. Interference problems related to the outside surface of the worm wheel are then discussed and load analysis of worm gears is performed, defining the thrust characteristics on shafts and bearings and determining their average efficiency. Further consideration on the worm and worm wheel sizing are made and short notes on double-enveloping worm gear pairs are provided. Finally, standard and non-standard worm drives and special worm drives are described.

11.1 Introduction

Worm gear pairs are gears with crossed axes, constituted by a *worm* (or *worm screw*) of cylindrical or toroidal shape that meshes with a *worm wheel* having tooth flanks capable of a line contact with the flanks of the worm *threads*. Usually, these gears are used for non-parallel, non-intersecting, right-angle crossed axes (more rarely, for crossed axes with shaft angle $\Sigma \neq \pi/2$), where high center distances, and high reduction ratios are required, although in many practical applications they are also employed for low and medium reduction ratios. In fact, worm gear pairs allow reaching large reduction ratios, and therefore a large multiplication of torque, using only one pair of mechanical geared members. This would not be possible, or would not be convenient, using gear pairs such as those so far studied and analyzed in the previous chapters.

© Springer Nature Switzerland AG 2020
V. Vullo, *Gears*, Springer Series in Solid and Structural Mechanics 10,
https://doi.org/10.1007/978-3-030-36502-8_11

As we have seen in the previous chapter, worm gears are a special case of crossed helical gears. However, unlike the latter gears, which are characterized by a point contact, the worm gears are characterized by a line contact. Given their extensive use for interesting practical applications, they are discussed in the most classical textbooks on gears (see Buckingham [2], Merritt [39], Dudley [13], Giovannozzi [17], Pollone [48], Henriot [19], Niemann and Winter [44], Colbourne [7], Townsend [65], Maitra [37], Jelaska [25], Radzevich [51, 52]). Here the discussion on this important topic is carried out not only in accordance with the classical one, available in more or less extensive form in the traditional aforementioned textbooks, but also using the most advanced numerical developments based on rigorous theoretical analysis that uses differential geometry methods (see Litvin and Fuentes [33]). It should be noted that the theoretical method described here is so general that it can be used for all gears covered in this textbook, including hypoid gears.

Since the instantaneous contact between the two members of a worm gear is a line contact, contrary to what we highlighted for crossed helical gears, it is able to transmit loads significantly greater. The meshing contact between the worm threads and worm wheel teeth is a combination of rolling and sliding, with the latter prevailing at higher reduction ratios. How best we will specify in the following sections, the relative motion of the worm with respect to worm wheel is a *screw motion* about an axis lying in a plane that is perpendicular to the line of shortest distance between the two axes.

From the point of view of kinematics, the worm gear pairs are known to be characterized by only one correct relative position between the axes of the two members, which is theoretically the only one that can guarantee optimal operating conditions. The cutting process and, more generally, the production and assembly processes of the two members of the worm gear must be extremely accurate. The worm wheel has curvilinear teeth, which are obtained by a generation cutting process by means of a hob having the same geometric characteristics of the worm.

To produce the two members of a worm gear, we encounter many difficulties, which include the construction of the worm as well as the tool with which the worm wheel is to be generated. This cutting tool must have the same characteristics of the worm. In fact, the hob is made of tool steel (see Davis [9], Radzevich [50]) and, due to the necessary hardening heat treatment, undergoes considerable deformations, also due to notches to create the cutting edges. Such deformations must be eliminated by a subsequent finishing operation, generally consisting of a precision grinding. The accuracy of this finishing operation of the worm thread and hob strongly affects the correct operating conditions of the worm gear.

Worm gear pairs are also sensitive to assembly errors, among which the variations of shaft angle and center distance, and axial displacements of the worm wheel play an important role. These errors result in a shift of the bearing contact to the edge, and determine a piecewise almost-linear function of transmission errors, which have the same frequency as the meshing cycle between the worm threads and worm wheel teeth. A proper mismatch between the worm and hob surfaces results in a more favorable function of transmission errors, and a more stable bearing contact of worm gear pair [33].

The worm can have one or more threads, and these threads may or may not have an involute profile, depending on their cutting process. The shape of the worm threads defines the shape of the worm wheel teeth, as the hob used to generate them is essentially a duplicate of the same worm. In gear terminology, the *number of threads* of the worm is called the *number of starts*. Thus, we can have *single-start worms* or *multi-start worms*. This number of starts can be, at most, equal to 2 or 3 (rarely 4 and, even more rarely, greater than 4), while the worm wheel may have several tens of teeth. The high value of the gear ratio allowed by a worm gear pair finds its justification: in fact, it is the ratio between the number of teeth, z_2, of the worm wheel and the number of starts, z_1, of the worm.

The worm can have a dual shape, so we have:

- a *cylindrical worm* or simply *worm*, when it is configured as a cylindrical helical pinion;
- an *enveloping worm*, when it is configured as a pinion having tip and root surfaces that consist of parts of toroid coaxial with the worm wheel, and radius of the mean circle equal to the center distance, or shortest distance, of the worm gear pair at which the pinion is intended.

Even the worm wheel can have a dual shape, for which we have: a *cylindrical worm wheel* or simply *worm wheel*, and an *enveloping worm wheel*.

Worm gear pairs can be configured according to three different coupling types:

- a cylindrical worm coupled with an enveloping worm wheel (Fig. 11.1);
- an enveloping worm coupled with a cylindrical worm wheel;
- an enveloping worm coupled with an enveloping worm wheel; in this case, we have a *double- enveloping worm gear pair.*

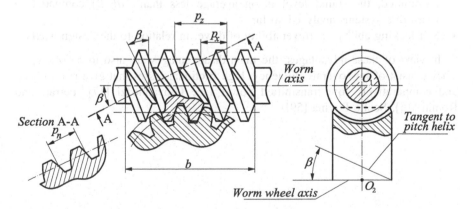

Fig. 11.1 Single-enveloping worm gear pair with crossed orthogonal axes, consisting of a cylindrical two-start worm and an enveloping worm wheel, and its characteristic quantities

At least one of the two members of a worm gear pair is therefore an enveloping member. Later we will give motivation for this configuration. Worm gear pairs most commonly used are still the first two, which we can call *single-enveloping worm gear pairs*.

The mutual operating position of the two members of a worm gear pair can vary, depending on the needs of connection with the types of driving machine (electric motor or internal combustion engine), and driven machine. In a worm-drive unit, the worm is generally the driving member, but not always the case: there are cases in which the worm wheel is the driving member. As a rule, when possible, it is convenient to have the worm, with its horizontal axis, below that of the worm wheel, also with a horizontal axis, and not vice versa. Several cases exist, however, in which the worm has horizontal axis and worm wheel has vertical axis, or the worm has vertical axis and the worm wheel has horizontal axis [17]. Of course, we are talking about the usual, conventional applications, which therefore do not concern the countless variety of worm-drive systems that have been developed, with characteristics entirely new and innovative, for special applications. These types of worm gear drives are beyond the purpose of this textbook. Nonetheless, the reader will find a brief reference to this topic in the Sect. 11.14.

The main working features of the worm gear pairs are summarized as follows:

- High gear ratios, u, with comparatively smaller overall dimensions and space requirements and, consequently, lower weight and lower cost. For a speed reducing unit, the range of variability of u is commonly $5 \leq u \leq 70$, mainly $15 \leq u \leq 50$, and for low power to be transmitted up to $u = 1000$.
- Possibility to drive the worm from both ends of its shaft, as well as to drive continuous shafts that carry several worms in series, which in turn drive as many driven devices.
- Silent and anti-vibration operating conditions, due to the high sliding and low speed of impact between the thread flanks of the worm and mating tooth flanks of the worm wheel. With equal number of turns of the worm and torque transmitted, the sound level is on average less than $7\,dB(B)$ compared to gear-drive systems analyzed so far.
- Self-locking ability or irreversibility of drive, in relation to the design needs.

In view of these advantages, the main disadvantages are also to be observed. They consist of low overall efficiency, high frictional losses that give rise to heat, and comparatively low transmitted power (see Giovannozzi [17], Ferrari and Romiti [15], Scotto Lavina [59]).

11.2   Geometry of Worm Thread Profiles and Worm Wheel Teeth, and Short Notes on Their Cutting Methods

In the worm-drive systems, cylindrical single-start or multi-start worms are usually employed, with threads that have, in axial section, trapezoidal shape, with straight or curved flank profiles, according to the generation process used. In many cases, the thread flanks are ruled helical surfaces, generated by the tangents of a cylinder of given diameter, in correspondence to the successive points of a helix with a constant lead, traced on it, and forming a constant angle with the tangents to the helix at these points. The sections of these helical surfaces (helicoids) with transverse planes, i.e. planes normal to the axis of the cylinder, are spirals, the shape of which varies with the diameter of the cylinder and with the angle that the generation straight line forms with the tangent to the helix.

In other cases, the thread flanks are not ruled surfaces, but envelopes to families of generating surfaces that perform a screw motion about the axis of the worm. These generation surfaces can be or conical surfaces, or surfaces of revolution whose axial sections are arcs of a circle.

The shape adopted for the thread flanks of the worm is entirely arbitrary, since, as we have already said, a cutting tool having the same shape of the worm generates the teeth of the worm wheel. Thus, the choice depends on the manufacturing process, which is convenient or desirable to use. Many worms have been made with straight-sided profiles on the axial section or the normal section, but usually we use involute profiles because of the comparative ease with which they can be profile-grounded (see Octrue and Denis [46], Predki [49], Zimmer et al. [73]).

In relation to the manufacturing process involved, thread profiles of the under mentioned five most common types of cylindrical worms are generally used in practical applications. The letters A, C, I, K and N, in accordance with the Technical Report ISO/TR 10828:1997 (E) [24], designate them. By definition, the five standardized types of cylindrical worms are the ones described below.

11.2.1   Type A Worm, with Straight-Sided Axial Profile

The thread flanks of this type of worm are generated as envelopes of straight lines in axial planes, which are inclined at a constant angle $[(\pi/2) - \alpha_{0t}]$ to the axis. Thus, this type of worm is the limit case in which the above-mentioned cylinder has diameter, d, equal to zero, and the tangent to the helix and the same helix coincide with the axis of the worm. The generation straight line of the helicoid then intersects the axis of the worm, forms with it a constant angle $[(\pi/2) - \alpha_{0t}]$, and moves along it with a *screw motion* or *helical motion* with lead, p_z; this motion consists of simultaneous uniform rotation about and translation along the axis. Therefore, we get a worm whose axial section is a common rack with trapezoidal teeth.

The worm flank surface is a *ruled surface*, which is the locus of the successive positions of the generation straight line during its screw motion with respect to the worm axis. The section of the helicoid with a transverse plane is an *Archimedean spiral*, whence the name of *Archimedean spiral worm* that we give commonly to this type of worm. In a normal section, the thread profiles are convex curves.

As Fig. 11.2 shows, threads may be cut on a lathe, with a cutting tool having straight edges, the cutting plane of which lies in an axial plane of the worm. Both flanks of a *thread space* may be cut simultaneously by using a cutter of trapezoidal shape. Type *A* worms can be also cut by using an involute shaper to produce the desired straight rack-type profile in an axial plane of the worm; its cutting face must lie in that axial plane. It is also necessary that the pitch circle of the shaper rolls without sliding on the datum line of the rack profile, which coincides with a straight-line generatrix of the worm pitch cylinder. This second cutting method of the worm is essentially an inversion of the process of cutting a helical gear wheel with a rack-type cutter.

It is to be noted that the type *A* worms may be grounded only using suitable profiled grinding wheels. This is because, in normal sections, the flank profiles of the threads are convex curves, that is curvilinear profiles. For more in-depth knowledge of the cutting and grinding methods of the various types of worms and related worm wheels, we refer the reader to specialized textbooks (see Galassini [16], Townsend [65], Radzevich [50]).

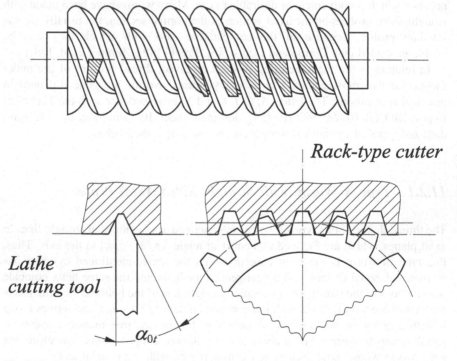

Fig. 11.2 Type *A* cylindrical worm and machining methods by turning or shaping

11.2.2   Type I Worm, with Involute Helicoid Flanks, and Generation Straight Line in a Plane Tangent to the Base Cylinder

The thread flanks of this type of worm (Fig. 11.3) are involute helicoidal surfaces, generated by the screw motion with lead, p_z, of a generation straight line not passing through the worm axis. This generation straight line forms with worm axis a constant angle β_b (the base helix angle), and is constantly tangent to the base

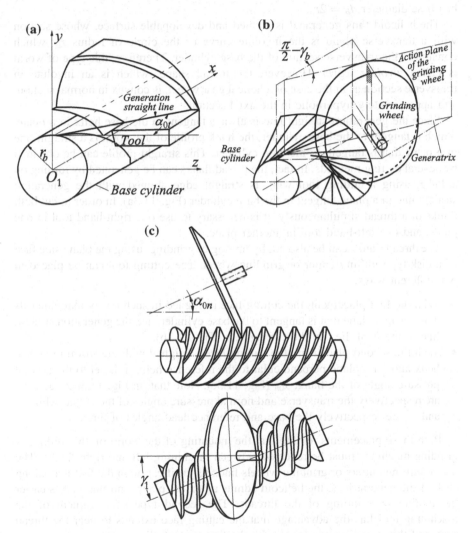

Fig. 11.3 Type I cylindrical worm and machining methods by: **a** turning; **b** milling or grinding with a first placement of the cutter; **c** milling or grinding with a second placement of the cutter

cylinder of radius r_b. These quantities are related between them by the following relationship:

$$p_z = 2\pi r_b \cot \beta_b. \tag{11.1}$$

With reference to the general case mentioned at the beginning of this section, in this case the generation straight line of the helicoid is tangent to the base helix. Therefore, the constant angle between the tangents to the base cylinder, in correspondence of successive points of the base helix, and tangents to this helix in these points is equal to zero. The base helix lies on the base cylinder of the worm, which has base diameter $d_b = 2r_b$.

The helicoid thus generated is a ruled and developable surface, whose section with a transverse plane is the involute curve of the circle of radius r_b, which constitutes the transverse section of the base cylinder. Therefore, this type of worm is called *involute worm*. However, the thread shape, which is an involute in transverse section as in the case of a helical gear wheel, is convex in normal section, and approximately hyperbolic in the axial section.

Since the generation straight line is always tangent to the base helix in a plane, which is tangent to the base cylinder, the flank profile of the worm is a straight line in a plane that is tangent to the base cylinder. This straight profile can be obtained by several cutting methods. In fact, the thread flanks can be generated by turning on a lathe, using a cutting tool with its straight edge aligned with the generation straight line in a plane tangent to the base cylinder (Fig. 11.3a). In order to cut both flanks of a thread simultaneously, it is necessary to use one right-hand tool in one plane, and one left-hand tool in another plane.

The thread flanks can be also cut by milling or grinding, using the plane side face of a disk-type milling cutter or grinding wheel. The cutting tool can be placed in two different ways:

- With the first placement, the cutting face is aligned in such a way that either its axis lies in a plane that is tangent to the base cylinder, and the generation straight line of the flank lies on the cutting face (Fig. 11.3b).
- With the second placement, the cutting face is aligned with the worm reference helix and, in a plane perpendicular to the reference helix, it is set to the normal pressure angle of the flank, α_{0n} (Fig. 11.3c). Note that, in Fig. 11.3, α_{0t} and α_{0n} are respectively the transverse and normal pressure angles of the cutter, while γ_b and γ_1 are respectively the base and reference lead angles of threads.

Both these placements require that the mounting of the worm on the milling or grinding machines must be reversed between machining left and right flanks. The use of milling cutters or grinding wheels finds its justification in the fact that, along each of the generatrices, the helicoid admits a tangent plane, and this makes easier the milling or grinding of the threads. The second type of alignment of the machining tool has the advantage that the cutting face extends to near the thread root, and this is not simple to do with the first type of alignment.

11.2.3   Type N Worm, with Straight Sided Normal Profile

The type N worm (Fig. 11.4) is similar to type A worm, with the variation that the thread profile is of trapezoidal shape, i.e. straight sided, in the normal section, rather than in the axial section. It represents the more general case of ruled helicoid, in the configuration of the thread flanks, which we discussed at the beginning of this section. The thread profile is slightly curved in the axial section. The generation straight line lies in a plane passing through the perpendicular to the worm axis, and forming the angle γ_1 (the reference lead angle, i.e. the lead angle on the reference pitch cylinder of the worm) with the worm axis. The transverse section of the worm thread, i.e. the transverse worm profile, is an extended involute.

As Fig. 11.4 shows, the machining methods are several. The threads can be cut in a lathe with a tool having trapezoidal shape and cutting edges placed in the cutting plane, which match the profile of the thread space in a plane normal to the reference helix of the thread space (Fig. 11.4a). The threads can be machined, in a more or less appropriate way, in a milling or grinding machine, using a biconical milling cutter or grinding wheel of small diameter, as Fig. 11.4b shows (when the diameter of the grinding wheel becomes large, the type N worm approaches the type K worm), or a small conical gear milling cutter or grinding wheel (Fig. 11.4c). These last two methods determine profiles that are approximate, because the effects due to the change of helix with the change of thread depth. Deviations and alterations with respect to the proper profile are much larger, the bigger the diameter of the milling cutter or grinding wheel and the lead of the worm.

11.2.4   Type K Worm, with Convex Thread Profiles in Axial Plane, and Helicoid Generated by Biconical Grinding Wheel or Milling Cutter

The thread profiles of this type of worm are concave in the normal section. This worm is frequently used for some advantage offered by it, inherent to the cutting process. Unlike those of types A, I, and N, the thread flanks of type K worms do not have a generation straight line, and their surfaces are not ruled surfaces, but envelopes to a family of cone surfaces. The thread spaces of these worms are generated with biconical grinding wheels or disk-type milling cutters having straight cone generatrices, and performing a screw motion about the worm axis.

As Fig. 11.5 shows, the common perpendicular to the tool spindle and worm axes lies on the line of intersection, l, of the middle plane of the tool and a transverse plane of the worm; the same figure shows, below, the trace point of this intersection line. The angle between these two planes is equal to the worm lead angle, γ_1. The straight generatrix of each conical side of the tool and the middle plane of the same tool forms an angle equal to the normal pressure angle α_{0n}. The worm is turned uniformly with simultaneous axial translation of threads, so that a

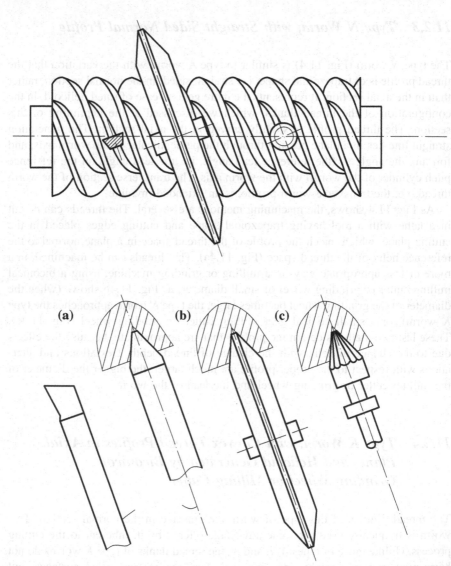

Fig. 11.4 Type N cylindrical worm and machining methods by turning (**a**), or by milling or grinding (**b, c**)

point on the common perpendicular, distant r_1 from the worm axis (r_1 is the reference radius of the worm), describes the reference helix. The conical sides of the tool generate the helicoidal flanks of the worm. The profile shape is affected by the change of helix angle with change of the thread depth, and points of the tool flanks, which contact the worm threads, lie on a curve and not on any one-cone generatrix.

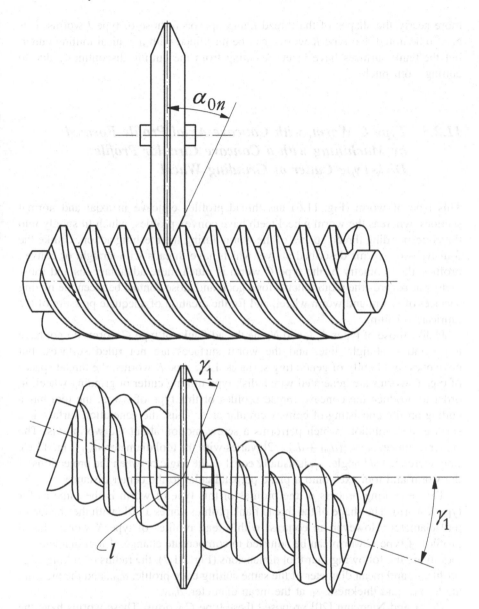

Fig. 11.5 Type K cylindrical worm and machining methods by grinding or milling

This type of worm has the advantage that the two flanks of a thread space can be machined simultaneously, but it has the disadvantage that the shape of the thread flanks varies with tool diameter, so the reproducibility is approximate. It should be noted that the smaller the tool diameter, the more nearly the normal profiles of the thread spaces approach those of type N worms, and the larger the tool diameter the more nearly the shapes of the thread flanks approach those of type I worms. It is also to be noted that type K worms can be machined with a conical milling cutter, but the flank surfaces have facets resulting from the cutting discontinuity due to cutting tooth pitch.

11.2.5 Type C Worm, with Concave Axial Profile Formed by Machining with a Concave Circular Profile Disk-Type Cutter or Grinding Wheel

This type of worm (Fig. 11.6) has thread profiles concave in axial and normal sections, whereas the worm wheel teeth have convex profiles, which fit snugly into the corresponding thread spaces of the worm during the meshing action. Unlike the four types of worms described above, which have threads with straight or convex profiles, the geometry of the type C worm is much more advantageous and satisfactory as regards the requirements for a good and close contact between the mating surfaces of worm and worm wheel, and for the creation of adequate pressure of the lubricant oil film.

Unlike those of types A, I, and N, the thread flanks of type C worms do not have a generation straight line, and the worm surfaces are not ruled surfaces, but envelopes to a family of generating surfaces. Like type K worms, the thread spaces of type C worms are generated with a disk-type milling cutter or grinding wheel. In order to produce the concave thread profiles of this type of worm, the tool has a cutting profile consisting of convex circular arcs. Thus, the generation surface is a surface of revolution, which performs a screw motion about the worm axis. The center distance, $a_0 = [(d_{m1} + d_{m0})/2]$, varies with the tool diameter, d_{m0} (Fig. 11.6). The reference lead angle, γ_1, is usually equal to the angle between the projections of the worm and tool axes onto a plane perpendicular to the center distance.

The generating process of the profiles of this type of worm is the same as for type K worm. The shape of the thread flank profiles varies a little with the change of tool diameter. However, in contrast with thread profiles of type K worm, thread profiles of type C worm can be adjusted to compensate change of tool diameter by modifying the following four tool dimensions (Fig. 11.7): the radius of cutting edge profile, ρ, and mean diameter of the same cutting edge profile, d_{m0}, and the pressure angle, α_{0n}, and thickness, s_0 at the mean diameter, d_{m0}.

Heyer and Niemann [20] proposed these type C worms. These worms have the great advantage of improving the conditions of lubrication, by virtue of the favorable shape of lines of contact between the worm and worm wheel mating surfaces. Consequently, they are characterized by a continuous maintenance of lubrication oil film and high hydrodynamic lubricant pressure, which reduces output losses and wear, as well as by a greater load carrying capacity, lesser frictional losses, and greater impact damping properties and noiselessness.

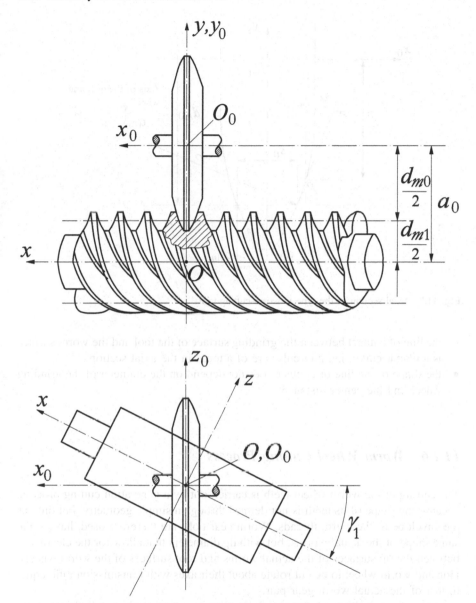

Fig. 11.6 Type C cylindrical worm and machining methods by milling or grinding

Litvin [30] proposed a variant of this type of worm, which is generated using the same cutting tool described above, but with a different cutting position with respect to the worm. This difference, which concerns only the setting parameters of the cutting machine, determines certain advantages over the aforementioned worm proposed by Heyer and Niemann, summarized as follows by the same Litvin:

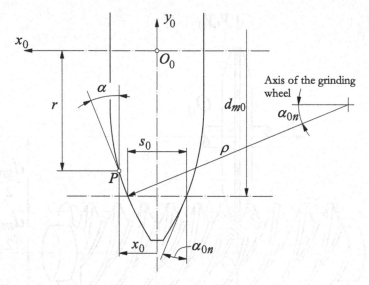

Fig. 11.7 Axial section of the type C worm-cutting tool

- the line of contact between the grinding surface of the tool and the worm surface is a planar curve, i.e. a circular arc of a torus in the axial section;
- the shape of the line of contact does not depend on the diameter of the grinding wheel and the center distance.

11.2.6 Worm Wheel Cutting Process

The cutting of the worm wheel teeth is carried out by a generation cutting process, because the shape of its teeth is not defined through a simple geometry, but only as the envelope of the worm threads. Worm gear hobs are therefore used, having the same shape of the actual worms, but with tip diameters that allow for the clearances between the tip surfaces of the actual worms and root surfaces of the worm wheels. Hob and worm wheel to be cut rotate about their axes with transmission ratio equal to that of the actual worm gear pair.

In order to cut the tooth progressively over its entire depth, a cylindrical worm gear hob can have a radial approach motion towards the worm wheel, until the desired depth (Fig. 11.8a). This radial approach, however, produces cutting interferences and thus undercut, which determine the removal of parts of the usable flank profile. The use of this type of cutting process is therefore limited to values of the helix angle, β, not exceeding $(6 \div 8)°$. For higher values of β, but also in general, conical worm gear hobs can be used, equipped with axial motion in

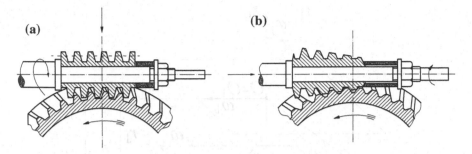

Fig. 11.8 Generation cutting process of worm wheels by: **a** cylindrical worm gear hob; **b** conical worm gear pair

addition to motion of rotation about their axes, in so that the final tooth depth is achieved gradually, as the hob deeper threads are progressively involved in the cutting process (Fig. 11.8b).

11.3 Coordinate Systems and Main Geometric Quantities of Worm and Worm Wheel

In the more general case, worm and worm wheel rotate about axes z_1 and z_2 that form, in a plane perpendicular to the shortest distance or center distance, a, a shaft angle $\Sigma \gtrless \pi/2$. Figure 10.8, which refers to any crossed helical gear pair, can also be used to show the meshing position between the two members of a worm gear pair with $\Sigma < \pi/2$. As coordinate systems we consider three Cartesian coordinate systems $O_1(x_1, y_1, z_1)$, $O_2(x_2, y_2, z_2)$ and $O_0(x_0, y_0, z_0)$, rigidly connected respectively to worm (gear 1), worm wheel (gear 2) and frame or housing (Fig. 11.9). In some cases, for convenience of calculation, it can be useful to introduce other auxiliary coordinate systems that will be described from time to time.

Though orientations having other shaft angles are possible, in most practical applications the members of a worm gear pair are generally mounted on crossed shafts with shaft angle $\Sigma = \pi/2$. For this reason, we sometimes consider a worm gear pair consisting of a cylindrical worm, having right-hand threads, and an enveloping worm wheel, positioned under the worm, whose axes are arranged with 90° shaft angle (Fig. 11.10). As coordinate system $O_1(x_1, y_1, z_1)$, we choose the Cartesian coordinate system with z_1-axis coinciding with the worm axis and origin at point O_1 on the worm axis, x_1-axis in the middle plane of the worm wheel and in its initial position coinciding with the shortest distance (i.e. with the common perpendicular to the worm and worm wheel axes), and y_1-axis that completes the direct coordinate system. Figure 11.10 shows this right-hand coordinate system for a worm gear pair with type A worm.

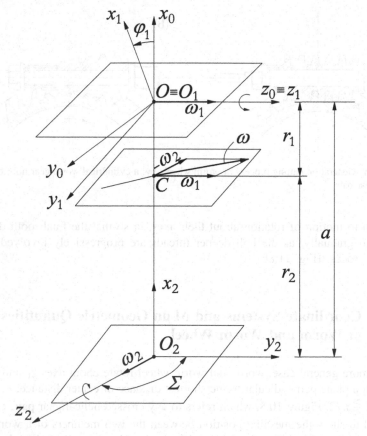

Fig. 11.9 Coordinate systems $O_1(x_1, y_1, z_1)$, $O_2(x_2, y_2, z_2)$ and $O_0(x_0, y_0, z_0)$

The study of kinematics of a worm gear pair can be carried out based on observation that the teeth shape of the worm wheel, when the thread profile of the worm is known, depends only on an axial advance motion of the same worm, which is precisely the motion that obliges the worm wheel rotates. In fact, the motion of the worm, which is a rotation about its own axis, can be considered as the difference of two motions: a helical motion with pitch equal to the lead, and the aforementioned axial advance motion. The first of these two motion components has no effect on the worm wheel, that is, it does not cause the rotation of the worm wheel, since in such a motion the thread of the worm simply slides on itself, and therefore the only consequence of such a motion is a power loss by friction.

Worm and worm wheel, by transmitting the motion, rotate about their two crossed axes, and their relative motion is a screw motion, that is an instantaneous helical motion. Therefore, for the worm gear pair, we cannot talk of pitch surfaces in the strict sense (see Sect. 10.3). However, with reference to the transmission of motion, since the worm behaves like a rack of endless length, which drags in

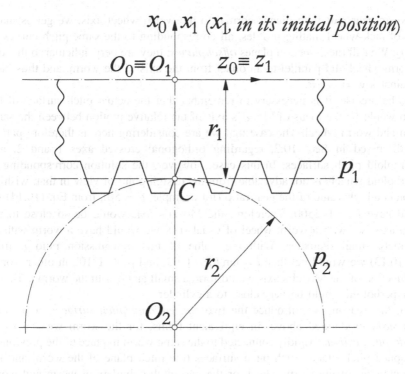

Fig. 11.10 Coordinate systems $O_1(x_1, y_1, z_1)$, and $O_0(x_0, y_0, z_0)$, and geometric quantities related to kinematics

rotation the worm wheel, we can recognize the existence of a pitch plane for the rack, and a pitch cylinder for the worm wheel. The relative motion between worm and worm wheel is therefore defined by a plane, parallel to the worm and worm wheel axes (and thus perpendicular to the shortest distance), and rigidly connected to the worm. This plane rolls without sliding over a cylinder, having as axis the axis of the worm wheel, and rigidly connected to the latter. Conventionally, we call this plane rigidly connected to the worm and the cylinder rigidly connected to the worm wheel *pitch plane of the worm* and, respectively, *pitch cylinder of the worm wheel*. The pitch plane, also called *pitch surface of the worm*, is given as the locus of the successive positions of the instantaneous axis of rotation in the relative motion of worm wheel to the worm threads, and is parallel to the worm wheel axis.

As shown in Fig. 11.10, the intersections of the above defined pitch plane and pitch cylinder with the middle plane of the worm wheel (this plane passes through the axis of the worm and is normal to the axis of the worm wheel) are respectively the straight-line p_1 and the circle p_2, which are tangent at the pitch point C. They are respectively the *pitch line* and the *pitch circle* of the rack-wheel pair, which we get sectioning the worm gear pair with the middle plane of the worm wheel. This plane coincides with the (x_0, z_0)-plane of the Cartesian coordinate system $O_0(x_0, y_0, z_0)$, that is with the plane having coordinate $y_0 = 0$. Performing a number of sections of the worm gear pair with planes normal to the worm wheel axis, we get as many pairs of rack-wheel mating profiles, all corresponding to the same pitch curves p_1 and p_2. We call these section planes *offset planes*; they are perpendicular to the axis of worm wheel and parallel to an offset from the axis of the worm, and thus have coordinates $y_0 = const$.

To be precise, it is necessary to remember that the actual pitch surface of the worm would be the locus of Mozzi's axis of the relative motion between the same worm and worm wheel. The case that we are considering here is therefore part of that discussed in Sect. 10.2, regarding orthogonal crossed axes 1 and 2, and hyperboloid pitch surfaces. In this case, however, the solution corresponding to hyperboloid wheels is not advisable from the design point of view in that, wishing to obtain a high value of the gear ratio (for example, $u = 50$), from Eq. (10.14) we would have $i^2 = 1/2500$. Therefore, the Mozzi's axis would be so close to the worm axis so, with a worm wheel of usual size, we would have a worm with an extremely small diameter. With the value of this transmission ratio i, from Eq. (10.13) we would get then $i = \tan \beta_1^* = 0.02$, and $\beta_1^* \cong 1°10'$; in other words, the direction of the Mozzi's axis would form a small angle with the worm axis, and the hyperboloid would be very close to a cylinder.

For this reason, we introduce the two *conventional pitch surfaces* or *pseudo-pitch surfaces* defined above. In equivalent terms, for the worm we can use a *pseudo-pitch cylinder* rigidly connected to the same worm in place of the previously mentioned pitch plane. Such pitch surfaces (the pitch plane of the worm and the pitch cylinder of the worm wheel, or the two pitch cylinders of worm and worm wheel, with radii r_1 and r_2 respectively) provide the same main point of contact C, coinciding with the point of tangency of the pitch surfaces considered in the middle plane of the worm wheel. Of course, as for other types of gears, we differentiate between *reference pitch surfaces* and *operating pitch surfaces*.

In the more general case in which the shaft angle is different from $\pi/2$ ($\Sigma \neq \pi/2$), the characteristic quantities of the worm gear pair are those defined in Sects. 10.4 and 10.5 for crossed helical gears. In the most common case, the shaft angle Σ is equal to $\pi/2$ and, in this case, the characteristic quantities of the worm gear pair can be deduced as follows.

During the apparent motion of the worm, which is identified with the translation motion of the rack to which the worm is reduced along its axis, all its points have the same value of axial velocity, v. If we denote by ω_1 and p_z respectively the angular velocity and lead (i.e. the axial distance between two consecutive corresponding profiles of the same worm thread) of the worm, the absolute value of vector, v, will be given by:

$$v = \omega_1 \frac{p_z}{2\pi}. \tag{11.2}$$

However, the tangential velocity on the pitch cylinder of the worm wheel must be equal to the translation velocity of the pitch plane of the worm-rack; so, if we denote by ω_2 and r_2 respectively the angular velocity and radius of the pitch cylinder of the worm wheel, this last will have to be equal to:

$$r_2 = \frac{\omega_1}{\omega_2} \frac{p_z}{2\pi}. \tag{11.3}$$

The cylinder of radius r_2 and the plane tangent to it and perpendicular to the shortest distance or, in equivalent terms, the same cylinder of radius r_2 and the cylinder of radius r_1, coaxial with the worm and rigidly connected to the latter, are therefore the pitch surfaces of the worm gear pair. Equation (11.3) shows that the worm wheel pitch radius, r_2, is determined when the transmission ratio $i = (\omega_1/\omega_2)$ and lead p_z of the worm are given. This deduction is valid regardless of the pitch diameter $d_1 = 2r_1$ of the worm, as it is kinematically possible to obtain the desired transmission ratio whatever the pitch radius r_1 of the worm. The shortest distance between the axes, that is the center distance, varies with change of this radius, the determination of which also depends on the geometric quantities of the thread and sliding velocity at the start of meshing, which affect the efficiency of the worm gear pair.

Figure 11.11 shows the axial section and front view of a four-start cylindrical worm $(z_1 = 4)$, with right-hand threads. The thread helical surface, i.e. the helicoid that constitutes the thread flanks, is comprised between two cylindrical surfaces, coaxial with the helicoid, which are the tip and root cylinders. The radial distance between these cylinders is the thread depth, $h = (h_a + h_f)$. The *nominal size* of the worm is given by the *reference* or *middle cylinder* diameter, d_1, whose transverse section is the *middle circle* or *reference pitch circle*. Usually, the value of d_1 lies between $(25 \div 60)\%$ of the center distance, a. The *axial pitch*, p_x, is related to the

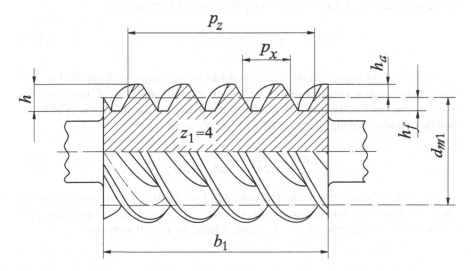

Fig. 11.11 Main quantities of a four-start cylindrical worm, with right-hand threads

axial module, m_x, by the relationship $p_x = \pi m_x$. In the case of worm gear pairs with driving worm, m_x is the *reference module* on which all calculations and specifications are based. For this reason, it is customary to denote this module simply by m, i.e. without any subscript.

The *axial pitch* p_x of the worm is the distance, measured along a generatrix of its pseudo-pitch cylinder, from a point of a given thread to the corresponding point of the adjacent thread (Fig. 11.11). It corresponds to the axial distance between two neighboring straight lines which represent the helices of two adjacent threads of the worm when we develop the pitch cylinder on a plane (Fig. 8.11). The *lead* p_z of the worm is the axial distance between two consecutive points of the same worm-thread when the thread helix makes a complete turn about the axis (Fig. 8.9). The axial pitch and lead of the worm are related as follows:

$$p_z = z_1 p_x. \tag{11.4}$$

Thus, the axial pitch is equal to the quotient of the lead divided by the number of threads (or number of starts). Of course, if the number of starts is one, we have $p_z = p_x$. The lead of the worm is an integer multiple of the axial pitch of the same worm. The number of starts usually does not exceed 4 (exceptionally 6). Only single-start worms are used when self-locking or irreversibility of the worm drive system is desired. Note that the worm axial pitch p_x does not change whatever the coaxial cylinder of the worm between the tip and root cylinders.

All the intersections of the helicoid with these coaxial cylinders are helices having the same *lead*, but different *lead angle*, γ (Fig. 8.9). This is the angle subtended between a tangent to the helix under consideration and a transverse plane of the worm. If r is the radius of any coaxial cylinder between tip and root cylinders, and $p_z = const$ the lead of the worm, the lead angle of the helix obtained as the intersection of this cylinder with the helicoid is given by:

$$\tan \gamma = \frac{p_z}{2\pi r} = \frac{p_z}{\pi d}. \tag{11.5}$$

Thus, the lead angle γ of the helix decreases with increasing diameter $d = 2r$ of the cylinder under consideration. As *reference lead angle* γ_1 of the worm we assume, conventionally, the one related to the helix on the reference cylinder, or middle cylinder, having diameter $d_1 = d_{m1}$; so, we have:

$$\tan \gamma_1 = \frac{p_z}{2\pi r_1} = \frac{z_1 p_x}{\pi d_1} = \frac{z_1}{d_1 P_{ax}}, \tag{11.6}$$

where P_{ax} is the diametral pitch. The reference lead angle γ_1 on the worm reference pitch cylinder, and the *operating lead angle* γ_{w1} on the worm operating pitch cylinder, having diameter $d_{w1} = 2r_{w1}$, are related by the relationship:

$$r_1 \tan \gamma_1 = r_{w1} \tan \gamma_{w1} = \frac{p_z}{2\pi} = p, \tag{11.7}$$

where $p = (p_z/2\pi)$ is the *screw parameter* (attention must be paid to the fact that here p has a different meaning that it has so far attributed to this symbol).

From Eqs. (11.6) and (11.7) we get:

$$\tan \gamma_{w1} = \frac{r_1}{r_{w1}} \tan \gamma_1 = \frac{z_1 p_x}{2\pi r_{w1}} = \frac{z_1}{2 r_{w1} P_{ax}}, \tag{11.8}$$

where r_{w1} is the radius of the chosen operating pitch cylinder. Note that the difference between r_{w1} and r_1 affects the shape of the lines of contact between the active flank surfaces of the worm threads and worm wheel teeth.

The helix angle, β, reference helix angle, β_1, and operating helix angle, β_{w1}, of the worm are complementary respectively to the lead angle, γ, reference lead angle, γ_1, and operating lead angle, γ_{w1}. Thus, we have: $\beta = [(\pi/2) - \gamma]$, $\beta_1 = [(\pi/2) - \gamma_1]$, and $\beta_{w1} = [(\pi/2) - \gamma_{w1}]$. Therefore, a worm can be thought as a helical gear wheel whose teeth make a complete revolution about the pitch cylinder.

In order to obtain a good efficiency of the worm gear pair, it is necessary to have a lead angle of the worm sufficiently high. The values of γ_{w1} can reach $(30 \div 40)^\circ$, for which the lead of the worm is great. If we wanted to use a single-start worm, with equal values of the thread thickness and thread space, to ensure good continuity of the transmission of motion, we should choose threads with depth too large. They would give rise to excessive sliding at the beginning of the meshing, and to cutting interferences during the generation of the worm wheel teeth. To overcome these drawbacks, we employ multi-start worms, with less thread depth; the pitch of their generating rack will be a sub-multiple of the lead of the worm.

We now consider the quantities of the worm related to its operating pitch cylinder. The normal module, m_n, and normal pitch, $p_n = \pi m_n$, are related respectively to the axial module, $m_x = m$, and axial pitch, p_x, by the relationships:

$$m_n = m_x \cos \gamma_{w1} = m \cos \gamma_{w1}, \tag{11.9}$$

$$p_n = p_x \cos \gamma_{w1}. \tag{11.10}$$

As indicative value, we can assume $(d_{w1}/15 \leq m \leq d_{w1}/6)$, with an average value $m \cong 0.1 d_{w1}$.

From Eqs. (11.8) and (11.10), we infer:

$$\sin \gamma_{w1} = \frac{z_1 p_n}{2\pi r_{w1}} = \frac{z_1 p_n}{\pi d_{w1}}. \tag{11.11}$$

The ratio between the reference pitch cylinder diameter, d_1, and the axial module, $m_x = m$, defines the *diameter quotient* z_F of the worm, also called *form number of the worm*, i.e.:

$$z_F = \frac{d_1}{m}. \tag{11.12}$$

This quantity is an important parameter of the worm sizing, as it determines the worm shape and, consequently, the bending load capacity of the worm. The worm reference pitch diameter is chosen as $d_1 = z_F m$. Keeping other parameters of the worm gear pair constant (for example, the transmission ratio, i, and the center distance, a), for a single-start worm $(z_1 = 1)$ we infer that, the smaller the value of z_F, the smaller is the worm reference diameter, the greater the lead angle and the maximum deflection of the worm shaft, and the smaller the tangential velocity, and vice versa. In practical applications, we have $(6 \leq z_F \leq 15)$, with an average value $z_F = 10$.

For small values of the operating lead angle $(\gamma_{w1} \leq 15°)$, the addendum, dedendum, and thread depth are chosen with reference to the axial module, $m_x = m$. Equation (11.9) shows that, for the same normal module, when the operating lead angle increases, also the axial module increases, and so does the thread depth. For larger values of the operating lead angle $(\gamma_{w1} > 15°)$, the addendum, dedendum, and thread depth are chosen with reference to the normal module, as in this way certain unfavorable consequences (for example, peaked teeth on hob, and peaked threads on worm and worm wheel) can be avoided.

The diameters of the tip and root reference cylinders of the worm are given respectively by:

$$d_{a1} = d_1 + 2h_{a1} \quad d_{f1} = d_{a1} - 2h_1, \tag{11.13}$$

where $h_1 = (h_{a1} + h_{f1})$ is the worm thread depth. Depending on the manufacturing process used, the variability range of the bottom clearance, c, is $(0.167m \leq c \leq 0.300m)$, with a preferred value between $(0.20 \div 0.25)m$. The bottom clearance should be as small as possible.

The *worm face width*, i.e. the length of the worm, b_1 (see Fig. 11.11), can be assumed approximately equal to:

$$b_1 \cong 2.5m\sqrt{z_2 + 1}; \tag{11.14}$$

as a broad approximation, we can choose $b_1 \cong 5p_x$.

On the reference pitch cylinder, the thread thickness of the worm in a normal section is given by:

$$s_n = \frac{\pi m_n}{2} = \frac{\pi m}{2} \cos \gamma_1. \tag{11.15}$$

The values of the pressure angles used in worm gear pairs depend on the values of the lead angles. They must be large enough to avoid undercut of the threads on the flank where the contact ends. If the lead angle increases, the cutting conditions of the worm wheel cutter become unfavorable. For manufacturing reasons, greater pressure angles are chosen for larger values of the lead angles. In the normal, axial and transverse sections of the worm, the reference normal, axial and transverse pressure angles α_n, α_x and α_t are notoriously related by the following relationships:

$$\tan \alpha_x = \frac{\tan \alpha_n}{\cos \gamma_1} \qquad \tan \alpha_t = \frac{\tan \alpha_n}{\sin \gamma_1}; \tag{11.16}$$

therefore, we have:

$$\tan \alpha_n = \tan \alpha_x \cos \gamma_1 = \tan \alpha_t \sin \gamma_1. \tag{11.17}$$

This equation relates the pressure angles in normal, axial and transverse sections, and the lead angle of the helix at the reference pitch cylinder. *Mutatis mutandis*, similar equations to Eqs. (11.16) and (11.17) can be written for any other coaxial cylinder between the tip and root cylinders of the worm: of course, profile angles in normal, axial and transverse sections, and the lead angle of the helix at the cylinder under consideration are involved.

In the particular case of an involute worm, we have:

$$r_{b1} = r_1 \cos \alpha_t. \tag{11.18}$$

From Fig. 8.9, where we put $r = r_1$ and $\gamma = \gamma_1$, and taking into account Eq. (11.18), we get:

$$\frac{r_{b1}}{r_1} = \frac{\tan \gamma_1}{\tan \gamma_{b1}} = \cos \alpha_t. \tag{11.19}$$

Equation (11.17) yields:

$$\tan \alpha_t = \frac{\tan \alpha_x}{\tan \gamma_1}. \tag{11.20}$$

The radius of the base cylinder (see Fig. 8.9), can be written as follows:

$$r_{b1} = \frac{p_z}{2\pi \tan \gamma_{b1}} = \frac{p}{\tan \gamma_{b1}} = \frac{p \cos \alpha_t}{\tan \gamma_1} = \frac{p}{\tan \gamma_1 (1 + \tan^2 \alpha_t)^{1/2}}. \tag{11.21}$$

Finally, Eqs. (11.21) and (11.20) yield:

$$r_{b1} = \frac{p}{(\tan^2 \gamma_1 + \tan^2 \alpha_x)^{1/2}}. \tag{11.22}$$

This equation expresses the base cylinder radius of an involute worm as a function of the screw parameter, p, reference lead angle, γ_1, ad axial pressure angle, α_x.

Figure 11.12 shows the transverse and axial sections of a single enveloping worm wheel for a shaft angle $\Sigma = \pi/2$. Many of the quantities that define the geometry of a worm wheel refer to its mid-plane, i.e. the plane perpendicular to its axis and containing the axis of the mating worm. The toothing is between the *root toroid*, i.e. the toroidal surface tangent to the root surface of the teeth, and the *outside surface*. This surface is formed by the *throat form surface*, in the region that straddles the mid-plane (the throat is the portion of the outside surface having toroidal geometry), and the *outside cylinder*, in the two outer regions (the outside cylinder is the cylindrical part of the outside surface). The *reference circle* of the worm wheel is the inner circle of intersection of the *reference toroid* and the mid-plane. For more details on the worm gear geometry, we refer the reader to ISO 1122-2: 1999 (E/F) [23].

The diameter of the reference circle of the worm wheel is given by:

$$d_2 = z_2 m, \tag{11.23}$$

while its *throat diameter*, i.e. the diameter of the throat circle at mid-plane, is given by:

$$d_{a2} = d_2 + 2h_{a2}. \tag{11.24}$$

Of course, Eqs. (11.23) and (11.24) refer to worm wheels with zero-shifted toothing.

The *throat form radius*, also called *radius of worm wheel face*, is the radius of the circle that surrounds the axial section of the throat; it is given by:

$$r_{th} = a - \frac{d_{a2}}{2}. \tag{11.25}$$

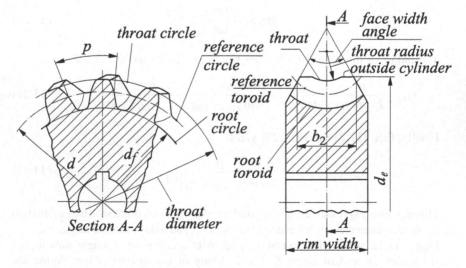

Fig. 11.12 Geometry and main characteristics of an enveloping worm wheel

The diameter of the outside cylinder, i.e. the outside diameter, d_e, of the worm wheel, is a function of its helix angle, β_2, and depending on whether $\beta_2 \leq 15°$ or $\beta_2 > 15°$, is given respectively by:

$$d_e = d_{a2} + m \tag{11.26}$$

$$d_e = d_{a2} + m_n. \tag{11.27}$$

The diameter of the root circle of worm wheel is given by:

$$d_{f2} = d_{a2} - 2h_2. \tag{11.28}$$

We define as *face width angle* (Fig. 11.12) the angle at the center that, in the generation circle of the reference toroid, is between the points of intersection of this circle with the end faces of the teeth. The face width, b_2, of the worm wheel is the distance between two planes perpendicular to the axis, which contain the circles of intersection of the reference toroid with the end faces of the teeth. In the most frequent case in which the teeth are symmetrical with respect to the mid-plane, the face width is the length of the chord (parallel to the axis) of the generating circle of the reference toroid, between the points of intersection of this circle with the end faces of the teeth. This quantity is defined using, as a usual guideline, the following relationship:

$$b_2 \cong 2m\left(0.5 + \sqrt{z_F + 1}\right). \tag{11.29}$$

We define as *wheel rim* and *rim width* (Fig. 11.12) the rim that contains the worm wheel teeth and, respectively, the maximum axial dimension of the rim. In the case of a worm gear pair with crossed orthogonal axes ($\Sigma = \pi/2$), in order to have a proper meshing, the helix angle of the worm wheel must satisfy the following conditions:

- The lead angle of the worm must be equal to the helix angle of the worm wheel.
- The axial pitch of the worm must be equal to the transverse pitch of the worm wheel.

11.4 Gear Ratio and Interdependences Between Worm and Worm Wheel Quantities

Let us consider the general case of a worm gear pair with shaft angle $\Sigma \neq \pi/2$. Figure 11.13a, b shows respectively the operating pitch cylinders and triangle of velocity of a right-hand worm gear pair. As Fig. 11.13a shows, the worm is placed above the worm wheel, and the operating pitch cylinders, having respectively radii

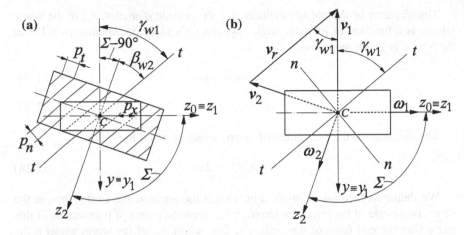

Fig. 11.13 Right-hand worm gear pair: **a** operating pitch cylinders; **b** triangle of velocity

r_{w1} and r_{w2}, are tangent at pitch point C. In the plane through this pitch point and normal to the shortest distance, both helices obtained as the intersections of the operating pitch cylinders with the worm thread surface and worm wheel tooth surface have a common tangent, the straight-line t–t.

We want to determine first the transmission ratio $i = \omega_1/\omega_2$ of the worm gear pair, considering r_{w1}, r_{w2}, γ_{w1} and Σ as input data (Σ is measured clockwise from $z_0 \equiv z_1$ to z_2), and assuming as given the direction and magnitude of the angular velocity vector, ω_1. We must first determine the direction and magnitude of the worm wheel angular velocity vector, ω_2, considering that the line of application of vector ω_2 is the z_2-axis. As Fig. 11.13b shows, in the plane through the pitch point C and normal to the center distance, the sliding velocity vector or relative velocity vector, $v_r = (v_1 - v_2)$, is collinear to the straight line t–t. In addition, the components of the velocity vectors v_1 and v_2 in the direction of the normal n–n to the tangent t–t must be equal, that is they must have the same direction and magnitude, for which we can write:

$$\omega_1 r_{w1} \sin\gamma_{w1} = \omega_2 r_{w2} \sin(\Sigma - \gamma_{w1});\qquad(11.30)$$

this is why $v_1 = \omega_1 r_{w1}$ and $v_2 = \omega_2 r_{w2}$.

From this equation we infer that, for $\Sigma > \gamma_{w1}$, ω_2 is positive, i.e. the vector ω_2 has the same direction of the positive direction of the z_2-axis, while for $\Sigma < \gamma_{w1}$, ω_2 is negative, and thus the vector ω_2 has opposite direction with respect to the positive direction of the z_2-axis. Note that, for $\Sigma = \gamma_{w1}$, Eq. (11.30) is not satisfied, because the helix on the operating pitch cylinder of the worm wheel becomes a circle, and the components of the velocity vectors v_1 and v_2 along the straight line $n - n$ are different.

From the same Eq. (11.30) we also infer the following relationship that gives the transmission ratio, i, as a function of r_{w1}, r_{w2}, γ_{w1} and Σ, with the condition $\Sigma \neq \gamma_{w1}$:

$$i = \frac{\omega_1}{|\omega_2|} = \pm \frac{r_{w2} \sin(\Sigma - \gamma_{w1})}{r_{w1} \sin \gamma_{w1}}, \tag{11.31}$$

where the upper and lower signs correspond respectively to the cases in which $\Sigma > \gamma_{w1}$, and $\Sigma < \gamma_{w1}$.

Figure 11.14a, b shows respectively the operating pitch cylinders and triangle of velocity of a left-hand worm gear pair. In this case, by using the same procedure described above, we obtain:

$$i = \frac{r_{w2} \sin(\Sigma + \gamma_{w1})}{r_{w1} \sin \gamma_{w1}}, \tag{11.32}$$

From this equation, in which the magnitude of γ_{w1} is considered as a positive value, we deduce that, for the chosen direction of the vector ω_1, the vector ω_2 has direction opposite to the positive direction of the z_2-axis.

In the case where $\Sigma = 90°$, which is what we find usually in practical applications, from both Eqs. (11.31) and (11.32) we obtain:

$$i = \frac{r_{w2}}{r_{w1}} \cot \gamma_{w1}. \tag{11.33}$$

The normal and axial pitches of the worm are related by Eq. (11.10), but the normal pitch is the same for the worm and worm wheel. From Fig. 11.13a and

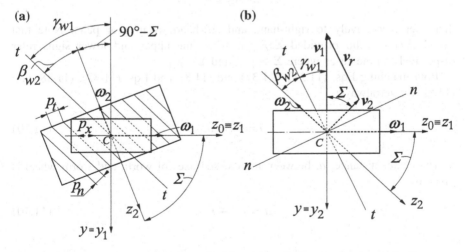

Fig. 11.14 Left-hand worm gear pair: **a** operating pitch cylinders; **b** triangle of velocity

taking into account Eq. (11.10), we obtain the following relationship of the transverse pitch, p_t, of the worm wheel:

$$p_t = \frac{p_n}{\cos \beta_{w2}} = \frac{p_x \cos \gamma_{w1}}{\cos[(\pi/2) \pm (\gamma_{w1} - \Sigma)]} = \pm \frac{p_x \cos \gamma_{w1}}{\sin(\Sigma - \gamma_{w1})}, \tag{11.34}$$

where β_{w2} is the helix angle on the operating pitch cylinder of the worm wheel. In this equation, which is valid provided $\Sigma \neq \gamma_{w1}$, the upper and lower signs refer respectively to the cases for which $\Sigma > \gamma_{w1}$, and $\Sigma < \gamma_{w1}$; the equation gives a positive sign for p_t.

By using the same procedure and with reference to Fig. 11.14a, for a left-hand worm gear pair we obtain:

$$p_t = \frac{p_x \cos \gamma_{w1}}{\sin(\Sigma + \gamma_{w1})}. \tag{11.35}$$

In the case where $\Sigma = 90°$, from both Eqs. (11.34) and (11.35) we obtain $p_t = p_x$.

Since

$$p_t z_2 = 2\pi r_{w2}, \tag{11.36}$$

where z_2 is the number of teeth of the worm wheel (attention must be made here to the different meaning of z_2), from Eqs. (11.34) and (11.35) we obtain:

$$r_{w2} = \pm \frac{p_x z_2 \cos \gamma_{w1}}{2\pi \sin(\Sigma - \gamma_{w1})}, \tag{11.37}$$

$$r_{w2} = \frac{p_x z_2 \cos \gamma_{w1}}{2\pi \sin(\Sigma + \gamma_{w1})}, \tag{11.38}$$

that refer respectively to right-hand and left-hand worm gear pairs. Note that Eq. (11.37) is valid provided $\Sigma \neq \gamma_{w1}$, while the upper and lower signs refer respectively to the cases where $\Sigma > \gamma_{w1}$, and $\Sigma < \gamma_{w1}$.

Both matching Eqs. (11.31), (11.37) and (11.8), and Eqs. (11.32), (11.38) and (11.8), we get always:

$$i = \frac{z_2}{z_1} = u. \tag{11.39}$$

The center distance, a, between the crossed axes of worm and worm wheel is given by:

$$a = r_{w1} + r_{w2}, \tag{11.40}$$

where r_{w1} is given by [see Eq. (11.8)]:

$$r_{w1} = \frac{z_1 p_x}{2\pi \tan \gamma_{w1}},$$ (11.41)

while r_{w2} is expressed by Eqs. (11.37) or (11.38), depending on whether the worm has right-hand or left-hand helix.

In the case where $\Sigma = 90°$ and the operating pitch cylinders coincide with the reference pitch cylinders, the center distance is given by:

$$a = \frac{p_x}{2\pi}\left(\frac{z_1}{\tan \gamma_{w1}} + z_2\right).$$ (11.42)

11.5 Elements of Differential Geometry of Surfaces

In order to choose some geometrical quantities of a worm gear pair, or to determine whether it is appropriate or not to use profile-shifted toothing, the worm gear designer needs to define beforehand the surface of contact of the gear in its operating conditions. To determine this surface by means of analytical or analytical-numerical methods, we should recall here briefly some fundamental concepts of differential geometry of surfaces (see Kreiszig [27], Louis [34], Stoker [62], Smirnov [60]). As we have already mentioned in the introduction to this chapter, these concepts are so general that they can be used for all the gears covered in this textbook, including hypoid gears.

A surface σ in a three-dimensional space can be defined as a locus of points whose *position vector*, r, directed from the origin O of any fixed Cartesian coordinate system $O(x, y, z)$ to a current point P on the surface, is a function of two independent variable parameters, u and v (Fig. 11.15a). The *parametric representation* of the surface can be given or in *vector form*, as follows:

$$r(u, v) = x(u, v)\mathbf{i} + y(u, v)\mathbf{j} + z(u, v)\mathbf{k},$$ (11.43)

where $r(u, v)$ is the position vector, and $\mathbf{i}, \mathbf{j}, \mathbf{k}$ are the *unit vectors* along the x-, y-, and z-axes, respectively, or in *scalar form*, as follows:

$$x = x(u, v) \quad y = y(u, v), \quad z = z(u, v),$$ (11.44)

where (x, y, z) are the Cartesian coordinates of points of the surface, and $x(u, v), y(u, v)$ and $z(u, v)$ are some definite, continuous and single-valued functions of the variable parameters u and v. If we eliminate the parameters u and v from

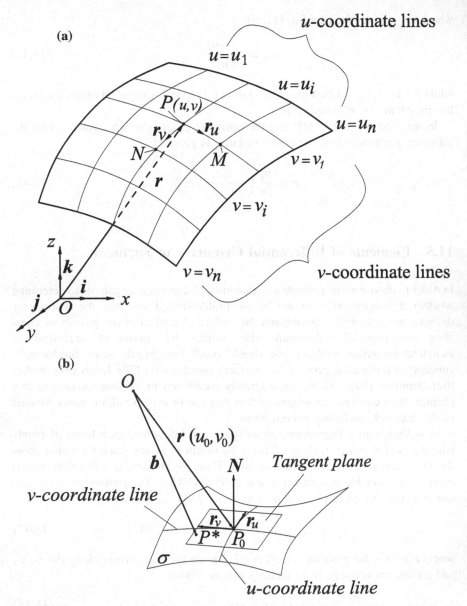

Fig. 11.15 **a** Position vector and coordinate lines of a surface; **b** tangent plane to a surface at a regular point

Eqs. (11.44), we obtain the following Cartesian equation of the surface, in an implicit form:

$$F(x, y, z) = 0, \tag{11.45}$$

which can also be written, in explicit form, as follows:

$$z = f(x, y). \tag{11.46}$$

Generally, one-to-one correspondence between the pairs of parameters (u, v) belonging to the plane of parameters and points of the surface σ is not guaranteed. It may happen that a given point $r(u, v)$ of the surface corresponds to more than one point of the plane of parameters. Here we assume that there is a one-to-one correspondence between the pairs of numbers (u, v) belonging to the plane of parameters and points of the surface. In this case, the surface does not have points of self-intersection, and then it is called a *simple surface*. Parameters (u, v) are called *curvilinear coordinates* or *Gaussian coordinates* on the surface. In this type of simple surface, it is immediate to define a family of ∞^1 *coordinate lines*, each corresponding to straight lines parallel to the coordinate axes of the plane (u, v). These are named u-coordinate lines (or u-lines) or v-coordinate lines (or v-lines), depending on whether they correspond to straight lines parallel to u-axis (these coordinate lines have equation $v = const$), or to straight lines parallel to v-axis (these coordinate lines have equation $u = const$).

The coordinate lines having equations $u = const$ and $v = const$ represent two families of parametric curves on the surface. Thus, a surface can be completely described by a double infinite set of such parametric curves where the position of any point on the surface is determined by the values of u and v. If the u- and v-coordinate lines are mutually perpendicular at all points on a surface (in this case, the angles between the tangents to these lines are equal to $\pi/2$), the curvilinear coordinates are said to be orthogonal. Here we will use extensively these *orthogonal curvilinear coordinates*.

The partial derivatives of the position vector $r(u, v)$ with respect to the curvilinear coordinates (u, v), here written with the notations

$$r_u = \frac{\partial r}{\partial u} \quad r_v = \frac{\partial r}{\partial v}, \tag{11.47}$$

are the *tangent vectors* at any point of the surface σ to the u- and v-coordinate lines, respectively. The directions of vectors r_u and r_v coincide, respectively, with directions of the vectors dr_u and dr_v, because u and v are scalar quantities. These vectors represent the chords joining two neighboring points P and M on u-coordinate line ($v = const$), having coordinates u and $u + \Delta u$, and two neighboring points P and N on v-coordinate line ($u = const$), having coordinates v and $v + \Delta v$. In the limit, as $\Delta u \to 0$ and $\Delta v \to 0$, these chords approach the tangents at a point

$P(u,v)$ along the coordinate lines u and v, respectively. Since u and v are assumed to be orthogonal, then the scalar product $r_u \cdot r_v$ must be equal to zero, i.e.:

$$r_u \cdot r_v = 0. \tag{11.48}$$

Consider now a small region of a *smooth surface* σ near a current point $P(u,v)$. As smooth surface we define a surface that is continuous, i.e. the function $F(x,y,z)$ and its first derivatives are continuous, and the latter are not simultaneously equal to zero; in addition, it contains no discontinuity of slope, i.e. *singular points*, as creases or vertices. It is to be noted that, in the case of creases, two branches of the surface have a common line, called *edge of regression*, and only a half-plane exists that is limited with the tangent line drawn at any point of this edge of regression, while in the case of vertices the tangent plane does not exist. If we draw various curves on the surface σ through a *regular point* P, the tangents to these curves are placed on one plane called the *tangent plane* to the surface at point P. A line that is perpendicular to the tangent plane and passes through point P is called the *normal* to the surface at point P, and is denoted by N. Since the surface is smooth, the tangent plane and, hence, the normal to the surface at point P are uniquely determined (see also Litvin and Fuentes [33]).

Points of a surface at which a tangent plane does not exist are called singular points. The tangent plane to a surface σ at a *regular point* P (the point is a regular point when the tangent plane exists) is determined by a pair of vectors r_u and r_v that are tangent to the u-coordinate line and v-coordinate line, respectively (Fig. 11.15b). We assume that $r_u \neq 0$ and $r_v \neq 0$ and that vectors r_u and r_v are not collinear. The tangent plane at point $P_0(u_0, v_0)$ of a smooth surface (point P_0 is defined by the position vector $r(u_0, v_0)$) is given by the *scalar triple product* (also called *mixed* or *box product*):

$$a \cdot (r_u \times r_v) = 0, \tag{11.49}$$

where $(r_u \times r_v)$ is the *cross product* of vectors r_u and r_v, and

$$a = b - r(u_0, v_0); \tag{11.50}$$

position vector b is drawn from the same origin of the position vector $r(u_0, v_0)$ to an arbitrary point P^* on the tangent plane (Fig. 11.15b). Equation (11.49) indicates that vector $a = \overline{P_0 P^*}$, applied to point P_0 defined by the position vector $r(u_0, v_0)$, lies in the plane drawn through vectors r_u and r_v.

The normal N to the surface σ at point P, and therefore perpendicular to the tangent plane to the surface at same point P, is given by the cross product:

$$N = r_u \times r_v; \tag{11.51}$$

its direction depends on the order of factors in the cross product. The normal N may be written in terms of projections on the coordinate axes, by the relationship:

$$N = \begin{vmatrix} i & j & k \\ x_u & y_u & z_u \\ x_v & y_v & z_v \end{vmatrix} = \begin{vmatrix} y_u & z_u \\ y_v & z_v \end{vmatrix} i + \begin{vmatrix} z_u & x_u \\ z_v & x_v \end{vmatrix} j + \begin{vmatrix} x_u & y_u \\ x_v & y_v \end{vmatrix} k. \tag{11.52}$$

The unit normal is given by:

$$n = \frac{N}{|N|} = \frac{N_x}{|N|} i + \frac{N_y}{|N|} j + \frac{N_z}{|N|} k, \tag{11.53}$$

provided $|N| = \left(N_x^2 + N_y^2 + N_z^2 \right)^{1/2} \neq 0$.

The point $r(u_0, v_0)$ of the surface is a singular point if

$$r_u \times r_v = 0, \tag{11.54}$$

and this happens when at least one of the two vectors r_u and r_v is equal to zero, or when the two vectors are collinear.

11.6 Parametric Equations of a Helicoid

A helicoid is a ruled or not-ruled surface, generated by a planar curve, which performs a screw motion about an axis perpendicular to the plane at which the curve itself belongs. For a worm, this generating curve is nothing more than the transverse profile of the helicoid. With reference to the Cartesian coordinate system $O_a(x_a, y_a, z_a)$, rigidly connected to the generation curve, this planar curve (Fig. 11.16a) can be described by the following parametric equation in vector form:

$$r_a(\vartheta) = r_a(\vartheta) \cos \vartheta i_a + r_a(\vartheta) \sin \vartheta j_a, \tag{11.55}$$

where $r_a(\vartheta)$ is the position vector of a current point P on the generation curve, i.e. the polar equation of the generation curve in the plane $z_a = 0$, $r_a(\vartheta) \cos \vartheta$ and $r_a(\vartheta) \sin \vartheta$ are the components of the position vector along the x_a-axis and y_a-axis, ϑ is the independent variable, which coincides with the angle between the x_a-axis and position vector, while i_a and j_a are the unit vectors of the x_a-axis and y_a-axis. It should be noted that, as Fig. 11.16b shows, the Cartesian coordinate system $O_a(x_a, y_a, z_a)$ can be regarded as an auxiliary and movable system. This auxiliary system, with respect to the fixed coordinate system $O_1(x_1, y_1, z_1)$, rigidly connected to the worm, has the z_a-axis that moves uniformly along the z_1-axis width velocity $p\zeta$, while the x_a-axis and y_a-axis are rotated of the angles ζ with respect to the

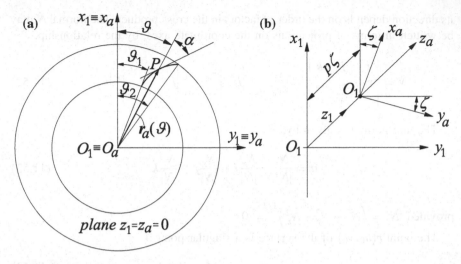

Fig. 11.16 **a** Generating transverse profile of a helicoid in the Cartesian coordinate system $O_a(x_a, y_a, z_a)$; **b** Cartesian coordinate systems $O_1(x_1, y_1, z_1)$ and $O_a(x_a, y_a, z_a)$

x_1-axis and y_1-axis. The movable coordinate system $O_a(x_a, y_a, z_a)$ coincides with the coordinate system $O_1(x_1, y_1, z_1)$ at the moment when the screw motion of the first with respect to the second begins.

In this way, the generation transverse profile of the helicoid, which has a given position in the plane with coordinate z_1, will have a position rotated of the angle ζ with respect to this position in the plane with coordinate $(z_1 + p\zeta)$.

In the same auxiliary coordinate system $O_a(x_a, y_a, z_a)$, the parametric equations in scalar form of the current point P on the generation curve of helicoid can be written as follows:

$$x_a = r_a(\vartheta) \cos \vartheta \quad y_a = r_a(\vartheta) \sin \vartheta \quad z_a = 0, \tag{11.56}$$

where $r_a(\vartheta)$ is the absolute value of the distance of point P from the origin O_a. As Fig. 11.16a shows, the angle, ϑ, between the x_a-axis and position vector $r_a(\vartheta)$ has a variability range $(\vartheta_1 \leq \vartheta \leq \vartheta_2)$. The transverse pressure angle, α, at a point, i.e. the angle between the position vector $r_a(\vartheta)$ and the tangent to the generation curve at current point P (Fig. 11.16a) is given by:

$$\alpha = \arctan\left(\frac{r_a(\vartheta)}{r_\vartheta}\right), \tag{11.57}$$

where $r_\vartheta = dr_a(\vartheta)/d\vartheta$.

The z_1-axis is the axis of the screw motion of the generating curve (Fig. 11.16b). This motion is defined by the angle of rotation ζ of a plane (x_a, y_a) about the

z_1-axis, and the axial displacement of the z_a-axis along the fixed z_1-axis with uniform axial velocity $p\zeta$. Thus, this axial velocity is proportional to ζ, the coefficient of proportionality being the screw parameter p, which represents the displacement along the z_1-axis corresponding to an angle of rotation ζ equal to one radian. Of course, it will be $p = p_z/2\pi$ [see Eq. (11.7)], since the lead p_z corresponds to one complete revolution. The sign of p is positive or negative for a right-hand or left-hand screw motions, respectively.

In the fixed Cartesian coordinate system $O_1(x_1, y_1, z_1)$, the parametric equations in scalar form of a current point P of the helicoid surface thus generated (this helicoid is the thread flank surface of the worm) can be written as follows:

$$x_1 = r_a(\vartheta)\cos(\vartheta + \zeta) \quad y_1 = r_a(\vartheta)\sin(\vartheta + \zeta) \quad z_1 = p\zeta, \tag{11.58}$$

where $(\vartheta_1 \leq \vartheta \leq \vartheta_2)$ and $(0 \leq \zeta \leq 2\pi)$. These equations are obtained taking into account the Eqs. (11.56) and the following matrix equation, where the coordinate transformation matrix M_{1a} in the transition from auxiliary coordinate system $O_a(x_a, y_a, z_a)$ to the fixed coordinate system $O_1(x_1, y_1, z_1)$ appears:

$$\begin{Bmatrix} x_1 \\ y_1 \\ z_1 \\ 1 \end{Bmatrix} = M_{1a} \begin{Bmatrix} x_a \\ y_a \\ z_a \\ 1 \end{Bmatrix} = \begin{bmatrix} \cos\zeta & -\sin\zeta & 0 & 0 \\ \sin\zeta & \cos\zeta & 0 & 0 \\ 0 & 0 & 1 & p\zeta \\ 0 & 0 & 0 & 1 \end{bmatrix} \begin{Bmatrix} x_a \\ y_a \\ z_a \\ 1 \end{Bmatrix}. \tag{11.59}$$

The same helicoid surface can be described, in equivalent terms, by the following matrix equation:

$$r_1(\vartheta, \zeta) = M_{1a}(\vartheta)r_a(\vartheta). \tag{11.60}$$

Using Eqs. (11.55) and (11.60), and taking account of the coordinate transformation matrix, M_{1a}, we can write:

$$r_1(\vartheta, \zeta) = r_a(\vartheta)\cos(\vartheta + \zeta)i_1 + r_a(\vartheta)\sin(\vartheta + \zeta)j_1 + p\zeta k_1, \tag{11.61}$$

where i_1, j_1, k_1 are the unit vectors of the x_1-axis, y_1-axis, and z_1-axis, and $p = (p_z/2\pi)$ is the screw parameter. This equation is the parametric equation of the helicoid surface in vector form. It should be noted that the planes $\vartheta = const$ and $\zeta = const$ are the axial and transverse planes of the worm, respectively.

The normal to the helicoid surface in its current point P, i.e. the perpendicular to the tangent plane to the same surface at point P, is given by the cross product:

$$N_1 = \frac{\partial r_1}{\partial \vartheta} \times \frac{\partial r_1}{\partial \zeta} = \frac{r_a(\vartheta)}{\sin\alpha}[p\sin(\vartheta + \zeta + \alpha)i_1 - p\cos(\vartheta + \zeta + \alpha)j_1 + r_a(\vartheta)\cos\alpha k_1].$$

$$\tag{11.62}$$

The unit normal to helicoid surface is given by the equation:

$$n_1 = \frac{N_1}{|N_1|} = \frac{1}{\left[p^2 + r_a^2(\vartheta)\cos^2\alpha\right]^{1/2}} \left[p\sin(\vartheta + \zeta + \alpha)i_1 - p\cos(\vartheta + \zeta + \alpha)j_1\right.$$
$$\left. + r_a(\vartheta)\cos\alpha k_1\right].$$

(11.63)

A helicoid with ruled surface is generated by a screw motion of a generation straight line which may intersect the z_1-axis of the screw motion, or it may form with the latter a crossed angle. In the first case (type A worm, Fig. 11.2), the cutting edge of the cutter is placed in an axial plane, where it intersects the worm axis, and has a transverse pressure angle α_{0t}. In the second case (type I worm, Fig. 11.3), the cutting edge of the cutter does not pass through the worm axis, but it is constantly tangent to the base cylinder of radius r_b, and has a transverse pressure angle α_{0t} (Fig. 11.3a). For both worms, p_z is the lead, and γ_1 is the lead angle. For type A and type I worms, the lengths, l, of the generation straight lines are given, respectively, with reference to the axial and normal planes, and are related to the thread depth by means of the pressure angles referred to the same planes.

Here we consider useful to analyze the two cases with a general model, which includes both. Results regarding the two cases above are then obtained as special cases of the general model. To this end, we consider a cylinder of z_1-axis and radius r_b, and two coplanar straight lines PT and $PQ = l$, both converging at point P, and rigidly connected with each other (Fig. 11.17). The plane through PT and PQ is parallel to the z_1-axis, and tangent to the cylinder along the generatrix through point P, whose distance from z_1-axis is the radius r_b. In this tangent plane, the lines PT and PQ are inclined with respect to the transverse plane through point P by the angles γ_1 and α_1, respectively.

We designate by ζ and $p\zeta$, respectively, the angle of rotation and axial displacement of the screw motion; $p = (p_z/2\pi)$ is the parameter of the screw motion. During the screw motion of the line PQ about the z_1-axis, point P generates a helix on the cylinder, whose tangent is PT, while the generation line PQ generates the helicoid. Compared to the transverse plane through point P, the two straight lines PT and PQ form the angles γ_1 and α_1, which are respectively the lead angle and the transverse pressure angle of the cutter.

Using suitable auxiliary coordinate systems (see also Litvin and Fuentes [33]), the parametric equation in vector form of a helicoid with ruled surface can be written as follows:

$$r_1 = (r_b\cos\zeta - l\cos\alpha_1\sin\zeta)i_1 + (r_b\sin\zeta + l\cos\alpha_1\cos\zeta)j_1 + (p\zeta - l\sin\alpha_1)k_1.$$

(11.64)

Fig. 11.17 Helix on the base cylinder, tangent PT to helix, and generation straight line PQ of the helicoid

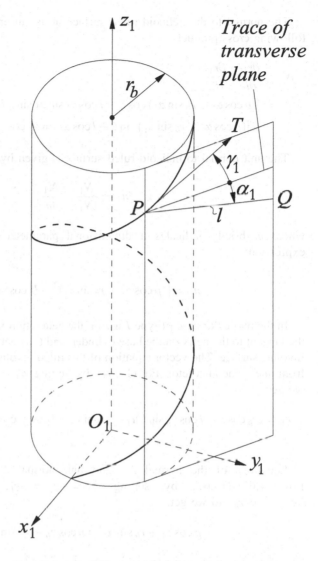

This general vector equation represents a ruled surface, where the l-lines (lines with $\zeta = const$) are straight lines, and ζ-lines (lines with $l = const$) are helices. It can be used not only for type I and type A worms, but also for type N worm, whose helicoid is a ruled surface generated by a generating planar curve lying on a plane that passes through the perpendicular to the worm axis and forms an angle γ_1 with the worm axis.

The normal to the helicoid ruled surface in its current point P is given by the following cross product:

$$
\begin{aligned}
N_1 &= \frac{\partial r_1}{\partial l} \times \frac{\partial r_1}{\partial \zeta} \\
&= [(p \cos \alpha_1 + r_b \sin \alpha_1) \cos \zeta - l \cos \alpha_1 \sin \alpha_1 \sin \zeta] i_1 \\
&\quad + [(p \cos \alpha_1 + r_b \sin \alpha_1) \sin \zeta + l \cos \alpha_1 \sin \alpha_1 \cos \zeta] j_1 + l \cos^2 \alpha_1 k_1.
\end{aligned}
\tag{11.65}
$$

The unit normal to helicoid ruled surface is given by:

$$
n_1 = \frac{N_1}{|N_1|} = \frac{N_1}{m},
\tag{11.66}
$$

where m briefly indicates a dimensional parameter given by the following expression:

$$
m = \left[(p \cos \alpha_1 + r_b \sin \alpha_1)^2 + l^2 \cos^2 \alpha_1 \right]^{1/2}.
\tag{11.67}
$$

In the particular case of type I worm, the generation straight line coincides with the tangent to the helix on the base cylinder, and thus generates a helicoid that is an involute surface. The vector equation of this ruled involute surface can be obtained from the general vector Eq. (11.64) by setting $\alpha_1 = -\gamma_1 = -\gamma_b$ (Fig. 11.17); we get:

$$
r_1 = (r_b \cos \zeta - l \cos \gamma_b \sin \zeta) i_1 + (r_b \sin \zeta + l \cos \gamma_b \cos \zeta) j_1 + (p\zeta + l \sin \gamma_b) k_1.
\tag{11.68}
$$

Equations of the normal N_1 and unit normal n_1 can be obtained from Eqs. (11.65)–(11.67) by setting $\alpha_1 = -\gamma_1 = -\gamma_b$, and considering that $\tan \gamma_b = p/r_b$. So we get:

$$
p \cos \alpha_1 + r_b \sin \alpha_1 = p \cos \gamma_b - r_b \sin \gamma_b = 0
\tag{11.69}
$$

$$
m^2 = (p \cos \alpha_1 + r_b \sin \alpha_1)^2 + l^2 \cos^2 \alpha_1 = l^2 \cos^2 \gamma_b
\tag{11.70}
$$

$$
N_1 = l \cos \gamma_b (\sin \gamma_b \sin \zeta i_1 - \sin \gamma_b \cos \zeta j_1 + \cos \gamma_b k_1).
\tag{11.71}
$$

If $l \cos \gamma_b \neq 0$, all points of the helicoid involute surface are regular points, and the equation of the unit normal at these points can be written as follows:

$$
n_1 = \cos \gamma_b (\sin \zeta i_1 - \cos \zeta j_1) + \cos \gamma_b k_1.
\tag{11.72}
$$

Thus, the direction of this unit normal does not depend on the surface parameter, l. Therefore, these unit normals have the same direction for all points of the generation straight line PQ, and the helicoid involute surface is a ruled developable surface.

Instead, in the particular case of type A worm (the Archimedean spiral worm), the generation straight line intersects the worm axis, that is the axis of the screw motion. The vector equation of ruled surface of this worm type (the equation is valid also for worms that are cut by straight-edged cutters) can be obtained from the general vector Eq. (11.64) by setting $r_b = 0$. So we get:

$$r_1 = l \cos \alpha_1 (- \sin \zeta i_1 + \cos \zeta j_1) + (p\zeta - l \sin \alpha_1) k_1. \tag{11.73}$$

Equation of the normal N_1 can be obtained from Eqs. (11.65), by setting $r_b = 0$, and dividing by a common factor $\cos \alpha_1$ (with the assumption that $\alpha_1 \neq 90°$). So we get:

$$N_1 = (p \cos \zeta - l \sin \alpha_1 \sin \zeta) i_1 + (p \sin \zeta + l \sin \alpha_1 \cos \zeta) j_1 + l \cos \alpha_1 k_1. \tag{11.74}$$

Equation of the unit normal n_1 is the following:

$$n_1 = \frac{N_1}{m}, \tag{11.75}$$

where m, from Eq. (11.70) by setting $r_b = 0$, is given by:

$$m = (p^2 + l^2)^{1/2}. \tag{11.76}$$

Directions of the normal N_1 and unit normal n_1, for the Archimedean spiral worm, depend on the location of the point on the generation straight line, and thus they depend on l. Therefore, the helicoid surface of this worm type is a ruled surface, but not a developable surface.

In the general case of a helicoid surface given by the vector Eq. (11.64), its profiles in transverse sections can be obtained by cutting the helicoid surface with transverse planes having equation $z_1 = c = const$. Since the transverse sections corresponding to the transverse planes $z_1 = 0$ and $z_1 = c$ give the same planar curve, i.e. the same profile, in two different positions, to simplify transformations we may consider only the transverse plane $z_1 = 0$. Therefore, the transverse sections may be obtained from the one corresponding to the transverse plane $z_1 = 0$, after rotation about the z_1-axis through an angle $\zeta = c/p$. From Eq. (11.64), by setting $z_1 = 0$, for which

$$l = \frac{p\zeta}{\sin \alpha_1}, \tag{11.77}$$

we obtain the following parametric equations in scalar form of the transverse profile of the helicoid corresponding to the transverse plane having equation $z_1 = 0$:

$$x_1 = r_b \cos \zeta - p\zeta \sin \zeta \cot \alpha_1 \quad y_1 = r_b \sin \zeta + p\zeta \cos \zeta \cot \alpha_1 \quad z_1 = 0. \quad (11.78)$$

The polar equation of the same transverse profile can be written as follows:

$$r_a(\vartheta) = \sqrt{x_1^2 + y_1^2} = \sqrt{\left[r_b^2 + (p\zeta \cot \alpha_1)^2\right]}, \quad (11.79)$$

with

$$\tan \vartheta = \frac{y_1}{x_1} = \frac{r_b \tan \zeta + p\zeta \cot \alpha_1}{r_b - p\zeta \tan \zeta \cot \alpha_1}, \quad (11.80)$$

where ϑ is the angle between the position vector $r_a(\vartheta)$ and the x_1-axis (Fig. 11.16a). In this general case, the profile in the transverse section is an extended involute.

11.7 Relative Velocity and Coordinate Transformation

In Sect. 11.3 we introduced three Cartesian coordinate systems $O_1(x_1, y_1, z_1)$, $O_2(x_2, y_2, z_2)$ and $O_0(x_0, y_0, z_0)$, rigidly connected to worm (gear 1), worm wheel (gear 2), and frame, respectively. Here, in order to limit the number of subscripts, which would be required to indicate that a quantity refers to the gear 1 and gear 2, or that it is given in the one or in the other coordinate system, we introduce a new notation. With this notation, subscripts refer to the coordinate system, and indices refer to gear 1 or gear 2, or to their mutual interactions. It is to be noted that this new notation is used here and in other sections of this chapter to avoid misunderstandings, while in the remainder of this textbook we preferred to use the usual notation, which is, however, consistent with that used by ISO Standards.

To derive the equation of meshing, it is convenient to write the *relative velocity*, already introduced in Sect. 10.6, in a vector form that is most suitable to the goal to be reached. Figure 11.18 shows the more general case of a worm gear pair with center distance, a, and shaft angle, $\Sigma \neq \pi/2$, where the worm and worm wheel rotate with angular velocities $\omega^{(1)}$ and $\omega^{(2)}$ about the axes $z_1 \equiv z_0$ and z_2. Point P is a common point to both surfaces σ_1 and σ_2 of the two rotating gear members, and the relative velocity vector $v^{(12)}$ of point P_1 (this is supposed rigidly connected to

Fig. 11.18 General case of
rotation about crossed axes
($\Sigma \neq \pi/2$), and coordinate
transformation

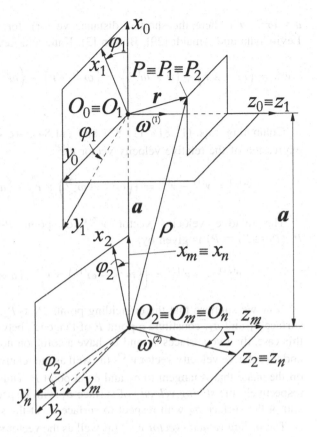

the worm surface σ_1) with respect to point $P_2 \equiv P_1$ (point P_2 is supposed rigidly
connected to the worm wheel surface σ_2) is given by:

$$v^{(12)} = v^{(1)} - v^{(2)}, \tag{11.81}$$

with

$$v^{(1)} = \omega^{(1)} \times r \tag{11.82}$$

$$v^{(2)} = \omega^{(2)} \times \rho, \tag{11.83}$$

where r and ρ are the position vectors drawn to point $P \equiv P_1 \equiv P_2$ from two
arbitrary points respectively on the lines of application of angular velocity vectors
$\omega^{(1)}$ and $\omega^{(2)}$. Here we choose these two points respectively coinciding with the
origins of coordinate systems $O_1(x_1, y_1, z_1) \equiv O_0(x_0, y_0, z_0)$ and $O_2(x_2, y_2, z_2)$.

By virtue of the *parallel axis theorem*, also known as *Huygens-Steiner theorem*,
we can express $v^{(2)}$ in an equivalent form to Eq. (11.83), replacing $\omega^{(2)}$ with
an equal vector applied to point $O_1 \equiv O_0$ and adding to it the *vector moment*

$a \times \omega^{(2)}$ (a is here the shortest distance vector), for which we shall have (see Levi-Civita and Amaldi [29], Burton [3], Kane and Levinson [26]):

$$v^{(2)} = \left(\omega^{(2)} \times r\right) + \left(a \times \omega^{(2)}\right) = \left(\omega^{(2)} \times r\right) - \left(\omega^{(2)} \times a\right) = \omega^{(2)} \times (r - a).$$

$$(11.84)$$

Combining Eqs. (11.81), (11.82) and (11.84), we obtain the following final expression of the relative velocity vector $v^{(12)}$:

$$v^{(12)} = v^{(1)} - v^{(2)} = \left[\left(\omega^{(1)} - \omega^{(2)}\right) \times r\right] - \left(a \times \omega^{(2)}\right). \qquad (11.85)$$

The relative velocity vector $v^{(21)}$ of point P_2 with respect to point P_1 ($P_1 \equiv P_2 \equiv P$) is given by:

$$v^{(21)} = -v^{(12)} = \left[\left(\omega^{(2)} - \omega^{(1)}\right) \times r\right] + \left(a \times \omega^{(2)}\right). \qquad (11.86)$$

Now we suppose that the coinciding points $P_1 \equiv P_2$, belonging respectively to surface σ_1 and σ_2, constitute a point P of tangency between these two surfaces. In this case, the two surfaces σ_1 and σ_2 have a common normal at point of contact P, and the relative velocity vector $v^{(12)}$ (as well as the relative velocity vector $v^{(21)}$) lies on the plane that is tangent to σ_1 and σ_2 at point P. Thus vectors $v^{(12)}$ and $v^{(21)}$ are respectively the *sliding velocity* of surface σ_1 with respect to surface σ_2 at point P, and of the surface σ_2 with respect to surface σ_1 at the same point P.

The *sliding velocity vector $v^{(12)}$* (as well as the vector $v^{(21)}$) can be referred to any of the three coordinate systems $O_1(x_1, y_1, z_1)$, $O_2(x_2, y_2, z_2)$ and $O_0(x_0, y_0, z_0)$. The subscript $i = (1, 2, 0)$ indicates that the vector is expressed in the coordinate system $O_i(x_i, y_i, z_i)$. In the cases where misunderstandings are not possible, the subscript 0, which indicates the fixed reference system $O_0(x_0, y_0, z_0)$, can be omitted. With reference to the coordinate system $O_i(x_i, y_i, z_i)$, Eq. (11.85) becomes:

$$v_i^{(12)} = \left[\left(\omega_i^{(1)} - \omega_i^{(2)}\right) \times r_i\right] - \left(a_i \times \omega_i^{(2)}\right). \qquad (11.87)$$

The sliding velocity vector $v^{(12)}$ in the fixed reference system $O_0 \equiv O_0(x_0, y_0, z_0)$ is as follows:

$$\begin{aligned}
v_{x0}^{(12)} &= -\left(\omega^{(1)} - \omega^{(2)} \cos \Sigma\right) y_0 - \left(\omega^{(2)} \sin \Sigma\right) z_0 \\
v_{y0}^{(12)} &= \left(\omega^{(1)} - \omega^{(2)} \cos \Sigma\right) x_0 - a\omega^{(2)} \cos \Sigma \\
v_{z0}^{(12)} &= (a + x_0)\omega^{(2)} \sin \Sigma.
\end{aligned} \qquad (11.88)$$

The same sliding velocity vector $v^{(12)}$ in the reference system $O_1(x_1, y_1, z_1)$ rigidly connected to the worm is as follows:

$$v_{x1}^{(12)} = -\left(\omega^{(1)} - \omega^{(2)} \cos \Sigma\right)y_1 - \omega^{(2)} \sin \Sigma \cos \varphi_1 z_1 - a\omega^{(2)} \cos \Sigma \sin \varphi_1$$

$$v_{y1}^{(12)} = \left(\omega^{(1)} - \omega^{(2)} \cos \Sigma\right)x_1 + \left(\omega^{(2)} \sin \Sigma \sin \varphi_1\right)z_1 - a\omega^{(2)} \cos \Sigma \cos \varphi_1$$

$$v_{z1}^{(12)} = (a + x_1 \cos \varphi_1 - y_1 \sin \varphi_1)\omega^{(2)} \sin \Sigma.$$

$$(11.89)$$

Often it is necessary to express the position vector r_1 given in the reference system $O_1(x_1, y_1, z_1)$, or any other vector given in this coordinate system, in the reference system $O_2(x_2, y_2, z_2)$, and vice versa. For transition from $O_1(x_1, y_1, z_1)$ to $O_2(x_2, y_2, z_2)$, the matrix equation of the coordinate transformation is the following:

$$r_2 = M_{21}r_1 = M_{2n}M_{nm}M_{m0}M_{01}r_1. \tag{11.90}$$

where

$$M_{21} = \begin{bmatrix} (\cos \varphi_1 \cos \varphi_2 + \sin \varphi_1 \sin \varphi_2 \cos \Sigma) & -(\sin \varphi_1 \cos \varphi_2 - \cos \varphi_1 \sin \varphi_2 \cos \Sigma) & -\sin \varphi_2 \sin \Sigma & a\cos \varphi_2 \\ -(\cos \varphi_1 \sin \varphi_2 - \sin \varphi_1 \cos \varphi_2 \cos \Sigma) & +(\sin \varphi_1 \sin \varphi_2 + \cos \varphi_1 \cos \varphi_2 \cos \Sigma) & -\cos \varphi_2 \sin \Sigma & -a\sin \varphi_2 \\ \sin \varphi_1 \sin \Sigma & \cos \varphi_1 \sin \Sigma & \cos \Sigma & 0 \\ 0 & 0 & 0 & 1 \end{bmatrix},$$

$$(11.91)$$

while M_{01} is the rotational matrix about the z_0-axis, M_{m0} is the translational matrix from $O_0(x_0, y_0, z_0)$ to $O_m(x_m, y_m, z_m)$, M_{nm} is the rotational matrix about the x_m-axis to obtain the shaft angle Σ, and M_{2n} is the rotational matrix about z_n-axis. $O_m \equiv O_m(x_m, y_m, z_m)$ and $O_n = O_n(x_n, y_n, z_n)$ are two auxiliary systems, both rigidly connected to the frame, which we introduce to make the coordinate transformation easier and more evident (Fig. 11.18). In the case where $\Sigma = \pi/2$, the matrix M_{21} is simplified very much, since $\cos(\pi/2) = 0$ and $\sin(\pi/2) = 1$.

From Eqs. (11.90) and (11.91) we obtain:

$$x_2 = (\cos \varphi_1 \cos \varphi_2 + \sin \varphi_1 \sin \varphi_2 \cos \Sigma)x_1 - (\sin \varphi_1 \cos \varphi_2 - \cos \varphi_1 \sin \varphi_2 \cos \Sigma)y_1$$
$$- (\sin \varphi_2 \sin \Sigma)z_1 + a\cos \varphi_2$$
$$y_2 = -(\cos \varphi_1 \sin \varphi_2 - \sin \varphi_1 \cos \varphi_2 \cos \Sigma)x_1$$
$$+ (\sin \varphi_1 \sin \varphi_2 + \cos \varphi_1 \cos \varphi_2 \cos \Sigma)y_1 - (\cos \varphi_2 \sin \Sigma)z_1 - a\sin \varphi_2$$
$$z_2 = (\sin \varphi_1 \sin \Sigma)x_1 + (\cos \varphi_1 \sin \Sigma)y_1 + (\cos \Sigma)z_1.$$

$$(11.92)$$

The inverse matrix $M_{12} = M_{21}^{-1}$ is as follows:

$$M_{12} = \begin{bmatrix} (\cos\varphi_1\cos\varphi_2 + \sin\varphi_1\sin\varphi_2\cos\Sigma) & -(\cos\varphi_1\sin\varphi_2 - \sin\varphi_1\cos\varphi_2\cos\Sigma) & \sin\varphi_1\sin\Sigma & -a\cos\varphi_1 \\ -(\sin\varphi_1\cos\varphi_2 - \cos\varphi_1\sin\varphi_2\cos\Sigma) & +(\sin\varphi_1\sin\varphi_2 + \cos\varphi_1\cos\varphi_2\cos\Sigma) & \cos\varphi_1\sin\Sigma & a\sin\varphi_1 \\ -\sin\varphi_2\sin\Sigma & -\cos\varphi_2\sin\Sigma & \cos\Sigma & 0 \\ 0 & 0 & 0 & 1 \end{bmatrix},$$

$$\tag{11.93}$$

From equation $r_1 = M_{12}r_2 = M_{21}^{-1}r_2$, we obtain:

$$\begin{aligned} x_1 &= (\cos\varphi_1\cos\varphi_2 + \sin\varphi_1\sin\varphi_2\cos\Sigma)x_2 - (\cos\varphi_1\sin\varphi_2 - \sin\varphi_1\cos\varphi_2\cos\Sigma)y_2 \\ &\quad + (\sin\varphi_1\sin\Sigma)z_2 - a\cos\varphi_1 \\ y_1 &= -(\sin\varphi_1\cos\varphi_2 - \cos\varphi_1\sin\varphi_2\cos\Sigma)x_2 \\ &\quad + (\sin\varphi_1\sin\varphi_2 + \cos\varphi_1\cos\varphi_2\cos\Sigma)y_2 + (\cos\varphi_1\sin\Sigma)z_2 + a\sin\varphi_1 \\ z_1 &= -(\sin\varphi_2\sin\Sigma)x_2 - (\cos\varphi_2\sin\Sigma)y_2 + (\cos\Sigma)z_2. \end{aligned}$$

$$\tag{11.94}$$

Equations (11.92) and (11.94) are respectively the Cartesian coordinate transformation in transition from reference system $O_1(x_1, y_1, z_1)$ to reference system $O_2(x_2, y_2, z_2)$, and vice versa.

11.8 Worm and Worm Wheel Meshing, and Lines of Contact

To be able to define completely the geometry of the worm gear pair, it is necessary to determine the *surface of contact*, i.e. the geometrical surface defined by contact points between the worm and worm wheel. The successive *lines of contact* during gear meshing constitute this surface, whose axial length is the sum of the length of approach and the length of recess. When the gear is working as speed reducing unit, the length of approach is the axial distance between the first point of contact of threads, and the instantaneous axis of rotation, while the length of recess is the axial distance between the last point of contact of threads, on withdrawal, and the instantaneous axis of rotation. Worm face width, and face width and rim width of the worm wheel (definitions are those of the ISO 1122-2:1999 [23]; see also Figs. 11.1 and 11.12) strongly depend on the extension of the surface of contact, which therefore needs to be evaluated as accurately as possible. Moreover, the choice of a worm face width larger than necessary does not give rise to disadvantages; indeed, it gives a certain freedom in adjusting its axial position relative to the worm wheel.

In this framework, it is necessary to write the equation of meshing, that relates parameters of the helicoid surface σ_1 of the worm and its angle of rotation φ_1 about the z_1-axis with the corresponding parameters of the worm wheel surface σ_2. The surface σ_1 of the worm thread flanks and surface σ_2 of the worm wheel tooth flanks are in contact along curved lines at every instant during the meshing cycle. The determination of these instantaneous curves of contact is based on the condition that the normal N to any point P of this curve and the relative velocity vector v_r, whose line of application is the common tangent at the same point, are perpendicular vectors.

In fact, when we consider the screw motion of a helicoid, the screw parameter p in this motion and the screw parameter of helicoid coincide. Any point of the helicoid traces out a helix, and the relative velocity vector v_r in the screw motion is tangent to the helix. Since the helix belongs to the helicoid and the relative velocity vector v_r (it is designed v, to avoid using too many subscripts) is tangent to helicoid, the following condition must be satisfied:

$$n \cdot v = N \cdot v = 0. \tag{11.95}$$

In general terms, the relative velocity vector, i.e. the velocity vector in screw motion, is given by:

$$v = (\omega \times r) + p\omega = \begin{vmatrix} i & j & k \\ 0 & 0 & \omega \\ x & y & z \end{vmatrix} + p\omega k = \omega(-yi + xj + pk), \tag{11.96}$$

while the normal N and unit normal n to the surface σ are given by:

$$N = N_x i + N_y j + N_z k \tag{11.97}$$

$$n = n_x i + n_y j + n_z k. \tag{11.98}$$

The equation of meshing (11.95), written in relation to the reference coordinate system of interest here, is the following:

$$N_i \cdot v_i = 0. \tag{11.99}$$

where $v_i = v_{i1} - v_{i2} = v_i^{(12)}$ is the relative velocity vector of surface σ_1 with respect to surface σ_2, N_i is the normal to the worm surface σ_1, while the subscripts $(i = 1, 2, 0)$ designate coordinate systems $O_1(x_1, y_1, z_1)$, $O_2(x_2, y_2, z_2)$, and $O_0(x_0, y_0, z_0)$, which are rigidly connected to the worm, worm wheel and housing-frame.

Taking into account that the worm thread surface σ_1 is a helicoid, the equation of meshing can be simplified in one of the following two relationships proposed by Litvin [30, 31]:

$$y_1 N_{x1} - x_1 N_{y1} - p N_{z1} = y_1 n_{x1} - x_1 n_{y1} - p n_{z1} = 0 \tag{11.100}$$

$$y_0 N_{x0} - x_0 N_{y0} - p N_{z0} = y_0 n_{x0} - x_0 n_{y0} - p n_{z0} = 0 \tag{11.101}$$

respectively valid in the coordinate systems $O_1(x_1, y_1, z_1)$ and $O_0(x_0, y_0, z_0)$. In these equations, (N_{x1}, N_{y1}, N_{z1}), (N_{x0}, N_{y0}, N_{z0}), (n_{x1}, n_{y1}, n_{z1}), and (n_{x0}, n_{y0}, n_{z0}), are the projections of the normal N_1 and unit normal n_1 on the axes of the two coordinate systems above.

Vector v_i in Eq. (11.99), with $(i = 1, 2, 0)$, can be also written as follows:

$$v_i = v_{xi}\boldsymbol{i} + v_{yi}\boldsymbol{j} + v_{zi}\boldsymbol{k}, \tag{11.102}$$

where (v_{xi}, v_{yi}, v_{zi}) are its projections on x-, y-and z-axes.

From Eqs. (11.99), (11.97) and (11.102) written in the coordinate system $O_1(x_1, y_1, z_1)$, we obtain:

$$N_1 \cdot v_1 = N_{x1}v_{x1} + N_{y1}v_{y1} + N_{z1}v_{z1} = 0. \tag{11.103}$$

Introducing Eq. (11.89) in the equation above, and dividing by $-(\omega_2 \sin \Sigma)$ all members of the equation thus obtained, we get:

$$(z_1 \cos \varphi_1 + a \cot \Sigma \sin \varphi_1)N_{x1} + y_1 \frac{u - \cos \Sigma}{\sin \Sigma} N_{x1} + (-z_1 \sin \varphi_1 + a \cot \Sigma \cos \varphi_1)N_{y1}$$

$$- x_1 \frac{u - \cos \Sigma}{\sin \Sigma} N_{y1} - (x_1 \cos \varphi_1 - y_1 \sin \varphi_1 + a)N_{z1} = 0. \tag{11.104}$$

Multiplying by $[(u - \cos \Sigma)/ \sin \Sigma]$ all terms of the left-hand side of Eq. (11.100), we infer that:

$$y_1 \frac{u - \cos \Sigma}{\sin \Sigma} N_{x1} - x_1 \frac{u - \cos \Sigma}{\sin \Sigma} N_{y1} = p \frac{u - \cos \Sigma}{\sin \Sigma} N_{z1}, \tag{11.105}$$

so, from Eq. (11.104) we obtain:

$$(z_1 \cos \varphi_1 + a \cot \Sigma \sin \varphi_1)N_{x1} + (-z_1 \sin \varphi_1 + a \cot \Sigma \cos \varphi_1)N_{y1}$$

$$- \left[(x_1 \cos \varphi_1 - y_1 \sin \varphi_1 + a) - p \frac{u - \cos \Sigma}{\sin \Sigma} \right] N_{z1} = 0. \tag{11.106}$$

This is the equation that expresses the meshing condition in the coordinate system $O_1(x_1, y_1, z_1)$. In this equation, (N_{x1}, N_{y1}, N_{z1}) are the projections of the normal N_1 to surface σ_1 on the x_1-, y_1-and z_1-axes, Σ is the shaft angle, a is the shortest distance or center distance, and $u = z_2/z_1$ is the gear ratio.

From Eqs. (11.99), (11.97) and (11.102) written in the coordinate system $O_0(x_0, y_0, z_0)$, we obtain:

$$N_0 \cdot v_0 = N_{x0} v_{x0} + N_{y0} v_{y0} + N_{z0} v_{z0} = 0. \tag{11.107}$$

Introducing Eqs. (11.88) in the equation above, and dividing by $-(\omega_2 \sin \Sigma)$ all terms of the equation thus obtained, we get:

$$-z_0 N_{x0} - \frac{u - \cos \Sigma}{\sin \Sigma} (y_0 N_{x0} - x_0 N_{y0}) - a N_{y0} \cot \Sigma + (x_0 + a) N_{z0} = 0. \tag{11.108}$$

From the first Eq. (11.101), we infer

$$y_0 N_{x0} - x_0 N_{y0} = p N_{z0}, \tag{11.109}$$

for which Eq. (11.108) can be written as follows:

$$z_0 N_{x0} + a N_{y0} \cot \Sigma - \left(x_0 + a - p \frac{u - \cos \Sigma}{\sin \Sigma} \right) N_{z0} = 0. \tag{11.110}$$

This equation expresses the meshing condition in the coordinate system $O_0(x_0, y_0, z_0)$. In this equation, still remaining the meaning of the other symbols, (N_{x0}, N_{y0}, N_{z0}) are the projections of the normal N_0 to surface σ_1 on the x_0-, y_0-and z_0-axes.

If we want to specialize the Eq. (11.110) in the case in which the worm surface is represented as a generalized helicoid, it is first necessary to derive expressions of N_{x0}, N_{y0} and N_{z0}. Equation of the normal N_0 can be obtained from Eq. (11.62) by means of the relationship:

$$N_0 = M_{01} \cdot N_1 = \begin{bmatrix} \cos \varphi_1 & -\sin \varphi_1 & 0 & 0 \\ \sin \varphi_1 & \cos \varphi_1 & 0 & 0 \\ 0 & 0 & 1 & 0 \\ 0 & 0 & 0 & 1 \end{bmatrix} N_1, \tag{11.111}$$

where $M_{01} = M_{10}^{-1}$ is the rotational matrix about the z_0-axis for the coordinate transformation in transition from coordinate system $O_1(x_1, y_1, z_1)$ to coordinate system $O_0(x_0, y_0, z_0)$. Thus, we have:

$$N_0 = \frac{r_a}{\cos \alpha} [p \sin(\vartheta + \zeta + \alpha + \varphi_1) i_0 - p \cos(\vartheta + \zeta + \alpha + \varphi_1) j_0 + r_a \cos \alpha k_0], \tag{11.112}$$

where $r_a = r_a(\vartheta)$ is the magnitude of the position vector of the current point of the worm transverse profile, and φ_1 is the rotation angle about the z_0-axis of the reference system $O_1(x_1, y_1, z_1)$ with respect to reference system $O_0(x_0, y_0, z_0)$.

Introducing into Eq. (11.110) the expressions of N_{x0}, N_{y0} and N_{z0} obtained from Eq. (11.112), we get:

$$r_a \cos \alpha \left(x_0 + a - p \frac{u - \cos \Sigma}{\sin \Sigma} \right) + ap \cot \Sigma \cos \psi = pz_0 \sin \psi = p^2 \zeta \sin \psi,$$

$$(11.113)$$

where ψ is an auxiliary angle given by $\psi = (\vartheta + \zeta + \alpha + \varphi_1)$ and $x_0 = r_a \cos(\vartheta + \zeta + \varphi_1)$. In the coordinate system $O_0(x_0, y_0, z_0)$, the coordinates of a current point of contact can be expressed by the relationships:

$$x_0 = r_a \cos(\vartheta + \zeta + \varphi_1) \quad y_0 = r_a \sin(\vartheta + \zeta + \varphi_1) \quad z_0 = p\zeta. \qquad (11.114)$$

Any of the three Eqs. (11.106), (11.110) and (11.113) expresses the relationship existing between the worm surface parameters (ϑ, ζ) and the angle of rotation φ_1 of the worm, i.e. the function:

$$f(\vartheta, \zeta, \varphi_1) = 0. \qquad (11.115)$$

The pair of equations

$$r_1 = r_1(\vartheta, \zeta) \quad f(\vartheta, \zeta, \varphi_1) = 0, \qquad (11.116)$$

where $r_1 = r_1(\vartheta, \zeta)$ represents the helicoid surface σ_1 of the worm, describe in the coordinate system $O_1(x_1, y_1, z_1)$ the family of lines of contact on surface σ_1. A given family of lines of contact corresponds to a given value of the angle of rotation φ_1 of the worm. Thus φ_1 is the parameter of the family of lines of contact.

The following pair of equations determine the lines of contact on the worm wheel surface σ_2:

$$r_2(\vartheta, \zeta, \varphi_1) = M_{21} r_1(\vartheta, \zeta) \quad f(\vartheta, \zeta, \varphi_1) = 0, \qquad (11.117)$$

where $M_{21} = M_{12}^{-1}$, given by relationship (11.91), is the matrix which describes the coordinate transformation from coordinate system $O_1(x_1, y_1, z_1)$, rigidly connected to the worm, to coordinate system $O_2(x_2, y_2, z_2)$, rigidly connected to the worm wheel.

Figure 11.19a, b shows the lines of contact on the surface σ_1 of a type A worm and, respectively, on the surface σ_2 of the mating worm wheel. Both the figures refer to an Archimedean worm gear pair having: $z_1 = 2$; $z_2 = 30$; $m_x = m = 8$ mm; $\Sigma = 90°$; $a = 176$ mm. Figure 11.19a also shows the envelope of the lines of contact on the generating surface, which is the worm thread surface, σ_1.

In this regard, it should be noted that, usually, the lines of contact on the generation surface σ_1 cover the entire working part of the surface. However, in some case, although not so rare, the lines of contact on the generation surface have an envelope line that divides the generation surface in two parts: a part covered by the lines of contact, which therefore contains the lines of contact and their envelope line, and the remaining part that is instead free of lines of contact. Towards the root of the worm threads, an edge line that generates the fillet surrounds the generation

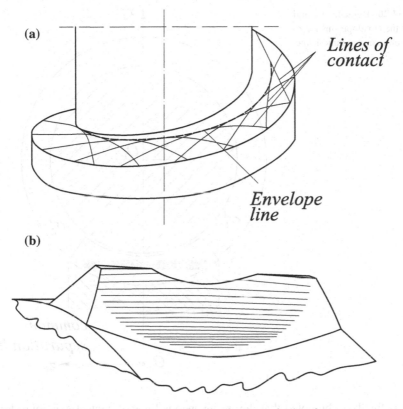

Fig. 11.19 Lines of contact on: **a** worm surface, σ_1; **b** worm wheel surface, σ_2

surface. Near the envelope line, the condition of lubrication and heat transfer become unfavorable. For this reason, this envelope line should be excluded from meshing, by choosing appropriate design parameters of the worm gear pair.

On the basis of a theorem on the necessary and sufficient conditions of existence of the envelope line on the generation surface, proposed by Litvin et al. [32], it is possible to demonstrate that, in the case of a conventional worm gear pair with type *I* worm, the lines of contact on the worm thread surface, σ_1, have an envelope line. Since each current line of contact has two branches that are recognized by the sign of the first derivative $f_{\varphi_1} = df/d\varphi_1$ with respect to the variable φ_1 of the function $f(\vartheta, \zeta, \varphi_1) = 0$ given by the second Eq. (11.116), the worm wheel tooth surface, σ_2, which is the envelope of the worm thread surface, σ_1, is formed by two branches, designed by σ_2^1 and σ_2^2 (Fig. 11.20). These two branches, on the worm wheel tooth surface, σ_2, are divided by a common separation line, which is the analogous of the envelope line on the worm thread surface, σ_1.

Fig. 11.20 Branches σ_2^1 and σ_2^2 of the envelope surface σ_2, and common separation line

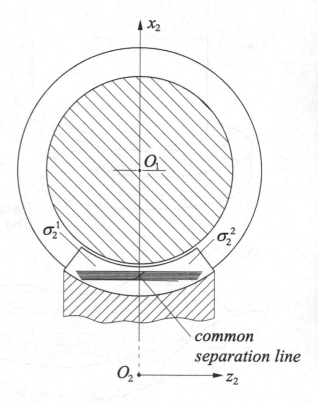

It is finally to be noted that the instantaneous lines of contact exist only for an ideal worm gear pair, without errors of manufacturing and misalignments. Actually, the contact between the surfaces σ_1 and σ_2 is an instantaneous point of contact that, due to the aforementioned errors, can suffer from hazardous shifts toward the edge (this is the bearing edge contact), and can adversely affect the function that expresses the transmission errors. Such transmission errors cause vibration during the meshing cycle.

Therefore, it is necessary to minimize the errors of manufacturing and misalignments. To this end, it is necessary to define as accurately as possible the shape and geometry of the bearing contact between the two mating surfaces σ_1 and σ_2, by reasonable averaging the results achievable with the use of theoretical and theoretical-numerical models, but not forgetting to treasure, when available, to any previous experience accumulated on this topic. The discussion described above, concerning the determination of the instantaneous lines of contact, is in any case basilar.

11.9 Surface of Contact: General Concepts and Determination by Analytical Methods

The *surface of contact*, also called *zone of contact* or *meshing surface*, is the active part of the *surface of action*. This is the surface swept out by the lines of contact during the meshing cycle; thus, it consists of the set of instantaneous lines of contact between the active flanks of the worm threads and worm wheel teeth. The determination of the meshing surface is a very important aspect of the worm gear pair design, since it determines the operating worm and worm wheel face widths, and the worm wheel rim width.

To avoid confusion and misunderstandings, we clarify first some basic concepts. To this end, we consider the orthogonal and right-hand worm gear pair shown schematically in Fig. 11.21. Assuming that the worm is the driving member of the gear pair, and considering the directions of rotation shown in this figure, we call *advancing* and *receding sides* (*front* and *rear parts*, or *entering* and *leaving sides*, are respectively synonyms) of the worm and worm wheel the sides (or parts) designed with *a.s.* and *r.s.* in the same figure. The contact starts between the

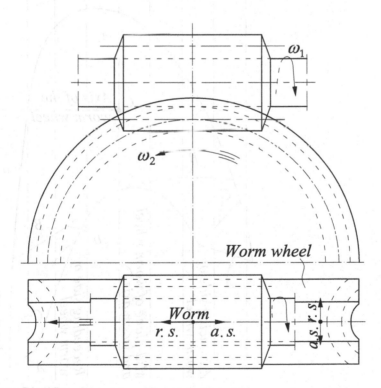

Fig. 11.21 Advancing (*a.s.*) and receding (*r.s.*) sides of the contact between the worm threads and worm wheel teeth

advancing side of the worm and the receding side of the worm wheel, then it also extends to the advancing part of the worm wheel and, finally, ends between the receding sides of the worm and worm wheel.

Figure 11.22 shows the projection of the surface of contact of an orthogonal worm gear pair with type A worm, on the pitch plane of the same worm, or on any plane parallel to it, and therefore perpendicular to the shortest distance line; one of these planes is the plane (y_0, z_0). Such a projection, which has roughly the shape of a horse-shoe, is bounded by the following two lines:

- The line, a, which is a smooth curve, and represents the end of the contact.
- The line, b, which is often a smooth curve at times, and represents the beginning of the contact; the geometry of the latter curve varies depending on the shape of the outside surface of the worm wheel.

Fig. 11.22 Projection of the surface of contact on the plane (y_0, z_0)

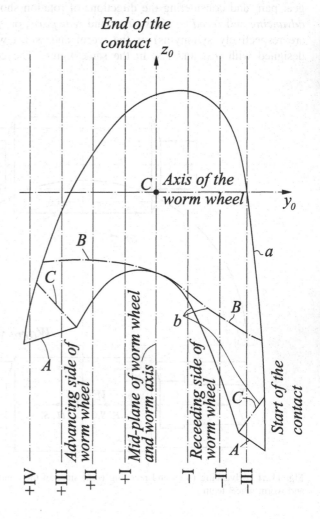

Figure 11.22 also shows that the horse-shoe shaped surface is approximately symmetric with respect to a straight line with direction normal to the worm thread at radius of the pitch cylinder. Therefore, with respect to the trace of the worm wheel mid-plane, it is the more asymmetric the greater the helix angle, β The contact starts at the beginning of the line, b, which turns its concavity towards the outside of the surface of contact, and ends at the top of the line, a, that instead turns outwardly its convexity.

The position and shape of the line, b, depend on the geometry of the outside surface of the worm wheel. From this point of view, we can distinguish the three following different geometries of the line, b:

- The type-A line, which corresponds to an outside surface of the worm wheel made, in the central part straddles the mid-plane, from the throat surface, and in the two side parts, made from two cylindrical surfaces (Fig. 11.23c). The two almost linear ends of the line, b, correspond to these cylindrical side parts of the worm wheel toothing.
- The type-B line, which corresponds to an outside surface of the worm wheel made of a single cylindrical surface, whereby the throat surface is missing (Fig. 11.23d).
- The type-C line, which corresponds to an outside surface of the worm wheel made, in the central part straddles the mid-plane, from the throat surface, and in the two side parts, made by two conical surfaces, whose generatrices are symmetric with respect to the mid-plane, and pass through the worm axis, as well as the end points of the throat surface (Fig. 11.23e).

The same Fig. 11.23 shows the projections of the instantaneous lines of contact between a single thread of the worm and a single tooth of the worm wheel on a transverse plane of the worm (Fig. 11.23a), and the projections of the surfaces of contact on the pitch plane of the worm (Fig. 11.23b).

After these necessary clarifications, we face the problem of determination of the surface of contact, which constitutes the subject of this section. Notoriously, this determination can be performed using analytical-numerical methods, graphical methods, and mixed analytical-numerical-graphical methods. Here we use an analytical-numerical method. Among the analytical methods described in the scientific literature, we accord our preference to that proposed by Litvin and Fuentes [33]. This method has very general validity, because it can be applied to the various types of worms described in Sect. 11.2. It assumes that the usable surface of the worm threads is represented as a generalized helicoid, that is, as a surface generated by a planar curve in transverse section, $r = r(\vartheta)$, which is performing a screw motion about the worm axis.

Figure 11.24a, b shows, in schematic form, an orthogonal worm gear pair and, respectively, its surface of contact, projected on the plane (y_0, z_0), where it is bounded by the curves a and b defined above. This surface, represented in the fixed coordinate system $O_0(x_0, y_0, z_0)$, which has the worm axis as its z_0-axis, does not

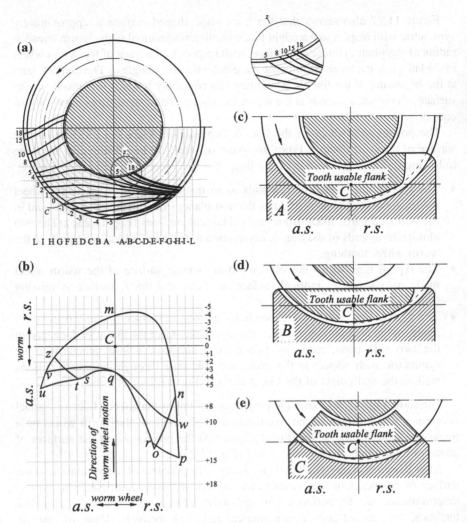

Fig. 11.23 Right-hand worm gear pair, with Archimedean spiral worm: **a** projections of the instantaneous lines of contact on a transverse plane of the worm; **b** projections of the surfaces of contact on the pitch plane of the worm; **c** type-*A* worm wheel; **d** type-*B* worm wheel; **e** type-*C* worm wheel

change in any plane parallel to the plane (y_0, z_0), including the pitch plane of the worm. As we said above, line *b* consists of the point corresponding to the entry into meshing of those points of the worm wheel tooth flanks belonging to the worm wheel addendum cylinder; line *a* consists instead of the points corresponding to the output from the meshing of those points of the thread flanks belonging to the worm addendum cylinder.

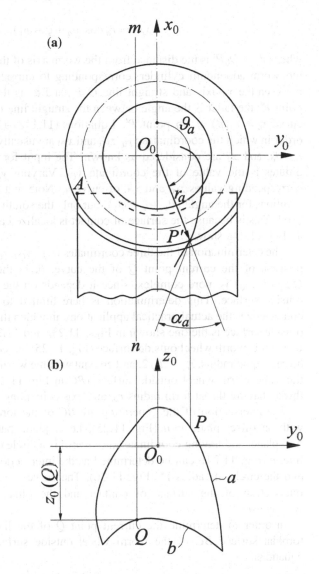

Fig. 11.24 a Generating planar curve of helicoid; **b** schematic projection of surface of contact on the plane (y_0, z_0)

The three coordinates (x_{P0}, y_{P0}, z_{P0}) that uniquely define the position of the current point P of the line a in the fixed coordinate system $O_0(x_0, y_0, z_0)$ are as follows:

$$y_{P0} = r_a \sin(\vartheta_a + \zeta + \varphi_1) \tag{11.118}$$

$$z_{P0} = \frac{[r_a \cos(\vartheta_a + \zeta + \varphi_1) + a - pu]}{p \sin[(\vartheta_a + \zeta + \varphi_1) + \alpha_a]} r_a \cos \alpha_a \tag{11.119}$$

$$x_{P0} = r_a \cos(\vartheta_a + \zeta + \varphi_1),\tag{11.120}$$

where $r_a = O_0 P'$ is the distance from the worm axis of the point P' (Fig. 11.24a) on the worm addendum cylinder, corresponding to current point P, ϑ_a is the angle between the x_0-axis and straight line $O_0 P'$, and α_a is the pressure angle at the tip point P'; the latter is the angle between the straight line $O_0 P'$ and the tangent to the curve $r_a = r_a(\vartheta)$ at tip point P'. Equations (11.118)–(11.120) are written in the order in which the coordinates y_{P0}, z_{P0} and x_{P0} are usually calculated. The quantities r_a, ϑ_a, and α_a are considered as known. The input for calculation of these coordinates is the value of the coordinate y_{P0}. Varying y_{P0}, we can determine the corresponding values z_{P0} and x_{P0} of curve, a. Note that Eq. (11.118) provides two solutions for the angle $(\vartheta_a + \zeta + \varphi_1)$, but only the solution for which $x_{P0} < 0$ can be used. This is because the surface of contact is localized all below the plane (y_0, z_0), as Fig. 11.24a shows.

The determination of the three coordinates (x_{Q0}, y_{Q0}, z_{Q0}) that uniquely define the position of the current point Q of the curve, b, in the fixed coordinate system $O_0(x_0, y_0, z_0)$ is more complex, since it depends on the shape of the worm wheel outside surface. This determination is here limited to the case that most often characterizes the actual practical applications, in which the shape of the worm wheel outside surface is the one shown in Figs. 11.23c and 11.24a. The side parts AB and CD of this worm wheel outside surface (Fig. 11.25) are constituted by two cylinders having equal radius, $r_e = d_e/2$, and coaxial with the worm wheel. The central part of the same worm wheel outside surface (BC in Fig. 11.25) is an arc of toroid, the throat, having throat form radius r_{th} and axis coinciding with the worm axis.

The intersection of the generating arc BC of the toroidal surface of the throat with an offset plane m-n (Fig. 11.25), i.e. a plane parallel to the worm wheel mid-plane, and having coordinate $y_0 = const$, is a circle of radius r_e^*. Point Q of the line, b (Fig. 11.24), can be determined as the intersection of the curve $z_0 = z_0(x_0)$ and the circle of radius r_e^* (Fig. 11.26). The curve $z_0 = z_0(x_0)$ is obtained as the intersection of the surface of contact and the offset plane having coordinate $y_0 = const$.

In order to determine the current point Q of the line b corresponding to the toroidal surface BC of the worm wheel outside surface, we use the following equations:

$$y_{Q0} - r(\vartheta) \sin(\vartheta + \zeta + \varphi_1) = f_1(\vartheta, (\zeta + \varphi_1)) = 0\tag{11.121}$$

$$z_{Q0} = \frac{[r(\vartheta) \cos(\vartheta, (\zeta + \varphi_1)) + a - pu]}{p \sin[\alpha(\vartheta) + (\vartheta, (\zeta + \varphi_1))]} r(\vartheta) \cos\alpha(\vartheta) = f_2(\vartheta, (\zeta + \varphi_1)) = 0$$

$$\tag{11.122}$$

$$x_{Q0} - r(\vartheta) \cos(\vartheta, (\zeta + \varphi_1)) = f_3(\vartheta, (\zeta + \varphi_1)) = 0\tag{11.123}$$

Fig. 11.25 Worm gear pair with type-A worm wheel, and offset plane m-n

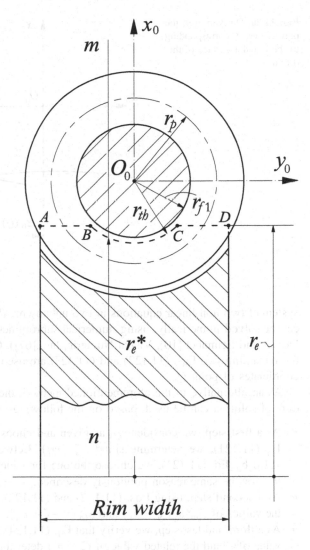

$$\left\{[a + x_{Q0}(\vartheta, ((\zeta + \varphi_1)))]^2 + z_{Q0}^2(\vartheta, ((\zeta + \varphi_1)))\right\}^{1/2} - a + \left[r_{th}^2 - y_{Q0}^2(\vartheta, ((\zeta + \varphi_1)))\right]^{1/2}$$
$$= f_4(\vartheta, ((\zeta + \varphi_1))) = 0.$$

(11.124)

Here $r_{th} = r_{f1} + c$, where r_{f1} is the radius of the worm root cylinder, and c is the clearance (usually, $c = 0.25m$). In addition, $\tan \alpha = [r(\vartheta)/(dr(\vartheta)/d\vartheta)]$. Coordinate y_{Q0} is considered as the input data for solution procedure, while Eqs. (11.122) and (11.123) are used for determination of $z_{Q0}(\vartheta, ((\zeta + \varphi_1)))$ and $x_{Q0}(\vartheta, ((\zeta + \varphi_1)))$ that appear in Eq. (11.124). Equations (11.121) and (11.124) can be considered as a

Fig. 11.26 Derivation of the part of line, b, corresponding to the toroidal surface of the throat

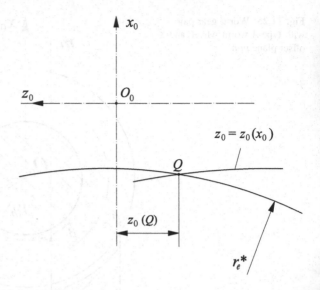

system of two non-linear equations in two unknowns, ϑ and $(\zeta + \varphi_1)$. This system can be solved numerically, using numerical subroutines (see Moré et al. [41, 42], Dennis and Schnabel [10], Visual Numeric, Inc. [67]). Once this system is solved, the remaining two Eqs. (11.122) and (11.123) are used to define fully the triad of coordinates of point Q.

As an alternative to the aforementioned method, the following iterative procedure of solution can be used, based on the following steps:

- As a first step, we consider y_{Q0} as given and choose a value of ϑ. Using then Eq. (11.121), we determine $\sin(\vartheta + \zeta + \varphi_1)$. Between the two solutions provided by Eq. (11.121), we choose the one for which $y_{Q0}(\vartheta + \zeta + \varphi_1) < 0$, and this for the same reason previously described.
- As a second step, using Eqs. (11.122) and (11.123), we determine respectively the values of $z_{Q0}(\vartheta + \zeta + \varphi_1)$ and $x_{Q0}(\vartheta + \zeta + \varphi_1)$.
- As a third and last step, we verify that Eq. (11.124) is satisfied with the chosen value of ϑ and the related value of $(\zeta + \varphi_1)$ determined by Eq. (11.121). In the case where the checking is not satisfied, it is necessary to perform a new iteration with a new value of ϑ, until convergence.

Instead, to determine the current point Q of the line b corresponding to the cylindrical parts AB and CD of the worm wheel outside surface, we use an equation system formed by the same Eqs. (11.121)–(11.123), and the following equation:

$$r_e - \left[(a + x_{Q0})^2 + z_{Q0}^2 \right] = f_5(\vartheta, (\zeta + \varphi_1)) = 0, \qquad (11.125)$$

which replaces Eq. (11.124).

The procedure of solution of these equations does not change compared to that previously described. This time we consider, as a system of two non-linear equations in the two unknowns ϑ and $(\zeta + \varphi_1)$, the system consisting of the two Eqs. (11.121) and (11.125), while the Eqs. (11.122) and (11.123) are used for determination of coordinates z_{Q0} and x_{Q0} for current point Q.

The determination of the surface of contact allows us to define the operating worm and worm wheel face widths, and the worm wheel rim width.

Example 1 In order to better clarify the conclusion that can be drawn based on the analysis of the surface of contact between the worm and worm wheel, as a first example, we consider (Fig. 11.23) a worm gear pair with a type A right-hand worm, i.e. a right-hand Archimedean spiral worm. This worm gear pair has the following characteristics: number of threads, $z_1 = 4$; nominal helix angle, $\beta_1 = 30°$; nominal module of the generating rack, $m = 9$ mm; pressure angle of the generating rack, $\alpha = 30°$; lead, $p_z = 113.10$ mm; nominal pitch diameter, $d_1 = 2r_1 = 62.35$ mm; tip diameter, $d_{a1} = 80.35$ mm; root diameter, $d_{f1} = 41.35$ mm. Instead, the worm wheel has the following characteristics: number of teeth, $z_2 = 28$; nominal module, $m = 9$ mm; nominal pitch diameter, $d_2 = z_2 m = 252$ mm; outside surface configured according to the three different shapes, A, B, and C, as Fig. 11.23c–e shows. The center distance is $a = 167.17$ mm.

Figure 11.23b shows the projections on the worm pitch plane of the surfaces of contact related to the three types of outside surfaces described above, and specifically:

- The surface of contact bounded by the mixed-line curve *umprqsu*, that corresponds to the worm wheel outside surface type-A.
- The surface of contact bounded by the mixed-line curve *vmwqv*, that corresponds to the worm wheel outside surface type-B.
- The surface of contact bounded by the mixed-line curve *zmnoqtz*, that corresponds to the worm wheel outside surface type-C.

Figure 11.23a shows the projections of the instantaneous lines of contact between a single thread of the worm and a single tooth of the worm wheel on a transverse plane of the worm.

Figure 11.23b and, albeit in schematic form, Fig. 11.22 shows that the extension of the surface of contact varies with the shape of the outside surface of the worm wheel. From the analysis of the three surfaces of contact shown in these figures, we infer that the meshing surface is more extensive in the advancing side of the worm with respect to what happens in its receding side, and that the outside surface of the worm wheel influences the extension of the contact in the advancing side of the worm. By increasing the number of teeth of the worm wheel, z_2, the surface of contact lengthens. In addition, we infer that the outside surface of the type-C worm wheel (Fig. 11.23d and mixed-line curve C-C in Fig. 11.22) is not very favorable, since the initial contacts take place according to instantaneous lines of small length, which therefore carry a very low contribution to the load bearing capacity at the start of the meshing. Furthermore, the sharp edge that is generated in the tooth of this type of worm wheel generally causes scoring and poor lubrication.

For this reason, worm wheels with outside surfaces type-A and type-B are usually used. The type-B outside surface of the worm wheel (Fig. 11.23d) involves the minimum surface of contact (see the mixed-line curve B-B in Fig. 11.22). The contact, just started, extends once to the worm wheel face width, so the gradualness of the contact fails. The best solution is therefore to use a worm wheel with type-A outside surface. In fact, it is characterized by the maximum extension of the surface of contact, as well as by a gradual contact at the beginning of meshing. We also note that, according to the length of the surface of contact, for worm wheel with type-A outside surface, three teeth are simultaneously in meshing, while for worm wheel with type-B outside surface only two teeth are simultaneously in meshing. The load-bearing capacity is, however, almost equal (the differences are small), because the main contribution is due to contacts in the vicinity of the mid-plane of the gear pair, i.e. near the pitch point C.

The worm must have the advancing and receding sides of equal lengths because, when the direction of rotation is reversed, the advancing and receding sides of the worm and worm wheel are exchanged between them. Thus, the usable worm face width (i.e. the length of the threaded part of the worm that is actually usable, which is denoted by its flanks being completely formed) must be at least twice compared to that of the advancing side of contact. It is often desirable to have worm wheels with type-B outside surface, as the corresponding worm is shorter, and therefore the worm is less sensitive to unavoidable small errors in the pitch. To establish definitively what type of worm wheel outside surface to be adopted, we must also keep in mind that the side parts of the teeth of the worm wheels with type-A and type-C outside surfaces may be too thin, in some cases in which the pressure angle is large.

Finally, a consideration must be made regarding the Archimedean spiral worm. In this case, if we choose as the generatrix of the pitch cylinder a straight line that does not divide the tooth depth of the rack according to the standard sizing, but instead it is closest to its axis, the extension of the surface of contact in the receding side of the worm increases. This position of the generatrix closest to the worm axis is only possible with large pressure angles and with high numbers of teeth, to avoid interference. In the projection on a transverse section plane of the worm, the instantaneous lines of contact will have greater inclinations compared to the concentric circles in the same transverse section (Fig. 11.23a).

Example 2 Let us now analyze, as a second example, the surface of contact between the worm and worm wheel of a worm gear pair with involute worm having the following characteristics: type I right-hand worm; number of threads, $z_1 = 4$; base helix angle, $\beta_b = 30°$; nominal module of the generating rack, $m = 9$ mm; nominal pressure angle of the generating rack, $\alpha = 20°58'$; lead, $p_z = 113.10$ mm; base cylinder diameter, $d_b = 62.35$ mm; tip diameter, $d_{a1} = 101.35$ mm; root diameter, $d_{f1} = 62.35$ mm; pitch diameter, $d_1 = 83.35$ mm; nominal helix angle, $\beta_1 = 23°21'30''$. Instead, the worm wheel has the following characteristics: number

Fig. 11.27 Right-hand worm gear pair, with involute worm: **a** projections of the instantaneous lines of contact on a transverse plane of the worm; **b** projection of the surface of contact on the pitch plane of the worm

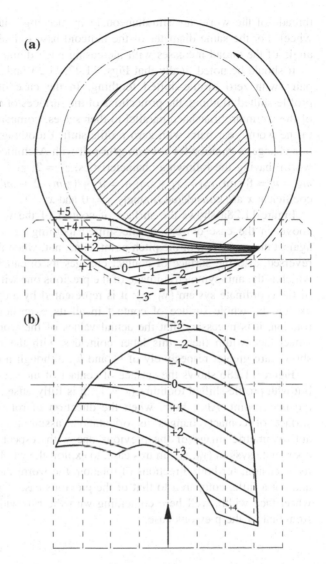

of teeth, $z_2 = 28$; nominal module, $m = 9\,\text{mm}$; nominal pitch diameter, $d_2 = 252\,\text{mm}$. Figure 11.27 shows the projections of the instantaneous lines of contact on a transverse plane of the worm, and the projection of the surface of contact on the pitch plane of the same worm.

In the same figure, a dashed line represents the outside surface of the worm wheel, which is a type-*A* outside surface. Instead, the projections of the instantaneous lines of contact, and projections of the mixed-line curves that surround the surface of contact are shown with continuous lines. From this figure, it is clear that the length of the surface of contact is about equal to $(0.66\,p_z)$, for which three threads of the worm are simultaneously in meshing with the teeth of the worm wheel. For the same diameter of the helicoid base cylinder, the nominal pressure angle of the thread increases with increasing pitch diameter of the worm.

It should be noted finally that Figs. 11.22, 11.23 and 11.27 refer to worm gear pairs with zero profile-shifted toothing. In the case of worm gear pairs with profile-shifted toothing, the projections of the surfaces of contact on the pitch plane of the worm are translated parallel to themselves, compared to the position related to the worm gear pairs with zero profile-shifted toothing. Figure 11.28 highlights this change of position for a worm gear pair with Archimedean spiral worm (type A worm) having the following characteristics: $z_1 = 2$; $z_2 = 30$; $r_1 = d_1/2 = 46$ mm; $m_x = m = 8$ mm; $r_{w1} = r_1 + xm$; $a = [r_1 + (z_2 m)/2 + xm]$. Two values of the shift coefficient x are considered, namely $x = 0$ and $x = 1$.

Figure 11.28a shows the surface of contact of the worm gear pair considered above, in the case of zero profile-shifted toothing ($x = 0$). It is quite centered, against the projection of the pitch point. Note that, when the direction of rotation is reversed, also the surface of contact changes its orientation, assuming a position which is the anti-symmetric image of the previous one with respect to the origin O_0 of the coordinate system (y_0, z_0); it is represented by a dashed line. Obviously, to exploit the whole surface of contact in all its extension, for both directions of rotation, it is necessary that the actual values of the worm face width and worm wheel face width (here, this latter coincides with the rim width, as Fig. 11.25 shows) are greater respectively of b_1^* and b_2^*, although not by much.

Figure 11.28b shows the surface of contact of the same worm gear pair above, but with profile-shifted toothing ($x = 1$). It is fully offset, against the projection of the pitch point. Also, here, when the direction of rotation is reversed, also the surface of contact changes its orientation, assuming a position which is the anti-symmetric image of the previous one with respect to the origin O_0 of the coordinate system (y_0, z_0). In this case, to exploit the whole surface of contact in all its extension, for both directions of rotation, the worm face width must be greater, and not a little, compared to that of the previous case. On the contrary, the worm wheel face width, well here coinciding with the rim width, does not differ much from that of the previous case.

11.10 Surface of Contact: Determination by Graphic-Analytical Methods

As we mentioned in the previous section, the determination of the instantaneous lines of contact and surface of contact between the worm and worm wheel can also be done using graphic methods and graphic-analytical methods. The purely graphic methods include, for example, those proposed by Stribeck [63], Ernst [14], Pohl [47], and Niemann and Weber [43]. Here we will not mention these methods.

Fig. 11.28 Surface of
contact of a worm gear pair
with type A worm: **a** worm
gear pair with zero
profile-shifted toothing
($x = 0$); **b** worm gear pair
with profile-shifted toothing
($x = 1$)

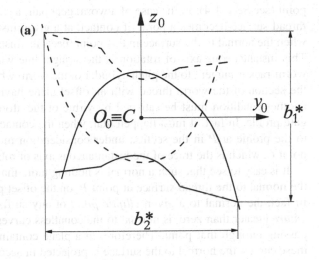

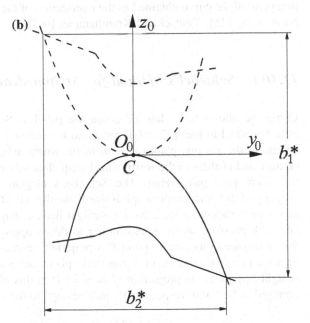

The mixed graphic-analytical methods include those proposed by Schiebel [56],
Ingrisch [22], Maschmeier [38], Altman [1], Buckingham [2] and Merritt [39]. Here
we will briefly describe two ingenious methods respectively proposed by Schiebel,
for the type A worm, and Ingrisch, for the type I worm.

Both methods use the basic law of conjugate gear-tooth action, which states that
as the gear members rotate, the common normal to the surface at the point of
contact must always intersect the line of centers at the same point, C, i.e. the pitch

point (see Sect. 1.4). In the case of a worm gear pair, a current point P on the worm thread surface becomes a point of contact, at a given instant of the meshing cycle, when the normal to the surface in this point meets the instantaneous axis of rotation. This instantaneous axis of rotation is the straight line where the pitch plane of the worm-rack is tangent to the pitch cylinder of the worm wheel. In fact, if we consider the section of the worm thread with an offset plane having coordinate $y_1 = const$, another condition must be satisfied by virtue of the aforementioned law of conjugate profile. In fact, it must happen that, when the contact is at point P, the normal to the profile at P in the section under consideration must pass through the pitch point C, which is the trace of the instantaneous axis of rotation in this section plane.

It is easy to see that such a normal is nothing more than the normal projection of the normal to the thread surface at point P, on the offset plane under consideration. In fact, the normal to a given *elliptic point* of any surface, having *Gaussian curvature* greater than zero, is normal to the countless curves lying on the surface and passing through that point. Therefore, in a plane containing the tangent to one of these curves, the normal to the surface is projected in accordance with the normal at that point of the curve obtained as the intersection of the surface with that plane (see Novozhilov [45], Ventsel and Krauthammer [66]).

11.10.1   Schiebel's Method for Archimedean Spiral Worms

Using the above basic law of conjugate profiles, Schiebel (see Schiebel [56], Schiebel and Lindner [57, 58]) proposed the following method, which allows to obtain, in the various planes normal to the worm axis, the instantaneous lines of contact, and profiles of the worm wheel teeth, thus solving the kinematic problem of the worm gear pair sizing. The Schiebel's elegant method also uses another property of the Archimedean spiral worm, the threads of which, in the axial section, configure a rack with teeth having straight flanks, shaped as an isosceles trapeze, and with pressure angle α. According to this property, the normal to the worm thread surface in its current point P is projected, regardless of whether this point is or not a point of contact, in a transverse plane, according to a straight line. This straight line joins the projection P' of point P in this plane with point A, which is arranged at $90°$ with respect to P', and belongs to the circle of radius e, given by:

$$e = (p_z/2\pi) \cot \alpha = p \cot \alpha = const; \tag{11.126}$$

the radius of this circle is a constant for all points of the surface (Fig. 11.29).

In fact, given any current point P on the thread surface (in Fig. 11.29, this point is localized in the section belonging to the plane of the same figure), at a distance r from the worm axis, the normal to the surface in this point must satisfy the following conditions:

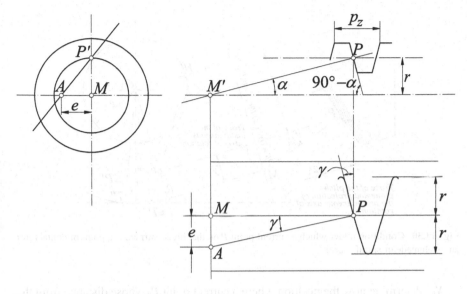

Fig. 11.29 Determination of distance, e

- It must be normal to the generation straight line passing through the point P, and thus contained in the plane normal to this straight line passing through this point; this plane has trace PM' in Fig. 11.29.
- It must be normal to the helix with lead p_z and helix angle γ, passing through point P, and thus having, in the plane view, projection PA inclined by an angle γ with respect to the projection of worm axis.

The point A of intersection of the normal to the helicoid surface with the plane passing through the worm axis, and forming an angle of $\pi/2$ compared to that passing through point P', is distant from the worm axis of the quantity, e, given by:

$$e = \overline{PM} \tan \gamma = r \cot \alpha \tan \gamma. \qquad (11.127)$$

As

$$\tan \gamma = \frac{p_z}{2\pi r} = \frac{p}{r}, \qquad (11.128)$$

we infer:

$$e = r \cot \alpha \frac{p_z}{2\pi r} = p \cot \alpha = const. \qquad (11.129)$$

It should be noted that the point A would be localized, in Fig. 11.29, in the position symmetrical with respect to the point M, if the worm had left-hand helix, rather than right-hand helix, such as that in Fig. 11.29.

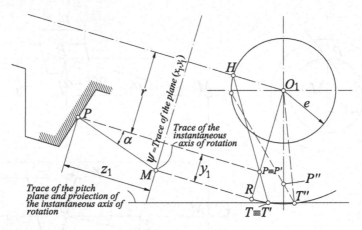

Fig. 11.30 Condition under which a current point P on the thread surface is a point of contact, for an Archimedean spiral worm

We determine now the position where a current point P, whose distance from the worm axis is r, and belonging to the worm transverse plane having coordinate $z_1 = const$, becomes a point of contact. Figure 11.30 shows, on the left, the thread axial section in a given axial plane of the worm (a left-hand worm is considered), and the trace, ψ, of the plane (x_1, y_1) passing through the worm wheel axis and the instantaneous axis of rotation (see Fig. 11.10). The same figure shows, on the right, what happens in the transverse section of the worm having coordinate $z_1 = const$, where the point P under consideration is localized. This transverse section is parallel to the plane (x_1, y_1).

Let us consider the axial plane passing through point P, and therefore containing the worm axis. This plane, which has trace O_1R in Fig. 11.30, will cut the thread surface according to a straight side rack, as it is shown in the same figure, on the left, in which this plane is represented revolved in the relevant view after rotating about its intersection with the plane (x_1, y_1). Given the above-described properties, in this plane, the projection of the normal PM is nothing more than the normal to the tooth straight flank. Moreover, the intersection of the normal to the surface with the instantaneous axis of rotation is projected into the point M, in which the normal to tooth straight flank at point P cuts the trace of the plane (x_1, y_1), coinciding with the projection of the instantaneous axis of rotation. It follows that, in this plane revolved in the relevant view, when point P is a point of contact, the intersection M of the normal to the tooth profile with the projection of the instantaneous axis of rotation must be away from point P of the amount (Fig. 11.30):

$$y_1 = z_1 \tan \alpha. \tag{11.130}$$

It should be noted explicitly that, for all the points for which z_1 is a constant, i.e. for all the points belonging to a given worm transverse plane, y_1 is also a constant.

It is now easy to find the position for which a current point P is a point of contact. In fact, once z_1, r, and $e = p \cot \alpha$ are known, we find the point of intersection H of the circle having radius e and center O_1, with the normal drawn from point O_1 to the straight line $O_1 P$. From point P, on the straight line $O_1 P$, we bring a segment $PR = y_1 = z_1 \tan \alpha$. Successively, we join point H with point P, and then we extend the segment HP until it meets at point T the normal drawn from point R to the straight line $O_1 P$. Let us now rotate rigidly about point O_1 the quadrilateral $O_1 HTR$, until the point T is located at point T', on the projection of the instantaneous axis of rotation. For simplicity, in Fig. 11.30, point P has been drawn in the position in which it is a point of contact, for which T coincides with T', and P coincides with P'. It is evident that, since the circumference of center O_1 and radius $O_1 T$ cuts in two points, T' and T'', the projection of the instantaneous axis of rotation (Fig. 11.30), we have correspondingly two possible positions, P' and P'', for which P can become a point of contact.

Maintaining $z_1 = const$, and thus $y_1 = const$, and varying r, it is easy to find, point by point, the projection of the line of contact on the plane $z_1 = const$, i.e. the intersection of the surface of contact, that is the locus of all points of contact, with the plane $z_1 = const$. If we consider a sufficient number of these sections, we can determine the entire surface of contact with the desired accuracy. To make the calculation procedure more streamlined and convenient, it should consider a number of planes $z_1 = const$ equally spaced of the same fraction of the product $p_z \cot \alpha$ (for example, the planes: $z_1 = 0$; $z_1 = \pm 0.1 p_z \cot \alpha$; $z_1 = \pm 0.2 p_z \cot \alpha$; etc.). Correspondingly, we will have the offset planes defined by coordinates $y_1 = const$, as well equidistant of a same amount (for the same example, the offset planes: $y_1 = 0$; $y_1 = \pm 0.1 p_z$; $y_1 = \pm 0.2 p_z$; etc.). It is noteworthy that z_1-coordinates are to be considered positive in the advancing side of the contact, and negative in the receding side of the contact; correspondingly, y_1-coordinates are positive towards the worm wheel, and negative towards the worm.

We do not consider here to further deepen the operating procedure of the Schiebel's method in its graphical part and, in this regard, we refer the reader directly to the already mentioned works of Schiebel or other scholars (for example, Giovannozzi [17]). However, operating with the previously described procedure, we can trace, point by point, the curves corresponding to various values of z_1, that is the sections of the surface of contact with the planes having coordinate $z_1 = const$. Figures 11.31 and 11.22 summarize the results obtained using the above procedure, for a two-start left-hand worm. In particular, Fig. 11.31a shows the curves indicated by 0, ± 1, ± 2,..., corresponding respectively to the intersections of the surface of contact with the planes having coordinates $z_1 = 0$, $z_1 = \pm 0.1 p_z \cot \alpha$, $z_1 = \pm 0.2 p_z \cot \alpha$, etc.

From the curves above, it is easy to get, point by point, the intersection curves of surface of contact with the offset planes, i.e. with the planes parallel to the worm wheel mid-plane, having coordinates $y_1 = const$. For example, the plane having the straight line $+ \mathrm{II}$ as a trace (Fig. 11.31a) cuts the curves $z_1 = const$ previously obtained in points marked with circles, the abscissas of which are the corresponding coordinates z_1. The curve obtained as intersection of the surface of contact with the

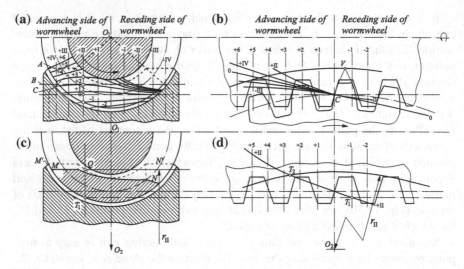

Fig. 11.31 Two-start left-hand worm gear pair, with Archimedean spiral worm: **a** projections of the instantaneous lines of contact on a transverse plane of the worm; **b** projections of the instantaneous lines of contact on an offset plane; **c** projection of the surface of contact on a worm transverse plane; **d** usable portion of the line $+\mathrm{II}$

plane $+\mathrm{II}$ is shown in Fig. 11.31b, as well as other curves obtained as intersections of the surface of contact with other planes parallel to the plane $+\mathrm{II}$, the traces of which are indicated with Roman numerals.

The extreme points T_1 and T_2 of the usable portion of this curve are defined by the following ordinates, x_1:

- the ordinate of the intersection of the trace of plane $+\mathrm{II}$ with the outside surface of the worm, for the point T_1 (Fig. 11.31c);
- the ordinate of the intersection of the line of contact with the circumference of radius r_{II}, coaxial with the worm wheel and passing through the point Q (Fig. 11.31c), for the point T_2; point Q is the intersection of the trace of the plane $+\mathrm{II}$, with the outside surface (the throat or outside cylinder) of the worm wheel.

Repeating the procedure for a number of offset planes parallel to the worm wheel mid-plane (Fig. 11.31b), we get a series of curves all passing through the pitch point C, which constitute the lines of contact related to the planar conjugate profiles obtained by intersecting the worm gear pair with these planes. In Fig. 11.31b, only some of these curves are shown, in order not to weight it down too much. The straight line parallel to the worm axis through the pitch point C, and the circle tangent to it, having center O_2, always constitute the pitch curves.

Using the usual methods of drawing conjugate profiles, we infer easily that, in different offset planes parallel to the worm wheel mid-plane, the teeth profiles of the worm wheel corresponding, as mating profiles, at the intersections of these planes with the worm thread, have a shape much more pointed, as their plane is far from the worm wheel mid-plane.

Finally, to get a clear idea on which to base the geometric-kinematic sizing of the worm gear pair, it is useful to analyze the projections of the surface of contact both on a transverse plane of the worm, and on the pitch plane of the worm. Figure 11.31c shows the projection, on a transverse plane of the worm, of the surface of contact of the worm gear pair under consideration here, when the worm wheel has a type-A outside surface. Figure 11.22 shows the projection of the surface of contact of the same worm gear pair on the pitch plane of the worm; in this regard, we refer to what we said in the Sect. 11.9.

It is to be noted that, in order that the contact is actually possible at all points of the surface obtained using the previously described procedure, it is necessary that this surface is contained within the following two well-defined limits:

- The first limit is constituted by the curve τ which, in different offset planes parallel to the worm wheel mid-plane (Fig. 11.31b), represents the locus of the extreme points at which the contact can get there, without the interference from occurring. As for involute cylindrical gears, on the lines of contact above, these points are those of minimum distance from the trace of the worm wheel axis O_2 (Fig. 11.31d). This curve τ is localized on the side of the top of the line a (Figs. 11.22 and 11.24b) of the surface of contact having a horseshoe shape; therefore, the danger of interference depends on the greater or lesser value of the worm addendum.

- The second limit is constituted by the curve σ, which is the locus of the extreme points that would be obtained on the lines of contact when the tooth flanks in the various sections had to touch, that is, when the teeth were pointed. In the various offset planes parallel to the worm wheel mid-plane, these extreme points are the intersections of the circles having center O_2 and passing through the apex V (Figs. 11.31b and 2.12) of the worm wheel teeth in that section, with the corresponding lines of contact.

The determination of the curve σ can be performed in an approximate way with the following procedure, which however provides reliable results from the application point of view. In the worm wheel mid-plane, containing the worm axis, the worm wheel tooth profile is an involute curve corresponding to the worm-rack and pitch curves p_1 and p_2 (Fig. 11.10); therefore, it can be drawn with well known procedures. The apex V (Fig. 11.31b) in which the extensions of the profiles of the two tooth flanks intersect, can thus be determined. The circle having the center on the worm wheel axis and passing through point V cuts the lines of contact on the various offset planes parallel to the worm wheel mid-plane at points that are, with a good approximation, the limiting points that would occur if in these planes the tooth profiles of the worm wheel had a pointed shape. Projecting these points on the pitch plane of the worm, we find the curve σ.

The line of contact in different offset planes parallel to the worm wheel mid-plane have variable slope, namely the higher in the receding side of the worm wheel, and the lower in its advancing side. This fact is unfavorable because, as in the parallel cylindrical spur gears, a variation of inclination of the force exchanged between the teeth, at equal tangential load to be transmitted, involves a variation of magnitude of this force, and therefore less smoothness of operation. The slope differences between the various lines of contact increase with the increasing of e, that is, according to Eq. (11.127), they increase with the decreasing of α, and the increasing of γ. Therefore, to reduce these differences for worms with strong lead angle of the thread, γ, it is convenient to adopt the highest values of the pressure angle, α.

The simultaneous lines of contact are the instantaneous lines of contact that are obtained intersecting the surface of contact with the worm thread surfaces, in a given position of the worm. For the determination of these simultaneous lines of contact, it is therefore necessary to consider a number of angular positions of the worm. Each of these angular positions of the worm is defined by the screw motion, and therefore by the position of the coordinate system $O_1(x_1, y_1, z_1)$, rigidly connected to worm, with respect to the coordinate system $O_0(x_0, y_0, z_0)$, rigidly connected to the frame. Of course, it is necessary to repeat the above procedure for each of the angular positions under consideration. These simultaneous lines of contact are projected, on the pitch plane of the worm and within the boundary of the surface of contact having a horseshoe shape, as approximately straight lines, parallel to the worm threads.

Recalling that, in the parallel cylindrical spur gears, it is most appropriate not to use the lines of contact in points too close to the limit points of tangency of the path of contact with the base circle, even now the use of these limit lines is very doubtful. In the parts of the surface of contact more distant from the instantaneous axis of rotation, the sliding is maximum, and the effects of errors of manufacturing and assembly are more consistent. Therefore, to a more extended use of the surface of contact, a higher manufacturing and assembly precision must correspond.

11.10.2   Ingrisch's Method for Involute Worm

Before describing the peculiarities of the Ingrisch's method for determining the surface of contact of an involute worm gear pair, it is necessary to premise a few fundamental relationships typical of these worm gears, in addition to those set out in Sect. 11.3. With symbols already introduced in that section, we have the following relationships:

$$p_z = 2\pi r_{b1} \tan \gamma_{b1} = 2\pi r_{b1} \cot \beta_{b1} \tag{11.131}$$

$$p_z = 2\pi r_1 \tan \gamma_1 = 2\pi r_1 \cot \beta_1 \tag{11.132}$$

$$r_{b1} = \frac{p_z}{2\pi \tan \gamma_{b1}} = \frac{p_z \tan \beta_{b1}}{2\pi} = r_1 \frac{\tan \gamma_1}{\tan \gamma_{b1}} = r_1 \frac{\cot \beta_1}{\cot \beta_{b1}} = r_1 \frac{\tan \beta_{b1}}{\tan \beta_1}. \qquad (11.133)$$

The pressure angle α_t in a transverse section of the worm is given by Eq. (11.18); therefore, it can be written as:

$$\cos \alpha_t = \frac{r_{b1}}{r_1} = \frac{\tan \gamma_1}{\tan \gamma_{b1}} = \frac{\tan \beta_{b1}}{\tan \beta_1}. \qquad (11.134)$$

The pressure angle α_n in a normal section of the worm thread is given by Eq. (8.17), shown here together with an equivalent relationship:

$$\tan \alpha_n = \tan \alpha_t \cos \beta_1 = \tan \alpha_t \sin \gamma_1. \qquad (11.135)$$

Considering Eq. (8.52), the same normal pressure angle α_n can be obtained using the following equation:

$$\cos \alpha_n = \frac{\sin \beta_{b1}}{\sin \beta_1} = \frac{\cos \gamma_{b1}}{\cos \gamma_1}. \qquad (11.136)$$

The pressure angle α_x in an axial section of the worm is given by the first of Eqs. (11.17), for which we have:

$$\tan \alpha_x = \frac{\tan \alpha_n}{\cos \gamma_1} = \frac{\tan \alpha_n}{\sin \beta_1}. \qquad (11.137)$$

From Eqs. (11.135) and (11.137), we get:

$$\tan \alpha_t = \frac{\tan \alpha_x}{\tan \gamma_1} = \frac{\tan \alpha_x}{\cot \beta_1}. \qquad (11.138)$$

The determination of the surface of contact of an involute worm gear pair according to Ingrisch's method is carried out following a procedure analogous to that described in the previous section for an Archimedean spiral worm. Therefore, we repeat what we have already described for the Archimedean spiral worm, but we consider now the planes tangent to the base cylinder, instead of the worm axial planes (Fig. 11.32). In so doing, we find easily that, in the front view, a current point P on the worm thread flank (Fig. 11.32, right), at a distance r from the worm axis and belonging to a plane having coordinate $z_1 =$ const, becomes a point of contact when the following condition is satisfied. The tangent PM through P to the base circle, consisting of the circumference obtained by sectioning the base cylinder with the plane $z_1 = const$, must intersect the projection of the instantaneous axis of rotation at a point T, whose radial distance, y_1, from P satisfies the condition given by:

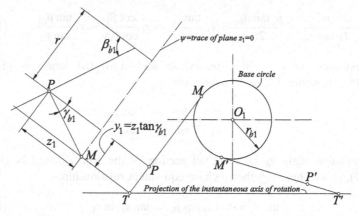

Fig. 11.32 Condition under which a current point P on the thread surface is a point of contact, for an involute worm

$$y_1 = PT = z_1 \tan \gamma_{b1} = z_1 \cot \beta_{b1}. \tag{11.139}$$

Therefore, once r and z_1 are known, the corresponding position of the point P, when it is a point of contact, is the one for which the circle of center O_1 and radius r, and the projection of the instantaneous axis of rotation intercept a segment of length $z_1 \tan \gamma_{b1} = z_1 \cot \beta_{b1}$ on the tangent to the base circle, passing through point P. As we saw for the Archimedean spiral worm, even now we have, generally, two positions of contact for point P. Therefore, to get the lines of contact in different planes $z_1 = const$, it is sufficient to trace a set of straight lines that are tangent to the base circle at points equidistant from it. Then we bring, starting from the intersections of these lines with the instantaneous axis of rotation, segments $(z_1 \cot \beta_{b1})$ in the same direction for each of the curves to be obtained. By joining the points corresponding to a given value of $(z_1 \cot \beta_{b1})$, we find the lines of contact, i.e. the curves of intersection of the surface of contact with the various planes $z_1 = const$.

However, if the distance $y_1 = z_1 \tan \gamma_{b1}$ is equal to $\sqrt{(r_1^2 - r_{b1}^2)}$ (Fig. 11.33), in that plane $z_1 = const$, there is only one position of the point of contact P, on the base circle having radius r_{b1}. This point coincides with the point in which the tangent to the base circle, passing through the pitch point C and corresponding to the direction of the thread surface, touches the same base circle. In this regard, the direction to be considered, of course depending on the fact that the worm is right-hand or left-hand, must be understood as the tangents PT and $P'T'$ shown in Fig. 11.32.

If $z_1 \tan \gamma_{b1} > \sqrt{(r_1^2 - r_{b1}^2)}$, in the aforementioned plane $z_1 = const$, two distinct points of contact P' and P'' exist on the base circle of radius r_{b1}. As Fig. 11.33

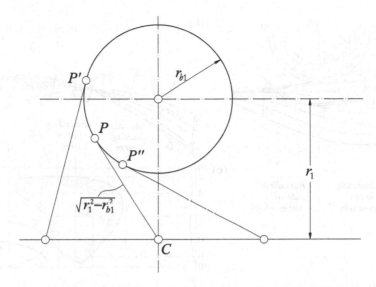

Fig. 11.33 Condition for which only one point of contact on the base circle exists

shows, an arc of finite length separates these two points, whereby the line of contact is not continuous, but has a gap. In other terms, the line of contact is composed of two separated and detached portions. As well as for an Archimedean spiral worm, also the usable part of the surface of contact of an involute worm is projected on the worm pitch plane according to a closed curve, having a horseshoe shape. Figure 11.34c shows this projection, for a five-start involute worm. On the same pitch plane of the worm, the points P' and P'' shown in Fig. 11.33, in which the two branches of the lines of contact in different planes $z_1 = const$ begin, are projected according to a curve τ', with respect to which the projection of the surface of contact must remain outside. The apex of this curve τ' corresponds to point P shown in Fig. 11.33, and approximately to the curve indicated by $+3$ in Fig. 11.34a.

Using the same identical procedure employed for the Archimedean spiral worm, we determine the intersections of the surface of contact with the different offset planes, parallel to the worm wheel mid-plane (Fig. 11.34b), the projections of the surface of contact and curves σ and τ on the worm pitch plane, the worm wheel tooth profiles, etc. In addition, to determine the instantaneous lines of contact, we proceed as indicated for the Archimedean spiral worm, with the only difference that, in the case of an involute worm, the sections of the surface of contact with the various planes $z_1 = const$ are involute curves of the base circle, instead of Archimedean spirals. Figure 11.27b shows the projections of these instantaneous lines of contact on the worm pitch plane, for a right-hand worm gear pair with an involute worm.

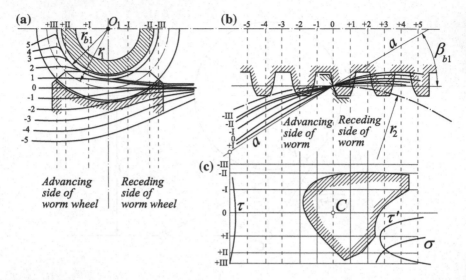

Fig. 11.34 Five-start worm gear pair, with involute worm: **a** projections of the instantaneous lines of contact on a transverse plane of the worm; **b** projections of the instantaneous lines of contact on an offset plane; **c** projections of the surface of contact, and curves τ, τ' and σ on the worm pitch plane

From what we have said previously, we can deduce that the intersection of the surface of contact with any plane tangent to the pitch cylinder of the worm is a straight line. Therefore, the surface of contact is a ruled surface. In fact, in a plane that has as a trace, for example the straight line MT (Fig. 11.32), the points P have coordinates y_1 which are proportional to the abscissas z_1 in the direction of the worm axis.

The surface of contact has, as limiting line, the straight line a-a (Fig. 11.34b) in the plane tangent to the base cylinder, and parallel to the worm wheel mid-plane (the plane indicated with $+\mathrm{III}$ in Fig. 11.34a). Therefore, the surface of contact has a shape much more asymmetric (thus it is worse usable), the more r_{b1} is large compared to r_1. It is therefore convenient to reduce the ratio $(r_{b1}/r_1) = \cos\alpha_t$, compatibly with the need not excessively to increase α_t, which is related to α_x by the Eq. (11.138).

For a good design of the worm gear pair, it is necessary to keep in mind that the outside surface bounding the worm threads and worm wheel teeth must be defined in such a way that the surface of contact does not go beyond the limiting lines τ, σ and τ' (Fig. 11.34c). However, at the same time, these outside surfaces must be drawn to ensure that the largest possible part of the lines of contact is used (see, in this regard, the following section).

11.11 Outside Surface of the Worm Wheel and Related Interference Problems

In Sect. 11.9 we have seen that the position and shape of the boundary line b of the projection of the surface of contact on the worm pitch plane (see Figs. 11.22 and 11.23), and consequently the extension of the same surface of contact, depend on the geometry of the outside surface of the worm wheel. In the same section we have examined what are, from the point of view of the worm gear design, the effects of three different shapes of the worm wheel outside surface (type-A, type-B, and type-C), and we calculated, in that regard, that the best outside surface of the worm wheel is the one called type-A.

With reference to this shape of the worm wheel outside surface, which is the best, it is to be noted that, once the radius r_e of the two side cylindrical parts has been defined (Fig. 11.26), the projection of the surface of contact on a transverse plane of the worm (Fig. 11.31c) results limited laterally by points M and N. Therefore, these side cylindrical parts can be chamfered with conical surfaces, usually passing through the worm axis, and arranged beyond the corresponding points M' and N', thus saving a useless work during the worm wheel cutting process. The slight conical chamfer shown in Fig. 11.31a, c responds to this concept. It is to be observed that, on the right side (Fig. 11.31c), the conical chamfer could start at point N, thus obtaining a worm wheel asymmetrical with respect to its mid-plane. Usually, we make the worm wheel symmetrical with respect to its mid-plane.

It should be borne in mind that, if the surface of contact does go beyond the limiting curve τ, that is, if the intersection of the surface of contact with the tip cylinder of the worm goes beyond the line τ, the interference with the worm wheel teeth occurs. Therefore, since during the cutting of the worm wheel teeth, a cutting tool (a hob or other suitable tools) replaces the worm, a part of the worm wheel tooth close to its root is removed, due to undercut. The result is a decrease in the usable part of the surface of contact.

For involute worms having nominal sizing of their threads, the curve τ is very far from the surface of contact, whereby the risk of interference due to undercut does not occur. This factual situation is very evident in Fig. 11.34c. Instead, for Archimedean spiral worms having mean values of the lead angle γ_1, the interference occurs for $z_2 < 36$ with $\alpha = 15°$, and for $z_2 < 20$ with $\alpha = 20°$.

Actually, the cutting interference is also higher, since the tool used to cut the worm wheel teeth is sized in such a way as to ensure the suitable clearance between the tip cylinder of the worm and the root surface of the worm wheel. Therefore, the cutting tool of the worm wheel cuts the surface of action in accordance with the dashed curve shown in Fig. 11.35. This curve is external with respect to that corresponding to the actual worm, and is obtained using a procedure similar to that described in the two previous sections.

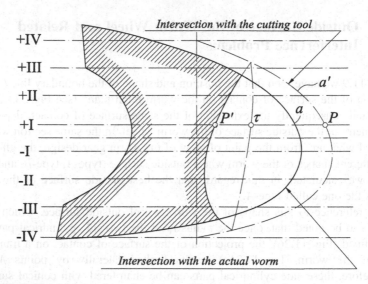

Fig. 11.35 Portion of the surface of contact lost for the cutting interference

Using an approximate method proposed by Schiebel (see Schiebel and Lindner [58]), the limit of the usable part of the surface of contact is obtained by drawing, starting from the line τ, and in direction of the worm axis, segments having a length $\tau P'$ equal to half of the segments $P\tau$, i.e. $\tau P' = (1/2)P\tau$. As Fig. 11.35 shows, in the cutting condition, the boundary line a of the surface of contact is replaced by the line a', outside the line a. The loss of a part of the surface of contact is therefore not limited only to the part intersected by the line τ but is much wider.

As for parallel cylindrical spur gears, to avoid the interference in a worm gear pair with Archimedean spiral worm, we can take, for cutting the worm wheel, a profile shift xm, where x is the profile shift coefficient, and $m = m_x = (p_x/\pi) = (p_z/\pi z_1)$ is the axial module of the worm. This profile shift does not involve any change in the shape of the worm, which is equal to that of the hob used to cut the worm wheel. Instead, the teeth of the worm wheel, as we saw in Chap. 6, and as shown in Fig. 11.36a, assume a more pointed shape. In the axial section of the worm that contains the worm wheel mid-plane, on the operating pitch cylinders having radii r_{w1} and r_{w2}, the thread and tooth thicknesses are respectively given by $[(\pi m/2) + 2xm \tan \alpha_x]$ and $[(\pi m/2) - 2xm \tan \alpha_x]$, where $\alpha_x = \alpha_{wx}$ is the operating axial pressure angle of the worm.

For an axial pressure angle $\alpha_x = 15°$, Wolff [68] calculated the values of the profile shift coefficients, x, which are necessary to avoid the interference in the worm gears with Archimedean spiral worm. These values are a function of the number of threads, z_1, number of teeth, z_2, and average value of the lead angle, γ_1. According to Wolff, for a worm gear pair having nominal sizing $[h_1 = h_{a1} + h_{f1} = (13/6)m]$, axial pressure angle $\alpha_x = 15°$, and cut with a hob with rounded edges and fillet

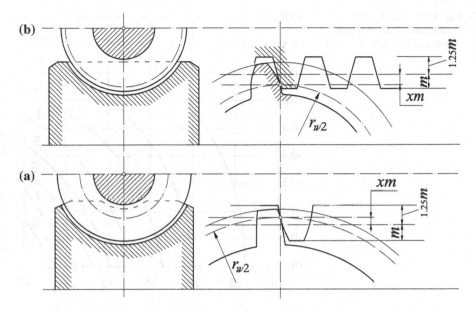

Fig. 11.36 Worm gear pairs with profile-shifted toothing: **a** Archimedean spiral worm, with positive profile shift; **b** involute worm, with negative profile shift

radius equal to 0.2 m, the profile shift coefficient can be calculated by the following relationship:

$$x = \pi z_1 \left[\frac{0.15}{z_1} - q \frac{0.01}{(z_1/z_2)} \right], \tag{11.140}$$

where q is a factor that is a function of γ_1 and z_1, as Table 11.1 shows. It should be noted that, by putting $x = 0$ in Eq. (11.140), we obtain the minimum number of teeth, z_2, for which the profile shift is not necessary. Profile shift is not even necessary, when we obtain negative values of the profile shift coefficient, x, from Eq. (11.140).

In worm gears with involute worm, interference may occur if the surface of contact cuts the line τ', beyond which contacts are not possible, for the reasons described in Sect. 11.10.2. In this case, the addendum of the worm wheel teeth must be reduced, and this is achieved by cutting the worm wheel with a negative profile

Table 11.1 Values of factor q according to Wolff

γ_1	6°	8°	10°	12°	14°	16°	18°	20°	22°	24°	26°
$z_1 = 1$	0.65	0.55	0.43	0.30	–	–	–	–	–	–	–
$z_1 = 2$	0.83	0.83	0.78	0.73	0.68	0.63	0.50	0.31	–	–	–
$z_1 = 3$	1.00	1.00	1.00	1.00	0.96	0.89	0.83	0.72	0.63	0.46	0.31

Fig. 11.37 Profile shift coefficient x of an involute worm as a function of γ_1 and z_1

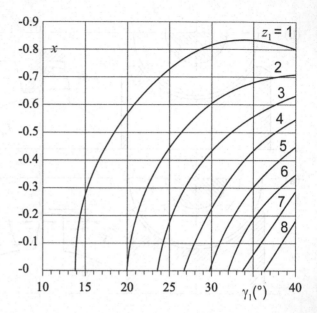

shift coefficient, x. In this regard, the abacus shown in Fig. 11.37 allows to calculate this profile shift coefficient, as a function of γ_1 and z_1, for an axial pressure angle $\alpha_x = 15°$. Here too, the thread of the worm remains unchanged, while the worm wheel teeth become stockier, as the example in Fig. 11.36b shows.

Due to the aforementioned negative profile shift, the usable surface of contact tends to move towards the line τ, becoming more centered with respect to the pitch point C (Fig. 11.34c). Therefore, excessive values of profile shift can cause interferences related to a high value of the addendum of the worm threads. For this reason, for type A, type N, type I, and type K worms, the trend today is to choose values of the profile shift coefficient included in the range $(-0.50 \leq x \leq +0.50)$, and preferably $x = 0$, or slightly positive. In doing so, we avoid the base circle going to land in the field of active flank (see Jelaska [25]).

For worm gears with type C worm, it is convenient to choose values of the profile shift coefficients included in the range $(0 \leq x \leq 1)$, with mean values preferably equal to $x = 0.50$. The lower values of this range of variability are to be preferred for high loads $(z_2 \geq 40)$, while the highest values are more suitable for high operating speed, which guarantee a high efficiency, but require a high precision. Other authors (see Jelaska [25]), for the type C worm, indicate a variability range of profile shift coefficients between $(0.50 \leq x \leq 1.50)$.

11.12 Load Analysis of Worm Gears, Thrust Characteristics on Shafts and Bearings, and Efficiency

The load analysis of worm gear pairs can be done in a similar way as in the case of crossed helical gears. Here this analysis is performed considering the most frequent case in practical applications, which is that of a worm-drive unit with shaft angle $\Sigma = \pi/2$. Similar to what we did for crossed helical gears, we consider only the power losses due to sliding in the direction tangent to the helix obtained intersecting the threads with the worm pitch cylinder, while still we neglect the power losses due to sliding in the direction of the thread depth.

For simplicity and according to what is usually made and accepted for this type of gear-drive unit, here we determine the forces acting on the worm and worm wheel shafts under the hypothesis that all the force that worm and worm wheel transmit to each other passes through the pitch point C (Fig. 11.38). We still remember that, at this pitch point, located in the worm wheel mid-plane, the worm pitch plane and worm wheel pitch cylinder are tangent, and thus they are touching. For the sake of simplifying the calculation, we assume also that the transmission of this total force takes place only at the pitch point C.

Fig. 11.38 Components F_n and $F_\mu = \mu F_n$ of the total force that worm and worm wheel transmit to each other

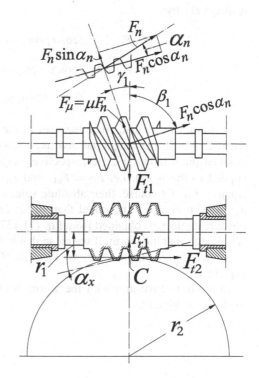

The total force that worm and worm wheel transmit to each other has two components:

- A component F_n normal to the two mating surfaces at the pitch point C.
- A friction component $F_\mu = \mu F_n$ directed along the tangent at point C to the pitch helix, having helix angle, β_1, and lead angle, $\gamma_1 = [(\pi/2) - \beta_1]$.

We must here necessarily take into account the latter component, because it, unlike what happens for parallel cylindrical spur and helical gears, generally has a non-negligible intensity, especially in relation to the effects of the calculation of the tangential force acting on the worm.

Considering the normal section of the worm thread through the pitch point C (Fig. 11.38), we see that the force F_n acts perpendicular to the thread profile, and is inclined of the normal pressure angle, α_n. This force can be resolved into three mutually perpendicular components, i.e. the tangential component, $F_{nt} = F_n \cos \alpha_n \sin \gamma_1$, the axial component, $F_{na} = F_n \cos \alpha_n \cos \gamma_1$, and the radial component $F_{nr} = F_n \sin \alpha_n$. The friction force $F_\mu = \mu F_n$, where $\mu = \tan \varphi$ is the coefficient of friction, and φ is the angle of friction, can be resolved into two components: the tangential component $F_{\mu t} = \mu F_n \cos \gamma_1$, acting in the same direction as the force F_{nt}, and the axial component $F_{\mu a} = \mu F_n \sin \gamma_1$, acting in the opposite direction to the axial component F_{na}.

Therefore, for a worm driving the worm wheel, the three resultant components of the effective total force acting on the worm at its nominal pitch radius, r_1, are given respectively by:

$$F_{t1} = F_n(\cos \alpha_n \sin \gamma_1 + \mu \cos \gamma_1) = F_{a2} \qquad (11.141)$$

$$F_{a1} = F_n(\cos \alpha_n \cos \gamma_1 - \mu \sin \gamma_1) = F_{t2} \qquad (11.142)$$

$$F_{r1} = F_n \sin \alpha_n = F_{r2}. \qquad (11.143)$$

It is noteworthy that, since the axes of the worm and worm wheel form a shaft angle $\Sigma = \pi/2$, the tangential and axial forces applied to the worm wheel at its nominal pitch radius, r_2, are respectively equal to the axial and tangential forces applied to the worm, i.e. $F_{t2} = F_{a1}$, and $F_{a2} = F_{t1}$, while the radial force F_{r2} is equal to F_{r1}. Of course, their absolute values are the same, but their directions are opposite. It is also to be noted that, in the case of an involute worm, the normal pressure angle α_n is obtained from Eq. (11.137), which also applies in the case of an Archimedean spiral worm, if we mean that α_x is the axial pressure angle of this type of worm. In the latter case, in fact, the Eq. (11.137) coincides with first of Eq. (11.16).

In a worm-drive unit with the worm driving the worm wheel, the transmitted load, F_{t2}, is given by:

$$F_{t2} = T_2/r_2, \tag{11.144}$$

where T_2 is the output torque available at the worm wheel shaft. The input torque, T_1, to the worm is given by:

$$T_1 = F_{t1}r_1. \tag{11.145}$$

Obtaining F_n as a function of F_{t2} from Eq. (11.142), and substituting the expression found in Eqs. (11.141) and (11.143), we get the following relationships that express F_{a2} and F_{r2} as a function of F_{t2}, which is generally one of the design data:

$$F_{a2} = F_{t2} \frac{\cos \alpha_n \sin \gamma_1 + \mu \cos \gamma_1}{\cos \alpha_n \cos \gamma_1 - \mu \sin \gamma_1} = F_{t2} \frac{\cos \alpha_n \tan \gamma_1 + \mu}{\cos \alpha_n - \mu \tan \gamma_1} \tag{11.146}$$

$$F_{r2} = F_{t2} \frac{\sin \alpha_n}{\cos \alpha_n \cos \gamma_1 - \mu \sin \gamma_1}. \tag{11.147}$$

If we consider that $F_{t1} = F_{a2}$, and take into account the Eq. (11.146), the input torque to the worm, T_1, given by Eq. (11.145), can also be expressed as follows:

$$T_1 = F_{t1}r_1 = F_{t2}r_1 \frac{\cos \alpha_n \tan \gamma_1 + \mu}{\cos \alpha_n - \mu \tan \gamma_1} \cong F_{t2}r_1 \tan(\gamma_1 + \varphi); \tag{11.148}$$

this is because $\cos \alpha_n \cong 1$, and $\mu = \tan \varphi$.

With the same approximation $\cos \alpha_n \cong 1$, and recalling that, according to Eq. (11.6), $\tan \gamma_1 = (p_z/2\pi r_1)$, Eq. (11.148) becomes:

$$T_1 = F_{t2}r_1 \frac{p_z + 2\pi\mu r_1}{2\pi r_1 - \mu p_z}. \tag{11.149}$$

The same Eq. (11.148) shows that, for a gear-drive unit with the worm driving worm wheel, the tangential forces F_{t1} and F_{t2} are related by the equation:

$$F_{t1} \cong F_{t2} \tan(\gamma_1 + \varphi). \tag{11.150}$$

With the worm wheel driving the worm, the above equation becomes:

$$F_{t1} \cong F_{t2} \tan(\gamma_1 - \varphi). \tag{11.151}$$

With the above approximate formulas, it is possible to express the efficiency of a worm gear pair in a simple way, as the quotient of the input power to the driven machine divided by the output power of the driving machine. With the worm driving the worm wheel, we have:

$$\eta = \frac{T_2\omega_2}{T_1\omega_1} = \frac{F_{t2}r_2\omega_2}{F_{t1}r_1\omega_1} = \frac{F_{t2}r_2\omega_2}{F_{t2}\tan(\gamma_1 + \varphi)r_1\omega_1} = \frac{\tan\gamma_1}{\tan(\gamma_1 + \varphi)}; \qquad (11.152)$$

this is because $(r_2\omega_2/r_1\omega_1) = \tan\gamma_1$, as we can infer from Eq. (11.30), written for the nominal pitch cylinders, and with $\Sigma = 90°$ (for further theoretical details, see Ferrari and Romiti [15]).

The diagram shown in Fig. 11.39 provides the curves that allow to calculate the efficiency given by Eq. (11.152) as a function of the lead angle, γ_1, and coefficient of friction $\mu = \tan\varphi$. From the theoretical point of view, from Eq. (11.152) we infer that, for a given value of the coefficient of friction μ, the efficiency increases with increasing of γ_1, up to reach a maximum value for $\gamma_1 = [(\pi/4) - (\varphi/2)]$. The curves in Fig. 11.39 show that the efficiency depends on the value of the coefficient of friction, and that, for values of this coefficient not higher than 0.1 that we have in practical applications, the efficiency practically reaches its maximum value already for $\gamma_1 = (15 \div 20)°$. This factual circumstance is advantageous because, with threads too inclined, the cutting is difficult, and in any case the related surfaces of contact have unfavorable asymmetrical shapes and are reduced in size. In practice, we use lead angles in the range $(15° \leq \gamma_1 \leq 30°)$.

The efficiency of a worm gear pair is rather low compared to other types of gear-drive systems with similar power transmission capacity. The main cause of the low efficiency is the large amount of frictional losses due to the relative sliding between the mating surfaces of the worm threads and worm wheel teeth. The efficiency of the worm gear pairs is a function of many factors, such as the lead angle, load, speed, coefficient of friction (and thus surface finishing, and type and

Fig. 11.39 Efficiency of a worm-drive unit with worm driving the worm wheel, as a function of γ_1 and μ

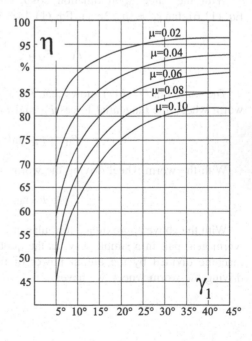

condition of lubrication), type of design of the gear-drive unit, and other factors. Since the efficiency increases with increasing lead angle, a higher efficiency can be achieved by using multi-start worms with small diameter. Good results can be also obtained by using rigid worms, with smooth, grounded or polished thread flanks.

If lubrication is appropriate and the gear-drive unit consists of a case-hardened and ground worm, meshing with an accurately machined worm wheel, then the efficiency will depend mainly on the coefficient of friction, μ, and lead angle, γ_1, according to Eq. (11.152). This equation expresses a purely theoretical condition, according to which the efficiency is not dependent on the load or speed. In reality, it instead depends on these two quantities, since the coefficient of friction, μ, is a coefficient of friction mediated, as between the worm and worm wheel we always have a good lubrication. In this regard, the results of researches done in this field confirm that a greater tangential velocity and a smaller load ensure ideal conditions of lubrication in the best way, so that the efficiency increases.

The speed increase leads, however, a greater heating of the worm gear pair, and a more rapid wear, so that the tangential velocity should not exceed 3 m/s for worm gears with steel worm and cast iron worm wheel, and 10 m/s for worm gears with steel worm and worm wheel made of phosphorus bronze. However, it is necessary that, as speed increases, the load must be correspondingly reduced. The experience has also shown that the efficiency and good operating conditions of the gear-drive unit are highly dependent on the accuracy of manufacturing and assembly. The operating conditions of this worm gear pair are in fact correct from a kinematic point of view only for one relative position of the worm and worm wheel axes.

With the worm wheel driving the worm, the efficiency is given by:

$$\eta' = \frac{\tan(\gamma_1 - \varphi)}{\tan \gamma_1}. \tag{11.153}$$

Equations (11.152) and (11.153) can be written as follows:

$$\eta = \frac{1 - \mu \tan \gamma_1}{1 + \frac{\mu}{\tan \gamma_1}} \tag{11.154}$$

$$\eta' = \frac{1 + \frac{\mu}{\tan \gamma_1}}{1 + \mu \tan \gamma_1}. \tag{11.155}$$

From Eqs. (11.154) and (11.155), we obtain:

$$\eta' = \frac{2 - \frac{1 - \mu \tan \gamma_1}{\eta}}{1 + \mu \tan \gamma_1} \cong 2 - \frac{1}{\eta}; \tag{11.156}$$

this is because $\mu \tan \gamma_1$ can be neglected compared to unit. It follows that, if the efficiency, η, is equal or less than 0.5, it is not possible for the worm wheel to drive the worm. This means that the gear-drive unit is irreversible or self-locking, i.e. the worm can drive the worm wheel, but the reverse drive is not possible. We use this property when the design requires the irreversibility condition as a categorical imperative. However, since the fulfillment of this condition depends on the coefficient of friction, μ, which for various reasons is very likely to change during the service, the irreversibility is not always automatically ensured, even if the above conditions are satisfied, and the proper coefficient of friction was initially chosen. In this framework, prudence requires to have always the availability of a brake in such a gear-drive unit.

Where irreversibility is required (see, for example, certain crane systems), the following values of the lead angle, γ_1, may be taken as guideline values for roughing calculation: $\gamma_1 \leq 5°$ and, respectively, $\gamma_1 \leq 6°$, when anti-friction bearings and journal bearings are used. The practical guideline values of the coefficient of friction, μ, are as follows:

- $\mu = 0.10$, for worm made of heat-treated steel, worm wheel made of cast-iron, and operating conditions characterized by smooth finished surfaces of threads and teeth, and grease lubrication;
- $\mu = (0.08 \div 0.09)$, for worm made of heat-treated steel, worm wheel made of phosphorus bronze, and operating conditions characterized by surfaces of threads and teeth without smooth finishing, and grease lubrication;
- $\mu = (0.06 \div 0.07)$, for worm made of heat-treated steel, worm wheel made of phosphorus bronze, and operating conditions characterized by machined surfaces of threads and teeth, and oil lubrication;
- $\mu = (0.05 \div 0.06)$, for worm made of case-hardened and ground steel, worm wheel made of phosphorus bronze, and operating conditions characterized by machined surfaces of threads and teeth, and oil lubrication.

Figure 11.39 shows that, for a given value of the coefficient of friction, the efficiency of a worm gear pair varies markedly with the lead angle. Since the efficiency increases with increasing lead angle, it follows that large lead angles are desirable. However, with increasing lead angle, the nominal pitch radius $r_1 = (z_1 m / 2 \tan \gamma_1)$ of the worm decreases correspondingly, as Eq. (11.6) shows. Consequently, even the root diameter of the worm decreases and, with it, its resistant area, while the worm wheel face width becomes narrower. From the same relationships it follows that, for a given value of the module, m, it is necessary to increase the number of threads, z_1, so that the radius r_1 is not too small, and the worm too weak. For further details on calculation of efficiency of worm gear drives, and experimental investigations, we refer the reader to specialized literature (see, for example: Magyar and Sauer [36], Stahl et al. [61]).

Since generally, in practice, the radius r_1 is assumed equal to $(2 \div 4)$ time the axial pitch $p_x = \pi m$, the maximum value of γ_1, corresponding to the minimum value of $r_1 = 2p_x$, should not exceed that we get from the relationship:

$$\tan \gamma_1 = \frac{z_1}{4\pi}. \tag{11.157}$$

For this reason, the values of γ_1 that are adopted are generally the following: $\gamma_1 = 9°$, for one-start worms; $\gamma_1 = (17 \div 18)°$, for two-start worms; $\gamma_1 = 25°$, for three-start worms. From the above considerations, it is clear that the choice of the lead angle cannot be made based on efficiency alone, and that the designer must make a reasonable balance between the factors governing the efficiency and strength.

In a worm-drive unit, the worm shaft is subjected to deflections due to bending and shear loads. In addition to strength considerations, the design of this worm-drive system involves stiffness considerations. From the point of view of the strength design, we have to determine the dimensions of the gear members to withstand the given design loads. From the stiffness point of view, it is necessary to prevent excessive deformations, such as large deflections of the worm shaft that might interfere with its performance. In fact, it is well known that the maximum deflection of the worm shaft must not exceed a predetermined limit value, compatible with the proper kinematics of the worm gear pair.

Response analysis and design analysis of the worm shaft in terms of strength and stiffness require the preliminary determination of the stress resultants, i.e. the bending moments, torques, axial and shear forces. The latter here are neglected, because their influence on the deflections is known to be small compared to that due to bending moments. For the determination of these stress resultants, we assume that the worm is symmetrically positioned relative to the bearings that support its shaft, and that the total force that worm and worm wheel transmit to each other passes through the pitch point C, located in the worm transverse plane, which is equidistant from the bearing mid-planes (Fig. 11.40). We also assume that the input power enters from the left side, and the thrust bearing is the one on the right in Fig. 11.40. With this arrangement, the stress resultants acting on the worm shaft are as follows:

- A bending moment acting in the worm wheel mid-plane, due to the total axial component, F_{a1}, and distributed as shown in Fig. 11.40a.
- A bending moment acting in the same plane as above, due to the radial component, F_{r1}, and distributed as shown in Fig. 11.40b.
- A bending moment acting in the plane orthogonal to the worm wheel mid-plane, due to the total axial component, F_{a2}, and distributed as shown in Fig. 11.40c.
- A torque $T_1 \cong F_{t2}r_1 \tan(\gamma_1 + \varphi)$ acting on the shaft part that goes from the worm mid-plane to the driving machine, due to the total tangential component, F_{t1}, and distributed as shown in Fig. 11.40d (in case the input power entered from the right side, this torque would be applied in the right half instead of the left half of the worm shaft).

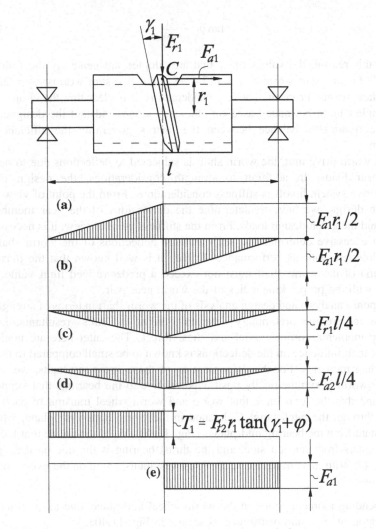

Fig. 11.40 Diagrams of the stress resultants acting on the worm shaft

- A compressive force, equal to the total axial component, F_{a1}, acting on the shaft part between the worm mid-plane and the thrust bearing, and distributed as shown in Fig. 11.40e (in case the thrust bearing was the one on the left, we would have a tensile force acting on the left half instead of the right half of the worm shaft).

The considerations described above and the relationships that follow apply in the theoretical case of carefully manufactured toothing and precise bearing mounting. Obviously, these theoretical conditions are not practically achievable. Therefore, under actual service conditions, some amount of deformations, errors and deviations from the theoretical conditions should be accepted. The magnitude of these deformations, deviations and errors is entrusted to the experience of the designer, and the judicious use of empirical formulas developed by industry experts. In any case, to minimize the bending deflections of the worm shaft, in consistency with the other design parameters, it is preferable that the shaft diameter is as large as possible, and that the distance between the bearings that support the same shaft is the minimum possible.

11.13 Worm and Worm Wheel Sizing: Further Considerations

In addition to what we said in Sect. 11.3, it should be noted that, generally, the worm sizing, that is the definition of the addendum and dedendum of the worm, is made in relation to the module, m, of the generating rack, which coincides with the worm axial module, i.e. $m = m_x$. The nominal pressure angles of the worm axial profiles, i.e. the pressure angles related to the pitch cylinder, are chosen according to the worm wheel number of teeth, z_2, and, with the same number of teeth, they are assumed greater than those adopted for the parallel cylindrical spur gears.

By doing so, we avoid the interference phenomena, which are manifested to a greater extent at the beginning of the worm threads meshing with the parts of the worm wheel teeth that are more distant from the worm wheel mid-plane. Therefore, we adopt angles α up to $30°$ for worm wheels having a minimum of 18 teeth ($z_2 = 18$), and we can get off to $\alpha = 15°$ for worm wheels with more than 60 teeth ($z_2 = 60$).

For the Archimedean spiral worms, the pressure angle of the generating axial profile does not change when the diameter of the cylinder we choose as nominal pitch cylinder of the worm changes. Instead, for the involute worms, such pressure angle changes when the chosen pitch cylinder of the worm changes. In fact, from Eq. (11.19) we obtain:

$$\tan \gamma_1 = \tan \gamma_{b1} \frac{r_{b1}}{r_1} = \tan \gamma_{b1} \frac{d_{b1}}{d_1}, \tag{11.158}$$

whereby the pressure angle of the generating pitch profile on the pitch cylinder is given by:

$$\tan \alpha_1 = \frac{\tan \gamma_{b1}}{\tan \gamma_1} \sqrt{1 - \frac{\tan^2 \gamma_1}{\tan^2 \gamma_{b1}}} = \frac{r_1}{r_{b1}} \sqrt{1 - \left(\frac{r_{b1}}{r_1}\right)^2}. \tag{11.159}$$

Instead, the pressure angle is zero on the base cylinder.

The values of the nominal lead angles of worm threads range from $\gamma_1 = (5 \div 6)°$ when the worm gear pair must have irreversible operating characteristics, to $\gamma_1 = (30 \div 40)°$ when we want to get worm gear pairs with high efficiency.

To reduce as much as possible the maximum deflection of the worm shaft, it is appropriate to have a comparatively large diameter of the resistant cross-sectional area of the worm shaft, and a distance between the bearings as small as possible in relation to the other design parameters. In this regard, the form number of the worm, given by Eq. (11.12), plays a very important role; therefore, it must be chosen with care. The diameter of the resistant cross-sectional area of the worm shaft, which is necessary for calculating the section modulus of the worm shaft, must obviously be less than the root circle diameter of the worm. It is determined on the basis of strength or stiffness considerations, or on the basis of manufacturing economy (for example, convenience or not to do the worm integral with the shaft, or to have the worm press-fitted or shrink-fitted on the shaft).

Once the module m has been chosen, the nominal pitch diameter $d_1 = 2r_1$ remains determined. Therefore, from Eq. (11.6) we can deduce that, in order to obtain large lead angles, and therefore high levels of efficiency, it is necessary to have a high number of threads, that is, it is necessary that the worm is a multi-start worm. The number of starts is also a function of the gear ratio $u = z_2/z_1$. As a guideline for selecting the number of starts of the worm as a function of u, the following indications can be used: $z_1 = 1$ for $u \geq 30$; $z_1 = 2$ for $u = (15 \div 29)$; $z_1 = 3$ for $u = (10 \div 14)$; $z_1 = 4$ for $u = (6 \div 9)$.

The pressure angles used in worm gear pairs depend on the lead angles. They must be sufficiently high, to avoid undercut of the worm wheel teeth on the side where the contact ends. With increasing lead angle, the cutting conditions of the worm wheel become unfavorable for which, for manufacturing reasons, greater pressure angles are chosen for larger lead angles. As a guideline for selecting the pressure angle as a function of the lead angle, the following indications can be used: $\alpha_1 = 20°$ for γ_1 up to $15°$; $\alpha_1 = 22.5°$ for γ_1 over $15°$ up to $25°$; $\alpha_1 = 25°$ for γ_1 over $25°$ up to $35°$; $\alpha_1 = 30°$ for γ_1 over $35°$. It is to be noted that, by choosing in this way the appropriate lead angle, a satisfactory tooth depth is also obtained.

In the case of involute worm, the root circle diameter of the worm, d_{f1}, must be greater than or at least equal to the base circle diameter, $d_{b1} = 2r_{b1}$. For lead angles γ_1 greater than $45°$ (in this case the worm wheel is the driving member of the worm gear pair), while maintaining the total depth of the thread unchanged, an increased addendum is chosen, and this increase is commensurate to the lead angle (for example, $h_{a1} = 1.4m \cos \gamma_1$).

The operating pitch diameter of the worm may be different from its nominal pitch diameter. This happens in the case of a worm gear pair with profile-shifted toothing. It is obvious that, in this case, for generating the toothing of the worm wheel, we will use a hob whose operating pitch cylinder is characterized by the same profile shift of actual worm.

The definition of the worm wheel face width, b_2, must take into account not only the guideline Eq. (11.29), but also the following relationship:

$$b_2 = \sqrt{d_{a1}^2 - d_1^2},\qquad\qquad(11.160)$$

according to which b_2 depends on the diameters of the tip and reference cylinders of the worm.

It is however to be noted that, in the case where the worm threads have high thickness and large lead angle, with the value of the face width determined using the previously mentioned procedure, the interference may occur between the side parts of the worm wheel toothing and the worm. In the cutting conditions, this type of interference leads to the removal of more or less substantial portions of the tooth flank by undercut. The determination of the maximum usable face width of the worm wheel therefore represents a critical point, and it must be made with great care using the known methods of mating profiles (see Pollone [48], Henriot [19]).

We have already pointed out that, from a theoretical point of view, only one relative position of the two members of a worm gear exists for its proper operating condition. Any deviation from this proper relative position, due to design and cutting errors or incorrect assembly, results in malfunction, abnormal tooth contact, loss of efficiency, and reduction of load bearing capacity. To overcome these drawbacks, it is good to foresee the use of crowned worm wheels. Indeed, this crowning not only eliminates the above drawbacks, but also promotes the formation of an appropriate oil film, which exalts the positive effects of lubrication.

As we have already said, a worm wheel is obtained by using a hob that is the copy of the worm. In other words, the hob has the same pitch diameter as that of the worm, and differs only for a greater addendum, which serves to ensure the necessary clearance for the actual worm gear pair. If, to cut the worm wheel, we use a so-sized hob, we get a worm wheel without crowning. To have crowned worm wheels, the following four methods can be used:

- Cut the worm wheel directly using a hob whose pitch diameter is slightly larger than that of the worm. This is a relatively simple but effective method, since it not only provides a localized contact at the mid-plane, but also a sufficient gap for the lubricant oil film formation.
- After cutting the worm wheel with a hob placed at the standard center distance, recut and finish it with the same hob, whose axis is shifted parallel to the worm wheel axis by $\pm\Delta z$. So the same crowing effect of the previous method is obtained, but with a higher cost.
- The crowned teeth of the worm wheel can be obtained by a different hob orientation, compared to its standard position to cut a crownless worm wheel. In fact, to obtain the crowning effect on the worm wheel teeth, the hob axis is slightly rotated by a $\Delta\vartheta$ angle about the shortest distance straight line, clockwise or counterclockwise.

- Use a worm with a larger pressure angle than the worm wheel. This is the only one of the four possible methods that involves modifications that do not affect the worm wheel (like the three previous methods), but the worm. It is also a very complex method, both theoretically and practically, because it is necessary to change the pressure angle and the axial pitch of the worm without changing the pitch line parallel to the axis, in accordance with the relationship $p_x \cos \alpha_x = p_{x,i} \cos \alpha_{x,i}$. The equations that allow us to calculate the final values, p_x and α_x, of the two variables from their initial values, $p_{x,i}$ and $\alpha_{x,i}$, are very complex. The subject, though very interesting, goes beyond the purpose of this textbook, so we refer the reader to more specialized treatises.

11.14   Double-Enveloping Worm Gear Pairs

A cylindrical worm and a worm wheel that is throated, i.e. configured in such a way that it can wrap partially and envelope the worm, constitute the worm gear pairs studied in the previous sections. These worm gear pairs are therefore called *single-enveloping worm gears*. In some practical applications, *double-enveloping worm gears* have been used, which are characterized by the fact that both the worm and worm wheel are throated. Therefore, the worm has globoidal shape, so that it is curved longitudinally to fit the curvature of the worm wheel (Fig. 11.41). These worm gears are also called *globoidal worm gear pairs*, and, sometimes, *hourglass worms*, because of their peculiar shape.

The double-enveloping gear concept is very ancient. It dates back to Renaissance. The development of the worm gearing principle progressed along conventional lines from the era of Archimedes, which is universally accredited as the inventor of worm gears, until the 15th century, when Leonardo da Vinci evolved the double-enveloping gearing concept (see Loveless [35], Dudas [12], Chen et al. [6]). In fact, among the hundreds of sketches from its notebooks, not only drawings of worm gears were found, but also drawings of hourglass worm gears.

We call *globoids* those solid bodies delimited externally by a surface generated by a curved line rotating about an axis. The globoid shown in Fig. 11.42 is delimited, at its outside periphery, from the surface generated by the rotation of the circular arc DE about the z-axis. If, during the rotation of the globoid about its axis, a tool ABC with cutting edge trapezoidal in shape, contained in the axial plane of the same globoid, rotates about an axis normal to that plane and passing through the center O of the circular arc DE (a_0 is the shortest distance between the globoid and cutting tool axes, in the cutting conditions), the cutting tool generates the globoidal worm. All this as long as the angular velocity ω_0 of the cutting tool is in a constant relationship with that of the globoid, ω_1. The sides of the cutting edge of trapezoidal shape can be straight or curved.

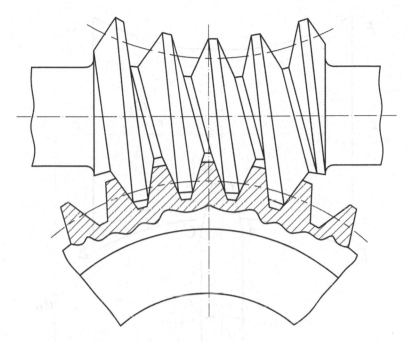

Fig. 11.41 Double-enveloping worm gear pair

Consider, as the pitch surface, that generated by the circular arc FG which bisects the depth of the worm thread, on which the angular width of the tool is γ (Fig. 11.42). If, for each revolution of globoid, the tool rotates by an angle 2γ, we have a single-start globoidal worm; instead, if the tool rotates by an angle $2z_1\gamma$, we will have a z_1-start globoidal worm. In the axial section of the globoid, the thread profile is that of the generating cutting tool. The helix angle on the pitch surface changes with continuity as we move from the transverse mid-plane to the transverse end planes of the worm. The circular pitch (Fig. 11.41) is constant, but the axial pitch is not constant from thread to thread. In addition, the profile of the worm thread is continuously variable, and so does the lead angle. Globoidal worms are also made with thread flanks consisting of helicoids having threads with curved flanks in axial section.

When, for the generation of the globoidal worm, a tool with cutting edge trapezoidal in shape is used, the generating lines in the process of generation keep the direction of tangents to the circle with center O and radius r_0. This circle is obtained by placing the tool with its mid-plane coinciding with the mid-plane of the worm, extending the two cutting edges, which are symmetrical with respect to the mid-plane, and finally drawing the circle tangent to the extensions thus obtained (Fig. 11.42). The generation of the worm wheel is obtained by using a hob identical to the generated worm.

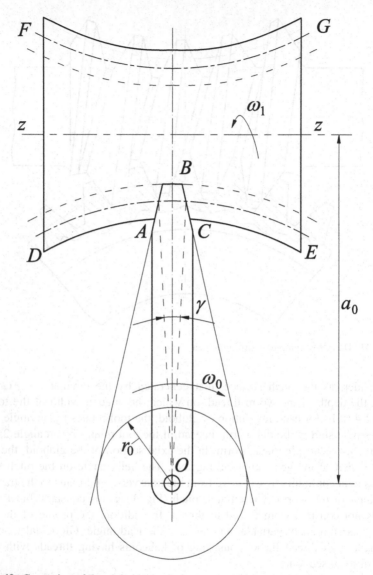

Fig. 11.42 Generation of the globoid worm

The generating process of globoidal worms has evolved over time, staring from the one introduced, in 1765, by the famous English clockmaker Henry Hindley (see Loveless [35]). With this process, which is considered the first process historically introduced in this specific field, the worm is cut by means of a trapezoidal-edged cutting blade, while the worm wheel is cut by a worm-type cutter which is identical

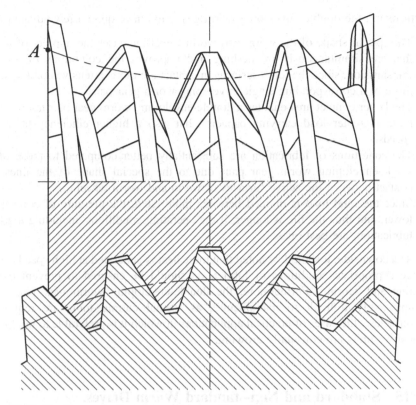

Fig. 11.43 Double-enveloping worm gear pair, with a single-start worm

to the Hindley's worm. Today, several cutting processes are used, with different gear cutting machines and different cutters.

All these processes, whatever the technology used, present considerable difficulties, both for the worm and for the worm wheel. Depending on the generating process used, both the worm and worm wheel differ, more or less, from the theoretical shape, as the undercut phenomena cannot be completely eliminated.

Figure 11.43 shows a double-enveloping worm gear pair, with a single-start worm. It highlights the fact that the worm face width cannot extend beyond the point A for which the tooth flank of the worm wheel is parallel to the shortest distance between the axes of rotation. In fact, in the case where the worm face width extends beyond that point, it would be not possible to either assemble the worm with the worm wheel, or mill the worm wheel with the cutting tool. Compared to a single-enveloping worm gear pair, the double-enveloping worm gear pairs require greater care in assembly. In fact, in addition to verifying that the shaft angle and the shortest distance are exactly those of the generating and manufacturing processes, it is necessary to ensure that the mid-plane of the globoidal worm contains the axis of rotation of the worm wheel.

In return, the double-enveloping worm gear pairs have quite a few advantages:

- The special shape of the worm and worm wheel increases the number of teeth that are simultaneously in meshing and improves the conditions of torque transmission, since the tooth surfaces in contact are more extensive compared to those of a corresponding single-enveloping worm gear pair.
- The larger contact area enables a double-enveloping worm gear pair not only to have a greater load carrying capacity, but also a higher efficiency at usual speeds.
- The conditions of lubrication are substantially better compared to those of a single-enveloping worm gear pair, due to the special shape of the lines of contact between the active surfaces of worm and worm wheel.
- Since the total load is divided between more teeth, the pressure of contact is lower, and thus the life is longer, if the operating conditions, including a good lubrication, are best ones.

As a downside, it should be noted that more heat is generated at high speeds, and hence appropriate conditions of lubrication should be ensured to prevent over-heating and scoring. Furthermore, the double-enveloping worm gear pairs are comparatively difficult to manufacture, as special cutting tools are required. Nevertheless, the cases of practical applications in which it is preferable to use these gear-drive units are constantly growing.

11.15   Standard and Non-standard Worm Drives, and Special Worm Drives

In Sect. 11.2 we described the five types of standardized worms by ISO, which characterize as many types of worm drives. However, types of worm drives, which include standard and non-standard worm drives and special worm drives, are much more numerous. From a practical application point of view, the various types of worm drives can be divided into power worm drives and precision worm drives. Power worm drives are mainly used in the heavy industry (e.g., mining, petro-chemical, metallurgy, mechanical, freight, etc.), so the problems to be solved are those related to the improvement of their load carrying capacity, which impose the use of double-enveloping hourglass worm drives. Precision worm drives are mainly used in fine mechanics (e.g., measuring equipment, scientific devices, fine robotics, indexing devices, actuators for orientation of antennas and solar panels, etc.), so the problems to be solved are those related to improving the precision of relative movements between worm and worm wheel, which impose the use of dual-lead cylindricalworm drives.

Power worm drives, which have been developed to meet stringent requirements for load carrying capacity, efficiency, and transmission accuracy, include at least the following eight types of worm drives (see Chen et al. [6]):

(1) *Cylindrical worm drives*, characterized by cylindrical worms with thread flanks having straight lines as generatrixes. Therefore, the worms of these worm drives are type *A*, type *I*, and type *N* worms.

(2) *Hindley worm drives*, characterized by double-enveloping gear pairs (see Loveless [35], Dudas [12], Crosher [8]). Since a trapezoidal-edged cutting blade cuts the Hindley globoidal worm, its thread flanks have straight lines as generatrixes. Therefore, this type of worm drive represents a generalization of the concept of type *N* cylindrical worm. A worm-type hob cutter instead cuts the worm wheel. This hob cutter, with the appropriate variations already mentioned at the end of Sect. 11.2, is identical to the worm. The surfaces of the worm thread flank cannot be grounded, and thus the worm cannot be subjected to surface hardening. Its surface hardness is usually within the range $(30 \div 40)$ HRC, so wear in operating conditions is remarkable.

(3) *Wildhaber worm drives*: an enveloping worm and a cylindrical worm wheel characterize these drives, proposed by Wildhaber, chief engineer of Gleason, in 1922 (see Dudas [12], Chen et al. [6]). The hourglass or enveloping worm is generated by a pinion-type cutter that, with the appropriate variations already highlighted above and at the end of Sect. 11.2, has cutting edges identical to the normal tooth profiles of the worm wheel. Thus, the thread flank surfaces of the enveloping worm can be easily grounded, with all the advantages that can be achieved.

(4) *Niemann worm drives*: a cylindrical worm, whose threads have concave axial profiles, and a cylindrical worm wheel, whose teeth have convex axial profiles, characterize these drives, developed by Niemann in 1935 (see Niemann and Winter [44], Chen et al. [6]). The worm, which originated the type *C* worms, is machined with a convex circular profile disk-type cutter or grinding wheel. Depending on the kinematic conditions between the worm to be cut and cutter, even a circular axial profile can be obtained. As usual, the worm wheel is cut with a worm-type hob cutter that, with the appropriate variations highlighted above, is identical to the worm. The advantages of the worm drives obtained by combining these two members are better lubrication conditions between the mating active surfaces, resulting in greater efficiency, and a substantial reduction in surface stresses due to the fact that the load is transmitted from a concave surface to the convex mating surface.

(5) *Spiroid worm drives*, which can be characterized by *circular cone worms* (these are the *spiroid worms* in the strict sense) or by cylindrical worms (in this case we are talking about *helicon worms*). In both cases, the mating worm wheels have related teeth distributed over their end surfaces, for which they resemble face-gear wheels with curved teeth. A good percentage of these teeth (from 10% to 15%) is simultaneously in meshing with threads of the spiroid worm. Therefore, these worm drives allow reducing noise and providing good load

carrying capacity. They are widely used in particular areas such as drive systems of astronomical telescopes, launcher missiles, and naval shipbuilding platforms as well as machine tool equipment and instrumentation (see Goldfard [18]).

(6) *Toroidal involute worm drives*, or shortly *TI worm drives*, which are characterized by an involute cylindrical worm wheel in meshing with an enveloping hourglass worm, the latter generated by the first (see Duan et al. [11]), with the usual modes described above. Theoretical analysis, TSC (Tooth Surface Contact) analysis as well as prototype experimental tests have shown that these worm drives have an appropriate load distribution, multi-tooth line contact, and high reliability. However, the performance of these worm drives is still better with lower loads, while wear increases and efficiency decreases with increasing load (see Sun et al. [64]).

(7) *Double-enveloping worm drives*: an hourglass worm, which can be cut either using a milling cutter or a grinding wheel with conical surface, and an enveloping worm wheel, generated by a hob-type cutter, with the usual modes described above, characterize these drives. The hourglass worm is in fact the generalization of the type K cylindrical worm, which is generated by a disk-type milling cutter having straight cone generatrixes or by a biconical grinding wheel. As in all worm gears, even in these worm drives a line contact occurs, but the meshing region is wider. Various types of these worm drives are described in scientific and technical literature, some of which stand out for improved efficiency, extended life, and higher transmission power compared to the Hindley worm drives described above, which are similar to those discussed here only in some aspects. These more remarkable features are essentially due to double-line contacts on the thread surface, resulting in a wider length of the contact line (see Sakai et al. [53], Zhang and Tan [70], Zhang et al. [71], Zhao and Zhang [72]).

(8) *Internal meshing worm drives*, which are characterized by an internal crown worm wheel and an enveloping barrel worm, generally generated by the same internal crown worm wheel, with the usual modes described above. These types of worm drives add to the benefits of hourglass worm drives (i.e., multi-tooth line contact, good lubrication conditions, long service life, and high load carrying capacity), the benefits of internal meshing, such as compacter structure, smaller volume, and lighter weight. Various types of these worm drives are described in scientific and technical literature (see Hoyashita [21], Chen [5], Chen et al. [4]). Theoretical and numerical analyses and experimental tests have substantially confirmed the aforementioned advantages. Therefore, the perspectives for their practical application in aerospace and other heavy-load areas with volume and weight restrictions are extremely promising.

As we have already pointed out, each of the eight types of worm drives described above includes one or more subtypes that fall into the same category. So the number of possible configurations of worm drives becomes much larger. It is then further expanded when precision worm drives are considered, which are growing applications in several modern technical areas due to their features, such as relative motion accuracy, adjustable backlash, and ability to compensate geometric deviations due to wear. In fact, to achieve these performances, not only the center distance can be changed (in this case, the worm drive configuration remains essentially unchanged, with the only variation being able to adjust the center distance), but also special solutions can be used, which configure new types of worm drives.

These new types of precision worm drives include (see Chen et al. [6]):

(a) *Dual-lead worm drives*, which allow an accurate backlash adjustment or wear compensation by axially moving the worm, which is manufactured with two leads, so that the thickness from one of the worm threads to another is increased.

(b) *Worm drives with a split worm*, composed of a half-shank worm and a half-hallow worm, and a worm wheel with modified tooth flank surfaces, to fit the worm split design. With these worm drives, the backlash adjustment or wear compensation are obtained by fixing the half-shank worm and rotating the half-hallow worm to ensure that the two sides of the worm are meshing with the worm wheel.

(c) *Non-backlash double-roller enveloping hourglass worm drives*, with the worm wheel composed of two half worm wheels and rollers distributed over the circumference of each half worm wheel, and the hourglass worm generated by the rollers. With the same worm drives, the backlash adjustment and wear compensation are obtained by rotating one of the half worm wheel.

(d) *Worm drives composed of a backlash-adjustable planar worm wheel* with variable tooth thickness enveloping a hourglass worm. The backlash adjustment or wear compensation are obtained by moving the worm wheel along its axis, thanks to the variable tooth thickness that depends on the fact that both tooth flanks of the worm wheel are manufactured with different inclination angles. The precision worm drives, composed of an involute beveloid worm wheel enveloping a hourglass worm generated by the involute beveloid worm wheel surface, can be considered a variation of these worm drives, as they share the same operating principle with them. However, these drives belong to single-enveloping hourglass worm drives, with multi-tooth line contact and high load carrying capacity.

Even these new types of precision worm drives are characterized by several variations, which contribute to further enhancing the already large family of worm drives (see Yang et al. [69]). In addition, in some cases we need to have worm drives with high-precision and heavy-load capacity, so the problems to be solved become more numerous, being the sum of those concerning both the power worm drives and the precision worm drives. These cases include, for example, industrial robots, indexing devices for gear cutting machines, speed reducing worm drives for high-speed hoisting machines, precision artillery systems such as gun-laying radars, ranging radars, quick-fire guns, etc.

Chen et al. [6] attempted to classify worm drives, taking into account the following three factors:

- Shape of generating body.
- Tooth position and meshing area.
- Shape of generating surface.

As for the first factor, it broadens the perspective compared to the assumption that a worm-type hob cutter generates the worm wheel; this cutter, with the appropriate variations already mentioned at the end of Sect. 11.2, is identical to the worm. With this usual assumption, which we have done so far, the generating body and related generating surfaces of the worm drive are those of the worm, whereby the generated body and corresponding generated surfaces are those of the worm wheel.

However, both theoretically and practically speaking, to get a worm drive, both the worm and the worm wheel can be used as the generating body. The spectrum of possibilities so widens, also because the geometric shape of the worm or worm wheel, when one of these two members of the worm drive is assumed as the generating body, can be a cylinder, a cone, a convex toroidal rotor or a concave toroidal rotor. However, the differences are very marked for the worms, for which we have cylindrical worms, spiral bevel worms or spiroid worms, barrel worms, and hourglass worms; instead, the differences are almost entirely evanescent for the worm wheels, so the four possible shapes actually form a single group. Therefore, from the point of view of the shape of generating body, worm drives can be divided into five types. In fact, we can have worm drives with cylindrical-, spiroid-, barrel- and hourglass-worm enveloping the worm wheel, and with worm wheel enveloping the worm.

From the point of view of the tooth (or thread) position and meshing area, five types of worm drives can be also identified. In fact, the tooth (or thread) of the generating body can be located on the outside circumference, end-face or inside circumference, while different generated bodies can be obtained by choosing different contact lines, and positioning the tooth of the generated body on the outside circumference, end-face or inside circumference. Therefore, we may have worm drives with a normal worm meshing a normal worm wheel, or with an end-face or internal worm meshing a worm wheel, or with an end-face or internal worm wheel meshing a worm.

Finally, the shape of generation surface (it can be a surface obtained as a path of a curve, a surface obtained as an envelope of another surface, or a surface obtained by a rotating tooth) directly affects the thread profile and meshing performance of the worm drives. From this point of view, Chen et al. [6] identify fourteen types of worm drives.

Considering simultaneously the three aforementioned points of view, it turns out that the possible types of worm drives amount to 350. These numerous sets of worm drives include not only all the types of worm drives that are currently in use, developed to meet the most different needs of today's practical applications, but also types not known, yet to be developed by researchers and designers working in this area. This shows that gears are still an ever-living and vital search area, with a

360-degree horizon open to the development of scientific and technological knowledge. In fact, the family of possible worm drives would be even bigger, as the 350 types mentioned above do not include the special worm drives on which we want to make a quick hint here.

These special worm gear drives essentially contain ball worm drives, whose operation is based on the combination of two mechanical principles, the one of the worm drives so far described, and the one of the recirculating ball screws (see Kulkarni et al. [28]). With this combination, the surfaces of worm threads and worm wheel teeth are no longer in direct contact, and motion and power are transmitted through the balls rolling between them. The key feature that can determine the success of such mechanism is the replacement of the sliding friction between the meshing surfaces of the worm threads and worm wheel teeth by the rolling friction of spherical balls complemented by a smooth recirculation path implemented into these two members.

Unfortunately, however, despite studies and research on these special worm drives, which have resulted in the release of numerous international patents, nobody has even been able to implement and fine-tuned a *double-enveloping recirculating ball worm drive* capable of ensuring correct operating conditions. These conditions include a proper kinematic coupling between worm and worm wheel, mediated by balls, as well as the smooth ball recirculation in the appropriate raceways, and the appropriate and continuous contact between the same balls and between the balls and side surfaces of the raceways.

For a traditional worm drive, where contact between the active mating surfaces is a direct sliding contact, the cutting process of two worm drive members are those described in the previous sections. When contact between the active worm and worm wheel surfaces is an indirect rolling contact, because it is mediated by the interposition of rolling balls, especially when worm and worm wheel have both hourglass shape, these cutting processes are much more complicated. This is because, in the axial plane of the worm, for the worm, and in the transverse plane of the worm wheel, for the worm wheel, the sections of the raceways cannot be simple circumference arcs for both these two worm drive members. In other words, the traditional cutting motions for the generation of a helicoid thread on a toroidal surface using a spherical head-cutting tool (rotating motion about the worm axis and simultaneous rotation of the cutter about the torus axis) do not generate an appropriate envelope. In fact, the surface obtained by envelope do not have geometry such to avoid both the interference (this would result in balls jamming inside the raceways), and excessive backlash (this would result in loss of contact). The appropriate coupling on all contact surfaces between balls, raceway on the worm, and raceway on the worm wheel are not guaranteed.

In the framework of a multi-annual consultancy provided to an important custom gear drive industry, since 1995 this author has had the chance to tackle and study the issues associated with the conceptual design and development of these innovative types of worm drives, an example of which is shown in Fig. 11.44. In-depth studies and theoretical analyses have been carried out in this regard, in conjunction with his research team at the University of Rome "Tor Vergata". These researches showed that the current scientific literature, which is still very poor, as well as the

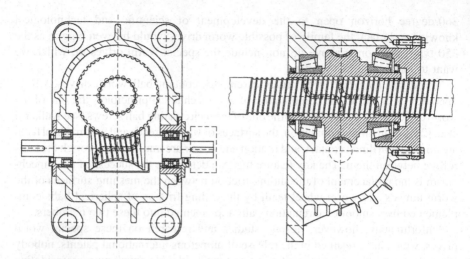

Fig. 11.44 Speed jack with double-enveloping and recirculating ball worm drive and recirculating ball screw

patents issued on this very advanced subject do not highlight the significant problems associated with the working interference or excessive backlash, which may occur between the balls and corresponding raceways.

The worm wheel and worm reference sections are notoriously the transverse mid-plane of the worm wheel and the transverse mid-plane of the worm, the latter containing the worm wheel axis. From the analysis of patent literature, it is apparent that nobody has studied the coupling and contact conditions between balls and corresponding raceways on the two members of the recirculating ball worm drive in sections other than the reference ones. The study of contact, which we performed in three-dimensional geometric conditions, has pointed out that interference or backlash are even more pronounced in sections parallel to the reference ones, the more the sections under consideration are far from the reference ones.

In fact, the cutting of the raceways on the two hourglass members of a recirculating ball worm drive, carried out with well-known traditional methods, results in a circumferential shift of the two centers of curvature of the two semi-raceways, leading to their displacement along the corresponding circumferential directrix. This can result in the impossibility of moving the balls, which would be blocked by ball jamming, or excessive backlash, resulting in loss of contact.

Figure 11.45a, b shows, respectively, what is happening in the reference transverse section of the worm wheel and in any transverse section of the same worm wheel, which is parallel to the ones assumed as reference. It is evident that, in the reference transverse section of the worm wheel, the centers of curvature of the two semi-raceways coincide, while they are shifted of a certain amount in any transverse section parallel to the reference one. The calculations show that the shift amount increases as the transverse section under consideration moves away from the reference one.

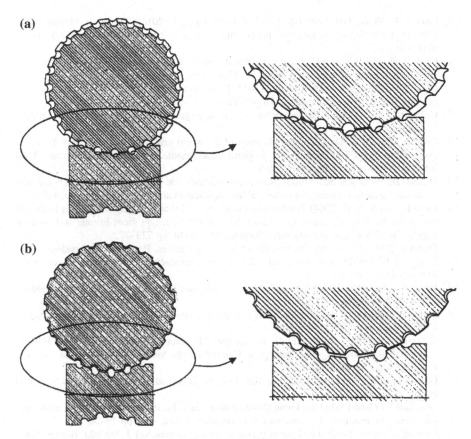

Fig. 11.45 a Reference transverse section of the worm wheel, where the centers of curvature of the two semi-raceways coincide; **b** any transverse section of the worm wheel parallel to the reference one, where these two centers of curvature are shifted

According to these studies and calculations, possible worm and worm wheel cutting technologies have been identified, which remedy the drawbacks of the traditional methods mentioned above. The peculiarities of these technological solutions are out of the content of this textbook. They are described in Salvini et al. [54, 55] in the two patents RM 2004A000138 (2004) and PCT/IB2005/050898 (2005), and in Minotti et al. [40], to which the author refers the reader.

References

1. Altmann FG (1932) Schraubengetriebe. VDI-Verlag, Berlin
2. Buckingham E (1949) Analytical mechanics of gears. McGraw-Hill Book Company Inc., New York
3. Burton P (1979) Kinematics and dynamics of planar machinery. Prentice-Hall, Upper Saddle River, NJ, USA

4. Chen Y-H, Zhang GH, Chen BK, Lou WJ, Li FJ, Chen Y (2013) A novel enveloping worm pair via employing the conjugating planar internal gear as counterpart. Mech Mach Theory 67:17–31
5. Chen Y-H (2013) Theoretical and experimental investigations on planar tooth internal gear enveloping crown worm drive. Chongqing University, Chongqing
6. Chen Y-H, Chen Y, Luo W, Zhang G (2015) Development and classification of worm drives. In: 14th IFToMM world congress, Taipei, Taiwan, 25–30 Oct 2015
7. Colbourne JR (1987) The geometry of involute gears. Springer, New York, Berlin. Heidelberg
8. Crosher WP (2002) Design and application of the worm gear. ASME Press, New York
9. Davis JR (ed) (2005) Gear materials, properties, and manufacture. Davis & Associates, ASM International, Materials Park OH, USA
10. Dennis JE, Schnabel RB (1966) Numerical methods for unconstrained optimization and nonlinear equations. Society of Industrial and Applied Mathematics, Philadelphia, USA
11. Duan L, Sun Y, Bi Q (2004) Tooth contact analysis of Ti worm gearing considering boundary condition. In: Yan XT, Jiang CY, Juster NP (eds) Perspectives from Europe and Asia on engineering design and manufacture. Springer, Dordrecht, pp 713–722
12. Dudas I (2000) The theory and practice of worm gear drives. Penton Press, London
13. Dudley DW (1962) Gear handbook. the design, manufacture, and application of gears. McGraw-Hill, New York
14. Ernst A (1902) Die Hebezeuge auf der Industrie- und Gewerbeausstellung in Düsseldorf. Z Ver deutsch Ing S. 1551 u.f
15. Ferrari C, Romiti A (1966) Meccanica Applicata alle Macchine. Unione Tipografica – Editrice Torinese (UTET), Torino
16. Galassini A (1962) Elementi di Tecnologia Meccanica, Macchine Utensili, Principi Funzionali e Costruttivi, loro Impiego e Controllo della loro Precisione, reprint 9th edn. Editore Ulrico Hoepli, Milano
17. Giovannozzi R (1965) Costruzione di Macchine, vol II, 4th edn. Casa Editrice Prof. Riccardo Pàtron, Bologna
18. Goldfard VI (2006) What we know about spiroid gears. In: Proceedings of the international conference on mechanical transmissions, Chongqing, China, sept, pp 19–26
19. Henriot G (1979) Traité théorique et pratique des engrenages, vol 1, 6th edn. Bordas, Paris
20. Heyer E, Niemann G (1953) Versuche an Zylinderschneckengetrieben, In: Braunschweig: Vieweg, VDI 95, pp 141–157
21. Hoyashita S (1996) Barrel worm-shaped tool with conjugate cutting-edge profile generated from tooth profile of internal gear. Jpn Soc Mech Eng 62(593):284–290
22. Ingrisch (1927) Untersuchung der Eingriffsverhältnisse bei der Evolventenschnecke, Dissertation an der deutschen technischen Hochschule in Prag
23. ISO 1122-2:1999(E/F) Vocabulary of gear terms - Part 2: Definitions related to worm gear geometry
24. ISO/TR 10828:1997(E), Worm gears—geometry of worm profiles
25. Jelaska DT (2012) Gears and gear drives. Wiley, New York
26. Kane TR, Levinson DA (2005) Dynamics, theory and applications. McGraw-Hill, New York
27. Kreyszig E (1959) Differential geometry. University of Toronto, Toronto
28. Kulkarni S, Kajale P, Patil DU (2015) Recirculating ball screw. Int J Eng Res Sci Technol 4 (2):252–257
29. Levi-Civita T, Amaldi U (1929) Lezioni di Meccanica Razionale – vol 1: Cinematica – Principi e Statica, 2nd edn., Reprint 1974. Zanichelli, Bologna
30. Litvin FL (1968) Theory of gearing, 2nd edn. Nauka, Moscow (in Russian)
31. Litvin FL (1989) Theory of gearing. NASA RP-1212, AVS COM 99-C-C035, Washington, DC
32. Litvin FL, Demenego A, Vecchiato D (2001) Formation of branches of envelope to parametric families of surfaces and curves. Comput Methods Appl Mech Eng 190:4587–4608

33. Litvin FL, Fuentes A (2004) Gear geometry and applied theory, 2nd edn. Cambridge University Press, Cambridge
34. Louis A (1967) Differential geometry. Harper and Row, New York
35. Loveless WG (1984), Cone drive double enveloping worm gearing design & manufacturing. Gear Technol 12–16 and 45
36. Magyar B, Sauer B (2015) Calculation of the efficiency of worm gear drives. Power Trans. Eng.
37. Maitra GM (1994) Handbook of gear design, 2nd edn. Tata McGraw-Hill Publishing Company Ltd., New Delhi
38. Maschmeier G (1930) Untersuchung an Zylinder- und Globoidschneckentrieben. Dissertation an der Technischen Hochschule su Berlin
39. Merritt HE (1954) Gears, 3th edn. Sir Isaac Pitman & Sons Ltd., London
40. Minotti M, Salvini P, Vivio F, Vullo V (2007) Metodo di taglio di viti e ruote in un riduttore a viti e ruota con ricircolazione di sfere, In: Atti XXXVI Convegno Nazionale AIAS, Ischia, 04–07 Sept 2007
41. Moré JJ, Garbow BS, Hillstrom KE (1980) User guide for MINIPACK-1. Argonne national laboratory report ANL-80-74, Argonne, Ill
42. Moré JJ, Sorensen DC, Hillstrom KE, Garbow BS, The MINIPACK (1984) Project. In: Cowel WJ (ed) Sources and development of mathematical software. Prentice-Hall, NJ, USA, pp 88–111
43. Niemann G, Weber C (1942) Schneckengetriebe mit flüssiger Reibung. VDI-Forshungsh 412, Berlin
44. Niemann G, Winter H (1983) Maschinen-Element Band III:Schraubrad-, Kegelrad-, Schneckn-, Ketten-, Riemen-, Reibradgetriebe, Kupplungen, Bremsen, Freiläufe. Springer, Berlin, Heidelberg
45. Novozhilov VV (1970) In: Radok JRM (ed) Thin shell theory. Wolters-Noordhoff Publishing, Groningen (NE)
46. Octrue M, Denis M (1982) Note technique CETIM no. 22, Géométrie des roues et vis tangentes. Ed. CETIM!, Senlis, France
47. Pohl WM (1933) Graphical determination of worm gear contact. Am Mach Eur Ed. 77:130E
48. Pollone G (1970) Il Veicolo. Libreria Editrice Universitaria Levrotto & Bella, Torino
49. Predki W (1985) Berechnung von Schneckenflankengeometrien verschiedener. Schneckentypen Antriebstechnik 24(2):54–85
50. Radzevich SP (2010) Gear cutting tools: fundamentals of design and computation. CRC Press, Taylor & Francis Group, Boca Raton, FL
51. Radzevich SP (2013) Theory of gearing: kinematics, geometry, and synthesis. CRC Press, Taylor & Francis Group, Boca Raton, FL
52. Radzevich SP (2016) Dudley's hadbook of practical gear design and manufacture, 3rd edn. CRC Press, Taylor & Francis Group, Boca Raton, FL
53. Sakai T, Maki P, Uesugi S, Horiuchi A (1978) A study on hourglass worm gearing with developable tooth surface. J Mech Des 100(3):451–459
54. Salvini P, Serpella D, Vivio F, Vullo V (2004) Metodo di taglio di vite e ruota in un riduttore a vite e ruota con ricircolazione di sfere e relativi utensili di taglio. Brevetto no. RM2004A000138, Italy, Università degli Studi di Roma "Tor Vergata"
55. Salvini P, Serpella D, Vivio F, Vullo V (2005) Method for cutting worm and worm wheel in a worm-gear reduction unit with circulation of bearing balls and related cutting tools, International Patent no. PCT/IB 2005/050898, Università degli Studi di Roma "Tor Vergata"
56. Schiebel A (1913) Zahnräder, II Teil, Räder mit schrägen Zähnen. Springer, Berlin
57. Schiebel A, Lindner W (1954), Zahnräder, Band 1: Stirn-und Kegelräder mit geraden Zähnen. Springer, Berlin, Heidelberg
58. Schiebel A, Lindner W (1957) Zahnräder, Band 2: Stirn-und Kegelräder mit schrägen Zähnen Schraubgetriebe. Springer, Berlin, Heidelberg

59. Scotto Lavina G (1990) Riassunto delle Lezioni di Meccanica Applicata alle Macchine: Cinematica Applicata, Dinamica Applicata - Resistenze Passive - Coppie Inferiori, Coppie Superiori (Ingranaggi – Flessibili – Freni). Edizioni Scientifiche SIDEREA, Roma

60. Smirnov V (1970) Cours de Mathématiques Supérieures, tome II. Éditions MIR, Moscou

61. Stahl K, Mautner E-M, Sigmund W, Stemplinger J-P (2016) Investigations on the efficiency of worm gear drives. Gear Solutions

62. Stoker JJ (1969) Differential geometry. Wiley, New York

63. Stribeck R (1898) Versuche mit Schneckengetrieben. Z Ver deutsch Ing S. 1156

64. Sun YH, Lu HW, Yan WY, Li GY (2011) A study on manufacture and experiment of hardened Ti worm gearing. Chin J Mech Eng 47(9):182–186

65. Townsend DP (1991) Dudley's gear handbook. McGraw-Hill, New York

66. Ventsel E, Krauthammer T (2001) Thin plates and shells: theory, analysis, and applications. Marcel Dekker Inc., New York

67. Visual Numerics Inc. (1998) ISML Fortran 90 MP library user's guide, vol 4.0. Houston, Texas, USA

68. Wolff W (1923) Über die Erzielung günstiger Eingriffsverhältnisse an Schneckengetrieben. Dissertation Technischen Hochschule, Aachen

69. Yang Z, Shang J, Luo Z, Wang X, Yu N, (2013) Nonlinear dynamics modeling and analysis of torsional spring-loaded antibacklash gear with time-varying meshing stiffness and friction. Adv Mech Eng 1–17

70. Zhang GH, Tan JP (1988) Theory investigation of sphere re-enveloping hourglass worm drive. J Chongqing Univ 10(1):42–49

71. Zhang GH, Zhang TP, Lou WJ (2007) Selecting and optimizing of the parameters on quasi-plane double-enveloping hourglass worm drive. Mech Trans 31(2):5–10

72. Zhao YP, Zhang Z (2010) Computer aided analysis on the meshing behavior of a height-modified dual-torus double-enveloping toroidal worm drive. Comput Aided Des 42:1232–1240

73. Zimmer M, Otto M, Stahl K (2016) Homogeneous geometry calculation of arbitrary tooth shape: mathematical approach and practical applications, Power Trans Eng 36–45

Chapter 12
Spiral Bevel Gears and Hypoid Gears

Abstract In this chapter, the geometry of the main types of spiral bevel gears is first defined and considerations about the spiral angle are made. The corresponding cutting processes are then briefly described, and the generation of active flank surfaces of the teeth is defined. The main geometrical quantities of these gears are then determined as well as those concerning the equivalent cylindrical gears obtained using the Tredgold approximation. The load analysis of spiral bevel gears is then performed, also defining the thrust characteristic on shafts and bearings. Subsequently, the concepts concerning these gears are extended to the most general gearing case represented by the hypoid gears. The fundamentals of these gears are then provided, based on the kinematics already described for the hyperboloid gears. Two approaches to theoretical analysis are described, the first of more limited validity, and the second more general, but both capable to provide reliable results in terms of geometric and kinematic characteristics of these types of gears. The load analysis is extended to these types of gears, and some indications on the design choices inherent to them to improve their efficiency are provided. Finally, the unified ISO procedure, which allow us to calculate the geometric quantities of spiral bevel and hypoid gears is summarized.

12.1 Introduction

In Chap. 9 we studied the straight bevel gears, which are a special case of power transmission through intersecting axes. In this chapter, we focus our attention on the *curved-toothed bevel gears*. These gears find broad application in the automotive and truck industry, helicopter industry and, more generally, in the industry of gear power transmissions through intersecting and crossed axes (see Buckingham [5], Merritt [37], Henriot [18], Niemann and Winter [42], Townsend [66], Stadtfeld [61], Maitra [36], Litvin and Fuentes [32], Radzevich [48], Klingelnberg [21]).

A curved-toothed bevel gear can be obtained by an intuitive procedure similar to that described in Sect. 8.1, which allows to obtain a cylindrical helical gear wheel, starting from a cylindrical spur gear wheel, divided into more and more segments,

© Springer Nature Switzerland AG 2020   569
V. Vullo, *Gears*, Springer Series in Solid and Structural Mechanics 10,
https://doi.org/10.1007/978-3-030-36502-8_12

when an infinite number of segments with infinitesimal small rotations along the face width is considered. By applying this same procedure to a conventional straight bevel gear wheel, we can consider a finite or infinite number of component gear wheels having correspondingly finite or infinitesimal thickness, all obtained by sectioning the initial straight bevel gear wheel with planes perpendicular to its axis. Then we can rotate a given angle in the tangential direction one of the gear wheels so obtained with respect to the adjacent one. In this way, if the number of component gear wheels is finite, and the relative rotation angle $\Delta \varphi$ of one with respect to the next is discrete and constant, we obtain *stepped bevel gears* (Fig. 12.1a). Instead, if the number of component gear wheels is infinite, and the relative rotation of one with respect to the next is infinitesimal and proportional to the coordinate z along the axis, we obtain *curved-toothed bevel gears* (Fig. 12.1b). This last figure shows, from below upwards, the next steps that allow to obtain a curved-toothed bevel gear starting from a simple bevel gear wheel (see Vol.3, Section 5.2).

In doing so, it is evident that, unlike what happens for straight bevel gears that have conical conjugate surfaces, the mating surfaces of the curved-toothed bevel gears are no longer conical surface. These curved-toothed bevel gears can be divided into two families:

- the gears with intersecting axes, called *spiral bevel gears*;
- the gears with crossed axes, called *hypoid gears*.

Like the straight bevel gears, even the spiral bevel gears are characterized by the fact that the axes of the pitch cones of the two members of the gear pair and the axis of their common generation crown wheel converge at the same pitch cone apex, O, which coincides with the crossing point, $O_c \equiv O$ (Fig. 12.2a). The common generation crown wheel meshes properly with each of these two members, considered separately. Instead, the two members of a hypoid gear pair have crossed axes, which intersect the pitch plane of the generation crown wheel in two different points, O_1 and O_2, which in this plane are offset by a certain *hypoid offset*, a, whose amount is generally not large (Fig. 12.2b). In this case, the crossing point, O_c, is the apparent point of intersection of the axes of the two members of the hypoid gear pair, when they are projected on a plane parallel to both (Fig. 12.2b). This figure shows that any point on the shortest distance, including its end points, can be taken as crossing point, O_c (see Niemann and Winter [42], ISO 23509:2016 (E) [20]).

It is to be noted that, strictly speaking, the hypoid gears (and so a few other gears that seam spiral bevel gears) do not fall within the family of the spiral bevel gears. Nevertheless, we considered, like other authors, to include in this chapter both the families of the spiral bevel gears and hypoid gears. This is because they have the common feature of not having straight teeth, but curved teeth, and are produced using entirely similar cutting processes based on the same technologies (see Micheletti [38], Galassini [13], Rossi [51], Henriot [17]; Litvin [29], Lin et al. [28], Suh et al. [65], Stadtfeld [63], Klingelnberg [21]). Substantial differences between these two gear families are present (see Wildhaber [72], Poritsky and Dudley [45],

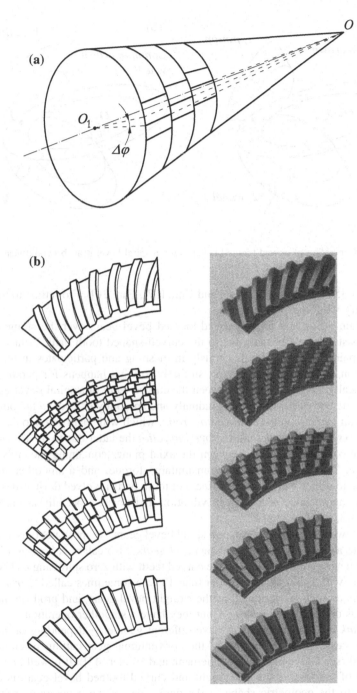

Fig. 12.1 a stepped bevel gear; **b** intuitive procedure to obtain a curved-toothed bevel gear wheel starting from a straight bevel gear wheel

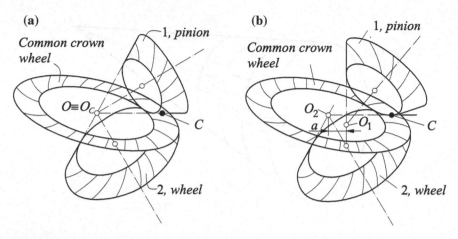

Fig. 12.2 Generation of curved-toothed bevel gears: **a** spiral bevel gear; **b** hypoid gear

Spear et al. [59], Pollone [44], Tsai and Chin [67], Fan [10]). From time to time, we will specify what these differences.

The main advantage of the curved-toothed bevel gears over the straight bevel gears consists in the fact that, due to the curved-shaped tooth flank surfaces, more than one pair of teeth is simultaneously in meshing and participates in the power transmission at any time. Moreover, similarly to what happens for parallel cylindrical helical gears, the contact between the curved teeth of a spiral bevel gear pair begins at one end of the tooth and gradually and smoothly extends to the other end. It follows an increase in the *total contact ratio*, which will result equal to the sum of the *transverse contact ratio* and the *overlap ratio*, the latter due to the *overlap arc*, i.e. the arc of the pitch circle between the axial planes containing the ends of one tooth trace. The result is a large transmittable torque, and a smoother meshing action, due to minor variations in stiffness of tooth parts involved during meshing. It follows an appreciable reduction in vibration and noise, especially at high speed [7, 8, 26, 43, 55, 57, 58, 74].

It is noteworthy that, with the term spiral bevel gears, we mean not only the spiral bevel gears as such, but also the *skew bevel gears*, also called *helical bevel gears*, and *Zerol bevel gears*. The latter have curved teeth with zero spiral angle. The spiral bevel gears with spiral angles smaller than 10° are sometimes called "*zerol*". Zerol bevel gears as such are mounted as the straight bevel gears, and produce the same thrust loads on shafts and bearings, but they have a smoother operation.

The *flank lines* or *tooth traces* have different geometry, depending on the generation processes that are typical of the conventional gear cutting machines currently used (see Giovannozzi [14], Niemann and Winter [42]). It is well known that the shape of the tooth flank of straight and curved-toothed bevel gears is defined according to the geometric shape of the flank lines of the generation crown gear wheel, which simulates the cutting motion of the cutter. Indeed, during rolling without sliding of the pitch plane of this crown gear on the pitch cone of the bevel gear to be generated, such a flank line, which is a planar curve, is wound on the pitch cone, univocally determining the shape of the tooth flank of the bevel gear that we want to get.

In practical applications, the flank lines actually used are those that are better suited to the generation process of the toothing, using conventional gear cutting machines, i.e. without Computerized Numerical Control, CNC [28]. Among them, those shown in Fig. 12.3 deserve specific mention. They are as follows:

- Straight lines on the pitch plane of the crown wheel, passing through the apex (Fig. 12.3a), which generate straight bevel gears as those studied in Chap. 9. These straight bevel gears may be used, for their intrinsic limitations (the engagement along tooth face width is not gradual, and therefore they are very noisy), for tangential speeds not very high, i.e. less than 10 m/s; if they are machined by grinding, tangential speeds can be a little more than 10 m/s.

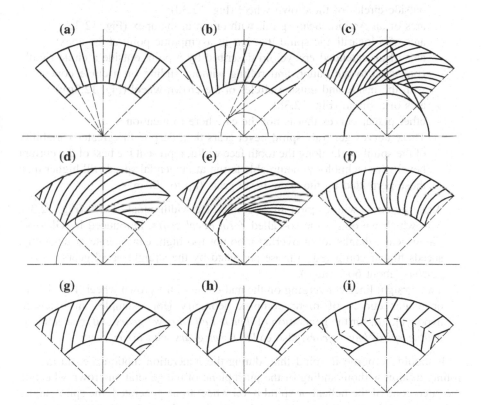

Fig. 12.3 Flank lines on the generation crown wheel: **a** straight lines passing through the apex; **b** straight lines not passing through the apex, but tangent to a concentric circle of given radius; **c** arcs of circle with centers on a concentric circle; **d** involute curves of a given concentric circle; **e** arcs of an Archimedean spiral; **f** arcs of circle with mean spiral angles equal to zero; **g** arcs of epicycloid; **h** arcs of sinoid; **i** two straight lines converging on the middle circle (double helical bevel gears)

- Straight lines on the pitch plane of the crown wheel, not passing through the apex (Fig. 12.3b), and tangent to a circle of given radius, concentric with the inside and outside circles of the same crown wheel, which generate *skew bevel gears*, also known as *helical bevel gears* or *oblique spiral bevel gears*. These helical bevel gears, whose tooth traces are non-cylindrical helices, may be used for tangential speeds much higher (up to 50 m/s, if machined by grinding). This is due to the gradualness of the engagement along the tooth face width, smoother meshing action between the mating tooth pairs as well as a greater total arc of transmission.
- Arcs of various geometrical curves on the pitch plane of the crown wheel, such as:

  - arcs of circle of a given radius, with centers on a circle concentric with the inside and outside circles of the crown wheel (Fig. 12.3c);
  - arcs of an involute curve of a given circle, concentric with the inside and outside circles of the crown wheel (Fig. 12.3d);
  - arcs of an Archimedean spiral, with origin at the apex (Fig. 12.3e);
  - arcs of a logarithmic spiral that has its asymptotic point at the apex;
  - arcs of an extended epicycloid described by a point outside a circle rigidly connected to the cutter, and rolling without sliding on a circle concentric with the inside and outside circles of the crown wheel (Fig. 12.3g);
  - arcs of a sinoid (Fig. 12.3h);
  - other planar curves that is not the case here to mention.

    These curves generate spiral bevel gears that, despite the greater variability of the spiral angle along the tooth face width, represent the best of the current bevel gear technology, being able to run at tangential speed still higher than those permitted by the skew bevel gears (up to 100 m/s).

- Arcs of a circle with a spiral angle at mid-face width equal to zero (Fig. 12.3f), with which we obtain the so-called *Zerol bevel gears*, introduced by Gleason. These gears, thanks to an overlap ratio not too high, can operate at tangential speeds higher compared to those permitted by the straight bevel gears (*ceteris paribus*, about 60% more).
- Two straight lines converging on the mid-circle of the crown wheel (Fig. 12.3i), or planar curves of more complex geometry (for example, an extended hypocycloid), for the generation of *double helical bevel gears* and, respectively, *double spiral bevel gears*, also called *bevel gears with arrow teeth*.

It should be borne in mind that, during the generation motion, i.e. during the rolling motion without sliding of the pitch plane of the generation crown wheel on the pitch cone of the helical or spiral bevel wheel to be generated, the spiral angles are preserved. In fact, these spiral angles remain unchanged, and the various generation curves characterizing the flank lines, described above, are wrapped according to other curves, which can be determined case by case, according to the basic laws of the mating profiles [14]. On this subject, we will return in the next section.

Similarly to what happens for parallel cylindrical helical gears in comparison with the parallel cylindrical spur gears, also the spiral bevel gears have considerable advantages compared to straight bevel gears. The main advantages are summarized as follows:

- Greater contact ratio, because the total contact ratio is the sum of the transverse contact ratio and overlap ratio.
- Gradual and progressive meshing action over the whole face width of the toothing.
- Longer tooth-engagement time due to the simultaneous meshing of several tooth pairs.
- Noise level considerably smaller.
- Lowest minimum number of teeth to avoid the cutting interference, and thus undercut, and therefore greater gear ratio for the same space requirements.
- Greater load carrying capacity of the teeth both in terms of surface durability, and in terms of tooth root strength.
- Comparatively higher transmission ratio.
- Sensitivity to load fluctuations not excessive, for which these gears can remedy, at least partially, to misalignment errors in mounting and bearing systems.

The tooth shape of the spiral bevel gears and hypoid gears depends on the manufacturing process and cutter profile used. In accordance with the cutting method and cutter employed, these gears may have a constant tooth depth along the face width, or the tooth depth may decrease progressively from the heel to the toe, i.e. from the back cone to the inner cone. More in detail, spiral bevel and hypoid gears have a parallel-depth along the face width if they are face-hobbed, i.e. manufactured by a continuous indexing process (all tooth spaces are cut progressively together, with a continuous cutting process). Instead, they have a tapered-depth profile along the face width when they are face-milled, i.e. manufactured using a single indexing process (the tooth spaces are cut one at time, with a discontinuous cutting process). For more details on this subject, we refer the reader to Sect. 12.11 (see also: Handschuh et al. [16]; ISO 23509:2016 (E) [20]; Wang and Fong [71]; Müller [40]).

12.2 Considerations on the Spiral Angle

The spiral angle at any point of the tooth flank of a spiral bevel gear is the angle between the cone generatrix and the tangent to the tooth trace at that point, measured on the tangent plane to the reference cone. Usually, the *mean spiral angle*, β_m, i.e. the *spiral angle at mid-face width*, is specified.

With reference to the generation method of these gears, however, it is convenient to consider what happens on the pitch plane of the generation crown wheel. On this pitch plane, whatever the geometry of the flank line, which is the generation curve, the spiral angle, β, is the angle between the tangent to this generation curve at any of its current points P, and the radius passing through the current point under consideration (Fig. 12.4). This angle is a function of the point under consideration, and therefore varies from point to point, passing from the *inner spiral angle*, β_i, at the *inner radius*, R_i, to the *outer spiral angle*, β_e, at the *outer radius*, R_e.

The spiral angle has a constant value only in the case in which the flank line on the pitch plane of the generation crown wheel is a *logarithmic spiral*. To demonstrate this, we assume, in the pitch plane of this crown wheel, a polar coordinate system $O(R, \varphi)$, such as that shown in Fig. 12.4. Considering a generic flank line, the condition $\beta = const$ leads to write the following relationship:

$$\tan\beta = \frac{Rd\varphi}{dR} = \frac{1}{k} = const, \tag{12.1}$$

where k is a constant. From this equation, separating the variables and integrating, we get:

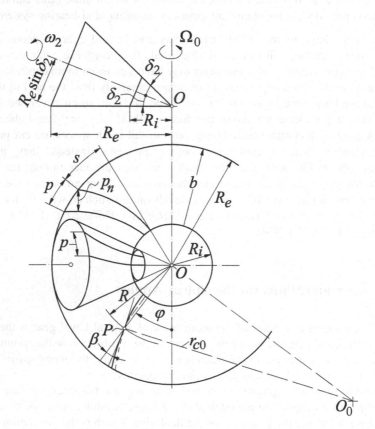

Fig. 12.4 Generation of a spiral bevel gear wheel by generation crown wheel

$$\ln R = k\varphi + const. \tag{12.2}$$

Then, passing from the logarithm function to the function, and measuring the angles starting from the straight radial line passing through the point having radial coordinate $R = R_i$, we obtain the following relationship:

$$R = R_i e^{k\varphi}, \tag{12.3}$$

which is the equation of the logarithmic spiral. This spiral has as its asymptotic point the pole O.

During the rolling without sliding of the pitch plane of the generation crown wheel on the pitch cone of the spiral bevel gear wheel to be generated, the logarithmic spiral is wound according to a *conical helix*, which cuts under a constant angle the generatrixes of the pitch cone. From the point of view of the cutting method, the condition $\beta = const$ cannot be realized with accuracy by mechanical means, i.e. simulating the motion of the cutting tool with a mechanism. Of course, this limitation does not exist for the current Computerized Numerical Control gear cutting machines (CNC-gear cutting machines). For further details on the geometry of planar (but also spherical) logarithmic spiral, on its kinematic properties as well as on some aspect concerning the technological cutting processes to obtain logarithmic spiral bevel gears, (see Li et al. [27], Figliolini et al. [12], Stachel et al. [60]).

The difficulty to realize mechanically, with precision, the logarithmic spiral on the generation crown wheel, led to its approximate simulation, by means of curves geometrically simple and easily achievable with the conventional cutting methods. The logarithmic spiral was therefore replaced with an arc of the osculating circle at the point P of the same spiral located on the pitch plane of the generation crown wheel, in correspondence of the mid-face width (Fig. 12.4). Gleason Works were the first to introduce and implement this fundamental concept, developing a face-milling process that uses a milling cutter, having cutter radius r_{c0} and center O_0 (see next section); theoretically, this center O_0 is obtained as the intersection of the normal to the logarithmic spiral at point P with the normal to the radius OP at point O (Fig. 12.4).

The logarithmic spiral can be realized, with a good approximation, by the *sinoid*. As it is well known, the sinoid is a spiral curve given by the following equation, written in a polar coordinate system $O(R, \varphi)$:

$$R = k + k'\cos(\lambda\varphi), \tag{12.4}$$

where k, k', and λ are positive constants. As shown in Fig. 12.5, the sinoid has a geometric profile fluctuating within the circular crown having inside and outside radii given by $(k \pm k')$. If we assume $\lambda = m/n$, with m and n integers and prime numbers between them, the spiral curve is closed for the first time after performing n oscillations over m angular revolutions. Figure 12.5 refers to the case $m = 3$, and $n = 10$.

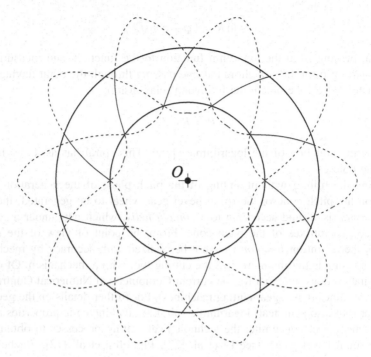

Fig. 12.5 Spiral curves obtained as sinoid arcs on the pitch plane of the crown wheel

Spiral bevel gears with teeth shaped like an arc of sinoid are those obtained with the *Brandenberger method* [14]. For each forward and back travel of the cutting tool in the radial direction, the wheel to be generated rotates to an integer number of revolutions, m, which is prime number with the number of teeth of the wheel, $z = n$. In this way, the sinoid closes only after it has machined all the teeth. So, we have a simultaneous machining of all the teeth, which avoids the inevitable errors of the universal indexing heads.

If we have the foresight to choose appropriately the values of k, k' and λ, we can get sinoids of a shape such that, in the part actually used, the angle β changes very little, $\pm(2 \div 3)°$, and the tooth size varies approximately in proportion to the distance R from the center O. The method leads to the generation of gearing characterized by very regular operating conditions. Technologically, the motion expressed by Eq. (12.4) is achievable by combining a uniform rotational speed of the wheel with a cutting tool speed in the direction of the generatrixes, variable according to a sinusoidal function. This is obtained, with a good approximation, by connecting the cutting tool to the connecting rod small end of a crank mechanism, with the connecting rod long enough with respect to the crank radius.

Before we tackle the topic regarding the various types of spiral bevel gears, which are produced today and find practical applications, we return to the generation methods by a generation crown wheel. With reference to Fig. 12.4, we denote by Ω_0 and ω_2 the angular velocities of the generation crown wheel and, respectively, spiral bevel gear wheel to be generated. Since $\delta_0 = 90°$, from Eq. (9.40) we infer the following relationship:

$$\Omega_0 = \omega_2 \sin\delta_2. \tag{12.5}$$

The tangential velocity at a current point P on the flank line of the generation crown wheel, which is at a distance R from the center O, is given by:

$$v_{OP} = R\Omega_0. \tag{12.6}$$

To ensure the desired spiral angle of the tooth of the generation crown wheel at any current point P of the flank line, it is necessary that the cutting tool and generation crown wheel have suitable motions. In this regard, the cutting tool must move radially, in the direction of the center O, with an instantaneous linear speed v_0 at point P, while the generation crown wheel must rotate with an instantaneous angular velocity Ω_0, and thus with a tangential velocity $R\Omega_0$ at the same point P. If β is the spiral angle at point P, the following relationship must be satisfied:

$$v_{OP} = \frac{R\Omega_0}{\tan\beta}. \tag{12.7}$$

Since v_0 is also the instantaneous linear speed of the cutting tool in the direction of the pitch cone apex of the spiral bevel gear wheel to be generated, to obtain the thread of the latter also the following relationship must be satisfied:

$$v_0 = \frac{R\omega_2 \sin\delta_2}{\tan\beta}. \tag{12.8}$$

This relationship, taking into account Eq. (12.5), coincides with Eq. (12.7).

Since the gear ratio $u = z_2/Z_0$ is equal to the ratio Ω_0/ω_2, and $\delta_0 = 90°$, from Eq. (9.15) we infer:

$$z_2 = Z_0 \sin\delta_2, \tag{12.9}$$

where Z_0 is the number of teeth of the fictitious generation crown wheel. The transverse pitch of the two wheels, measured on the reference circles, i.e. the outside circles having radii $R\sin\delta_2$ for the spiral bevel gear wheel to be generated, and R_e for the generation crown wheel (Fig. 12.4), is given by:

$$p = \frac{2\pi R_e \sin\delta_2}{z_2} = \frac{2\pi R_e}{Z_0}. \tag{12.10}$$

Due to the spiral angle, the *arc of transmission*, compared to what happens for straight bevel gears having the same profile, is increased by the *face advance, s*, i.e. the arc length corresponding to the angular displacement of the tooth profiles between the toe and heel. Therefore, the *total arc of transmission* is the sum of the *transverse arc of transmission* and *overlap arc*. The overlap angles are equal to (s/R_e) and $s/(R_e \sin\delta_2)$, respectively for the spiral bevel gear wheel to be generated, and for generation crown wheel. Obviously, the *total angle of transmission* is the sum of the *transverse angle of transmission* and *overlap angle*.

Generally, however, always with reference to the flank line geometry of the generation crown wheel, the spiral angle changes from point to point. We define as *mean spiral angle*, β_m, or simply *spiral angle* in common usage, the spiral angle referred to the point of intersection between the flank line and the middle circle having radius R_m. Since the spiral angle varies as a function of the distance R from the center O, and R varies in the range $R_i \leq R \leq R_e$ (Fig. 12.4), β may also vary in the range $\beta_i \leq \beta \leq \beta_e$, where β_i and β_e are respectively the *inner spiral angle* and *outer spiral angle*, i.e. the spiral angles at the inside and outside circles of the generation crown wheel.

Since during the generation motion these spiral angles are preserved, they also characterize the spiral bevel gear wheel that is generated. However, using as generation curves on the pitch plane of the generation crown wheel the usual flank lines, which are characterized by a spiral angle that is variable as a function of the radius, we obtain spiral bevel gear wheels whose flank lines are conical helices with variable helix angle. As it is well known, the usual flank lines are the arc of a circle, the arc of an involute curve, and the arc of an epicycloid, respectively shown in Fig. 12.3c, d, g.

Still more generally, in relation to the geometry of the flank lines on the pitch plane of the generation crown wheel, we can get spiral bevel gear wheels whose flank lines are: conical helices with constant helix angle and variable lead; conical helices with variable helix angle and constant lead; conical helices with helix angle and lead both variable.

In a curved-toothed bevel gear (spiral bevel gear and hypoid gear), the hands of the spirals of the two members are always opposite, that is one right-hand and the other left-hand, or vice versa. In the common language of gear technology, usually we indicate a spiral bevel gear pair, or hypoid gear pair, with reference to the spiral hand of the pinion, which is normally the driving member. Therefore, the gear pair is a right-hand pair or a left-hand pair depending on whether the pinion is a right-hand pinion or a left-hand pinion.

Right-hand or left-hand teeth are respectively the teeth whose successive profiles show clockwise or counterclockwise displacement with increasing distance from an observer looking along the straight line generatrixes of the reference cone. The spirals are thus right-hand spirals or left-hand spirals for an observer looking along the generatrixes of the reference cone, regardless of the fact that he is being positioned by the side of the cone apex, or from the opposite side of the apex.

Another way of defining the hand of the spiral is that of an observer placed on the wheel axis, which looks like a tracing point moves along the tooth spiral. The spiral is a right-hand spiral or a left-hand spiral, if the observer sees the tracing point move in the clockwise direction or in the counterclockwise direction, moving away from himself. When an algebraic sign is associated with the spiral angle, this second convention may be usefully applied to spiral gears in which contact occurs at a point of the common normal. In fact, in these cases, the algebraic sum of the spiral angles is equal to the shaft angle.

According to ISO 23509:2016 (E) [20], a right-hand (or left-hand) spiral bevel wheel is one in which the outer half of a tooth is inclined in the clockwise (or counterclockwise) direction from the axial plane through the mid-point of the tooth, as viewed by an observer looking at the face of the gear.

It should be noted that the rule that the spiral angles of a spiral bevel gear, at any point where the mating tooth spirals are touching each other, must be equal and opposite, must be considered carefully when meshing with the generation crown wheel is analyzed. In fact, the hand of the generation crown wheels of the two members of a spiral bevel gear must be equal and opposite, respectively, compared to the hands of the pinion and wheel. Therefore, if a pair of spiral bevel gear is described as right-hand pair or left-hand pair, it must be understood, as we said above, that the description applies to the pinion.

12.3 Geometry and Cutting Process of the Main Types of Spiral Bevel and Hypoid Gears

The geometry of spiral bevel and hypoid gears and, more generally, the geometry of the curved-toothed bevel gears, depends on the cutting process used. It is however to be noted that the cutting process of these gears constitutes a rather complex kinematic and technological problem. In this regard, many studies have been carried out, and various solutions have been devised, none of which, however, is to be considered strictly exact, understood as the potential of ensuring tooth profiles having the shape of theoretical spherical involutes [19]. The reasons are to be found in the fact that, for the cutting of spiral bevel and hypoid gears, other approximations are required, in addition to those already described regarding the cutting process of straight bevel gears.

Obviously, however, as it happens for straight bevel gears, two wheels cut using the same generation cutting process and, therefore, the same gear cutting machine and the same cutting tool, are able to mesh between them correctly. This provided that the generation pitch cones constitute also the working pitch cones (see Schiebel and Lindner [52, 53], Giovannozzi [14], Niemann and Winter [42]).

However, the detailed study of the various solutions to the problem of the generation of spiral bevel and hypoid gears, and various geometries of teeth that can be achieved comes from the limits of this textbook (see Handschuh and Litvin [15], Litvin and Fuentes [32]). Here we limit ourselves to do a summary of the main types of curved-toothed bevel gears used in most practical applications, which are those described in the following sections.

12.3.1 Gleason Spiral Bevel Gears

Gleason spiral bevel gears have as generation curves on the pitch plane of the generation crown wheel arcs of circle. These gears are made with spiral bevel gear cutting machines developed by Gleason, which use circular milling cutters with inserted blades. These cutters could be used for both right-hand and left-hand spiral teeth. The flank lines on the pitch plane of the generation crown wheel are shaped like arcs of circle of a given radius, r_{c0}, with centers on a circle concentric with the inside and outside circles of the crown wheel. These two circles constitute the contours of the fictitious circular crown wheel which defines the face width of the to-be-generated spiral bevel gear wheel (see Henriot [17, 18], Niemann and Winter [42], Stadtfeld [63]).

The original Gleason process is referred as *face-milling process*, as the manufacturing is done using the *single-indexing process*, where each tooth space is generated separately. The process of generation is interrupted when generation of a given tooth space is ended. Then the workspace is indexed to the next tooth space, and the process of generation is repeated, and so on until the cutting process of the to-be-generated gear wheel is completed. The curved-toothed bevel gears obtained with this process have a tapered-depth profile along the face width.

As Fig. 12.6 shows, this face-milling process uses a milling cutter having cutter radius, r_{c0}, and axis parallel to that of the generation crown wheel. The head-cutter is mounted on a cradle held at rest, and its axis is placed an *offset*, ρ_{P0} (this offset is also called *crown gear to cutter center*), with respect to the apex, O, where the axes of the generation crown wheel and spiral bevel wheel to be generated converge. Therefore, the center of curvature at the mean point and cutter center as well as the radius of curvature of the flank line at the mean point and the cutter radius coincide.

The generation surface of the toothing is a conical surface, whose generatrixes have as directrix curve the arc of circle AB, and are inclined with respect to the cutter axis passing through point O_0 of an angle equal to the pressure angle, α. Otherwise placing the head-cutter with respect to the axis of the generation crown wheel, it is possible to get the desired value and direction of the spiral angle, β_m. The theoretical location of the center O_0, described in the previous section, is therefore not respected. It follows that the active flank of the tooth is a conical surface, with apex V on the normal to the pitch plane of the generation crown wheel, passing through the point O_0. The surface of the other tooth flank, i.e. the flank that becomes active when the direction of rotation is reversed, is also a conical surface, whose generatrixes converge at apex, V^*, which is the symmetrical point of V with respect to the above-mentioned pitch plane.

This cutting process is also referred as *formate-cut process* (*spread blade process*, *fixed setting process*, and *five-cut process* are synonymous). The head-cutter, which is mounted on the cradle held at rest, is rotated about its axis with the suitable cutting velocity, and generates the tooth surfaces between one tooth space as the copy of the surfaces of its blades. During this cutting process, the to-be-generated gear wheel does not perform any rotation about its own axis and with respect to the cradle.

Fig. 12.6 Operating mode of
the Gleason spiral bevel gear
cutting machine

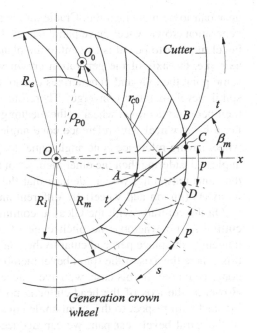

*Generation crown
wheel*

   Gleason has also developed a cutting process of spiral bevel gears referred as
face-hobbing process. With this process, the to-be-generated gear wheels are
manufactured in a continuous indexing process, and have a parallel-depth profile
along the face width. In this case, the head-cutter installed on the cradle performs a
planetary motion, consisting of a rotation in transfer motion with the cradle about
the cradle axis, and a rotation in relative motion with respect to the cradle about the
cutter axis. The to-be-generated spiral bevel wheel is placed with spiral angle, β_m,
with respect to the pitch plane of the generation crown wheel, which is simulated
and materialized by the head-cutter, and rotates about its axis, dragged in rolling
without sliding by the generation crown wheel. Rotations of the cradle, head-cutter
and spiral bevel wheel to be generated are therefore related in a timed
relationship. The spiral angle, β_m, of the spiral bevel wheel to be generated represents the setting angle of the cutting machine.
   The generation curves on the pitch plane of the generation crown wheel are arcs
of extended epicycloid. The milling head-cutter is provided with generation surfaces formed by a number of trapezoidal blades with dual cutting edges, profiling
respectively the concave and convex flanks of the teeth, or by a number of alternate
blades with single cutting edge, profiling alternatively the aforementioned tooth
flanks. Each tooth space is thus generated simultaneously, with a continuous
motion. Therefore, at any instant, all the teeth are in the same generation stage.
   It is noteworthy that, with the Gleason single-indexing process, which is the one
usually used, since it allows a fast and easy set-up of the cutting machine, a
common generation crown wheel for pinion and wheel to be cut cannot exist. To
convince ourselves of this, we consider any of the two members of the spiral bevel

gear pair to be manufactured. Cradle and head-cutter that, together, materialize the generation crown wheel, have parallel axes. Furthermore, the root cone of the spiral bevel member to be cut is tangent to the plane passing through a point of the cradle axis (i.e. the axis of the generation crown wheel) and perpendicular to it. In this same point the axis and generatrixes of root and pitch cones of the to-be-generated spiral bevel member converge. Therefore, the generation crown wheel of this member is not a crown wheel in the meaning that we have given so far to this term, i.e. a bevel wheel with reference cone angle of 90°. Instead, it is actually a bevel wheel having reference cone angle equal to $(90° − \vartheta_f)$, where ϑ_f is the dedendum angle. Considering then the other member of the spiral bevel gear pair and repeating the same reasoning, we can deduce that the generation crown wheels for the two members (pinion and wheel) are different and have different axes.

On the contrary, with the Gleason continuous-indexing process, it is possible a cutting set-up whereby the pitch cone of the spiral bevel member to be cut is tangent to the plane perpendicular to the cradle axis, and passing through a point of this, where the axis of the spiral bevel member and generatrixes of its pitch and root cones converge. In this case, we have a generation crown wheel in the strict sense. However, the axis of the head-cutter is no longer parallel to the cradle axis, but inclined with respect to this at an angle equal to ϑ_f. Considering the two members of the spiral bevel gear pair, we can deduce that the generation crown wheels for pinion and wheel to be generated are equal and have the same axis.

With reference to the Gleason single-indexing process, it is to be noted that the most used set-up of the cutting machine has the drawback of not ensuring the correct meshing conditions along the whole face width of the spiral bevel gear pair, but only at the mean point. This follows from the fact that the two fictitious and improper generation crown wheels related to the two members of the gear pair to be cut are different from each other, having respectively reference cone angles equal to $(90° − \vartheta_{f1})$ and $(90° − \vartheta_{f2})$. Only at the mean point, where these two fictitious reference conical surfaces intersect, and therefore the related pitch circles are tangent, the meshing conditions are correct. In all other points, the generated teeth are not strictly conjugate. Furthermore, we are faced with a relative error of pressure angle, which changes sign, passing from one side to the other with respect to the midpoint, and that leads to the so-called *bias-in contact* and *bias-out contact*. It is possible to remedy, at least partially, this drawback, by means of appropriate profile modifications (*profile crowning*, *lengthwise crowning*, and *flank twist*). However, the deepening of this topic is beyond the scope of this textbook, and we refer the reader to more specialized literature.

Depending on the set-up of the cutting machine, with the Gleason single-indexing process we can get: *Zerol bevel gears*, having spiral angle equal to zero $(\beta_m = 0°)$; *Gleason spiral bevel gears* themselves, with spiral angles up to 45° (usually, $\beta_m = 35°$). The toothing geometry is characterized by transverse tooth thickness, tooth depth, and tooth space width tapered from the heel to toe. However, there are Gleason spiral bevel gears with constant tooth depth.

12.3.2 Modul-Kurvex Spiral Bevel Gears

Modul-Kurvex spiral bevel gears have as generation curves on the pitch plane of the generation crown wheel arcs of circle. Special milling head-cutters with inserted blades in two halves generate these gears. With these special milling head-cutters it is possible to generate both the two members (pinion and wheel) of the desired gear. Depending on the set-up of the cutting machine, we can obtain spiral angles ranging from $\beta_m = 25°$ to $\beta_m = 45°$. Modul-Kurvex spiral bevel gears are similar to Gleason spiral bevel gears, except that the tooth depth generally remains a constant throughout. With a proper set-up of the cutting machine, it is also possible to obtain tooth depth tapered from the heel to toe.

12.3.3 Oerlikon-Spiromatic Spiral Bevel Gears

Oerlikon-Spiromatic spiral bevel gears have as generation curves on the pitch plane of the generation crown wheel arcs of epicycloid or hypocycloid. These gears, also called *eloid spiral bevel gears*, are generated by milling head-cutters with inserted blades, which are different for right-hand spiral and left-hand spiral. The Oerlikon process is referred as a *face-hobbing process*, since the manufacturing is done using a continuous indexing process. This is obtained with the cradle axis, working axis and head-cutter axis that roll together in a timed relationship, as it is shown below. Gears obtained with this cutting process have a parallel-depth profile along the face width (see Henriot [17, 18], Niemann and Winter [42]).

As Fig. 12.7 shows, z_0-groups of equal blades, arranged alternatively inwardly and outwardly of a circle having radius r_{c0}, form the milling head-cutter with inserted blades (five groups of blades in the case shown in Fig. 12.7). The fictitious generation crown wheel, whose mean radius is equal to R_m, rotates uniformly about its axis. It, however, after the head-cutter has carried out $(1/z_0)$ turns, rotates further of an angular pitch $\tau_0 = (2\pi/Z_0)$, where Z_0 is its virtual number of teeth. Of course (see Sect. 9.8), we have $Z_0 = (2R_m/m_0)$, where m_0 is the module.

The ratio between the angular velocity of the generation crown wheel, Ω_0, and angular velocity of the head-cutter, ω_0, is given by:

$$\frac{\Omega_0}{\omega_0} = \frac{z_0}{Z_0}. \tag{12.11}$$

This type of motion can be reduced to the pure rolling (i.e. rolling without sliding) of the centrode circle having radius ρ_c and center H, rigidly connected to the head-cutter, on the centrode circle having radius ρ_b and center O, rigidly connected to the generation crown wheel. This last centrode circle is then the *epicycloid base circle*, and ρ_b is its radius. Therefore, we infer that the generation curves on the pitch plane of the crown wheel are arcs of extended epicycloid, as

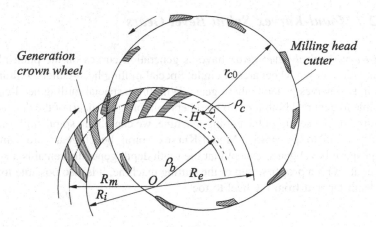

Fig. 12.7 Operating mode of the Oerlikon spiral bevel gear cutting machine

they are described by the cutting edges of the cutter blades, which are arranged at the outside of the rolling centrode circle having radius ρ_c. In fact, we have:

$$\frac{\rho_b}{\rho_c} = \frac{\omega_0}{\Omega_0} = \frac{Z_0}{z_0}. \tag{12.12}$$

For the direction of rotation shown in Fig. 12.7, the flank line on the pitch plane of the generation crown wheel is an *extended epicycloid* or *conical cycloid*, also called *eloid* (see Lawrence [24]). A continuous division process, resulting from the combination of three motions, realizes it. These motions are as follows: rotation of the cutting tool, rotation of the spiral bevel wheel to be cut, and motion for the tooth profile generation. Figure 12.7 shows the combination of motions of the gear wheel to be cut and the milling head-cutter. This last is composed of a number of identical groups of blades. Each group has at least one blade, whose inside cutting edge cuts the convex side of the teeth, and one blade whose outside cutting edge cuts the concave side of the teeth. Each subsequent group of blades penetrates the following tooth space between the teeth. Blades for roughing the tooth spaces can characterize the milling cutter. If one of the directions of rotation shown in Fig. 12.7 is changed, the flank lines on the pitch plane of the generation crown wheel are *extended hypocycloids*.

The rotational speed of the milling head-cutter is uniform. Therefore, since a group of blades must penetrate into a tooth space of the toothing to be cut, while the next group of blades must penetrate into the next tooth space, the so-called *division ratio* must be respected. It is defined as the quotient of the angular velocity of the spiral bevel wheel to be cut divided by the angular velocity of the milling head-cutter, given by Z_0/z, where z is the number of teeth of the gear wheel to be cut. The angular velocities of the milling head-cutter and spiral bevel gear to be cut allow to materialize the fictitious generation crown wheel, which rotates about an axis coinciding with the cradle axis.

To get the toothing of the spiral bevel wheel by generation motion, it is necessary that the pitch plane of the fictitious generation crown wheel rolls without sliding on the pitch cone of the spiral bevel wheel to be generated. Without this motion, the milling cutter would cut the same tooth spaces. Therefore, it is necessary that the cradle on which the milling cutter is mounted, and then the fictitious generation crown wheel, are equipped with an additional angular velocity. It follows a corresponding additional angular velocity of the gear wheel to be generated. It is to be add to that due to the *division ratio* above. This additional rotational speed is expressed by the so-called *generation ratio*, defined as the quotient of the additional angular velocity of the spiral bevel wheel to be generated divided by the angular velocity of the cradle, given by z_0/z. This additional rotational speed of the gear wheel to be cut is obtained by a differential device arranged in the kinematic chain of the gear cutting machine.

Through the combination of the three motions above-mentioned, the Eq. (12.12) is satisfied. Apparently, the spiral bevel gears cut with Oerlikon process have the same appearance of those obtained with the Gleason process. Instead, they are fundamentally different, at least for the following two aspects:

- The Oerlikon toothing has a constant depth, while usually a taper depth characterizes the Gleason toothing.
- The axes of milling cutters used to cut the pinion and wheel are perpendicular to the generatrixes of the pitch cones. Therefore, both milling head-cutters materialize a common generation crown wheel, in the strict meaning of the term, so that the cutting process complies with the theoretical conditions for conjugate toothing. Thus, the problems regarding the contact distribution between the active flanks of the teeth, which are typical of the Gleason process, are overcome.

Depending on the adjustment of the Oerlikon cutting machine, which may be performed with reference to a point between the inside and outside circles of the generation crown wheel, not necessarily coinciding with the mean point, it is possible to obtain two types of toothing. The first type consists of N-type Oerlikon-Spiromatic spiral bevel gears, with spiral angle ranging from $\beta_m = 30°$ to $\beta_m = 50°$, and maximum normal module at the mean point which decreases towards the tooth sides. The second type consists of G-type Oerlikon-Spiromatic spiral bevel gears, with spiral angle ranging from $\beta_m = 0°$ to $\beta_m = 50°$.

12.3.4 Klingelnberg-Ziclo-Palloid Spiral Bevel Gears

Klingelnberg-Ziclo-Palloid spiral bevel gears have as generation curves on the pitch plane of the generation crown wheel arcs of epicycloid or hypocycloid. These gears, which are similar to the Oerlikon-Spiromaric spiral bevel gears, are generated by milling head-cutters with inserted blades in two parts, which can be used for right-hand spiral or left-hand spiral, with replacement of the blades. Depending on the set-up of the cutting machine, spiral angles ranging from $\beta_m = 0°$ to $\beta_m = 45°$ can be obtained. Geometry of the toothing is characterized by a constant tooth depth, or by a normal module and normal pitch that, depending on the spiral angle, are tapered from the heel to toe, to become almost a constant (see Niemann and Winter [42]).

12.3.5 Klingelnberg-Palloid Spiral Bevel Gears

Klingelnberg-Palloid spiral bevel gears have as generation curves on the pitch plane of the generation crown wheel arcs of an involute of circle. These gears are made with spiral bevel gear cutting machines developed by Klingelnberg, which use conical (or cylindrical) hobs with left-hand cutting threads for right-hand spiral bevel wheels, and vice versa. The flank lines on the pitch plane of the generation crown wheel are shaped like arcs of involute of a base circle concentric with the inside and outside circles of the generation crown wheel and having radius $r_b \leq R_i$ (Fig. 12.3d). The Klingelnberg-Palloid process is referred also as a face-hobbing process, since the manufacturing is done using a continuous-indexing process. This is obtained with the cradle axis, working axis and head-cutter axis that roll together in a timed relationship. Gears obtained with this process have a parallel-depth profile along the face width (see Galassini [13], Giovannozzi [14], Henriot [17], Niemann and Winter [42], Radzevich [47], Klingelnberg [21]).

As Fig. 12.8 shows, the conical hob, whose transverse pitch is a constant, is placed in such a way that its cutting pitch cone (usually it is coinciding with the pitch cone) is tangent along the straight line, t, to the pitch plane of the generation crown wheel, as well as to the base circle at point T. The generation surface of the toothing is also here a conical surface, whose generatrixes have, as directrix curve, the arc AB of the involute of the base circle having radius, r_b. These generatrixes are inclined with respect to the axis of the generation crown wheel of an angle equal to the pressure angle, α. Therefore, the generation surface is an involute helicoid, the base cylinder of which has the base circle as transverse section. All transverse sections of this involute helicoid are arcs of an involute of the base circle.

In relation to the generation process, we consider first the planetary rotation of the conical hob about the crown wheel axis, with an angular velocity, Ω. During this planetary rotation, the straight line t continues to remain tangent to the base circle. Any of points of this straight line t describes an involute curve, provided the conical hob rotates simultaneously about its own axis with an angular velocity such that the linear velocity with which the point P under consideration moves along a generatrix of the cone is equal to $r_b\Omega$. This linear velocity is of course equal to the product of the pitch measured along the cone generatrix by the number of revo-lutions per second. Therefore, through these two combined motions, the conical hob covers and describes all the involute curves on the generation crown wheel, con-sidered non-rotating. All this happens because the involute of a circle can be

Fig. 12.8 Operating mode of the Klingelnberg-Palloid spiral bevel gear cutting machine

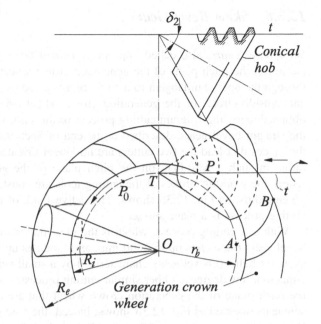

generated in a manner different, but equivalent, to what we have seen in Sect. 2.1. In fact an involute can be generated by a point P on a tangent line to the base circle (tangent line and base circle rotate rigidly connected, i.e. without relative rolling), when the point P moves on the tangent line with a linear velocity equal to the tangential velocity on the same base circle.

To obtain a given spiral bevel gear with this generation process, we place first the gear wheel to be generated in such a way that its cutting pitch cone, having a pitch angle δ_2, is tangent to the pitch plane of the generation crown wheel (Fig. 12.8). Then we give to the conical hob an additional rotation ω_0 about its axis. Since the conical hob is a single-start hob, and the generation crown wheel has a fictitious number of teeth Z_0, this additional rotation causes a rotation $\Omega_0 = \omega_0/Z_0$ of the generation crown wheel considered in meshing with the hob, and thus a rotation $\omega_2 = \Omega_0/\sin\delta_2 = \omega_0/(Z_0\sin\delta_2)$ of the spiral bevel wheel to be generated.

By virtue of the three rotations considered above, the conical hob, while drags in rotation the fictitious generation crown wheel, and with it the spiral bevel wheel to be cut, moves sweeping all the generation crown wheel, which also rotates, and cuts progressively the spiral bevel wheel to be generated over all its face width.

In practical applications, the teeth are shaped like shortened involutes, for reasons on which it is no need to dwell here. The spiral angle, β_m, is usually equal to 35°, but it may be up to 38°. Teeth of constant depth characterize the geometry of the toothing. The pitch and tooth thickness are the nominal ones. The surface finishing is not high, and the tooth flank surfaces are scaly, due to the conical hob.

12.3.6 Skew Bevel Gears

Skew bevel gears, also called improperly *helical bevel gears*, have as generation curves on the pitch plane of the generation crown wheel straight lines not passing through the apex, but tangent to a circle of given radius, concentric with the inside and outside circles of the generation crown wheel (Fig. 12.3b). These gears are obtained using the Bilgram cutting process or the same Gleason cutting process of the straight bevel gears, described at the end of Sect. 9.4. Obviously, in this case, the guides that lead the two cutters are no longer oriented in a radial direction, but rather in such a way that, on the pitch plane of the generating fictitious crown wheel, the working strokes of the cutting tools are constantly tangent to a circle of given radius, as Fig. 12.3b shows. The active flank of the tooth obtained by this cutting process is a plane surface.

With this cutting process, which is the most important for large curved-toothed bevel gears (diameters up to 2500 mm, and modules up to 20 mm), we can obtain skew (or helical) bevel gears characterized by a small value of the spiral angle (β_m equal to a few degrees). Other similar cutting processes are based on flank lines on the pitch plane of the generation crown wheel that are only virtually straight-line segments oriented as Fig. 12.3b shows; indeed, they are obtained as arcs of sinoid. With an appropriate sizing of this sinoid, deviations from the straight line are of the order of micrometer (thousandth of a millimeter). With this cutting process, the teeth are also obtained with a continuous and simultaneous generation motion [14].

12.3.7 Archimedean Spiral Bevel Gears

Archimedean spiral bevel gears have as generation curves on the pitch plane of the generation crown wheel arcs of Archimedean spirals. These gears are not much used in practical applications, since the cutting process with which they are generated is not able to ensure an accuracy grade comparable to the one typical of the previously described processes.

12.3.8 Branderberger Spiral Bevel Gears

Branderberger spiral bevel gears have as generation curves on the pitch plane of the generation crown wheel arcs of sinoid (Fig. 12.5). We have already described the cutting process of these gears, and its potential, in the previous section, to which we refer the reader.

12.3.9 Double-Helical Bevel Gears

Double-helical bevel gears or *double spiral bevel gears* have as generation curves on the pitch plane of the generation crown wheel two straight lines of opposite hand, converging on the mean circle (Fig. 12.3i). These gears have a spiral tooth in two portions of opposite hand, but are uncommon. They in fact have been developed and made with the aim of eliminating the end thrust. However, the results have been disappointing so far, since the maldistribution of the load on the meshing teeth constitutes a limit, which is difficult to overcome.

These gears are generated using a variation of the Reinecker-Böttcher gear cutting machine with which the double-helicoid cylindrical gears are produced. In reality, the denomination of double-helical bevel gears is somewhat improper, since the generation curves on the pitch plane of the generation crown wheel are not two straight lines of opposite hand, converging on the mean circle, but rather a continuous curve in the shape of an arc of extended hypocycloid. The more correct name would be *spiral bevel gears with arrow-shaped teeth*.

In addition to the above-mentioned main families of spiral bevel gears, other families exist that take their trade names from the cutting processes with which they are generated. This is not the place to describe their characteristics that, essentially, are attributable to the generation processes previously highlighted [13].

12.4 Generation Process of the Tooth Active Flank of Spiral Bevel Gears

A general process of generation of the tooth active flank of spiral bevel gears is described here. It is however be borne in mind that the active surfaces of the tooth flanks obtained with the process made by means of the existing gear cutting machines are not perfect spherical involutes, but present with respect to the latter small differences. Therefore, the properties of the mating surfaces theoretically obtained may be considered sufficiently representative of those actually realized with the current gear cutting machines. They are therefore to be considered reliable and can be used for calculation purposes. Thus, we are faced with the same type of approximation between the spherical involute teeth and octoid teeth we talked about the straight bevel gears (see Poritsky and Dudley [45], Romiti [49, 50], Ferrari and Romiti [11]).

Without loss of generality of the discussion, we here describe the generation process considering a skew bevel wheel, i.e. a curved-toothed bevel wheel that has, as generating flank line on the pitch plane of the generation crown wheel, a straight line not passing through the apex (Fig. 12.3b). With reference to Fig. 9.5, which shows the generation of spherical involutes and tooth profiles of straight bevel gears, let's consider the base cone $(O; c_1)$ of the curved-toothed bevel wheel 1, and the plane π, tangent to it, as well as any straight-line segment, l_1, not passing through the apex, O (Fig. 12.9).

Fig. 12.9 Generation process
of the tooth active surface of a
skew bevel wheel

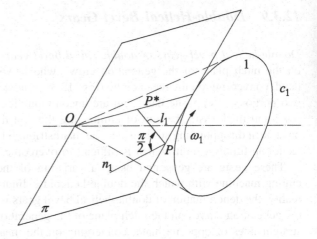

We then roll without sliding the plane, π, on the base cone. Each point, P, of the straight line, l_1, describes, on the sphere having radius OP and center O, a spherical involute of the base circle obtained as the intersection of the base cone with the afore-mentioned sphere. In this rolling motion without sliding, the successive positions of the straight line, l_1, define a surface, which we assume as the active surface of the wheel tooth. This surface, σ_1, is a ruled surface, whose generation lines, l_1, are tangent to the base cone under consideration. The locus, λ_1, of points of tangency of straight lines, l_1, with the base cone is a geodesic on the base cone, i.e. a curve of shortest path between two points on the cone surface. It is a developable curve, in the sense that, with a reverse operation, that is by rolling without sliding the base cone on the plane π, the geodesic, λ_1, develops in the straight line, l_1.

In each point P^* of the geodesic, the tangent to it lies on the plane π tangent to the base cone, which also includes the generation line l_1 of surface σ_1 passing through point P^*. Since then the tangent to geodesic and generation line l_1 of surface σ_1 have the same inclination with respect to the generatrix of the base cone passing through the same point P^*, they obviously coincide. The surface σ_1 thus proves to be the ruled surface, defined as the locus of tangents to the geodesic; it is therefore a developable surface.

From this generation process of surface σ_1, we can deduce the following theoretical properties, which concern the same surface σ_1 and the engagement between two mating curved-toothed bevel wheels, whose active tooth surfaces are obtained by the same generation process.

1. The normal n_1 to surface σ_1 at any current point P is the straight line on the plane π (the plane tangent at point P to the base cone), and is perpendicular to the generation line l_1 of surface σ_1 passing through the same point (Fig. 12.9). In fact, this normal must be perpendicular to the tooth profile at point P. This profile is, however, the path described by point P during the rolling motion of plane π about the generatrix of contact of the same plane π with the base cone; therefore, the tooth profile is perpendicular to the plane π. Consequently, the

normal n_1 must belong to plane π, and must be as well perpendicular to the generation line l_1 of surface σ_1, lying on plane π.

2. Let's consider two mating curved-toothed bevel wheels, such as those shown in schematic form in Fig. 9.5, and assume that their active surfaces σ_i (with $i = 1, 2$) are obtained with the generation process described above. Suppose that they are brought into contact in a common point of tangency P, and consider the plane π passing through point P and tangent to the two base cones of the two wheels; with reference to Fig. 9.5, this plane is evidently the plane containing the maximum circle, a. Given the properties described in the previous point 1, both the generation lines of surfaces σ_i, and the normals n_i (with $i = 1, 2$) to these surfaces, which are straight lines of the plane π perpendicular to the generation lines (Fig. 12.9), belong to this plane. Given that in the rolling motion of plane π on the two base cones, the same generation line is used to generate the two mating surfaces, σ_i, in this common tangent plane π, the two generation lines and the two normals coincide, i.e. $l_1 \equiv l_2$ and $n_1 \equiv n_2$. Therefore, we infer that the two surfaces σ_i in contact at point P are tangent to each other along the whole generation line passing through point P.

To make possible the meshing, it is however necessary that the contact between the two surfaces is maintained during rotation of the two wheels, to which these surfaces are rigidly connected. Now, with reference to Fig. 9.5, we infer that, while the wheel 1 rotates about its own axis with an angular velocity ω_1, the intersection of the surface σ_1 with the plane π rotates about the normal to the same plane passing through point O with an angular velocity $\Omega_1 = \omega_1 \sin \delta_{b1}$ (see Eq. 9.39). In fact, the angular velocity ω_1 of surface σ_1 about the axis of the wheel 1 can be resolved into two components: the angular velocity $\Omega_1 = \omega_1 \sin \delta_{b1}$, just defined, and the angular velocity $\Omega_1' = \omega_1 \cos \delta_{b1}$ about the generatrix OT_1 of the base cone along which this cone is tangent to the plane π (Fig. 9.5). This last component, which represents the rolling motion of plane π on the base cone (this motion generates the surface σ_1), does not alter the intersection of surface σ_1 on plane π, while the first component defines the motion on plane π of that intersection, which is however always a generation line of the same surface σ_1.

Similarly, we can deduce that the intersection of the surface σ_2 with the plane π rotates about the normal to the same plane passing through point O with an angular velocity $\Omega_2 = \omega_2 \sin \delta_{b2}$. Keeping in mind the Eqs. (9.42) and (9.41), we can write $\omega_2 \sin \delta_{b2} = \omega_2 \sin \delta_2 \cos \alpha = \omega_1 \sin \delta_1 \cos \alpha = \omega_1 \sin \delta_{b1}$, for which it is $\Omega_1 = \Omega_2 = \Omega$. It follows that the intersections of the surfaces σ_i (with $i = 1, 2$) with the common plane π tangent to the two base cones rotate about the normal to the same plane passing through point O with the same angular velocity. Therefore, since these intersections are coinciding at location of point P under consideration, they continue to remain coinciding, whereby the meshing is possible.

3. As a result of the property described in the previous point 2, we deduce immediately that the meshing surface of the curved-toothed bevel pair under

consideration belongs to the plane of action, which is the common tangent plane π to the two base cones of the same bevel pair. The *meshing surface* is the region of this plane of action having the shape of a circular ring sector between (Fig. 12.10).

- the two circles K_e and K_i, obtained as intersections of the plane of action with the two spheres that surround the face width of each tooth, having radii R_e and R_i respectively (Figs. 9.2 and 9.3);
- the two radii that project the points E and A from the point O, where the axes of the two wheels intersect (points E and A are the intersections of the circle K_e with the tip circles of the two wheels on the sphere having radius R_e).

At every instant, the active surfaces of the teeth, σ_1 and σ_2, are touching along a generation line segment belonging to the aforementioned region. This segment appears to rotate about the normal to the plane of action passing through point O with the angular velocity $\Omega_1 = \Omega_2 = \Omega$, defined above, while wheels 1 and 2 rotate with angular velocities ω_1 and ω_2 (Fig. 12.10). The positions on the circle K_e of points E and A as well those of points of tangency T_2 and T_1 of circle K_e with circles c_1 and c_2 (Fig. 9.5) on the sphere having radius R_e can be easily obtained by considering the spherical triangles $(O_1 T_1 C)$, $(O_1 T_1 E)$, $(O_2 T_2 C)$, and $(O_2 T_2 A)$, where O_1 and O_2 are the poles of the pitch surfaces. Figure 12.11 shows the first two spherical triangles, which refer to wheel 1. *Mutatis mutandis*, the other two spherical triangles are obtained in a similar way. Considering these spherical triangles, which are right angled at point T_1 or T_2, we get respectively the following relationships:

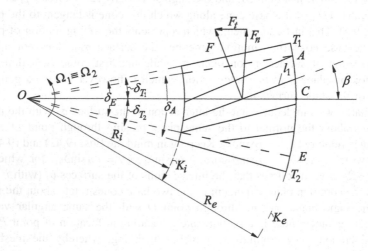

Fig. 12.10 Meshing surface of a skew bevel wheel

$$\cos\delta_1 = \cos\delta_{b1}\cos\delta_{T1} \quad \cos\left(\delta_1 + \frac{m}{R_e}\right) = \cos\delta_{b1}\cos\delta_E$$
$$\cos\delta_2 = \cos\delta_{b2}\cos\delta_{T2} \quad \cos\left(\delta_2 + \frac{m}{R_e}\right) = \cos\delta_{b2}\cos\delta_A$$

$$(12.13)$$

where (Fig. 12.10): δ_{T1} is the angle between the straight lines OC and OT_1; δ_A is the angle between the straight lines OA and OT_2; δ_{T2} is the angle between the straight lines OC and OT_2; δ_E is the angle between the straight lines OE and OT_1; m is the arc shown in Fig. 12.11. Equations (12.13) allow to obtain the angles δ_{T1}, δ_{T2}, δ_E and δ_A and the relative positions of the various corresponding points on the circle K_e.

For each position of the generation line of contact, and thus for each relative position of the bevel pair, it is easy to determine the mutual force transmitted between two teeth that touch each other along the same generation line, under the assumption that only one pair of teeth are in meshing. To this end, we consider the generation line of contact, l_1, in the position where it is divided into two equal halves by the segment of straight line OC between the circles K_i and K_e (Fig. 12.10). We assume also that the force, F, that the wheel 2 exerts on wheel 1 is applied at point of intersection between the generation line, l_1, and the straight line, OC (this is the mid-point, usually used by the technical standards), and that friction is negligible, for which it is directed perpendicularly to the same generation line, l_1. Finally, we denote by β the angle between the generation line, l_1, and straight line, OC.

As Fig. 12.10 shows, the force F has a component normal to the straight line OC, $F_n = F\cos\beta$, and a component directed along the line OC, $F_t = F\sin\beta$. The force F as well as its two components F_n and F_t are vectors that lie on the meshing surface, which is a portion of the pitch plane of the generation crown

Fig. 12.11 Spherical triangles to calculate δ_{T1} and δ_E (Fig. 12.10)

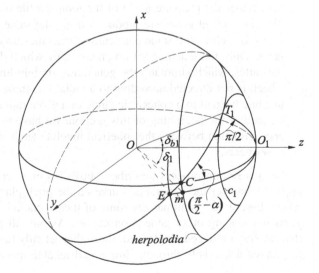

wheel. By using the coordinate system $O(x, y, z)$ shown in Fig. 12.11, the components of this force F are given by the following relationships:

$$F_x = F_n \cos\alpha = F\cos\beta\cos\alpha$$
$$F_y = -F(\cos\beta\sin\alpha\cos\delta_1 + \sin\beta\sin\delta_1) \qquad (12.14)$$
$$F_z = F\sin\beta(\sin\alpha\sin\delta_1 - \cos\delta_1).$$

In the case in which the pinion 1 is the driving wheel, and the driving torque T_1 is applied to it, for steady state operating conditions the following relationship must be true:

$$F_x\left(\frac{R_{e1} + R_{i1}}{2}\right) = T_1, \qquad (12.15)$$

from which we get:

$$F = \frac{2T_1}{\cos\beta\cos\alpha(R_{e1} + R_{i1})}. \qquad (12.16)$$

4. The above-described properties of the surface obtained as successive positions of the generating straight lines on the pitch plane of the generation crown wheel, not passing through its apex, are theoretical properties related to the spherical involute. In practical reality, the conventional gear cutting machines actually used to implement the generation process of skew bevel gears (for example, Bilgram cutting machines, Gleason cutting machines, and similar cutting machines with which these types of gears are cut) are not able to achieve exactly this theoretical generation process. However, they are able to emulate it, with a greater approximation, the smaller the tooth depth on the sphere of motion, and the smaller the pressure angle of the tooth profile on the same sphere.

The skew bevel gears are produced using the same tools with straight cutting edges as well as the same gear cutting machines used for cutting straight bevel gears. Only the machine set-up changes, by which the reciprocating motion of the cutter, which simulates the generating straight line on the generation crown wheel, is not directed according to a radial line through the apex, but according to a line tangent to a concentric circle of a given radius, as Fig. 12.3b shows. As we said at the beginning of this section, we have to do with the same kind of approximation between the spherical involute teeth and octoid teeth for straight bevel gears.

The generation process described above is quite general, and therefore can be applied whatever the generation curve on the pitch plane of the generation crown wheel. Figure 12.3 summarizes some of these generation curves, but nothing prevents introducing other generation curves. Among all possible generation curves that can be introduced or planned, those that actually had practical applications are the curves achievable through simple and reliable mechanisms. The conventional gear cutting machines enjoying greater success today use, as generation lines, the circular arc (Fig. 12.3c), the arc of epicycloid (Fig. 12.3g), and the involute arc (Fig. 12.3d).

Figure 12.12 shows a generation crown wheel on whose pitch plane generation lines are traced, having geometry in the form of an arc of circle. They constitute a special case of curved generation lines. In this case, as well as in the general case of generation lines of any geometry, repeating the procedure described above, we can derive the active surface of the tooth flank of the to-be-generated curved-toothed wheel, whose properties are quite similar to those obtained above. In particular, at each instant, the mating gear wheels are touching along the points of the generation line, l_1, in the common plane π tangent to the two base cones of the same wheels. This generation line appears rotate in this plane with an angular velocity Ω previously defined, while the two wheels rotate about their axes with angular velocities ω_1 and ω_2. Also the mutual force, F, that the teeth exchange each other, in the assumption that only one pair of teeth is in meshing, can be obtained in the same way.

Finally, it should be noted that the meshing between the curved-toothed bevel gears obtained with the above described generation process remains kinematically correct with respect to small changes of the shaft angle. Instead, it is no longer kinematically correct when displacements are such that the axes of the two members of the gear pair are no longer converging, so that the gear becomes a hypoid gear.

Fig. 12.12 Meshing surface of a curved-toothed bevel gear with generation line in the form of an arc of circle

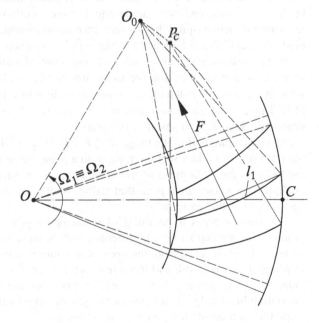

12.5 Spiral Bevel Gears: Main Quantities and Equivalent Cylindrical Gears

Various types of sizing are used for spiral bevel gears: some of them provide teeth of constant depth, while others require geometric parameters related to those we have described in Chap. 9 about straight bevel gears. Frequently, spiral bevel gears with profile-shifted toothing are used, most often characterized by profile shift coefficients equal and opposite for the two members of the gear pair $(x_2 = -x_1)$. These coefficients are related to the thickness factors given by Eq. (9.111). This discussion on the sizing of the curved-toothed bevel gears as well as that of the hypoid gears will still be deepened in Sect. 12.11.

Nominal profiles of the teeth are those related to the back cone. The tooth profiles obtained as sections of the same teeth with countless cones between the back and inner cones (from a theoretical point of view, they are infinite), are similar to each other, with dimensions linearly proportional to the distances from the apices of the pitch cones.

The Tredgold method (see Buchanan [4]), already described for straight bevel gears in Sect. 9.5, can be used also for spiral bevel gears. Depending on the geometric shape of the tooth trace (i.e., the line of intersection of the tooth flank with the reference pitch cone), we will have, generally, virtual equivalent cylindrical wheels with straight, skew (or improperly helical) and spiral teeth, which are used for the analysis of contact. In the case of spiral bevel gears of interest here, often we are content to reduce the actual gear wheel to a cylindrical helical wheel, equivalent to the spiral bevel wheel under consideration, having virtual number of teeth $z_v = z/\cos\delta$, and helix angle equal to the spiral angle, β_m. Obviously, if the teeth of the actual spiral bevel gear pair are crowned, also those of the virtual equivalent cylindrical helical gear pair will be crowned.

If the teeth are cut with a cutter having nominal standard sizing in the normal section, we can assume that the minimum number of teeth, z_{min}, that can be cut without the risk of undercut, decreases according to a factor equal to $\cos\delta\cos^3\beta_m$ [14]. This extends the considerations already made for cylindrical involute helical gears (see Sect. 8.5) to spiral bevel gears.

At a given radius R in the range $R_i \leq R \leq R_e$ (Fig. 12.4), the normal pitch, p_n, i.e. the pitch measured in a section made with a plane perpendicular to the flank line of the tooth on the pitch plane of the generation crown wheel, is equal to the circular pitch, or transverse pitch, p, at that radius, multiplied by $\cos\beta$. The angle β is the spiral angle at the same radius.

The various types of toothing of spiral bevel gears, generated by the cutting processes developed so far, are different, both because they have a spiral angle that varies with the distance from the apex of the reference pitch cone, both for the way in which the tooth depth and tooth transverse thickness change as a function of the radius, R. Once again it is to remember that only the logarithmic spiral used as generation line on the pitch plane of the generation crown wheel determines a spiral angle that is a constant when the radius changes.

About the operating conditions, it is to point out that, by usual convention, the concave pinion flank in meshing with the convex wheel flank is generally the *drive-side*, i.e. the driving flank. Otherwise we would have a greater spiral angle on the inner cone, β_i (it corresponds to the inside radius R_i on the pitch plane of the generation crown wheel in Fig. 12.4) and, consequently, a pointed tooth. This applies in the recommended case in which the directions of the spiral and rotation are the same. Nevertheless, there are cases where the pinion-driving flank is the *coast-side*. In these cases, the convex pinion flank is in meshing with the concave wheel flank.

In addition, it is to be noted that the use of the drive-side as main load transmission direction is for spiral bevel gears (but also and especially for hypoid gears) a rather binding rule. Transmission of speed and torque and the additional lengthwise sliding forces would lead on the coast-side to a deflection of pinion towards the gear wheel, consequently leading to a reduction of the backlash, in extreme cases to zero backlash. This occurs already under moderate torque, compromising the lubrication conditions. Surface damages followed by tooth fracture can occur.

The curvature of the flank line on the pitch plane of the generation crown wheel is that resulting from the radius of the inserted blade milling head-cutter used for cutting as well as the type of gear cutting machine employed. Gear cutting processes can be those already described in Sect. 12.3 (Gleason, Modul-Kurvex, Oerlikon-Spiromatic, Klingelnberg-Ziclo-Palloid, Klingelnberg-Palloid, Branderberger, etc.) or others. Instead, the spiral angle, β_m, depends on the set-up and adjustment of the gear cutting machine. Prudence demands that the generation of these gears is carried out following the instructions of the manufacturers of gear cutting machines used for this purpose. As a broad indication, it is appropriate that the spiral angle is sufficiently high so that the overlap ratio ε_β is greater than 1.5. This in order to put a remedy to the disadvantages resulting from a surface of contact that is limited in depth and length (see Niemann and Winter [42], Handschuh and Litvin [15], Su et al. [64]).

The spiral angle also significantly affects the axial thrust on the shafts, which must be supported by suitable bearings. In any case, the spiral angle and direction of rotation should be selected in such a way that the axial thrust on the shafts of the pinion and wheel has direction such as to move both members of the spiral bevel gear pair out of mesh when this gear pair is operating in the predominant working conditions. Otherwise, the pinion would tend to be sucked from the wheel, resulting in the above-mentioned significant reduction or total loss of backlash, and correlated risk of scuffing. To avoid this, it is appropriate that both members of a spiral bevel gear pair (or hypoid gear pair) are held against axial movement in both directions. Often the hand of spiral is imposed by the mounting conditions.

The *outer transverse module*, m_{et}, i.e. the transverse module on the *back cone*, and the *mean normal module*, m_{mn}, i.e. the normal module on the mean cone, are given respectively by the relationships:

$$m = m_{et} = \frac{d_1}{z_1} = \frac{d_2}{z_2} = \frac{m_{en}}{\cos\beta_2} \tag{12.17}$$

$$m_{mn} = m_{mt}\cos\beta_m \tag{12.18}$$

where remaining unchanged the meaning of other symbols, m_{en} and m_{mt} are respectively the normal module on the back cone, and the transverse module on the mean cone (it is to be noted that, to avoid misunderstandings, we used the double-subscript et to indicate the nominal module, m).

In addition to those obtained in Sect. 9.3, we have to consider other quantities that are typical of the spiral bevel gears, the main of which are described below. The *mean pitch diameters* of the pinion and wheel are given by:

$$d_{m1} = d_1 - b\sin\delta_1 = \frac{m_{mn}z_1}{\cos\beta_m} \tag{12.19}$$

$$d_{m2} = d_2 - b\sin\delta_2 = ud_{m1}, \tag{12.20}$$

where $u = z_2/z_1$ is the gear ratio, and b is the face width.

Furthermore, the following relationship is valid:

$$\tan\alpha_t = \frac{\tan\alpha_n}{\cos\beta}, \tag{12.21}$$

where α_t and α_n are respectively the transverse and normal pressure angles. To have teeth not too tapered, it is necessary that the following condition is met: $(b/R_e) \le (0.25 \div 0.35)$.

In the case of teeth tapered from the heel towards the toe (Fig. 9.11a), but assuming that the nominal clearance, c, is a constant (in this case the apex of the tip cone will be located at the inside of the reference pitch cone), the reference quantity is m_{et}, whereby we have the following relationships:

$$\delta_{a1} = \delta_1 + \vartheta_{f2} \quad \delta_{f1} = \delta_1 - \vartheta_{f1} \tag{12.22}$$

$$\delta_{a2} = \delta_2 + \vartheta_{f1} \quad \delta_{f2} = \delta_2 - \vartheta_{f2} \tag{12.23}$$

and $\vartheta_{a1} = \vartheta_{f2}$, and $\vartheta_{a2} = \vartheta_{f1}$ by design choice. We also have:

$$\tan\vartheta_{f1} = \tfrac{h_{fe1}}{r_e} \quad \tan\vartheta_{f2} = \tfrac{h_{fe2}}{r_e} \tag{12.24}$$

$$h_{ae1} = m_{et}(1 + x_{he}) = h_{am1} + \frac{b\tan\vartheta_{a1}}{2}; \quad h_{fe1} = m_{et}(1 - c - x_{he}) \tag{12.25}$$

$$h_{ae2} = m_{et}(1 - x_{he}) = h_{am2} + \frac{b\tan\vartheta_{a2}}{2}; \quad h_{fe2} = m_{et}(1 - c + x_{he}) \tag{12.26}$$

We must keep in mind that, with tapered teeth from heel to toe, it is possible to have a greater contact ratio, without the danger of meshing interference is manifested at the tooth toe. The nominal clearance, c, is generally assumed equal to $(0.1 \div 0.3)m$.

In the case of teeth having constant depth (Fig. 9.11b), the reference quantity is m_{mn}, whereby we have the following relationships:

$$\delta_{a1} = \delta_{f1} = \delta_1 \tag{12.27}$$

$$\delta_{a2} = \delta_{f2} = \delta_2 \tag{12.28}$$

$$\vartheta_a = \vartheta_f = 0 \tag{12.29}$$

$$h_{am1} = m_{mn}(1 + x_{hm}) \quad h_{fm1} = m_{mn}(1 + c - x_{hm}) \tag{12.30}$$

$$h_{am2} = m_{mn}(1 - x_{hm}) \quad h_{fm2} = m_{mn}(1 + c + x_{hm}). \tag{12.31}$$

As already mentioned previously, the virtual equivalent cylindrical helical wheels of interest here are those that correspond, according to the Tredgold method, to the mean cones. The relationships from Eqs. (9.44) to (9.52) are still valid. They concern quantities related to this equivalent cylindrical helical gear pair. The following relationships are to be added to them.

Transverse pressure angle, α_{vt}:

$$\tan\alpha_{vt} = \tan\alpha_{mt} = \frac{\tan\alpha_n}{\cos\beta_m} \quad (\text{with } \alpha_{vn} = \alpha_n). \tag{12.32}$$

Base helix angle, β_{bv}:

$$\sin\beta_{bv} = \sin\beta_m\cos\alpha_n \quad (\text{with } \beta_{vm} = \beta_m). \tag{12.33}$$

Virtual number of teeth in normal section, z_{vn}:

$$z_{vn} = \frac{z_v}{\cos^2\beta_{bv}\cos\beta_m}. \tag{12.34}$$

We will have then: $m_{vt} = m_{mt} = d_{m1}/z_1 = d_{v1}/z_{v1} = d_{v2}/z_{v2}$, and $m_{vn} = m_{mn} = m_{vmt}\cos\beta_m$, where m_{vt} and m_{vn} are respectively the transverse and normal modules of the virtual gear pair. The profile shift coefficients are given by:

$$x_{hm1,2} = \frac{(h_{am1,2} - h_{am2,1})}{2m_{mn}}. \tag{12.35}$$

Length of path of contact, $g_{v\alpha}$:

$$g_{v\alpha} = \frac{1}{2}\left[\left(d_{va1}^2 - d_{vb1}^2\right)^{1/2} + \left(d_{va2}^2 - d_{vb2}^2\right)^{1/2}\right] - a_v\sin\alpha_{vt}. \tag{12.36}$$

Transverse contact ratio, $\varepsilon_{v\alpha}$:

$$\varepsilon_{v\alpha} = \frac{g_{v\alpha}\cos\beta_m}{\pi m_{mn}\cos\alpha_{vt}}. \tag{12.37}$$

Transverse contact ratio in normal section, $\varepsilon_{v\alpha_n}$:

$$\varepsilon_{v\alpha_n} = \frac{\varepsilon_{v\alpha}}{\cos^2\beta_b}. \tag{12.38}$$

Overlap ratio, $\varepsilon_{v\beta}$:

$$\varepsilon_{v\beta} = \frac{b\sin\beta_m}{\pi m_{mn}}\frac{b_{eH}}{b}. \tag{12.39}$$

Total contact ratio, $\varepsilon_{v\gamma}$:

$$\varepsilon_{v\gamma} = \varepsilon_{v\alpha} + \varepsilon_{v\beta}. \tag{12.40}$$

Partial contact ratio of the pinion, ε_{v1}:

$$\varepsilon_{v1} = \frac{z_{v1}}{2\pi}\left\{\left[\left(\frac{d_{va1}}{d_{vb1}}\right)^2 - 1\right]^{1/2} - \tan\alpha_{vt}\right\}. \tag{12.41}$$

Partial contact ratio of the wheel, ε_{v2}:

$$\varepsilon_{v2} = \frac{z_{v2}}{2\pi}\left\{\left[\left(\frac{d_{va2}}{d_{vb2}}\right)^2 - 1\right]^{1/2} - \tan\alpha_{vt}\right\}. \tag{12.42}$$

With reference to the overlap contact ratio and total contact ratio, we must keep in mind that, when calculating the load carrying capacity of the gear pair, we usually assume an effective face width, b_{eH}, equal to $0.85b$, i.e. $b_{eH} \cong 0.85b$. Of course, the transverse module and pitch of bevel gears change along the face width. As already we mentioned, depending on the problem to be talked, we use or the outer transverse module, m_{et}, or the mean normal module, m_{mn}. Although it is not necessary, as a series of cutters cover a large range of modules, these are generally chosen with reference to the standard set of modules for parallel cylindrical gears.

12.6 Load Analysis for Spiral Bevel Gears and Thrust Characteristics on Shafts and Bearings

Load analysis for spiral bevel gears can be made with the same procedure used for parallel cylindrical spur and helical gears and for straight bevel gears, i.e. neglecting the contribution of the friction forces. These forces are in fact not very significant from the design point of view. To simplify this load analysis and evaluate the thrust characteristics on shafts and bearings, we assume that the total force, $F = F_n$, exchanged between the meshing teeth, acts along the direction of the normal to the tooth surface. We also assume that this force is applied at the mean point, M, of the tooth, and thus in the middle face width, in the tooth position where that point is located in the plane containing the axes of the two members of the gear (see Giovannozzi [14]; Brown [3]; Su et al. [64]).

With reference to Fig. 12.13, we consider the normal section of the tooth, passing through its mean point, M, and suppose that this point lies in the plane of the figure (Fig. 12.13a). The total tooth force, $F = F_n$, acting on the tooth, lies in the plane of the normal section to the tooth, and is inclined, with respect to the perpendicular to the plane of Fig. 12.13a through point M, of the pressure angle, α_n. Therefore, the force F_n can be resolved into two orthogonal components, $F_n \sin\alpha_n$ and $F_n \cos\alpha_n$ (Fig. 12.13c), the first of which acts in the axial plane of the gear wheel under consideration in the direction perpendicular to the generatrix of the pitch cone (Fig. 12.13a), while the second in the front view of the same gear wheel shown in Fig. 12.13b is inclined of the angle β_m with respect to the perpendicular through point M to the projection of the straight line OM. From the same figure, which also shows the axial view of the gear wheel (Fig. 12.13b), as well as the various components of the total tooth force, $F = F_n$, we can infer the following relationships:

$$F_t = F_n \cos\alpha_n \cos\beta_m \qquad (12.43)$$

$$F_a = F_n(\sin\alpha_n \sin\delta + \cos\alpha_n \sin\beta_m \cos\delta) \qquad (12.44)$$

$$F_r = F_n(\sin\alpha_n \cos\delta - \cos\alpha_n \sin\beta_m \sin\delta), \qquad (12.45)$$

where F_t, F_a, and F_r are respectively the tangential, axial, and radial components (or tangential, axial, and radial forces).

If we obtain F_n from Eq. (12.43), and we replace it in Eqs. (12.44) and (12.45), we get the following relationships, which express, with the signs on the top, the axial and radial components, F_a and F_r, as a function of the tangential force, F_t:

$$F_a = F_t \left(\frac{\tan\alpha_n \sin\delta}{\cos\beta_m} \pm \tan\beta_m \cos\delta \right) \qquad (12.46)$$

Fig. 12.13 Total tooth force $F = F_n$ exchanged between the meshing curved teeth, and its components

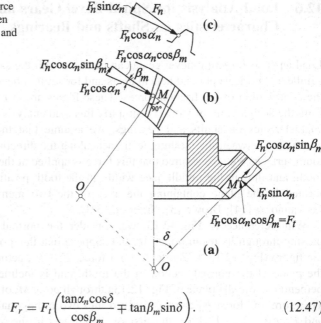

$$F_r = F_t \left(\frac{\tan\alpha_n \cos\delta}{\cos\beta_m} \mp \tan\beta_m \sin\delta \right). \tag{12.47}$$

Conventionally, in the equations above, the axial component, F_a, also called *axial thrust*, is considered as positive when it tends to distance the gear wheel from the apex of the pitch cone, and negative in the opposite case. Therefore, a positive axial force tends to move the two members of the gear pair out of meshing, while a negative axial force moves the two members towards each other. Instead, the radial component, F_r, is considered positive when it is directed from the mean pitch circle (i.e. the intersection of the pitch cone with the mean cone) to the axis of the gear wheel, and negative in the opposite case.

Whatever the directions of the tooth spiral and rotation of the gear wheel, the component $F_n \sin\alpha_n = F_t(\tan\alpha_n/\cos\beta_m)$ always retains the same direction, the one that goes from the pitch cone towards the axis of the gear wheel (see Fig. 12.13a). Instead, the component $F_n \cos\alpha_n \sin\beta_m$, whose application line coincides with a generatrix of the pitch cone, changes its direction when the direction of the tooth spiral or direction of rotation of the gear wheel change.

We agree now to assume as a direction of rotation of the gear wheel that seen by an observer positioned on the side of the pitch circle, i.e. the circle of intersection of the pitch cone with the back cone. Thus, this observer sees the same gear wheel interposed between himself and the apex of the pitch cone. The possible cases that we can have are as follows:

– If the gear wheel is driving, in Eqs. (12.46) and (12.47) the signs at the top are valid for left-hand spiral and clockwise rotation, as well as for right-hand spiral and counterclockwise rotation. These are the cases (a) and (b) shown in Fig. 12.14, and the case shown in Fig. 12.13.

– If the gear wheel is driving like above, in Eqs. (12.46) and (12.47) the signs
   below are valid in the other two cases shown in Fig. 12.14, i.e. the cases (c) and
   (d).

If the gear wheel is driven, the signs opposite to those aforementioned are valid.

We must also keep in mind that the direction of the axial thrust in a spiral bevel
gear (and in a hypoid gear) depends on the direction of rotation, hand of the spiral
angle, relative position of the driving and driven members, and pitch cone angle.
Furthermore, it depends on the fact that the gear wheel is the driving or driven
member. All these factors determine the condition for which the axial component,
F_a, will make the gear wheel move away or towards the cone apex.

The following fundamental difference exists between a straight bevel gear and a
spiral bevel gear or a hypoid gear. In a straight bevel gear as well as in a Zerol bevel
gear, the axial components F_{a1} and F_{a2} always tend to force the pinion and the
wheel out meshing. On the contrary, in a spiral bevel gear as well as in a hypoid
gear, the direction to which the axial components act may be either towards or away
from the cone apex, depending on the algebraic sign appearing in Eq. (12.46).

Since in the case of spiral bevel gears and hypoid gears the axial thrust may act
in both directions, because of the aforementioned factors, it follows that the choice
of the direction and magnitude of the spiral angle constitutes a fundamental point on
which the designer has to focus his attention. The members of the spiral bevel and
hypoid gear pairs can move away or move towards each other. In the first case, an
unnecessary large backlash is created between them. In the second case, a tight
mesh is to be determined, and this may result in jamming or scuffing.

The proper running and operating conditions of a spiral bevel or hypoid gear need a
right kind of support, which can prevent possible axial displacements of both members
of the gear. To this end, meticulous design calculations of the axial thrusts are necessary, in order to choose the appropriate types of support and the constructive and

Fig. 12.14 Possible cases for
the choice of the signs in
Eqs. (12.46) and (12.47)

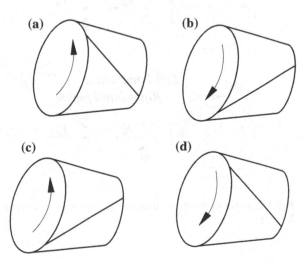

assembly solutions related to them. Figure 12.15 shows the direction of the axial and radial forces in a spiral bevel gear pair with shaft angle $\Sigma = 90°$; it highlights the influence of the hand of spiral and direction of rotation on the direction of the axial thrust.

It is to be noted that, in the curved-toothed bevel gears, the hand of the spiral and the direction of rotation of the two members are opposite. Furthermore, in the case where the shaft angle is equal to $90°(\Sigma = \Pi/2)$, the axial thrust acting on one of the two members is equal, in absolute value and sign, to the radial force acting on the other member, and vice versa. It is also to be kept in mind that the axial thrust, F_a, is a vector parallel to the axis of the gear wheel, which is applied at a distance from the axis equal to the mean pitch radius, r_m. Therefore, this axial force, F_a, causes a concentrated bending moment, $M_b = F_a r_m$, which in turn determines related reaction forces on the bearings.

Figure 12.15 also shows the radial force acting on the two members of the gear: it has positive sign when it is directed towards the wheel axis, while it has negative sign when it is directed away from the wheel axis, i.e. towards the mating gear wheel.

Therefore, we can deduce that, in the case of a spiral bevel gear characterized by a shaft angle $\Sigma = 90°$ (this is the most frequent case), if F_{a1} and F_{r1} are the axial and radial forces acting on the pinion, respectively calculated using Eqs. (12.46) and (12.47), the axial and radial forces $F_{a2} = F_{r1}$ and $F_{r2} = F_{a1}$ acting on the mating wheel are those obtained using respectively Eqs. (12.47) and (12.46).

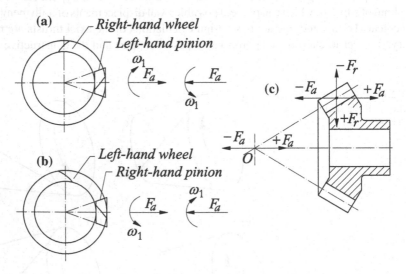

Fig. 12.15 Direction of the axial and radial forces in a spiral bevel gear pair with shaft angle $\Sigma = 90°$

12.7 Hypoid Gears: Basic Concepts

In Sects. 10.1 and 10.2, we saw that, for a crossed helical gear pair with axes 1 and 2, shaft angle $\Sigma = (\beta_1^* + \beta_2^*)$, center distance $a = (r_1^* + r_2^*) = O_1O_2$, and angular velocities ω_1 and ω_2, the relative motion is an instantaneous helical motion, or screw motion, about an axis, m, passing through the point O lying between the end points O_1 and O_2 of the center distance O_1O_2 (Fig. 10.1). In Sect. 10.7 we determined (see Fig. 10.13): the angle, β_1^*, between the axis of the screw motion, m, and the axis 1 of the first gear wheel, given by Eqs. (10.54); the distances $r_1^* = OO_1$ and $r_2^* = OO_2$ of the same instantaneous axis from the axes of the two wheels, measured along the center distance and given by Eqs. (10.56); the characteristics of the instantaneous helical motion, i.e. the relative angular velocity, $\omega_{r1,2}$ and the relative velocity vector, $v_{r1,2}$, whose absolute values are given respectively by Eqs. (10.57) and (10.58).

Assuming that this instantaneous axis is rigidly connected first with the axis 1 of the first wheel, which is rotating, and subsequently with the axis 2 of the second wheel, also rotating, it generates a pair of one-sheeted double ruled hyperboloids (Fig. 10.2). These hyperboloids can be used as surfaces of two friction wheels that can mutually transmit the motion according to the transmission ratio $i = \omega_1/\omega_2$ when they touch along a common generatrix and are pressed one against the other. Obviously, the motion would be transmitted with greater safety and efficiency in the case in which the two hyperboloids are provided with teeth. These two hyperboloids of revolution are the gear axodes that are in tangency along the instantaneous axis of screw motion, and perform in relative motion rotation about and translation along this instantaneous screw axis.

In the transmission of motion between two crossed axes, we have a special case when the shaft angle Σ between the directions of the vectors ω_1 and $-\omega_2$ (Fig. 10.5) is greater than $90°$, and β_2^* is equal to $90°$. In this case, $\Sigma = [(\pi/2) + \beta_1^*]$, for which, from Eq. (10.9), we obtain:

$$\sin\beta_1^* = \frac{\omega_2}{\omega_1} = \frac{z_1}{z_2} = \frac{1}{i} \qquad (12.48)$$

while from Eqs. (10.5) and (10.6), since $\tan\beta_2^* = \tan 90° = \infty$, we get:

$$r_1^* = OO_1 = 0 \quad r_2^* = OO_2 = a. \qquad (12.49)$$

Therefore, in this case, the Mozzi's axis, m (i.e. the axis of the instantaneous relative helical motion, see Sect. 10.1), lies in the plane perpendicular to the shortest distance O_1O_2 containing the first axis 1, intersects the axis 1 at point O_1, and is normal to the projection of the second axis 2 on this plane. Thus, m' is normal to $2'$ as Fig. 10.5 shows (see Mozzi del Garbo [39], Scotto Lavina [54]). Consequently, the pitch hyperboloid of the first wheel, having as axis the axis 1, degenerates into a cone with apex O_1 and cone angle $\beta_1^* = (\Sigma - \pi/2)$. Instead, the pitch hyperboloid of the second wheel, having as axis the axis 2, degenerates into a ruled plane according to the tangent lines to the circle having axis 2 and radius equal to the center distance, a (these tangent lines coincide with the successive positions of the Mozzi's axis). This second wheel is the so-called *hyperboloid crown wheel*. Figure 10.5 also shows the successive positions of the Mozzi's axis for a transmission of motion between two crossed axes with $\Sigma = \left(\beta_1^* + \beta_2^*\right) > 90°$ and $\beta_2^* = 90°$, and hyperboloid crown wheel.

We can achieve the same result starting from Fig. 10.4b, and considering what happens when the distance of the axis of instantaneous screw motion, m (remember that it generates the pair of double pseudo-bevel hyperboloids of one sheet), from the axis of the pinion 1 is cancelled (such distance, in the general case, is equal to the radius of the throat). In this case, the hyperboloid 1 is transformed into a cone. In addition, if such instantaneous axis forms a right angle with the axis of the wheel 2, the hyperboloid 2 is transformed into a planar hyperboloid, which can be taken as a hyperboloid crown wheel.

The transmission ratio $i = \omega_1/\omega_2$ is equal to $\left(1/\sin\beta_1^*\right)$, while the sliding velocity at the points of contact line m (the Mozzi's axis), according to Eq. (10.8), will be equal to $v_r = \omega_2 a$. Equipping of appropriate sets of teeth the pitch surface thus defined, we can implement the transmission of motion between crossed axes with a gear pair whose members are, respectively, the first a straight bevel wheel and, the second, a crown wheel with curved teeth. It is however to be noted that such a gear pair cannot be used to transmit the motion between orthogonal crossed axes, since the shaft angle between these axes cannot be equal to $\pi/2$ (in this case should be $\beta_2^* = \pi/2$ and $\beta_1^* = 0$).

However, since the hyperboloid crown wheel defined above actually coincides with the crown wheel, the hypoid gear pairs can be generated using as generation wheel the hyperboloid crown wheel. According to this generation method, the pinion is generated by placing it in such a way that its axis is positioned with a crossed axis with respect to that of the generation hyperboloid crown wheel, while the wheel is generated by positioning it in such a way that its axis is coplanar with that of the generation hyperboloid crown wheel.

The reference pitch surfaces of the two members of the hypoid gear pair thus obtained are two conical surfaces, each of which has a common straight line with the generation hyperboloid crown wheel. The two straight lines do not coincide when the two-hypoid gear wheels are in contact meshing. In fact, the pitch cone of the hypoid pinion touches the pitch plane of the generation hyperboloid crown wheel along the straight line O_1A (see Fig. 12.16, where $\beta_2^* = \beta_2 = \pi/2$ and $\beta_1^* = \beta_1$), while the pitch cone of the hypoid wheel touches the pitch plane of the generation hyperboloid crown wheel along a radius.

Figure 12.17 shows schematically, in three orthogonal projections, a hypoid gear pair characterized by a shaft angle $\Sigma = 90°$, and having a gear ratio $u = z_2/z_1 = 1$. The shortest or center distance $a = O_1O_2$ (Fig. 12.17a), between the two crossed axes of wheels 1 and 2, is called *hypoid offset*. In all three projections, the boundary line of the hyperboloid crown wheel in its position relative to the two conical pitch

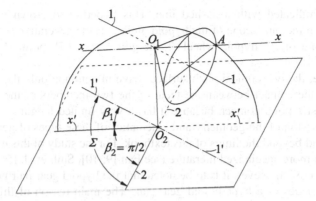

Fig. 12.16 Position of the pitch cone of the hypoid pinion with respect to the pitch plane of the generation hyperboloid crown wheel

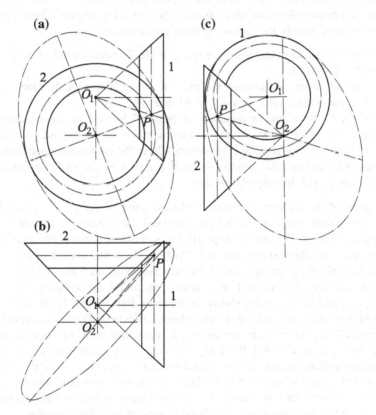

Fig. 12.17 Hypoid gear pair with $\Sigma = 90°$ and $u = 1$, in three orthogonal projections

surfaces is indicated with a dashed line. This hyperboloid crown wheel has in common with the pitch cone of the hypoid pinion, 1, the generatrix $O_1 P$, and with the pitch cone of the hypoid wheel, 2, the generatrix $O_2 P$ (point P is the mean point).

Therefore, the two conical pitch surfaces have in common only the point P. For this reason, the contact between the teeth of the two members of the hypoid gear pair is almost a point contact, because it has a very limited length.

The discussion of the geometry and kinematics of these types of gearing is very laborious, and beyond the limits of this textbook. For the study of this case, we refer the reader to more specialized literature (see Fan [9, 10], Shih et al. [56], Vimercati [68]). As of now, however, it is to be noted that the hypoid gear pair no longer has the characteristics of a hyperboloid gear pair. The main reason of this is that the location of the axodes is out of the meshing zone of hypoid gears, for which the concept of axodes of hypoid gears cannot be used for practical applications; it is used only for visualizing the relative velocity.

In this framework, the design of blanks of hypoid gears is aimed at determining the operating pitch cones instead of the hypoid gear axodes, the latter coinciding with the two hyperboloids of revolution of the hypoid gear pair. These operating pitch cones must satisfy the following main requirements:

- Their axes must form the design crossing angle, Σ, between the axes of rotation (usually, $\Sigma = 90°$).
- The hypoid offset, a, between the axes of the operating pitch cones must be equal to the design value of the hypoid gear set.
- These pitch cones are in tangency at point P located in the meshing zone between the active flank surfaces of two members of the hypoid gear pair.
- The sliding velocity at point P is directed along the common tangent to the two curves obtained as intersections of pitch cones with the tooth surfaces; these curves are called, improperly, helices.

The generation method of mating surfaces, corresponding to a given law of motion, for power transmission between crossed axes, deserves further attention with respect to what we said in Chap. 10. Let's consider a transmission drive with shaft angle Σ and shortest distance $a = O_1 O_2$, such as that shown in Fig. 12.18a. We obtain firstly the pitch surfaces of the gear pair, assuming that the transmission ratio is a constant. To this end, we assume, as coordinate systems, the systems $O_1(x_1, y_1, z_1)$ and $O_2(x_2, y_2, z_2)$ shown in the same Fig. 12.18a. In the first system, z_1 coincides with the axis of the driving wheel 1 (the pinion) and is directed as the corresponding angular velocity vector ω_1, x_1 coincides with the shortest distance straight line and is directed from O_1 to O_2, and y_1 completes the right-hand Cartesian coordinate system. In the second system, z_2 coincides with the axis of the driven wheel 2 (the wheel) and is directed as the corresponding angular velocity vector ω_2, x_2 coincides with the shortest distance straight line, but is directed from O_2 to O_1, and y_2 completes the right-hand Cartesian coordinate system.

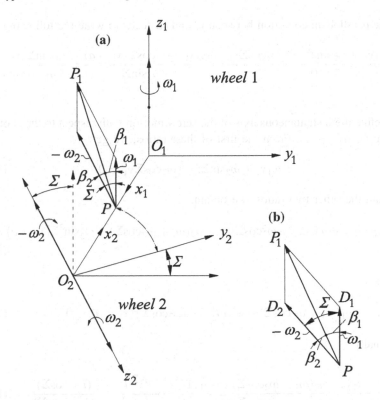

Fig. 12.18 **a** Axis of the instantaneous screw motion in the relative motion of a gear pair with crossed axes; **b** parallelogram of composition of vectors ω_1 and $-\omega_2$

During the relative motion of the pinion with respect to the wheel, the coordinate system $O_1(x_1, y_1, z_1)$ rotates about the z_2-axis with angular velocity $-\omega_2$, while the motion of member 1 with respect to member 2 of the gear pair is a screw motion which is obtained by composing the two angular velocity vectors ω_1 and $-\omega_2$. At any instant of time t, the axis of the instantaneous screw motion is defined by the condition that the relative velocity vector v_r at any point P is directed as the same instantaneous axis, i.e. as the instantaneous relative angular velocity vector $\omega_r = (\omega_1 - \omega_2)$. In the coordinate system $O_1(x_1, y_1, z_1)$, the axes of which have as unit vectors respectively (i_1, j_1, k_1), we can write $\omega_1 = \omega_1 k_1$ and $-\omega_2 = (-\omega_2 \sin\Sigma j_1 + \omega_2 \cos\Sigma k_1)$. Therefore, we get:

$$v_r = \omega_1 \times (P - O_1) - \omega_2 \times (P - O_1)$$
$$= \begin{vmatrix} i_1 & j_1 & k_1 \\ 0 & 0 & \omega_1 \\ x_1 & y_1 & z_1 \end{vmatrix} + \begin{vmatrix} i_1 & j_1 & k_1 \\ 0 & -\omega_2 \sin\Sigma & \omega_2 \sin\Sigma \\ (x_1 - a) & y_1 & z_1 \end{vmatrix} \qquad (12.50)$$

$$\omega_r = (\omega_1 - \omega_2) = -\omega_2 \sin\Sigma j_1 + (\omega_1 + \omega_2 \cos\Sigma) k_1. \qquad (12.51)$$

The parallelism condition between v_r and ω_r leads to write the following ratios:

$$\frac{-\omega_1 y_1 - \omega_2 \sin\Sigma z_1 - \omega_2 \cos\Sigma y_1}{0} = \frac{\omega_1 x_1 + \omega_2 \cos\Sigma (x_1 - a)}{-\omega_2 \sin\Sigma} = \frac{\omega_2 \sin\Sigma (x_1 - a)}{\omega_1 + \omega_2 \cos\Sigma},$$
$$(12.52)$$

that define the instantaneous axis of the screw motion with respect to the coordinate system $O_1(x_1, y_1, z_1)$. From the first of these ratios, we get:

$$\omega_1 y_1 + \omega_2 \sin\Sigma z_1 + \omega_2 \cos\Sigma y_1 = 0. \qquad (12.53)$$

From the other two ratios, we obtain:

$$\omega_1 x_1 (\omega_1 + \omega_2 \cos\Sigma) + \omega_2 \cos\Sigma (x_1 - a)(\omega_1 + \omega_2 \cos\Sigma) + \omega_2^2 \sin^2\Sigma (x_1 - a) = 0,$$
$$(12.54)$$

and then

$$\omega_1 (\omega_1 + \omega_2 \cos\Sigma) x_1 + (\omega_2^2 + \omega_1 \omega_2 \cos\Sigma)(x_1 - a) = 0, \qquad (12.55)$$

and finally

$$\frac{(a - x_1)}{x_1} = \frac{\omega_1 (\omega_1 + \omega_2 \cos\Sigma)}{\omega_2 (\omega_2 + \omega_1 \cos\Sigma)} = \frac{\omega_1}{\omega_2} \frac{\left(\frac{\omega_1}{\omega_2} + \cos\Sigma\right)}{\left(1 + \frac{\omega_1}{\omega_2}\cos\Sigma\right)} = i \frac{(i + \cos\Sigma)}{(1 + i\cos\Sigma)}. \qquad (12.56)$$

To determine the point C of intersection of the instantaneous axis of the screw motion with the plane (x_1, y_1), just put in Eq. (12.53) $z_1 = 0$, for which we get $y_1 = 0$. Therefore, we infer that the instantaneous axis of the screw motion intersects the straight line to which the center distance between the axes of the two wheels belongs. However, it is still necessary to determine the location of point C along the shortest distance $a = O_1 O_2$. To do this, from the parallelogram of composition of angular velocity vectors ω_1 and $-\omega_2$, we obtain firstly the relationship:

$$\frac{\omega_1}{\omega_2} = \frac{\sin\beta_2}{\sin\beta_1}, \qquad (12.57)$$

where β_1 and β_2 are the angles between the relative angular velocity vector ω_r and, respectively, the vectors ω_1 and $-\omega_2$ or, what is the same, z_1 and $-z_2$. Then, introducing Eq. (12.57) into Eq. (12.56), and considering that $\Sigma = (\beta_1 + \beta_2)$, we obtain:

$$\frac{(a - x_1)}{x_1} = \frac{\sin\beta_2 \left[\sin\beta_2 + \sin\beta_1 \cos(\beta_1 + \beta_2)\right]}{\sin\beta_1 \left[\sin\beta_1 + \sin\beta_2 \cos(\beta_1 + \beta_2)\right]}. \tag{12.58}$$

This equation can be write as follows, since we have $[\sin\beta_2 + \sin\beta_1 \cos(\beta_1 + \beta_2)]$ $= \cos\beta_1 \sin(\beta_1 + \beta_2)$ and $[\sin\beta_1 + \sin\beta_2 \cos(\beta_1 + \beta_2)] = \cos\beta_2 \sin(\beta_1 + \beta_2)$:

$$\frac{(a - x_1)}{x_1} = \frac{\sin\beta_2 \cos\beta_1 \sin(\beta_1 + \beta_2)}{\sin\beta_1 \cos\beta_2 \sin(\beta_1 + \beta_2)} = \frac{\tan\beta_2}{\tan\beta_1}. \tag{12.59}$$

Thus, we infer that the instantaneous axis of the screw motion divides the center distance $a = O_1 O_2$ between the axes of the two wheels in parts that are directly proportional to the tangents of angles between the instantaneous relative angular velocity vector ω_r and the axes of rotation of the two wheels. Bearing in mind that $\Sigma = (\beta_1 + \beta_2)$, from Eq. (12.57), we get:

$$i = \frac{\omega_1}{\omega_2} = \frac{\sin\beta_2}{\sin\beta_1} = \frac{\sin(\Sigma - \beta_1)}{\sin\beta_1} = \frac{\sin\beta_2}{\sin(\Sigma - \beta_2)}. \tag{12.60}$$

Therefore, if the transmission ratio $i = \omega_1/\omega_2$ is a constant, as we have assumed, also angles β_1 and β_2 are constants, and consequently the abscissa x_1 defined by Eq. (12.59) will be a constant. In conclusion, we infer that the pitch surface of the driving wheel 1 is a ruled surface, obtained as the locus of straight lines that form with the z_1-axis of the same wheel an angle β_1 that is constant, and pass through the points of a given circle in the plane $z_1 = 0$. This locus is the surface described by the straight line PP_1, crossed with respect to the z_1-axis, during the motion of rotation about this axis. Therefore, it is the round ruled hyperboloid having z_1 as its axis, the successive positions of the straight line PP_1 as generatrixes, and throat radius defined by Eq. (12.59). In a similar way, we deduce that the pitch surface of the driven wheel 2 is the round ruled hyperboloid having z_2 as its axis, the successive positions of the straight line PP_1 during the rotation about this axis as generatrices, and throat radius corresponding to Eq. (12.59).

Hypoid gears are obtained by taking, as members 1 and 2 of the gear pair, two cones, the axes of which coincide with the axes of the two wheels, and arranged so as to have a common point of tangency, P. The pitch plane of the generation crown wheel of both members of the gear pair is the plane, π, passing through the apices O_1 and O_2 of the two cones and point P. Therefore, it is tangent to both cones, which are the cutting pitch cones of the two wheels. The generation surface of the teeth, σ_0, is a conical surface rigidly connected to the fictitious generation crown wheel. The axis of this surface is perpendicular to plane π, while its directrix curve, i.e. its intersection with plane π, is a straight-line segment or curved arc as those shown in Fig. 12.3 or still other (Fig. 12.19).

While the two wheels 1 and 2 rotate about their axes with angular velocities ω_1 and ω_2, we move the generation crown wheel so that its motion relative to the

Fig. 12.19 Schematic
representation of generation
of a hypoid gear pair

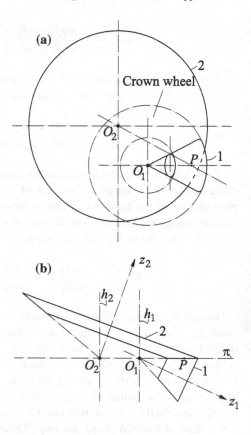

wheel 1 corresponds to rolling without sliding of plane π on the pitch cone of
wheel 1. Clearly, this relative motion corresponds to meshing of the wheel 1 with
the fictitious crown wheel, whose pitch plane is π. Therefore, in accordance with
the generation process described in Sect. 12.4, the absolute motion of the crown
wheel is a rotation about an axis perpendicular to plane π and passing through point
O_1, with angular velocity $\Omega = \omega_1 \sin\delta_1$, where δ_1 is the cone angle of wheel 1.

Consequently, the motion of the generation crown wheel relative to the wheel 2
is a motion composed by two rotations, the first about its axis, with angular velocity
Ω_2, and the second about the perpendicular to plane π passing through point O_1,
with angular velocity $\Omega = \omega_1 \sin\delta_1$. This motion has, as pitch surfaces, two double
ruled hyperboloids of one sheet, which are completely defined by the problem data,
as mentioned above. It follows that the surface σ_1, which is the envelope of successive positions of surface σ_0 in the motion of the generation crown wheel relative
to the wheel 1, coincides with the active surface of the teeth of hypoid pinion 1,
conjugated with the generation crown wheel. Furthermore, in accordance with the
properties described in Sect. 12.4 , the two surfaces σ_0 and σ_1 are touching, in every
instant during meshing, along points of a curved line, s_{10}, which is a characteristic
of the envelope.

The envelope of the same surface σ_0, in the motion of the generation crown wheel relative to the wheel 2, is instead another surface σ_2, which touches the generation surface σ_0 in points of a curved line, s_{20}, different from s_{10}, which can be determined analytically. The contact between the active surfaces σ_1 and σ_2 is therefore a point contact, in a point coinciding with the common point to the two characteristic curved lines s_{10} and s_{20}.

Generally, the larger wheel between the two members of the hypoid gear pair is the driven wheel; its cone angle δ_2 is thus somewhat greater than that of the other member of the gear pair, which is the driving pinion ($\delta_2 > \delta_1$). From the generation process shown in Fig. 12.19 and described above, it appears that the pinion 1 is a usual spiral bevel wheel. The hypoid wheel 2 is generated with the same generation crown wheel used for pinion, i.e. cutting its teeth with the same cutter used to cut the pinion teeth, with the difference however that the apex of wheel 2 has a different position compared to that corresponding to the apex of the wheel 1. Furthermore, the cutter feed motion is now corresponding to the simultaneous rolling and sliding of two round ruled hyperboloids, instead of the pure rolling of two cones of revolution.

The contact between the two members of the hypoid gear pair is a point contact. It is correct from a kinematic point of view only for a given position of the two members of the gear pair, and thus for given values of the shaft angle and shortest distance. Therefore, the hypoid gear pair requires a very accurate assembly. However, hypoid gears have the considerable advantages that we summarized in Sect. 12.1, which usually overshadow the limitations above. Due to these advantages, in many practical applications it is preferable to use hypoid gears even when it would possible to realize the transmission of motion with straight bevel and spiral bevel gears. The most practical applications of these gears are still those characterized by a shaft angle $\Sigma = 90°$.

Finally, it should be noted that the generation of the hypoid gear pair described above, related to the diagram shown in Fig. 12.19, is only an example to understand the generation process. We can consider the dual situation with respect to the example of Fig. 12.19, where the wheel 2 is a usual spiral bevel wheel, while the pinion is a hypoid wheel. This is done for the generation process of the hypoid gear pair made with the most of the Gleason gear cutting machines.

The method developed by Gleason for cutting hypoid gears pairs, which is the most used method, employs the hyperboloid crown wheel as a virtual generation cutter. The hypoid wheel is generated as a common spiral bevel wheel, i.e. by arranging the to-be-generated wheel and the generation hyperboloid crown wheel simulating the cutting tool in such a way that the apices of the two cutting pitch cones coincide. Instead, the pinion is generated by placing it in such a way that the axis of rotation of the generation hyperboloid crown wheel and that of the pinion to be generated are crossed axes. The tooth flank lines on the pitch plane of the generation hyperboloid crown wheel may be straight lines not converging on the axis (Fig. 10.5) or curved lines. Of course, it is also possible to cut hypoid gear pairs whose members are both hypoid gear wheels. In this case, both pinion and wheel are generated by placing their blanks in such a way that the axis of rotation of the generation hyperboloid crown wheel and those of to-be-generated pinion and wheel are crossed axes.

12.8 Approximate Analysis of Hypoid Gears

We describe here an approximate analysis of a hypoid gear, with which it is possible to determine some of its geometric and kinematic parameters. The approximation is because we replace the pitch ruled hyperboloids of screw motion with the pitch cones, and the generation hyperboloid crown wheel with a common crown wheel. This approximation, universally adopted, however leads to reliable results from the point of view of practical applications. Two analytical methods are described here, the first of more limited validity, and the second more general.

12.8.1 First Analytical Method

Let's consider the pitch cones of the two members 1 (the pinion) and 2 (the wheel) of a hypoid gear, that are tangent to the plane π, and assume that pinion and wheel are located above and, respectively, below the pitch plane π, as Fig. 12.19b shows. In this condition, $O_1 P = l_1$ and $O_2 P = l_2$ are the generatrixes of contact of the two pitch cones with plane π; they intersect each other at point P. We can appreciate the ability to transmit motion and torque, imagining that a fictitious generation crown wheel is associated with each of the two pitch cones, and precisely the crown wheel W_{01} conjugated with wheel 1, and the crown wheel W_{02} conjugated with wheel 2. The pitch plane of both crown wheels coincides with the plane π, while their axes of rotation are different. In fact, the crown wheel W_{01} has as its axis the straight line h_1, perpendicular to plane π and passing through point O_1, while the crown wheel W_{02} has as its axis the straight line h_2, perpendicular to plane π and passing through point O_2 (Fig. 12.19b).

We now rotate the wheel 1 about its axis, with angular velocity ω_1; the crown wheel W_{01} will rotate about its axis h_1 with angular velocity $\Omega_{01} = \omega_1 \sin\delta_1$. We then rotate the wheel 2 about its axis, with angular velocity ω_2; the crown wheel W_{02} will rotate about its axis h_2 with angular velocity $\Omega_{02} = \omega_2 \sin\delta_2$. If the crown wheel W_{02} is driven in its motion by the crown wheel W_{01}, to make possible the transmission of motion between the two crown wheels, it is necessary and sufficient that the relative motion of W_{01} compared to W_{02} induces, in the point of contact, velocity vectors which must be tangent to the mating surfaces touching at point P (see Ferrari and Romiti [11]). As point of contact, P, we consider the mean point.

Let u_1 and u_2 be the unit vectors of axes of the two wheels, directed in accordance with the respective angular velocity vectors, ω_1 and ω_2, as Fig. 12.20a shows. Let's consider the usual case in which u_1 and u_2, and thus the axes of the two wheels, are orthogonal ($\Sigma = 90°$). Then consider the Cartesian coordinate system $P(x_1, y_1, z_1)$ with origin at point P, x_1-axis coinciding with the generatrix of contact of the pitch cone of wheel 1 with plane π, y_1-axis in this plane and normal to x_1-axis, and z_1-axis normal to plane π; these axes are directed as Fig. 12.20a shows. From this figure, we can easily derive the following relationships, where $\zeta_{mp} < \pi/2$ is the angle between O_1P and O_2P (this is the *pinion offset angle in pitch plane*, according to ISO standards), while (i_1, j_1, k_1) are the unit vectors of x_1-, y_1-, and z_1-axes, respectively:

$$u_1 = -\cos\delta_1 i_1 - \sin\delta_1 k_1$$
$$u_2 = \cos\delta_2\cos\zeta_{mp} i_1 - \cos\delta_2\sin\zeta_{mp} j_1 - \sin\alpha_2 k_1. \tag{12.61}$$

Since we have assumed $\Sigma = 90°$, i.e. the axes of the two wheels are orthogonal, the dot product of unit vectors u_1 and u_2 must be equal to zero ($u_1 \cdot u_2 = 0$), for which, by imposing this condition, from Eqs. (12.61) we get:

$$-\cos\delta_1\cos\delta_2\cos\zeta_{mp} + \sin\delta_1\sin\delta_2 = 0, \tag{12.62}$$

and then

$$\cos\zeta_{mp} = \tan\delta_1\tan\delta_2. \tag{12.63}$$

Based on geometric elements indicated above, it is also possible to derive the center distance, a, in a simple way. Consider the v-plane (Fig. 12.20b), which contains the unit vector u_1 and the shortest distance straight line, and then the end points O_{c1} and O_{c2} of the center distance. From Fig. 12.20b, it becomes obvious

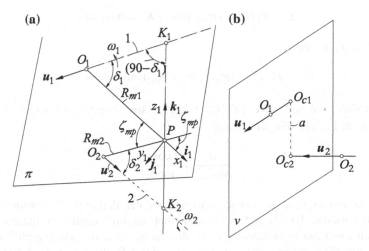

Fig. 12.20 **a** Axes of wheels 1 and 2, and generatrices of contact of their pitch cones with plane π; **b** v-plane through the axis of wheel 1 and shortest distance straight line

that $(-a)$ represents the bending arm of the unit vector $\boldsymbol{u}_2$ applied to point O_2 with respect to the line of action of the unit vector $\boldsymbol{u}_1$ passing through point O_1, so we can write the following equation, formed by a dot product of cross product:

$$-a = (O_2 - O_1) \times \boldsymbol{u}_2 \cdot \boldsymbol{u}_1, \qquad (12.64)$$

where $(O_2 - O_1)$ is the position vector of point O_2 with respect to point O_1. In the coordinate system $P(x_1, y_1, z_1)$, the position vector of points O_1 and O_2 with respect to point P are respectively: $(O_1 - P) = R_{m1}\boldsymbol{i}_1$ and $(O_2 - P) = R_{m2}\cos\zeta_{mp}\boldsymbol{i}_1 + R_{m2}\sin\zeta_{mp}\boldsymbol{j}_1$, where R_{m1} and R_{m2} are the *mean cone distances*. Therefore, from Eq. (12.64) we have:

$$
\begin{aligned}
a &= \begin{vmatrix} R_{m2}\cos\zeta_{mp} - R_{m1} & R_{m2}\sin\zeta_{mp} & 0 \\ \cos\delta_2\cos\zeta_{mp} & \cos\delta_2\sin\zeta_{mp} & \sin\delta_2 \\ \cos\delta_1 & 0 & -\sin\delta_1 \end{vmatrix} \\
&= (R_{m2}\cos\delta_1\sin\delta_2 + R_{m1}\sin\delta_1\cos\delta_2)\sin\zeta_{mp} \\
&= (R_{m1}\tan\delta_1 + R_{m2}\tan\delta_2)\cos\delta_1\cos\delta_2\sin\zeta_{mp}.
\end{aligned}
\qquad (12.65)
$$

Let K_1 and K_2 be the intersections of the normal to plane π through point P, which coincides with the z_1-axis, with axes of wheels 1 and 2 (Fig. 12.20a), and O_c^* be the point of intersection of axis of wheel 1 with the plane through the axis of wheel 2 and containing the shortest distance straight line (Fig. 12.21a). The distance L between points K_1 and O_c^* is called *taper ratio* from Gleason. It can be expressed simply by means of the geometric parameters previously considered, taking into account that $L = K_1 O_c^* = K_1 K_2 \sin\delta_1$. Since then $K_1 K_2 = (K_1 P + K_2 P) = (R_{m1}\tan\delta_1 + R_{m2}\tan\delta_2)$, as Fig. 12.20a shows, we get:

$$L = K_1 O_c^* = (R_{m1}\tan\delta_1 + R_{m2}\tan\delta_2)\sin\delta_1. \qquad (12.66)$$

Similarly, we get:

$$O_c^* K_2 = (R_{m1}\tan\delta_1 + R_{m2}\tan\delta_2)\sin\delta_2. \qquad (12.67)$$

Therefore, we will have $(K_1 O_c^* / O_c^* K_2) = (\sin\delta_1 / \sin\delta_2)$, for which, by comparing with Eq. (12.65), we obtain:

$$\frac{a}{\sin\zeta_{mp}} = \frac{L\cos\delta_2}{\tan\delta_1}. \qquad (12.68)$$

In addition to these geometric equations, we can derive the kinematic relationships that follow. To this end, we assume that the tooth depths are quite small, so that each tooth can be reduced to its tooth-axis on the corresponding pitch cone. If the mating teeth are in contact at the mean point P, the two curved lines representing the tooth-axes will be tangent to each other at point P. Consider two

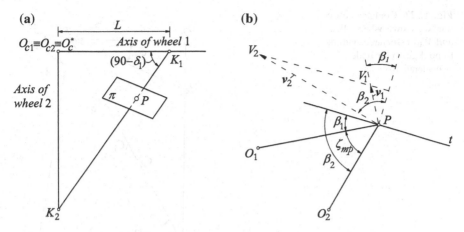

Fig. 12.21 **a** Taper ratio; **b** calculation diagram of the relative velocity at mean point P on the π-plane

differential elements ds_1 and ds_2 of these curved lines, which can be assumed as linear, as well as belonging to the plane π. Let t be the straight line of plane π, to which ds_1 and ds_2 belong.

We indicate with $v_2 = (V_2 - P)$ and $v_1 = (V_1 - P)$ the velocity vectors at point P of the wheels 1 and 2, corresponding to their angular velocity vectors ω_1 and ω_2 about the respective axes. The condition that the relative velocity vector v_r at point of contact P has component equal to zero in the direction of the common normal of contact between the mating surfaces involves the condition that this relative velocity vector is directed along the straight line t (Fig. 12.21b). Since the absolute values of two vectors, v_2 and v_1, are given respectively by $v_2 = PV_2 = \omega_2 R_{m2} \sin\delta_2$, and $v_1 = PV_1 = \omega_1 R_{m1} \sin\delta_1$, the following equality must necessarily be satisfied:

$$\omega_2 R_{m2} \sin\delta_2 \cos\beta_2 = \omega_1 R_{m1} \sin\delta_1 \cos\beta_1, \qquad (12.69)$$

where β_1 and β_2 are the spiral angles of the two wheels, i.e. the angles between the straight line t and, respectively, the straight lines $O_1 P$ and $O_2 P$.

Suppose now that each of the wheels 1 and 2 is in meshing, at point P, with the associated generation crown wheels, W_{01} and W_{02}, in the manner set out above. The angles β_1 and β_2 coincide with the spiral angles of the fictitious teeth of these generation crown wheels, while the tooth-axes of the same crown wheels, at the instant under consideration, are tangent to each other at point P. We denote these tooth-axes with λ_1 and λ_2.

To determine the centers of curvature of λ_1 and λ_2 at point P, we first observe that the motion of the crown wheel W_{02} relative to the crown wheel W_{01} is a planar motion, and that the center of the instantaneous rotation C of this motion, at the instant in which λ_1 and λ_2 are touching at point P, belongs to the extended straight line $O_1 O_2$ (Fig. 12.22). On the other hand, point C must also belong to the common

Fig. 12.22 Contacts between
the two crown wheels W_{01}
and W_{02} associated with the
hypoid gear pair under
consideration

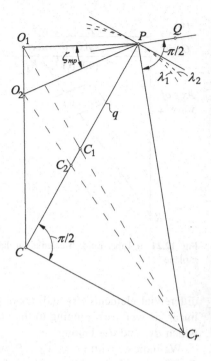

normal to λ_1 and λ_2, which are always tangent to each other, and therefore it
coincides with the point of intersection of these two straight lines.

After an infinitesimal time dt, the point of contact between the meshing teeth
moves from point P to point Q, which can be considered belonging to the π-plane.
Thus point Q will be the new tangency point between λ_1 and λ_2. For ease of
understanding, we materialize the common normal to λ_1 and λ_2 through point
P with a semi-infinite rod, q, whose end point on the finished side coincides at every
instant with the point of contact between λ_1 and λ_2. The instantaneous center of
motion of this rod on π-plane must be on the normal to path of this extreme point,
that is the normal to the straight line PQ through point P. It is then to be observed
that, since the angular velocities of rotation of the crown wheels W_{01} and W_{02} are
respectively $\Omega_1 = \omega_1 \sin\delta_1$ and $\Omega_2 = \omega_2 \sin\delta_2$, the point C, which is the instantaneous center of rotation of their relative motion, will have to satisfy the relationship:

$$\frac{CO_1}{CO_2} = \frac{\omega_2 \sin\delta_2}{\omega_1 \sin\delta_1}. \tag{12.70}$$

Since $i = (\omega_1/\omega_2) = const$, and positions of points O_1 and O_2 do not vary, from
this latter relationship we infer that also the position of point C on the π-plane
remains in a fixed position during the mutual engagement of the two crown wheels.
Therefore, the common normal to λ_1 and λ_2 must constantly pass through this point,

so that its motion can be considered, at every instant, as consisting of a rotation about point C, and a translation movement in the direction of axis of the afore-mentioned rod, i.e. in the direction of the straight line CP. It follows that the instantaneous center of this motion also belongs to the normal to the straight line CP passing through point C, for which it is the intersection, C_r, of the two straight lines just defined.

However, the motion of the rod q can be also regarded as composed of the motion of q with respect to the crown wheel W_{01}, which is a rotation about the center of curvature C_1 of λ_1 at point P, and the rotation in transfer motion with the crown wheel W_{01} about the normal to π-plane through point O_1. The center of curvature C_1 belongs simultaneously to the straight line PC, normal to λ_1 at point P, and to the straight line C_rO_1, for which it is the intersection of the two straight lines C_rO_1 and CP. In a similar way, we infer that the center of curvature C_2 of λ_2 is the intersection of the straight lines O_2C_r and CP.

According to the discussion above, it follows that, for a Gleason hypoid gear with face-milled teeth, the pinion is conjugate to the crown wheel, whose teeth have axes in the shape of arc of a circle, with radius C_1P. Instead, the wheel is conjugate to the crown wheel, whose teeth have axes in the shape of arc of a circle, with radius C_2P. All this, of course, within the framework of the approximations cor-responding to the assumptions indicated above.

By means of the previously established relationships, the following equations can be obtained. These new equations allow determining other basic geometric parameters of the hypoid gear pair, once the problem data are known. From Eq. (12.69) we obtain the following expression of the transmission ratio:

$$i = \frac{\omega_1}{\omega_2} = \frac{R_{m2}\sin\delta_2\cos\beta_2}{R_{m1}\sin\delta_1\cos\beta_1} = \frac{d_{m2}\cos\beta_2}{d_{m1}\cos\beta_1} = \frac{1}{K}\frac{d_{m2}}{d_{m1}}, \tag{12.71}$$

where

$$d_{m1} = 2R_{m1}\sin\delta_1 \quad d_{m2} = 2R_{m2}\sin\delta_2 \tag{12.72}$$

are the *mean pitch diameters* of the two gear members, while $K = (\cos\beta_1/\cos\beta_2)$ is a ratio whose value Gleason assumes within the range $(1.3 \div 1.5)$. Therefore, once the value of K was chosen, since the transmission ratio i is an input datum of the gear design, Eq. (12.71) allows to derive one of the two mean pitch diameters when the other is fixed. Correspondingly, Gleason recommended that the taper ratio was determined using the following relationship:

$$L = \frac{3}{8}d_{m1}\left(\frac{z_1}{z_2} + K\frac{z_2}{z_1}\right), \tag{12.73}$$

where the subscripts 1 and 2 indicate respectively the driving and driven members of the gear pair.

From Eqs. (12.66) and (12.72), we obtain:

$$\frac{d_{m1}}{\cos\delta_1} + \frac{d_{m2}}{\cos\delta_2} = \frac{2L}{\sin\delta_1}.$$ (12.74)

From Eqs. (12.65) and (12.66), we get:

$$\frac{L}{\sqrt{a^2+L^2}} = \frac{\sin\delta_1}{\sqrt{\sin^2\delta_1 + \cos^2\delta_1\cos^2\delta_2\sin^2\zeta_{mp}}};$$ (12.75)

since then, taking into account Eq. (12.63), it can be shown that $(\sin^2\delta_1 + \cos^2\delta_1\cos^2\delta_2\sin^2\zeta_{mp}) = \cos^2\delta_2$, Eq. (12.75) becomes:

$$\frac{L}{\sqrt{a^2+L^2}} = \frac{\sin\delta_1}{\cos\delta_2}.$$ (12.76)

Combining Eqs. (12.74) and (12.76), we get:

$$\frac{L}{\sqrt{a^2+L^2}} = \frac{2L}{d_{m2}} - \frac{d_{m1}}{d_{m2}}\tan\delta_1.$$ (12.77)

With this equation, we calculate δ_1, since the other quantities that appear in it are now known. By Eq. (12.76), we calculate δ_2. After that, in order to determine ζ_{mp}, R_{m1} and R_{m2}, just use Eqs. (12.63), (12.65) and (12.66). Finally, taking into account that $(\cos\beta_2/\cos\beta_1) = (1/K)$ and $(\beta_2 - \beta_1) = \zeta_{mp}$, we obtain:

$$\tan\beta_1 = \frac{\cos\zeta_{mp} - (1/K)}{\sin\zeta_{mp}},$$ (12.78)

from which we determine β_1 and then $\beta_2 = \beta_1 + \zeta_{mp}$.

12.8.2   Second Analytical Method

The aforementioned quantities as well as the other previously described quantities concerning the hypoid gear pair, characterized by a shaft angle $\Sigma = 90°$, can be obtained by following a more general procedure, due to Litvin [30]. It is to remember that Litvin, after Baxter [2], is the researcher who has most contributed with his followers to elaborating the theoretical basis for calculating hypoid gears (see Litvin et al. [34], Litvin and Gutman [33], Litvin et al. [31], Litvin [30], Litvin et al. [35], Litvin and Fuentes [32]). In the coordinate system $O_i(x_i, y_i, z_i)$, the parametric equations of the operating pitch cones of the two wheels, having pitch angles δ_i, are as follows (Fig. 12.23).

Fig. 12.23 Reference system, operating pitch cone and its geometry

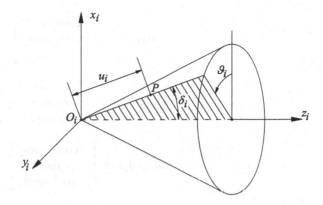

$$x_i = u_i\sin\delta_i\cos\vartheta_i$$
$$y_i = u_i\sin\delta_i\sin\vartheta_i \qquad\qquad (12.79)$$
$$z_i = u_i\cos\delta_i,$$

where u_i an ϑ_i are the Gaussian coordinates (see Sect. 11.5), while $i = (1,2)$. According to Eqs. (11.51) and (11.53), the surface normal N_i and unit normal n_i are given by the equations:

$$N_i = \frac{\partial r_i}{\partial u_i} \times \frac{\partial r_i}{\partial \vartheta_i} \quad n_i = \frac{N_i}{|N_i|}, \qquad\qquad (12.80)$$

where $r_i = [x_i, y_i, z_i]^T$ is the position vector represented in coordinate system $O_i(x_i, y_i, z_i)$. Provided that $u_i\sin\delta_i \neq 0$, from Eqs. (12.79) and (12.80) we get:

$$n_i = [\cos\delta_i\cos\vartheta_i \quad \cos\delta_i\sin\vartheta_i \quad -\sin\delta_i]^T. \qquad\qquad (12.81)$$

To define the equations of tangency of the operating pitch cones at mean point P, which is the pitch point, it is convenient to represent these pitch cones in a fixed coordinate system. To this end, we consider, in addition to the two coordinate systems, $O_1(x_1, y_1, z_1)$ and $O_2(x_2, y_2, z_2)$, rigidly connected to the two members 1 and 2 of the hypoid gear pair, also the coordinate system $O_0(x_0, y_0, z_0)$, rigidly connected to the frame. Figure 12.24 shows the position and orientation of the two systems $O_1(x_1, y_1, z_1)$ and $O_2(x_2, y_2, z_2)$, one with respect to the other, as well as that of these two systems with respect to the fixed system $O_0(x_0, y_0, z_0)$. By means of a coordinate transformation from $O_1(x_1, y_1, z_1)$ and $O_2(x_2, y_2, z_2)$ to $O_0(x_0, y_0, z_0)$, we obtain the following vector functions:

$$r_{1,0}(u_1, \vartheta_1) = \begin{bmatrix} r_1\cos\vartheta_1 \\ r_1\sin\vartheta_1 \\ r_1\cot\delta_1 - l_1 \end{bmatrix} \qquad\qquad (12.82)$$

$$n_{1,0}(\vartheta_1) = \begin{bmatrix} \cos\delta_1\cos\vartheta_1 \\ \cos\delta_1\sin\vartheta_1 \\ -\sin\delta_1 \end{bmatrix} \tag{12.83}$$

$$r_{2,0}(u_2,\vartheta_2) = \begin{bmatrix} -r_2\cos\vartheta_2 + a \\ -r_2\cot\delta_2 + l_2 \\ r_2\sin\vartheta_2 \end{bmatrix} \tag{12.84}$$

$$n_{2,0}(\vartheta_2) = \begin{bmatrix} -\cos\delta_2\cos\vartheta_2 \\ -\sin\delta_2 \\ -\cos\delta_2\sin\vartheta_2 \end{bmatrix}, \tag{12.85}$$

which express, in the fixed coordinate system $O_0(x_0, y_0, z_0)$, the pitch cones of pinion and wheel and their unit normals. It is to be noted that the first of the two subscript refers to the member 1 or 2 of the hypoid gear pair, while the second refers to the fixed reference system, $O_0(x_0, y_0, z_0)$. In these vector functions, the meaning of symbols already introduced remains unchanged. Instead, the new symbols represent:

- $r_i = u_i\sin\delta_i$ (with $i = 1, 2$), is the radius of the cross section at point P of each of the two pitch cones (being P the mean point, r_i is the mean pitch radius, that is $r_i = r_{mi} = d_{mi}/2$; the use of a single subscript is here preferred for the sake of brevity, but the indication with double subscript will be used in the final part of this section);
- l_i (with $i = 1, 2$), indicates the location of the apex of each of the two pitch cones (Fig. 12.24);
- a, is the shortest distance.

Fig. 12.24 Reference systems $O_1(x_1, y_1, z_1)$, $O_2(x_2, y_2, z_2)$, and $O_0(x_0, y_0, z_0)$, and pitch plane

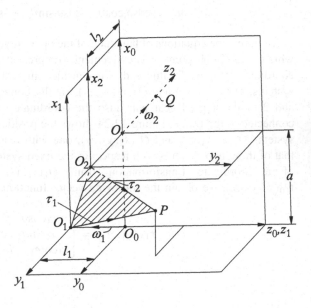

The equations of tangency of the operating pitch cones at the pitch point P are given by:

$$r_{1,0}(u_1, \vartheta_1) = r_{2,0}(u_2, \vartheta_2) = r_{P,0} \tag{12.86}$$

$$n_{1,0}(\vartheta_1) = n_{2,0}(\vartheta_2) = n_{P,0}, \tag{12.87}$$

where $r_{P,0}$ and $n_{P,0}$ are respectively the position vector and the common normal to the pitch cones at pitch point P. Since the mating pitch cones are located above and below the pitch plane (see also Fig. 12.19), their surface unit normals at point P have opposite directions. Therefore, the coincidence of the surface unit normal is obtained with a negative sign in Eq. (12.87). Vector Eqs. (12.86) and (12.87) give the following system of six scalar equations:

$$
\begin{aligned}
r_1\cos\vartheta_1 &= -r_2\cos\vartheta_2 + a = x_{P,0} \\
r_1\sin\vartheta_1 &= -r_2\cot\delta_2 + l_2 = y_{P,0} \\
r_1\cot\delta_1 - l_1 &= r_2\sin\vartheta_2 = z_{P,0} \\
\cos\delta_1\cos\vartheta_1 &= -\cos\delta_2\cos\vartheta_2 = n_{x,P,0} \\
\cos\delta_1\sin\vartheta_1 &= -\sin\delta_2 = n_{y,P,0} \\
-\sin\delta_1 &= -\cos\delta_2\sin\vartheta_2 = n_{z,P,0}.
\end{aligned}
\tag{12.88}
$$

It should be noted that only two of the last three equations of the system (12.88) are independent, because it must be $|n_{1,0}| = |n_{2,0}| = 1$.

Making some transformations and eliminating the trigonometric functions containing the variable ϑ_i, from Eqs. (12.88) we obtain the following relationships, which are the basis for design of hypoid pitch cones:

$$\frac{r_1}{r_2} = \frac{(a/r_2)\cos\delta_1}{\sqrt{\cos^2\delta_1 - \sin^2\delta_2}} - \frac{\cos\delta_1}{\cos\delta_2} \tag{12.89}$$

$$l_1 = -\frac{r_2}{\sin\delta_1\cos\delta_2} + \frac{a\cos\delta_1\cot\delta_1}{\sqrt{\cos^2\delta_1 - \sin^2\delta_2}} \tag{12.90}$$

$$l_2 = \frac{r_2}{\sin\delta_2\cos\delta_2} - \frac{a\sin\delta_2}{\sqrt{\cos^2\delta_1 - \sin^2\delta_2}} \tag{12.91}$$

$$x_{P,0} = a - \frac{r_2\sqrt{\cos^2\delta_1 - \sin^2\delta_2}}{\cos\delta_2} \tag{12.92}$$

$$y_{P,0} = r_2\tan\delta_2 - \frac{a\sin\delta_2}{\sqrt{\cos^2\delta_1 - \sin^2\delta_2}} \tag{12.93}$$

$$z_{P,0} = \frac{r_2\sin\delta_1}{\cos\delta_2} \tag{12.94}$$

$$\begin{cases} n_{P,x_0} = \sqrt{\cos^2\delta_2 - \sin^2\delta_1} = \sqrt{\cos^2\delta_1 - \sin^2\delta_2} \\ \qquad n_{P,y_0} = -\sin\delta_2 \\ \qquad n_{P,z_0} = -\sin\delta_1. \end{cases} \tag{12.95}$$

It is to be noted that, in Eqs. (12.95), two subscripts are used: the first indicates the point at which the normal refers (in this case, point P); the second, composed of a letter with numeric subscript, indicates the component of the unit normal in the direction of the corresponding axis of the coordinate system $O_0(x_0, y_0, z_0)$.

We indicate now with τ_1 and τ_2 the unit vectors of the generatrices of contact of the pitch cones with the pitch plane, which intersect each other at pitch point P, and are directed as Fig. 12.24 shows. Consider then the tooth traces λ_1 and λ_2 (Fig. 12.22) of the mating tooth flanks at point P on the pitch plane, i.e. the intersections of the tooth flank surfaces with the pitch plane. These tooth traces or flank lines, improperly called spirals or helices, are curves depending on the geometry of the teeth, and therefore on the cutting process used to obtain the toothing. Obviously, these flank lines have a common tangent at point P on the plane, π. The so-called mean spiral angles, $\beta_{m1} = \beta_1$ and $\beta_{m2} = \beta_2$, in the pitch plane are the angles between the common tangent to the tooth traces λ_1 and λ_2 and the generatrices of the respective pitch cones passing through point P (Fig. 12.25).

The pinion offset angle in pitch plane, ζ_{mp} (Fig. 12.25) is given by the following dot product:

$$\cos\zeta_{mp} = \tau_1 \cdot \tau_2. \tag{12.96}$$

Considering that the unit vectors τ_1 and τ_2 can be expressed by the following relationships:

$$\tau_1 = \frac{\overline{O_1P}}{|\overline{O_1P}|} = \frac{\partial r_{1.0}/\partial u_1}{|\partial r_{1.0}/\partial u_1|} = [\sin\delta_1\cos\vartheta_1 \quad \sin\delta_1\sin\vartheta_1 \quad \cos\delta_1]^T$$

$$\tau_2 = \frac{\overline{O_2P}}{|\overline{O_2P}|} = \frac{\partial r_{2.0}/\partial u_2}{|\partial r_{2.0}/\partial u_2|} = [\sin\delta_2\cos\vartheta_2 \quad -\sin\delta_2 \quad \sin\delta_2\sin\vartheta_2]^T. \tag{12.97}$$

From Eqs. (12.96) and (12.97), we obtain:

$$\cos\zeta_{mp} = \cos(\beta_{m1} - \beta_{m2}) = \cos(\beta_1 - \beta_2) = \tan\delta_1\tan\delta_2. \tag{12.98}$$

We have thus obtained the same Eq. (12.63), previously deduced otherwise.

The relative velocity vector or sliding velocity vector of the pinion with respect to the wheel at point P (Fig. 12.25) can be expressed as (see also Eq. 11.85):

$$v_r = v_{1,2} = v_1 - v_2 = [(\omega_1 - \omega_2) \times r_P] - (a \times \omega_2), \tag{12.99}$$

where $r_P = \overline{O_0P}$ is the position vector of pitch point P in the coordinate system $O_0(x_0, y_0, z_0)$, and a is the position vector of an arbitrary point Q of the line of

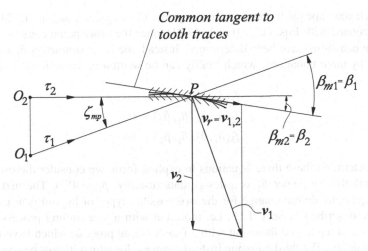

Fig. 12.25 Pinion offset angle in pitch plane, mean spiral angles, and relative velocity

action of vector ω_2 with respect to the origin O_0 of the same coordinate system, i.e.
$\overline{O_0 Q}$ (Fig. 12.24). The angular velocity vector ω_1 is a vector passing through the
origin of coordinate system $O_0(x_0, y_0, z_0)$. For point P, vectors v_1 and v_2 lie in the
pitch plane, and are perpendicular to the generatrices of contact of the pitch cones
with this plane. By means of some transformations (see Sect. 11.7), from
Eq. (12.99) we obtain:

$$v_{1,2} = -\omega_1 r_1 (\tan\beta_1 - \tan\beta_2)\cos\beta_1 \begin{bmatrix} 0 \\ \sin\beta_1 \\ \cos\beta_1 \end{bmatrix} \qquad (12.100)$$

$$i = \frac{\omega_1}{\omega_2} = \frac{r_2\cos\beta_2}{r_1\cos\beta_1} = \frac{z_2}{z_1}. \qquad (12.101)$$

The relative velocity vector $v_r = v_{1,2}$ given by Eq. (12.100) is related to the
coordinate system $O_e(e_1, e_2, e_3)$, where e_1 is the unit vector normal to the pitch
plane, $e_3 = \tau_1$ is the unit vector corresponding to the generatrix of contact of the
pinion pitch cone with the pitch plane, and $e_2 = e_3 \times e_1$. From Eqs. (12.100) and
(12.101) we get:

$$\tan\beta_1 = \frac{ir_1 - r_2\cos\zeta_{mp}}{r_2\sin\zeta_{mp}} \qquad (12.102)$$

$$\tan\beta_2 = \frac{r_1\cos\zeta_{mp} - (r_2/i)}{r_1\sin\zeta_{mp}}. \qquad (12.103)$$

The quantities β_i, δ_i and l_i (width $i = 1, 2$) are the basic design parameters of the
hypoid gear pair under consideration. Parameters l_i, which determine the location of

the pitch cone apexes in the coordinate system $O_0(x_0, y_0, z_0)$, as Fig. 12.24 shows, are calculated with Eqs. (12.90) and (12.91), once the other parameters that appear in these equations have been determined. Instead, the four parameters β_i and δ_i are related by three equations, which briefly can be written in the following form:

$$
\begin{aligned}
f_1(\delta_1, \delta_2, \beta_1) &= 0 \\
f_2(\delta_1, \delta_2, \beta_1, \beta_2) &= 0 \\
f_3(\delta_1, \delta_2, \beta_1, \beta_2) &= 0.
\end{aligned}
\tag{12.104}
$$

To determine these three equations in explicit form, we consider the usual case for which the parameter β_1 is a design data (usually, $\beta_1 = 45°$). The first two of these equations do not change for the two possible types of hypoid gear pairs that we have described in Sect. 12.3, i.e. those cut with a face-milling process, which have tapered teeth, and those cut with a face-hobbing process, which have teeth of uniform depth. The third equation instead changes, for which it must be specifically derived for each of these two types of hypoid gear pairs. This is because the face-milled tapered teeth are generated by a surface (the cone surface of the head-cutter), while the face-hobbed teeth having uniform depth are generated by a straight-line segment materialized by the blade edge.

For derivation of the first two Eq. (12.104) in explicit form, first we obtain from Eqs. (12.89) and (12.101) the following relationship:

$$
\frac{(a/r_2)\cos\delta_1}{\sqrt{\cos^2\delta_1 - \sin^2\delta_2}} - \frac{\cos\delta_1}{\cos\delta_2} = \frac{z_1\cos\beta_2}{z_2\cos\beta_1},
\tag{12.105}
$$

which, in compact form, can be written as follows:

$$
\cos\beta_2 = \frac{\cos\beta_1}{p},
\tag{12.106}
$$

where p is a parameter that collects all the other quantities appearing in Eq. (12.105), and therefore given by the following relationship:

$$
p = \frac{z_1\cos\delta_2\sqrt{\cos^2\delta_1 - \sin^2\delta_2}}{z_2\cos\delta_1\left[(a/r_2)\cos\delta_2 - \sqrt{\cos^2\delta_1 - \sin^2\delta_2}\right]}.
\tag{12.107}
$$

Then, from Eq. (12.98), we get:

$$
\cos(\beta_1 - \beta_2) = \cos\beta_1\cos\beta_2 + \sin\beta_1\sin\beta_2 = \tan\delta_1\tan\delta_2 = q.
\tag{12.108}
$$

From Eqs. (12.106) and (12.108), we obtain:

$$\left(\frac{\cos^2\beta_1}{p} - q\right)^2 = (-\sin\beta_1\sin\beta_2)^2 = (1 - \cos^2\beta_1)(1 - \cos^2\beta_2)$$

$$= (1 - \cos^2\beta_1)\left(1 - \frac{\cos^2\beta_1}{p}\right), \tag{12.109}$$

which, after further processing, gives:

$$f_1(\delta_1, \delta_2, \beta_1) = \cos^2\beta_1 - \frac{(1 - q^2)b^2}{(1 + p^2 - 2pq)} = 0. \tag{12.110}$$

Thus the first of Eqs. (12.104) has been obtained. In fact, q is a function of δ_1 and δ_2 (see Eq. 12.108), but also p is a function of δ_1 and δ_2, because the quantities z_1, z_2, a and r_2 appearing in Eq. (12.107) are considered as known.

The second of Eqs. (12.104) written in explicit form has been already obtained. In fact, it consists of the Eq. (12.98), which can be written as follows:

$$f_2(\delta_1, \delta_2, \beta_1, \beta_2) = \cos(\beta_1 - \beta_2) - \tan\delta_1\tan\delta_2 = 0. \tag{12.111}$$

The derivation of the third of Eqs. (12.104) for hypoid gear pairs with face-milled teeth is based on the *limit contact normal*, introduced for the first time by [73], and applied by Gleason to design of these types of gears. With reference to the Wildhaber's concept of limit contact normal and associated *limit pressure angle*, we refer the reader directly to the Wildhaber's paper [73]. However, it is worth noting that, in the Gleason approach, by a procedure on which it is not the place to dwell, it is possible to express the limit pressure angle α_{lim}, i.e. the angle between the limit contact normal to the hypoid wheel tooth surface at point P and the pitch plane, by the following relationship:

$$\tan\alpha_{lim} = \frac{\frac{r_2}{\sin\delta_2}\sin\beta_2 - \frac{r_1}{\sin\delta_1}\sin\beta_1}{\frac{r_2}{\sin\delta_2} + \frac{r_1}{\sin\delta_1}}. \tag{12.112}$$

From this equation, negative values of the normal pressure angle can be derived $(\alpha_{lim} < 0)$, which have no physical meaning. It follows that the limit pressure angle, α_{lim}, can assume the value $\alpha_{lim} = 0$ as the minimum value (see also: Litvin [30]; Litvin and Fuentes [32]).

Figure 12.26 shows both tooth profiles of the hypoid wheel, in the normal section passing through the pitch point P, which are not symmetrical with respect to the normal to the pitch plane passing through the same pitch point P. In this figure, n is the unit vector of the limit contact normal, while n_1 and n_2 are the unit normals to the concave and convex tooth sides, whose lines of action form with the line of action of n the same angle. It follows that the normal pressure angles α_{n1} and α_{n2} of the corresponding profiles must satisfy the following relationship:

Fig. 12.26 Limit unit normal
and surface unit normals to
the concave and convex tooth
sides of the hypoid wheel

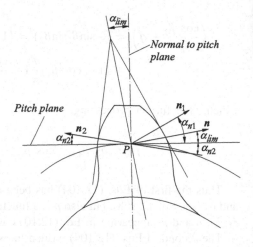

$$\alpha_{n1} - |\alpha_{lim}| = \alpha_{n2} + |\alpha_{lim}|. \tag{12.113}$$

It is to be noted that α_{n1} is different from α_{n2}, and in accordance with the Gleason approach, the pressure angle α_{n1} on the concave side is larger than the pressure angle α_{n2} on the convex side, i.e. $\alpha_{n1} = \alpha_{n2} + 2|\alpha_{lim}|$.

Another equation can be derived that correlates the limit pressure angle with the design parameters of the pitch cones, and is independent from the previous one. The derivation of this equation is based on the consideration that a hypoid gear pair with face-milled teeth (and then obtained by means of a formate-cut process) consists of a generated pinion (which does not interest here), and a non-generated wheel whose tooth flank surfaces coincide with the head-cutter surface. In accordance with what we said in Sect. 12.3 , the intersection of the wheel tooth surface with the pitch plane is an arc of circle having radius r_{c0}, which coincides with the mean radius of the head-cutter; it is given by the following relationship:

$$r_{c0} = \frac{\tan\beta_1 - \tan\beta_2}{\dfrac{\sin\delta_1}{r_1\cos\beta_1} - \dfrac{\sin\delta_2}{r_2\cos\beta_2} - \left(\dfrac{\tan\beta_1\cos\delta_1}{r_1} + \dfrac{\tan\beta_2\cos\delta_2}{r_2}\right)\tan\alpha_{lim}}. \tag{12.114}$$

The Eqs. (12.112) and (12.114) taken together make it possible to obtain the explicit form of the requested third of Eqs. (12.104), valid for hypoid gear pairs with face-milled teeth.

Instead, the derivation of the third of Eqs. (12.104) for hypoid gear pairs with face-hobbed teeth is based on the specific location of the head-cutter for the generation of these types of gears. Figure 12.27 shows the generatrices of contact O_1P and O_2P of the two pitch cones with the pitch plane (the plane passing through points O_1, O_2, and P), the pitch circles that are tangent each other at point P, and the unit vectors $\tau_1 = \overline{O_1P}$ and $\tau_2 = \overline{O_2P}$ of the pitch cone generatrices, whose lines of action converge at point P. We find now the point C of intersection of the

Fig. 12.27 a Location of the head-cutter axis in the pitch plane, and instantaneous center of rotation in its relative motion with respect to the crown wheel; **b** diagram for derivation of Eq. (12.119)

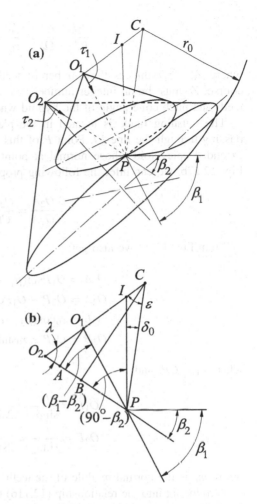

head-cutter with the pitch plane, assuming that it belongs to the extended line O_1O_2, in accordance with the set-up of the gear cutting machine, comprised of head-cutter. We denote then with z_0 the number of finishing blade groups of this head-cutter.

Let's consider now the imaginary generation crown wheel, simultaneously in meshing with pinion and wheel of the hypoid gear pair, whose axode is the aforementioned pitch plane, which can be regarded as a circular cone. Assume also that, while the head-cutter rotates about its axis with angular velocity ω_0, the generation crown wheel rotates about its axis with the angular velocity Ω_0, remembering that both these axes are perpendicular to the pitch plane, and pass respectively through points C and O_2. If I is the instantaneous center of rotation of the head-cutter in its relative motion with respect to the generation crown wheel, its position on the extended line O_1O_2 will be given by the following relationship:

$$\frac{O_2 I}{IC} = \frac{\omega_0}{\Omega_0} = \frac{Z_0}{z_0} = \frac{z_2}{z_0 \sin\delta_2}, \tag{12.115}$$

where $Z_0 = z_2/\sin\delta_2$ is the number of teeth of the imaginary crown wheel (of course, Z_0 must be an integer number); z_2 and δ_2 are respectively the number of teeth and pitch cone angle of the hypoid wheel.

The finishing blade is located in the plane perpendicular to the pitch plane, passing through the line PI. Point P of this blade generates in the pitch plane an extended epicycloid whose normal at point P coincides with the line PI. From Fig. 12.27b, we can infer the following proportion:

$$\frac{O_2 A}{O_1 A} = \frac{O_2 B}{CB}. \tag{12.116}$$

From Fig. 12.27 we also get:

$$\begin{aligned} O_1 A &= O_1 P \sin(\beta_1 - \beta_2) \\ O_2 A &= O_2 P - O_1 P \cos(\beta_1 - \beta_2) \\ CB &= r_0 \cos(\beta_2 - \delta_0) \\ O_2 B &= O_2 P - r_0 \sin(\beta_2 - \delta_0), \end{aligned} \tag{12.117}$$

where $r_0 = CP$, and

$$\begin{aligned} O_1 P &= \frac{r_1}{\sin\delta_1} = \frac{z_1 m_n}{2\sin\delta_1 \cos\beta_1} \\ O_2 P &= \frac{r_2}{\sin\delta_2} = \frac{z_2 m_n}{2\sin\delta_2 \cos\beta_2}, \end{aligned} \tag{12.118}$$

where m_n is the normal module of the teeth.

Introducing into the relationship (12.116) the Eqs. (12.117), taking into account Eqs. (12.118), we obtain the following explicit form of the requested third of Eqs. (12.104), valid for hypoid gear pairs with face-hobbed teeth:

$$\frac{r_0 \cos(\beta_2 - \delta_0)}{r_2 - r_0 \sin\delta_2 \sin(\beta_2 - \delta_0)} - \frac{z_1 \cos\beta_2 \sin(\beta_1 - \beta_2)}{z_2 \sin\delta_1 \cos\beta_1 - z_1 \sin\delta_2 \cos\beta_2 \cos(\beta_1 - \beta_2)} = 0, \tag{12.119}$$

where the angle $\delta_0 = \widehat{IPC}$ is given by (see Fig. 12.27b):

$$\sin\delta_0 = \frac{z_0 r_2 \cos\beta_2}{z_2 r_0}. \tag{12.120}$$

The derivation of this last equation is based on the following three relationships, as inferred from Fig. 12.27b:

$$\frac{CI}{PI} = \frac{\sin\delta_0}{\sin\varepsilon}; \quad \frac{O_2 I}{PI} = \frac{\cos\beta_2}{\sin\lambda}; \quad \frac{O_2 P}{CP} = \frac{\sin\varepsilon}{\sin\lambda}, \tag{12.121}$$

where ε and λ are respectively the angles $\widehat{O_2 CP}$ and $\widehat{CO_2 P}$.

Once the formulation of the three Eqs. (12.104) in explicit form is completed for both types of hypoid gear pairs (those with face-milled gears, and those for face-hobbed gears), it is possible to make the determination of the three quantities δ_1, δ_2 and β_2 (note yet that the parameter β_1 is considered as a design datum). Since we have to solve a system of three nonlinear equations in both cases, the computational procedure to be used is *a fortiori* an iterative procedure. The input data to implement this procedure are β_1, r_2, a, z_1 and z_2, for the case of face-milled gears. For the case of face-hobbed gears, z_0 is to be added to these input data. In the case where the pitch outside radius, $r_{e2} = d_{e2}/2$, and face width, b, are given instead of the mean pitch radius $r_2 = r_{m2} = d_{m2}/2$, the following relationship between r_{e2}, b, and r_{m2} must be used at each iteration:

$$r_2 = r_{m2} = r_{e2} - \frac{b\sin\delta_2}{2}. \tag{12.122}$$

About the iterative procedure of solution, at each iteration we can consider the equations represented in echelon form, so that they can be solved separately in the case in which one of the unknowns (for example, δ_2) is regarded as given. Then the third nonlinear equation is used for checking of the iterative procedure. The determination of the three unknowns δ_1, δ_2 and β_2 the appear in the equation system (12.104), written in explicit form, involves the use of computer-aided programs, with appropriate subroutines for the numerical solution of nonlinear equations. However, when these subroutines are used, calculations must be completed by requiring that the following requirements be met:

$$\tan\delta_1 \tan\delta_2 < 1; \quad \cos^2\delta_1 - \sin^2\delta_2 > 0. \tag{12.123}$$

To begin the iteration procedure, it is recommended to choose an initial value of δ_2 such that $\delta_2 < \tan^{-1}(z_1/z_2)$.

12.9 Main Characteristics of the Hypoid Gears, and Some Indications of Design Choices

In the two previous sections, we have seen that the hypoid gears are gear wheel pairs of conical or approximately conical shape, which mesh with their axes crossed and offset. We can say that the hypoid gears are special spiral bevel gears that transmit motion and torque between crossed axes, arranged with a given offset each other. Therefore, the axis of the pinion does not intersect the axis of the wheel, and has a hypoid offset, a, with respect to it. Generally, this offset has a not high value.

With a suitable value of this offset, it is possible to have the axis of the pinion below that of the wheel. Thus, it becomes possible to arrange the bearings straddling the toothed part. Typical advantages of this relative position between the two axes that transmit the motion are manifold.

The first great advantage of the hypoid gear pairs is that they allow us to have crossed axes of the two gear members, and therefore to support their shafts with bearings arranged at the two sides of the same gear wheels. Thus, the cantilever mounting of the pinion, inevitable for gear pairs with limited space, is avoided.

The second great advantage of these gears is related to the fact that the shaft of the pinion, which is the driving member, occupies a lower position with respect to the shaft of the wheel. This relative position of the two members of the gear pair is adopted by many automobile manufacturers, and used in the differential for the rear axle drive of the cars (see Powell and Barton [46], Pollone [44], Lechner and Naunheimer [25], Naunheimer et al. [41]). This arrangement with a lowered drive shaft, and consequently lowered transmission shaft, allows in fact having a lower-floored body of the vehicle. It follows a lowering of the center of gravity of the transmission system and, therefore, of that of the whole vehicle, and thus a greater stability of the same vehicle.

A third advantage achievable with hypoid gears, which are characterized by sufficient offset between the two shafts, concerns the possibility of supplying power to several machines, using a single input shaft on which several hypoid pinions are mounted. This advantageous and useful feature is used in various industrial applications (e.g., textile machines and plants, machine tools, etc.).

However, the kinematics of the hypoid gear pair is characterized by sliding, not only in the direction of the tooth depth, but also along the tooth trace, and the latter sliding increases with the increasing of the offset. Therefore, the fourth advantage of the silent operating conditions, typical of this gear and related to the longitudinal or lengthwise sliding, is associated with other significant disadvantages, such as lower efficiency, more high wear, and higher operating temperatures, with consequent risk of scuffing (see Coleman [6], Kolivand et al. [23], Stadtfeld [62]).

At first sight, the hypoid gears can be confused with the spiral bevel gears, as the shape of their teeth is very similar to that of the latter. A substantial difference exists however between the two types of gears: in fact, the axodes, i.e. the pitch surfaces, are two cones in the spiral bevel gears, while they are two hyperboloids of revolution in hypoid gears. Hypoid gears transmit roughly the same power of spiral bevel gears, having equal nominal pitch diameters and equal transmission ratio.

The *hypoid offset* $a = O_1 O_2$ between the axes of the two members of a hypoid gear pair may reach 30% of the mean pitch diameter of the hypoid wheel, d_{m2}, when the gear ratio is equal to $1(u = 1)$, by must not exceed 20% of the same mean pitch diameter if $u > 3$. For gear ratios included in the range $(1 \leq u \leq 3)$, the hypoid offset can be made to vary linearly between the two aforementioned values. Other limitations exist in regard to the magnitude of the hypoid offset. Indeed, it should

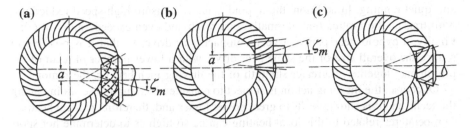

Fig. 12.28 Hypoid gear pairs with: **a** positive hypoid offset; **b** negative hypoid offset; **c** without hypoid offset (spiral bevel gear pair)

not exceed 40% of the outer cone distance, R_{e2}, of the hypoid wheel and, for heavy duty equipment, it should be nearer 20% of the same outer cone distance.

The hypoid offset can be positive or negative, depending on whether the axis of the hypoid pinion is placed below or above the axis of the hypoid wheel. Figure 12.28 shows three hypoid gear pairs, respectively with positive hypoid offset (Fig. 12.28a), negative hypoid offset (Fig. 12.28b), and without hypoid offset (Fig. 12.28c). In this latter case, the hypoid gear becomes a normal spiral bevel gear, which can therefore be considered a special case of the more general family of hypoid gears (see also Niemann and Winter [42]). Figures 12.28a and 12.28b also highlight the pinion offset angle in axial plane of the same pinion, ζ_m (see also Fig. 12.36b).

With a positive hypoid offset, the hypoid pinion has a larger *mean spiral angle*, β_{m1}, than that of the hypoid wheel, for which $\beta_{m2} < \beta_{m1}$. With reference to the *mean cones*, we have in fact $\beta_{m1} = (\beta_{m2} + \zeta_{mp})$, where ζ_{mp} is the *pinion offset angle in pitch plane*, i.e. measured in the pitch plane of the common generation hyperboloid *crown wheel*. With the same *mean pitch diameter*, d_{m2}, of the hypoid wheel and the same transmission ratio, i, the mean pitch diameter, d_{m1}, and pitch angle, δ_1, of the hypoid pinion are larger than the values that define the corresponding spiral bevel gear pair. Also, the *outer transverse module*, m_{et}, will be increased, as a result of the increase of β_{m1}. At the same time, the overlap ratio increases, and so does the axial force with respect to a spiral bevel gear pair, i.e. a gear pair without offset. The greater diameter of the pinion also makes possible an increase of the diameter of its shaft. This is the preferred solution in the automotive industry. In borderline cases, $\beta_{m1} = 0$ (straight bevel and Zerol bevel hypoid pinion), and $\beta_{m2} = 0$ (straight bevel or Zerol bevel hypoid wheel).

Conversely, with a negative hypoid offset, the hypoid pinion has a smaller mean spiral angle than that of the hypoid wheel: thus $\beta_{m1} = \beta_{m2} - \zeta_{mp}$. With the same mean pitch diameter of the hypoid wheel and the same transmission ratio, the mean pitch diameter, pitch angle, overlap ratio, and axial force are smaller than the values that define the corresponding spiral bevel gear pair without offset. In the extreme case, the hypoid pinion assumes a cylindrical shape, becoming a cylindrical pinion.

Besides the advantages mentioned above, the hypoid gears are more suitable than the ordinary spiral bevel gears in industrial applications that require a smooth and quiet running. In addition, the hypoid gears can ensure high-speed reductions, with transmission ratios that, normally, can reach and even exceed values of 60:1. Finally, a hypoid gear pair is a very compact gear drive, so that it is possible to reduce the overall size of the same gear drive, with a lower number of teeth of the pinion and, together, a greater strength of the driving member, i.e. the pinion.

Of course, the meshing action is subject to considerable lengthwise sliding along the teeth, and this sliding leads to great power losses and, therefore, to local heating. Temperatures related to this local heating can be so high as to determine hot spots and consequent micro-welding. To inhibit the tendency to the generation of these local micro-welding, as well as to withstand the high pressure of contact, anti-scuff lubricants (the so-called *EP*, Extreme Pressure) are needed.

Finally, it is to emphasize the need to use bearings able to withstand the axial thrusts, and an adequate lubrication circuit, which also has the task of dissipating the heat generated due to the high power losses. In the hypoid gears, these power losses, and the consequent efficiency, are highly dependent on many parameters, such as the hypoid offset, mean spiral angles, load in the design operating conditions, finishing of the tooth flank surfaces, rotational speed, characteristics of the lubricant, etc.

Only the study and most recent analyses have shown the great influence of the load exerted on the hypoid gear efficiency. This predominant influence as well as that exerted by the coefficient of friction have led to the proposal of approximate empirical formulas, other than those previously developed and proposed, which were characterized by the specific parameters of the tooth geometry. The disappearance of these parameters in the more recent formulas is due to the fact that many of these geometric parameters have an influence entirely negligible in front of the coarse approximation that inevitably characterizes the evaluation of the coefficient of friction, μ.

To improve the efficiency of the hypoid gears, as well as that of the spiral bevel gears, now the margins are not very high, since the manufacturing technologies and assembly techniques of these gears are so advanced as not to allow further large gains. From the design point of view, it is to be remembered that the efficiency of the hypoid gears increases, reducing the hypoid offset, pressure angle, and the difference between the mean spiral angles, increasing the operating load and rotational speed, improving the finishing quality of the tooth flank surfaces, and using the best lubricants.

One of the approximate empirical formulas, mainly used for the calculation of the hypoid gear efficiency, is that developed and proposed by Gleason; we can write this formula in the following form:

$$\eta = \frac{100}{1 + \sqrt{\frac{T_{max}}{T}} \left[\mu \sec\alpha_{eD} \sqrt{(\tan\beta_{m1} - \tan\beta_{m2})^2 + k_1} + k_2 \right]}, \qquad (12.124)$$

where: T_{max} is the *maximum torque*, to be taken equal to 2.75 times the torque T_{Flim} corresponding to the nominal stress number for bending σ_{Flim} $(T_{max} = 2.75T_{flim})$; T is the operating torque; μ is the coefficient of friction; α_{eD} is the effective pressure angle on drive side of the driving member of the hypoid gear pair; $\sec\alpha_{eD} = (1/\cos\alpha_{eD})$; β_{m1} and β_{m2} are respectively the mean spiral angles of the pinion and wheel; k_1 and k_2 are two numerical dimensionless factors that take into account the profile sliding losses and, respectively, the bearing losses. The expression $(\tan\beta_{m1} - \tan\beta_{m2})^2$ appearing in this formula takes into account the lengthwise sliding losses.

To take into account the profile sliding losses, Gleason recommended to take $k_1 = 2.25 \times 10^{-2}$. For the most frequent case in which the shafts of pinion and wheel are supported by anti-friction journal bearings, Gleason recommended to take $k_2 = 0.01$. When other types of bearings are used, this value of k_2 must be increased, determining it preferably by means of suitable experimental measurements. The coefficient of friction, μ, varies mainly depending on the lubricant used and its viscosity, rolling and sliding velocities between the tooth mating surfaces, finishing quality of these surfaces, and transmitted load. The experimental evidence in this regard shows that it is included in the range $(0.01 < \mu < 0.12)$. Usually, a value $\mu = 0.05$ is considered as a reasonable average value to be taken in the efficiency calculations.

The curves shown in Fig. 12.29 allow us to immediately see the great influence exerted by the ratios (a/d_{m2}) and (T/T_{max}) on the efficiency of the spiral bevel gears and hypoid gears. These curves were obtained with $\alpha_{eD} = 20°$, $\mu = 0.05$, and with certain values of β_{m1} and β_{m2} (of course, $\beta_{m1} = \beta_{m2}$ for spiral bevel gears), considered to be representative of mean operating conditions. The same curves

Fig. 12.29 Efficiency of hypoid and spiral bevel gears (for $\alpha_{eD} = 20°$; $\mu = 0.05$; and given values of β_{m1} and β_{m2})

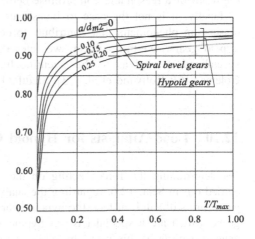

show that, *ceteris paribus*, the hypoid gears have a lower efficiency than that of spiral bevel gears, with much higher differences, the greater the hypoid offset.

The values of the efficiency of straight bevel, Zerol, and spiral bevel gears are almost equivalent. They are very much influenced by the characteristics of the bearings and assembly, and by the finishing quality of the tooth mating surfaces. For $(T/T_{max}) = 1$, that is for transmission of torques equal to T_{max} defined above, efficiency values included in the range $(0.98 \le \eta \le 0.99)$ can be achieved with these types of gears, when they have a high quality accuracy, and are properly mounted and supported by anti-friction journal bearings. For lower loads, the efficiency decreases, assuming values of about 0.96, for torques that cause root bending stresses equal to the nominal stress number for bending, σ_{Flim}.

Due to the lengthwise sliding losses, hypoid gears are characterized by much lower efficiency values than those of the aforementioned gears. In fact, with the exception of gear pairs characterized by high transmission ratio ($i = 10$, and greater), the values of efficiency of these gears are included in the range $(0.90 \le \eta \le 0.98)$, for loads for which the ratio T/T_{max} is equal to 1. For torque values which cause root bending stresses equal to the nominal stress number for bending, σ_{Flim}, the efficiency of these gears decreases again to values in the range $(0.86 \le \eta \le 0.97)$, depending on the transmission ratio and pinion hypoid offset.

Of course, it is to consider that the above-mentioned need to reduce the hypoid offset, difference between the mean spiral angles, and pressure angle, in order to increase the efficiency of these gears, contrasts with the current conveniences in the design of automotive transmissions. In fact, for these applications, it is required to have centers of gravity at a lower level (to this end, a larger hypoid offset is required), low noise (to this end, a larger difference between the mean spiral angles is required), and greater tooth strength (to this end, a larger value of the pressure angle is required). However, we know that the design of any mechanical system is the result of a reasonable compromise between conflicting requirements.

Finally, it is not the case to investigate what the contribution of the various dissipation phenomena that contribute to determine the total efficiency, η, of the hypoid gears. On this subject, we refer the reader to textbooks that are more specialized as well as to the scientific literature references (see, for example: Coleman [6]; Kolivand et al. [23]; Stadtfeld [62]; Artoni et al. [1]).

12.10 Load Analysis for Hypoid Gears

For determining the forces acting on the hypoid gear pairs, the considerations carried out in Sect. 12.6, regarding the analysis of loads acting on the spiral bevel gears, are still valid. In fact, the hypoid gears are the generalization of spiral bevel gears, which are a special case of hypoid gears. We focus our attention on the general aspects of this problem, while for the more particular ones we refer the reader to specialized scientific literature (see, for example: Vimercati and Piazza [69]; Wang et al. [70]; Kolivand and Kahraman [22]).

The tangential force, F_{mt2} (in N), at the mean pitch diameter d_{m2} (in mm) of the hypoid wheel is given by:

$$F_{mt2} = \frac{2000T_2}{d_{m2}},$$ (12.125)

where T_2 (in Nm) is the torque transmitted by the hypoid wheel. The tangential force, F_{mt1} (in N), at the mean pitch diameter d_{m1} (in mm) of the mating hypoid pinion is instead given by:

$$F_{mt1} = \frac{2000T_1}{d_{m1}} = \frac{F_{mt2}\cos\beta_{m1}}{\cos\beta_{m2}},$$ (12.126)

where T_1 (in Nm) is the torque transmitted by the hypoid pinion, and β_{m1} and β_{m2} are the mean spiral angles of pinion and wheel.

The Eqs. from (12.43) to (12.45), which express respectively the tangential, axial, and radial forces acting on the meshing teeth of spiral bevel gears, as a function of the total tooth force, normal pressure angle, pitch angle and spiral angle on the mean cone, and Eqs. (12.46) and (12.47) resulting from the equations above, are also valid for hypoid gears. Of course, when they are written with reference to the hypoid pinion or hypoid wheel, we add respectively additional subscripts 1 and 2.

The axial force, F_a, can be a repulsive force (in this case, the two members of the hypoid gear pair tend to move away each other, and the axial force is considered positive), or a suction force (in this case, the two members of the hypoid gear pair tend to approach each other, and the axial force is considered negative). For signs that appear in Eqs. (12.46) and (12.47), the same conventions as described in Sect. 12.5 are valid.

Figure 12.30 highlights, in a synoptic view, as the direction of the axial thrust, F_{a1}, acting on the pinion of a hypoid gear pair changes with the hand of the spiral, direction of rotation and position of the hypoid pinion with respect to the hypoid wheel (positive or negative hypoid offset). The four possible cases are considered, namely: pinion with right-hand spiral (RH-spiral) and clockwise rotation (Fig. 12.30a); pinion with right-hand spiral (RH-spiral) and counterclockwise rotation (Fig. 12.30b); pinion with left-hand (LH) spiral and clockwise rotation (Fig. 12.30c); pinion with left-hand (LH-spiral) and counterclockwise rotation (Fig. 12.30d).

Figure 12.31 shows as the direction of the total tooth force, and that of its axial and tangential components, acting on the pinion of a hypoid gear pair, change with the hand of the spirals, direction of rotation of the two members, and position of the hypoid pinion with respect to the hypoid wheel (positive or negative offset).

It is necessary to note that the hand of the spiral must be such that the spiral angle of the pinion at mid-face width is greater than that of the hypoid wheel, i.e. $\beta_{m1} > \beta_{m2}$. In addition, if the axial thrust acting on the pinion is a repulsive force, the active flank of the pinion is the concave flank, while, if the axial thrust acting on the pinion is a suction force, the active flank of the pinion is the convex flank.

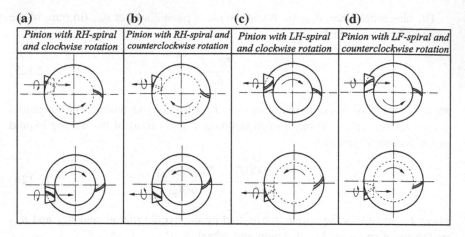

(a)	(b)	(c)	(d)
Pinion with RH-spiral and clockwise rotation	*Pinion with RH-spiral and counterclockwise rotation*	*Pinion with LH-spiral and clockwise rotation*	*Pinion with LF-spiral and counterclockwise rotation*

Fig. 12.30 Direction of the axial thrust acting on the pinion as a function of the hand of the spiral (RH and LH), direction of rotation, and positive or negative offset

To avoid errors and misunderstandings in the determination of forces and moments, which act on shafts and bearings, which are the result of the tangential, axial and radial components of the gear tooth forces, here we need to give explicitly expressions of the axial and radial components, which depend on the curvature of the loaded tooth flank. Table 12.1 allows determining the loaded tooth flank as a function of the driver hand of spiral and direction of rotation of driver.

For *drive side flank loading*, the pinion axial force, $F_{ax1,D}$, and wheel axial force, $F_{ax2,D}$, are given respectively by:

$$F_{ax1,D} = F_{mt1}\left(\tan\alpha_{nD}\frac{\sin\delta_1}{\cos\beta_{m1}} + \tan\beta_{m1}\cos\delta_1 \right) \tag{12.127}$$

$$F_{ax2,D} = F_{mt2}\left(\tan\alpha_{nD}\frac{\sin\delta_2}{\cos\beta_{m2}} - \tan\beta_{m2}\cos\delta_2 \right), \tag{12.128}$$

while for *coast side flank loading*, the pinion axial force, $F_{ax1,C}$, and wheel axial force, $F_{ax2,C}$, are given respectively by:

$$F_{ax1,C} = F_{mt1}\left(\tan\alpha_{nC}\frac{\sin\delta_1}{\cos\beta_{m1}} - \tan\beta_{m1}\cos\delta_1 \right) \tag{12.129}$$

$$F_{ax2,C} = F_{mt2}\left(\tan\alpha_{nC}\frac{\sin\delta_2}{\cos\beta_{m2}} + \tan\beta_{m2}\cos\delta_2 \right). \tag{12.130}$$

In these equations, α_{nD} and α_{nC} are the *generated pressure angles* on drive and, respectively, coast side, while the meaning of other symbols already introduced remains unchanged. It is also to be noted that the positive sign (+) indicates

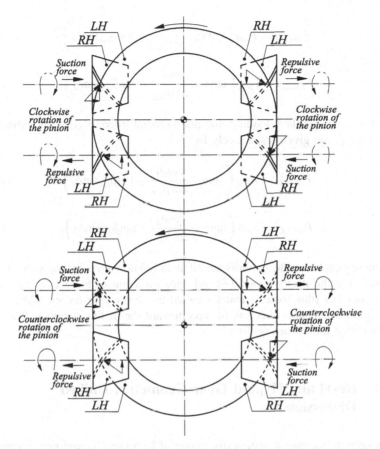

Fig. 12.31 Direction of the total tooth force and its axial and tangential components, acting on the pinion as a function of the hand of the spirals (RH or LH), direction of rotation, and positive or negative offset

Table 12.1 Loaded tooth flank	Driver hand of spiral	Rotation of driver	Loaded flank	
			Driver	Driven
	Right	Clockwise	Convex	Concave
		Counterclockwise	Concave	Convex
	Left	Clockwise	Concave	Convex
		Counterclockwise	Convex	Concave

direction of thrust is away from pitch apex, while the negative sign (−) indicates direction of thrust is toward pitch apex.

For drive side flank loading, the pinion radial force, $F_{rad1,D}$, and wheel radial force, $F_{rad2,D}$, are given respectively by:

$$F_{rad1,D} = F_{mt1} \left(\tan\alpha_{nD} \frac{\cos\delta_1}{\cos\beta_{m1}} - \tan\beta_{m1}\sin\delta_1 \right) \qquad (12.131)$$

$$F_{rad2,D} = F_{mt2} \left(\tan\alpha_{nD} \frac{\cos\delta_2}{\cos\beta_{m2}} + \tan\beta_{m2}\sin\delta_2 \right), \qquad (12.132)$$

while for coast side flank loading, the pinion radial force, $F_{rad1,C}$, and wheel radial force, $F_{rad2,C}$, are given respectively by:

$$F_{rad1,C} = F_{mt1} \left(\tan\alpha_{nC} \frac{\cos\delta_1}{\cos\beta_{m1}} + \tan\beta_{m1}\sin\delta_1 \right) \qquad (12.133)$$

$$F_{rad2,C} = F_{mt2} \left(\tan\alpha_{nC} \frac{\cos\delta_2}{\cos\beta_{m2}} - \tan\beta_{m2}\sin\delta_2 \right). \qquad (12.134)$$

In these equations, the symbols are all already known. As for the signs, it should be noted that the positive sign (+) indicates direction of force is away from the mating member (this force is usually called the *separating force*), while the negative sign (−) indicates direction of force toward the mating member (this force is usually called the *attracting force*).

12.11 Bevel and Hypoid Gear Geometry: Unified Discussion

In this section, we take a unified discussion of bevel and hypoid gear geometry, in accordance with ISO 23509:2016 (E) [20]. Therefore, from now on, the term bevel gear(s) is to be understood in its more general meaning, i.e. including straight, skew (or helical) and spiral bevel, Zerol and hypoid gear(s). Thus, symbols and relationships that gradually we introduce, unless noted otherwise, are applicable to all these types of gears. This unified treatment is also compliant with the introduction of universal, multi-axes, CNC-gear cutting machines that, in principle, are capable of producing nearly all types of gearing, including the different curved-toothed bevel gears previously described as well as hypoid gears.

Hypoid gear pairs, which consist of pinion and wheel with skew and non-intersecting axes, and with curved teeth in the lengthwise direction, can be considered the most general type of gear pair. All other types of gear pairs are subsets of the hypoid gear pairs. So spiral bevel gears, including skew (or helical) bevel gears and Zerol bevel gears, are hypoid gears with zero hypoid offset, straight bevel gears are hypoid gears with zero hypoid offset and zero tooth curvature, and parallel cylindrical helical gears are hypoid gears with zero shaft angle and zero tooth curvature.

Figure 12.32 shows a hypoid gear pair with shaft angle $\Sigma = 90°$, and highlights some of its geometrical quantities. In hypoid gears, axes do not intersect, so we call conventionally, as a crossing point, O_c, the apparent point of intersection of these axes, when they are projected on a plane parallel to both. Given the high flexibility allowed by the current hypoid gear cutting machines, apices of face (or tip), pitch, and root cones of a member of the hypoid gear pair may be beyond or before of the crossing point, which is the trace of the centerline of mate on projection plane above. Conventionally, the distances between these apices and the crossing point are considered positive or negative, depending on whether the apices are beyond the crossing point or before the crossing point. Therefore, according to this convention, for the hypoid pinion, t_{zF1} (*face apex beyond crossing point*), t_{zR1} (*root apex beyond crossing point*), and t_{z1} (*pitch point beyond crossing point*) are all positive, while for hypoid wheel the corresponding quantities t_{zF2}, t_{zR2}, and t_{z2} are all negative (these quantities are not shown in Fig. 12.32). Obviously, all these quantities would be zero in the case in which the apices of both members of the hypoid gear pair coincide with the crossing point.

Fig. 12.32 Some geometrical quantities of a hypoid gear pair

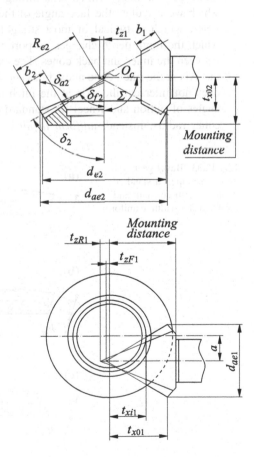

In addition to quantities already defined in the previous sections, as well as in Chap. 9, Fig. 12.32 shows: the *crown to crossing point*, t_{x01} and t_{x02}, which define the distances between the crossing point and crown for hypoid pinion and wheel; the *front crown to crossing point*, t_{xi1}, which defines the distance between the crossing point and the front crown of the hypoid pinion; the *face angle of blank*, δ_{a2}, of the hypoid wheel.

Notoriously, the geometry of bevel and hypoid gears is strongly influenced by the cutting process used, which is defined by the type of cutting machine, its set-up, as well as the type of cutting tool with which teeth are machined (face mill cutters, face hob cutters, planer tools, cup-shaped grinding wheels, etc.). The final tooth proportions, and the size and shape of the blank depend on various tapers that characterize the teeth.

Deepening what we have already said in Sect. 9.5, we can have the following basic types of taper, often interrelated with each other:

– *Depth taper* (Figs. 9.11a and 12.33a), i.e. the change in tooth depth along the face width, measured perpendicularly to the pitch cone. This taper directly influences the blank dimensions through its effect on the dedendum angle by which we calculate the face angle of the mating member. The *standard depth taper*, which is typical of most straight bevel gears, is the configuration for which the tooth depth changes proportionally to the cone distance at any cone between the inner and back cones. The extension of the tooth root line intersects the axis at the pitch cone apex, but generally the extension of the tooth face line does not intersect the axis at the pitch cone apex. The sum of the dedendum angles of pinion and wheel for standard depth taper, $\Sigma\vartheta_{fS}$, does not depend on cutter radius. Instead, *uniform depth* is the configuration for which the tooth

Fig. 12.33 Bevel gear depthwise taper: **a** standard depth taper; **b** constant and modified slot width; **c** uniform depth

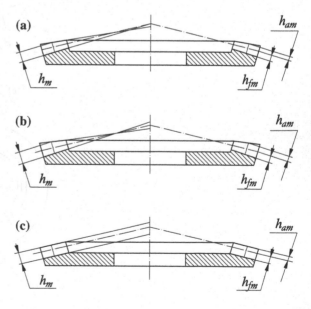

depth is a constant along the face width (Figs. 9.11b and 12.33c), regardless of cutter radius. The sum of the dedendum angles of pinion and wheel for uniform depth, $\Sigma\vartheta_{fU}$, is equal to zero. In this case, the cutter radius, r_{co}, should be greater than $R_{m2}\sin\beta_{m2}$, but less than $1.5R_{m2}\sin\beta_{m2}$. In this way, the variation in normal circular thickness along the face width is a minimum for both the pinion and wheel. In the case in which a narrow inner topland occurs on the pinion, a small tooth tip chamfer may be provided; the length measured along the face width, and the angle between the chamfer generatrix and pinion axis define its geometry (see Fig. 12.34).

- *Space width taper* and *thickness taper*, i.e. respectively the change in the space width and tooth thickness along the face width, both measured generally in the pitch plane.
- *Slot width taper*, i.e. the change in the point width identified by a V-shaped cutting tool with nominal pressure angle, whose top is tangent to the root cone and whose sides are tangent to the two sides of the tooth space, along the face width. This taper is of primary consideration for production, since the width of the slot at its narrowest point defines the point width of the cutting tool, and limits the edge radius of the cutter blade. It depends on the dedendum angle and lengthwise curvature, and for straight bevel gears can be changed by varying the depth taper, i.e. by tilting the root line (Fig. 9.11c). This rotation is generally carried out about the mid-section at the pitch line, in order to maintain the desired working depth at the tooth mean section. For spiral bevel and hypoid gears, the amount of the root line tilting is further dependent on a number of geometric characteristics including the cutter radius. The wheel and pinion root line can be rotated about the mean point, with a resulting *dedendum angle modification* or *tilting*, normally ranging between $-5°$ and $+5°$ (Fig. 9.11c); this is done to avoid cutting interference with a hub or shoulder.

From the point of view of the *tooth depth configuration*, we can have bevel gear wheels with *constant slot width*, and bevel gear wheels with *modified slot width* (Fig. 12.33b). The gear wheels with constant slot width have zero width slot taper on both members of the gear pair; therefore, this taper is related to a tilt of the root line such that the slot width is constant while maintaining the proper

Fig. 12.34 Tooth tip chamfer on the pinion

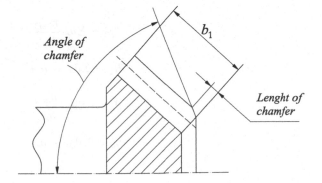

space width taper. Instead, the gear wheels with modified slot width have a slot width taper characterized by a root line tilted about the mean point of an intermediate amount between zero and the value related to the constant slot width. In this case, the slot width of the wheel is constant along the face width, while the pinion does not have any slot width taper.

The formulas for calculating the sum of the dedendum angles, $\Sigma\vartheta_f$ (in degrees), of the pinion and wheel, for the above-mentioned four possible cases of depthwise taper that are chosen in the accordance with the cutting method, are as follows:

$$\Sigma\vartheta_{fS} = \arctan\left(\frac{h_{fm1}}{R_{m2}}\right) + \arctan\left(\frac{h_{fm2}}{R_{m2}}\right) \tag{12.135}$$

$$\Sigma\vartheta_{fU} = 0 \tag{12.136}$$

$$\Sigma\vartheta_{fC} = \left(\frac{90m_{et}}{R_{e2}\tan\alpha_n\cos\beta_m}\right)\left(1 - \frac{R_{m2}\sin\beta_{m2}}{r_{c0}}\right) \tag{12.137}$$

$$\Sigma\vartheta_{fM} = \Sigma\vartheta_{fC} \quad \text{or} \quad \Sigma\vartheta_{fM} = 1.3\Sigma\vartheta_{fS}, \quad \text{whichever is smaller,} \tag{12.138}$$

which apply respectively to standard depth taper, uniform depth, constant slot width and modified slot width. It is to be noted that subscripts S, U, C and M for quantities appearing in Eqs. (12.135) to (12.138) refer respectively to standard depth taper, uniform depth, constant slot width, and modified slot width depthwise taper.

Instead, the formulas that allow us to calculate the addendum and dedendum angles of the wheel, ϑ_{a2} and ϑ_{f2} (both in degrees), apportioning the sum of the dedendum angles between pinion and wheel as a function of the desired depthwise typer, are as follows:

$$\vartheta_{a2} = \arctan\left(\frac{h_{fm1}}{R_{m2}}\right) \tag{12.139}$$

$$\vartheta_{f2} = \Sigma\vartheta_{fS} - \vartheta_{a2} \tag{12.140}$$

$$\vartheta_{a2} = \vartheta_{f2} = 0 \tag{12.141}$$

$$\vartheta_{a2} = \Sigma\vartheta_{fC}\frac{h_{am2}}{h_{mw}} \tag{12.142}$$

$$\vartheta_{f2} = \Sigma\vartheta_{fC} - \vartheta_{a2} \tag{12.143}$$

$$\vartheta_{a2} = \Sigma\vartheta_{fM}\frac{h_{am2}}{h_{mw}} \tag{12.144}$$

$$\vartheta_{f2} = \Sigma\vartheta_{fM} - \vartheta_{a2} \tag{12.145}$$

It is to be noted that: Eqs. (12.139) and (12.140) apply to standard depth taper; Eq. (12.141) applies to uniform depth; Eqs. (12.142) and (12.143) apply to constant slot depth; Eqs. (12.144) and (12.145) apply to modified slot width. It is also to be noted that the quantities and related symbols in Eqs. (12.135) to (12.145) are all known, but we feel it opportune to specify that the subscript, m, refers to the mean cone quantities, and that the mean working depth, h_{mw}, is the depth of engagement of two members of the hypoid gear pair at mean cone distance.

For constant slot width, Eq. (12.137) shows that the sum of dedendum angles, $\Sigma \vartheta_{fC}$, is strongly influenced by the cutter radius, r_{c0}, as this has a significant effect on the amount by which the root line is tilted. It is to be noted that, for a given design, a large cutter radius and a small cutter radius increases and, respectively, decreases the sum of dedendum angles. However, the cutter radius should be neither too large nor too small: it must be within the limits $(1.1 R_{m2} \sin \beta_{m2} \leq r_{c0} \leq R_{m2})$. In fact, if the cutter radius is too large (beyond the upper limit, i.e. $r_{c0} > R_{m2}$), the resultant depthwise taper could adversely influence the tooth depth at both ends, with a too shallow tooth at inner end for a proper tooth contact, and a too depth tooth at the outer end that may cause undercut and narrow toplands. Then, if the cutter radius is equal to $R_{m2} \sin \beta_{m2}$, $\Sigma \vartheta_{fC}$ becomes zero. Therefore, we would get uniform depth teeth. If r_{c0} is less than $R_{m2} \sin \beta_{m2}$, we would get a reverse depthwise taper, and the teeth would be deeper at the inner end than at the outer end. In order to not have teeth with excessive depth at the inner end, with the risk of undercut and narrow toplands, it is appropriate that r_{c0} is at least equal to the minimum value mentioned above. In addition, it is to be noted that standard taper is the norm for gears cut with planer tools; in this case, the cutter center is considered to be at infinity, and root lines are not tilted.

For modified slot width, the sum of dedendum angles, $\Sigma \vartheta_{fM}$, must not exceed either 1.3 times the sum of the dedendum angles for standard depth taper, $\Sigma \vartheta_{fS}$, nor the sum of the dedendum angles for constant slot width taper, $\Sigma \vartheta_{fC}$. In practice, therefore, the smaller of the values $1.3 \Sigma \vartheta_{fS}$ or $\Sigma \vartheta_{fC}$ is used (see Eq. 12.138).

Since the geometry of bevel and hypoid gears is a function of the cutting method used, as well as the setting of the cutting machine that implements it, an infinite number of pitch surfaces will exist for any hypoid gear pair. However, once the initial data related to a given method have been defined, we will have one pitch surface for each method. If we limit our attention to the cutting method of spiral bevel gears, and the three cutting methods of hypoid gears (Gleason, Oerlikon, and Klingelnberg), we can identify the following four design procedures, referred as Method 0, Method 1, Method 2, and Method 3 by ISO 23509:2016 (E) [20]. The fields of use of these four methods are as follows:

- Method 0 is used for spiral bevel gears, i.e. non-hypoid gears (curved-toothed bevel gears, without hypoid offset).
- Method 1 is used for hypoid gears manufactured by the face-milling process. The pitch surfaces of these gears are chosen in such a way that the hypoid radius

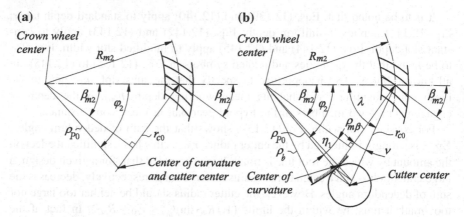

Fig. 12.35 Geometry of the main cutting processes of spiral bevel gears: **a** face-milling process; **b** face-hobbing process

of curvature matches the radius of curvature of the cutter at the mean point for the to-be-generated gear wheels (so the mean radius of curvature of the tooth is equal to the cutter radius, i.e. $\rho_{m\beta} = r_{c0}$ as Fig. 12.35a shows).

- Method 2 is used for hypoid gears manufactured by the face-hobbing process. The pitch surfaces of these gears are chosen in such a way that the hypoid radius of curvature matches the mean epicycloid curvature at the mean point, with the condition that the wheel pitch apex, pinion pitch apex and cutter center lie on a straight line.
- Method 3 is also used for hypoid gears manufactured by the face-hobbing process. The pitch surfaces of these gears are chosen in such a way that the hypoid radius of curvature matches the mean epicycloid curvature at the mean point for gear wheel to be generated, without the condition described for Method 2 (Fig. 12.35b).

For spiral bevel gears, Method 0 has to be used, and face width factor is set to be $c_{be2} = 0.5$; for initial data other than those shown in Table 12.2, the formulas can be easily converted. With Method 1, which is used by Gleason, it is necessary to determine the face width factor c_{be2}, since the calculation point is not in the middle of the wheel face width. The pitch cone parameters obtained with this method have similar values compared to those obtained using Method 3, which is the method used by Klingelnberg. Method 2 is the method used by Oerlikon.

The quantities shown in Fig. 12.35, and related symbols are as follows: mean cone distance, R_{m2}; spiral angle, β_{m2}; intermediate angle, φ_2; crown gear to cutter center, ρ_{P0}; cutter radius, r_{c0}; lengthwise tooth mean radius of curvature, $\rho_{m\beta}$; first auxiliary angle, λ; second auxiliary angle, η_1; lead angle of the cutter, ν; epicycloid base circle radius, ρ_b.

To define in more detail the geometry of the hypoid gears, which represent the most general case of gears, it is necessary to consider not only the three main views,

Table 12.2 Initial data for calculation of the pitch cone parameters

Symbol	Description	Method 0	Method 1	Method 2	Method 3
Σ	Shaft angle	x	x	x	x
a	Hypoid offset	0	x	x	x
$z_{1,2}$	Number of teeth	x	x	x	x
d_{m2}	Mean pitch diameter of wheel	–	–	x	–
d_{e2}	Outer pitch diameter of wheel	x	x	–	x
b_2	Wheel face width	x	x	x	x
β_{m1}	Mean spiral angle of pinion	–	x	–	–
β_{m2}	Mean spiral angle of wheel	x	–	x	x
r_{c0}	Cutter radius	x	x	x	x
z_0	Number of blade groups (only face-hobbing)	x	x	x	x

but also three appropriate sections, as shown in Fig. 12.36, where: O_1 and O_2 are the *pinion pitch apex* and respectively the *wheel pitch apex*; O_c is the *crossing point*, i.e. the apparent point of intersection of axes, when it is projected on a plane parallel to both axes (therefore, it is also the trace of the common normal to pinion and wheel axes through crossing point); P is the *mean point*; M and N are the intersection points of contact normal through mean point at pinion axis and respectively at wheel axis; F and G are the crossing points at pinion axis and respectively at wheel axis; l is the distance along wheel axis between crossing point and intersection of contact normal.

The other quantities shown in Fig. 12.36, and related symbols are as follows: hypoid offset, a; pinion pitch angle and *wheel pitch angle*, δ_1 and, respectively, δ_2; *pinion spiral angle* and *wheel spiral angle*, β_{m1} and, respectively, β_{m2}; *pinion offset angle in axial plane* and *wheel offset angle in axial plane*, ζ_m and, respectively, η; pinion mean cone distance and wheel mean cone distance, R_{m1} and, respectively, R_{m2}; pinion offset angle in pitch plane, ζ_{mp}; *pinion pitch apex beyond crossing point* and *wheel pitch apex beyond crossing point*, t_{z1} and, respectively, t_{z2}; crossing point to mean point along pinion axis and crossing point to mean point along wheel axis, t_{zm1} and, respectively, t_{zm2}; tangent to tooth trace at mean point, t.

Calculations of bevel and hypoid gears are divided into the two main steps described in the following two subsections. The calculation development basically follows that from ISO 23509:2016 (E) [20], which uses relationships obtainable according to the theoretical treatment described in the previous sections. However, we have tried to eliminate repeated equations, while retaining some repetitions so as not overload the logic flow with references to previous calculation steps.

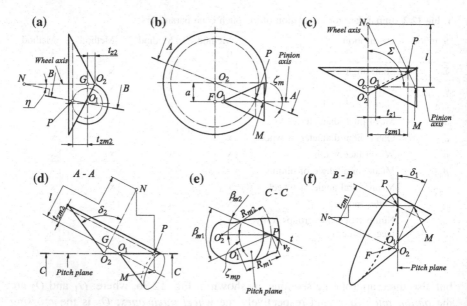

Fig. 12.36 Main angles and quantities of a hypoid gear pair with $\Sigma = 90°$: **a** side view looking along the pinion axis; **b** side view looking along the wheel axis; **c** top view showing the shaft angle; **d** view of the wheel section along the plane making the offset angle, ζ_m, in the pinion axial plane; **e** view of the pitch plane of the hypoid gear pair; **f** view of the pinion section along the plane making the offset angle, η, in the wheel axial plane

12.11.1   First Step of Calculation

In this first step of calculation, the pitch cone parameters are determined from the initial data. This determination is carried out using a specific set of formulas for each of the four methods described above. For spiral bevel gears (Method 0), it is possible a simple determination of the pitch cone parameters. For hypoid gears, this is not possible; in this case, a procedure of successive approximation or iteration must be used. To start the calculation procedure related to each of the four methods, it is necessary to have a series of initial data, which are summarized in Table 12.2. However, it is to be noted that the pitch cone parameters of spiral bevel gears also can be determined with different initial data as given in Table 12.2 (see Sect. 12.5).

The shaft angle, Σ, and in most cases also hypoid offset, a, are imposed by the practical application involved. A positive pinion hypoid offset (Figs. 12.30 and 12.31) is recommended because of the increasing diameter of the pinion, higher pitting load carrying capacity, and higher face contact ratio. It is however necessary to check the scuffing load capacity due to additional lengthwise sliding. For this reason, the pinion hypoid offset should not exceed $0.25d_{e2}$ and, for heavy-duty applications, it should be limited to half of the above value.

The calculation of the outer pitch diameter of wheel, d_{e2}, involves the preliminary determination of the outer pitch diameter of pinion, d_{e1}, which is correlated to the pinion torque, T_1 (in Nm), given by:

$$T_1 = \frac{P}{\omega_1} = \frac{60P}{2\pi n_1}, \tag{12.146}$$

where P is the power (in W), ω_1 is the pinion angular velocity (in rad/s), and n_1 is the pinion rotational speed (in min^{-1}, revolutions per minute). When the load is not constant, the pinion torque will vary, and its operating value must be calculated considering the power and speed values at which the expected operating cycle of the driven machine is carried out. When peak loads are present, if their total duration exceeds ten million cycles during the total expected life of the gear, the

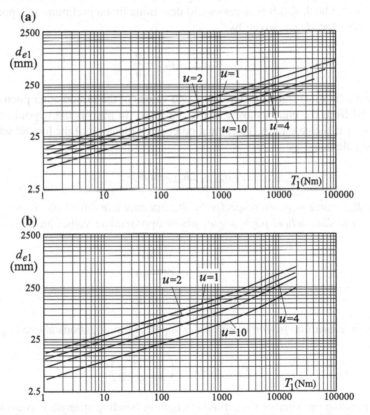

Fig. 12.37 Outer pitch diameter of pinion, d_{e1}, as a function of pinion torque, T_1, for: **a** pitting strength; **b** bending strength

determination of the gear size should be made according to peak load. If, however, their total duration is less than ten million cycles, the evaluation of the gear size should be started based on the greater of the two values corresponding to the highest sustained load or half of the peak load.

For spiral bevel gears with $\Sigma = 90°$, the charts shown in Fig. 12.37 enable us to determine the outer pitch diameter of commercial quality spiral bevel pinions of case-hardened steel, at 55 minimum HRC, as a function of the pinion torque, respectively for pitting and bending strengths. For shaft angles $\Sigma \neq 90°$, the charts give preliminary values of d_{e1} less accurate, and this could require additional adjustments of design choices. For straight bevel and Zerol bevel gears, the values of d_{e1} obtained by these charts are to be multiplied respectively by 1.2 and 1.3 (for Zerol bevel gears, this is due to a face width limitation). For hypoid gears, the values of d_{e1} obtained by the same charts are to be considered as equivalent pinion outer pitch diameters.

In the more general case of hypoid gears, to calculate the outer pitch diameter of the hypoid wheel, d_{e2}, it is necessary to determine first a preliminary hypoid pinion pitch diameter, d_{eplm1}, given by:

$$d_{eplm1} = d_{e1} - \frac{a}{u}, \tag{12.147}$$

where d_{e1} (in mm) is the greater of the two values of the pinion outer pitch diameter obtained from the charts shown in Fig. 12.37, a (in mm) is the hypoid offset, and $u = z_2/z_1$ is the gear ratio. The actual outer pitch diameter of the hypoid wheel, d_{e2}, is then calculated as:

$$d_{e2} = 2R_{eint2}\sin\delta_{int2}, \tag{12.148}$$

where R_{eint2} and δ_{int2} are respectively the *intermediate wheel outer cone distance* and *intermediate wheel pitch angle*, whose approximate values are given by:

$$R_{eint2} = \frac{d_{eplm1}}{2\sin\delta_{int1}}, \tag{12.149}$$

$$\delta_{int2} = \Sigma - \delta_{int1}. \tag{12.150}$$

The *intermediate pinion pitch angle*, appearing in equations above is given by:

$$\delta_{int1} = \arctan\left(\frac{\sin\Sigma}{u + \cos\Sigma}\right). \tag{12.151}$$

Depending on whether the pitting strength or bending strength is considered, we have two values of d_{e1}, and thus two values of the preliminary hypoid pinion pitch diameter, d_{eplm1}. For calculations, we have to choose the highest value between the two. This choice must be respected, even for precision-finished gears. In this case, the pitting load carrying capacity is increased, whereby the value of d_{e1} determined

Table 12.3 Material factor, k_M

Gear set materials				
Wheel material and hardness		Pinion material and hardness		k_M
Material	Hardness	Material	Hardness	
Case-hardened steel	58 HRC min.	Case-hardened steel	60 HRC min.	0.85
Case-hardened steel	55 HRC min.	Case-hardened steel	55 HRC min.	1.00
Flame-hardened steel	50 HRC min.	Case-hardened steel	55 HRC min.	1.05
Flame-hardened steel	50 HRC min.	Flame-hardened steel	50 HRC min.	1.05
Oil-hardened steel	375 HB–425 HB	Oil-hardened steel	375 HB–425 HB	1.20
Heat-treated steel	250 HB–300 HB	Case-hardened steel	55 HRC min.	1.45
Heat-treated steel	210 HB–245 HB	Case-hardened steel	55 HRC min.	1.45
Cast iron	–	Case-hardened steel	55 HRC min.	1.95
Cast iron	–	Flame-hardened steel	50 HRC min.	2.00
Cast iron	–	Annealed steel	160 HB–200 HB	2.10
Cast iron	–	Cast iron	–	3.10

from the charts shown in Fig. 12.37a, and the value of d_{eplm1} calculated by Eq. (12.147) are to be multiplied by 0.8. Furthermore, for materials other than case-hardened steel at 55 minimum HRC, the values of d_{e1} obtained from the charts shown in Fig. 12.37, and the values of d_{eplm1} determined by Eq. (12.147) are to be multiplied by a *material factor*, k_M, given in Table 12.3.

For statically loaded gears, only the bending strength is considered. In fact, the load conditions related to bending strength are more restrictive than those related to the pitting resistance. Depending on whether or not the gears are subjected to vibration, the values of the outer pitch diameter of pinion taken from the charts shown in Fig. 12.37b or calculated by Eqs. (12.147) are multiplied by 0.7 or respectively by 0.6.

Theoretically, the choice of the numbers of teeth z_1 and z_2 can be made in an arbitrary manner, as long as the assigned gear ratio $u = z_2/z_1$ is respected. In reality, however, for general applications, it is appropriate that the number of pinion teeth, z_1, is made using the two charts shown in Figs. 12.38a and 12.38b, respectively valid for spiral bevel and hypoid gears, and for straight bevel and Zerol bevel gears; these two charts provide approximate but reliable values of the number of teeth of pinion, as a function of d_{ei} and u. For automotive applications, the pinion of spiral bevel gears and hypoid gears often has a fewer number of teeth, as shown in Table 12.4. In order to achieve an acceptable contact ratio without undercut, straight bevel gears and Zerol bevel gears are respectively designed with 12 teeth and higher, and with 13 teeth and higher.

Spiral bevel gears and hypoid gears can have a fewer numbers of teeth (see Table 12.6), due to the additional overlap ratio resulting from oblique teeth, which allows the teeth to be stubbed to avoid undercut and still maintain an acceptable contact ratio. A careful analysis of undercut must however be done.

Table 12.4 Suggested minimum numbers of pinion teeth for spiral bevel and hypoid gears

Approximate ratio, u	Minimum number of pinion teeth, z_1
$1.00 \leq u \leq 1.50$	13
$1.50 < u \leq 1.75$	12
$1.75 < u \leq 2.00$	11
$2.00 < u \leq 2.50$	10
$2.50 < u \leq 3.00$	9
$3.00 < u \leq 3.50$	9
$3.50 < u \leq 4.00$	9
$4.00 < u \leq 4.50$	8
$4.50 < u \leq 5.00$	7
$5.00 < u \leq 6.00$	6
$6.00 < u \leq 7.50$	5
$7.50 < u \leq 10.0$	5

The wheel face width, b_2 (in mm), of spiral bevel gears with $\Sigma = 90°$ can be determined as a function of the outer pitch diameter of pinion, d_{e1} (in mm), and gear ratio, u, as Fig. 12.39 shows. This figure refers to face widths corresponding to 30% of the outer cone distance. For shaft angle less than or greater than 90°, respectively a face width larger or smaller than that given in Fig. 12.39 can be used. Generally, the face width should not exceed 30% of the outer cone distance or $10m_{et2}$, where m_{et2} is the outer transverse module of the wheel. The face width of Zerol bevel gears should not exceed 25% of the outer cone distance, and should be determined by multiplying by 0.83 the value read in the chart of Fig. 12.39. The face width of hypoid gear wheels is also determined with the chart in this figure; the hypoid pinion face width is generally greater than that of hypoid wheel (see Sect. 12.12.2.5).

The outer transverse module of the wheel is obtained by dividing the outer pitch diameter by the number of teeth, i.e. $m_{et2} = d_{e2}/z_2$. It is not necessary that this module be an integer, since the cutting tools for bevel gears are not standardized according to the module. The determination of the mean pitch diameter of wheel, d_{m2}, is carried out using the above defined quantities and those that follows.

The mean spiral angles of pinion and wheel, β_{m1} and β_{m2}, are chosen so as to achieve a face contact ratio, ε_β, approximately equal to 2.0. For maximum smoothness and quietness and high-speed applications, it is appropriate to have $\varepsilon_\beta > 2$, although values of ε_β less than 2.0 are allowed. The chart shown in Fig. 12.40 is a guideline for choosing the spiral angles of spiral bevel gears with face width equal to 30% of the outer cone distance, i.e. $b/R_e = 0.3$. For any value of b/R_e, the face contact ratio, ε_β, is given by:

$$\varepsilon_\beta = \left(K_z \tan\beta_m - \frac{K_z^3}{3} \tan^3\beta_m \right) \frac{R_e}{\pi m_{et}}, \tag{12.152}$$

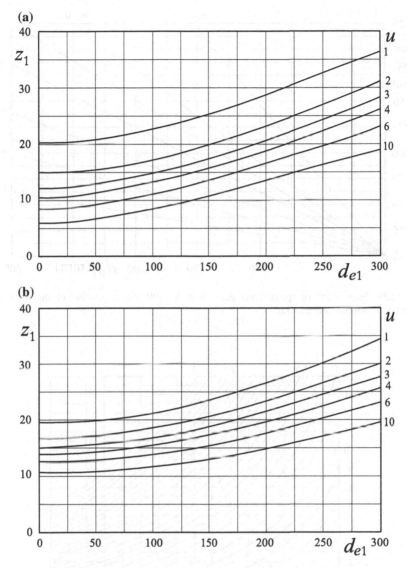

Fig. 12.38 Approximate number of pinion teeth, for: **a** spiral bevel and hypoid gears; **b** straight bevel and Zerol bevel gears

where β_m is the mean spiral angle at pitch surface, quantities R_e, m_{et}, and b are given in mm, and K_z is a dimensionless factor depending on the ratio b/R_e, given by:

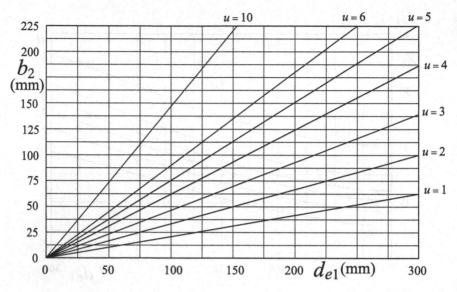

Fig. 12.39 Face width of spiral bevel gears with $\Sigma = 90°$, as a function of the outer pitch diameter of pinion, and gear ratio

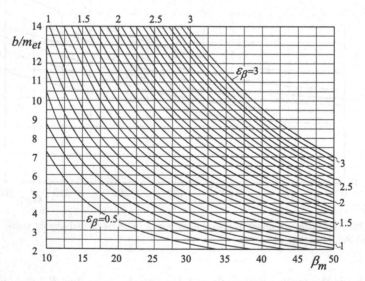

Fig. 12.40 Face contact ratio for spiral bevel gears as a function of spiral angle and ratio (b/m_{et})

$$K_z = \frac{b}{2R_e} \frac{\left(2 - \frac{b}{R_e}\right)}{\left(1 - \frac{b}{R_e}\right)}. \tag{12.153}$$

For the use of chart shown in Fig. 12.40, where $b/R_e = 0.3$, we have $\varepsilon_\beta = [(0.3885\tan\beta_m - 0.0171\tan^3\beta_m)(b/m_{et})]$.

The mean spiral angle of the pinion of a hypoid gear pair is determined by the relationship:

$$\beta_{m1} = 25 + 5\sqrt{u} + 90\frac{a}{d_{e2}}, \tag{12.154}$$

with a and d_{e2} in mm. The mean spiral angle β_{m2} of the wheel depends on the hypoid geometry, and is calculated as shown in Sect. 12.12.1.

Design and manufacture of bevel and hypoid gears also depend, for face-milled gears, on the cutter radius, r_{c0}, and number of blade groups, z_0, and, for face-hobbed gears, only on the cutter radius. For both cutting processes, Table 12.5 provides a data list of standard cutters. For the face-milling process, the cutter diameter $2r_{c0}$ is given in inches, for $2r_{c0} < 500$ mm, and in millimeters, for $2r_{c0} \geq 500$ mm.

By using the above-described initial data as well as the appropriate equations in the next section, we can determine the pitch cone parameters $R_{m1}, R_{m2}, \delta_1, \delta_2, \beta_{m1}, \beta_{m2}$, and c_{be2}, by which a schematic diagram of a spiral bevel or hypoid gear can be drawn, like the one shown in Fig. 12.41.

The parameter $c_{be2} = (R_{e2} - R_{m2})/b_2$, i.e. the *face width factor*, must necessarily be considered for Method 1, as in this case the calculation point does not always coincide with the mean point P of the wheel face width. For Methods 0, 2 and 3, the calculation point coincides with the mean point, for which $c_{be2} = 0.5$.

The schematic diagram of Fig. 12.41 shows a common tangential plane, T, between both pitch cones, with mean pitch diameters d_{m1} and d_{m2}, which are in contact each other at the mean point, P. Besides, both pitch cones contact with the tangential plane, T, along two straight lines, which are the mean cone distances, R_{m1} and R_{m2}, and include the offset angle, ζ_{mp}. The normal straight line to the plane, T, through the mean point, P, intersects the pinion axis x_P at point N_P, and the gear wheel axis x_G at point N_G. The straight line $N_P - N_G$ represents the center distance, a_v, of the equivalent mean virtual cylindrical gear.

In doing so, we applied to hypoid gears the concepts we have described in Chap. 9, concerning straight bevel gears. However, contrary to what happens in the last gears, the pinion and wheel axes of a hypoid gear pair are not in the same plane. Therefore, to have virtual cylindrical gears with parallel axes, an approximation is made according to which both axes are arranged in the direction that divides the offset angle ζ_{mp} into half. This does not mean that the thus-defined virtual cylindrical gears have the same meshing conditions such as hypoid gears. This goal is achieved afterwards, by introducing several appropriate correction factors such as the hypoid factor, Z_{HyP}, which takes into account the influence of the lengthwise sliding of

Table 12.5 Nominal cutter radius, r_{c0}, and number of blade groups, z_0

Face hobbing						Face milling	
Two-part cutter (two divided cutter parts for inner and outer blades)		Two-blade cutter (outer, and inner blade per group)		Three-blade cutter (rougher, outer, and inner blade per group)			
r_{c0} (mm)	z_0	r_{c0} (mm)	z_0	r_{c0} (mm)	z_0	Cutter diameter, $2r_{c0}$	
						inch	mm
25	1	30	7	39	5	2.50	500
25	2	51	7	49	7	3.25	640
30	3	64	11	62	5	3.50	800
40	3	64	13	74	11	3.75	1000
55	5	76	7	88	7	4.375	
75	5	76	13	88	13	5	
100	5	76	17	110	9	6	
135	5	88	11	140	11	7.50	
170	5	88	17	150	12	9	
210	5	88	19	160	13	10.50	
260	5	100	5	181	13	12	
270	3	105	13			14	
350	3	105	19			16	
450	3	125	13			18	
		150	17				
		175	19				

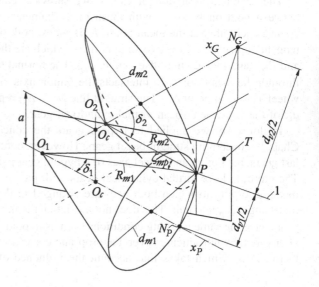

Fig. 12.41 Schematic
diagram of a hypoid gear

hypoid gear teeth. Nevertheless, virtual cylindrical gears supply the required geometrical basis to achieve a practicable rating system for all types of bevel gears.

12.11.2 Second Step of Calculation

In this second step of calculation, the gear dimensions are determined, starting from the pitch cone parameters, and introducing the set of additional data presented in Table 12.6. These additional data can be given in terms of European standards (data type I) or in terms of AGMA standards (data type II). For example, European standards describe the *gear tooth proportions* with an addendum factor, k_{hap}, a dedendum factor, k_{hfp}, a profile shift coefficient, x_{hm}, and a thickness modification coefficient, x_{smn}, while AGMA standards describe the same tooth proportions with a *depth factor*, k_d, a *clearance factor*, k_c, a *mean addendum factor* of wheel, c_{ham}, and a *thickness factor*, k_t or *wheel mean slot width*, W_{m2}. European factors and AGMA factors are related to each other (see Table 12.7), and both lead to the same result of tooth geometry. This geometry, contrary to pitch cone parameters, is expressed by only one set of formulas for bevel and hypoid gears, no matter which method was chosen. All formulas for hypoid gears, with the hypoid offset, a, set to zero, also apply to spiral bevel gears. Figure 12.42 highlights some of the above-mentioned quantities, with reference to the basic rack tooth profile of the wheel.

As Table 12.6 shows, some additional data for calculation of gear dimensions are common to European and AGMA standards. Other additional data are rather different. In this regard, in this textbook, we refer to data type I, and we leave to the reader the ability to turn them into data type II, according to the correlations reported in Table 10.7. In any case, whatever the standards, European or AGMA,

Table 12.6 Additional data for calculation of gear dimensions

Data type I (European standards)		Data type II (AGMA standards)	
Symbol	Description	Symbol	Description
α_{dD}	Nominal design pressure angle on drive side		
α_{dC}	Nominal design pressure angle on coast side		
$f_{\alpha lim}$	Influence factor of limit pressure angle		
x_{hm1}	Profile shift coefficient	c_{ham}	Mean addendum factor of wheel
k_{hap}	Basic crown gear addendum factor	k_d	Depth factor
k_{hfp}	Basic crown gear dedendum factor	k_c	Clearance factor
x_{smn}	Thickness modification coefficient	k_t W_{m2}	Thickness factor or wheel mean slot width
j_{mn}, j_{mt2} j_{en}, j_{et2}	Backlash (choice of four)		
ϑ_{a2}	Addendum angle of wheel		
ϑ_{f2}	Dedendum angle of wheel		

the choice of these additional data should be done carefully, to ensure optimal performance of the gear to be designed.

With regard to the normal pressure angles, it is to distinguish between:

- Nominal design pressure angle, α_d (α_{dD} on drive side, and α_{dC} on coast side), which is the start value for the calculation, and may be half of the sum of pressure angles or different on drive and coast sides.
- Generated pressure angle, α_n (α_{nD} on drive side, and α_{nC} on coast side), which is the pressure angle of the generation crown wheel, and characterizes the tooth flank in the mean normal section.
- Effective pressure angle, α_e (α_{eD} on drive side, and α_{eC} on coast side), which is a calculated value as reported below.

Generally, the nominal design pressure angles on drive side, α_{dD}, and on coast side, α_{dC}, are balanced, but in some optimized applications these angles are unbalanced. For bevel gears, the most commonly used design pressure angle is $\alpha_d = 20°$. This pressure angle greatly influences the gear design. In fact, lower generated pressure angles reduce the axial and separating forces, and increase the transverse contact ratio as well as toplands and slot widths, while higher generated pressure angles cause adverse effects. In addition, lower effective pressure angles increase the risk of undercut.

For hypoid gears, in many practical applications, it would be appropriate to have unequal generated pressure angles on the drive and coast sides, to obtain mesh balanced conditions. In the case in which full balanced mesh conditions are required, the effective pressure angle on drive side, α_{eD}, has a different value compared to that of the effective pressure angle on the coast side, α_{eC}, and the *influence factor of limit pressure angle*, $f_{\alpha lim}$, is set equal to the unit. Thus, the generated normal pressure angles on the drive side, α_{nD}, and on the coast side, α_{nC}, are obtained from the corresponding design pressure angles, α_{dD} and α_{dC}, by adding and respectively subtracting the limit pressure angle, α_{lim}, which is given by:

$$\alpha_{lim} = -\arctan\left[\frac{\tan\delta_1 \tan\delta_2}{\cos\zeta_{mp}}\left(\frac{R_{m1}\sin\beta_{m1} - R_{m2}\sin\beta_{m2}}{R_{m1}\tan\delta_1 + R_{m2}\tan\delta_2}\right)\right]. \qquad (12.155)$$

In the general case of balanced mesh conditions, the generated normal pressure angles on drive side, α_{nD}, and on coast side, α_{nC}, are given respectively by the following relationships:

$$\alpha_{nD} = \alpha_{dD} + f_{\alpha lim}\alpha_{lim} \qquad (12.156)$$

$$\alpha_{nC} = \alpha_{dC} - f_{\alpha lim}\alpha_{lim}, \qquad (12.157)$$

while the effective pressure angles on drive side, α_{eD}, and on coast side, α_{eC}, are given respectively by:

$$\alpha_{eD} = \alpha_{nD} - \alpha_{lim} \tag{12.158}$$

$$\alpha_{eC} = \alpha_{nC} + \alpha_{lim}. \tag{12.159}$$

Reducing the generated pressure angles on drive side, several benefits can be obtained in terms of contact ratio, contact stress, and axial and radial forces. However, we must not go down to values lower than $(9 \div 10)°$, due to inherent limitations of the cutting tools, as well as to the risk of undercut. In all cases, however, the effective pressure angles are calculated with Eqs. (12.158) and (12.159).

For not-hypoid gears, i.e. straight, Zerol, and spiral bevel gears, the limit pressure angle is always equal to zero, for which the nominal design pressure angles and generated pressure angles have the same values. If then the effective pressure angles have the same values, the mesh conditions on drive side and coast side are equal.

The following guidelines are to be observed for the choice of the nominal pressure angle, in order to prevent undercut:

- For straight bevel gears, $\alpha_d = 20°$ or higher for pinions with 14 to 16 teeth, and $\alpha_d = 25°$ for pinions with 12 or 13 teeth.
- For Zerol bevel gears with high transmission ratios or low tooth numbers, or both, $\alpha_d = 22.5°$ for pinions with 14 to 16 teeth, and $\alpha_d = 25°$ for pinions with 13 teeth.
- For spiral bevel gears, $\alpha_d = 20°$ or higher for pinions with 12 or fewer teeth.
- For hypoid gears, $\alpha_d = 18°$ or $\alpha_d = 20°$ for light-duty drives, and $\alpha_d = 22.5°$ or $\alpha_d = 25°$ for heavy-duty drives (it is to be noted that, to balance the mesh conditions on drive and coast sides, it should be $f_{\alpha lim} = 1$, but for use of standard cutting tools, the value of $f_{\alpha lim}$ can be different from unity).

As regard the tooth depth components, it should be noted that, in terms of European standards, in common cases, the basic crown wheel addendum factor, k_{hap}, and basic crown wheel dedendum factor, k_{hfp}, are chosen respectively equal to 1.00 and 1.25. To prevent then undercut, the profile shift coefficients have to be included within specified ranges (see Sect. 12.12.3). AGMA standards, however, provide more detailed data type II, which can be used to calculate the corresponding data type I, with the help of equations collected in Table 12.7.

Table 12.7 Relations between data type I and data type II

$x_{hm1} = k_d\left(\frac{1}{2} - c_{ham}\right)$	$c_{ham} = \frac{1}{2}\left(1 - \frac{x_{hm1}}{k_{hap}}\right)$
$k_{hap} = k_d/2$	$k_d = 2k_{hap}$
$k_{hfp} = k_d\left(k_c + \frac{1}{2}\right)$	$k_c = \frac{1}{2}\left(\frac{k_{hfp}}{k_{hap}} - 1\right)$
$x_{smn} = \frac{k_t}{2} = \frac{1}{2}\left[\frac{W_{m2}}{m_{mn}} + k_d\left(k_c + \frac{1}{2}\right)(\tan\alpha_{nD} + \tan\alpha_{nC}) - \frac{\pi}{2}\right]$	$k_t = 2x_{smn}$

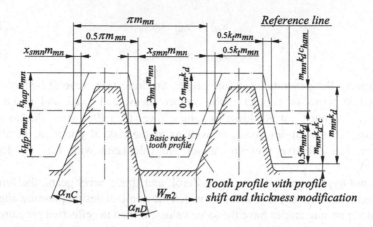

Fig. 12.42 Basic rack tooth profile of wheel, and wheel tooth profile with profile shift and thickness modification

Table 12.8 Suggested values of depth factor, k_d

Type of gear	Depth factor	Number of pinion teeth
Straight bevel	2.000	12 or more
Spiral bevel	2.000	12 or more
	1.995	11
	1.975	10
	1.940	9
	1.895	8
	1.835	7
	1.765	6
Zerol bevel	2.000	13 or more
Hypoid	2.000	11 or more
	1.950	10
	1.900	9
	1.850	8
	1.800	7
	1.750	6

Usually, a depth factor $k_d = 2$ is used to determine the mean working depth, h_{mw}, but it can be changed to obtain other specified design requirements. Table 12.8 gives a guideline to choose the depth factor values as a function of type of gear and number of pinion teeth.

For constant clearance along the whole tooth depth, the calculation is made at mean point, and a clearance factor $k_c = 0.125$ is used, but also this value can be changed to obtain other specified design requirements. For fine pitch gearing ($m_{et} = 1.27$ and finer), and for teeth to be finished in a secondary machining operation, the corresponding values of the clearance must be increased by 0.051 mm.

The mean addendum factor, c_{ham}, apportions the mean working depth, h_{mw}, between the pinion and wheel addendum. To prevent undercut, the pinion addendum is usually greater than the wheel addendum, with the exception of the case in which the numbers of teeth are equal. Table 12.9 gives a guideline to choose the mean addendum factor, c_{ham}, for $\Sigma = 90°$, as a function of type of gear, number of pinion teeth, and equivalent ratio, u_a, defined as:

$$u_a = \sqrt{\frac{\cos\delta_1 \tan\delta_2 \cos\eta}{\cos\delta_2}}. \tag{12.160}$$

where

$$\eta = a\sin(\sin\zeta_m \cos\delta_2) \tag{12.161}$$

is the wheel offset angle in axial plane.

As for the tooth thickness components, it is to be remembered that, in terms of European standards, the values of the thickness modification coefficient, x_{smn}, can be found by imposing the bending strength balancing between pinion and wheel, and choosing consequently the cutting process that allows to get it. Even here, the AGMA standards (they calculate the mean normal circular tooth thickness, s_{mn}, at the mean point) are more detailed. In fact, these standards allow to calculate the circular thickness factor, $k_t = 2x_{smn}$, as a function of the gear ratio, u, and number of teeth of the pinion, z_1. In the case of bending strength balancing between pinion

Table 12.9 Mean addendum factor, c_{ham}, for shaft angle $\Sigma = 90°$

Type of gear	Mean addendum factor	Number of pinion teeth
Straight bevel	$0.210 + 0.290/u_a^2$	12 or more
Zerol bevel	$0.210 + 0.290/u_a^2$	13 or more
Spiral bevel and hypoid	$0.210 + 0.290/u_a^2$	12 or more
	$0.210 + 0.280/u_a^2$	11
	$0.175 + 0.260/u_a^2$	10
	$0.145 + 0.235/u_a^2$	9
	$0.130 + 0.195/u_a^2$	8
	$0.110 + 0.160/u_a^2$	7
	$0.100 + 0.115/u_a^2$	6

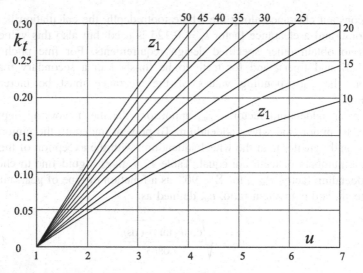

Fig. 12.43 Circular thickness factor, k_t, as a function of u and z_1

and wheel, the curves shown in Fig. 12.43 may be used. These curves have been obtained with the following relationship:

$$k_t = -0.088 + 0.092u - 0.004u^2 + 1.6 \cdot 10^{-3}(z_1 - 30)(u - 1). \qquad (12.162)$$

Of course, nothing prevents from using other values of k_t, when a different bending strength balancing is desired.

Finally, Table 12.10 provides a useful guideline for choosing the minimum values of the outer normal backlash, j_{en} (in mm). From this table, we can deduce

Table 12.10 Typical minimum normal backlash measured at outer cone

Outer transverse module, m_{et}	Minimum normal backlash, j_{en} (in mm)	
	ISO accuracy grades, 4–7	ISO accuracy grades, 8–12
25.00–20.00	0.61	0.81
20.00–16.00	0.51	0.69
16.00–12.00	0.38	0.51
12.00–10.00	0.30	0.41
10.00–8.00	0.25	0.33
8.00–6.00	0.20	0.25
6.00–5.00	0.15	0.20
5.00–4.00	0.13	0.15
4.00–3.00	0.10	0.13
3.00–2.50	0.08	0.10
2.50–2.00	0.05	0.08
2.00–1.50	0.05	0.08
1.50–1.25	0.03	0.05
1.25–1.00	0.03	0.05

that the backlash allowance is proportional to the outer transverse module m_{et} (in mm), and depends on the ISO accuracy grades, which are divided into two ranges of values.

12.12 Calculation of Spiral Bevel and Hypoid Gears

In the previous section, we said that the calculation of spiral bevel and hypoid gears is divided in two main steps, which respectively relate to the determination of the pitch cone parameters, and gear dimensions. Figure 12.44 shows the flow-chart of the aforementioned calculation structure, which deserves to be described here in more detail.

12.12.1 Determination of the Pitch Cone Parameters

Both the flow chart in Fig. 12.44 and Table 12.2 show the initial data needed to make the first step of calculation, the one concerning the determination of the pitch cone parameters. For all four methods (Method 0, Method 1, Method 2 and Method 3), these initial data include the numbers of teeth of pinion and wheel, z_1 and z_2, and therefore the gear ratio $u = z_2/z_1$. However, the calculation procedure is different

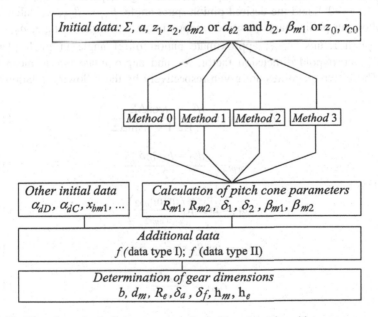

Fig. 12.44 Flow-chart of calculation structure of spiral bevel and hypoid gears

for each of these four methods. It is very simple for spiral bevel gears (Method 0), while it is an iterative procedure for the other three methods, respectively used by Gleason (Method 1), Oerlikon (Method 2), and Klingelnberg (Method 3). Therefore, the four methods here are differentiated from each other.

12.12.1.1 Method 0

The pinion pitch angle, δ_1, and wheel pitch angle, δ_2, are given by:

$$\delta_1 = \arctan\left(\frac{\sin\Sigma}{\cos\Sigma + u}\right) \quad \delta_2 = \Sigma - \delta_1. \tag{12.163}$$

The outer cone distance, R_{ei}, and mean cone distance, R_{mi}, are given by:

$$R_{ei} = \frac{d_{ei}}{2\sin\delta_i} \quad R_{mi} = R_{ei} - \frac{b_i}{2}, \tag{12.164}$$

with $i = (1.2)$.

Finally, the mean spiral angles of pinion and wheel are the same $(\beta_{m1} = \beta_{m2})$, and the face width factor of the wheel is assumed equal to 0.5 $(c_{be2} = 0.5)$.

12.12.1.2 Method 1

Using the initial data, and according to the shaft angle departure from 90°, given by $\Delta\Sigma = (\Sigma - 90°)$, and the desired pinion spiral angle, $\beta_{\Delta1} = \beta_{m1}$, we calculate first the following preliminary quantities: approximate wheel pitch angle, δ_{int2}; wheel mean pitch radius, r_{mpt2}; approximate pinion offset angle in pitch plane, ε_1'; approximate hypoid dimension factor, K_1; and approximate pinion mean radius, r_{mn1}. These five quantities are given respectively by the following relationships:

$$\delta_{int2} = \arctan\left(\frac{u\cos\Delta\Sigma}{1.2(1 - u\sin\Delta\Sigma)}\right) \tag{12.165}$$

$$r_{mpt2} = \frac{d_{e2} - b_2\sin\delta_{int2}}{2} \tag{12.166}$$

$$\varepsilon_1' = \arcsin\left(\frac{a\sin\delta_{int2}}{r_{mpt2}}\right) \tag{12.167}$$

$$K_1 = \tan\beta_{\Delta1}\sin\varepsilon_1' + \cos\varepsilon_1' \tag{12.168}$$

$$r_{mn1} = \frac{r_{mpt2}K_1}{u}. \tag{12.169}$$

Now we start the iterative calculation procedure, and for the purpose we introduce, in succession, the quantities defined below, some of which are common to
gears obtained with face-hobbing process and with face-milling process, while
others are specific of each of these two cutting processes. The quantities common to
the two cutting processes are those described below:

- Wheel offset angle in axial plane, η, with which the iterative calculation procedure begins; it is given by:

$$\eta = \arctan \left[\frac{a}{r_{mpt2}(\tan\delta_{int2}\cos\Delta\Sigma - \sin\Delta\Sigma) + r_{mn1}} \right]. \tag{12.170}$$

- Intermediate wheel offset angle in axial plane, ε_2, intermediate pinion pitch
  angle, δ_{int1}, intermediate wheel offset angle in pitch plane, ε_2', and intermediate
  pinion mean spiral angle, β_{mint1}, which are given respectively by the following
  relationships:

$$\varepsilon_2 = \arcsin \left(\frac{a - r_{mn1}\sin\eta}{r_{mpt2}} \right) \tag{12.171}$$

$$\delta_{int1} = \arctan \left(\frac{\sin\eta}{\tan\varepsilon_2\cos\Delta\Sigma} + \tan\Delta\Sigma\cos\eta \right) \tag{12.172}$$

$$\varepsilon_2' = \arcsin \left(\frac{\sin\varepsilon_2\cos\Delta\Sigma}{\cos\delta_{int1}} \right) \tag{12.173}$$

$$\beta_{mint1} = \arctan \left(\frac{K_1 - \cos\varepsilon_2'}{\sin\varepsilon_2'} \right). \tag{12.174}$$

- Increment in hypoid dimension factor, ΔK, pinion mean radius increment,
  Δr_{mpt1}, pinion offset angle in axial plane, ε_1, and pinion pitch angle, δ_1, which
  are given respectively by the following relationships:

$$\Delta K = \sin\varepsilon_2'(\tan\beta_{\Delta 1} - \tan\beta_{mint1}) \tag{12.175}$$

$$\Delta r_{mpt1} = r_{mpt2}(\Delta K/u) \tag{12.176}$$

$$\varepsilon_1 = \arcsin \left(\sin\varepsilon_2 - \frac{\Delta r_{mpt1}}{r_{mpt2}}\sin\eta \right) \tag{12.177}$$

$$\delta_1 = \arctan\left(\frac{\sin\eta}{\tan\varepsilon_1 \cos\Delta\Sigma} + \tan\Delta\Sigma \cos\eta\right). \tag{12.178}$$

– Pinion offset angle in pitch plane, ε_1', pinion spiral angle, β_{m1}, wheel spiral angle, β_{m2}, and wheel pitch angle, δ_2, which are given respectively by the following relationships:

$$\varepsilon_1' = \arcsin\left(\frac{\sin\varepsilon_1 \cos\Delta\Sigma}{\cos\delta_1}\right) \tag{12.179}$$

$$\beta_{m1} = \arctan\left(\frac{K_1 + \Delta K - \cos\varepsilon_1'}{\sin\varepsilon_1'}\right). \tag{12.180}$$

$$\beta_{m2} = \beta_{m1} - \varepsilon_1' \tag{12.181}$$

$$\delta_2 = \arctan\left(\frac{\sin\varepsilon_1}{\tan\eta \cos\Delta\Sigma} + \tan\Delta\Sigma \cos\varepsilon_1\right). \tag{12.182}$$

– Mean cone distance of pinion, R_{m1}, mean cone distance of wheel, R_{m2}, and pinion mean pitch radius, r_{mpt1}, which are given respectively by the following relationships:

$$R_{m1} = \frac{r_{mn1} + \Delta r_{mpt1}}{\sin\delta_1} \tag{12.183}$$

$$R_{m2} = \frac{r_{mpt2}}{\sin\delta_2} \tag{12.184}$$

$$r_{mpt1} = R_{m1}\sin\delta_1. \tag{12.185}$$

– Limit pressure angle, α_{lim}, and limit radius of curvature, ρ_{lim}, which are given respectively by:

$$\alpha_{lim} = \arctan\left[-\frac{\tan\delta_1 \tan\delta_2}{\cos\varepsilon_1'}\left(\frac{R_{m1}\sin\beta_{m1} - R_{m2}\sin\beta_{m2}}{R_{m1}\tan\delta_1 + R_{m2}\tan\delta_2}\right)\right] \tag{12.186}$$

$$\rho_{lim} = \frac{\sec\alpha_{lim}(\tan\beta_{m1} - \tan\beta_{m2})}{\left[-\tan\alpha_{lim}\left(\frac{\tan\beta_{m1}}{R_{m1}\tan\delta_1} + \frac{\tan\beta_{m2}}{R_{m2}\tan\delta_2}\right) + \frac{1}{R_{m1}\cos\beta_{m1}} - \frac{1}{R_{m2}\cos\beta_{m2}}\right]}. \tag{12.187}$$

The quantities specific of the face-hobbed gears are the following: number of crown gear teeth, z_p, lead angle of cutter, v, first auxiliary angle, λ, crown gear to cutter center distance, ρ_{P0}, second auxiliary angle, η_1, and lengthwise tooth mean radius of curvature, $\rho_{m\beta}$. These quantities are given respectively by the following relationships:

$$z_p = \frac{z_2}{\sin\delta_2} \tag{12.188}$$

$$v = \arcsin\left(\frac{R_{m2}z_0}{r_{c0}z_p}\cos\beta_{m2}\right) \tag{12.189}$$

$$\lambda = 90° - \beta_{m2} + v \tag{12.190}$$

$$\rho_{P0} = \sqrt{R_{m2}^2 + r_{c0}^2 - 2R_{m2}r_{c0}\cos\lambda} \tag{12.191}$$

$$\eta_1 = \arccos\left[\frac{R_{m2}\cos\beta_{m2}}{\rho_{P0}z_p}(z_p + z_0)\right] \tag{12.192}$$

$$\rho_{m\beta} = R_{m2}\cos\beta_{m2}\left[\tan\beta_{m2} + \frac{\tan\eta_1}{1 + \tan v(\tan\beta_{m2} + \tan\eta_1)}\right], \tag{12.193}$$

where r_{c0} and z_0 are respectively the cutter radius and the number of blade groups.

The specific quantity of face-milled gears is only the lengthwise tooth mean radius of curvature, $\rho_{m\beta}$, which is given by:

$$\rho_{m\beta} = r_{c0}. \tag{12.194}$$

The iterative procedure involves several calculation steps, the first of which involves the determination of all the above-mentioned quantities. With the second step, the initial value of the offset wheel angle in axial plane, η, given by Eq. (12.170), is changed, and consequently all the quantities given by Eqs. (12.171) to (12.187) are recalculated, and so on, until the following condition is satisfied:

$$\left(\frac{\rho_{m\beta}}{\rho_{lim}} - 1\right) \leq 0.01. \tag{12.195}$$

At this point, the iterative procedure is stopped, and the face width factor, c_{be2}, is calculated as follows:

$$c_{be2} = \frac{1}{b_2}\left(\frac{d_{e2}}{2\sin\delta_2} - R_{m2}\right). \tag{12.196}$$

12.12.1.3 Method 2

Using the initial data, we calculate first the following preliminary quantities:

− Lead angle of cutter, v, first auxiliary angle, λ, first approximate pinion pitch angle, δ_{1app}, first approximate wheel pitch angle, δ_{2app}, and first approximate pinion offset angle in axial plane, ζ_{mapp}, which are given respectively by the following relationships:

$$v = \arcsin\left(\frac{z_0 d_{m2}\cos\beta_{m2}}{2 r_{c0} z_2}\right) \tag{12.197}$$

$$\lambda = 90° - \beta_{m2} + v \tag{12.198}$$

$$\delta_{1app} = \arctan\left(\frac{\sin\Sigma}{\cos\Sigma + u}\right) \tag{12.199}$$

$$\delta_{2app} = \Sigma - \delta_{1app} \tag{12.200}$$

$$\zeta_{mapp} = \arcsin\left[\frac{(2a/d_{m2})}{\left(1 + \frac{\cos\delta_{2app}}{u\cos\delta_{1app}}\right)}\right]. \tag{12.201}$$

• Approximate hypoid dimension factor, F_{app}, approximate pinion mean pitch diameter, d_{m1app}, intermediate angle, φ_2, approximate mean radius of crown gear, R_{mapp}, and second auxiliary angle η_1, which are given respectively by the following relationships:

$$F_{app} = \frac{\cos\beta_{m2}}{\cos(\beta_{m2} + \zeta_{mapp})} \tag{12.202}$$

$$d_{m1app} = (F_{app} d_{m2})/u \tag{12.203}$$

$$\varphi_2 = \arctan\left[\frac{u\cos\zeta_{mapp}}{\frac{u}{\tan\delta_{2app}} + (F_{app} - 1)\sin\Sigma}\right] \tag{12.204}$$

$$R_{mapp} = d_{m2}/(2\sin\varphi_2) \tag{12.205}$$

$$\eta_1 = \arctan\left(\frac{r_{c0}\cos v - R_{mapp}\sin\beta_{m2}}{r_{c0}\sin v + R_{mapp}\cos\beta_{m2}}\right). \tag{12.206}$$

− Approximate angle, φ_3, second approximate pinion pitch angle, δ_1'', and approximate wheel pitch angle, δ_2'', projected into pinion axial plane along the common pitch plane (see Fig. 12.36), which are given respectively by the following relationships:

$$\varphi_3 = \arctan\left[\frac{\tan(\beta_{m2} + \eta_1)}{\sin\varphi_2}\right] \tag{12.207}$$

$$\delta_1'' = \arctan\left[\frac{d_{m1app}\sin\Sigma}{d_{m2}\cos\zeta_{mapp} + d_{m1app}\cos\Sigma - \dfrac{2a}{\tan(\varphi_3 + \zeta_{mapp})}}\right] \tag{12.208}$$

$$\delta_2'' = \Sigma - \delta_1''. \tag{12.209}$$

Now we start the iterative calculation procedure, and for this purpose we introduce, in succession, the quantities defined as follows:

- Improved wheel pitch angle, δ_{2imp} (the iterative procedure begins with this first quantity), auxiliary angle, η_p, approximate wheel offset angle, η_{app}, improved pinion offset angle in axial plane, ζ_{mimp}, and improved pinion offset angle in pitch plane, ζ_{mpimp}, which are given respectively by the following relationships:

$$\delta_{2imp} = \arctan\left(\tan\delta_2''\cos\zeta_{mapp}\right) \tag{12.210}$$

$$\eta_p = \arctan\left[\frac{\sin\zeta_{mapp}\cos\delta_{2imp}}{\cos(\Sigma - \delta_{2imp})}\right] \tag{12.211}$$

$$\eta_{app} = \arctan\left[\frac{2a}{d_{m2}\tan\delta_{2imp} + d_{m1app}\dfrac{\cos\eta_p\sin(\beta_{m2}+\eta_1)}{\cos(\Sigma-\delta_{2imp})}}\right] \tag{12.212}$$

$$\zeta_{mimp} = \arcsin\left[\frac{2a}{d_{m2}} - \frac{F_{app}\tan\eta_{app}\sin\delta_{2imp}\cos\eta_p}{u\cos(\Sigma - \delta_{2imp})}\right] \tag{12.213}$$

$$\zeta_{mpimp} = \arctan\left[\frac{\tan\zeta_{mimp}\sin\Sigma}{\cos(\Sigma - \delta_{2imp})}\right]. \tag{12.214}$$

- Hypoid dimension factor, F, pinion mean pitch diameter, d_{m1}, intermediate angle, φ_4, improved pinion pitch angle, δ_{1imp}'', improved wheel pitch angle, projected into pinion axial plane along the common pitch plane, δ_{2imp}'', wheel pitch angle, δ_2, and intermediate angle, φ_5, which are given respectively by the following relationships:

$$F = \frac{\cos\beta_{m2}}{\cos\left(\beta_{m2} + \zeta_{mpimp}\right)} \tag{12.215}$$

$$d_{m1} = (Fd_{m2})/u \tag{12.216}$$

$$\varphi_4 = \arctan\left[\frac{\sin\lambda\sin\Sigma}{(d_{m2}/2r_{c0}) - \cos\lambda\sin\delta_{2imp}}\right] \tag{12.217}$$

$$\delta''_{1imp} = \arctan\left[\frac{d_{m1}\sin\Sigma}{d_{m2}\cos\zeta_{mimp} + d_{m1}\cos\Sigma\cos\eta_p - \dfrac{2a}{\tan\left(\varphi_4 + \zeta_{mimp}\right)}}\right] \tag{12.218}$$

$$\delta''_{2imp} = \Sigma - \delta''_{1imp} \tag{12.219}$$

$$\delta_2 = \arctan\left(\tan\delta''_{2imp}\cos\zeta_{mpimp}\right) \tag{12.220}$$

$$\varphi_5 = \arctan\left(\frac{\tan\delta_2}{\cos\zeta_{mimp}}\right). \tag{12.221}$$

– Improved auxiliary angle, η_{pimp}, wheel offset angle in axial plane, η, pinion
  offset angle in axial plane, ζ_m, pinion offset angle in pitch plane, ζ_{mp}, pinion
  spiral angle, β_{m1}, pinion mean pitch diameter, d_{m1}, and auxiliary angle, ξ (this
  quantity is expressed by two different equations, depending on whether $\Sigma \neq 90°$
  or $\Sigma = 90°$), which are given respectively by the following relationships: the
  two Eqs. (12.228) are respectively valid for $\Sigma \neq 90°$ and $\Sigma = 90°$.

$$\eta_{pimp} = \arctan\left[\frac{\tan\eta_{app}\sin\varphi_5}{\cos(\Sigma - \varphi_5)}\right] \tag{12.222}$$

$$\eta = \arctan\left[\frac{2a}{d_{m2}\tan\delta_2 + d_{m1}\dfrac{\cos\eta_{pimp}\sin\varphi_5}{\cos(\Sigma - \varphi_5)}}\right] \tag{12.223}$$

$$\zeta_m = \arcsin(\tan\delta_2\tan\eta) \tag{12.224}$$

$$\zeta_{mp} = \arctan\left[\frac{\tan\zeta_m\sin\Sigma}{\cos(\Sigma - \delta_2)}\right] \tag{12.225}$$

$$\beta_{m1} = \beta_{m2} + \zeta_{mp} \tag{12.226}$$

$$d_{m1} = \frac{d_{m2}\cos\beta_{m2}}{u\cos\beta_{m1}} \qquad (12.227)$$

$$\xi = [\arctan(\tan\Sigma\cos\zeta_m) - \delta_2] \quad \text{or} \quad \xi = (90° - \delta_2); \qquad (12.228)$$

– Pinion pitch angle, δ_1, mean cone distance of pinion, R_{m1}, mean cone distance of wheel, R_{m2}, crown gear to cutter center distance, ρ_{P0}, intermediate angle, φ_6, complementary angle, φ_{comp}, and checking variable, R_{mcheck}, which are given respectively by the following relationships:

$$\delta_1 = \arctan\left(\tan\xi\cos\zeta_{mp}\right) \qquad (12.229)$$

$$R_{m1} = d_{m1}/(2\sin\delta_1) \qquad (12.230)$$

$$R_{m2} = d_{m2}/(2\sin\delta_2) \qquad (12.231)$$

$$\rho_{P0} = \sqrt{R_{m2}^2 + r_{c0}^2 - 2R_{m2}r_{c0}\cos\lambda} \qquad (12.232)$$

$$\varphi_6 = \arcsin\left(\frac{r_{c0}\sin\lambda}{\rho_{P0}}\right) \qquad (12.233)$$

$$\varphi_{comp} = 180° - \zeta_{mp} - \varphi_6 \qquad (12.234)$$

$$R_{mcheck} = \frac{R_{m2}\sin\varphi_6}{\sin\varphi_{comp}}. \qquad (12.235)$$

The iterative procedure involves also here several calculation steps, the first of which involves the determination of all the quantities above. With the second step, the initial value of the improved wheel pitch angle, δ_{2imp}, given by Eq. (12.210), is changed, and consequently all the quantities given by Eqs. (12.211) to (12.235) are recalculate, and so on, until the following condition is satisfied (note that δ_{2imp} must be increased if $R_{m1} < R_{mcheck}$, and vice versa):

$$\left|\frac{R_{m1}}{R_{mcheck}} - 1\right| \leq 0.01. \qquad (12.236)$$

At this point, the iterative procedure is stopped, and the face width factor c_{be2} is taken equal to 0.5 ($c_{be2} = 0.5$).

12.12.1.4   Method 3

Using the initial data, we calculate first the wheel pitch angle, δ_2, and pinion pitch angle, δ_1, which are to be considered as two preliminary quantities; they are given respectively by the following relationships:

$$\delta_2 = \arctan\left(\frac{\sin\Sigma}{(F/u) + \cos\Sigma}\right) \tag{12.237}$$

$$\delta_1 = \Sigma - \delta_2, \tag{12.238}$$

where F is the hypoid dimension factor, to be assumed equal to 1 $(F = 1)$ for iteration that follows.

Now we start the iterative calculation procedure, and for this purpose we introduce, in succession, the quantities defined below, which are:

– Wheel mean pitch diameter, d_{m2} (this is the quantity with which the iterative procedure begins), pinion offset angle in axial plane, ζ_m, pinion pitch angle, δ_1, pinion offset angle in pitch plane, ζ_{mp}, mean normal module, m_{mn}, and spiral angle of pinion, β_{m1}, which are given respectively by the following relationships:

$$d_{m2} = d_{e2} - b_2\sin\delta_2 \tag{12.239}$$

$$\zeta_m = \arcsin\left[\frac{2a}{d_{m2}\left(1 + \frac{F\cos\delta_2}{u\cos\delta_1}\right)}\right] \tag{12.240}$$

$$\delta_1 = \arcsin(\cos\zeta_m\sin\Sigma\cos\delta_2 - \cos\Sigma\sin\delta_2) \tag{12.241}$$

$$\zeta_{mp} = \arcsin\left(\frac{\sin\zeta_m\sin\Sigma}{\cos\delta_1}\right) \tag{12.242}$$

$$m_{mn} = (d_{m2}\cos\beta_{m2})/z_2 \tag{12.243}$$

$$\beta_{m1} = \beta_{m2} + \zeta_{mp}. \tag{12.244}$$

– Hypoid dimension factor, F, pinion mean pitch diameter, d_{m1}, mean cone distance of pinion, R_{m1}, mean cone distance of wheel, R_{m2}, lead angle of cutter, v, and auxiliary angle, ϑ_m, which are given respectively by the following relationships:

$$F = (\cos\beta_{m2}/\cos\beta_{m1}) \tag{12.245}$$

$$d_{m1} = (Fd_{m2})/u \tag{12.246}$$

$$R_{m1} = d_{m1}/(2\sin\delta_1) \tag{12.247}$$

$$R_{m2} = d_{m2}/(2\sin\delta_2) \tag{12.248}$$

$$v = \arcsin\left(\frac{z_0 m_{mn}}{2r_{c0}}\right) \tag{12.249}$$

$$\vartheta_m = \arctan(\sin\delta_2\tan\zeta_m). \tag{12.250}$$

- Intermediate variables A_3, A_4, A_5, A_6, A_7, and R_{mint}, pinion pitch angle, δ_1, and wheel pitch angle, δ_2, which are given respectively by the following relationships:

$$A_3 = r_{c0}\cos^2(\beta_{m2} - v) \tag{12.251}$$

$$A_4 = R_{m2}\cos(\beta_{m2} + \vartheta_m)\cos\beta_{m2} \tag{12.252}$$

$$A_5 = \sin\zeta_{mp}\cos\vartheta_m\cos v \tag{12.253}$$

$$A_6 = R_{m2}\cos\beta_{m2} + r_{c0}\sin v \tag{12.254}$$

$$A_7 - \cos\beta_{m1}\cos(\beta_{m2} + \vartheta_m) - \frac{\sin(\beta_{m2} + \vartheta_m - v)\sin\zeta_{mp}}{\cos(\beta_{m2} - v)} \tag{12.255}$$

$$R_{mint} = \frac{A_3 A_4}{A_5 A_6 + A_3 A_7} \tag{12.256}$$

$$\delta_1 = \arcsin\left(\frac{d_{m1}}{2R_{mint}}\right) \tag{12.257}$$

$$\delta_2 = \arccos\left(\frac{\sin\delta_1\cos\zeta_m\sin\Sigma + \cos\delta_1\cos\zeta_{mp}\cos\Sigma}{1 - \sin^2\Sigma\sin^2\zeta_m}\right). \tag{12.258}$$

Here also, the iterative procedure involves several calculation steps, the first of which involves the determination of all quantities above. With the second step, the initial value of the wheel mean pitch diameter, d_{m2}, given by Eq. (12.239), is changed, and consequently all the quantities given by Eqs. (12.240) to (12.258) are recalculated, and so on, until the following condition is satisfied:

$$|R_{\text{mint}} - R_{m1}| \leq 1 \cdot 10^{-4} R_{m1}. \tag{12.259}$$

At this point, the iterative procedure is stopped, and the face width factor c_{be2} is taken equal to 0.5 ($c_{be2} = 0.5$).

12.12.2   Determination of the Gear Dimensions

Once the pitch cone parameters are known, for determination of the gear dimensions, which constitutes the second main step of the calculation procedure of spiral bevel and hypoid gears (see flow chart shown in Fig. 12.44), the set of additional data summarized in Table 12.6 is required. Some of the gear dimensions to be determined are common to the four calculation methods described in the two previous sections, other instead change from method to method, and other, for a given method, are different depending on whether the gears under consideration are face-hobbed gears or face-milled gears. Whenever differences are found in this regard, we will care to highlight them on time.

12.12.2.1   Determination of the Basic Data

Using the pitch cone parameters and additional data, we calculate successively the basic data, to be used at least partially in further calculation, as follows:

- Pinion mean pitch diameter, d_{m1}, wheel mean pitch diameter, d_{m2} and shaft angle departure from 90°, $\Delta\Sigma$, which are given respectively by the following relationships:

$$d_{m1} = 2R_{m1}\sin\delta_1 \tag{12.260}$$

$$d_{m2} = 2R_{m2}\sin\delta_2 \tag{12.261}$$

$$\Delta\Sigma = \Sigma - 90°. \tag{12.262}$$

- Offset angle in pinion axial plane, ζ_m, and offset angle in pitch plane, ζ_{mp}, which are given respectively by the following relationships:

$$\zeta_m = \arcsin\left(\frac{2a\cos\delta_1}{d_{m2}\cos\delta_1 + d_{m1}\cos\delta_2}\right) \tag{12.263}$$

$$\zeta_{mp} = \arcsin\left(\frac{\sin\zeta_m\sin\Sigma}{\cos\delta_1}\right). \tag{12.264}$$

- Offset in pitch plane, a_p, mean normal module, m_{mn}, and limit pressure angle, α_{lim}, which are given respectively by the following relationships:

$$a_p = R_{m2}\sin\zeta_{mp} \tag{12.265}$$

$$m_{mn} = \frac{2R_{m2}\sin\delta_2\cos\beta_{m2}}{z_2} \tag{12.266}$$

$$\alpha_{lim} = -\arctan\left[\frac{\tan\delta_1\tan\delta_2}{\cos\zeta_{mp}}\left(\frac{R_{m1}\sin\beta_{m1} - R_{m2}\sin\beta_{m2}}{R_{m1}\tan\delta_1 + R_{m2}\tan\delta_2}\right)\right]. \tag{12.267}$$

- Generated normal pressure angles on drive side, α_{nD} and on coast side, α_{nC}, and effective pressure angles on drive side, α_{eD}, and on coast side, α_{eC}, which are given respectively by the following relationships:

$$\alpha_{nD} = \alpha_{dD} + f_{\alpha lim}\alpha_{lim} \tag{12.268}$$

$$\alpha_{nC} = \alpha_{dC} - f_{\alpha lim}\alpha_{lim} \tag{12.269}$$

$$\alpha_{eD} = \alpha_{nD} - \alpha_{lim} \tag{12.270}$$

$$\alpha_{eC} = \alpha_{nC} + \alpha_{lim}. \tag{12.271}$$

- Outer and inner pitch cone distances of wheel, R_{e2} and R_{i2}, outer and inner pitch diameters of wheel, d_{e2} and d_{i2}, and outer transverse module, m_{et2}, which are given respectively by the following relationships:

$$R_{e2} = R_{m2} + c_{be2}b_2 \tag{12.272}$$

$$R_{i2} = R_{e2} - b_2 \tag{12.273}$$

$$d_{e2} = 2R_{e2}\sin\delta_2 \tag{12.274}$$

$$d_{i2} = 2R_{i2}\sin\delta_2 \tag{12.275}$$

$$m_{et2} = d_{e2}/z_2. \tag{12.276}$$

- Wheel face width from calculation point to outside, b_{e2}, wheel face width from calculation point to inside, b_{i2}, crossing point to calculation point along wheel axis, t_{zm2}, crossing point to calculation point along pinion axis, t_{zm1}, and pitch apex beyond crossing point along axis, $t_{z1,2}$, which are given respectively by the following relationships:

$$b_{e2} = R_{e2} - R_{m2} \tag{12.277}$$

$$b_{i2} = R_{m2} - R_{i2} \tag{12.278}$$

$$t_{zm2} = \frac{d_{m1}\sin\delta_2}{2\cos\delta_1} - \frac{a\tan\Delta\Sigma}{\tan\zeta_m} \tag{12.279}$$

$$t_{zm1} = \frac{d_{m2}\cos\zeta_m\cos\Delta\Sigma}{2} - t_{zm2}\sin\Delta\Sigma \tag{12.280}$$

$$t_{z1,2} = R_{m1,2}\cos\delta_{1,2} - t_{zm1,2}. \tag{12.281}$$

12.12.2.2 Determination of the Tooth Depth at Calculation Point

The quantities concerning the tooth depth are the mean working depth, h_{mw}, mean addendum of wheel, h_{am2}, mean dedendum of wheel, h_{fm2}, mean addendum of pinion, h_{am1}, mean dedendum of pinion, h_{fm1}, clearance, c, and mean whole depth, h_m. These quantities are given respectively by the following relationships (see Table 12.8 for other symbols):

$$h_{mw} = 2m_{mn}k_{hap} \tag{12.282}$$

$$h_{am2} = m_{mn}\left(k_{hap} - x_{hm1}\right) \tag{12.283}$$

$$h_{fm2} = m_{mn}\left(k_{hfp} + x_{hm1}\right) \tag{12.284}$$

$$h_{am1} = m_{mn}\left(k_{hap} + x_{hm1}\right) \tag{12.285}$$

$$h_{fm1} = m_{mn}\left(k_{hfp} - x_{hm1}\right) \tag{12.286}$$

$$c = m_{mn}\left(k_{hfp} - k_{hap}\right) \tag{12.287}$$

$$h_m = h_{am1,2} + h_{fm1,2} = m_{mn}\left(k_{hap} + k_{hfp}\right) \tag{12.288}$$

12.12.2.3 Determination of the Root Angles and Face Angles

The quantities concerning the root angles and face angles are, in ordered sequence, as follows:

- Face and root angles of wheel, δ_{a2} and δ_{f2}, auxiliary angle for calculating pinion offset angle in root plane, φ_R and auxiliary angle for calculating pinion offset angle in face plane, φ_0, which are given respectively by the following relationships:

$$\delta_{a2} = \delta_2 + \vartheta_{a2} \tag{12.289}$$

$$\delta_{f2} = \delta_2 - \vartheta_{f2} \tag{12.290}$$

$$\varphi_R = \arctan\left[\frac{a(\tan \Delta\Sigma \cos \delta_{f2})}{R_{m2} \cos \vartheta_{f2} - t_{z2} \cos \delta_{f2}}\right] \tag{12.291}$$

$$\varphi_0 = \arctan\left[\frac{a(\tan\Delta\Sigma\cos\delta_{a2})}{R_{m2}\cos\vartheta_{a2} - t_{z2}\cos\delta_{a2}}\right]. \tag{12.292}$$

- Pinion offset angle in root plane, ζ_R, pinion offset angle in face plane, ζ_0, face angle of pinion, δ_{a1}, root angle of pinion, δ_{f1}, addendum angle of pinion, ϑ_{a1}, and dedendum angle of pinion, ϑ_{f1}, which are given respectively by the following relationships:

$$\zeta_R = \arcsin\left[\frac{a(\cos\varphi_R\sin\delta_{f2})}{R_{m2}\cos\vartheta_{f2} - t_{z2}\cos\delta_{f2}}\right] - \varphi_R \tag{12.293}$$

$$\zeta_0 = \arcsin\left[\frac{a(\cos\varphi_0\sin\delta_{a2})}{R_{m2}\cos\vartheta_{a2} - t_{z2}\cos\delta_{a2}}\right] - \varphi_0 \tag{12.294}$$

$$\delta_{a1} = \arcsin\left(\sin\Delta\Sigma\sin\delta_{f2} + \cos\Delta\Sigma\cos\delta_{f2}\cos\zeta_R\right) \tag{12.295}$$

$$\delta_{f1} = \arcsin(\sin\Delta\Sigma\sin\delta_{a2} + \cos\Delta\Sigma\cos\delta_{a2}\cos\zeta_0) \tag{12.296}$$

$$\vartheta_{a1} = \delta_{a1} - \delta_1 \tag{12.297}$$

$$\vartheta_{f1} = \delta_1 - \delta_{f1}. \tag{12.298}$$

– Wheel face apex beyond crossing point along wheel axis, t_{zF2}, wheel root apex beyond crossing point along wheel axis, t_{zR2}, pinion face apex beyond crossing point along pinion axis, t_{zF1}, and pinion root apex beyond crossing point along pinion axis,t_{zR1}, which are given respectively by the following relationships:

$$t_{zF2} = t_{z2} - \frac{R_{m2}\cos\vartheta_{a2} - h_{am2}\cos\vartheta_{a2}}{\sin\delta_{a2}} \tag{12.299}$$

$$t_{zR2} = t_{z2} + \frac{R_{m2}\sin\vartheta_{f2} - h_{fm2}\cos\vartheta_{f2}}{\sin\delta_{f2}} \tag{12.300}$$

$$t_{zF1} = \frac{a\sin\zeta_R\cos\delta_{f2} - t_{zR2}\sin\delta_{f2} - c}{\sin\delta_{a1}} \tag{12.301}$$

$$t_{zR1} = \frac{a\sin\zeta_0\cos\delta_{a2} - t_{zF2}\sin\delta_{a2} - c}{\sin\delta_{f1}}. \tag{12.302}$$

12.12.2.4 Determination of the Pinion Face Width

For determination of the pinion face width, b_1, we calculate first the pinion face width in pitch plane, b_{p1}, and pinion face width from calculation point to front crown, b_{1A}, using the following relationships:

$$b_{p1} = \sqrt{R_{e2}^2 - a_p^2} - \sqrt{R_{i2}^2 - a_p^2} \tag{12.303}$$

$$b_{1A} = \sqrt{R_{m2}^2 - a_p^2} - \sqrt{R_{i2}^2 - a_p^2}. \tag{12.304}$$

Once this is done, we differentiate the calculation for each of the above-mentioned four methods.

For Method 0, the quantities involved are the pinion face width, b_1, pinion face width from calculation point to outside, b_{e1}, and pinion face width from calculation point to inside, b_{i1} (see Fig. 12.45a), which are given respectively by the following relationships:

$$b_1 = b_2; \quad b_{e1} = c_{be2}b_1; \quad b_{i1} = b_1 - b_{e1}. \tag{12.305}$$

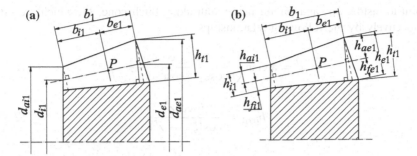

Fig. 12.45 Pinion geometry: **a** inner and outer diameters, and face width; **b** tooth depth, and face width

For Method 1, the quantities involved are the auxiliary angle, λ', pinion face width in pitch plane, b_{reri1} (see Fig. 12.46), pinion face width increment along pinion axis, Δb_{x1}, increment along pinion axis from calculation point to outside, Δg_{xe}, increment along pinion axis from calculation point to inside, Δg_{xi}, pinion face width from calculation point to outside, b_{e1}, pinion face width from calculation

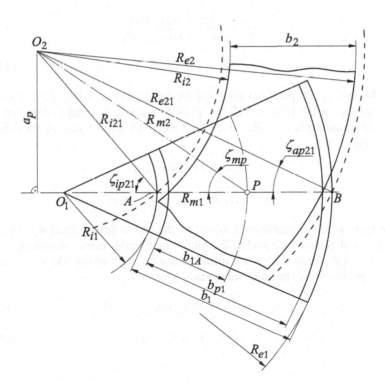

Fig. 12.46 Quantities in pitch plane for determination of pinion spiral angle

point to inside, b_{i1}, and pinion face width along pitch cone, b_1, which are given respectively by the following relationships:

$$\lambda' = \arctan\left(\frac{\sin\zeta_{mp}\cos\delta_2}{u\cos\delta_1 + \cos\delta_2\cos\zeta_{mp}}\right) \tag{12.306}$$

$$b_{reri1} = \frac{b_2\cos\lambda'}{\cos(\zeta_{mp} - \lambda')} \tag{12.307}$$

$$\Delta b_{x1} = h_{mw}\sin\zeta_R\left(1 - \frac{1}{u}\right) \tag{12.308}$$

$$\Delta g_{xe} = \frac{c_{be2}b_{reri1}}{\cos\vartheta_{a1}}\cos\delta_{a1} + \Delta b_{x1} - (h_{fm2} - c)\sin\delta_1 \tag{12.309}$$

$$\Delta g_{xi} = \frac{(1 - c_{be2})b_{reri1}}{\cos\vartheta_{a1}}\cos\delta_{a1} + \Delta b_{x1} + (h_{fm2} - c)\sin\delta_1 \tag{12.310}$$

$$b_{e1} = \frac{\Delta g_{xe} + h_{am1}\sin\delta_1}{\cos\delta_{a1}}\cos\vartheta_{a1} \tag{12.311}$$

$$b_{i1} = \frac{\Delta g_{xi} - h_{am1}\sin\delta_1}{\cos\delta_1 - \tan\vartheta_{a1}\sin\delta_1} \tag{12.312}$$

$$b_1 = b_{i1} + b_{e1}. \tag{12.313}$$

For Method 2, the quantities involved are the pinion face width along pitch cone, b_1, pinion face width from calculation point to outside, b_{e1}, and pinion face width from calculation point to inside, b_{i1}, which are given respectively by the following relationships:

$$b_1 = b_2\left(1 + \tan^2\zeta_{mp}\right) \tag{12.314}$$

$$b_{e1} = c_{be2}b_1; \quad b_{i1} = b_1 - b_{e1}. \tag{12.315}$$

For Method 3, the quantities involved are the pinion face width along pitch cone, b_1, additional pinion face width, b_x, pinion face width from calculation point to inside, b_{i1}, and pinion face width from calculation point to outside, b_{e1}, which are given respectively by the following relationships:

$$b_1 = \text{int}\left(b_{p1} + 3m_{mn}\tan|\zeta_{mp}| + 1\right) \tag{12.316}$$

$$b_x = (b_1 - b_{p1})/2 \tag{12.317}$$

$$b_{i1} = b_{1A} + b_x; \quad b_{e1} = b_1 - b_{i1}. \tag{12.318}$$

From what has been said above, we realize that each method involves their specific formulas, which are related to the corresponding calculation procedures of the pitch cone parameters. For spiral bevel gears, the pinion and wheel face widths are equal. The calculation point is located in the middle of the pinion face width for hypoid gears calculated with Method 2; consequently, the inner and outer pinion face widths, b_{i1} and b_{e1}, at the pitch cone are equal. For hypoid gears calculated with Method 1 and 3, the inner and outer pinion face widths, b_{i1} and b_{e1}, are instead different. With Method 3, an integer value of the pinion face width along the pitch cone, b_{i1}, is chosen. Figure 12.45a shows the quantities of interest as well as how to get b_{i1} and b_{e1}, which represent the actual distances from calculation point to inside and outside, measured along the pitch cone generatrix. In accordance with what has been done for pinion, also the inner and outer wheel face widths, b_{i1} and b_{e1}, can be determined. Once b_{i1}, b_{e1}, b_{i2} and b_{e2} have been calculated, it is possible to determine the inner and outer diameters of pinion and wheel, d_{ai1}, d_{ae1}, d_{ai2}, and d_{ae2}, for all methods with the same equations.

12.12.2.5   Determination of the Inner and Outer Spiral Angles

For the determination of the inner and outer spiral angles, we must distinguish between pinion and wheel, because they are calculated with different formulas for face-hobbed and face-milled wheels, while the previous quantities have been calculated using the same formulas both for face-hobbed and face-milled wheels.

The spiral angles of the pinion are determined at its front and rear crowns, i.e. at the inner and outer cone distances, R_{i1} and R_{e1} (Fig. 12.46). To determine the values of these spiral angles, it is first necessary to define the spiral angles of the wheel at boundary points of the hypoid gear pair in pitch plane (points A and B in Fig. 12.46). At these points, the corresponding cone distances, R_{i21} and R_{e21}, of the wheel may be smaller/larger than the inner/outer wheel cone distances, R_{i2} and R_{e2}; this is due to the overlap of the pinion face over the wheel face. The circles with center O_2 and radii R_{i21} and R_{e21} are shown by dashed lines in Fig. 12.46, which also highlights the pinion offset angles in pitch plane, ζ_{ip21} and ζ_{ep21}, at inner boundary point, A, and at outer boundary point, B. These angles are used to determine the inner and outer spiral angles, β_{i1} and β_{e1}, of pinion. However, it is noteworthy that, to arrive at common formulas valid for both face-hobbed and face-milled pinions, a different set of intermediate formulas is used for face-hobbing and face-milling.

For the determination of the inner and outer spiral angles of pinion, we first calculate the wheel cone distance of outer pinion boundary point, R_{e21} (it may be larger than R_{e2}), and the wheel cone distance of inner boundary point, R_{i21} (it may be smaller than R_{i2}), using the following equations:

$$R_{e21} = \sqrt{R_{m2}^2 + b_{e1}^2 + 2R_{m2}b_{e1}\cos\zeta_{mp}} \qquad (12.319)$$

$$R_{i21} = \sqrt{R_{m2}^2 + b_{i1}^2 - 2R_{m2}b_{i1}\cos\zeta_{mp}}. \qquad (12.320)$$

For face-bobbed pinions, the intermediate quantities involved are the lead angle of cutter, v, crown gear to cutter center distance, ρ_{P0}, epicycloid base circle radius, ρ_b, auxiliary angle, φ_{e21}, auxiliary angle, φ_{i21}, wheel spiral angle at outer boundary point, β_{e21}, and wheel spiral angle at inner boundary point, β_{i21}, which are given respectively by the following relationships:

$$v = \arcsin\left(\frac{z_0 m_{mn}}{2r_{c0}}\right) \qquad (12.321)$$

$$\rho_{P0} = \sqrt{R_{m2}^2 + r_{c0}^2 - 2R_{m2}r_{c0}\sin(\beta_{m2} - v)} \qquad (12.322)$$

$$\rho_b = \frac{\rho_{P0}}{[1 + (z_0/z_2)\sin\delta_2]} \qquad (12.323)$$

$$\varphi_{e21} = \arccos\left(\frac{R_{e21}^2 + \rho_{P0}^2 - r_{c0}^2}{2R_{e21}\rho_{P0}}\right) \qquad (12.324)$$

$$\varphi_{i21} = \arccos\left(\frac{R_{i21}^2 + \rho_{P0}^2 - r_{c0}^2}{2R_{i21}\rho_{P0}}\right) \qquad (12.325)$$

$$\beta_{e21} = \arctan\left(\frac{R_{e21} - \rho_b\cos\varphi_{e21}}{\rho_b\sin\varphi_{e21}}\right) \qquad (12.326)$$

$$\beta_{i21} = \arctan\left(\frac{R_{i21} - \rho_b\cos\varphi_{i21}}{\rho_b\sin\varphi_{i21}}\right). \qquad (12.327)$$

For face-milled pinions, the intermediate quantities involved are the wheel spiral angle at outer boundary point, β_{e21}, and wheel spiral angle at inner boundary point, β_{i21}, which are given respectively by the following relationships:

$$\beta_{e21} = \arcsin\left(\frac{2R_{m2}r_{c0}\sin\beta_{m2} - R_{m2}^2 + R_{e21}^2}{2R_{e21}r_{c0}}\right) \qquad (12.328)$$

$$\beta_{i21} = \arcsin\left(\frac{2R_{m2}r_{c0}\sin\beta_{m2} - R_{m2}^2 + R_{i21}^2}{2R_{i21}r_{c0}}\right). \qquad (12.329)$$

The following formulas are common to face-hobbled and face-milled pinions. The quantities involved in these formulas are the pinion offset angle in pitch plane at outer boundary point, ζ_{ep21}, pinion offset angle in pitch plane at inner boundary point, ζ_{ip21}, outer pinion spiral angle, β_{e1}, and inner pinion spiral angle, β_{ei1}, which are given respectively by the following relationships:

$$\zeta_{ep21} = \arcsin\left(\frac{a_p}{R_{e21}}\right) \tag{12.330}$$

$$\zeta_{ip21} = \arcsin\left(\frac{a_p}{R_{i21}}\right) \tag{12.331}$$

$$\beta_{e1} = \beta_{e21} + \zeta_{ep21} \tag{12.332}$$

$$\beta_{i1} = \beta_{i21} + \zeta_{ip21}. \tag{12.333}$$

Of course, despite the final formulas are the same, the results will be different, since the quantities appearing in Eqs. (12.332) and (12.333) are different for face-hobbed and face-milled pinions.

For the determination of the inner and outer spiral angles of wheel, we have to use different formulas for face-hobbing and face-milling.

For face-hobbed wheels, the quantities involved are the auxiliary angle, φ_{e2}, auxiliary angle, φ_{i2}, outer wheel spiral angle, β_{e2}, and inner wheel spiral angle, β_{i2}, which are given respectively by the following relationships:

$$\varphi_{e2} = \arccos\left(\frac{R_{e2}^2 + \rho_{P0}^2 - r_{c0}^2}{2R_{e2}\rho_{P0}}\right) \tag{12.334}$$

$$\varphi_{i2} = \arccos\left(\frac{R_{i2}^2 + \rho_{P0}^2 - r_{c0}^2}{2R_{i2}\rho_{P0}}\right) \tag{12.335}$$

$$\beta_{e2} = \arctan\left(\frac{R_{e2} \quad \rho_b\cos\varphi_{e2}}{\rho_b\sin\varphi_{e2}}\right) \tag{12.336}$$

$$\beta_{i2} = \arctan\left(\frac{R_{i2} - \rho_b\cos\varphi_{i2}}{\rho_b\sin\varphi_{i2}}\right). \tag{12.337}$$

For face-milled wheels, the quantities involved are the outer wheel spiral angle, β_{e2}, and inner wheel spiral angle, β_{i2}, which are given respectively by the following relationships:

$$\beta_{e2} = \arcsin\left(\frac{2R_{m2}r_{c0}\sin\beta_{m2} - R_{m2}^2 + R_{e2}^2}{2R_{e2}r_{c0}}\right) \tag{12.338}$$

$$\beta_{i2} = \arcsin\left(\frac{2R_{m2}r_{c0}\sin\beta_{m2} - R_{m2}^2 + R_{i2}^2}{2R_{i2}r_{c0}}\right). \tag{12.339}$$

12.12.2.6   Determination of the Tooth Depth

Figure 12.45b shows the quantities concerning the inner and outer tooth depth of pinion. It is to be noted that quantities h_{ai1}, h_{fi1}, h_{am1}, h_{fm1}, h_{ae1}, and h_{fe1} are measured along the normal to the pitch cone generatrix, while quantity h_{t1} is measured along the normal to the root cone generatrix. The inner and outer addendum and dedendum can obtained easily taking into account inner and outer face widths of the pinion.

Here, the quantities that interest us are the outer addendum, h_{ae}, outer dedendum, h_{fe}, outer whole depth, h_e, inner addendum, h_{ai}, inner dedendum, h_{fi}, and inner whole depth, h_i, which are given respectively by the following relationships:

$$h_{ae1,2} = h_{am1,2} + b_{e1,2}\tan\vartheta_{a1,2} \tag{12.340}$$

$$h_{af1,2} = h_{fm1,2} + b_{e1,2}\tan\vartheta_{f1,2} \tag{12.341}$$

$$h_{e1,2} = h_{ae1,2} + h_{fe1,2} \tag{12.342}$$

$$h_{ai1,2} = h_{am1,2} - b_{i1,2}\tan\vartheta_{a1,2} \tag{12.343}$$

$$h_{fi1,2} = h_{fm1,2} - b_{i1,2}\tan\vartheta_{f1,2} \tag{12.344}$$

$$h_{i1,2} = h_{ai1,2} + h_{fi1,2}. \tag{12.345}$$

12.12.2.7   Determination of the Tooth Thickness

In order to determine the tooth thickness of pinion and wheel, the outer normal backlash, j_{en}, normal backlash at calculation point, j_{mn}, transverse backlash at calculation point, j_{mt2}, and outer transverse backlash, j_{et2}, are first to be calculated. As Fig. 12.42 shows, the thickness modification coefficient, x_{smn}, which is included among type-I data, is a theoretical value, and does not take into account the backlash. To take into account the backlash, the thickness modification coefficients, x_{sm1} and x_{sm2}, are to be determined. The backlash has to be set equal to zero only when the theoretical tooth thickness is considered.

For the determination of the tooth thickness of pinion and wheel, it is necessary first to calculate the mean normal pressure angle, α_n, using the following relationship:

$$\alpha_n = \frac{1}{2}(\alpha_{nD} + \alpha_{nC}). \tag{12.346}$$

For the pinion, it is necessary to calculate the thickness modification coefficient, x_{sm1} , using one of the following four equations:

$$x_{sm1} = x_{smn} - \frac{j_{en}}{4m_{mn}\cos\alpha_n} \frac{R_{m2}\cos\beta_{m2}}{R_{e2}\cos\beta_{e2}} = x_{sm1} - \frac{j_{et2}R_{m2}\cos\beta_{m2}}{4m_{mn}R_{e2}}$$

$$= x_{smn} - \frac{j_{mn}}{4m_{mn}\cos\alpha_n} = x_{smn} - \frac{j_{mt2}\cos\beta_{m2}}{4m_{mn}}. \tag{12.347}$$

These equations are equivalent to each other, and we choose that for which the backlash that appears in their expressions is one of the additional data summarized in Table 12.6.

The mean normal circular tooth thickness of pinion, s_{mn1}, is given by:

$$s_{mn1} = 0.5\pi m_{mn} + 2m_{mn}(x_{sm1} + x_{hm1}\tan\alpha_n). \tag{12.348}$$

For the wheel, it is necessary to calculate the thickness modification coefficient, x_{sm2}, using one of the following four equations:

$$x_{sm2} = -x_{smn} - \frac{j_{en}}{4m_{mn}\cos\alpha_n} \frac{R_{m2}\cos\beta_{m2}}{R_{e2}\cos\beta_{e2}} = -x_{smn} - \frac{j_{et2}R_{m2}\cos\beta_{m2}}{4m_{mn}R_{e2}}$$

$$= -x_{smn} - \frac{j_{mn}}{4m_{mn}\cos\alpha_n} = -x_{smn} - \frac{j_{mt2}\cos\beta_{m2}}{4m_{mn}}. \tag{12.349}$$

These four equations are also equivalent to each other, and we choose that for which the backlash appearing in their expressions is one of the additional data shown in Table 12.6.

The mean normal circular tooth thickness of wheel, s_{mn2}, is given by:

$$s_{mn2} = 0.5\pi m_{mn} + 2m_{mn}(x_{sm2} - x_{hm1}\tan\alpha_n). \tag{12.350}$$

Finally, the mean transverse circular thickness, s_{mt}, mean normal diameter, d_{mn}, mean normal chordal tooth thickness, s_{mnc}, and mean chordal addendum, h_{amc}, are given respectively by the following expressions:

$$s_{mt1,2} = s_{mn1,2}/\cos\beta_{m1,2} \tag{12.351}$$

$$d_{mn1,2} = \frac{d_{m1,2}}{(1 - \sin^2\beta_{m1,2}\cos^2\alpha_n)\cos\beta_{m1,2}\cos\delta_{1,2}} \tag{12.352}$$

$$s_{mnc1,2} = d_{mn1,2}\sin(s_{mn1,2}/d_{mn1,2}) \tag{12.353}$$

$$h_{amc1,2} = h_{am1,2} + 0.5d_{mn1,2}\cos\delta_{1,2}\left[1 - \cos\left(\frac{s_{mn1,2}}{d_{mn1,2}}\right)\right]. \tag{12.354}$$

12.12.2.8   Determination of the Remaining Dimensions

The remaining dimensions to be determined are the outer pitch cone distance of pinion, R_{e1}, inner pitch cone distance of pinion, R_{i1}, outer and inner pitch diameters of pinion, d_{e1} and d_{i1}, outside diameter, d_{ae}, diameters d_{fe}, d_{ai} and d_{fi}, crossing point to crown along axis, $t_{xo1,2}$, crossing point to front crown along axis, $t_{xi1,2}$, and pinion whole depth perpendicular to the root cone, h_{t1}, which are given by:

$$R_{e1} = R_{m1} + b_{e1} \tag{12.355}$$

$$R_{i1} = R_{m1} - b_{i1} \tag{12.356}$$

$$d_{e1} = 2R_{e1}\sin\delta_1 \tag{12.357}$$

$$d_{i1} = 2R_{i1}\sin\delta_1 \tag{12.358}$$

$$d_{ae1,2} = d_{e1,2} + 2h_{ae1,2}\cos\delta_{1,2} \tag{12.359}$$

$$d_{fe1,2} = d_{e1,2} - 2h_{fe1,2}\cos\delta_{1,2} \tag{12.360}$$

$$d_{ai1,2} = d_{i1,2} + 2h_{ai1,2}\cos\delta_{1,2} \tag{12.361}$$

$$d_{fi1,2} = d_{i1,2} - 2h_{fi1,2}\cos\delta_{1,2} \tag{12.362}$$

$$t_{xo1,2} = t_{zm1,2} + b_{e1,2}\cos\delta_{1,2} - h_{ae1,2}\sin\delta_{1,2} \tag{12.363}$$

$$t_{xi1,2} = t_{zm1,2} - b_{i1,2}\cos\delta_{1,2} - h_{ai1,2}\sin\delta_{1,2} \tag{12.364}$$

$$h_{t1} = \frac{t_{zF1} + t_{xo1}}{\cos\delta_{a1}}\sin(\vartheta_{a1} + \vartheta_{f1}) - (t_{zR1} - t_{zF1})\sin\delta_{f1}. \tag{12.365}$$

12.12.3   Undercut Check

The equations in this section are used to check if undercut occurs or not. They are valid for generated gears with non-constant and constant tooth depth, and can be applied selecting any point on the pinion or wheel face widths. These equations are based on the concept of the common generation crown wheel, which is able to mesh with pinion and wheel at the same time; this common generation crown wheel has already been described in detail in the previous sections.

To make things even clearer, we can add to what we have already said that the action of the blades in the cutter-head simulates the one of a tooth of the generation crown wheel. The axis of rotation of this generation crown gear coincides with the generation cradle axis of the gear-cutting machine. The gear wheel to be generated (e.g., the pinion) rolls without sliding with the imaginary generating and mating crown gear, and in this motion its tooth spaces and flanks are manufactured. To generate the mating gear wheel to be fit properly with this pinion, a mirrored arrangement is used, i.e. this mating gear wheel is cut on the backside of the same generation crown wheel. This is a general principle, which applies independently of the lengthwise tooth form (circular, epicycloid or involute arcs).

For the undercut check of pinion, we first choose the pinion cone distance of the point to be checked, R_{x1}, obviously within the range ($R_{i1} \leq R_{x1} \leq R_{e1}$), and then we define the wheel cone distance of the appropriate pinion boundary point, R_{x2} (as Fig. 12.46 shows, this pinion boundary point may be smaller than R_{i2} and larger than R_{e2}), using the relationship:

$$R_{x2} = \sqrt{R_{m2}^2 + (R_{m1} - R_{x1})^2 - 2R_{m2}(R_{m1} - R_{x1})\cos\zeta_{mp}}. \qquad (12.366)$$

For face-hobbed pinion, we determine in sequence the auxiliary angle, φ_{x2}, and wheel spiral angle at checkpoint, β_{x2}, with the equations:

$$\varphi_{x2} = \arccos\left(\frac{R_{x2}^2 + \rho_{P0}^2 - r_{c0}^2}{2R_{x2}\rho_{P0}}\right) \qquad (12.367)$$

$$\beta_{x2} = \arctan\left(\frac{R_{x2} - \rho_b\cos\varphi_{x2}}{\rho_b\sin\varphi_{x2}}\right), \qquad (12.368)$$

while for face-milled pinion we determine the wheel spiral angle at checkpoint, β_{x2}, with the equation:

$$\beta_{x2} = \arcsin\left(\frac{2R_{m2}r_{c0}\sin\beta_{m2} - R_{m2}^2 + R_{x2}^2}{2R_{x2}r_{c0}}\right). \qquad (12.369)$$

The following equations are common to face-hobbed and face-milled pinions. The quantities involved in these equations are the pinion offset angle in pitch plane at checkpoint, ζ_{xp2}, pinion spiral angle at checkpoint, β_{x1}, pinion and wheel pitch diameters at checkpoint, d_{x1} and d_{x2}, normal module at checkpoint, m_{xn}, effective diameter at checkpoint of pinion, d_{Ex1}, appropriate cone distance, R_{Ex1}, intermediate value, z_{nx1}, limit pressure angle at checkpoint, α_{limx}, and effective pressure angles at checkpoint on drive side and coast side, α_{eDx} and α_{eCx}. These quantities are given respectively by the following relationships:

$$\zeta_{xp2} = \arcsin\left(\frac{a_p}{R_{x2}}\right) \qquad (12.370)$$

$$\beta_{x1} = \beta_{x2} + \zeta_{xp2} \qquad (12.371)$$

$$d_{x1} = 2R_{x1}\sin\delta_1 \tag{12.372}$$

$$d_{x2} = 2R_{x2}\sin\delta_2 \tag{12.373}$$

$$m_{xn} = (d_{x2}/z_2)\cos\beta_{x2} \tag{12.374}$$

$$d_{Ex1} = d_{x2}\frac{z_1\cos\beta_{x2}}{z_2\cos\beta_{x1}} \tag{12.375}$$

$$R_{Ex1} = \frac{d_{Ex1}}{2\sin\delta_1} \tag{12.376}$$

$$z_{nx1} = \frac{z_1}{\left(1 - \sin^2\beta_{x1}\cos^2\alpha_n\right)\cos\beta_{x1}\cos\delta_1} \tag{12.377}$$

$$\alpha_{limx} = -\arctan\left[\frac{\tan\delta_1\tan\delta_2}{\cos\zeta_{mp}}\left(\frac{R_{Ex1}\sin\beta_{x1} - R_{x2}\sin\beta_{x2}}{R_{Ex1}\tan\delta_1 + R_{x2}\tan\delta_2}\right)\right] \tag{12.378}$$

$$\alpha_{eDx} = \alpha_{nD} - \alpha_{limx} \tag{12.379}$$

$$\alpha_{eCx} = \alpha_{nC} + \alpha_{limx}. \tag{12.380}$$

It is to be noted that, for further calculations, the smaller effective pressure angle must be chosen, i.e. $\alpha_{eminx} = \alpha_{eCx}$, if $\alpha_{eCx} < \alpha_{eDx}$, and $\alpha_{eminx} = \alpha_{eDx}$, if $\alpha_{eCx} \geq \alpha_{eDx}$.

Now the minimum profile shift coefficient of pinion at calculation point, $x_{hmminx1}$, should be determined. The quantities involved are the working tool addendum at checkpoint, k_{hapx}, minimum profile shift coefficient of pinion at checkpoint, x_{hx1}, and minimum profile shift coefficient of pinion at calculation point, $x_{hmminx1}$, which are given respectively by the following relationships:

$$k_{hapx} = k_{hap} + \frac{(R_{x2} - R_{m2})\tan\vartheta_{a2}}{m_{mn}} \tag{12.381}$$

$$x_{hx1} = 1.1k_{hapx} - \frac{z_{nx1}x_{xn}\sin^2\alpha_{eminx}}{2m_{mn}} \tag{12.382}$$

$$x_{hm\,min\,x1} = x_{hx1} + \frac{(d_{Ex1} - d_{x1})\cos\delta_1}{2m_{mn}}. \tag{12.383}$$

The undercut of pinion at checkpoint is avoided, if $x_{hm1} > x_{hmminx1}$.

For the undercut check of the wheel, we also choose firstly the wheel cone distance of the point to be checked, R_{x2}, within the range $(R_{i2} \leq R_{x2} \leq R_{e2})$.

For face-hobbed wheel, we determine in sequence the auxiliary angle, φ_{x2}, and wheel spiral angle at checkpoint, β_{x2}, with the same Eqs. (12.367) and (12.368). For face-milled wheel, we determine the wheel spiral angle at checkpoint, β_{x2}, using Eq. (12.369). The other equations are common for face-hobbed and face-milled wheels. We calculate the wheel pitch diameter at checkpoint, d_{x2}, and normal module at checkpoint, m_{xn}, using Eqs. (12.373) and (12.374). Then we calculate the intermediate value, z_{nx2}, with the equation:

$$z_{nx2} = \frac{z_2}{\left(1 - \sin^2\beta_{x2}\cos^2\alpha_n\right)\cos\beta_{x2}\cos\delta_2}, \tag{12.384}$$

and we note that, for further calculations, the smaller effective pressure angle must be chosen, i.e. $\alpha_{eminx} = \alpha_{nC}$, if $\alpha_{nC} < \alpha_{nD}$, and $\alpha_{eminx} = \alpha_{nD}$, if $\alpha_{nC} \geq \alpha_{nD}$.

Now the maximum profile shift coefficient of pinion at calculation point, $x_{hmmaxx1}$, should be determined. The quantities involved are the working tool addendum at checkpoint, k_{hapx}, and maximum profile shift coefficient of pinion at calculation point, $x_{hmmaxx1}$, which are given respectively by the following relationships:

$$k_{hapx} = k_{hap} + \frac{(R_{x2} - R_{m2})\tan\vartheta_{f2}}{m_{mn}} \tag{12.385}$$

$$x_{hmmaxx1} = -\left(1.1k_{hapx} - \frac{z_{nx2}m_{xn}\sin^2\alpha_{eminx}}{2m_{mn}}\right) \tag{12.386}$$

The undercut of wheel at checkpoint is avoided, if $x_{hm1} > x_{hmmaxx1}$.

The author refers the reader to Appendix F of the aforementioned ISO 23509:2016 (E) [20], in which four examples of calculations are carried out clearly and comprehensively. They concern: the first, a spiral bevel gear set (Method 0); the second, a face-milled hypoid gear set (Method 1); the third, a face-hobbed hypoid gear set (Method 2); the fourth, another face-hobbed hypoid gear set (Method 3).

References

1. Artoni A, Gabiccini M, Guiggiani M, Kahraman A (2011) Multi-objective ease-off optimization of hypoid gears for their efficiency, noise, and durability performances. J Mech Des 133(12):267–277
2. Baxter ML (1961) Basic geometry and tooth contact of hypoid gears. Ind Math 11(2):19–42
3. Brown MD (2009) Design and analysis of a spiral bevel gear. Master of Engineering in Mechanical Engineering, Rensselaer Polytechnic Institute, Hartford, Connecticut, USA
4. Buchanan R (1823) Practical essays on mill work and other machinary, with notes and additional articles, containing new researches on various mechanical subjects by Thomas Tredgold. J Taylor, London
5. Buckingham E (1949) Analytical mechanics of gears. McGraw-Hill Book Company Inc, New York
6. Coleman W (1975) Computing efficiency for bevel and hypoid gears. Mach Des 47:64–65
7. Coleman W, Lehmann EP, Mellis DW, Peel DM (1969) Advancement of straight and spiral bevel gear technology, USAAVLABS Technical Report 69-75. U.S. Army Aviation Material Laboratories, Fort Eustis, VA (Virginia), pp 1–267

8. Falah B, Gosselin C, Cloutier L (1998) Experimental and numerical investigation of the meshing cycle and contact ratio in spiral bevel gears. Mech Mach Theory 33(1/2):21–37
9. Fan Q (2006) Computerized modeling and simulation of spiral bevel and hypoid gears manufactured by Gleason face hobbing process. J Mech Des 128(6):1315–1327
10. Fan Q (2011) Advanced developments in computerized design and manufacturing of spiral bevel and hypoid drives. Appl Mech Mater 86:439–442
11. Ferrari C, Romiti A (1966) Meccanica Applicata alle Macchine. Unione Tipografica – Editrice Torinese (UTET), Torino
12. Figliolini G, Stachel H, Angeles J (2019) Kinematic properties of planar and spherical logarithmic spirals: applications to the synthesis of involute tooth profiles. Mech Mach Theory 136:14–26
13. Galassini A (1962) Elementi di Tecnologia Meccanica, Macchine Utensili, Principi Funzionali e Costruttivi, loro Impiego e Controllo della loro Precisione, reprint 9th edn. Editore Ulrico Hoepli, Milano
14. Giovannozzi R (1965) Costruzione di Macchine, vol II, 4th edn. Casa Editrice Prof. Riccardo Pàtron, Bologna
15. Handschuh RF, Litvin FL (1991) How to determine spiral bevel gear tooth geometry for finite element analysis. NASA Technical Memorandum 105150, AVSCOM Technical Report 91-C-018, pp 1–9
16. Handschuh RF, Nanlawala M, Hawkins JM, Mahan D (2001) Experimental comparison of face-milled and face-hobbed spiral bevel gears. NASA Technical Memorandum 2001-210940, ARL-Technical Report-1104, pp 1–8
17. Henriot G (1972) Traité théorique et pratique des engrenages. Fabrication, contrôle, lubrification, traitement thermique, vol 2. Dunod, Paris
18. Henriot G (1979) Traité théorique et pratique des engrenages, vol 1, 6th edn. Bordas, Paris
19. Huston RL, Coy JJ (1981) Ideal spiral bevel gears—a new approach to surface geometry. J Mech Des 103(1):127–132
20. ISO 23509:2016 (E) Bevel and hypoid gear geometry
21. Klingelnberg J (ed) (2016) Bevel gear, fundamentals and applications.Spinger, Berlin, Heidelberg
22. Kolivand M, Kahraman A (2009) A load distribution model for hypoid gears using ease-off topography and shell theory. Mech Mach Theory 44:1848–1865
23. Kolivand M, Li S, Kahraman A (2010) Prediction of mechanical gear mesh efficiency of hypoid gear pairs. Mech Mach Theory 45:1568–1582
24. Lawrence JD (2014) A catalog of special plane curves. Dover Publications, New York
25. Lechner G, Naunheimer H (1999) Automotive transmissions: fundamentals, selection, design and application. Springer, Berlin, Heidelberg
26. Li M, Hu HY (2003) Dynamic analysis of a spiral bevel-geared rotor-bearing system. J Sound Vib 259(3):605–624
27. Li Q, Jiang JF, Chang YL, Wang LT (2017) Modeling optimization and machining detection of logarithmic spiral bevel gears. Tool Technol 51(3):51–54
28. Lin C-Y, Tsay C-B, Fong Z-H (2001) Computer-aided manufacturing of spiral bevel and hypoid gears by applying optimization techniques. J Mater Process Technol 114:22–35
29. Litvin FL (1992) Development of gear technology and theory of gearing. NASA Reference Publication 1406, ARL-TR-1500, U.S. Army Research Laboratory, Cleveland. Lewis Research Center, Ohio
30. Litvin FL (1994) Gear geometry and applied theory. Englewood Cliffs, Prentice Hall Inc, New Jersey
31. Litvin FL, Chaing W-S, Lundy M, Tsung W-J (1990) Design of pitch cones for face-hobbed hypoid gears. ASME J Mech Des 112:413–418
32. Litvin FL, Fuentes A (2004) Gear geometry and applied theory, 2nd edn. Cambridge University Press, Cambridge
33. Litvin FL, Gutman Y (1981) Methods of synthesis and analysis for hypoid gear drives of formate and Helixform. Parts 1, 2 and 3. ASME J Mech Des 103(1):83–113

34. Litvin FL, Petrov KM, Ganshin VA (1974) The effect of geometrical parameters of hypoid and spiroid gears on their quality characteristics. J Eng Ind 96:330–334
35. Litvin FL, Wang AG, Handschuh RF (1998) Computerized generation and simulation of meshing and contact of spiral bevel gears with improved geometry. Comput Methods Appl Mech Eng 158:35–64
36. Maitra GM (1994) Handbook of gear design, 2nd edn. Tata McGraw-Hill Publishing Company Ltd., New Delhi
37. Merritt HE (1954) Gears, 3th edn. Sir Isaac Pitman&Sons, Ltd, London
38. Micheletti GF (1958) Lavorazioni dei Metalli ad Asportazione di Truciolo. Libreria Editrice Universitaria Levrotto&Bella, Torino
39. Mozzi del Garbo GG (1763) Discorso matematico sopra il rotolamento momentaneo dei corpi. Stamperia di Donato Campo, Napoli
40. Müller H (2007) Face-off: face hobbing vs. face milling. Gear Solution Magazine, 1 Sept 2007
41. Naunheimer H, Bertsche B, Ryborz J, Novak W (2011) Automotive transmissions: fundamentals, selection, design and application, 2nd edn. Springer, Berlin, Heidelberg
42. Niemann G, Winter H (1983) Maschinen-Elemente, Band III: Schraubrad-, Kegelrad-, Schnecken-, Ketten-, Rienem-, Reibradgetriebe, Kupplungen, Bremsen, Freiläufe. Springer, Berlin, Heidelberg
43. Pauline K, Irbe A, Torims T (2014) Spiral bevel gears with optimized tooth-end geometry. Procedia Eng 69:383–392
44. Pollone G (1970) Il Veicolo. Libreria Editrice Universitaria Levrotto & Bella, Torino
45. Poritsky H, Dudley DW (1948) Conjugate action of involute helical gears with parallel or inclined axes. Quaterly Appl Mathem VI(3)
46. Powell DE, Barton HR (1959) Analytical study of surface loading and sliding velocity of automotive hypoid gears. ASME Trans II(2):173–183
47. Radzevich SP (2010) Gear cutting tools: fundamentals of design and computation. CRC Press, Taylor&Francis Group, Boca Raton, Florida
48. Radzevich SP (2016) Dudley's handbook of practical gear design and manufacture, 3rd edn. CRC Press, Taylor&Francis Group, Boca Raton, Florida
49. Romiti A (1960) Sopra un tipo di ingranaggi conici per assi concorrenti e sghembi, vol XCIV. Atti dell'Accademia delle Scienze di Torino, Torino
50. Romiti A (1960) Problemi di dimensionamento, accoppiamento e fabbricazione di ingranaggi conici per uso universale, vol XCIV. Atti dell'Accademia delle Scienze di Torino, Torino)
51. Rossi M (1965) Macchine Utensili Moderne. Editore Ulrico Hoepli, Milano
52. Schiebel A, Lindner W (1954) Zahnräder, Band 1: Stirn-und Kegelräder mit geraden Zähnen. Springer, Berlin, Heidelberg
53. Schiebel A, Lindner W (1957) Zahnräder, Band 2: Stirn-und Kegelräder mit schrägen Zähnen Schraubgetriebe. Springer, Berlin, Heidelberg
54. Scotto Lavina G (1990) Riassunto delle Lezioni di Meccanica Applicata alle Macchine: Cinematica Applicata, Dinamica Applicata - Resistenze Passive - Coppie Inferiori, Coppie Superiori (Ingranaggi – Flessibili – Freni). Edizioni Scientifiche SIDEREA, Roma
55. Sheveleva GI, Volkov AE, Medvedev VI (2007) Algorithms for analysis of meshing and contact of spiral bevel gears. Mech Mach Theory 42:198–215
56. Shih YP, Fong ZH, Lin GC (2007) Mathematical model for a universal face hobbing hypoid gear generator. J Mech Des 129(1):38–47
57. Simon V (2007) Computer simulation of tooth contact analysis of mismatched spiral bevel gears. Mech Mach Theory 42:365–381
58. Simon V (2009) Design and manufacture of spiral bevel gears with reduced transmission errors. ASME J Mech Des 131(4):041007-1–041007-11
59. Spear GM, King CB, Baxter ML (1960) Helixform bevel and hypoid gears. ASME Trans 82-B(III):179–190
60. Stachel H, Figliolini G, Angeles J (2019) The logarithmic spiral and its spherical counterpart. J Ind Des Eng Graph 14(1):91–98

61. Stadtfeld HJ (1993) Handbook of bevel and hypoid gears: calculation. Rochester Institute of Technology, Manufacturing and Optimization, Rochester
62. Stadtfeld HJ (2011) Tribology aspects in angular transmission systems, Part VII: hypoid gears. Gear Technol 66–72
63. Stadtfeld HJ (2014) Gleason bevel gear technology. Gleason Works, Rochester
64. Su J, Fang Z, Cai X (2013) Design and analysis of spiral bevel gears with seventh-order function of transmission error. CSAA, Chin J Aeronaut 26(5):1310–1316
65. Suh S-H, Jung D-H, Lee E-S, Lee S-W (2003) Modeling, implementation, and manufacturing of spiral bevel gears with crown. Int J Adv Manuf Technol 21:775–786
66. Townsend DP (1991) Dudley's gear handbook. McGraw-Hill, New York
67. Tsai YC, Chin PC (1987) Surface geometry of straight and spiral bevel gears. J Mech Transmissions Autom Des 109(4):443–449
68. Vimercati M (2007) Mathematical model for tooth surfaces representation of face-bobbed hypoid gears and its application to contact analysis and stress calculation. Mech Mach Theory 42(6):668–690
69. Vimercati M, Piazza A (2005) Computerized design of face hobbed hypoid gears: tooth surface generation, contact analysis and stress calculation. AGMA Fall Technical Meeting, Paper N. 05FTM05
70. Wang J, Lim TC, Li M (2007) Dynamics of a hypoid gear pair considering the effects of time-varying mesh parameters and backlash nonlinearity. J Sound Vib 308:302–329
71. Wang P-Y, Fong Z-H (2005) Adjustability improvement of face-milling spiral bevel gears by modified radial motion (MRM) method. Mech Mach Theory 40:69–89
72. Wildhaber E (1946) Basic relationship of hypoid gears. Am Machinist, XC, February, pp 14, and 28, March, p 14, June, pp 6 and 20, July, p 18, and August, pp 1 and 15
73. Wildhaber E (1956) Surface curvature. Prod Eng 27:184–191
74. Zhang Y, Litvin FL, Handschuh RF (1995) Computerized design of low-noise face-milled spiral bevel gears. Mech Mach Theory 30(8):1171–1178

Chapter 13
Gear Trains and Planetary Gears

Abstract In this chapter, the general concepts of ordinary gear trains are first described, particularly those concerning their efficiency and the obtainable transmission ratios. The same concepts are then extended to planetary gear trains and various methods of calculating the transmissions ratios achievable with these types of gear drives are defined. Some problems related to simple planetary gear trains are then examined, and the main characteristics of some of these gear drives are described and discussed. Particular planetary gear trains are then dealt with, such as those concerning summarizing and differential gear trains. The analysis is then extended to multi-stage planetary gear trains and parallel mixed power trains consisting of ordinary and planetary gear trains, and the calculation procedure is described of the braking torque to be applied to the members of the planetary gear train that are held at rest. Finally, the problem of the efficiency of a planetary gear train is addressed and discussed in its general terms, also with reference to those particular planetary gear trains used to realize large transmission ratios.

13.1 Introduction

A *gear pair* is a mechanism consisting of two gears rotating about axes the relative position of which is fixed, and one gear turns the other by action of teeth successively in meshing contact. A *train of gears* or *gear train* is any combination of gear pairs, i.e. a combination of two or more gear pairs (or a combination of two or more gears), mounted on rotating shafts, to transmit power or torque. In common practical applications, a gear train acts as a *speed reducer*, but the industrial application where the gear train acts as a *speed increaser* are also frequent.

The more or less numerous gear pairs of a gear train can be arranged in an endless variety of ways to meet the particular shaft position, direction of rotation, and transmission ratio requirements (see: Merritt [37], Dudley [12], Ferrari and Romiti [14], Pollone [51], Maitra [36].

From the theoretical point of view, we can get a large transmission ratio using only two gears. In doing so, however, we would get a gear pair characterized by a

© Springer Nature Switzerland AG 2020
V. Vullo, *Gears*, Springer Series in Solid and Structural Mechanics 10,
https://doi.org/10.1007/978-3-030-36502-8_13

small pinion and an enormously large wheel, with obvious problems of overall dimensions and technical feasibility. Such a design choice is not technically and economically viable. The desired transmission ratio is then achieved by using a gear train, which may be constituted by different types of gears (e.g. parallel cylindrical spur and helical gears, cylindrical crossed helical gears, straight bevel, spiral bevel and hypoid gears, worm gears, etc.).

The gear trains can be classified into two main families:

- the family of the *ordinary gear trains*, in which all the gears rotate about axes which are fixed with respect to a housing, which is the reference frame;
- the family of the *epicyclic gear trains*, also called *planetary gears trains*, in which at least one gear rotates with respect to its axis, which in turn rotates with respect to the fixed frame.

Each of these two families of gear trains has its own characteristics. However, we can consider the family of the ordinary gear trains as a special case of the more general family of planetary gear trains, since the former can be obtained from the second when the moving axes of the planetary gear trains are restrained and held at rest.

Ordinary gear trains as well as planetary gear trains are widely used in practical geared transmissions, which are often irreplaceable in many technological areas (industrial plants, machine tools, textile machines, power trains for cars, aircrafts and helicopters, auxiliary devices for steam and gas turbines, mills and kilns for cement production, ceramic and brick industry, food industry, lifting and transport machines, wind generators, etc.) They are in fact used to mediate between the characteristics of the power delivered by the prime mover and those of the driven machine. For example, in automotive transmissions, ordinary and planetary gear trains have the function of adapting the available traction power to the required power, while also ensuring the desired performance [31, 45, 51].

13.2   Ordinary Gear Trains

An ordinary gear train is constituted by a set of gears, arranged to transmit power or torque between a driving shaft and a driven shaft. All the gears of the gear trains are mounted on rotating shafts around fixed axes, and the shafts are supported by a fixed housing. Figure 13.1 shows schematically some ordinary gear trains, in which the driving and driven shafts are not coaxial. Figure 13.2 shows instead other ordinary gear trains, in which the driving and driven shafts are coaxial. In these last types of ordinary gear trains, only one set of gears may be disposed between the first and the last gear of the gear train (Fig. 13.2a), or two or more sets of gears can be arranged, working in parallel (Fig. 13.2b–d). The conditions to be met in order to have two or more gears in parallel between the driving and driven gears will be examined by us later, when we discuss the epicyclic gear trains, where these schematic configurations are frequently used [51].

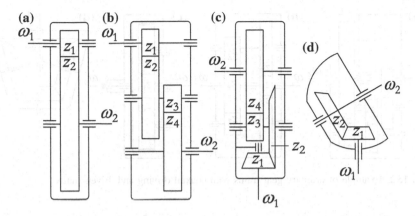

Fig. 13.1 Examples of ordinary gear trains with non-coaxial driving and driven shafts

An ordinary gear train may be simple or compound. We define as a *simple ordinary gear train*, a train of gears in which each shaft carries a single gear, which in turn transmits the motion to the gear mounted on the next shaft, and so on. The gears mounted on the intermediate shafts are *idle gears* (or simply *idlers*), as they serve the purpose of changing the direction of rotation, without affecting the gear ratio and the transmission ratio of the whole gear train. The idlers also serve to bridge the distance between the driving member (or *driver*) and driven member (or *follower*), thus helping to reduce the diameters of the gear wheels necessary to transmit motion. Only the transmission systems shown in Fig. 13.1a, d can be considered as simply ordinary gear trains, although in these systems idle gears are not present.

Instead, we define as a *compound ordinary gear train*, a train of gears in which some of the shafts carry two or more gear wheels. These transmission systems are very compact, and therefore have small size, while allowing to achieve large transmission ratios. Depending on the design technical requirements to be fulfilled, the driving and driven shafts may be arranged to be coaxial, if necessary (Fig. 13.2a, b), or not (Fig. 13.1b, c). Most of industrial transmission systems fall under the category of compound ordinary gear trains.

The *transmission ratio*, *i*, of an ordinary gear train is defined as the quotient of the angular speed of the first driving gear divided by the angular speed of last driven gear of the gear train. When necessary, a plus sign should be added to the transmission ratio when the directions of rotation are the same or a minus sign when they are opposite. For a simple ordinary gear train consisting of only two gear wheels (Fig. 13.1a, d), the transmission ratio is given by:

$$i = \frac{\omega_1}{\omega_2} = \frac{n_1}{n_2} = \frac{z_2}{z_1} \tag{13.1}$$

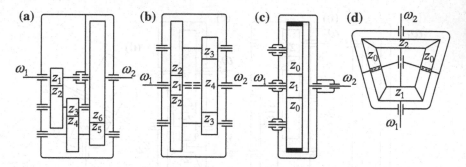

Fig. 13.2 Example of ordinary gear trains with coaxial driving and driven shafts

where ω_1 and ω_2 are the angular velocities of the driving and driven gears, expressed in rad/s, n_1 and n_2 are the rotational speeds of the same gears, expressed in $\min^{-1}$ (rpm, revolution per minute), and z_1 and z_2 are their numbers of teeth. In this case, the transmission ratio, i, coincides with the gear ratio, u, which is defined as the quotient of the number of teeth of the wheel divided by the number of teeth of the pinion.

For a compound ordinary gear train constituted by two gear pairs, arranged in series (Fig. 13.1b, c), the transmission ratio is given by:

$$i = \frac{\omega_1}{\omega_2} = \frac{n_1}{n_2} = \frac{z_2 z_4}{z_1 z_3}, \tag{13.2}$$

where z_1 and z_3 are the numbers of teeth of the two driving gears, while z_2 and z_4 are the numbers of teeth of the two driven gears.

Generally, for a compound ordinary gear train consisting of more than two gear pairs, arranged in series, the transmission ratio is equal to the quotient of the product of numbers of teeth of the driven gears divided by the product of numbers of teeth of the driving gears. It is therefore expressed by a relationship that is the generalization of Eq. (13.2).

If the gear axes of the compound ordinary gear train are all parallel, we can assign a sign to the transmission ratio, making it positive if the first and last gear wheels have the same direction of rotation and negative in the opposite case. It is therefore apparent that, if a gear wheel, W_i, mounted on any intermediate shaft, operates both as a driving and driven gear wheel (it then simultaneously engages with the proceeding gear wheel $(W_i - 1)$ and with the following gear wheel $(W_i + 1)$, it does not alter the absolute value of the transmission ratio, corresponding to the direct engagement of the gear wheels, $(W_i - 1)$, and $(W_i + 1)$, but reverses the direction of rotation that the gear wheel $(W_i + 1)$ would have in the case of its direct engagement. Thus, the effect of each idle gear is to reverse the direction of rotation, while the overall transmission ratio is not changed. Intermediate gear wheels can also be used in the case of gear trains including straight bevel and spiral bevel gears.

It should be remembered that a gearbox, however it is constituted, is an oscillatory system. Therefore, to avoid resonance problems (and consequent malfunctions and noise), it is advisable to choose the numbers of teeth of each gear pair of an ordinary gear train in order to avoid a common factor between them [56]; this choice is especially recommended if the gear train under consideration is a compound gear train. With this choice, another important goal is achieved: to have a tooth wear as uniform as possible on all teeth, rather than localized as would be the case when these number of teeth, instead of being prime numbers, were chosen with a common factor among them.

The ratio between the speeds of rotation of the driven and driving shafts of an ordinary gear train of the type of those described above is a constant, and this ratio is determined by the numbers of teeth of the gear wheels that constitute the gear train. When we want to obtain, for a given speed of rotation of the driving shaft, a given number of different speeds of rotation of the driven shaft, it is necessary to have a set of different transmission gear trains, between the two driving and driven shafts. In this way, it is possible to operate, at will, the appropriate gear train that realizes the desired transmission ratio.

A system of gear trains that is able to realize different transmission ratios between the driving and driven shafts is called *gearbox*. The exclusion from or the inclusion in the transmission of motion of one of the aforementioned gear trains may be done using gear wheels displaceable axially on their shafts. It is also possible to employ gear pairs with fixed positions with respect to their shafts, one of which is mounted idle and is made rigidly connected to its shaft by means of dog clutches or friction clutches.

The reduction gear units currently used in industrial applications are chosen based on the following main parameters or design criteria:

- overall speed reduction ratio;
- maximum allowable speed reduction in any of the stages that constitute the gear system;
- space requirements and constraints arising from links with the driving and driven machines;
- values of speed reduction from stage to stage, which are usually in geometrical progression, so as to avoid resonance phenomena;
- progressively increasing values of the gear tooth modules from stage to stage, proportionate to the torque that in the reduction units increases from the input shaft (or input end) to the output shaft (or output end).

With reference to the gear train for which the Eq. (13.1) is valid, the transmission ratio can be also defined as the quotient of the angle of rotation, φ_1, of the first driving gear divided by the angle of rotation, φ_2, of last driven gear of the gear train; this provided that the two angles of rotation are measured starting from a given initial meshing condition. We can therefore write:

$$i = \frac{\varphi_1}{\varphi_2}. \tag{13.3}$$

Then neglecting the power losses due to friction and other causes, by the resulting equality between the input power, $P_1 = T_1\omega_1$, and the output power, $P_2 = T_2\omega_2$, we deduce:

$$i = \frac{T_2}{T_1} = i_T, \tag{13.4}$$

where T_1 and T_2 are respectively the input and output torque, and i_T is the torque conversion factor defined in Sect. 1.1.

If we want to take into account the inevitable power losses due to various causes, we have to introduce the concept of *efficiency*, η, defined as the quotient of the output power from the last driven gear divided by the input power in the first driving gear of the gear train. Equation (13.4) is no longer valid because, due to the lost power $P_d = P_1 - P_2$, the efficiency is always less than one ($\eta < 1$). In this regard, we must keep in mind that it is necessary to consider at least two values of the efficiency, depending on whether the prime mover power flows from gear 1 to gear 2, or vice versa. In the first case, the efficiency will be η_1, while in the second case the efficiency will be η_2 [54].

In the first case, it is $\eta_1 = P_2/P_1$, for which the Eq. (13.4) should be replaced with the equation:

$$i\eta_1 = \frac{T_2}{T_1}, \tag{13.5}$$

while, in the second case, it is $\eta_2 = P_1/P_2$, for which the Eq. (13.4) should be replaced with the equation:

$$\frac{i}{\eta_2} = \frac{T_2}{T_1}. \tag{13.6}$$

The two ratios between the output and input torques given by Eqs. (13.5) and (13.6) are different from each other even when $\eta_1 = \eta_2$. Under the hypothesis that discontinuities do not exist passing from the static condition to a dynamic condition, we infer the rule according to which gears initially at rest continue to remain at rest in case where the ratio (T_2/T_1) is in the range:

$$i\eta_1 < T_2/T_1 < i/\eta_2, \tag{13.7}$$

while the motion takes place if this ratio comes out of aforementioned range.

When the transmission ratio is slightly different from the unit, as it occurs for spur and bevel gears, we can obtain very high efficiencies η_1 and η_2, which differ little from one another. In practice, we consider a single value of gear efficiency, given by $\eta = \eta_1 = \eta_2$. The parallel cylindrical spur gears can reach value of efficiency equal to $\eta = \eta_1 = \eta_2 = 0.98$, while bevel gears are characterized by a little lower efficiency. Frequently, due to manufacturing inaccuracies, the values of efficiency are lower.

Instead, when the transmission ratio is small or large, the efficiencies η_1 and η_2 generally get worse, and become very different from each other. Indeed, we find that the best efficiency is when the gear works as a speed reducer $(i > 1)$, i.e. when the driving gear wheel is the fast wheel, while the worst efficiency is when the gear works as a speed multiplier $(i < 1)$, i.e. when the driving gear wheel is the slow wheel. In the first case we speak of *direct motion*, while in the latter case we speak of *retrograde motion*.

A possible reason for this behavior can be given assuming that the lost power, $P_d = P_1 - P_2$, in the gear is proportional to the power, P_2, applied to the slower wheel, on which the most relevant torque acts. In direct motion, for which 1 and 2 are respectively the fast and slow gear wheels, we have:

$$\frac{P_d}{P_2} = \frac{P_1 - P_2}{P_2} = \frac{1}{\eta_1} - 1. \tag{13.8}$$

In retrograde motion, while maintaining the ratio (P_p/P_2) characterizing the direct motion, being

$$\eta_2 = \frac{P_1}{P_2} = 1 - \frac{P_d}{P_2}, \tag{13.9}$$

we get:

$$\eta_2 = 2 - \frac{1}{\eta_1}. \tag{13.10}$$

This equation shows that:

- η_1 and η_2 differ from each other increasingly when η_1 decreases;
- η_2 becomes equal to zero when $\eta_1 = 50\%$;
- for $\eta_1 < 50\%$, η_2 becomes negative, whereby the retrograde motion can not be obtained, resulting in spontaneous arrest, unless a driving torque is also applied on the shaft 1 (this torque is considered conventionally as negative).

Of course, the hypothesis that the ratio (P_d/P_2) is a constant, is somewhat arbitrary, for which the results described above have only qualitative value. It is undeniable that many gears with very small transmission ratio, and consequently with low efficiency η_1, can have a negative value of η_2. An example of this condition is made up of a worm gear, when we want that the wormwheel is the driving member.

To calculate the efficiency of an ordinary gear train, we must consider the fact that it is a typical example of a *machine*, consisting of the series connection of several mechanisms, i.e., mechanism linked so that the follower (or driven member) of any of the component mechanisms is the driving member of the next one. It is also an example of a machine that can be considered as *absolute regime machine*, i.e. capable of working, under normal operating conditions, so that its kinetic energy is maintained constant, so its variation in any time interval is equal to zero.

In fact, during the engagement of two gear wheels, even under steady-state conditions, a periodicity in operating conditions exist, correlated with the time needed to rotate each gear wheel of an angular pitch. The corresponding period is, however, so small, and so does the kinetic energy variation within the same period, so the effects of the related variations can generally be neglected. Now, the *absolute regime machines* resulting from a series connection of several mechanisms possess a well-known property: the one that the efficiency of the machine is the product of the efficiencies of the single mechanism that make up the same machine [14]. *Rebus sic stantibus*, it follows that the efficiency of an ordinary gear train, whatever it is, simple or compound, is equal to the product of the efficiencies of the single gear pairs, since they are arranged in series between them. Therefore, with product notation Π, we will have:

$$\eta = \prod_{i=1}^{n} \eta_i \qquad (13.11)$$

where η_i is the efficiency of the i-th gear pair that makes up the gear train considered. Therefore, with regard to efficiency, the idler gears (they, as we have already seen, do not affect the transmission ratio) are playing the same role as all the other gear wheels that make up the gear train.

In the case of more complex gear trains in which the input power first is divided in two or more flow lines, and then it rejoins downstream of the various gears through which the same power has flowed, the calculation of the efficiency is more complicated. In this case in fact the efficiency is no longer equal to the product of the efficiencies of each single gear pair. Therefore, Eq. (13.11) is no longer valid, since the efficiency of the single gears has a different weight, depending on the amount of power that passes through them (see Sects. 13.8 and 13.12).

13.3   Epicyclic or Planetary Gear Trains: Definitions and Generalities

Since classical antiquity, as demonstrated by both the Archimedes planetarium, dating back to the third century BC (see: Drachmann [11], Pastore [47]) and by the Antikythera Mechanism, dating back to the first century BC [60], planetary gear trains have drawn attention of the most valued scholars and scientists, who have faced complex problems that required their study and development. The topic is dealt with in almost all of the most famous gear textbooks and treatises (see: Merritt

[37], Dudley [12], Henriot [21], Lyndwander [35], Maitra [36], Litvin and Fuentes [32], Jelaska [24], Radzevich [53]); many textbooks, which deal with general machine design topics (see: Giovannozzi [18], Juvinall [25], Niemann and Winter [46], Budynas and Nisbett [3]); textbooks on automotive design and transmissions (see: Pollone [51], Morelli [43], Lechner and Naunheimer [31], Genta and Morello [17], Fisher et al. [15]); specific textbooks and literature (see: Kelley [28], Henriot [22], Pearson [49], Ruggieri [54], Litvin et al. [33], Cooley and Parker [5], Balbayev [1]). However, this topic continues to attract the attention of prominent scholars, which still make it an advanced research subject, especially with reference to new and innovative applications, such as gear drive systems for helicopters and hybrid electric vehicles. In this regard, in addition to the works referred by Litvin and Fuents [32], we limit ourselves to citing the following papers: Coy et al. [6], Simionescu [55], Miller [41], Chen and Angeles [4], Galvagno [16], Liu and Peng [34], Pennestri et al. [50], Davies et al. [7], Gupta and Remanarayanan [20], Dheaud and Pullen [9], Gupta et al. [19], Yang et al. [62], Kahraman et al. [27], Xue et al. [61], Essam and Isam [13].

According to ISO 1122-1:1998 [23], an *epicyclic gear train* (also called *planetary gear train*, or more simply *epicyclic gear* or *planetary gear*) is a combination of coaxial elements, of which one or more are *annulus gears* (or *ring gears*) and one or more are *planet carriers* (or *train arms* or *spiders*). Planet carriers rotate around the common axis and support one or more *planet gears* that mesh with the annulus gears and one or more *sun gears*. Usually, the planet gears are equally spaced, to balance the forces acting on sun gear, annulus gear and planet carrier. The first and last gear wheels of the planetary gear train, taken in the direction of the subsequent engagements of their teeth, are called *main wheels*. Generally, a planetary gear train can be derived from an ordinary gear train whose housing, instead of being held at rest, is rotated about the driving axis. In so doing, the speed of the driven axis becomes a function of speeds of the driving axis and housing.

However, it should be borne in mind that the above-mentioned ISO definition does not encompass all the possible configurations of planetary gear trains conceived and developed to solve real practical problems. In fact, by limiting attention to usual applications, the axes of the main wheels and the axis of the planet carrier are parallel, while the intermediate axes may have different directions. The family of planetary gear trains is therefore much larger, as it includes (see Ferrari and Romiti [14]:

(a) planetary gear trains whose end axes are fixed and coincide with the axis of the planet carrier;
(b) planetary gear trains with a mobile end wheel, whereby the last axis is distinct from the first, but parallel to it;
(c) planetary gear trains with a mobile end wheel, whereby the last axis is distinct from the first, but is directed according to any angle with respect to the latter, while the axis of the planet carrier coincides with the axis of the first main wheel;

(d) planetary gear trains in which also the axis of the planet carrier is mobile.

These mechanisms are systems with two degrees of freedom. When, however, they are considered in the framework of the machine to which they belong, the number of degrees of freedom is reduced to one by virtue of the constraints that immobilize the first (or the last) wheel or by virtue of the couplings that establish a determined correlation between rotations of the main wheels and rotation of the planet carrier. A relationship between these rotations then exist, which depends on the geometry of the gear wheels that make up the planetary gear train. This relationship depends only on the geometry of the gear system and can easily be obtained if the axes of the main wheels and planet carrier are parallel, and the axis of the planet carrier is held at rest (see next section).

The simplest of the planetary gear trains consists of two gears mounted so that the center of one gear revolves around the center of the other gear (Fig. 13.3). A *planet carrier* connects the centers of the two gears and rotates to carry the *planet gear* around the *sun gear*. The planet and sun gears mesh so that their pitch circles roll without sliding. When the sun gear is fixed and the planetary gear roll around the sun gear, a point on the pitch circle of the planet gear traces an epicycloid curve.

The simple epicyclic gear train described above may be housed at the inside of an annulus gear and assembled in such a way that the pitch circle of the planet gear rolls without sliding on the inside of the pitch circle of the annulus gear supposed fixed (Fig. 13.4). In this case, a point on the pitch circle of the planet gear traces a hypocycloid curve.

Generally, we call planetary gear train a mechanism consisting of the combination of all four coaxial component members shown in Fig. 13.4, i.e. the mechanism composed of at least one planet gear, one sun gear, one planet carrier, and one annulus gear. However, in most cases the terms planetary gear train and epicyclic gear train are used interchangeably. In the most frequent case, the annulus gear is usually fixed, the planet carrier is driving, and the sun gear is driven.

Albeit in a more streamlined shape, this kind of mechanism has an ancient history: in fact, it dates back to the 5th century BC, when the Greeks invented the idea of epicycles, i.e. of circle traveling on the circular orbits. According to the epicyclic theory, the Antikythera Mechanism (about the '80 BC) had gearing that was able to approximate the moon's elliptical path through the heavens, while

Fig. 13.3 Simple epicyclic gear train consisting of a planet gear, sun gear and planet carrier

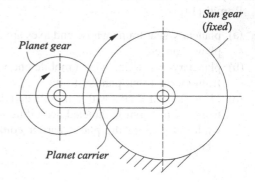

Fig. 13.4 Epicyclic gear
train consisting of one planet
gear, sun gear, planet carrier,
and annulus gear

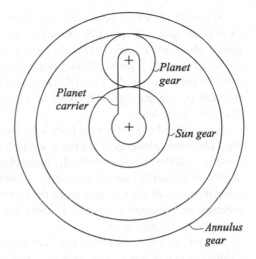

Claudius Ptolemy in the Almagest (in 148 AD) was able to predict planetary orbital
paths (Ptolemy, in translated edition, [52].

Compared to an ordinary gear train, which has only one degree of freedom
movement, a planetary gear train has two degrees of freedom of movement.
Therefore, the planetary gear trains, as they are characterized by three main component members (the sun gear, annulus gear, and planet carrier), can be used in
various ways to obtain combinations of rotary motions [48, 57].

A planetary gear system, even in its simplest configurations, possesses unique
characteristics, which are summarized as follows: compactness, also due to the
coaxial arrangement of the driving and driven shafts; high power-to-weight ratio;
large speed reduction possibilities compared to its overall size; possibilities of a
large number of combinations of driving and driven inputs and outputs; possibilities
of large torque conversion; different possibilities of orientation of the gear system.

Planetary gear trains can be typically classified as simple or compound planetary
gear trains. A *simple planetary gear train* has one sun gear, one annulus gear, one
planet carrier and one planet gear set. A *compound planetary gear train* involves
one or more of the following three types of substructures:

- *meshed-planet gears*, for which the compound planetary gear train is characterized by at least two more planet gear in mesh with each other;
- *stepped-planet gears*, for which the compound planetary gear train is characterized by a shaft connection between two planet gears;
- *multi-stage structures*, for which the compound planetary gear train contains
  two or more planet gear sets.

The already remarkable characteristics, which we highlighted above for the
simple planetary gear trains, are further enhanced in the compound planetary gear
trains. The latter in fact have the advantages of higher torque-to-weight ratio, larger
reduction ratio and more flexible configurations.

A simple planetary gear train may have different configurations. Usually, the axes of all gears are parallel. However, for special cases, such as *differentials* or *pencil sharpeners*, these axes can be placed at an angle, introducing members configured as bevel gears. Furthermore, the axes of the sun gear, planet carrier and annulus gear are usually coaxial. Depending on whether the axes are parallel (or coaxial) or not parallel, we speak of *planar planetary gear trains* or *spherical planetary gear trains*.

A compound planetary gear train with meshed-planet gears is characterized by at least two more planet gears in mesh with each other. One of these two planet gears meshes with the sun gear, while the other planet gear meshes with the annulus gear. With a compound planetary gear train so configured, the annulus gear rotates in the same direction of the sun gear when the planet carrier is held at rest; in this way, the mechanism provides a reversal in direction of rotation compared with a simple usual epicyclic gear train.

Instead, a compound planet gear, composed by two differently sized planet gears forming a single block with their common shaft, characterizes a compound planetary gear train with stepped-planet gears. The smaller planet gear meshes with the sun gear, while the larger planet gear meshes with the annulus gear. We use this type of compound planetary gear train when the overall package size is limited, and it is necessary to achieve smaller step changes in gear ratio. The speed hub of internally geared bicycle hubs constitutes an example of this type of compound epicyclic gear train. As a down side the problems posed by these types of compound epicyclic gear trains, which have a *relative gear mesh phase* (or *timing marks*), should not be underestimated. Moreover, their assembly conditions are more restrictive than those of a simple planetary gear train. In fact, these compound planetary gear trains with stepped-planet gears must be assembled in such a way that their initial relative position is correct, in order to prevent their teeth from simultaneously meshing the annulus and sun gear at opposite ends of the compound planet, as this condition would determine a very rough running and short life.

Finally, a multi-stage structure is a compound planetary gear system where more planet and sun gear units are placed in series into the same annulus gear housing. The various component units (or substructures) are assembled in such a way that the output shaft of the first stage become the input shaft of the next stage, and so on. As an example, consider a simple planetary gear train, where no component member is fixed, and the power comes from the sun gear and does go out from the planet carrier. In this case, the sun gear drives the planet gears assembled with the annulus gear to operate. The planet carrier, connected to the planet gears, revolves on its own axis and along the annulus gear, and the output shaft, connected to the planet carrier, rotates with a speed reduced with respect to the input speed. A higher reduction ratio can be obtained by doubling the multiple staged gears and planet carriers. A gear system so configured may provide a larger or smaller gear ratio, depending on the sizing of the various component substructures. By choosing a suitable combination of the component units, speed ratio of 10^4:1 can be easily obtained. Some automatic transmission of vehicles works in this way. The number of combinations in which the various component units may be arranged in the overall gear system can be of an endless variety.

In a simple planetary gear train, any one of the three main members (the sun gear, planet carrier and annulus gear) can be made to be the fixed member. Any of the two remaining main members can be used as the input or the output member for power transmission. Thus, a simple planetary gear train has six possible combinations of speed ratios. However, it can happen that no component member of a single planetary gear train is fixed. This happens for example, in the case of an *automobile differential*, which is a *spherical* or *bevel planetary gear train*.

In addition to the advantages described above, the planetary gear trains have other special characteristics, some of which deserve to be specifically mentioned here. In comparison to the ordinary gear trains with parallel axes, planetary gear trains provide a higher power density, volume reduction due to coaxial shafts, purely torsional reaction, and multiple kinematic combinations. The efficiency of a planetary gear train is about 97% for stage. Therefore, a high portion of the input energy is transmitted through the gearbox, and only a small portion of this energy is lost, due to mechanical losses inside the gearbox.

In a planetary gear train, loads are shared among multiple (two or more) planet gears, whereby torque capability is greatly increased. A great advantage of the planetary gear trains is the load distribution between the sun gear and annulus gear, which is known as *power branching*. The more planet gears in the planet gear train, the greater the load capacity and the higher the torque density. The planetary gear trains also provide great stability due to a uniform distribution of masses and body forces, and increased rotational stiffness. Forces applied radially on the gears of a planetary gear train are transferred radially, without longitudinal loads on the teeth. Furthermore, the use of two, three or more planet gears for load sharing reduces the specific load (load for length unit of face width) at each mesh, and eliminates the radial thrust on shafts.

It is also to be observed that in an ordinary gear train with parallel gears the driving force is transmitted through a small number of points of contact, where all loads are concentrated on a few contacting surfaces, making the gear to wear and crack. A planetary gear reducer has instead many gear contacting surfaces (the number of surfaces of contact depend of the number of planet gears) with a larger area on which the load is distributed evenly. Even impact loads arising from abrupt changes in driving or resistant torque are distributed uniformly. A greater resistance to the impact follows. These loads therefore do not damage the bearings and their housings.

In view of these advantages, the following disadvantages are to be observed. They included the inaccessibility of the gear unit, constant lubrication requirements, higher costs compared to those of the ordinary gear trains, greater accuracy of manufacture and assembly, high loads on pins that withstand the planet gears, and design complexity.

13.4 Transmission Ratio in a Planetary Gear Train

The transmission ratio in a planetary gear train can be obtained using several methods, among them equivalent, that lead to the same results. Here we describe only three of these methods, which are commonly used. In addition, we mention two other methods also usually used.

13.4.1 Algebraic Method and Willis Formula

Consider the most frequent case of planetary gear train, which is the one corresponding to the schematic drawing shown in Fig. 13.5 for a planar planetary gear train, or a similar schematic drawing for a spherical planetary gear train.

Let ω_a, ω_s, ω_c, and ω_p respectively be the angular velocities of the annulus gear, sun gear, planet carrier and planet gear. Of course, in place of these angular velocities (in rad/s), the equivalent rotational speeds n_a, n_s, n_c, and n_p (in $\min^{-1}$, i.e. rpm), can be considered. From what is known on the kinematics of mechanisms, it is obvious that the intercorrelation existing between the three angular velocities ω_a, ω_s and ω_c does not change, when we do rotate the whole planetary system with angular velocity $-\omega_c$ about the common axis of rotation. In doing so, the angular velocities of the annulus gear and sun gear are respectively equal to $(\omega_a - \omega_c)$ and $(\omega_s - \omega_c)$, while the angular velocity of the planet gear becomes equal to zero $(\omega_c - \omega_c = 0)$. In this condition, the planetary gear train is thus reduced to an ordinary gear train, with a driving shaft and a driven shaft, and with the axes of rotation of the planet gears fixed in three-dimensional space (see: Paul [48], Uicker et al. [57]).

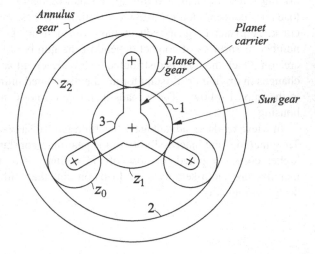

Fig. 13.5 Planetary gear train with three planet gear arranged at 120°

We denote by i_0 the transmission ratio (it is also called *characteristic ratio* or *basic ratio*) of the planetary gear train in the condition in which the planet carrier is held fixed, i.e. the condition in which the planet gear train has become an ordinary gear train. This characteristic ratio, i_0, is easily determined by knowing the number of teeth of the gear wheels involved in the power flow from the first to the last gear wheel of the planetary gear train under consideration. We attribute to it the sign, positive or negative, depending on whether, helding the planet carrier at rest, the directions of rotation of the first and last gear wheels are the same or opposite. As mentioned in Sect. 13.2, the absolute value of i_0 is given by the product of the ratios between the number of teeth of the individual gear pairs, which occur in the gear system.

In the case considered here, assuming that the driving and driven gears are respectively the sun gear (gear 1) and the annulus gear (gear 2), and that their numbers of teeth are $z_1 = z_s$ and $z_2 = z_a$, we have: $i_0 = -(z_2/z_1) = -(z_a/z_s)$. Usually, we design the gear system with a sizing such that i_0 is included in the range $(-9 < i_0 < -3/2)$. If we reverse between them the gear wheels 1 and 2, we will have: $(-2/3 < i_0 < -1/9)$.

However, when the planetary gear train as such performs its functions, the Eq. (13.1), which allows calculating the transmission ratio of an ordinary gear train, is not in general valid. It applies only when the planet carrier is held at rest, and the planetary gear train acts as an ordinary gear train. Moreover, when the planet carrier moves, the transmission ratio or does not make sense, or is still defined case by case, since the gear system no longer has a single degree of freedom, but two degree of freedom (see Dooner [10]).

Anyhow, a well-defined relationship between the angular velocities of the sun gear, annulus gear and planet carrier exists, in the aforementioned relative motion with respect to the planet carrier. In fact, once a common positive direction of rotation for the three members of the planetary gear train has fixed, assuming that the shaft of the sun gear is the driving shaft 1 (thus, $\omega_1 = \omega_s$), and that the shaft of the annulus gear is the driven shaft 2 (thus, $\omega_2 = \omega_a$), and indicating with 3 the shaft of the planet carrier (thus $\omega_3 = \omega_c$), it must be:

$$i_0 = \frac{\omega_{r1}}{\omega_{r2}} \tag{13.12}$$

where $\omega_{r1} = (\omega_1 - \omega_3)$ and $\omega_{r2} = (\omega_2 - \omega_3)$ are, in value and sign, the relative angular velocities of the first gear wheel and the last gear wheel with respect to the planet carrier. Introducing these two relative angular velocities in Eq. (13.12), we obtain:

$$i_0 = \frac{\omega_1 - \omega_3}{\omega_2 - \omega_3}. \tag{13.13}$$

This is the famous and well-known *Willis formula*, which correlates linearly between them the three angular velocities $\omega_1 = \omega_s$, $\omega_2 = \omega_a$, and $\omega_3 = \omega_c$. This formula allows to calculate one of the three angular velocities, once the values of the other two angular velocities, and the numbers of teeth of the gear wheels of the planetary gear train are known [59].

The Willis formula can also be written in terms of angles of rotation, in accordance with Eq. (13.3). Thus indicating respectively with φ_1, φ_2, and φ_3 the angles of rotation of the three main members of the planetary gear train (in the order, the sun gear, 1, annulus gear, 2, and planet carrier, 3), all three measured starting from a given initial position, we can write:

$$i_0 = \frac{\varphi_1 - \varphi_3}{\varphi_2 - \varphi_3}. \tag{13.14}$$

In the technical and scientific literature, the use of the reciprocal of the characteristic ratio, that is the quantity $\psi = 1/i_0$, is often preferred, for which the Willis formula is written in the following form:

$$\psi = \frac{1}{i_0} = \frac{\omega_2 - \omega_3}{\omega_1 - \omega_3}. \tag{13.15}$$

Like i_0, also ψ can be positive or negative, and the two sign conventions do not differ. Resolving the Eq. (13.15) with respect to ω_1, ω_2, and ω_3, we get the following three relationships:

$$\omega_1 = \frac{1}{\psi}[\omega_2 + \omega_3(\psi - 1)]; \quad \omega_2 = \psi\omega_1 + (1 - \psi)\omega_3; \quad \omega_3 = \frac{1}{\psi - 1}(\psi\omega_1 - \omega_2).$$

$$\tag{13.16}$$

These equations show that each of the three angular velocities ω_1, ω_2, and ω_3 is a linear function of the other two. From the second of Eq. (13.16), we infer the following three special cases:

(a) If $\psi = +1$, we have $\omega_2 = \omega_1$, whatever the value of ω_3. Therefore, the angular velocity ω_2 is independent of ω_3, and the planetary gear train behaves like a rigid coupling.

(b) If $\psi = +(1/2)$, we have:

$$\omega_2 = \frac{1}{2}(\omega_1 + \omega_3). \tag{13.17}$$

(c) If $\psi = -1$, we have:

$$\omega_3 = \frac{1}{2}(\omega_1 + \omega_2). \tag{13.18}$$

The planetary gear trains for which the Eqs. (13.17) and (13.18) are valid, are used in cases in which the speed of rotation of a shaft should be proportional to the sum of the speeds of rotation of the other two shafts.

Equations (13.16) allow us to easily handle the two particular cases in which one of the two main wheels, 1 or 2, is held at rest. In fact, if the first main wheel is held at rest, i.e. $\omega_1 = 0$, from the second of Eqs. (13.16) we find that the transmission ratio between the shafts of the second main wheel and the planet carrier is given by $\omega_2/\omega_3 = (1 - \psi)$. If instead the second main wheel is held at rest, i.e. $\omega_2 = 0$, from the first of Eqs. (13.16) we find that the transmission ratio between the shafts of the first main wheel and the planet carrier is given by $\omega_1/\omega_3 = (\psi - 1)/\psi$. Both of these two conditions allow us to obtain planetary gear trains capable to achieving very small transmission ratios between two parallel shafts, that of the planet carrier and that of the main wheel, which is rotating. In fact, from what we have said above, we infer that, to reach this interesting design goal, it is sufficient to design the gear wheels so that $i_0 = 1/\psi$ is close to the unit as much as is desired; consequently, the transmission ratio ω_1/ω_3 (or ω_2/ω_3) becomes close to zero as much as is desired.

The characteristic ratio, i_0, and its reciprocal, ψ, may be expressed as the quotient of the product of number of teeth of driving gears in a gear train divided by the product of number of teeth of driven gears in the same gear train, i.e. in the form

$$\psi = \frac{1}{i_0} = \frac{\omega_2 - \omega_3}{\omega_1 - \omega_3} = \pm \frac{z_1 z_3 \ldots}{z_2 z_4 \ldots}, \tag{13.19}$$

where z_1, z_3, ... are the numbers of teeth of driving gears, and z_2, z_4, ... are the numbers of teeth of the driven gears. In Eq. (13.19), we take the plus or minus sign depending on whether, with the planet carrier held at rest, the directions of rotation of the first gear wheel (the driving gear wheel, 1) and the last gear wheel (the driven gear wheel, 2) are the same or opposite.

Furthermore, it is to be noted that the angular velocities ω_1, ω_2, and ω_3 (and thus the corresponding rotational speeds n_1, n_2, and n_3) are absolute angular velocities, that is angular velocities in absolute motion of the three main members of the planetary gear train with respect to its housing, which constitutes the frame.

Finally, it should be noted that the above-described algebraic method, from which the Willis formula is obtained, is applicable to planetary gear trains of type (a) and (b), as defined in the previous Sect. 13.3. To solve the kinematic problems of planetary gear trains of type (c) and (d) defined in the same Sect. 13.3, other more elaborate methods must be used. For planetary gear trains of type (c), a method similar to the one described above may be used, with the addition of other conditions due to their geometric configuration (see Ferrari and Romiti [14]). For planetary gear trains of the type (d), a more complex method must be used, such as the one proposed and described by Ferrari and Romiti [14]. This method, on which we do not dwell here for brevity, is quite general, so it is applicable in all possible cases, including the three previous cases, though it is less easy to apply in these cases than the one above described.

13.4.2 Torque Balance on Single Members

This method involves the analysis of the balance of torques acting on the single members of the planetary gear train. For the description of the method, we consider the Fig. 13.6, where these members are shown separately, that is in exploded view that allows a free-body force analysis to be carried out (see: Juvinall [25], Juvinall and Marshek [26]). With the notations shown in this figure, we find that the distance of the axes of the planet gears from the axis of the gear system is equal to the sum of the radii of pitch circles of the sun gear and planet gear. This distance, equal to $(r_s + r_p)$, can be expressed as follows:

$$(r_s + r_p) = r_s + \frac{r_a - r_s}{2} = \frac{r_a + r_s}{2},$$ (13.20)

where r_s, r_p, and r_a are respectively the radii of the pitch circles of the sun gear, planet gear, and annulus gear.

Let us assume that the power losses are equal to zero, the sun gear is held at rest, the driving torque or input torque, T_i, is applied to the annulus gear, and the resistant torque or output torque, T_0, is applied to the planet carrier. To satisfy the conditions of balance of torques acting on the three members of the planetary gear train, through which the driving power flows (they are, in order, the annulus gear, planet gears, and planet carrier), we introduce the necessary forces acting at various radii. This procedure, summarized in Fig. 13.6, leads to the following final result:

$$\frac{\omega_i}{\omega_0} = \frac{T_0}{T_i} = 1 + \frac{r_s}{r_a}$$ (13.21)

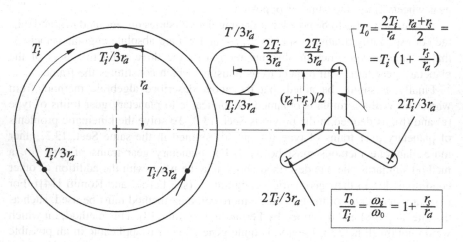

Fig. 13.6 Determination of the transmission ratio by free-body force analysis, and balance of torques (annulus gear as input, planet carrier as output, and sun gear as fixed member)

where ω_i and ω_0 are respectively the input and output angular velocities.

Under the same aforementioned conditions, using the second Eq. (13.16) in which we place $\omega_1 = 0$, $\omega_3 = \omega_i$, and $\omega_2 = \omega_0$, and bearing in mind that in this case $\psi = -(z_s/z_a) = -(r_s/r_a)$, we infer that the second Eqs. (13.16) and (13.21) give the same result.

13.4.3 Analysis of Tangential Velocity Vectors

This method involves the analysis of the tangential velocity vectors on the pitch circles of the single members of the planetary gear train, as well as on the circle passing through the axes of the planet gears. For a description of the method, we denote by v an arbitrary value of the tangential velocity at the point of contact between the pitch circle of the annulus gear and the pitch circle of any one of the planet gears of the planetary gear train examined (Fig. 13.7).

Even here we assume that the sun gear is held at rest. In this condition, the tangential velocity of point of contact between the pitch circles of the sun gear and the planet gear under consideration will obviously be zero. The tangential velocity of the planet carrier at point corresponding to the axis of the planet gear is evidently equal to $v/2$. The angular velocities of the annulus gear and the planet carrier can then be determined as the quotient, respectively, of the tangential velocities v and $v/2$ divided by the corresponding radii r_a and $(r_a + r_s)/2$. In doing so, in another way, we get the same Eq. (13.21).

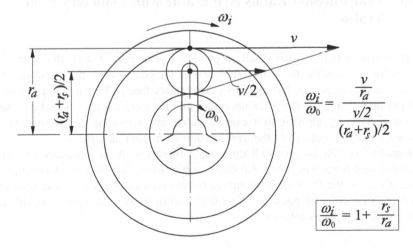

Fig. 13.7 Determination of the transmission ratio by tangential velocity vectors (annulus gear as input, planet carrier as output and sun gear as fixed number)

13.4.4 Other Methods

In addition to the three above described methods, other methods exist for the determination of the transmission ratio of a planetary gear train. We make here a brief reference to the *tabulation method* and *geometrical method*, illustrating the basic concepts on which they are based. For details, we refer the reader to more specialized textbooks (see: Müller [44], Lyndwander [35], Maitra [36], Jelaska [24]).

The tabulation method is a summation process in which the planetary gear train is first considered with all its members locked to one another (thus the whole gear system rotates as if it is a single block). Successively, the same planetary gear train is considered as an ordinary gear train (thus all the gears are free to rotate about their own axes, and planet carrier is held at rest). Results related to these two steps are then added and presented in a tabulation form. Compared with the algebraic method, which provides only the final transmission ratio of the gear system, this method has the advantage of giving a complete picture of the angular motions of the various members, and also to provide any intermediate transmission ratio related to them.

Ex nomine, the geometrical method is based on geometrical considerations. According to this method, starting from a given initial position, the pitch circle of any of the planet gears is rolled without sliding on the pitch circle of the sun gear, assumed as fixed. The angle between the initial and new positions of the planet carrier is then considered and, in correlation with it, the transmission ratio is evaluated.

13.5   Transmission Ratios Achievable with Planetary Gear Trains

If we interpose between two shafts a planetary gear train, and rigidly connect to each of them two of the three main members of the gear system, the transmission of motion cannot occur, as the third member rotates freely. To carry out the transmission of motion, it is therefore necessary that the third member is locked, that is held stationary; by applying to it a brake-clutch, and activating or deactivating this last, we can carry out or not the transmission of motion, at will.

Usually, we can assign to all three main members of the planetary gear train three different functions: input of motion, output of motion, and fixed member. By locking one of the three main members (of course one at a time), and then considering three different possible operating conditions, we can obtain six different transmission ratios, as follows.

(A) With the planet carrier locked ($\omega_3 = 0$), we have an ordinary gear train (all axes are fixed), and the sun gear and annulus gear (the one or the other can be indifferently driving or driven members) rotate in the opposite direction, realizing a reversal of the direction of rotation. From Eqs. (13.16) or (13.15) we get the following expression of the transmission ratio:

$$i = i_0 = \frac{\omega_1}{\omega_2} = \frac{1}{\psi}. \tag{13.22}$$

Therefore, depending on whether we take the sun gear or the annulus gear as driving gear 1 and driven gear 2, or vice versa, two different transmission ratios are possible.

(B) With the sun gear locked, the annulus gear and planet carrier (again one or the other can be indifferently driving or driven members) rotate in the same direction, but with different angular velocities. Since $\omega_1 = 0$, from Eqs. (13.16) or (13.15) we get the following expression of the transmission ratio:

$$i = \frac{\omega_2}{\omega_3} = 1 - \psi. \tag{13.23}$$

In addition, in this case, depending on whether we take the annulus gear or the planet carrier as driving gear 2 and driven gear 3, or vice versa, two different transmission ratios are possible.

(C) With the annulus gear locked, the sun gear and planet carrier (also here the one or the other can be indifferently driving or driven members) rotate in the same direction, but with different angular velocities, and with ratios between these velocities different from those obtained above for the case where the sun gear was locked. Since $\omega_2 = 0$, from Eqs. (13.16) or (13.15) we get the following expression of the transmission ratio:

$$i = \frac{\omega_1}{\omega_3} = 1 - \frac{1}{\psi}. \tag{13.24}$$

Also, in this case, depending on whether we take the sun gear or the planet carrier as driving gear 1 and driven gear 3, or vice versa, two different transmission ratios are possible.

We also have a seventh chance. In fact, if the gear system is characterized by a brake-clutch which allows to rigidly connect between them any pair formed by any two of the three main members of the planetary gear train, the whole gear system will rotate as a single block, coming to operate as a rigid coupling. We have therefore the possibility to realize the *direct drive*, i.e. the direct engagement between the driving and driven shafts ($i = 1$). This mode of operation is common to the three possible interlocks between any two of the three main members of a planetary gear train. The three related trivial states of motion, mentioned by someone, are reduced de facto to a single trivial state of motion in which the gear system rotates as a rigid coupling.

In summary, we can say that a planetary gear train provides seven combinations of possible states of motion (they obviously become nine when the three trivial states of motion are considered as distinct states of motion), which are derived from the fact that, in principle, the position of the annulus gear, planet carrier and sun gear can be locked, for which they act as a frame. The two remaining members of the planetary gear train can be used as input or output of the power to be transmitted. The ratios of the single states of motion cannot be selected independently of each other, but are defined by the number of teeth z_1 of the sun gear and z_2 of the annulus gear. Table 13.1 summarizes, as an example, the six non-trivial states of motion of a planetary gear train such as that shown in Fig. 13.5, and provides, for each of these states of motion, the operating conditions and transmission ratios obtainable.

Therefore, this type of gear system makes it possible to achieve different transmission ratios between the input and output shafts, depending on how the individual main members are linked together and which members are in a locked position. The members are linked together by clutches, while they are locked to the housing by brakes. The operating condition with the sun gear locked is the one commonly used. Depending on the relative sizes of the various members of the planetary gear train, the transmission ratio calculated by Eqs. (13.16) or (13.15) can take values between 1 and 2. With the annulus gear as a driving gear and planet carrier as output shaft, the planetary gear train is often used as a speed reducing gear drive in propeller planes. With the planet carrier as an input shaft and annulus gear as an output gear, the planetary gear train forms the basis of the *overdrive* used in automobiles [25, 43, 45].

If several planetary gear trains of the type shown in Fig. 13.5 are connected together, the result is a compound planetary gear train of the multi-stage structure type (see Sect. 13.3). This type of compound gear train makes it possible to obtain different transmission ratios between input and output, which depend on how individual transmission members are connected together, and which members are in a fixed position. The connection between the various members of the compound

Table 13.1 States of motion of a planetary gear train such as that shown in Fig. 13.5, operating conditions, and transmission ratio obtainable

State of motion	Operating condition	Input	Output	Frame	Transmission ratio
1	Ordinary gear train	1	2	3	$i = \frac{\omega_1}{\omega_2} = -\frac{z_2}{z_1}$
2		2	1		$i = \frac{\omega_2}{\omega_1} = -\frac{z_1}{z_2}$
3	Planetary gear train	1	3	2	$i = \frac{\omega_1}{\omega_3} = 1 + \frac{z_2}{z_1}$
4		3	1		$i = \frac{\omega_3}{\omega_1} = \frac{1}{1+\frac{z_2}{z_1}}$
5		2	3	1	$i = \frac{\omega_2}{\omega_3} = \frac{1}{1+\frac{z_1}{z_2}}$
6		3	2		$i = \frac{\omega_3}{\omega_1} = 1 + \frac{z_1}{z_2}$

planetary trains is made by means of clutches, while the locking of members to be stationary is achieved by means of brakes, which connect them to the housing. Thus, the aforementioned large variety of possible transmission ratios, which is typical of just one planetary gear train, is further substantially increased with these compound planetary gear trains, although not all the possible transmission ratios can be suitable for use in motor vehicle transmissions. With regard to this interesting topic, we refer the reader to specialized textbooks (see: Pollone [51], Lechner and Naunheimer [31], Naunheimer et al. [45]).

Traditional automatic transmission in motor vehicles are made up of the combination of several planetary gear trains, the member of which are differently connected to each other and with the housing, so as to obtain the desired number of transmission ratios. However, the ratios of individual gear steps cannot be freely selected independently of each other, as the same gearwheels are used for several gear steps. The first automatic transmission, which switched the optimum gear without driver intervention, except for starting off and going in reverse, was developed by General Motors, based on ideas and patents dating back to the first thirty years of the last century (see: Meyer [39], Birch [2], Warwick [58]). The same General Motors introduced the related technology in the 1940 Oldsmobile model, as Hydra-Matic Transmission. With the exception of CVTs-(Continuous Variable Transmissions), current automatic transmissions are essentially identical to this automatic transmission, as the few differences are limited to the use of more sophisticated hydraulic and electronic systems, which are responsible for their operation and control.

Of course, automatic transmissions were developed on the basis of manual or semi-automatic gearboxes, which used planetary gear trains. The Wilson gearbox, which relied on a number of epicyclic gear trains, coupled in an ingenious manner (it is not to be confused with the pre-selector or self-changing gearbox, due to the same inventor, Wilson, W.G.), can be considered to be the precursor of these automatic transmissions. It had a planetary gear train for each intermediate gear, with a cone clutch for the straight-through top gear, and a further planetary gear for going in reverse (see: Ferrari and Romiti [14], Lechner and Naunheimer [31], Naunheimer et al. [45], Meyer [39]).

13.6 Some Problems Related to Simple Planetary Gear Train

Let z_1, z_2 and z_0, respectively, the number of teeth of the sun gear, annulus gear, and planet gears. In a simple planetary gear train, a first problem arises in relation to these numbers of teeth, because it is possible to have n planet gears arranged at an angular distance $(2\pi/n)$ from one another, so that the centrifugal forces acting on them have zero resultant, thus being balanced.

In view of known geometrical properties, we can reduce the problem to the diagram shown in Fig. 13.8, in which the small circles indicate the intersections of the tooth profiles with the pitch circles. Let's consider one of the planet gears in the position in which a tooth of this planet gear touches at point A a tooth of the sun gear, and distinguish the case in which z_0 is an even number (Fig. 13.8a) from that in which z_0 is an odd number (Fig. 13.8b).

In the first case (z_0, even number), another tooth of the planet gear exists, which touches at point A' a tooth of the annulus gear. The pitch circle of the next planet gear has the center C' on the straight line OC', at an angular distance $(2\pi/n)$ from the straight line OC. The tooth B of the sun gear, and the tooth B' of the planet gear, which meshes with B, are located on the respective pitch circles of a number of circular pitches less of 1, and equal to $[(z_1/m) - Z]$, wherein Z is an integer (in this regard, we assume that B is the tooth that immediately precedes the straight line OC'). The same thing happens for the tooth B'' diametrically opposite with respect to B'. Between the tooth A' and the tooth B''' of the annulus gear, which must mesh with B'', an integer number of circular pitches should be, for which the following equation must be satisfied:

$$\frac{z_2}{n} + \left(\frac{z_1}{n} - Z\right) = Z. \tag{13.25}$$

The condition that must be met is therefore the following:

$$z_1 + z_2 = \text{multiple of } n. \tag{13.26}$$

In the second case (z_0, odd number), repeating the same reasoning with reference to the diagram shown in Fig. 13.8b, we find that the condition to be satisfied is as follows:

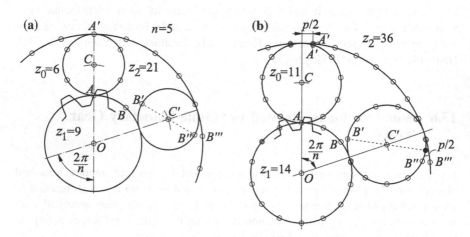

Fig. 13.8 Conditions between the numbers of teeth to be met for: **a** z_0 even number; **b** z_0 odd number

$$\frac{A'B'''}{p} = \frac{z_2}{n} - \frac{1}{2} + \left(\frac{z_1}{n} - Z\right) + \frac{1}{2} = Z, \tag{13.27}$$

from which we still gain the condition given by relationship (13.26). However, this necessary condition is not sufficient. The following condition must be added to it:

$$z_2 - z_1 = 2z_0 = Z, \tag{13.28}$$

where Z is an even number; this condition expresses that the numbers of teeth of the sun gear and annulus gear must be both even, or both odd.

Sometimes it happens that the planet gears are mounted on ball bearings. A second problem arises in this case: the one related to the determination of the number of revolutions according to which these bearings must be calculated, based on the indications of catalogs of the manufactures. This number of revolutions, which corresponds to the relative motion between the bearing and the pin, is calculated by impressing to the whole gear system an angular velocity $-\omega_3$, equal and opposite to the angular velocity of the planet carrier. We have already said that, by doing so, the planetary gear train becomes an ordinary gear train. The angular velocities of the sun gear and annulus gear pass from the values ω_1 and ω_2 to the values $(\omega_1 - \omega_3)$ and $(\omega_2 - \omega_3)$; therefore, the number of revolutions required is given by:

$$\frac{1}{2\pi}(\omega_1 - \omega_3)\frac{z_1}{z_0} = \frac{1}{2\pi}(\omega_2 - \omega_3)\frac{z_2}{z_0}. \tag{13.29}$$

The ball bearing of the planet gear is loaded by a force equal to the resultant of the two forces (they are approximately of equal value), exchanged between the sun gear and planet gear, and between the planet gear and annulus gear (see Fig. 13.6). This resultant force is about double of the tangential force acting on the planet gear. Therefore, often the need occurs to increase the size of the gear planet than would be required according to the strength calculation of the planet gear itself, so that it is possible to house the bearing inside the coaxial hole practiced in the center of the same planet gear.

13.7 Main Characteristics of Some Planetary Gear Trains

Some planetary gear trains are examined here as examples and, for each of them, the main characteristics are described, with reference to the conditions that must be met in terms of numbers of teeth of the toothed members, angular distance between the planet gears and obtainable transmission ratios. All the examples mentioned below are mainly derived from practical applications in vehicle transmissions.

13.7.1 Planetary Gear Train Type 1 (Example 1)

As Fig. 13.9 shows, this planetary gear train is composed of two coaxial sun gears having z_1 and z_4 teeth, and a given number n of planet gears, each of which consists of two gear wheels having z_2 and z_3 teeth, and between them rigidly connected. The various toothed pairs in meshing with each other have the same face width, and their mid-planes coincide. As usual, all teeth of the toothed members have equal module, m.

The condition that the numbers of teeth of the gear wheels must meet, resulting from the need that the two gear pairs have the same center distance, is as follows:

$$z_1 + z_2 = z_3 + z_4. \tag{13.30}$$

Since the sizing we are considering is the nominal one, the condition that two successive planet gears do not interfere with each other is expressed by the following relationship:

$$(z_1 + z_2) \sin \frac{\varphi}{2} > z_2 + 2, \tag{13.31}$$

where $\varphi = 2\pi/n = \tau$ is the angular pitch. This angular distance is then given by any of the following two equations, each of which is expressed in radians or degrees:

$$\varphi = Z \frac{2\pi}{z_4 - z_1 \frac{z_3}{z_2}} \text{(rad)} \qquad \varphi = Z \frac{360}{z_4 - z_1 \frac{z_3}{z_2}} (°) \tag{13.32}$$

$$\varphi = Z \frac{2\pi}{z_1 - z_4 \frac{z_2}{z_3}} \text{(rad)} \qquad \varphi = Z \frac{360}{z_1 - z_4 \frac{z_2}{z_3}} (°), \tag{13.33}$$

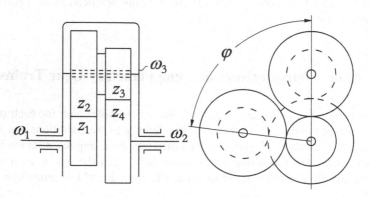

Fig. 13.9 Planetary gear train type 1

where Z is an integer selected so that the Eq. (13.31) is satisfied. Thus, we can deduce that φ must be a multiple according to an integer of the angle expressed by the fraction that appears in the above equations. The planet gears are arranged in pairs diametrically opposed, unless they are put at $\varphi = 120°$.

The determination of the transmission ratio of this planetary gear train (as well as the determination of the transmission ratios of the planetary gear trains described in the following sections) is performed using the Eqs. (13.23) and (13.24), in the case in which only two members are mobile, and with Eqs. (13.16) in the case in which all three members are mobile.

Putting $(z_2 - z_3) = x$, and taking account of eq. (13.30), it must also be $(z_4 - z_1) = x$, and therefore:

$$\psi = \frac{z_1 z_3}{z_2 z_4} = \frac{z_1(z_2 - x)}{z_2(z_1 + x)} = \frac{1 - (x/z_2)}{1 + (x/z_1)} < 1. \tag{13.34}$$

Therefore, we infer that a planetary gear train for which it is $x = 0$, is not interesting for practical applications. In the case in which only two members of the planetary gear train are mobile, in place of the transmission ratios given by Eqs. (13.22)–(13.24), we obtain the following transmission ratios:

$$\omega_2 = \omega_1 \psi = \omega_1 \frac{1 - (x/z_2)}{1 + (x/z_1)} \tag{13.35}$$

$$\omega_2 = \omega_3(1 - \psi) = \omega_3 \frac{x[(1/z_1) + (1/z_2)]}{1 + (x/z_1)} \tag{13.36}$$

$$\omega_1 = \omega_3\left(1 - \frac{1}{\psi}\right) = -\omega_3 \frac{x[(1/z_1) + (1/z_2)]}{1 - (x/z_2)}, \tag{13.37}$$

which are respectively valid for $\omega_3 = 0$, $\omega_1 = 0$, and $\omega_2 = 0$.

Wanting to realize large values of the reduction ratio, in cases where $\omega_1 = 0$ or $\omega_2 = 0$, small values of x must be adopted, and sun gears with a large number of teeth. The minimum value of x may be $x = 1$. Wanting to realize planetary gear trains not very bulky, that is, with gear wheels having not very large numbers of teeth, large reduction ratios can not be achieved.

With teeth having nominal sizing and pressure angle $\alpha = 20°$, limiting as much as possible the number of teeth, and then assuming $z_1 = 15$ and $z_2 = 20$, for $x = 1$ and $\omega_1 = 0$, from Eq. (13.36) we get: $\omega_2 = (7/64)\omega_3$. If m is the module of the teeth, the bulk of the planetary gear train, corresponding to the diameter enveloping the tip circles of the largest gear wheels of the planet gears, is given by: $(z_1 + 2z_2 + 2)m = 57m$.

A particular case of this planetary gear train is that in which $z_4 = (z_1 + 1)$ and $z_2 = z_3$. Since the two pairs of gear wheels having respectively numbers of teeth z_1 and z_2, z_3 and z_4, must have the same center distance, they must necessary be characterized by profile shifted toothing. In this case it is $\psi = z_1/(z_1 + 1)$, for which we will have: $\omega_2 = \omega_3[1/(z_1 + 1)]$ for $\omega_1 = 0$, and $\omega_1 = -\omega_3(1/z_1)$ for $\omega_2 = 0$. With the first gear pair having profile shifted toothing and $z_1 = 19$ and $z_2 = 17$, and the second gear pair having zero profile shifted toothing and $z_3 = 17$ and $z_4 = 20$, and with $\alpha = 20°$, the minimum gear ratio is given by $\omega_2 = \omega_3/20$, so that the bulk of the planetary gear train will be $(z_4 + 2z_3 + 2)m = 56m$.

13.7.2 Planetary Gear Train Type 2 (Example 2)

As shows Fig. 13.10, this planetary gear train is composed of two sun gears, having z_1 and z_6 teeth, and of p planet gears, each of which is composed of two rigidly connected pairs of gear wheels having z_2, z_3, z_4, and z_5 teeth. The relative position of the two gear wheels of each pair is identical to that determined for the planet gear of the planetary gear train described in the previous Sect. 13.7.1. All teeth of the toothed members have equal module, m.

It must be verified that the pins of the planet gear with z_2 and z_3 teeth do not interfere with the sun gear having z_6 teeth. The angular distance φ must correspond to one of the values given by the following relationships:

$$\varphi = Z\frac{2\pi}{z_6 + z_1 \frac{z_3 z_5}{z_2 z_4}}(\text{rad}) = Z\frac{360}{z_6 + z_1 \frac{z_3 z_5}{z_2 z_4}}(°) \qquad (13.38)$$

$$\varphi = Z\frac{2\pi}{z_1 + z_6 \frac{z_4 z_2}{z_5 z_3}}(\text{rad}) = Z\frac{360}{z_1 + z_6 \frac{z_4 z_2}{z_5 z_3}}(°), \qquad (13.39)$$

in which Z is an integer. However, the value of Z should be selected in such a way that the planet gears do not interfere with each other.

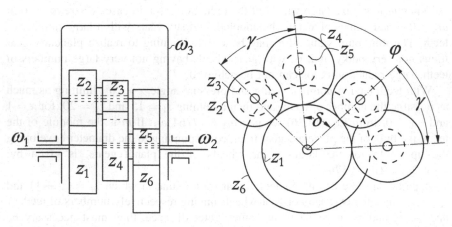

Fig. 13.10 Planetary gear train type 2

The numbers of teeth of the single pairs of the gear wheels that mesh between them should be chosen so as to avoid interference between the tooth flank profiles. The planet gears are arranged in opposed pairs along a diameter. The angle γ between the two radial straight lines passing through the axes of each of the two rigidly connected pairs of gear wheels of the planet gears is given by the following relationship:

$$\cos \gamma = \frac{(z_1 + z_2)^2 - (z_3 + z_4)^2 + (z_5 + z_6)^2}{2(z_1 + z_2)(z_5 + z_6)}. \tag{13.40}$$

This type of planetary gear train, for which it is:

$$\psi = -\frac{z_1 z_3 z_5}{z_2 z_4 z_6}, \tag{13.41}$$

is not very interesting from the point of view of practical applications, due to its constructive complexity. The case in which $z_1 = z_6$ and $z_2 = z_3 = z_4 = z_5$ otherwise makes exception; since in this case $\psi = -1$, these types of planetary gear trains with three mobile members are used as speed adders or as differential torque splitters into two equal parts.

13.7.3 Planetary Gear Train Type 3 (Example 3)

As Fig. 13.11 shows, this planetary gear train is composed of a sun gear having z_1 teeth, an annulus gear with z_2 teeth, and p planet gears, each of which is composed of a single gear wheel having z_0 teeth. All teeth of the toothed members have equal module, m.

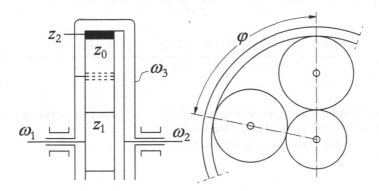

Fig. 13.11 Planetary gear train type 3 (example 3)

Considering that the pitch diameter of the annulus gear is equal to the sum of the pitch diameter of the sun gear plus twice the pitch diameter of the planet gears, we obtain the following relationship that correlates the number of teeth of the various toothed members:

$$z_2 - z_1 = 2z_0. \tag{13.42}$$

Since $2z_0$ is an even number, we infer that z_1 and z_2 must be both even or both odd. The planet gears need to be arranged at an angular distance, φ, given by:

$$\varphi = Z\frac{2\pi}{z_1 + z_2}\,(\text{rad}) = Z\frac{360}{z_1 + z_2}\,(°), \tag{13.43}$$

where Z is an integer. If the sum $(z_1 + z_2)$ is divisible by an integer, p, that is, if it is

$$\frac{z_1 + z_2}{Z} = p, \tag{13.44}$$

a number p of planet gears can be arranged at an angular distance, $(360°/p)$. In order that the planet gears do not interfere with each other and do not touch, the following relationship must be satisfied:

$$(z_1 + z_0)\sin\frac{\varphi}{2} > z_0 + 2. \tag{13.45}$$

Substituting z_0 in place of z_1 in Eqs. (5.10) and (5.11), and taking into account Eq. (13.42), for which it is $z_2 = (z_1 + 2z_0)$, we get the following two relationships that express, as a function of z_1, the minimum number of teeth, z_0, of the planet gears which does not cause interference with the annulus gear having z_2 teeth:

$$z_{0,min} = \frac{4 - z_1 \sin^2\alpha + \sqrt{16 - 12\sin^2\alpha + 4z_1\sin^2\alpha + z_1^2\sin^4\alpha}}{3\sin^2\alpha}. \tag{13.46}$$

$$z_{0,min} = \frac{4(h_1/m) - z_1^2\sin^2\alpha + \sqrt{\left(16 - 12\sin^2\alpha\right)\left(\frac{h_1}{m}\right)^2 + 4\frac{h_1}{m}z_1\sin^2\alpha + z_1^2\sin^4\alpha}}{3\sin^2\alpha}; \tag{13.47}$$

these relationships are respectively valid for $h_1 = m$ (standard or nominal sizing), and for $h_1 \neq m$ (in particular, for stub sizing).

Since for this planetary gear train we have

$$\psi = -\frac{z_1}{z_2}, \tag{13.48}$$

the second Eq. (13.16), which is valid in the case where its three main members are all mobile, becomes:

$$\omega_2 = -\frac{z_1}{z_2}\omega_1 + \left(1 + \frac{z_1}{z_2}\right)\omega_3. \tag{13.49}$$

Locking successively, one at a time, the three aforementioned members, we get the following transmission ratios:

$$\omega_2 = -\omega_1\frac{z_1}{z_2} \tag{13.50}$$

$$\omega_2 = \left(1 + \frac{z_1}{z_2}\right)\omega_3 \tag{13.51}$$

$$\omega_1 = \left(1 + \frac{z_2}{z_1}\right)\omega_3, \tag{13.52}$$

which are respectively valid for $\omega_3 = 0$, $\omega_1 = 0$, and $\omega_2 = 0$.

Wanting to take annulus gears not too bulky, with these planetary gear trains it is not possible to achieve large transmission ratios. Here also, with teeth having nominal sizing and pressure angle $\alpha = 20°$, and a sun gear with $z_1 = 15$, from Eq. (13.46) we derive that the minimum number of teeth of the sun gear is equal to $z_{0,min} = 21$, for which we shall have: $z_2 = (z_1 + 2z_0) = 57$ teeth. It is then possible to make the following transmission ratios: $\omega_2 = -(15/57)\omega_1$, for $\omega_3 = 0$; $\omega_2 = (72/57)\omega_3$, for $\omega_1 = 0$; $\omega_1 = (72/15)\omega_3$, for $\omega_2 = 0$.

This planetary gear train with $z_1 + z_2 = 72$ is achievable with 2, 3 or 4 planet gears, since 72 is divisible by such numbers. It is possible to increase the number of teeth of the planet gear, z_0, and therefore the number of teeth of the annulus gear, z_2, according to Eq. (13.42), without the interference between them occurs. However, the maximum number of teeth of the planet gear will be the one for which the interference occurs between it and the sun gear. From the diagram shown in Fig. 4.3, for $\alpha = 20°$ and $z_1 = 15$, we obtain that the maximum number of teeth of the planet gear is $z_0 = 43$. Therefore, it is $z_2 = 101$. Thus it is possible to make the following transmission ratios: $\omega_2 = -(15/101)\omega_1$, for $\omega_3 = 0$; $\omega_2 = (116/101)\omega_3$, for $\omega_1 = 0$; $\omega_1 = (116/15)\omega_3$, for $\omega_2 = 0$.

Considering the first of the two above examples, and taking, as bulk dimension, the root diameter of the annulus gear, if m is the module of the teeth, this bulk dimension will be given by $(z_2 + 2 \cdot 1.25)m = 59.5m$. Therefore, we see that, for equal overall dimensions, with this type of planetary gear train we can realize reduction ratios lower than those obtainable with the planetary gear trains previously examined.

13.7.4 Planetary Gear Train Type 4 (Example 4)

As Fig. 13.12 shows, this planetary gear train differs from that considered in the previous section only because each planet gear consists of two gear wheels rigidly connected, having z_2 and z_3 teeth. The sun gear and the annulus gear have respectively z_1 and z_4 teeth. All teeth of the toothed members have equal module, m.

Since the radii r_1, r_2, r_3, and r_4 of the four gear wheels are linked by the relationship $r_4 = r_1 + 2r_2 - (r_2 - r_3)$, we obtain:

$$z_4 = z_1 + z_2 + z_3, \tag{13.53}$$

for which z_1 and z_4 may be either odd or even. The angular distance between the straight lines passing through the axes of the planet gears is given by one of the two following relationships:

$$\varphi = Z \frac{2\pi}{z_4 + z_1 (z_3/z_2)} (\text{rad}) = Z \frac{360}{z_4 + z_1 (z_3/z_2)} (°) \tag{13.54}$$

$$\varphi = Z \frac{2\pi}{z_1 + z_4 (z_2/z_3)} (\text{rad}) = Z \frac{360}{z_1 + z_4 (z_2/z_3)} (°), \tag{13.55}$$

where Z is an integer. In order that the planet gears do not interfere with each other, Z must be selected so as to satisfy the following inequality:

$$(z_1 + z_2) \sin \frac{\varphi}{2} > z_2 + 2. \tag{13.56}$$

This type of planetary gear train is much more cumbersome than the previous one, without realizing much larger reduction ratios. For this gear train we have:

$$\psi = -\frac{z_1 z_3}{z_2 z_4}. \tag{13.57}$$

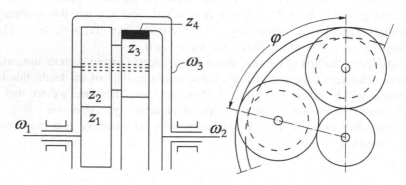

Fig. 13.12 Planetary gear train type 4

The various reduction ratios achievable are obtained by substituting this value of ψ in Eqs. (13.35), (13.36) and (13.37), which refer respectively to cases where $\omega_3 = 0$, $\omega_1 = 0$, and $\omega_2 = 0$.

13.7.5 Planetary Gear Train Type 5 (Example 5)

As Fig. 13.13 shows, this planetary gear train consists of two annulus gears having z_1 and z_4 teeth and a single planet gear, composed of two gear wheels having z_2 and z_3 teeth and rigidly connected to each other. All teeth of the toothed members have equal module, m.

For this planetary gear train, the first condition to be satisfied is given by the following relationship:

$$z_4 - z_3 = z_1 - z_2, \tag{13.58}$$

which indicates that the two internal gear pairs have the same center distance. The second condition to be met imposes that the choice of the numbers of teeth of the various gear wheels in meshing is carried out so as to avoid the meshing interference. For this planetary gear train, we have:

$$\psi = \frac{z_1 z_3}{z_2 z_4}, \tag{13.59}$$

and the various reduction ratios achievable are obtained by substituting this value of ψ in Eqs. (13.35), (13.36) and (13.37), which refer respectively to cases where $\omega_3 = 0$, $\omega_1 = 0$, and $\omega_2 = 0$. From Eqs. (13.36) and (13.37), we can deduce that,

Fig. 13.13 Planetary gear train type 5

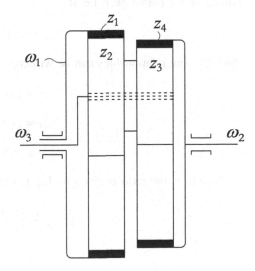

in order to obtain large reduction ratios in cases where $\omega_1 = 0$ and $\omega_2 = 0$, it is necessary that the value of ψ is as much as possible close to 1. It is therefore necessary that z_1 and z_4 differ as little as possible between them. In accordance with Eq. (13.58), the difference between z_2 and z_3 must be equal to the difference between z_1 and z_4. In addition, if we do not want to take too large values of z_1 and z_4, in order to limit the overall dimensions, we have to choose values of z_2 and z_3 as large as possible.

With a given annulus gear of z_4 teeth, the minimum difference $(z_4 - z_3)_{min}$, and therefore the maximum size of the gear wheel of z_3 teeth that we can take, without that the secondary interference occurs, is that for which the difference

$$z_4 - z_3 = y \tag{13.60}$$

is given by the diagram shown in Fig. 5.7.

By means of the relationships (5.10) and (5.11), however written in terms of z_4 and z_3 instead of z_2 and z_1, substituting in them the expression of z_4 given by Eq. (13.60), we obtain the minimum number of teeth z_3 of the planet gear that, differing by y from the number of teeth z_4 of the annulus gear, does not interfere with it. From the relationship (5.11), which is valid for standard sizing, we get:

$$z_{3,min} = \frac{2 - y\sin^2\alpha + \sqrt{(2 - y\sin^2\alpha)^2 + 4(y-1)\sin^2\alpha}}{\sin^2\alpha}. \tag{13.61}$$

Once we have chosen the value of y, using the diagram shown in Fig. 5.7, we determine $z_{3,min}$, and then $z_{4,min}$ by Eq. (13.60). The minimum values of z_3 and z_4 can be also read directly on the diagram of Fig. 5.4, since they correspond to the points that delimit, at the bottom right, the curves relating to the various pressure angles.

If we denote by x the difference between the numbers of teeth of the two gear wheels of the planet gear, i.e. if:

$$z_2 - z_3 = x, \tag{13.62}$$

the following relationships can be written:

$$z_2 = z_3 + x \tag{13.63}$$

$$z_4 = z_3 + y \tag{13.64}$$

$$z_1 = z_3 + x + y. \tag{13.65}$$

Therefore, the ratio ψ given by Eq. (13.59) can be written in the form

$$\psi = \frac{z_3(z_3 + x + y)}{(z_3 + x)(z_3 + y)}, \tag{13.66}$$

and it is obviously smaller than 1 ($\psi < 1$). Since y depends on the pressure angle, to achieve large reduction ratios, the value of x should be chosen as smaller as possible, that is $x = 1$. With reference to the angular velocities indicated in Fig. 13.13, the transmission ratios achievable are as follows:

$$\omega_2 = \psi \omega_1 \tag{13.67}$$

$$\omega_2 = (1 - \psi)\omega_3 \tag{13.68}$$

$$\omega_1 = \left(1 - \frac{1}{\psi}\right)\omega_3 = -\left(\frac{1}{\psi} - 1\right)\omega_3, \tag{13.69}$$

which are respectively valid for $\omega_3 = 0$, $\omega_1 = 0$, and $\omega_2 = 0$.

For example, with a pressure angle $\alpha = 20°$, and with $y = 8$, from Eq. (13.61) or Fig. 5.4 we get $z_3 = 27$, and from Eq. (13.60), $z_4 = 35$, while from Eqs. (13.62) and (13.65) we obtain $z_2 = 28$ and $z_1 = 36$. With the planetary gear train so dimensioned, we have: $\omega_2 = (243/245)\omega_1$, for $\omega_3 = 0$; $\omega_2 = (2/245)\omega_3$, for $\omega_1 = 0$; $\omega_1 = -(2/243)\omega_3$, for $\omega_2 = 0$. If m is the module, the root diameter of the bigger annulus gear is equal to 38.34 m. Thus, this planetary gear train allows large reduction ratios with small overall dimensions.

If we choose $z_1 = 54$, $z_2 = 46$, $z_3 = 45$ and $z_4 = 53$, we obtain: $\omega_2 = (1215/1219)\omega_1$, for $\omega_3 = 0$; $\omega_2 = (1/304, 75)\omega_3$, for $\omega_1 = 0$; $\omega_1 = -(1/303, 75)$ ω_3, for $\omega_2 = 0$. The overall dimension corresponding to the root diameter of the bigger annulus gear equals 56.34 m; therefore, it is comparable to that of the planetary gear trains examined in Sects. 13.7.1 and 13.7.3, but the reduction ratios obtained are much greater.

It is also to be noted that the reduction ratios would be greater in the case in which, with annulus gears having $z_1 = 54$ and $z_4 = 53$ teeth, we would choose for the two planet gears $z_3 = 21$ and $z_2 = 22$, that is equal to the minimum number of teeth for which the interference with the corresponding annulus gears is avoided. In this case, in fact, we would have: $\omega_2 = (527/563)\omega_1$, for $\omega_3 = 0$; $\omega_2 = (1/36, 4)\omega_3$, for $\omega_1 = 0$; $\omega_1 = -(1/35, 4)\ \omega_3$, for $\omega_2 = 0$.

Finally, it should be noted that all the examples considered in this section as well as in the previous ones relate to cases in which the pressure angle is equal to $\alpha = 20°$. Increasing the pressure angle, the minimum number of teeth of the planet gears and annulus gears can be reduced. Furthermore, with increasing the pressure angle, also the difference y between the numbers of the teeth of the annulus gears and planet gears, which is necessary to prevent the secondary interference, can be decreased.

13.7.6 Planetary Gear Train Type 6 (Example 6)

As Fig. 13.14 shows, this planetary gear train is composed of two sun bevel gears, having z_1 and z_2 teeth, and p planet gears made up of bevel gears with z_0 teeth. All teeth of the toothed members have equal module, m.

Among the numbers of teeth of the toothed gear wheels and their geometric quantities, the following relationships exist:

$$\Sigma = \delta_1 + \delta_0 \tag{13.70}$$

$$\tan \delta_0 = \frac{\sqrt{4z_0^2 - (z_2 - z_1)^2}}{z_1 + z_2} \tag{13.71}$$

$$\tan \delta_1 = \frac{z_1(z_1 + z_2)}{2z_0^2 + z_1(z_2 - z_1)} \tan \delta_0 \tag{13.72}$$

$$\tan \delta_2 = \frac{z_2(z_1 + z_2)}{2z_0^2 - z_2(z_2 - z_1)} \tan \delta_0 \tag{13.73}$$

The lengths of the segments AO, OB and $AB = (AO + OB)$ are given by:

$$AO = \frac{mz_1}{2 \tan \delta_1} \tag{13.74}$$

$$OB = \frac{mz_2}{2 \tan \delta_2} \tag{13.75}$$

$$AB = \frac{m}{2}\left(\frac{z_1}{\tan \delta_1} + \frac{z_2}{\tan \delta_2}\right) = \frac{m}{2}\sqrt{z_0^2 - \left(\frac{z_2 - z_1}{2}\right)^2}. \tag{13.76}$$

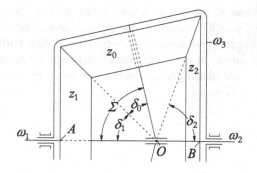

Fig. 13.14 Planetary gear train type 6

If we want the pitch cone angle of the largest sun bevel gear is less than $\pi/2(\delta_2 < \pi/2)$, due to Eq. (13.73) the following inequality must be satisfied:

$$2z_0^2 > z_2(z_2 - z_1), \tag{13.77}$$

that is

$$z_0 > \sqrt{\frac{z_2(z_2 - z_1)}{2}}. \tag{13.78}$$

We will choose, therefore, a number of teeth of the planet bevel gears z_0 that satisfies inequality (13.78). The angular distance φ at which the planet bevel gears are arranged is given by:

$$\varphi = Z\frac{2\pi}{z_1 + z_2}\,(\text{rad}) = Z\frac{360}{z_1 + z_2}\,(°), \tag{13.79}$$

where Z is a integer. If $(z_1 + z_2)/Z = p$ is an integer, p planet bevel gears can be arranged at an angular distance $(2\pi/p) = (360°/p)$.

For this planetary gear train, we have

$$\psi = -\frac{z_1}{z_2}. \tag{13.80}$$

The various reduction ratios achievable are obtained by substituting this values of ψ in Eqs. (13.35), (13.36), and 13.37). Thus, we have:

$$\omega_2 = \psi\omega_1 = -\frac{z_1}{z_2}\omega_1 \tag{13.81}$$

$$\omega_2 = (1 - \psi)\omega_3 = \left(1 + \frac{z_1}{z_2}\right)\omega_3 \tag{13.82}$$

$$\omega_1 = \left(1 - \frac{1}{\psi}\right)\omega_3 = \left(1 + \frac{z_2}{z_1}\right)\omega_3, \tag{13.83}$$

which are respectively valid for $\omega_3 = 0$, $\omega_1 = 0$, and $\omega_2 = 0$.

It is not convenient to adopt ratios $(z_1/z_2) < 1/2$, as it is appropriate that the numbers of teeth z_0 of the planet bevel gears is not too large. This type of planetary gear train is used: with the planet carrier locked ($\omega_3 = 0$), to obtain small reduction ratios; with the three main members all mobile, and the two sun bevel gears of equal size ($\psi = -1$), in the differentials that share the torque equally.

13.8   Summarizing and Differential Planetary Gear Trains

If none of the members of a planetary gear train is held at rest, it can become a *summarizing gearbox*, a *transfer gearbox*, a *differential drive* or a *differential gear* depending on how it is used. From this last point of view, we can classify the planetary gear trains into the following three distinct families (see also Ferrari and Romiti [14]):

- Planetary gear trains with only one input and one output; those described in the previous section belong to this family, provided the condition of one input and one output is respected.
- Planetary gear trains with two inputs and one output; these are the *summarizing planetary gear trains*.
- Planetary gear trains with one input and two outputs; these are the *transfer gearboxes* and the *differential gears*.

13.8.1   Summarizing Planetary Gear Trains

Figure 13.15 schematically shows the operating mode of a typical summarizing gearbox, which is driven by two engines, $E1$ and $E2$, for rotating the drum of a lift winch, W. The $E1$-engine drives the sun gear 1 that meshes with the planet gears 4, which are arranged symmetrically with respect to the axis of the sun gear, and have their axes carried by the planet carrier 3, rotating about the axis of the same sun gear. The planet gear teeth mesh with the inner teeth of the annulus gear 2; this is also equipped with external teeth, with which the teeth of the pinion 5 mesh. This pinion is driven by the $E2$-engine. Finally, the pinion 6, which is rigidly connected with the planet carrier 3, drives the mating gear wheel 7, which drags the drum of the lift winch. Figure 13.15 highlights all this in a schematic diagram.

Fig. 13.15 Schematic
diagram of a typical
summarizing gearbox

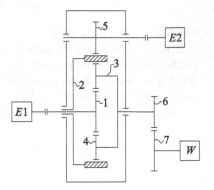

The angular velocities ω_1 and ω_5 of the sun gear, 1, and pinion, 5, are of course equal to those of the corresponding engines, $E1$ and $E2$. The angular velocities ω_2 and ω_5 of the annulus gear, 2, and pinion 5, are then correlated by the following relationship:

$$\omega_2 = -\omega_5(z_5/z_2), \qquad (13.84)$$

where the minus sign indicates that these two members of the planetary gear train rotate in opposite directions, while z_5 and z_2 are the numbers of teeth of the pinion, 5, and respectively of external and internal teeth of the annulus gear, 2.

If, for the planetary gear train under consideration, we assume 1 and 2 as input power members, and the planet carrier, 3, as output power member, by the third of the Eq. (13.16) we find the following relationship which express the angular velocity ω_3 of the planet carrier as function of ω_1 and ω_5:

$$\omega_3 = \frac{z_1/z_2}{(z_1/z_2)+1}\omega_1 - \frac{z_5/z_2}{(z_1/z_2)+1}\omega_5 = \frac{z_1}{z_1+z_2}\omega_1 - \frac{z_5}{z_1+z_2}\omega_5; \qquad (13.85)$$

this equation indicates how the planetary gear train under consideration allows to combine the angular velocities of the two engine shafts.

Now let's consider the mechanism described above in steady-state conditions, and assume that it works in ideal conditions, that is, without friction losses. In these conditions, the power balance leads to writing the following relationship:

$$T_1\omega_1 + T_2\omega_2 + T_r\omega_3 = 0, \qquad (13.86)$$

where: T_1 is input torque applied to the sun gear, 1; T_2 is the input torque applied to the annulus gear, 2, related to the force transmitted to it by the pinion, 5; T_r is the resistant torque, acting on the pinion 6 rigidly connected to the planet gear 3, and applied to it by the mating gear wheel 7.

On the other hand, the torque balance equation of the system constituted by members 1, 2 and 3 requires the following relationship to be satisfied:

$$T_1 + T_2 + T_r = 0. \qquad (13.87)$$

Taking into account the equations found before, which correlate the angular velocities, we can write the following relationship:

$$T_1\omega_1 + T_2\omega_2 = (T_1 + T_2)\left(\frac{z_1}{z_1+z_2}\omega_1 + \frac{z_2}{z_1+z_2}\omega_2\right), \qquad (13.88)$$

from which we get:

$$T_1 = T_2 \frac{z_1}{z_2}. \tag{13.89}$$

We will have:

$$T_2 = T_2' \frac{z_5}{z_2}, \tag{13.90}$$

where T_2' is the drive torque provided by engine $E2$. Therefore, assuming $z_1 = z_5$, it follows that $T_1 = T_2'$, that is, the drive torques applied to the two input shafts are identical in steady-state conditions.

Instead, in the actual operating conditions for which frictional power losses have to be considered, the power balance applied to the two engines, $E1$ and $E2$, and the planetary gear train leads to writing the following relationship:

$$T_1\omega_1 - (1 - \eta_\mathrm{I})T_1\omega_1 + T_2\omega_2 - (1 - \eta_\mathrm{II})T_2\omega_2 + T_r\omega_3 = 0, \tag{13.91}$$

where η_I is the efficiency of mechanism I, consisting of engine $E1$, sun gear 1, planet gears 4, annulus gear 2, and planet carrier 3, while η_II is the efficiency of mechanism II, consisting of engine $E2$, pinion 5, annulus gear 2, planet gears 4, and planet carrier 2.

Quantities $(1 - \eta_\mathrm{I})T_1\omega_1$ and $(1 - \eta_\mathrm{II})T_2\omega_2$ give the frictional power losses in such mechanisms. Therefore, we deduce that Eq. (13.86), which is valid under the ideal conditions without friction losses, under the actual operating conditions in replaced by the following relationship:

$$\eta_\mathrm{I}T_1\omega_1 + \eta_\mathrm{II}T_2\omega_2 + T_r\omega_3 = 0. \tag{13.92}$$

It follows that the efficiency of the compound mechanism, consisting of the coupling of the two mechanisms, I and II, is given by:

$$\eta = \frac{|T_r\omega_3|}{T_1\omega_1 + T_2\omega_2} = \frac{\eta_\mathrm{I}T_1\omega_1 + \eta_\mathrm{II}T_2\omega_2}{T_1\omega_1 + T_2\omega_2}. \tag{13.93}$$

This last relationship shows that the efficiency of the compound mechanism $(\mathrm{I} + \mathrm{II})$ is the weighted average of individual mechanism efficiencies.

The result thus obtained corresponds to a general property of mechanical systems resulting from the union of several component mechanisms, made in one of the two following ways:

- by linking together all the input members of the component mechanism, so as to have a mechanical system with just one input, and as many output as there are mechanisms;

- by connecting together all the output members, in order to obtain a mechanical system with just one output, and as many inputs as there are mechanisms.

In both of these cases, we say that the component mechanisms are connected in parallel.

13.8.2 Differential Planetary Gear Trains

Usually, in automotive engineering, the power delivered by the engine is fed to the driving wheels on one powered axle. In vehicles with more than one powered axle, the power must be distributed to the various powered axles. Especially in commercial vehicles, the need to provide power to auxiliary units can also arise. The mechanism that enables us to meet these needs is the *differential*.

In the automotive industry, a distinction is made between [31, 45]:

- the *interaxle differential* or *transferbox*, which is a mechanism intended to split the power to more than one powered axle, in the longitudinal direction, i.e. in the direction of travel of the vehicle;
- the *interwheel differential* or *differential gear unit*, which is a mechanism intended to split the power to the driving wheels of one axle, in the transverse direction with respect to the direction of travel;
- the *power take-offs*, which is a mechanism intended to split the power from actual power train to auxiliary units.

The *differentials* can be *spur gear planetary differentials* or *bevel gear differentials*. We give here the general concepts that apply to one and other type of differential. However, by way of example, we focus our attention primarily on the bevel gear differentials, which find a general use as *final drives* for vehicle transmissions. The following three types are commonly used: (*i*), spur gear final drives; (*ii*), bevel gear final drives, with helical bevel or hypoid gears; (*iii*), worn gear final drive (see Naunheimer et al. [45]). Therefore, with reference to the differential shown in Fig. 13.14, we denote by ω_3 and T_3 the angular velocity of the driving member (the planet carrier) and the torque acting on it, and by ω_1 and ω_2 the angular velocities of the two driven members, and T_1 and T_3 the torques acting on them.

To perform its design functions, a differential must meet the following three requirements:

1. For a given angular velocity, ω_3, of the driving member, the angular velocities ω_1, ω_2 and ω_3 of the three main members of the differential must be linked by a unique relationship, so that the difference $(\omega_1 - \omega_2)$ is indeterminate.
2. The driving torque, T_3, must be splitted into the torques T_1 and T_2 acting on the driven members, according to a constant ratio.
3. The two torque T_1 and T_2 acting on the driven members must have the same sign, that is they must operate in the same direction of travel.

The epicyclic gear train meet the first two requirements described above. In fact, if we denote by ω_1 and ω_2 the angular velocities of the two driven members, ω_3 the angular velocity of the planet carrier, which is the driving member, and $\psi = 1/i_0$ the reciprocal of the characteristic ratio of the ordinary gear train from which the planetary gear train comes, we get the relationship between the above three angular velocities. As we showed in Sect. 10.4, this relationship is given by any of the three Eqs. (13.16). Subtracting member to member, from identity $\omega_1 = \omega_1$, the two members of the second Eq. (13.16), we can deduce the following other equation:

$$\omega_1 - \omega_2 = (1 - \psi)(\omega_1 - \omega_3). \tag{13.94}$$

This equation shows that the difference $(\omega_1 - \omega_2)$ can take any value.

In a mechanical system in conditions of steady-state equilibrium, the sum of torque must be equal to zero (see Eq. 1.5); therefore, we have:

$$T_3 = T_1 + T_2. \tag{13.95}$$

Furthermore, in the ideal case in which there are no losses of efficiency ($\eta = 1$), and the system is in conditions of steady-state equilibrium, by virtue of the theorem of virtual work, also the algebraic sum of the powers that came into play must be equal to zero (see Eq. 1.12). Therefore, we can write the following relationship:

$$T_3\omega_3 = T_1\omega_1 + T_2\omega_2. \tag{13.96}$$

From Eqs. (13.16), (13.95) and (13.96), we obtain the following expressions:

$$T_1 = -T_3 \frac{\psi}{1 - \psi} \tag{13.97}$$

$$T_2 = T_3 \frac{1}{1 - \psi}. \tag{13.98}$$

From these equations, we can deduce that the ratio between the two torques T_1 and T_2 applied to the driven shafts is a constant, and equal to:

$$\frac{T_1}{T_2} = -\psi. \tag{13.99}$$

To satisfy the third condition, according to which the two torques T_1 and T_2 must have the same sign, the ratio ψ must be negative. So we come to the conclusion that the differentials are epicyclic gear trains with negative ratio, ψ.

If $\psi = -1$, from the previous equation we get

$$\omega_3 = \frac{\omega_1 + \omega_2}{2}, \tag{13.100}$$

and, in the case where there are no losses of efficiency ($\eta = 1$), we have:

$$T_1 = T_2 = \frac{T_3}{2}. \tag{13.101}$$

Therefore, a differential with ratio $\psi = -1$ splits equally the driving torque between the driven shafts. In other words, in this type of differential, the torques applied to the output shafts are the same. Furthermore, according to Eq. (13.100), the angular velocity of the planet carrier is equal to the half sum of the angular velocities of the two driven shafts. This differential is used to drive the two equal driving wheels of the same powered axle of a vehicle. From Eq. (13.100) we can deduce that, when the vehicles are in straight traveling, the angular velocities of the three main members of the differential are equal to each other, i.e. $\omega_1 = \omega_2 = \omega_3$. Therefore, relative movements between the members of the differential does not occur, and the differential behaves as a rigid coupling.

If $\psi = -\xi$, the second Eq. (13.16) becomes:

$$\omega_2 + \xi\omega_1 = (1 + \xi)\omega_3. \tag{13.102}$$

In the case where there are no losses of efficiency ($\eta = 1$), instead of Eqs. (13.97), (13.98) and (13.99), we have the following equations:

$$T_1 = T_3 \frac{\xi}{1 + \xi} \tag{13.103}$$

$$T_2 = T_3 \frac{1}{(1 + \xi)} \tag{13.104}$$

$$\frac{T_1}{T_2} = \xi. \tag{13.105}$$

This differential splits the driving torque in different parts between the driven shafts. Also, in this differential, when the driven shafts rotate at the same angular velocities ($\omega_1 = \omega_2$), the relative motions between the members of the differential do not occur, as Eq. (13.102) shows. This type of differential can be used in the transmissions of a vehicle with two powered axles, to split the driving torque between the two powered axles in proportion to the weights acting on each of them, that is according to the maximum forces of adhesion that are available.

All the equations written above are valid under the hypothesis that the losses of efficiency are equal to zero ($\eta = 1$). To take into account the losses of efficiency that inevitably occur in the real case, it is convenient to consider the motion of the planetary gear train, which has two degrees of freedom, as the sum of two elementary motions, which both satisfy Eq. (13.15). For example, we can consider the actual motion as the sum of two motions. A first motion in which the three main members of the planetary gear train rotate rigidly connected to each other, with the same angular velocity ω_3 of the planet carrier; this is the so-called dragging motion, which is characterized by $\omega_1 = \omega_2 = \omega_3$. A second motion, which is the relative motion with respect to the planet carrier, in which the other two members of the same planetary gear train rotate with angular velocities $\omega_{r1} = (\omega_1 - \omega_3)$ and $\omega_{r2} = (\omega_2 - \omega_3)$, linked together by Eqs. (13.13) or (13.15), while the planet carrier remains stationary.

We use the Eq. (13.96), which is derived from the theorem of virtual work and expresses the balance of power during the actual motion of the planetary gear train, in the case where the losses of efficiency are equal to zero ($\eta = 1$). This equation can be applied separately, only to the dragging motion as well as only to the relative motion, so we can write respectively:

$$T_3\omega_3 = T_1\omega_3 + T_2\omega_3 \qquad (13.106)$$

$$T_1\omega_{r1} + T_2\omega_{r2} = 0. \qquad (13.107)$$

Eliminating ω_3 from Eq. (13.106), we get Eq. (13.95). Instead, taking into account Eqs. (13.12) and (13.15), from Eq. (13.107) we obtain the Eq. (13.99). Therefore, the transmission of power between the three main members of the planetary gear train takes place through two different flow ways, namely, in part through the dragging motion, and in part through the relative motion.

For the determination of the actual efficiency of the gear system, it should be noted that the losses of efficiency occur only in the relative motion, in the meshing between the various toothed wheels that characterize the planetary gear train. Instead, the losses of power in the dragging motion are negligible, as similar to those that might occur in a rotating rigid coupling. Since during the relative motion the planet carrier is locked, in Eq. (13.107) we can consider as driving member the gear wheel 1 (in this case, the input power is given by $P_{ri} = T_1\omega_{r1} > 0$, while the output power is given by $P_{ro} = T_2\omega_{r2} < 0$), or the gear wheel 2 (in this case, the input power is given by $P_{ri} = T_2\omega_{r2} > 0$, while the output power is given by $P_{ro} = T_1\omega_{r1} < 0$). In both cases, the power dissipated for the various causes of loss, P_d, is given by:

$$P_d = (1 - \eta_0)P_{ri}, \qquad (13.108)$$

where η_0 is the *characteristic efficiency* or *basic efficiency* (see Sect. 13.12).

Once the value of P_d is known, the determination of the actual efficiency, η, of the planetary gear train, which is given by the quotient of the total output power divided by the total input power, can be done with the following relationship:

$$\eta = 1 - \frac{P_d}{P_i}, \qquad (13.109)$$

where P_i is the input power in the overall motion. Then we have:

$$\frac{1 - \eta}{1 - \eta_0} = \frac{P_{ri}}{P_i}, \tag{13.110}$$

from which we get

$$\eta = 1 - (1 - \eta_0)\frac{P_{ri}}{P_i}. \tag{13.111}$$

This conclusion anticipates and demonstrates a general concept, which is described in Sect. 13.12.

Of course, if we take into account the power dissipated, P_d, the balance of power in the relative motion is no longer given by Eq. (13.107). Instead, it is given by the relationship:

$$T_1\omega_{r1} + T_2\omega_{r2} = P_d. \tag{13.112}$$

Consequently, the Eq. (13.99) is no longer valid. It should be replaced by the following two equations that relate respectively to cases in which the driving member is the gear wheel 1, or the gear wheel 2:

$$\frac{T_1}{T_2} = -\frac{\psi}{\eta_0} \tag{13.113}$$

$$\frac{T_1}{T_0} = -\psi\eta_0. \tag{13.114}$$

An evaluation of first approximation of P_d can be obtained by determining the power flows in the ideal condition, in which the power losses are neglected, and for which the balance of power previously described is valid. In this case, the input power and output power (the latter coinciding with the resistant power) are equal to each other, so we speak simply of transmitted power. In particular, for the evaluation of the efficiency, we can use still Eq. (13.111) where, however, P_{ri} represent the power transmitted in relative motion (we can assume, either, $P_{ri} = |T_1\omega_{r1}|$ or $P_{ri} = |T_2\omega_{r2}|$), while P_i is the power transmitted (we can assume either the input power or the output power) in the overall motion.

We have so far focused our attention on bevel gear differentials, with which we are more familiar, as there are no passenger cars and commercial vehicles that can do without them. However, in many vehicles it is required that all four wheels be driven, permanently or only in poor traction conditions, which take place for a limited time. In the first case, all four wheels are permanently powered, while in the second case front axle wheels (or rear axle wheels) must be driven on request.

To meet these needs, transfer gearboxes with differential are used: they can be transfer gearboxes with bevel gear differential (Fig. 13.16a) or transfer gearboxes with spur gear planetary differential (Fig. 13.16b). Both of these transfer gearboxes make it possible to equalize the speed and forces between the powered axes. With a bevel gear differential, as we have already seen, the torque is split equally between the front and rear axles. Instead, with a spur gear planetary differential, the torque is split unevenly.

Here, it is not necessary to further investigate the topic concerning these transfer gearboxes with spur gear planetary differential, also because their discussion does not show any significant differences compared to transfer gearboxes with bevel gear differential described above. The reader can deepen the topic with specialized textbooks (see Naunheimer et al. [45]).

13.9   Multi-stage Planetary Gear Trains

According to the definition already given in Sect. 13.3, multi-stage planetary gear trains are compound structures containing two or more planetary gear trains arranged in series. These compound mechanisms are widely used in semi-automatic and fully automatic transmissions for passenger cars and commercial vehicles. To clarify the potential of these mechanisms, we would like to examine here the simplest case of multi-stage planetary gear train made up of two planetary gear trains like the one shown in Fig. 13.5, arranged in series.

With reference to Fig. 13.17, and with the same notations used in Sect. 13.4.1, the sun gears, annulus gears and planet carriers of the two component planetary gear trains are indicated by 1, 2, and 3. One superscript and two superscripts are used respectively to distinguish the first from the second of the two component planetary gear trains. The notation ω_i (with $i = 1, 2, 3$) is used to indicate the angular velocities of the three main members of the two planetary gear trains, and notation z_i (with $i = 0, 1, 2$) for number of teeth of the planet gears, sun gears, and annulus gears. Finally, the quantities $\psi_1 = 1/i_{01}$ and $\psi_2 = 1/i_{02}$, given by

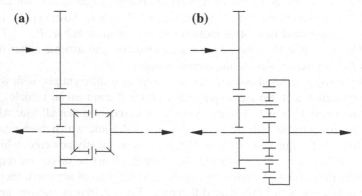

Fig. 13.16 Transfer gearboxes with: **a** bevel gear differential; **b** spur gear planetary differential

Eq. (13.15), indicate the reciprocal of the transmission ratios of the two planetary gear trains, which have become ordinary gear trains (i.e., with planet carriers held at rest).

Let us introduce the following quantities:

$$\psi_1' = z_1'/z_2'$$
$$\psi_2' = z_1''/z_2''. \tag{13.115}$$

The quantities ψ_1 and ψ_2, which are negative, are then given by:

$$\psi_1 = -\psi_1' = -z_1'/z_2'$$
$$\psi_2 = -\psi_2' = -z_1''/z_2''. \tag{13.116}$$

Therefore, for the two planetary gear trains under consideration, the second of the Eqs. (13.16) becomes respectively:

$$\omega_2' = -\psi_1'\omega_1' + (1 + \psi_1')\omega_3'$$
$$\omega_2'' = -\psi_2'\omega_1'' + (1 + \psi_2')\omega_3''. \tag{13.117}$$

If a member of the first planetary gear train is connected with a member of the second, so that these two members rotate at the same speed, a compound planetary gear train is obtained. The six rotational speeds of the members of this compound gear system are however correlated only by three equations. Therefore, in order to determine the relationship between the various rotational speeds, it is necessary to assign the speed values of three members not rigidly connected two-to-two between each other. In this way, any of the other three members can constitute the driven member, and its rotational speed is a linear combination of those of the three driving members whose speeds are assigned.

Fig. 13.17 Numbers of teeth of the toothed members of two planetary gear trains like the one shown in Fig. 13.5

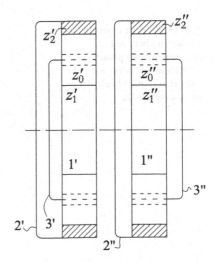

To reduce the independent variables to two, it is possible to make a second connection between another member of the first planetary gear train and another of the second. In this case, only the rotational speed of two members of the compound gear system, not connected to each other, will have to be assigned. The rotation speed of any other member, not directly connected with the driving members, is a linear combination of the rotation speeds of driving members. In particular, the rotational speed of one of the driving members of the compound gear system can be assumed equal to zero, since this member can be held at rest by means of a brake. If instead such a member is allowed to rotate, the motion is not transmitted to the driven member, which remains stationary.

Let us now look at two different examples of connection between the members of the two planetary gear trains arranged in series, based on the above described premises.

The *first example of connection* is that shown in Fig. 13.18, which highlights the following features: the two planet carriers, $3'$ and $3''$, are connected to each other; the annulus gear $2'$ of the first planetary gear train is connected to the sun gear $1''$ of the second planetary gear train; the annulus gear $2''$ of the second planetary gear train is held at rest by means of a brake; the sun gear $1'$ of the first planetary gear train is the driving member; the mechanical subset formed by the interconnection of members $3'$ and $3''$ is the driven member.

In this case, in addition to the two Eq. (13.117), the following equalities are valid:

$$
\begin{aligned}
\omega_3' &= \omega_3'' \\
\omega_2' &= \omega_1'' \\
\omega_2'' &= 0.
\end{aligned}
\tag{13.118}
$$

Fig. 13.18 First example of connection between the members of two planetary gear trains in series, and second example of calculation of braking torque

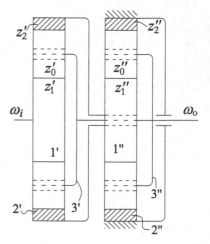

In addition, the angular velocity ω_1' of the driving sun gear $1'$ is known. From the above equations, we get the following relationship that gives the transmission ratio of the compound gears system under consideration:

$$\frac{\omega_1''}{\omega_3'} = -\frac{1 - \psi_1'\psi_2'}{\psi_1'\psi_2'}. \tag{13.119}$$

It should be noted that, when it is not possible to make the mechanical connection between two members that need to be braked, two separate brakes must be used for their braking.

The *second example of connection* is that shown in Fig. 13.19, which highlights the following characteristics: members $1'$ and $2''$ as well as members $3'$ and $1''$ are interconnected; member $2'$ is the driving member; member $3''$ is the driven member; with only one brake, it is possible to held at rest the mechanical subset constituted by the interconnection of the members $1'$ and $2''$. In this case, in addition to the Eq. (13.117), the following equalities are valid:

$$\omega_1' = \omega_2''$$
$$\omega_3' = \omega_1'' \tag{13.120}$$
$$\omega_1' = 0.$$

In addition, the angular velocity ω_2' of the driving annulus gear $2'$ is known. From the above equations, we get the following relationship that gives the transmission ratio of the compound gear system under consideration:

$$\frac{\omega_2''}{\omega_3'} = \frac{(1 + \psi_1')(1 + \psi_2')}{\psi_2'}. \tag{13.121}$$

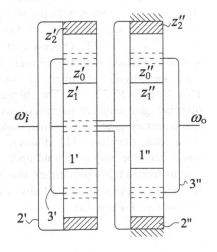

Fig. 13.19 Second example of connection between the members of two planetary gear trains in series, and third example of calculation of braking torque

Several special cases of connection between the members of the two aforementioned planetary gear trains are possible. Let us consider for example the case in which one of the two connections is made between two of the members of one of the two planetary gear trains. The planetary gear train where two of its main members are interconnected comes to operate as a rigid coupling, so its three main members rotate all at the same speed. The compound planetary gear train behaves as if this planetary gear train did not exist, so only the other planetary gear train is active.

As another example, we consider the case of two pairs of interconnected members between the two planetary gear trains, where each pair has a member belonging to the first planetary gear train and the other member belonging to the second gear train. We also assume that a pair of connected members is the driving pair, and the other is the driven pair or is held at rest. In this case, only that of the two planetary gear trains whose third member is respectively the driven member or the stationary member remains active with reference to power transmission.

Therefore, when operating conditions dictate that both planetary gear trains participate in power transmission, only one pair of connected members must be the driving pair, or driven pair, or held at rest. For example, if one of the two pairs of connected members is the driving pair, one of the two non-interconnected members of the planetary gear trains may be stationary member, and the other the driven member.

The number of transmission ratios achievable by a compound planetary gear train consisting of two planetary gear trains arranged in series varies depending on whether the interconnected members belong to one and the other of the two planetary gear trains or only to one of them. With a compound planetary gear train, with two members of the first planetary gear train separately connected with two members of the second, 108 different transmission ratios can be achieved, each of which can be obtained with multiple connections. All 108 transmission ratios are functions of the quantities ψ'_1 and ψ'_2, given by Eq. (13.115).

With the same compound planetary gear train where one of the connections is made between two members of the same planetary gear train, only 13 different transmission ratios can be obtained. Six of these ratios are only a function of ψ'_2, since they are obtained with the first planetary gear train rotating rigidly connected with the driving member of the second, due to one of the connections $1' - 3', 1' - 2', 2' - 3'$. Other six of these ratios are only a function of ψ'_1, since they are obtained with the second planetary gear train rotating rigidly connected with the driven member of the first, due to one of the connections $1'' - 3'', 1'' - 2'', 2'' - 3''$. The thirteenth ratio is unitary, as two members of each planetary gear train are interconnected, and a further connection exists between the two planetary gear trains, which makes them rotate rigidly connected.

In total, therefore, 121 different transmission ratios are achievable with the compound planetary gear train under consideration. Here, for brevity, it is not the case to deepen this topic further. For details, including the values in terms of ψ'_1 and ψ'_2 of the achievable transmission ratios, we refer the reader to specialized textbooks (e.g., Pollone [51]).

However, before leaving this subject, we have to point out that a compound planetary gear train consisting of two planetary gear trains arranged in series, automatically performs the two-way motion transmission, so it can be used as a summarizing planetary gear system, similar to those discussed in Sect. 13.8.1. In fact, in a compound planetary gear train in which two connections exist between the two component planetary gear trains, the drive power is transmitted through the two paths formed by these connections. The two examples described below clarify the salient aspects that are of interest.

As a first example, we here consider the compound planetary gear train shown in Fig. 13.18, and calculate the power distribution between the two paths. Let's denote with $T_1', T_2', T_3', T_1'', T_2'',$ and T_3'' the torques applied respectively to members $1', 2', 3', 1'', 2'',$ and $3''$ of the compound planetary gear train under the condition shown in Fig. 13.18, and with T_i and T_0 the driving and resistant torques respectively applied to members $1'$ and $3''$. The following relationships are valid between the above torques:

$$T_1' - T_2' = T_3'$$
$$T_1'' - T_2'' = T_3''. \tag{13.122}$$

Due to the connection between members $2'$ and $1''$, the following equality is also valid:

$$T_2' = T_1''. \tag{13.123}$$

Bearing in mind the values of ψ_1' and ψ_2' given by Eqs. (13.115), for the first planetary gear train we will have:

$$T_1' = -T_2'\psi_1', \tag{13.124}$$

while for the second planetary gear train we will have:

$$T_2'' = -\frac{T_1''}{\psi_2'} = -\frac{T_2'}{\psi_2'}. \tag{13.125}$$

Since the resistant or output torque, T_0, applied to the driven shaft is given by:

$$T_0 = T_3' + T_3'', \tag{13.126}$$

using the previous equations, we get:

$$\frac{T_3'}{T_0} = -\frac{(1+\psi_1')\psi_2'}{1-\psi_1'\psi_2'}$$
$$\frac{T_3''}{T_0} = \frac{1+\psi_2'}{1-\psi_1'\psi_2'}. \tag{13.127}$$

Since both members 3' and 3" are rigidly connected to the driven shaft, in the ideal case where the friction losses are zero, so the efficiency of the whole planetary gear train is equal to $1 (\eta = 1)$, the drive power, P_i, is divided between the two paths according to the two following fractions:

$$\frac{P'}{P_i} = \frac{T_3'}{T_0} = -\frac{(1+\psi_1')\psi_2'}{1-\psi_1'\psi_2'}$$

$$\frac{P''}{P_i} = \frac{T_3''}{T_0} = \frac{1+\psi_2'}{1-\psi_1'\psi_2'}.$$

(13.128)

It can easily be verified that the following equality exist:

$$T_3' = T_1'\frac{1+\psi_1'}{\psi_1'};$$

(13.129)

therefore, taking into account the first Eq. (13.127), we get:

$$T_0 = -T_1'\frac{1-\psi_1'\psi_2'}{\psi_1'\psi_2'} = -T_i\frac{1-\psi_1'\psi_2'}{\psi_1'\psi_2'},$$

(13.130)

as $T_1' = T_i$. We finally get the following expression of the transmission ratio:

$$\frac{\omega_i}{\omega_0} = \frac{T_0}{T_i} = -\frac{1-\psi_1'\psi_2'}{\psi_1'\psi_2'}.$$

(13.131)

As a second example, we consider the compound planetary gear train shown in Fig. 13.20, which highlights its operating condition. Here, for brevity, we limit ourselves to giving the final relationships. Using the same notations of the previous example, we will have:

$$\frac{T_1''}{T_i} = -\frac{(1+\psi_1')\psi_2'}{\psi_1'-\psi_2'}$$

$$\frac{T_1'}{T_i} = \frac{(1+\psi_2')\psi_1'}{\psi_1'-\psi_2'}.$$

(13.132)

In the ideal case where the friction losses are zero, and thus $\eta = 1$, the drive power is divided between the two paths according to the two following fractions:

$$\frac{P'}{P_i} = \frac{T_1''}{T_i} = -\frac{(1+\psi_1')\psi_2'}{\psi_1'-\psi_2'}$$

$$\frac{P''}{P_i} = \frac{T_1'}{T_i} = \frac{(1+\psi_2')\psi_1'}{\psi_1'-\psi_2'},$$

(13.133)

Fig. 13.20 Compound
planetary gear train in a given
operating condition

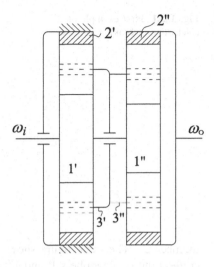

where T_i'' is the portion of driving torque directly transmitted to the member $1''$,
while T_1' is the part transmitted through the other path.

The transmission ratio is given by:

$$\frac{\omega_i}{\omega_0} = \frac{T_0}{T_i} = \frac{1 + \psi_1'}{\psi_1' - \psi_2'}. \tag{13.134}$$

13.10 Braking Torque on Members of a Planetary Gear System Held at Rest

The planetary gear systems discussed in the previous sections allow to transmit the
motion when some members of the compound planetary gear trains are held at rest
(note that in the previous figures, these members are highlighted by hatching).
Instead, when these members are free to rotate, transmission of motion cannot take
place. It is therefore necessary to equip them with brakes capable of transmitting to
them braking torques whose value depends on that of the drive torque to be
transmitted, and on the characteristics of the compound planetary gear train under
considerations. In other words, these braking torques must be of such magnitude as
to make the braking members rigidly connected with the housing. Four examples of
calculation of braking torque to be applied to the members to be held at rest are
described below.

The first example corresponds to the schematic diagram shown in Fig. 13.21,
which highlights that the two members $1'$ and $1''$ are interconnected and held at rest,
and that the driving and driven members of the compound planetary gear train in
the operating condition under consideration are respectively member $3'$ and

Fig. 13.21 First example of
calculation of braking torques

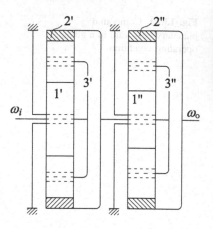

member 2″. The same figure shows that it is not possible to mechanically inter-
connect the two members 1′ and 1″, so to held them both at rest, it is necessary to
use two distinct brakes simultaneously.

As usual, we indicate with: $z_1', z_2', z_0', z_1'', z_2'', z_0''$, the number of teeth of the toothed
members of the two component planetary gear trains (here too we must pay
attention to the fact that z_0' and z_0'' indicate the numbers of teeth of the planet gears);
T_i and T_0, the driving and driven torques applied to the corresponding shafts, which
are respectively the ones of planet carrier of the first planetary gear train, and of the
annulus gear of the second planetary gear train; ψ_1' and ψ_2', the quantities given by
Eq. (13.115). The braking torque to be applied to member 1′ shall be:

$$T_1' = -T_i \frac{z_1'}{z_1' + z_2'} = -T_i \frac{\psi_1'}{1 + \psi_1'}. \qquad (13.135)$$

Therefore, the torque T_3'' applied to planet carrier 3″ will be given by:

$$T_3'' = T_i \frac{z_2'}{z_1' + z_2'} = T_i \frac{1}{1 + \psi_1'}, \qquad (13.136)$$

so, the breaking torque to be applied to member 1″ must be:

$$T_1'' = -T_3'' \frac{z_1''}{z_1'' + z_2''} = -T_3'' \frac{\psi_2'}{1 + \psi_2'} = -T_i \frac{\psi_2'}{(1 + \psi_1')(1 + \psi_2')}. \qquad (13.137)$$

The second example concerns the schematic diagram shown in Fig. 13.18,
which also highlights how the power transmission is carried out through two paths.
As we have seen in the previous section, the torque T_3'' that the planet carrier of the
second planetary gear train transmits to the driven shaft is given by the second of
Eq. (13.127), which here for convenience is rewritten in the from:

$$T_3'' = T_0 \frac{1 + \psi_2'}{1 - \psi_1' \psi_2'}. \tag{13.138}$$

Taking into account the transmission ratio, given by Eq. (13.131), the same torque T_3'' will also be given by:

$$T_3'' = -T_i \frac{1 + \psi_2'}{\psi_1' \psi_2'}. \tag{13.139}$$

Therefore, the braking torque to be applied to the annulus gear $2''$ of the second planetary gear train will have to be equal to:

$$T_2'' = T_3'' \frac{z_2''}{z_1'' + z_2''} = T_3'' \frac{1}{1 + \psi_2'} = -T_i \frac{1}{\psi_1' \psi_2'}. \tag{13.140}$$

The third example corresponds to the schematic diagram shown in Fig. 13.19, which highlights that the two members $1'$ and $2''$ are interconnected and held at rest. In this case, the mechanical interconnection between these two members is possible, so they can be braked with a single brake, which will have to develop a braking torque equal to the sum of the braking actions that the two members must exercise.

Since the driving torque T_i applied to the annulus gear $2'$ of the first planetary gear train is known, the braking torque to be applied to the sun gear $1'$ of the same planetary gear train will be given by:

$$T_1' = T_i \frac{z_1'}{z_2'} = T_i \psi_1'. \tag{13.141}$$

In the configuration of Fig. 13.19, the transmission ratio of the compound planetary gear train is given by eq. (13.121). Therefore, in the ideal case of zero friction losses, and therefore of unitary efficiency ($\eta = 1$), the torque applied to the driven shaft will be:

$$T_0 = T_i \frac{(1 + \psi_1')(1 + \psi_2')}{\psi_2'}, \tag{13.142}$$

while the braking torque to be applied to the annulus gear $2''$ of the second planetary gear train shall be equal to:

$$T_2'' = T_0 \frac{z_2''}{z_1'' + z_2''} = T_0 \frac{1}{1 + \psi_2'} = T_i \frac{1 + \psi_1'}{\psi_2'}. \tag{13.143}$$

Consequently, the total braking torque T to be applied to the pair of interconnected members $1' - 2''$ shall be equal to:

$$T = T_1' + T_2'' = T_i \frac{1 + \psi_1' + \psi_1' \psi_2'}{\psi_2'}, \tag{13.144}$$

The fourth and last example refers to the compound planetary gear train shown in schematic form in Fig. 13.20, whereby two-way motion transmission is evident. In the previous section, we found that the portion of the driving torque transmitted by the sun gear $1'$ of the first planetary gear train is given by the second Eq. (13.132), which here is rewritten for convenience in the form:

$$T_1' = T_i \frac{(1 + \psi_2') \psi_1'}{\psi_1' - \psi_2'}, \tag{13.145}$$

The braking torque to be applied to the annulus gear $2'$ must therefore be equal to:

$$T_2' = T_1' \frac{z_2'}{z_1'} = T_1' \frac{1}{\psi_1'} = T_i \frac{1 + \psi_2'}{\psi_1' - \psi_2'}. \tag{13.146}$$

13.11   Parallel Mixed Power Trains

Parallel mixed power trains are gear trains that transmit power between the driving and driven shafts through two parallel ordinary gear trains, which drive the driven shaft through a planetary gear train. In these mixed gear drives, torque and power are divided between two transmission paths. These mixed gear trains differ from the summarizing planetary gear trains, discussed in Sect. 13.8.1, both because they incorporate ordinary gear trains in addition to a planetary gear train, and because they are characterized by only one prime mover instead of two. Three types of these parallel mixed power trains are shown in schematic form in Fig. 13.22.

The schematic diagrams in Fig. 13.22 show that the driving shaft, connected to the prime mover (it is not shown in the figure), drives one of the two ends of two parallel transmission lines, consisting of as many ordinary gear trains. The other end of each of these two transmission lines drives one of the mobile members of a planetary gear train, which constitute the driving members of the latter.

Let ψ_1 and ψ_2 denote respectively the transmission ratios of the two ordinary transmission lines. They will be positive or negative, depending on whether the member of the planetary gear train operated by the corresponding transmission line and the driving shaft rotate in the same direction or in the opposite direction. We also introduce the following quantities: T_i, the driving torque applied to the driving shaft; T_1 and T_2, the two parts of driving torque, respectively operating the two transmission lines, I and II; T_1' and T_2', the torques applied to the two driving members of the planetary gear train; T_0, the driven torque applied to the driven shaft of the power train under consideration; $\omega_i = \omega$, the angular velocity of the driving

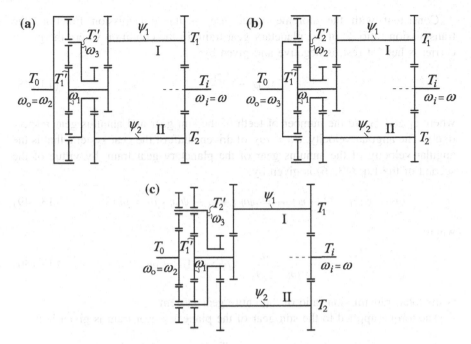

Fig. 13.22 Three types of parallel mixed power trains: **a** first example; **b** second example; **c** third example

shaft; ω_1, ω_2 and ω_3, the angular velocities of the two shafts and planet carrier of the planetary gear train; ψ, the transmission ratio of the planetary gear train, in the condition where the planet carrier is held at rest. It should be noted that, in this discussion, we are here using the transmission function $i_n = \psi$ defined at the end of Sect. 1.1, in place of the strictly defined transmission ratio.

Let us now examine how driving torque and driving power are shared, in the ideal condition of zero friction losses, and therefore efficiency equal to $1 (\eta = 1)$. To this end, the three examples shown in Fig. 13.22 are discussed separately.

In the first example, as shown in Fig. 13.22a, the parallel mixed power train uses a planetary gear train characterized by an annulus gear, as the one shown in Fig. 13.11 (planetary gear train type 3). Its sun gear is driven by the transmission line, I, and its angular velocities is ω_1; its planet carrier is driven by the transmission line, II, and its angular velocities is ω_3. Being $\omega_i = \omega$ the angular velocity of the driving shaft, and ψ_1 and ψ_2 the two above defined transmission ratios, we will have:

$$\omega_1 = \psi_1 \omega$$
$$\omega_3 = \psi_2 \omega. \tag{13.147}$$

Consistent with the aforementioned use of the transmission function, the transmission ratio ψ, of the planetary gear train, in the condition where the planet carrier is held at rest, is negative and given by:

$$\psi = -\frac{z_1}{z_2},$$ (13.148)

where z_1 and z_2 are the number of teeth of the sun gear and annulus gear respectively. The angular velocity, $\omega_0 = \omega_2$, of driven shaft of the gear system, that is the angular velocity of the annulus gear of the planetary gear train, by virtue of the second of the Eq. (13.16) is given by:

$$\omega_2 = \psi\omega_1 + (1 - \psi)\omega_3 = [\psi\psi_1 + (1 - \psi)\psi_2]\omega = \rho\omega,$$ (13.149)

where

$$\rho = \frac{\omega_0}{\omega_i} = \frac{\omega_2}{\omega} = \psi\psi_1 + (1 - \psi)\psi_2$$ (13.150)

is the total transmission ratio of the entire gear system.

The torque applied to the sun gear of the planetary gear train is given by:

$$T_1' = \frac{T_1}{\psi_1},$$ (13.151)

while the one applied to the planet carrier is given by:

$$T_2' = \frac{T_2}{\psi_2}.$$ (13.152)

In steady-state equilibrium conditions, the algebraic sum of torques applied to the planetary gear train must be zero, so we can write the following relationship:

$$T_1' + T_2' = T_0.$$ (13.153)

Under the assumed conditions of absence of friction losses, i.e. $\eta = 1$, for which $T_0\omega_0 = T_i\omega_i$, the following relationship between the input and output torques must be also be met:

$$T_0 = T_i\frac{\omega_i}{\omega_0} = T_i\frac{\omega}{\omega_2} = \frac{T_i}{\rho}.$$ (13.154)

For this type of planetary gear train, the following relationship is obtainable:

$$T_1' = T_2'\frac{\psi}{1 - \psi}.$$ (13.155)

Using Eqs. (13.151)–(13.155), we get the following two relationships:

$$T_1 = T_i \frac{\psi \psi_1}{\rho} = T_i \frac{\psi \psi_1}{\psi \psi_1 + (1 - \psi)\psi_2}$$

$$T_2 = T_i \frac{(1 - \psi)\psi_2}{\rho} = T_i - T_1 = T_i \left(1 - \frac{\psi \psi_1}{\rho}\right), \tag{13.156}$$

while from Eq. (13.150) we get:

$$\psi_2 = \frac{\rho - \psi \psi_1}{1 - \psi}. \tag{13.157}$$

The two parts T_1 and T_2 in which the driving torque is divided between the two transmission lines vary by varying the ratios ψ, ψ_1 and ψ_2, and therefore varying the total transmission ratio, ρ. In some cases, one of these two torques can assume higher values than the driving torque, T_i, and the other can become negative. Reaction torques are then born in the gear system, and the transmission lines are much more stressed than they would if each of them transmitted all the input power. This power is divided between the two transmission lines in the same proportions as the two torques T_1 and T_2.

In the second example, as shown in Fig. 13.22b, the parallel mixed power train also uses a planetary gear train with an annulus gear, such as the one of the first example. However, in this second example, while the sun gear is still driven by the transmission line I, it is the annulus gear that is driven by the transmission line II. Furthermore, the output shaft of the mixed power train is the one of the planet carrier of planetary gear train that rotates with angular velocity $\omega_0 = \omega_3$ and to which the resistant torque T_0 is applied. Using the same notations of the first example, some of which are shown in Fig. 13.22b, in this second example instead of Eq. (13.147) we will have:

$$\omega_1 = \psi_1 \omega$$
$$\omega_2 = \psi_2 \omega, \tag{13.158}$$

while the Eq. (13.148) will still be valid.

Bearing in mind the second of the Eq. (13.16), we will have:

$$\omega_3 = \omega \frac{\psi_2 - \psi \psi_1}{1 - \psi} = \omega \rho. \tag{13.159}$$

The total transmission ratio will then be given by:

$$\rho = \frac{\psi_2 - \psi \psi_1}{1 - \psi}. \tag{13.160}$$

Equations (13.151) and (13.152) are still valid, with the warning that T_2' is this time the torque applied to the annulus gear, which rotates at the angular velocity ω_2. In the usual steady-state equilibrium conditions, the algebraic sum of torques applied to the planetary gear train must be zero, so we can write the same Eq. (13.153), where however T_2' has the different meaning just highlighted. Therefore, given the mode of use of the planetary gear train, the following relationship can be written, which relates T_1' and T_2':

$$T_2' = -T_1' \frac{1}{\psi_1}. \tag{13.161}$$

Additionally, in the usual assumption of absence of friction losses, i.e. $\eta = 1$, the relationship (13.154), which links the input and output torques, is still valid.

Here, using all the equations described or recalled above, we get the following two relationships:

$$
\begin{aligned}
T_1 &= -\frac{\psi\psi_1}{(1-\psi)\rho} T_i = -\frac{\psi\psi_1}{\psi_2 - \psi\psi_1} T_i \\
T_2 &= \frac{\psi_2}{(1-\psi)\rho} T_i = \frac{\psi_2}{\psi_2 - \psi\psi_1} T_i = T_i - T_1,
\end{aligned}
\tag{13.162}
$$

while from Eq. (13.160) we get:

$$\psi_2 = (1-\psi)\rho + \psi\psi_1. \tag{13.163}$$

Here too the input power is shared between the two transmission lines in the same proportion as the input torque.

In the third example, as shown in Fig. 13.22c, the parallel mixed power train uses a planetary gear train characterized by only external gear wheels, whose transmission ratio with planet carrier held at rest is given by:

$$\psi = \frac{z_1 z_3}{z_2 z_4}, \tag{13.164}$$

where z_1, z_2, z_3, and z_4 are the numbers of teeth shown in Fig. 13.9.

Using the same procedure described for the first example, it is proved that the input torque is divided between the two transmission lines according to the relationship (13.156). In addition, the relationship (13.150), which gives the total transmission ratio, and the relationship (13.157), with which we calculate ψ_2 as a function of ρ, ψ, and ψ_1, are still valid.

From the comparison between the first (Fig. 13.22a) and the third example (Fig. 13.22c), it is clear that the transmission ratio ψ of the planetary gear train with planet carrier held at rest is negative in the first example, and positive in the third example. However, as we have shown above, in both cases the same relationships are valid that allow to calculate T_1 and T_2. This results in an obvious deduction:

with the same total transmission ratio, ρ, we will obtain the same torque distributions, in the event that for the gear system shown in Fig. 13.22c we adopt a transmissions ratio ψ_1 that has an opposite sign with respect to the one used for the gear system shown in Fig. 13.22a.

All the discussion above relates to some design solutions of speed variators in which the transmission ratio ψ_1 of one of the two transmission lines has a constant and predetermined value, while the other transmission ratio, ψ_2, is varied using a speed variator, to vary the speed of the driven shaft, in the belief that the speed variator can transmit only a small part of the drive power. This conviction is not always correct since, as we have just seen, it follows that, due to the reaction torques that arise between the two transmission lines, when the variation range of the total transmission ratio is wide, the speed variator may be in the condition of having to transmit powers, positive or negative, in absolute value greater than the drive power. It is therefore preferable to adopt a simpler type of transmission, with the speed variator directly connected to it.

For further details, we refer the reader to more specialized textbook (e.g., Naunheimer et al. [45]).

13.12   Efficiency of a Planetary Gear Train

In Sect. 13.2 we already introduced the concept of efficiency of an ordinary gear train. Among other things, we also highlighted that this efficiency depends on the direction of the power flow, which is different depending on whether one of the two end gear wheels of the gear drive is chosen as power input and the other as power output or vice versa. In Sect. 13.8.1 we also showed that the efficiency of a summarizing planetary gear train equals the weighted average of efficiencies of the individual gear mechanisms making up the gear drive under consideration. Finally, in Sect. 13.8.2, we introduced the basic efficiency or characteristic efficiency, η_0, of a planetary gear train, and we used this quantity to calculate the actual efficiency of the gear drive in its particular working condition.

It is now time to tackle this topic in as many general terms as possible, exploring its most interesting design aspects. In Sect. 13.4 we introduced the characteristic ratio $i_0 = 1/\psi$ that allow us to calculate, for different operating conditions, the transmission ratios, which can be speed reducing or speed increasing ratios. Similarly, to evaluate the efficiency of a planetary gear train, we already introduced the basic efficiency or characteristic efficiency, η_0, defined as the efficiency of the planetary gear train in that condition in which the planet carrier is held at rest, i.e. the condition of the planetary gear train became an ordinary gear train.

The evaluation of this basic efficiency can be made by means Eq. (13.11), i.e. by multiplying among themselves the various efficiencies that characterize the successive gear pairs that make up the planetary gear train in the condition where it has been reduced to an ordinary gear train. However, before proceeding to this topic, it is worth to dwell on the Eq. (13.11), which is usually used to evaluate the efficiency of an ordinary gear train. This equation is not the only one that is available to estimate the gear train efficiency with sufficient approximation. Other equations have been introduced, also because this subject has been studied for over a century, even though no simple and all-encompassing theory has so far been established. This is due to the fact that several sources of power loss arise in a power train, such as gear mesh losses, windage and churning losses, bearing and seal losses, and oil pump losses, and none of the so far proposed theory is able to consider all simultaneously.

We will focus our attention on gear mesh losses, as if they were the only losses in an ordinary gear train. As it is well known, the losses of each gear pair in mesh, and consequently the related efficiency, depend on the manufacturing and assembly accuracy, the lubrication conditions, as well as the number of teeth of each gear wheel. With reference to this last aspect, Fig. 13.23, which is based on data by Merritt [38], shows how the losses are so much higher (and therefore the efficiency is much lower), the more the number of teeth is low. The chart of Fig. 13.23 points out, however, that losses for a single gear pair are very small, usually about 2% or less.

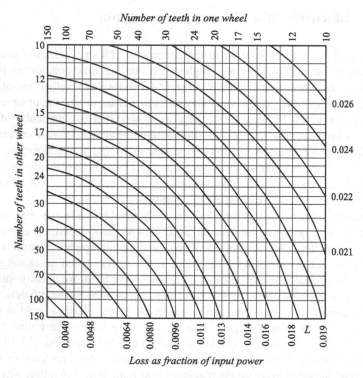

Fig. 13.23 Chart of percentage power loss for a single gear pair, as a function of the number of teeth of the two gear wheels

In accordance with the first equality of Eq. (1.4), the basic efficiency of a single gear pair is defined as the quotient of the output power, P_0, divided by the input power, P_i. Instead, in accordance with the last equality of the same Eq. (1.4), the basic efficiency is defined as $(1 - L)$, where $L = P_d/P_i$ is the percentage power loss, which can be estimated using the chart shown in Fig. 13.23. This is an initial estimate, which can be refined using the rules outlined below. With the notations that are relevant here ($\eta_t = \eta_0$ and L), Eq. (1.4) can be written in the following form:

$$\eta_0 = \frac{P_0}{P_i} = 1 - \frac{P_d}{P_i} = 1 - L. \qquad (13.165)$$

For an ordinary gear train made up of two or more gear pairs, the efficiency is given by the Eq. (13.11). Therefore, considering the simplest example of an ordinary gear train consisting only of two gear pairs, we can write the following relationship:

$$\eta_0 = \eta_1\eta_2 = (1 - L_1)(1 - L_2), \qquad (13.166)$$

which expresses the overall efficiency of the gear train under consideration as the product of the efficiencies of the individual component gear pairs, in terms of η and $(1 - L)$.

Expending the last equality of this equation, we get:

$$\eta_0 = 1 - L_1 - L_2 + L_1L_2. \qquad (13.167)$$

Since L_1 and L_2 are in the order of (2×10^{-2}) or less, product L_1L_2 is in the order of (4×10^{-4}), i.e. two orders of magnitude less; therefore, it is usually overlooked with respect to the other two terms, and thus omitted, also because this omission plays in favor of security. The above procedure is of a general nature, so it can be extended to any ordinary gear train consisting of n gear pairs. In this case, neglecting all the terms of a lower order of magnitude, the Eq. (13.167) becomes:

$$\eta_0 = 1 - L_1 - L_2 \cdots - L_i - \cdots - L_n. \qquad (13.168)$$

For a sufficiently approximate estimate of L_i (with $i = 1, 2, \ldots, n$), just keep in mind the following rules, suggested by Molian [42]:

- for external spur gears, just take the L_i values on the chart in Fig. 13.23;
- for an internal spur gear, the value of L_i must be calculated by multiplying the value obtained from the chart in Fig. 13.23, as if it was an external gear pair, by the ratio $[(u - 1)/(u + 1)]$, where $u = z_2/z_1$ is the gear ratio;
- for external helical gear pairs, the value L_i given by the chart must be multiplied by $(0.8 \cos \beta)$, where β is the helix angle;
- for internal helical gear pairs, L_i value must be modified by using the same rule of internal spur gear;

- for bevel gears, the number of teeth to be used to enter the chart in Fig. 13.23 are the virtual numbers of teeth z_v, calculated according to the Tredgold approximation.

It should be noted that, within the aforementioned approximation limits, the values of losses L, obtained using the chart in Fig. 13.23 and the above described rules, are the same, whichever of the two members of a gear pair is the driving wheel. Thus, the basic efficiency of an ordinary gear train does not depend on the direction of the power flow. In addition, with regard to a reasonably accurate bearing design, this type of calculation provides sufficient tolerance margin, which also includes bearing losses. It should not be forgotten that the values of L so deduced are approximate, and that the actual values depend on speeds, loads, gear materials and heat treatments, tooth flank surface finishing, shaft rigidity and alignment, lubricant properties, lubrication system, etc. This is why, at present, general theoretical models that can provide great accuracy in this regard are not available. These calculations are still very important, especially for comparing alternative proposed designs, while their validity is much more limited when they are used to develop existing gear drives; in this latter case, the efficiency tests in working conditions are unavoilable.

Let us now see how to calculate as easily as possible reliable values of the overall efficiency of a planetary gear train. The first step to achieve this important goal is in any case to calculate its basic efficiency, η_0. The subject has great design interest in this specific area, as the designer must try to design planetary gear trains with a higher efficiency than η_0, avoiding the real risk of planetary gear drives with an efficiency so low that they are unusable. It is worth remembering that the power losses of a planetary gear train can be surprising high, and therefore the efficiency is too low.

Several methods have so far been introduced and used to calculate more or less accurate values of the overall efficiency of a planetary gear train, such as: straightforward methods, more elaborate manual methods, and systematic computerized methods. To have a brief idea of these methods, we refer the reader to Del Castillo [8], who also proposed other methods.

We here only deal this peculiar subject in accordance with the procedure proposed by Molian [42], which allows to obtain sufficiently accurate values of the efficiency of a planetary gear drive. It is obvious that all the friction power losses in a planetary gear train occur only under conditions of relative motion among its main members, as they are due to the rubbing of the teeth between them. When all these members rotate like a rigid coupling, because two of its main members are rigidly connected to each other, the power loss is theoretically equal to zero (actually, the power loss is not equal to zero, but so small that it can to be considered equal to zero, and therefore neglected).

In Sect. 13.4.1 we have already introduced the relative angular velocities ω_{r1} and ω_{r2} of the first gear wheel and last gear wheel, 1 and 2, of a planetary gear train with respect to its planet carrier, 3; they are respectively given by:

$$\omega_{r1} = \omega_1 - \omega_3$$
$$\omega_{r2} = \omega_2 - \omega_3. \tag{13.169}$$

Mesh power losses due to rubbing of the teeth between them can only occur when these relative angular velocities are both nonzero. When the entire planetary gear train rotates as a rigid coupling, we will get $\omega_1 = \omega_2 = \omega_3$, and therefore $\omega_{r1} = \omega_{r2} = 0$. From all this, we infer that all friction power losses are proportional to the relative angular velocities, ω_{r1} and ω_{r2}, rather than to the absolute angular velocities, ω_1, ω_2 and ω_3. So, we assume that the power flow inside a planetary gear train and friction losses can only be affected by relative motions that cause rubbling between the tooth flank surfaces of mating teeth, but not by the gear drive rotation as a rigid coupling.

In order to derive the expression of efficiency, it is necessary to use the two fundamental Eqs. (1.5) and (1.2), which represent respectively the torque balance equation and the power balance equation. To avoid misunderstanding, considering a planetary gear train such as the one shown in Fig. 13.5, it is convenient to write the two equations above-mentioned in the following form:

$$T_1 + T_2 + T_3 = 0$$
$$T_1\omega_1 + T_2\omega_2 + T_3\omega_3 = 0. \tag{13.170}$$

The first of these equations, i.e. the torque balance equation, points out that the torques applied to the three main members of the planetary gear train cannot be all positive or all negative. The usual sign convention is then valid, according to which the torque is positive if it acts in the same direction (and thus has the same sign) of the corresponding angular velocity, while it is negative if it acts in the opposite direction (and hence has opposite sign) to that of the corresponding angular velocity. In the first case, the corresponding power involved is positive $(P > 0)$, and the shaft to which the torque is applied is a driving shaft, which is connected to the driving member. In the second case, the corresponding power involved is negative $(P < 0)$, and the shaft to which the torque (this is the resisting torque) is applied is a driven shaft, which is connected to the driven member. The second of the Eq. (13.170), i.e. the power balance equation, expresses energy conservation in the ideal condition where the total frictional losses are equal to zero, so the power entering the planetary gear drive is the same as the one that comes out of it.

In relation to the goal to be achieved, it is convenient to express the power balance equation (the second of Eq. 13.170) by showing in it the relative angular velocities given by Eq. (13.169). This is because, as mentioned above, friction power losses are only proportional to these relative angular velocities. Therefore, we get the following relationship:

$$T_1(\omega_{r1} + \omega_3) + T_2(\omega_{r2} + \omega_3) + T_3\omega_3 = 0, \tag{13.171}$$

which, rearranged, becomes:

$$T_1\omega_{r1} + T_2\omega_{r2} + (T_1 + T_2 + T_3)\omega_3 = 0. \tag{13.172}$$

However, by virtue of the first of the Eq. (13.170), the sum $(T_1 + T_2 + T_3)$ is equal to zero, so the Eq. (13.172) reduces to the following simpler form:

$$T_1\omega_{r1} + T_2\omega_{r2} = 0, \tag{13.173}$$

whose terms express respectively the power consumed by the shafts of members 1 and 2 of the planetary gear train under consideration.

Let us now consider, instead of the ideal condition of total losses equal to zero, and hence unitary efficiency, the actual condition of nonzero total losses. We also assume that the gear set is operating in steady-state conditions, that is at constant speed, or that at least the speed change so slowly that changes in internal kinetic energy of the same gear set can be ignored. Thus, while the first of Eq. (13.170) remains valid, instead of Eq. (13.173) we will have:

$$\eta_0 T_1\omega_{r1} + T_2\omega_{r2} = 0, \tag{13.174}$$

or

$$T_1\omega_{r1} + \eta_0 T_2\omega_{r2} = 0. \tag{13.175}$$

Equation (13.174) is to be used in the following two cases:

- the planet carrier, 3, constitutes the output member, while 1 is the input member and 2 is the fixed member or vice versa;
- the planet carrier, 3, is the member held at rest, while 1 is the input member and 2 is the output member.

Instead, the Eq. (13.175) is to be used in the following two cases:

- the planet carrier, 3, constitutes the input member, while 1 is the output member and 2 is the fixed member or vice versa;
- the planet carrier, 3, is the member held at rest, while 2 is the input member and 1 is the output member.

Apart from the aforementioned indications for their use, in any case to be well-attended, it should be noted that the Eqs. (13.174) and (13.175) are quite general. In fact, they come from the two Eq. (13.170), with the only variation to consider friction power losses; for the rest, we have not made any assumption on which of three main members of the planetary gear train under consideration is held at rest or not. Therefore, these equations can be used for any planetary gear set operation mode, provide we are able to identify the power flow path in relation to the planet carrier.

The approximate method of calculating the efficiency of a planetary gear train, proposed by Molian [42], supported by Kudryavtsev et al. [30]\with few significant variations, and endorsed by Michlin and Myunster [40] and Klebanov and Groper [29], is based on the use of the first of the Eq. (13.170) and one of the two Eqs. (13.174) or (13.175), depending on the cases highlighted above. It assumes that the geometry of the planetary gear train, whose efficiency we want to evaluate, has already been determined on the basis of other design considerations. Therefore, the number of teeth z_1, z_2, and z_3 of the sun gear, annulus gear and planet gears are already known, so the basic ratio, $i_0 = 1/\psi$, is also known.

As we have already mentioned in Sect. 13.4.1, this basic ratio should be considered as an algebraic quantity. Therefore, it is positive ($i_0 > 0$) if the directions of rotation of the input member and output member of the planetary gear train, which has become an ordinary gear train, coincide (see, for example, the planetary gear train type 1, shown in Fig. 13.9). Indeed, i_0 is negative ($i_0 < 0$) if these directions are opposite (see, for example, the planetary gear train shown in Fig. 13.5). The algebraic value of i_0 can be easily determined by Eq. (13.19).

We then choose and introduce a plausible value of the basic efficiency, η_0, determined by the above-described method for an ordinary gear train, assuming that power flows, and thus friction power losses, are equally divided between the different planet gears. Furthermore, we assume that the torque acting on the driven member of the planetary gear train under consideration is known. It is also to be stated that the T_i-torques (with $i = 1, 2, 3$) are to be regarded as positive if i is the driving member, whereas they are to be considered as negative if i is the driven member or the member held at rest. For the driving member, even the angular velocity ω_i (with $i = 1, 2, 3$) must be considered as positive, since it has the same direction of torque T_i, so also the power applied to its shaft, $P_i = T_i\omega_i$ is positive, that is $T_i\omega_i > 0$. Instead, for the driven member and for the member held at rest, the angular velocities ω_i must be considered as negative as they are in the opposite directions to the T_i-torques, so that the power P_i applied to the corresponding shafts are also negative, that is $T_i\omega_i < 0$.

The overall efficiency of a planetary gear train depends on the power flow path and, in turn, this is strongly influenced by the actual operating conditions of the planetary gear set. Therefore, for the determination of the efficiency of a planetary gear train, once its geometry is known, and the parameters dependent on it have been calculated, the next step is to define the power flow path in the operating conditions of the same planetary gear set.

For positive values of the basic ratio ($i_0 > 0$), six possible cases are identifiable. The corresponding results are shown in Table 13.2. In the columns of this table, the following indications or quantities are listed in order, from left to right: case under consideration, designed with a progressive number from 1 to 6; input member; fixed member; output member; transmission ratio, defined as the quotient of the angular velocity of the driving member divided by the angular velocity of the driven member, and expressed both in terms of i_0 and ψ; torques T_1, T_2, and T_3; efficiency

Table 13.2 Efficiencies in the six possible cases of a planetary gear train with $i_0 > 0$

Case	Input member	Fixed member	Output member	Transmission ratio, i	T_1	T_2	T_3	Efficiency η
1	1	2	3	$1 - i_0 = \frac{\psi - 1}{\psi}$	$\frac{T_3}{i_0\eta_0 - 1}$	$\frac{i_0\eta_0 T_3}{1 - i_0\eta_0}$	T_3	$\frac{i_0\eta_0 - 1}{i_0 - 1}$
2	2	1	3	$\frac{1 - i_0}{i_0} = \psi - 1$	$\frac{T_3}{i_0\eta_0 - 1}$	$\frac{i_0\eta_0 T_3}{1 - i_0\eta_0}$	T_3	$\frac{i_0\eta_0 - 1}{\eta_0(i_0 - 1)}$
3	3	1	2	$\frac{i_0}{i_0 - 1} = \frac{1}{1 - \psi}$	$-\frac{\eta_0}{i_0} T_2$	T_2	$\frac{\eta_0 - i_0}{i_0} T_2$	$\frac{i_0 - 1}{i_0 - \eta_0}$
4	3	2	1	$\frac{1}{1 - i_0} = \frac{\psi}{\psi - 1}$	T_1	$-\frac{i_0}{\eta_0} T_1$	$\frac{i_0 - \eta_0}{\eta_0} T_1$	$\frac{\eta_0(i_0 - 1)}{i_0 - \eta_0}$
5	1	3	2	$i_0 = \frac{1}{\psi}$	$-\frac{T_2}{i_0\eta_0}$	T_2	$\frac{1 - i_0\eta_0}{i_0\eta_0} T_2$	η_0
6	2	3	1	$\frac{1}{i_0} = \psi$	T_1	$-\frac{i_0}{\eta_0} T_1$	$\frac{i_0 - \eta_0}{\eta_0} T_1$	η_0

η, defined as the quotient of the output power, $P_0 = T_0\omega_0$, divided by the input power $P_i = T_i\omega_i$ (attention must be paid to the fact that here the subscript, i, indicates the input member).

The results shown for the cases 1, 2 and 5 in Table 13.2 are obtained using the first of the two Eqs. (13.170) and (13.174), while those shown for the cases 3, 4, and 6 in the same table are obtained using the first of the two Eqs. (13.170) and (13.175). All six possible cases correspond to a particular operating condition of the planetary gear train examined, where any of its main members 1, 2, or 3, is held at rest, while one of the remaining two members is the input member and the other is the output member, or vice versa. Therefore, for each of these six operating conditions, we know the sign of the torques applied to each of the three main members, in accordance with the above-mentioned sign convention. According to the kinematics, we also know the sign of the angular velocities of the rotating members in their absolute motion with respect to the frame of the planetary gear set, and the sign of the angular velocities relative to the shaft of the planet carrier, so we can identify the path of the power flow.

By way of example, let's examine in detail the three cases 1 (example 1), 3 (example 2), and 5 (example 3) of the Table 13.2.

Example 1 this example is characterized by an operating condition where 1, 2, and 3 are respectively the input, fixed and output members. In this case, T_3, η_0, i_0, ω_1 and ω_2 are known quantities, while the unknowns to be determined are T_1, T_2, ω_3, and the overall efficiency, η. Power flows from shaft of member 1 to shaft of member 3, so we have to solve the equation system consisting of the first of the two Eqs. (13.170) and (13.174). By solving this system, and taking into account the Eq. (13.12), we get:

$$T_1 = \frac{T_3}{i_0\eta_0 - 1}$$

$$T_2 = \frac{i_0\eta_0 T_3}{1 - i_0\eta_0},$$

(13.176)

which express the torques T_1 and T_2 as a function of the torque T_3 and quantities i_0 and η_0.

By definition, the overall efficiency, η, is given by Eq. (1.4), which we rewrite here in more complete terms, and neglecting the subscript, t:

$$\eta = \frac{P_0}{P_i} = \left| \frac{T_0 \omega_0}{T_i \omega_i} \right| = \left| \frac{T_3 \omega_3}{T_1 \omega_1} \right|. \tag{13.177}$$

The angular velocity ω_3 of the planet carrier can be determined from the third of the Eq. (13.16). From this equation, written in terms of $i_0 = 1/\psi$, we get the following relationship:

$$\omega_3 = \frac{\omega_1 - i_0 \omega_2}{1 - i_0}, \tag{13.178}$$

which expresses the angular velocity ω_3 as a function of ω_1, ω_2, and i_0, in the most general possible case. In the specific case examined here ($\omega_2 = 0$), this equation becomes:

$$\omega_3 = \frac{\omega_1}{1 - i_0}. \tag{13.179}$$

Finally, from Eq. (13.177), taking into account this last equation as well as the first of the two Eq. (13.176), we get:

$$\eta = \frac{i_0 \eta_0 - 1}{i_0 - 1}. \tag{13.180}$$

Example 2 this example is characterized by an operating condition where 3, 1, and 2 are respectively the input, fixed and output members. In this case T_2, η_0, i_0, ω_2 and ω_1 are known quantities, while the unknowns to be determined are T_1, T_3, ω_3, and η. Power flows from shaft of member 3 to shaft of member 2, so we have to solve the equation system consisting of the first of the two Eqs. (13.170) and (13.175). By solving this system, and taking into account the Eq. (13.12), we get:

$$T_1 = -\frac{\eta_0}{i_0} T_2$$
$$T_3 = \frac{\eta_0 - i_0}{i_0} T_2, \tag{13.181}$$

which express the torques T_1 and T_3 as a function of the torque T_2 and quantities i_0 and η_0.

The overall efficiency is given by:

$$\eta = \left| \frac{T_2 \omega_2}{T_3 \omega_3} \right|. \tag{13.182}$$

From this equation, taking into account the second of the two Eqs. (13.181) and (13.178) with $\omega_1 = 0$, we get:

$$\eta = \frac{i_0 - 1}{\eta_0 - i_0}. \tag{13.183}$$

Example 3 this last example is characterized by an operating condition where 1, 3, and 2 are respectively the input, fixed and output members. In this case T_2, η_0, i_0, ω_1, ω_3 and $\eta = \eta_0$ are known quantities, while the unknowns to be determined are only T_1, T_3, and ω_2. From Eq. (13.178), with $\omega_3 = 0$, we obtain $\omega_2 = \omega_1 / i_0$. Power flows from shaft of member 1 to shaft of member 2, so we have to solve the equation system consisting of the first of the two Eqs. (13.170) and (13.174). By solving this system, and taking into account the Eq. (13.12), we get:

$$T_1 = -\frac{T_2}{i_0 \eta_0}$$
$$T_3 = \frac{1 - i_0 \eta_0}{i_0 \eta_0} T_2. \tag{13.184}$$

We leave the reader the solution of the other three case in Table 13.2, here not explicitly performed.

For negative values of the basic ratio ($i_0 < 0$), six other possible cases are identifiable. The corresponding results are shown in Table 13.3. In the columns of this table, the same indications and quantities shown in Table 13.2 are listed, in the same order from left to right.

Here also the results shown for the cases 1, 2 and 5 of the above table are obtained using the first of the two Eqs. (13.170) and (13.174), while those shown for the cases 3, 4, and 6 of the same table are obtained using the first of the two Eqs. (13.170) and (13.175). It is not the case here to dwell on these results, concerning the negative values of the basic ratio, i_0, since the same considerations apply to them, which we have made for positive values of the basic ratio.

Instead, it is very important to draw design guidelines from the detailed examination of the results obtained above. To do this, we leave aside cases 5 and 6 shown in both tables, as they refer to conditions in which the planetary gear train became an ordinary gear train, as its planet carrier was held at rest. Instead, we focus our attention on the results concerning cases 1 to 4. To give a quantitative idea of how things go, let's consider two different values of the basic efficiency, η_0, and precisely $\eta_0 = 95\%$, and $\eta_0 = 80\%$.

Table 13.3 Efficiencies in the six possible cases of a planetary gear train with $i_0 < 0$

Case	Input member	Fixed member	Output member	Transmission ratio, i	T_1	T_2	T_3	Efficiency η
1	1	2	3	$1 - i_0 = \frac{\psi-1}{\psi}$	$\frac{T_3}{i_0\eta_0-1}$	$\frac{i_0\eta_0 T_3}{1-i_0\eta_0}$	T_3	$\frac{i_0\eta_0-1}{i_0-1}$
2	2	1	3	$\frac{i_0-1}{i_0} = 1-\psi$	$\frac{\eta_0 T_3}{i_0-1}$	$\frac{i_0 T_3}{\eta_0-i_0}$	T_3	$\frac{i_0-\eta_0}{i_0-1}$
3	3	1	2	$\frac{i_0}{i_0-1} = \frac{1}{1-\psi}$	$-\frac{T_2}{i_0\eta_0}$	T_2	$\frac{1-\eta_0 i_0}{i_0\eta_0} T_2$	$\frac{\eta_0(i_0-1)}{i_0\eta_0-1}$
4	3	2	1	$\frac{1}{1-i_0} = \frac{\psi}{\psi-1}$	T_1	$-\frac{i_0}{\eta_0} T_1$	$\frac{i_0-\eta_0}{\eta_0} T_1$	$\frac{\eta_0(i_0-1)}{i_0-\eta_0}$
5	1	3	2	$i_0 = \frac{1}{\psi}$	$-\frac{T_2}{i_0\eta_0}$	T_2	$\frac{1-i_0\eta_0}{i_0\eta_0} T_2$	η_0
6	2	3	1	$\frac{1}{i_0} = \psi$	T_1	$-\frac{i_0}{\eta_0} T_1$	$\frac{i_0-\eta_0}{\eta_0} T_1$	η_0

Figures 13.24a, b show, for the four cases above, the distribution curves of overall efficiency, respectively obtained for positive and negative values of the basic ratio, under the assumption that the basic efficiency is equal to 95% ($\eta_0 = 0.95$). The examination of these curves suggests that:

(a) Cases 1 and 4 are very similar to each other and are characterized by unacceptable low efficiencies for positive basic ratios, and relatively low efficiencies for negative basic ratios. Moreover, efficiency decreases drastically when the basic ratio decreases, for positive basic ratios, while it increases slightly when the absolute value of the basic ratio decreases, for negative basic ratios, until it reaches its maximum value when the absolute value of the basic ratio is equal to 1.

(b) Cases 2 and 3 are also very similar to each other and are characterized by a trend of efficiency distribution curves similar to that of cases 1 and 4, but with higher efficiency values for positive basic ratios. For negative basic ratios, however, the trend of the distribution curves shows an increase in efficiency when the absolute value of the basic ratio increases, starting from a minimum value corresponding to the unitary absolute value of the basic ratio.

It is, however, necessary that too hasty conclusions are not drawn, based on the above trends of the overall efficiency distribution curves. In fact, at first glance, and therefore without the necessary insights, it would seem that the designer should always accord his preference to cases 2 and 3 with a negative basic ratio, since they have the highest efficiency. However, a more comprehensive analysis should take into account that cases 2 and 3 do not allow large overall transmission ratios (see the fifth column in the Table 13.2). The usual compromise choices, which we have often talked about, can lead the designer to opt for design solutions based on cases 1 and 4, which instead provide the required high transmission ratios.

Figures 13.25a, b show, for the same four cases above, the distribution curves of overall efficiency, respectively obtained for positive and negative values of the basic ratio, under the assumption that the basic efficiency is equal to 80% ($\eta_0 = 0.80$). The examination of these curves leads to the same considerations we have made on the distribution curves shown in the previous figure, with the following two variations:

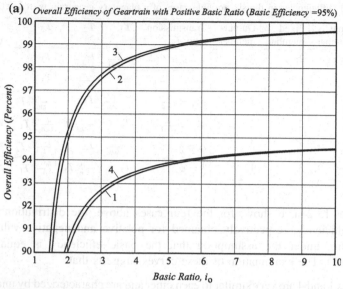

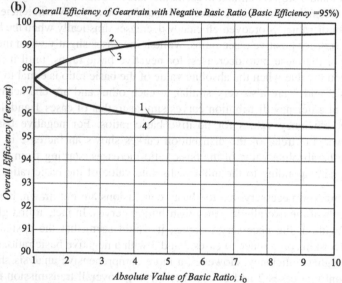

Fig. 13.24 Distribution curves of overall efficiency η (in %) as a function of the absolute value of basic ratio, for cases 1, 2, 3, and 4, with basic efficiency $\eta_0 = 0.95$: **a** positive basic ratio; **b** negative basic ratio

- *ceteris paribus*, due to the lower values of basic efficiency (80% instead of 95%), the overall efficiency values are much lower, so much that a different scale of the ordinates is needed;
- as before, here also cases 1 and 4, and cases 2 and 3, are two-to-two very similar, but the differences between the two curves of each pair are more pronounced, and overlaps are virtually nonexistent.

13.13   Efficiency of Planetary Gear Trains for Large Transmission Ratios

Here we want to analyse the efficiency of planetary gear trains capable of achieving large transmission ratios (see also Pollone [51]. To this end, let us consider the planetary gear train type 5 shown in Fig. 13.13. As we already showed in Sect. 13.7.5, this planetary gear train allows for obtaining large transmission ratios in cases where ψ is as close as possible to 1, and ω_1 or ω_2 are equal to zero. In this framework, let us assume that the operating conditions of this planetary gear set is as follows: second annulus gear held at rest by suitable braking torque, for which $\omega_2 = 0$; driving member is the planet carrier, whose angular velocity is ω_3; driven member is the first annulus gear, which rotates with angular velocity ω_1, and has T_0 as resisting torque applied to its shaft. Finally, let z_1, z_2, z_3, and z_4 be the numbers of teeth of toothed gear wheels, as shown in Fig. 13.13.

As we have shown in Sect. 13.7.5, the angular velocity ω_1 is correlated to the angular velocity ω_3 and ratio $\psi = 1/i_0$ by the Eq. (13.69), with ψ given by Eq. (13.59). The output power applied to the driven shaft is given by $P_0 = T_0\omega_1$. Since the torque, T_0, is applied to the rotating annulus gear and its relative velocity with respect to the planet carrier is given in absolute value by $(\omega_1 - \omega_3) = [\omega_1/(1 - \psi)]$, the gear pair with z_1 and z_2 teeth behaves as if transmitting the power $P_{1,2}$ given by:

$$P_{1,2} = \frac{P_0}{1 - \psi},\tag{13.185}$$

which is called *rolling power* by some scholars.

If we indicate with $\eta_{1,2}$ the efficiency of the above gear pair, the lost power will be given by:

$$P_{d1,2} = \left(\frac{1}{\eta_{1,2}} - 1\right)P_{1,2} = \left(\frac{1}{\eta_{1,2}} - 1\right)\frac{P_0}{1 - \psi};\tag{13.186}$$

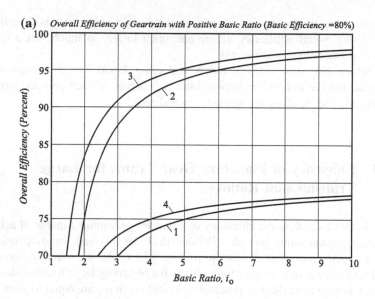

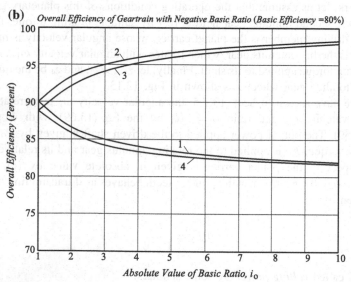

Fig. 13.25 Distribution curves of overall efficiency η (in %) as a function of the absolute value of basic ratio, for cases 1, 2, 3, and 4, with basic efficiency $\eta_0 = 0.80$: **a** positive basic ratio; **b** negative basic ratio

so, the input power applied to the planet gear with z_2 teeth is given by:

$$P_2 = P_0 \left[1 + \left(\frac{1}{\eta_{1,2}} - 1 \right) \frac{1}{1 - \psi} \right] = T_0 \omega_1 \left[1 + \left(\frac{1}{\eta_{1,2}} - 1 \right) \frac{1}{1 - \psi} \right]. \quad (13.187)$$

Thus, the torque applied to the annulus gear with z_4 teeth is as follows:

$$T_4 = T_0 \left[1 + \left(\frac{1}{\eta_{1,2}} - 1 \right) \frac{1}{1 - \psi} \right] \frac{z_2 z_4}{z_1 z_3} = T_0 \left[1 + \left(\frac{1}{\eta_{1,2}} - 1 \right) \frac{1}{1 - \psi} \right] \frac{1}{\psi}. \quad (13.188)$$

Since the angular velocity of the same annulus gear relative to the planet carrier is given by $(\omega_2 - \omega_3) = -\omega_3 = \omega_1 / [(1/\psi) - 1]$, the output rolling power will be given by:

$$
\begin{aligned}
P_{3,4} &= T_0 \omega_1 \frac{(1/\psi)}{[(1/\psi) - 1]} \left[1 + \left(\frac{1}{\eta_{1,2}} - 1 \right) \frac{1}{1 - \psi} \right] \\
&= \frac{P_0}{1 - \psi} \left[1 + \left(\frac{1}{\eta_{1,2}} - 1 \right) \frac{1}{1 - \psi} \right]. \quad (13.189)
\end{aligned}
$$

If, for simplicity of calculation, we assume that the efficiency of the gear pair with z_3 and z_4 teeth is equal to that of the gear pair with z_1 and z_2 teeth, that is $\eta_{3,4} = \eta_{1,2}$, we get the following relationship, which expresses the lost power in this gear pair:

$$
\begin{aligned}
P_{d3,4} &= P_0 \left(\frac{1}{\eta_{1,2}} - 1 \right) \frac{1}{1 - \psi} \left[1 + \left(\frac{1}{\eta_{1,2}} - 1 \right) \frac{1}{1 - \psi} \right] \\
&= P_0 \left\{ \left(\frac{1}{\eta_{1,2}} - 1 \right) \frac{1}{1 - \psi} + \left[\left(\frac{1}{\eta_{1,2}} - 1 \right) \frac{1}{1 - \psi} \right]^2 \right\}. \quad (13.190)
\end{aligned}
$$

Therefore, the input power, P_i, must be equal to:

$$P_i = P_0 + P_{d1,2} + P_{d3,4}; \quad (13.191)$$

so, the overall efficiency of the planetary gear train under consideration, in the operating conditions described above, will be given by:

$$\eta = \frac{P_0}{P_i} = \frac{1}{\left[1 + \left(\frac{1}{\eta_{1,2}} - 1 \right) \frac{1}{1 - \psi} \right]^2} = \left(\frac{1 - \psi}{\frac{1}{\eta_{1,2}} - 1} \right)^2. \quad (13.192)$$

To give an idea of the magnitude of this overall efficiency, we reconsider one of the examples discussed in Sect. 13.7.5, namely the planetary gear train shown in Fig. 13.13, and characterized by the following quantities: pressure angle $\alpha = 20°$; $y = 8$; $z_1 = 36$; $z_2 = 28$; $z_3 = 17$, and $z_4 = 35$. Thus, for this planetary gear train we will have $1/(1 - \psi) = 122,5$, and therefore $\psi = (121.5/122.5) = 972/980$. We also assume that the efficiency of each pair of gears is very high, that is $\eta_{1,2} = \eta_{3,4} = 0.99$ (this value of efficiency is difficult to achieve in practical applications, but it is deliberately chosen very high here to empathize the significant conclusion outlined below).

By performing the calculations based on the above input data, we get $\eta = 0.20$. It is evident that this is a very low efficiency value. If the same speed reduction was achieved using an ordinary gear train consisting of three gears, that is, three pairs of gear wheels, we would achieve a much higher overall efficiency. For example, with $\eta_{1,2} = \eta_{3,4} = \eta_{5,6} = 0.96$ (to be noticed: 0.96 vs. 0.99 above), we would get $\eta = \eta_{1,2}^3 = 0.88$.

A peremptory conclusion is to be drawn from what has been discussed above: the choice of design solutions based on the use of planetary gear trains necessary requires an in-depth preliminary assessment of their efficiency, before any other evaluation is made, such as load carrying capacity and similar. In fact, it is absolutely necessary to avoid the risk of too low efficiencies, which would make the planetary gear train completely useless, as it is not able to fulfill all its design requirements satisfactory.

References

1. Balbayev G (2015) New planetary gearbox: design and testing. LAP Lambert Academy Publishing, Riga, Latvia
2. Birch TW (2012) Automatic transmissions and transaxles. Pearson Education, Upper Saddle River, NJ
3. Budynas RG, Nisbett JK (2008) Shigley's mechanical engineering design, 8th edn. McGraw-Hill Book Companies Inc, New York
4. Chen C, Angeles J (2007) Virtual-power flow and mechanical gear-mesh power losses of epicyclic gear trains. J Mech Des 129(1):107–113
5. Cooley CG, Parker RG (2014) A review of planetary and epicyclic gear dynamics and vibrations research. ASME Appl Mech Rev 66:040804-1–040804-15
6. Coy JJ, Townsend DP, Zaretsky EV (1985) Gearing, NASA Reference Publication 1152, AVSCOM Technical Report 84-C-15
7. Davies K, Chen C, Chen BK (2012) Complete efficiency analysis of epicyclic gear train with two degree of freedom. J Mech Des 134(7):071006
8. De Castillo JM (2002) The analytical expression of the efficiency of planetary gear trains. Mech Mach Theor 37:187–214
9. Dhand A, Pullen K (2014) Analysis of continuously variable transmission for flywheel energy storage systems in vehicular application. In: Proceedings of Institution of Mechanical Engineers, Part. C: Journal of Mechanical Engineering Science, vol 229, issue 2
10. Dooner DB (2012) Kinematic geometry of gearing, 2nd edn. Wiley&Sons, New York

11. Drachmann AG (1963) The mechanical technology of Greek and Roman antiquity, a study of the literary sources. Munksgaard, København
12. Dudley DW (1962) Gear handbook. the design, manufacture, and applications of gears. McGraw-Hill, Book Companies, Inc., New York
13. Essam LE, Isam EI (2018) Influence of the operating conditions of two-degree-of-freedom planetary gear trains on tooth friction losses. J Mech Des 140(5)
14. Ferrari C, Romiti A (1966) Meccanica Applicata alle Macchine. Unione Tipografico-Editrice Torinese (UTET), Torino
15. Fisher R, Kücükay F, Jürgens G, Najork R, Pollak B (2015) The automotive transmission book. Springer International Publishing Switzerland, Cham, Heidelberg
16. Galvagno E (2010) Epicyclic gear train dynamics including mesh efficiency. Int J Mech Control 11(2):41–47
17. Genta G, Morello L (2007) L'Autoveicolo: Progetto dei componenti, vol 1. ATA-Associazione Tecnica dell'Automobile, Orbassano (TO)
18. Giovannozzi R (1965) Costruzioni di Macchine, vol II, 4th edn. Casa Editrice Prof. Riccardo Pàtron, Bologna
19. Gupta AK, Kartik V, Ramanarayanan CP (2015) Design, development and performance evolution of a light diesel hybrid electric pickup vehicle using a new parallel hybrid transmission system. In: Transportation Electrification Conference (ITEC), IEEE International, Chennai, India, 27–29 Aug 2015
20. Gupta AK, Ramanarayanan CP (2013) Analysis of circulating power within hybrid electric vehicle transmissions. Mech Mach Theor 64:131–143
21. Henriot G (1979) Traité théorique and pratique des engrenages, vol 1, 6th edn. Bordas, Paris
22. Henriot G (1990). Gears and planetary gear trains. Reggio Emilia: Brevini (ed)
23. ISO 1122-1:1998. Vocabulary of gear terms—Part. 1: Definition related to geometry
24. Jelaska DT (2012) Gears and gear drives. Wiley, New York
25. Juvinall RC (1983) Fundamentals of machine component design. Wiley, New York
26. Juvinall RC, Marshek KM (2012) Fundamentals of machine component design, 5th edn. Wiley, New York
27. Kahraman A, Hilty DR, Singh A (2015) An experimental investigation of spin power losses of a planetary gear set. Mech Mach Theory 86:48–61
28. Kelley OK (1973) Design of planetary gear trains, Chapter 9 of design practices-passenger car automatic transmissions. SAE-Society of Automotive Engineers, New York
29. Klebanov BM, Groper M (2016) Power mechanisms of rotational and cyclic motions. CRC Press, Taylor&Francis Group, Boca Raton, Florida
30. Kudryavtsev VN, Derzhavets YA, Gluharev EM (1994) Calculation and design of gear transmissions-handbook. Politeknika Publishing, St. Petersburg (in Russian)
31. Lechner G, Naunheimer H (1999) Automative transmissions: fundamentals, selection, design and applications. Springer, Heidelberg
32. Litvin FL, Fuentes A (2004) Gear geometry and applied theory, 2nd edn. Cambridge University Press, Cambridge, UK
33. Litvin FL, Fuentes A, Vecchiato D, Gonzalez-Perez I (2004) New design and improvement of planetary gear trains, NASA Technical Report, NASA/Cr-2004-213101, ARL-CR-0540
34. Liu J, Peng H (2010) A systematic design approach for two planetary gear split hybrid vehicles. Veh Syst Dyn 48(11):1395–1412
35. Lyndwander P (1983) Gear drive system design and application. Marcel Dekker Inc, New York
36. Maitra GM (1994) Handbook of gear design, 2nd edn. Tata McGraw-Hill Publishing Company Ltd, New Delhi
37. Merritt HE (1954) Gears, 3th edn. Sir Isaac Pitman&Sons, London
38. Merritt HE (1972) Gear engineering. Wiley, New York
39. Meyer PB (2011) The Wilson Preselector Gearbox, Armstrong-Siddley Type. pbm verlag, Seevetal, Hamburg, Germany

40. Michlin Y, Myunster V (2002) Determination of power losses in gear transmissions with rolling and sliding friction incorporated. Mech Mach Theor 37:167–174
41. Miller JM (2006) Hybrid electric vehicle propulsion system architectures of the e-CVT type. IEEE Trans Power Electron 21(3):756–767
42. Molian S (1982) Mechanism design: an introductory text. Cambridge University Press, Cambridge
43. Morelli A (1999) Progetto dell'autoveicolo: concetti di base. Celid, Torino
44. Müller HW (1982) Epicyclic drive trains: analysis, synthesis, and applications. Wayne State University Press, Detroit
45. Naunheimer H, Bertsche B, Ryborz J, Novak W (2011) Automatic transmission: fundamentals, selection, design and application, 2nd edn. Springer, Heidelberg
46. Niemann G, Winter H (1983) Maschinen-Element, Band II: Getriebe allgemein, Zahnradgetriebe-Grundlagen, Stinradgetriebe. Springer, Berlin
47. Pastore G (2013). Il Planetario di Archimede Ritrovato, and The Recovered Archimedes Planetarium. Pastore Ed, Rome
48. Paul B (1979) Kinematics and dynamics of planar machinery. Prentice Hall, New Jersey
49. Pearson T (1991) Planetary gear. Roof Books, New York
50. Pennestrì E, Mariti L, Valentini PP, Mucino VH (2012) Efficiency evaluation of gearboxes for parallel hybrid vehicles: theory and applications. Mech Mach Theory 49:157–176
51. Pollone G (1970) Il veicolo. Libreria Editrice Universitaria Levrotto&Bella, Torino
52. Ptolemy C (2014) The Almagest: introduction to the mathematics of the heavens, Translated by Bruce M. Perry, William H. Donahue (ed). Green Lion Press
53. Radzevich SP (2016) Dudley's handbook of practical gear design and manufacture, 3rd edn. CRC Press, Taylor&Francis Group, Boca Raton, Florida
54. Ruggieri G (2003) Rotismi Epicicloidali. McGraw-Hill Companies, srl, Milano
55. Simionescu PA (1998) A unified approach to the assembly condition of epicyclic gears. ASME J Mech Des 120(3):448–453
56. Smith JD (1983) Gears and their vibration: a basic approach to understanding gear noise. Marcel Dekker Inc, New York
57. Uicker JJ, Pennock GR, Shigley JE (2003) Theory of machines and mechanisms. Oxford University Press, New York
58. Warwick A (2014) Who invented the automatic gearbox. North West Transmissions Ltd., (ed). Retrieved 11 Oct
59. Willis R (1841) Principles of mechanism. Longmans-Green&C, London
60. Wright MT (2007) The Antikythera mechanism reconsidered. Interdisc Sci Rev 32(1):27–43
61. Xue HL, Liu G, Yang XH (2016) A review of graph theory approach research in gears. In: Proceedings of the Institution of Mechanical Engineers, Part. C: Journal of Mechanical Engineering Science, vol 230, issue 10
62. Yang F, Feng J, Zhang H (2015) Power flow and efficiency analysis of multiflow planetary gear trains. Mech Mach Theor 92:86–99

Chapter 14
Face Gear Drives

Abstract In this chapter, the ever-increasing importance of face gear drives is first underlined, and short notes on their cutting processes are provided. The generation processes and geometrical elements of a face gear pair are then defined, and the basic topics of face gear manufacturing and bearing contact are given. The basic equations of the tooth flank surfaces of these gears are then obtained using differential geometry methods, and the non-undercut conditions of the tooth surfaces and those for avoiding pointed teeth are also defined. The correct meshing and contact conditions between the two members of a face gear pair are then studied and analyzed using the same methods, and the point contact conditions are determined. Finally, the generation and grinding processes of face gears by worms are deepened, always using the aforementioned analytical procedures.

14.1 Introduction and General News

According to the definition of ISO 1122-1 [15], a *face gear*, or *contrate gear*, is a bevel gear with tip and root angles of 90° (Fig. 14.1). The same ISO defines as *face gear pair*, or *contrate gear pair*, a gear pair consisting of a face gear and its mating pinion, with either intersecting or crossed axes, with a shaft angle equal to 90°. Figure 14.2 shows a conventional face gear pair formed by a spur involute pinion (this can also be a cylindrical helical pinion) and a conjugate face gear (see Feng et al. [8], Kawasaki [16], Litvin et al [23], Townsend [35]).

Face gear drives are among the oldest types of gears of which we have historical memory, but also they can be considered as futuristic gears, because we do not know well yet all their features, and above all their potential in relation to possible practical applications, in competition with other types of gears. These types of gear drives, albeit in a rudimentary form, have been used by the Chinese (see, for example, the chariot compass, described in Vol. III, concerning *A concise history of gears*), as well as in ancient Egypt, Mesopotamia, and in the Roman world, for waterwheels, watermills and windmills. A face mill pair, consisting of a face gear

© Springer Nature Switzerland AG 2020   773
V. Vullo, *Gears*, Springer Series in Solid and Structural Mechanics 10,
https://doi.org/10.1007/978-3-030-36502-8_14

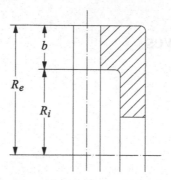

Fig. 14.1 Face gear (or contrate gear)

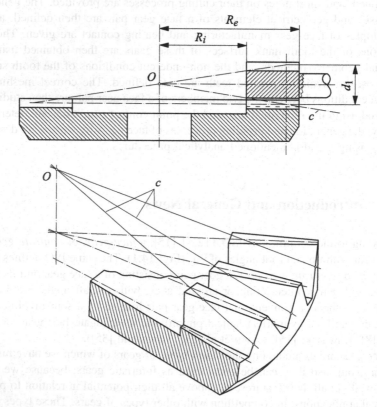

Fig. 14.2 Face gear pair formed by a spur involute pinion and a mating face gear, with intersecting axes and shaft angle $\Sigma = 90°$

with pins, and a lantern pinion, in fact characterized these ancient practical applications.

The revival of interest in these types of gears came from the middle of the past century [3, 5, 6, 9]. This interest was intensified towards the end of the same century, and the beginning of the 21th century [1, 2, 10–13, 17, 19, 20, 25, 29, 33, 40]. This interest is also witnessed by specific research programs, such as the program ART (Advanced Rotorcraft Transmission) of the US Army, USA, and BRITE EURAM FACET (development of face gears for use in aerospace transmission) in Europe.

The reasons of this delayed interest on these types of gears are to be found in the fact that the benefits of using face gears in the helicopter transmissions also began to be obvious with some delay. Early studies on the subject clearly showed that the use of face gear technology would have allowed a power density improvement and lower cost when applied to a helicopter transmission. In fact, the ability of face gear pairs to provide high reduction ratios as well as a self-adjusting torque splitting, allows to replace conventional multi-stage gearboxes with gear units requiring a lower number of stages. This leads to a better power/weight ratio, a small number of parts of the gearbox, and a reduction of volume. In addition, the split torque design of this type of transmission offers improved reliability and reduced costs against existing conventional gearing designs used in large horsepower applications.

A helicopter transmission must meet, among other things, the following two requirements: it must allow a split of torque from the engine shafts to the combining gear; it must be able to ensure a drastic speed reduction from about 25×10^3 rpm (~ 2.618 rad/s) of the engine shafts to about 300 rpm ($\sim 31,4$ rad/s) of the main rotor. The combining gear is rigidly connected with the input member, the sun gear, of a planetary gear train, whose output is the planet carrier, which drives the main rotor shaft. Figure 14.3 highlights the concept of split of torque. Figure 14.3a shows a more traditional design solution, which uses two spiral bevel pinions forming a single rigid body, which drive two mating spiral bevel gear wheels that transmit power to the combining gear. Figure 14.3b shows instead a far more advanced design solution, in which a single spur (or helical) pinion is in mesh with two face gears. In both figures, the transmission part made up of the planetary gear train, which drives the main rotor, is omitted; in this regard, see Litvin et al. [29].

The same Fig. 14.3 shows that the solution with face gears has at least two main advantages (with all that follows) over the one utilizing spiral bevel gears. On the one hand, the torque split configuration allows to transmit reduced loads on bearings that support the various shafts, and, on the other hand, the pinion is single and consists of a simple spur (or helical) pinion, compared to the double spiral bevel pinion, which in itself is much more complex. It results in greatly reduced weight and cost. This is another very important design goal, also because the transmission is called to support the weight of the entire aircraft, including that of several flight accessories.

Now, we focus our attention on the face gear pair. The axes of its two toothed members (pinion and face gear) generally, but not necessary, intersect under a right angle. When this configuration is made, the face gear pair is called *on-center face gear pair*; instead, if the axes do not intersect (this is a much rarer case in practical applications, also because it is much more complex), the face gear pair is called *off-set face gear pair*. We here limit the discussion only to face gear pairs with

Fig. 14.3 Split of torque for a helicopter transmission, with: **a** spiral bevel gears; **b** face gears

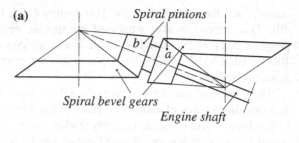

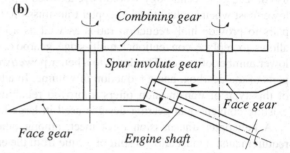

intersecting axes, that is to on-center face gear pairs. Below, in the absence of further details, for face gear pairs we mean those with intersecting axes.

The peculiarities of an on-center face gear pairs (as well as those of an off-set face gear pair) are closely related to the generation process of its two toothed members. We already know very well the generation process of the spur (or helical) pinion. For the generation of the mating face gear, a generation shaper is usually used. This shaper, with its relative motions compared to the face gear to be generated, must simulate the engagement of the actual pinion with the same face gear. In this generation process, the surfaces of the teeth of shaper and face gear are in line contact of all times, just as the tooth surfaces of the actual face gear pair (see Fig. 14.2). However, when the shaper is exactly identical to the actual pinion (with the obvious exception of the cutting edges), the face gear pair thus generated becomes sensitive to misalignment, hence an undesirable shift of the bearing contact and also separation of surfaces, i.e. absence of contact, may occur.

Therefore, to eliminate or decrease such a misalignment sensitivity, which is proper to an on-center face gear pair, and is due to line contact, it is necessary to make the instantaneous contact between the tooth surfaces of the pinion and mating face gear become a point contact. In fact, with a localized contact, theoretically limited to one point, the face gear pair is less sensitive to misalignment. This important goal is achieved by using a shaper with a number of teeth $z_0 > z_1$, where z_1 is the number of teeth of the pinion. To this end, it is generally enough that the difference $(z_0 - z_1)$ is equal to $(2 \div 3)$. For an off-set face gear pair, this attention is not necessary, since contact is for itself a point contact; in this case, however, the operating conditions are worse, because they are characterized by rolling and considerable sliding.

Figure 14.2 represents the theoretical operating condition of an involute face gear pair, consisting of a spur gear and a mating face gear with constant tooth depth and large profile modification along the face width. The same figure highlights the fact that, from the point of view of kinematics, this gear pair is a bevel gear pair, as the lines of contact, c, are the generatrixes of two pitch cones, whose axes intersect at a right angle.

Mutatis mutandis, the same figure represents the cutting conditions where the actual pinion is replaced by an involute shaper. It should be noted that, under these cutting conditions, the tooth flank surface of the face gear to be generated consists of two portions. The first portion is the working surface, formed by the successive positions of the instantaneous lines of tangency between the cutting pitch cones of the shaper and face gear. The second portion is the fillet surface, which is generated by the shaper top edge. These two portions of the face gear tooth surface are bounded by the common boundary line, L_{lim}, shown in Fig. 14.4a, where several instantaneous contact lines, L_{2s}, are also drawn. Figure 14.4b shows instead how the cross sections of the tooth surface of a face gear vary along the face width.

It should be noted that the working surface and related instantaneous contact lines, L_{2s}, relate to cutting conditions. In accordance with the convention adopted in this textbook, according to which with the subscript 0 we indicate the cutting tool or, more generally, the cutting conditions, we should designate these instantaneous contact lines with the symbol L_{20}. We have not done this (and we will not do this in all this chapter), to avoid confusion and misunderstanding as, as we shall see, we will introduce a second cutting tool, i.e. the cutting worm. Therefore, for face gears, depending on the cutting process used, we may have to do with two cutting tools, i.e. the shaper and the worm. To distinguish these two cutting tools, we will use the subscripts, s and w, respectively. We will use the same convention also for coordinate systems rigidly connected with them as well as for any other quantity that refers to them.

From Fig. 14.4, it is apparent that the face width, b, of a face gear is to be limited to avoiding the risk of undercut towards the inside (A, in Fig. 14.4a) and the danger of pointed tooth towards the outside (B, in Fig. 14.4b). These two risks must be kept in mind by the designer. To overcome these dangers, as an indicative guidelines, for $\alpha_n = 20°$, the following quantities can be chosen (see Dudley and Winter [7]):

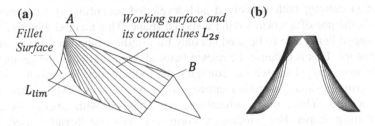

(a) *A* — *Working surface and* **(b)**
Fillet Surface *its contact lines L_{2s}*
B
L_{lim}

Fig. 14.4 Geometry of a face gear tooth: **a** fillet surface, and working surface, with its contact lines; **b** cross sections of a face gear tooth

$R_e = (1.10 \div 1.20)z_2 m_n/2$ and $R_i = (0.95 \div 1.05)z_2 m_n/2$, where the higher and lower values are to be selected, respectively, for low gear ratios (about $u = 1.5$), and for large gear ratios (about $u = 8$); $b = (R_e - R_i) \cong 0.07 z_2 m_n$. Other scholars express the permissible face width of the face gear as a function of the module, m, by the relationship $b = (R_e - R_i) = km$, where k is a coefficient that mainly depends on the gear ratio, u. Usually, k has values within the range $(8 \leq k \leq 15)$. If the pinion was realized with transverse crowning, the corresponding face gear pair would be insensitive to misalignment, but would have a load carrying capacity a little less.

14.2   About the Face Gear Cutting Process

As long as adequate cutting process were not available, face gears were not considered suitable for transmitting high power. The limited use of these types of gear drives, which lasted almost to the end of the last century, can be attributed both to the unavailability of technologically and economically viable manufacturing processes, capable of meeting the demanding requirements of high-power applications, and to the insufficient knowledge of three-dimensional geometry of their tooth surfaces, and their meshing properties. The grinding processes of face-gear teeth (which could have improved the accuracy grade by promoting their use for aerospace applications) were even more inadequate than cutting processes.

Of course, no difficulty exists for grinding the spur involute pinion. Significant difficulties existed, however, for grinding the face gear; it was therefore not possible to harden its tooth surfaces, which could have resulted in a substantial increase in permissible contact stresses. A first attempt to develop a face-gear manufacturing process is certainly that made by Miller [31], who proposed a method of generating a face gear by a hob. Due to its not a few limitations (see Litvin and Fuentes [24]), this method did not have much luck, but had the merit of drawing the attention of scholars and experts, and launching a research line on the subject.

These researches were intensified towards the end of the last century and the beginning of this century, and culminated with the substantial contribution of Litvin and his followers (see Litvin et al. [20, 22, 26–29]). This contribution is to be considered as a substantial step in development of the grinding and cutting processes of face gears. These processes use a worm for grinding, and a hob for generation cutting, both conceived and developed according to a common idea, based on the use of a worm having a special shape. This method therefore has the advantage of being able to be used not only for grinding, but also for the generation of face gears. This is because the related bases allow us to define the geometry of a suitable worm by which we can configure the tools needed to perform both operations, grinding and generation cutting, to be made respectively with grinding wheel or hob. Thus, many advantages are obtained with respect to conventional cutting shaper. For grinding, a worm with only one thread is used, but the number of turns of this thread has to be limited to avoid the appearance of singularities on the worm surface.

In the previous section we have already said that the conventional face-gear generation process is based on the use of an involute shaper, and on the meshing simulation of this shaper with the face gear to be generated. Here we will focus our attention on the concept of generation of face gears by grinding or cutting worm. The discussion is in line with the afore-mentioned method of generating face gears by a worm, which was development by Litvin and his followers, and therefore called from here on *Litvin's method*. This is a general method as it can be applied both for grinding the face gears, using a grinding worm, and for generating the face gear by a hob used as generation wheel. However, we cannot forget another major contribution, developed by Tan [34], which provides an improved method of dressing the grinding worm, which offers many unique features and advantages (see Heath et al. [13]). In the application of the Litvin's method, we will consider this important contribution.

Figure 14.5 shows in schematic form the grinding operation of a face gear according to the Litvin's method, which also incorporates the contribution of Tan. In this figure, an imaginary pinion, a face gear, and a grinding wheel are mutually overlapped with respect to each other in their meshing positions. The dressing (or truing) tool is also shown in the same figure. The grinding worm has a thread geometry that, in synchronous rotation with the face gear to be generated, must simulate the generation action of the imaginary pinion in rotation.

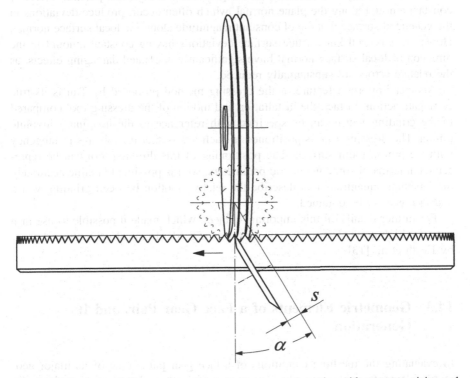

Fig. 14.5 Face gear grinding configuration, and dressing configuration with respect to pinion and face gear as references

The installation of this grinding worm on the face gear must therefore take into account the lead angle γ_w of its thread, which is given by:

$$\sin \gamma_w = \frac{z_w}{z_{v1}} \frac{d_{v1}}{d_w},$$

(14.1)

where z_w and z_{v1} are respectively the number of threads of the worm and the number of teeth of the imaginary pinion (usually $z_w = 1$), while d_{v1} and d_w are respectively the pitch diameter of the imaginary pinion and the diameter of the circle passing through the pitch point on the grinding wheel (see also next section).

The generation motion is relatively easy to get. Instead, it is much harder to obtain the correct shape of the grinding worm (and so the one of the generation hob). Often this step is the keystone and the biggest challenge in the development of face gears manufacturing processes. For dressing the grinding worm, Litvin uses a tool whose geometry follows that of the tooth space of the imaginary pinion. Tan has instead proposed a dressing tool configured to disk shape, having a plane as generation surface. This extremely simple geometry, besides being very easy to prepare with the desired high accuracy, has the inherent advantage of absorbing installation and motion errors. This is because any component of error contained in the generation plane has no effect on the generation action, and position errors of constant amount along the plane normal, which often occur, produce deviations in the generated surface that are of constant magnitude along the local surface normal. However, it is well known that surface deviations having constant amount in the direction of local surface normal have significantly alleviated damaging effects, as the relative errors are substantially reduced.

Another important feature of the dressing method proposed by Tan is its true conjugate action. In fact, the installation and motion of the dressing tool compared to the grinding worm can be specified with reference to the imaginary involute pinion. The dressing tool is positioned in such a way that it is always in tangency with the pinion tooth surface. The positioning of this dressing tool can be represented in terms of some geometric parameters, so it is possible to define accurately the kinematic quantities that describe the relative motion between grinding worm and face gear to be obtained.

For further details on this important subject, which made it possible to use face gears for high power transmissions, we refer the reader to the aforementioned paper by Heath et al. [13].

14.3 Geometric Elements of a Face Gear Pair, and Its Generation

In evaluating the meshing conditions of a face gear pair, some of its major geometric elements come into play, such as reference surfaces, pitch surfaces (or axodes), and pitch point. For a correct definition of these geometric elements, avoiding misunderstandings, we consider the general case of an intersecting face gear pair, characterized by any shaft angle, Σ, as shown in Fig. 14.6. Of course, the case that concerns us here, that of the face gear pair characterized by a shaft angle $\Sigma = 90°$, can be obtained from general case discussed below, placing in the formulas we will get $\Sigma = 90°$.

If the pinion was conical, instead of cylindrical, we would have a straight bevel gear pair such as that shown in Fig. 9.1a. Let's rotate the two members of this straight bevel gear pair about their intersecting axes OO_1 and OO_2, which form the shaft angle, Σ. The transmission ratio, i, which is equal to the gear ratio, u, will be given by:

$$i = \frac{\omega_1}{\omega_2} = \frac{\varphi_1}{\varphi_2} = \frac{z_2}{z_1} = u, \tag{14.2}$$

where ω_i and z_i (with $i = 1, 2$) are respectively the angular velocities and the number of teeth of the pinion ($i = 1$) and the bevel wheel ($i = 2$). The reference cones of these two members are the cones having cone angles δ_1 and δ_2, given by the Eqs. (9.20) and (9.21). These two cones touch each other along the common tangent line, OB, which is the instantaneous axis of rotation in their relative motion (Fig. 9.1). Therefore, they are also the pitch cones (or axodes) of the bevel gear drive. In fact, by definition, axodes are the *loci* of the instantaneous axes of rotation generated in coordinate system S_i (with $i = 1, 2$) rigidly connected with the pinion, 1, and wheel, 2. In other words, for a straight bevel gear pair with zero profile-shifted toothing, reference and pitch surfaces coincide.

However, in our case, the face gear pair consists of a cylindrical spur pinion and mating face gear. As shown in Fig. 14.6, the reference surfaces are, for the pinion, the cylinder of radius r_1, and, for the face gear, the cone having cone angle Σ, which coincides with the shaft angle. These reference surfaces touch each other along the common tangency line $O't'$, which is the *reference line*. In the case of intersecting orthogonal face gear pair, for which the shaft (or crossing) angle is $\Sigma = 90°$, the reference surface of the face gear is a plane. However, under the operating conditions, the two members of the face gear pair under consideration touch each other along the instantaneous line of contact Ot, which is the instantaneous axis of rotation in relative motion. Therefore, the pitch surface (or axodes)

Fig. 14.6 Geometric elements of a face gear pair with any shaft angle, Σ

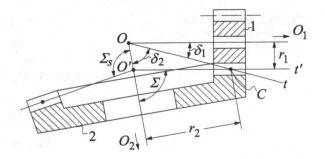

of the face gear pair are the two cones having cone angles δ_1 and δ_2, with $(\delta_1 + \delta_2) = \Sigma$. Thus, reference and pitch surfaces are no longer coincident, but different.

The point of intersection, C, of the reference line, $O't'$, with the instantaneous axis of rotation Ot, is what we call *pitch point*. The location of the pitch point along the instantaneous axis of rotation is of great importance as it affects the size of the meshing area as well as the hazards associated with the pointed teeth. At pitch point, the relative motion is pure rolling, while at any other point of the reference line, $O't'$, and instantaneous line of contact, Ot, the relative motion is a combined rolling and sliding motion.

As we have already anticipated in Sect. 14.1, the conventional method of generation of the face gear uses an involute shaper as a cutting tool, with a number of teeth, z_s, which exceeds two or three teeth the number of teeth, z_1, of the pinion. In this generation process, the shaper and face gear rotate about their respective intersecting axes with angular velocities ω_s and ω_2, which are correlated by the relationship (see also Fig. 14.7):

$$\frac{\omega_s}{\omega_2} = \frac{z_2}{z_s}, \tag{14.3}$$

where z_2 is the number of teeth of the face gear.

As shown in Fig. 14.7, the shaper also performs a reciprocating motion in the direction of the generatrix of the reference cone of the face gear, which is parallel to the shaper axis; this motion is the feed motion. Of course, the shaper and face gear axes intersect, forming the same shaft angle, Σ, which characterizes the actual face gear pair. The angle Σ_s is the supplementary angle of the shaft angle, i.e. $\Sigma_s = (\pi - \Sigma)$.

The tooth surface σ_2 of the face gear generated by an involute shaper is that shown in Fig. 14.4, and already partly commented. The L_{2s}-lines shown in Fig. 14.4a represent the instantaneous lines of tangency between the surface σ_2 and σ_s of the face gear being generated and the generation shaper; they are represented on the σ_2-surface, which consists of the sum of the fillet and working surfaces. Investigations made by Litvin et al. [25] have shown that the points of the face gear tooth flank surfaces are hyperbolic points, i.e. points where the product of principal curvatures, that is the Gaussian curvature, is negative. This means that the two

Fig. 14.7 Diagram of the generation process of a face gear by shaper

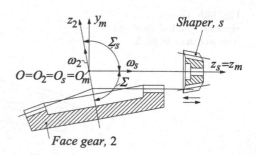

centers of curvature positioned on the principal direction straight lines lie on
opposite sides of the surface (see Novozhilov [32], Ventsel and Krauthammer [37]).
Figure 14.4b shows the various cross sections of the face gear tooth along the face
width; they show a remarkable tapering of the tooth from the inside to the outside.

The fillet of the tooth surface of face gear is generated by the shaper edge.
Figure 14.4a shows that the surface associated with the fillet thus generated is very
extensive. In the same paper above, the authors propose to generate the fillet by
means of a rounded edge of the shaper (see Fig. 14.8), as it results in a reduction of
about 10% of the bending stresses. In the case of a traditional shaper, without
rounded top edge, the fillet is generated by the active generatrix of the addendum
cylinder, whose intersections with the straight-line tooth profiles are points A or A'
in Fig. 14.8a. In the case of shaper with rounded top edge, the fillet is generated as
the envelope to the family of circles having radii $\rho = \rho_1 = \rho_2$ in Fig. 14.8.

It is here to emphasize that the ability of a face gear pair to withstand bending
stress is strongly dependent on the dimensionless coefficient $k = (R_e - R_i)/$
$m = b/m$, which we have already introduced in Sect. 14.1. Usually, for an involute
pinion that must transmit high powers, this coefficient is assumed equal to 10
($k = 10$). Higher values of this coefficient can be chosen for face gear pair with
higher gear ratios and pinions with larger number of teeth. Figure 14.9 is a good
guideline to choosing the appropriate k-value; it shows how this coefficient varies as
a function of the number of teeth z_1 of the pinion, for three different values of the
gear ratio, when the difference $(z_s - z_1) = 3$.

Fig. 14.8 Shaper tooth with
or without rounded top

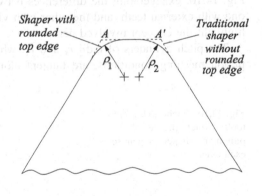

Fig. 14.9 Distribution curves
of coefficient k for involute
pinion

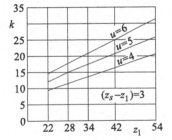

14.4 Basic Topics of Face Gear Manufacturing and Bearing Contact

If the face gear was generated by a shaper that, apart from the cutting edges, was an identical duplication of the pinion, its generation process would be an exact simulation of the meshing between pinion and face gear, so we would have a line contact at every instant. However, as we have already said, such a generation process cannot be practically used because of the sensitivity to the misalignment of the face gear pair made through it. This sensitivity is then exacerbated by assembly and manufacturing errors, which can cause separation of the contacting surfaces or undesirable contact at the edge.

We have also said that, to eliminate the above sensitivity to misalignment, the bearing contact between pinion and mating face gear must be a localized contact and this goal can only be achieved if the generation process is able to provide a point contact between the tooth flank surfaces of the two members that form the face gear pair. To achieve this goal, it is sufficient that the face gear is obtained as the envelope to the family of surfaces of an involute shaper having a number of teeth $z_s > z_1$, with the difference $(z_s - z_1)$ equal to 2 or 3.

To analyze the contact conditions occurring during the manufacturing process, made using an involute shaper, we introduce an imaginary internal gear pair consisting of the pinion, 1, whose pitch cylinder has radius r_1, and shaper, s, whose pitch cylinder has radius r_s. This imaginary internal gear pair is shown in Fig. 14.10, exaggerating the differences between the radii of these two members, one with external teeth, and the other with virtual internal teeth, in order to better highlight the concept involved here.

The pitch cylinders of radii r_s and r_1, which are the axodes in the shaper and pinion meshing conditions, are tangent along a common tangent, t_{s1}, which is

Fig. 14.10 Pitch circles and tooth involute profiles of pinion and shaper, tangent to each other

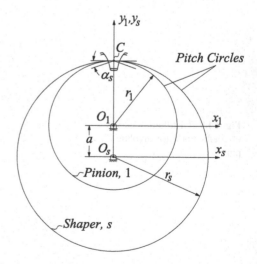

parallel to the respective axes of rotation. This common tangent is the pitch line, that is, the instantaneous axis of rotation in the relative motion of the shaper with respect to the pinion. The intersections of the pitch cylinder, pitch line, and axes of rotation with a transverse plane are respectively the pitch circles, pitch point, C, and centers O_s and O_1, shown in Fig. 14.10. This figure also shows the two coordinate systems, $O_s(x_s, y_s, z_s)$ and $O_1(x_1, y_1, z_1)$, rigidly connected to shaper and pinion respectively, whose z_i-axes coincide with their axes of rotation, and form right-hand coordinate systems with the corresponding axes x_i and y_i (with $i = s, 1$).

By generalizing the problem, we now consider not only the engagement of two members of the imaginary internal gear pair, as defined above, but also the engagements between the shaper and face gear to be generated, and between the pinion and generated face gear. By doing so, we come to consider three surfaces, σ_s, σ_1, and σ_2, which are respectively the tooth flank surfaces of shaper, pinion, and face gear. These three surfaces are in mesh simultaneously. During the process of generation of the face gear by shaper, surfaces σ_s and σ_2 are in line contact at every instant, along the instantaneous axis of rotation, t_{s2}. During the process of imaginary meshing between the shaper and pinion, surfaces σ_s and σ_1 are in line contact at every instant, along the instantaneous axis of rotation t_{s1}. Finally, in actual operating conditions, surface σ_1 and σ_2 are in point contact at every instant at point C (the pitch point), which belongs to the instantaneous axis of rotation, t_{12}, in the relative motion of the pinion with respect to the face gear.

Figure 14.11 summarizes everything we have just said. It shows the location of the three instantaneous axes of rotation during the two-to-two meshing of the three surfaces σ_s, σ_1, and σ_2. The instantaneous axis of rotation, t_{s1}, in the meshing of the imaginary internal gear pair, is parallel to the axes of rotation of shaper and pinion. The other two instantaneous axes of rotation, t_{s2} and t_{12}, are oriented with respect to the direction of t_{s1} by angles δ_s and δ_1. Angle δ_1 that is formed between the pinion axis and t_{12} is given by Eq. (9.20). Angle δ_s that is formed between the shaper axis and t_{s2} is given by the following equation:

Fig. 14.11 Location and orientation of instantaneous axes of rotation t_{s1}, t_{s2} and t_{12}

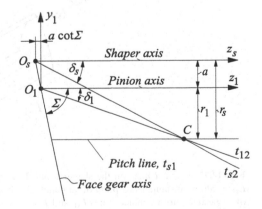

$$\tan \delta_s = \frac{\sin \Sigma}{u_{s2} + \cos \Sigma} = \frac{\sin \Sigma}{(z_2/z_s) + \cos \Sigma}, \tag{14.4}$$

which is obtained in the same way as Eq. (9.20).

As Fig. 14.11 shows, all three instantaneous axes of rotation t_{s1} (this axis coincides with the pitch line), t_{s2} and t_{12} intersect each other at pitch point C. The distance, a, between the axes of the shaper and pinion, which are parallel, is given by:

$$a = (r_s - r_1) = \frac{(z_2 - z_1)}{2} m. \tag{14.5}$$

It is very interesting to see how the contact lines are localized and oriented on the shaper tooth flank surface. Figure 14.12a shows the contact lines, L_{s2}, corresponding to the meshing of the shaper with the face gear; they are differently inclined with respect to the generatrixes of the shaper tooth surface. Figure 14.12b shows the contact lines, L_{s1}, corresponding to the meshing of the shaper with the pinion; these contact lines are all parallel to the generatrixes above. Finally, Fig. 14.12c shows the current instantaneous point of tangency, P, between surfaces σ_1 and σ_2, also represented on the shaper tooth surface. This point is the intersection point of respective current contact lines L_{s1} and L_{s2}.

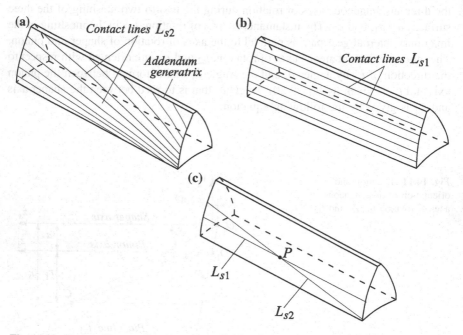

Fig. 14.12 Contact lines on the shaper tooth flank surface: **a** contact lines, L_{s2}; **b** contact lines, L_{s1}; **c** current instantaneous point of tangency of surfaces σ_1 and σ_2, obtained as intersection point of respective current contact lines L_{s1} and L_{s2}

Finally, it is necessary to determine the path of contact of surfaces, σ_1 and σ_2. To do this, just impose the know condition according to which the normal to the generation surface, σ_s, at the instantaneous point of contact, P, must pass through the pitch point, C. On this topic, we will return later (see Sect. 14.8).

14.5   Tooth Flank Surfaces of Face Gears, and Their Equations

The tooth flank surfaces of the face gears, which are of interest here, are those obtained under cutting conditions. Of course, these surfaces are strongly related to the shape of the cutting tool edge used in their manufacturing process and, more generally, to the geometry of tooth flank surfaces of the same cutting tool. This is because the face gear tooth surfaces are obtained as the envelope of the cutting tool surfaces, used for this purpose.

Here we will only consider the case of face gears generated by an involute shaper, following the theoretical development proposed by Litvin and Fuentes [24]. However, it is to be remembered that Litvin and his followers have further developed the theory, extending it to the face gears generated by shapers conjugate to parabolic rack cutters as well as face gears generated by involute helical shapers (see Litvin et al. [26–28]). We limit our discussion here, circumscribing it to the simplest case of face gears generated by an involute shaper, which is already quite complex for itself. The discussion is, however, of a very general nature, so that, with the proper variations, it can easely be extended to any geometry of the tooth profiling surface of shaper.

In this framework, we must first write the equation of the involute surface of the spur involute shaper. To this end, let us consider the shaper transverse section shown in Fig. 14.13, and assume, as coordinate system, the system $O_s(x_s, y_s, z_s)$ rigidly connected to it. This coordinate system has its origin at the crossing point $O \equiv O_s \equiv O_2$ between axes of shaper and face gear (see Fig. 14.7). Its z_s-axis is coinciding with the shaper axis, and directed as Fig. 14.7 shows, while its x_s-axis is coinciding with the straight line that symmetrically divides space width and is directed from the origin O_s to the pitch point C (this axis is the trace of the plane of equation $y_s = 0$, in Fig. 14.13). The y_s-axis completes the right-hand coordinate system.

In the transverse section shown in Fig. 14.13, i.e. in the Cartesian plane $O_s(x_s, y_s)$, the position of a current point P on the involute profile I is defined by the position vector $\overline{O_sP} = \overline{O_sT} + \overline{TP}$; the absolute value of this last vector is given by $|\overline{TP}| = \widehat{TP_0} = r_{bs}\vartheta_s$, where r_{bs} is the radius of the shaper base circle, and ϑ_s is the involute roll angle (see Sect. 2.1). Considering also the current point P' on the involute profile II, which is symmetric of point P with respect to the x_s-axis, we obtain the following system of Cartesian equations that define the position of these two points in the Cartesian plane $O_s(x_s, y_s)$:

Fig. 14.13 Transverse section of the shaper and involute tooth profiles

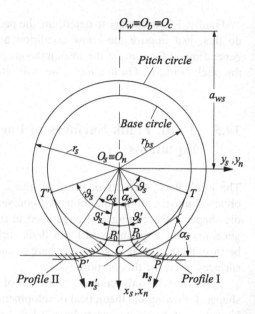

$$x_s = r_{bs}\left[\cos\left(\vartheta_s + \vartheta_s'\right) + \vartheta_s \sin\left(\vartheta_s + \vartheta_s'\right)\right]$$
$$y_s = r_{bs}\left[\pm \sin\left(\vartheta_s + \vartheta_s'\right) \mp \vartheta_s \cos\left(\vartheta_s + \vartheta_s'\right)\right], \tag{14.6}$$

where ϑ_s' is space width half angle on the base cylinder, which for a standard shaper is given by

$$\vartheta_s' = \frac{\pi}{2z_s} - \mathrm{inv}\alpha_s, \tag{14.7}$$

where z_s is the number of teeth of the shaper, and α_s is the pressure angle. The upper and lower signs in equation system (14.6) correspond respectively to involute profile I and II in Fig. 14.13.

When we move from the Cartesian plane $O_s(x_s, y_s)$ to Cartesian space $O_s(x_s, y_s, z_s)$, according to what we have seen in Sect. 11.5, for the parametric representation of the shaper involute tooth surface, σ_s, we must introduce two independent variable parameters, u and v. In this case, parameter v coincides with the involute roll angle, ϑ_s, while parameter $u = u_s$ represents the third coordinate in the direction of the z_s-axis of the current point P on the involute tooth surface in the three-dimensional space.

By limiting the discussion to the involute profile I in Fig. 14.13, and using homogeneous coordinates, in order to benefit from their known advantages for matrix calculus, we can represent the involute tooth surface of the shaper, σ_s, by the following vector function:

$$r_s(u_s, \vartheta_s) = \begin{bmatrix} r_{bs}\left[\cos\left(\vartheta_s + \vartheta_s'\right) + \vartheta_s \sin\left(\vartheta_s + \vartheta_s'\right)\right] \\ r_{bs}\left[\sin\left(\vartheta_s + \vartheta_s'\right) - \vartheta_s \cos\left(\vartheta_s + \vartheta_s'\right)\right] \\ u_s \\ 1 \end{bmatrix}. \tag{14.8}$$

According to what we have seen in Sect. 11.5, the unit normal n_s to the tooth flank surface of the shaper is given by:

$$n_s = \frac{N_s}{|N_s|} = \begin{bmatrix} \sin\left(\vartheta_s + \vartheta_s'\right) \\ -\cos\left(\vartheta_s + \vartheta_s'\right) \\ 0 \end{bmatrix}. \tag{14.9}$$

We can now determine the face gear tooth surface, σ_2, as the envelope to the family of the shaper tooth surface, σ_s. For this purpose, it is first necessary to define the relative positions of shaper and face gear to be generated under cutting conditions. To do this in the best and easiest way, we use the following coordinate systems:

- Coordinate system $O_s(x_s, y_s, z_s)$ rigidly connected to the shaper, already defined above (see Figs. 14.13, 14.14 and 14.7).
- Coordinate system $O_2(x_2, y_2, z_2)$ rigidly connected to the face gear (see Figs. 14.7 and 14.14a), with origin at the crossing point $O \equiv O_s = O_2$ between axes of shaper and face gear, z_2-axis coinciding with the face gear axis, and directed as Figs. 14.7 and Fig. 14.14 show, and axes x_2 and y_2 in the plane perpendicular to the z_2-axis and passing through the origin O_2, and directed as Fig. 14.14b shows.

Fig. 14.14 Coordinate systems for generation and determination of the face gear tooth surface, σ_2: **a** schematic drawing of the shaper/face gear pair; **b** other geometric and kinematic parameters

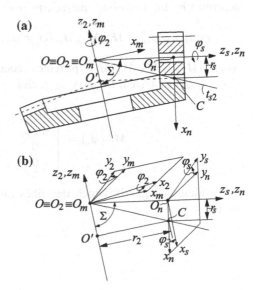

- Coordinate system $O_m(x_m, y_m, z_m)$ rigidly connected to the frame: this fixed system has its origin at the crossing point $O \equiv O_s \equiv O_2 \equiv O_m$, z_m-axis coinciding with the z_2-axis and equally directed, and axes x_m and y_m in the same plane of axes x_2 and y_2, as Fig. 14.14 shows;
- Coordinate system $O_n(x_n, y_n, z_n)$ rigidly connected to the frame: this fixed system has its origin at point O_n, z_n-axis coinciding with the z_s-axis and equally directed, and axes y_n and z_n in the plane perpendicular to the z_n-axis and passing through the origin, O_n (Fig. 14.14b). Point O_n is the point of intersection of z_s-axis with the plane perpendicular to it, passing through the pitch point, C, as shown in Fig. 14.14b.

Figure 14.14b also highlights other geometric and kinematic parameters, such as the radii, r_s and r_2, of the pitch circles of shaper and face gear respectively, as well as the angles of rotation, $\varphi_s = \omega_s dt$ and $\varphi_2 = \omega_2 dt$ of the shaper and face gear about their axes. These angles define the instantaneous positions of the mobile coordinate system $O_s(x_s, y_s, z_s)$ and $O_2(x_2, y_2, z_2)$, with respect to the fixed coordinate systems $O_n(x_n, y_n, z_n)$ and respectively $O_m(x_m, y_m, z_m)$.

The derivation of the face gear surface, σ_2, is based on the assumption that shaper and face gear perform rotations about intersecting axes, z_s and z_2, that form a non-orthogonal shaft angle, Σ, and that their angular velocities, ω_s and ω_2, are constant. Therefore, the angles of rotation of the shaper, φ_s, and face gear, φ_2, are related by the relationship:

$$\frac{\varphi_s}{\varphi_2} = \frac{\omega_s}{\omega_2} = \frac{z_2}{z_s}, \tag{14.10}$$

where z_2 and z_s are the numbers of teeth of the shaper and face gear, respectively.

In the coordinate system, $O_2(x_2, y_2, z_2)$, the family of shaper surfaces, σ_s, is described by the following matrix equation:

$$r_2(u_s, \vartheta_s, \varphi_s) = M_{2m}M_{mn}M_{ns}(\varphi_s)r_s(u_s, \vartheta_s) = M_{2s}(\varphi_s)r_s(u_s, \vartheta_s), \tag{14.11}$$

where $M_{2s} = M_{2m}M_{mn}M_{ns}$ provides coordinate transformation from system $O_s(x_s, y_s, z_s)$ to system $O_2(x_2, y_2, z_2)$, and

$$M_{ns}(\varphi_s) = \begin{bmatrix} \cos\varphi_s & -\sin\varphi_s & 0 & 0 \\ \sin\varphi_s & \cos\varphi_s & 0 & 0 \\ 0 & 0 & 1 & 0 \\ 0 & 0 & 0 & 1 \end{bmatrix} \tag{14.12}$$

is the rotational matrix which describes rotation about the z_n-axis with the unit vector,

$$
\boldsymbol{M}_{mn}(\varphi_s) =
\begin{bmatrix}
-\cos \Sigma & 0 & \sin \Sigma & r_2 \\
0 & 1 & 0 & 0 \\
-\sin \Sigma & 0 & -\cos \Sigma & -r_2 \cot \Sigma \\
0 & 0 & 0 & 1
\end{bmatrix}
\tag{14.13}
$$

is the translational matrix in transition from coordinate system $O_n(x_n, y_n, z_n)$ to coordinate system $O_m(x_m, y_m, z_m)$ (in this respect, it is to be noted that the shaper and face gear axes are non-orthogonal axes that form the shaft angle, Σ, as Fig. 14.14 shows), and

$$
\boldsymbol{M}_{2m} =
\begin{bmatrix}
\cos \varphi_2 & \sin \varphi_2 & 0 & 0 \\
-\sin \varphi_2 & \cos \varphi_2 & 0 & 0 \\
0 & 0 & 1 & 0 \\
0 & 0 & 0 & 1
\end{bmatrix}
\tag{14.14}
$$

is the rotational matrix, which describes rotation about the z_2-axis with the unit vector.

The meshing equation to be used is expressed by the dot product given by Eq. (11.95), which we can write here in the double form:

$$
\boldsymbol{n}_s \cdot \boldsymbol{v}_s^{(s2)} = \boldsymbol{N}_s \cdot \boldsymbol{v}_s^{(s2)} = 0,
\tag{14.15}
$$

where $\boldsymbol{n}_s$ and $\boldsymbol{N}_s$ are respectively the unit normal and normal to any point P of the shaper surface σ_s, and $\boldsymbol{v}_s^{(s2)}$ is the relative velocity, i.e. the velocity of a point on surface σ_s with respect to a point on surface σ_2. Of course, when these two points coincide with each other and the two surfaces σ_s and σ_2 are regular surfaces, they represent the common point of tangency of the two surfaces.

In this respect, it should be noted that the necessary condition of existence of surface σ_2 provides that, if it exists, it is in tangency with surface σ_s. Indeed, the sufficient condition of existence of surface σ_2 provides that it is not only in tangency with surface σ_s, but also a regular surface, i.e. a surface without singular points. It should also be emphasized that instantaneous line contacts are typical for cases where σ_s is the cutting tool surface and σ_2 is the generated surface, and this is our case.

Using the second of the Eq. (14.15) and remembering the expressions of the relative velocity, given in Sect. 11.7, we get the following meshing equation between the surfaces σ_s and σ_2, which is expressed as a dot product of cross product:

$$
\boldsymbol{n}_s \cdot \boldsymbol{v}_s^{(s2)} = f_{s2}(u_s, \vartheta_s, \varphi_s) = \left(\frac{\partial \boldsymbol{r}_2}{\partial u_s} \times \frac{\partial \boldsymbol{r}_2}{\partial \vartheta_s} \right) \cdot \frac{\partial \boldsymbol{r}_2}{\partial \varphi_s} = 0;
\tag{14.16}
$$

in fact, the product within the parenthesis is a cross product, while the other product is a dot product.

This last equation is called the meshing equation as it relates the curvilinear coordinates (u_s, ϑ_s) of surface σ_s with the generalized parameter of motion, φ_s. It represents the necessary condition of existence of the envelope to the family of surfaces defined by Eq. (14.11). If this equation is satisfied and the envelope really exists, we can proceed as indicate below. It should be noted, however, that vector $\partial r_2 / \partial \varphi_s$ has the same direction as the already defined vector $v_s^{(s2)}$. Therefore, substituting $v_s^{(s2)}$ instead of $\partial r_2 / \partial \varphi_s$ in the last expression of Eq. (14.16), and taking into account that the cross product represents the normal N_s, we get the first equality of the same equation, which is a dot product. The engineering approach to the problem discussed here uses this dot product, as it does not depend on the chosen coordinate system. Of course, normal N_s can be replaced by unit normal n_s.

Surface σ_2 of the face gear can be determined using Eq. (14.11), which describes the family of shaper surfaces σ_s in the coordinate system $O_2(x_2, y_2, z_2)$, and simultaneously the meshing Eq. (14.16). The matrix equation of $r_2(u_s, \vartheta_s, \varphi_s)$, given by Eq. (14.11), is expressed in a three-parameter form: the first two parameters that appear in this equation designate the surface parameter of the shaper, (u_s, ϑ_s), while the third parameter is the generalized parameter of motion, φ_s.

From the previous discussion it follows that the shaper surface σ_s is in line contact with the face gear tooth surface σ_2. Such a type of line contact is due to the fact that surface σ_2 is generated as the envelope of the shaper surface σ_s. We have already designated by L_{s2} the lines of tangency between surfaces σ_s and σ_2 in relative motion (see Fig. 14.12).

Surface σ_2 can be written in a two-parameter form, using the theorem of implicit function system existence (see Korn and Korn [18], Zalgaller [39]). To this end, following the procedure proposed by Zalgaller and Litvin [38], we assume that the meshing equation, $f_{s2}(u_s, \vartheta_s, \varphi_s) = 0$, is satisfied at a point $P^0(u_s^0, \vartheta_s^0, \varphi_s^0)$. We also assume that at this point the following condition is satisfied:

$$\frac{\partial f_{s2}}{\partial u_s} \neq 0. \tag{14.17}$$

Thus, equation $f_{s2}(u_s, \vartheta_s, \varphi_s) = 0$ can be solved in the neighborhood of point P^0 by a function of class C^1 as

$$u_s = u_s(\vartheta_s, \varphi_s) \in C^1 \tag{14.18}$$

and the face gear tooth surface, σ_2, can be determined locally as:

$$r_2(u_s(\vartheta_s, \varphi_s), \vartheta_s, \varphi_s) = R_2(\vartheta_s, \varphi_s). \tag{14.19}$$

Finally, lines of contact L_{s2} on the face gear tooth surface are determined by vector function $R_2(\vartheta_s, \varphi_s)$, taking $\varphi_s = const.$

14.6 Conditions of Non-undercut of Face Gear Tooth Surface

The appearance of singularities on the face gear tooth surface, σ_2, is the first symptom of its imminent undercut. As it is well known, in order to avoid the undercut of this surface, it must be a regular surface, i.e. a surface without singularities, which occur at points where a tangent plane does not exist, and therefore the local normal, N_2, becomes equal to zero. Now, if the generated surface σ_2 is expressed by Eq. (14.19), i.e. it is represented in a two-parameter form, the normal N_2 will be given by the following cross product:

$$N_2 = \frac{\partial R_2}{\partial \vartheta_s} \times \frac{\partial R_2}{\partial \varphi_s}. \tag{14.20}$$

Therefore, a singular point at this surface appears if at least one of the vectors in the cross product is equal to zero or if the two vectors are collinear. However, detection of the singular points of the face gear surface σ_2 through this method is very laborious and heavy from a computational point of view. For the determination of singularities of this surface, Litvin et al. [30] proposed a different approach which is described below. This other approach is preferable, since it is less costly computationally.

As it well known, the absolute velocity vector, v_i (with $i = s, 2$), of a point of contact between surfaces σ_s and σ_2, for each of the two members of the gear pair under consideration, is expressed as the vector sum of two components. The first is the component $v_{tr}^{(i)}$ in transfer motion with one of the two members of the gear pair. The second is the component $v_r^{(i)}$ in relative motion over the tooth surface σ_i (with $i = s, 2$). Since the point of contact belongs to both surfaces, for continuity the resulting velocities at a given point of contact must be the same for both gear members, so we get:

$$v_s = v_{tr}^{(s)} + v_r^{(s)} = v_2 = v_{tr}^{(2)} + v_r^{(2)}. \tag{14.21}$$

From this equation we obtain:

$$v_r^{(2)} = v_r^{(s)} + v_{tr}^{(s)} - v_{tr}^{(2)} = v_r^{(s)} + v_s^{(s2)} \tag{14.22}$$

where $v_s^{(s2)} = v_{tr}^{(s)} - v_{tr}^{(2)}$ is the sliding velocity at the point of tangency between the two surfaces, while $v_r^{(s)}$ is the velocity at a point of contact in its motion over surface σ_s of the shaper.

Litvin et al. [30] showed that the following condition must be satisfied at a singular point of the surface σ_2:

$$v_r^{(s)} + v_s^{(s2)} = 0. \tag{14.23}$$

This equation and the following differentiated meshing equation

$$\frac{d}{dt}[f_{s2}(u_s, \vartheta_s, \varphi_s)] = 0 \tag{14.24}$$

allow us to determine a line L_s on surface σ_s that generates singular points on surface σ_2. Therefore, limiting σ_s with line L_s, we can avoid the appearance of singular points on surface σ_2. Thus, it is necessary to determine this line. To this end, we first write in a different form Eq. (14.23), from which we get:

$$\frac{\partial \boldsymbol{r}_s}{\partial u_s}\frac{du_s}{dt} + \frac{\partial \boldsymbol{r}_s}{\partial \vartheta_s}\frac{d\vartheta_s}{dt} = -v_s^{(s2)} \tag{14.25}$$

where $\boldsymbol{r}_s(u_s, \vartheta_s)$ is given by Eq. (14.8). It is to be noted that $(\partial \boldsymbol{r}_s/\partial u_s)$, $(\partial \boldsymbol{r}_s/\partial \vartheta_s)$, and $v_s^{(s2)}$ are three- or two-dimensional vectors for spatial and planar gears, respectively, and are expressed in coordinate system $O_s(x_s, y_s, z_s)$.

Thus, from Eq. (14.24) we get:

$$\frac{\partial f_{s2}}{\partial u_s}\frac{du_s}{dt} + \frac{\partial f_{s2}}{\partial \vartheta_s}\frac{d\theta_s}{dt} = -\frac{\partial f_{s2}}{\partial \varphi_s}\frac{d\varphi_s}{dt}. \tag{14.26}$$

Equations (14.25) and (14.26) are a system of four linear equations in the two unknowns (du_s/dt) and $(d\vartheta_s/dt)$, since $(d\varphi_s/dt)$ is considered as given. This system has a well-defined solution for the unknowns only if the matrix:

$$\boldsymbol{M} = \begin{bmatrix} \frac{\partial \boldsymbol{r}_s}{\partial u_s} & \frac{\partial \boldsymbol{r}_s}{\partial \vartheta_s} & -v_s^{(s2)} \\ \frac{\partial f_{s2}}{\partial u_s} & \frac{\partial f_{s2}}{\partial \vartheta_s} & -\frac{\partial f_{s2}}{\partial \varphi_s}\frac{d\varphi_s}{dt} \end{bmatrix} \tag{14.27}$$

has the rank $r = 2$. This yields:

$$\Delta_1 = \begin{bmatrix} \frac{\partial x_s}{\partial u_s} & \frac{\partial x_s}{\partial \vartheta_s} & -v_{xs}^{(s2)} \\ \frac{\partial y_s}{\partial u_s} & \frac{\partial y_s}{\partial \vartheta_s} & -v_{ys}^{(s2)} \\ \frac{\partial f_{s2}}{\partial u_s} & \frac{\partial f_{s2}}{\partial \vartheta_s} & -\frac{\partial f_{s2}}{\partial \varphi_s}\frac{d\varphi_s}{dt} \end{bmatrix} = 0 \tag{14.28}$$

$$\Delta_2 = \begin{bmatrix} \frac{\partial x_s}{\partial u_s} & \frac{\partial x_s}{\partial \vartheta_s} & -v_{xs}^{(s2)} \\ \frac{\partial z_s}{\partial u_s} & \frac{\partial z_s}{\partial \vartheta_s} & -v_{zs}^{(s2)} \\ \frac{\partial f_{s2}}{\partial u_s} & \frac{\partial f_{s2}}{\partial \vartheta_s} & -\frac{\partial f_{s2}}{\partial \varphi_s}\frac{d\varphi_s}{dt} \end{bmatrix} = 0 \tag{14.29}$$

$$\Delta_3 = \begin{bmatrix} \dfrac{\partial y_s}{\partial u_s} & \dfrac{\partial y_s}{\partial \vartheta_s} & -v_{ys}^{(s2)} \\[2mm] \dfrac{\partial z_s}{\partial u_s} & \dfrac{\partial z_s}{\partial \vartheta_s} & -v_{zs}^{(s2)} \\[2mm] \dfrac{\partial f_{s2}}{\partial u_s} & \dfrac{\partial f_{s2}}{\partial \vartheta_s} & -\dfrac{\partial f_{s2}}{\partial \varphi_s}\dfrac{d\varphi_s}{dt} \end{bmatrix} = 0 \qquad (14.30)$$

$$\Delta_4 = \begin{bmatrix} \dfrac{\partial x_s}{\partial u_s} & \dfrac{\partial x_s}{\partial \vartheta_s} & -v_{xs}^{(s2)} \\[2mm] \dfrac{\partial y_s}{\partial u_s} & \dfrac{\partial y_s}{\partial \vartheta_s} & -v_{ys}^{(s2)} \\[2mm] \dfrac{\partial z_s}{\partial u_s} & \dfrac{\partial z_s}{\partial \vartheta_s} & -v_{zs}^{(s2)} \end{bmatrix} = 0 \qquad (14.31)$$

where Δ_i (with $i = 1, 2, 3$) are three determinants obtained from matrix M, while the fourth determinant Δ_4, when it is equal to zero, i.e. $\Delta_4 = 0$, expresses the meshing equation. In fact, equality (14.31) can be expressed in the form:

$$\left(\frac{\partial \boldsymbol{r}_s}{\partial u_s} \times \frac{\partial \boldsymbol{r}_s}{\partial \vartheta_s} \right) \cdot \boldsymbol{v}_s^{(s2)} = \boldsymbol{N}_s \cdot \boldsymbol{v}_s^{(s2)} = f_{s2}(u_s, \vartheta_s, \varphi_s) = 0, \qquad (14.32)$$

which is just the meshing equation. Therefore, this equation cannot be used to determine the condition of singularity for surface σ_2. To this end, only Eqs. (14.28) to (14.30) can be used.

The simultaneous fulfillment condition of the system consisting of the three Eqs. (14.28) to (14.30) can be expressed by the following vector equation, which can easily be obtained by combining the same three equations:

$$\boldsymbol{m} = \frac{\partial f_{s2}}{\partial u_s}\left[\frac{\partial \boldsymbol{r}_s}{\partial \vartheta_s} \times \boldsymbol{v}_s^{(s2)} \right] - \frac{\partial f_{s2}}{\partial u_s}\left[\frac{\partial \boldsymbol{r}_s}{\partial u_s} \times \boldsymbol{v}_s^{(s2)} \right] + \frac{\partial f_{s2}}{\partial \varphi_s}\frac{d\varphi_s}{dt}\left[\frac{\partial \boldsymbol{r}_s}{\partial u_s} \times \frac{\partial \boldsymbol{r}_s}{\partial \vartheta_s} \right] = 0. \quad (14.33)$$

This vector equation is satisfied when both of the following conditions are met: at least one Eqs. (14.28) to (14.30) is satisfied, and vector $\boldsymbol{m}$ is not perpendicular to any of the axes of the coordinate system $O_s(x_s, y_s, z_s)$.

A sufficient condition for detecting the singularities of the generated surface σ_2, and therefore the solutions to avoid undercut, is represented by the following simple equation:

$$\Delta_1^2 + \Delta_2^2 + \Delta_3^2 = F_{s2}(u_s, \vartheta_s, \varphi_s) = 0. \qquad (14.34)$$

In conclusion, the following three equations:

$$\begin{aligned} \boldsymbol{r}_s &= \boldsymbol{r}_s(u_s, \vartheta_s) \\ f_{s2}(u_s, \vartheta_s, \varphi_s) &= 0 \\ F_{s2}(u_s, \vartheta_s, \varphi_s) &= 0 \end{aligned} \qquad (14.35)$$

allow us to determine a limit line, L_s, on the generation tooth surface, σ_s, of the shaper, which generates singularities on the generated surface, σ_2. The detection of this limit line allows to limit the values of the three parameters u_s, ϑ_s, and φ_s, whose overcoming generates singularities of the face gear tooth surface, σ_2. From the point of view of the face gear design, it is therefore important to choose the appropriate settings for surface σ_s that generates the surface σ_2.

The procedure for detecting the aforementioned limit line implies, first, that the plane (u_s, ϑ_s) of parameters of the shaper tooth surface is considered, and that on this plane, the lines of tangency L_{s2} between shaper surface σ_s and face gear surface σ_2 are plotted, as well as the line M of the points $(u_s, \vartheta_s, \varphi_s)$ corresponding to the singular points of the surface σ_2. To do this first step, we use the second and third of Eq. (14.35). The result of this first step is therefore a representation of the plane (u_s, ϑ_s), where lines L_{s2} and line M are plotted, as shown in Fig. 14.15. The second decisive step is to completely remove line M of the space of parameters (u_s, ϑ_s). Since line M designates the shaper parameter measured along its axis, to avoid singularities of the surface σ_2 it is sufficient to eliminate the parts of the L_{s2}-lines to the left of the vertical line $u_s^* - K$, where u_s^* corresponds to the addendum of the shaper and K is the limiting point on the parameter plane defined by the coordinates (u_s^*, ϑ_s^*). For generation of face gear by a shaper, as the one shown in Fig. 14.15, singularities are avoided in the parameter sub-area for which $\vartheta_s < \vartheta_s^*$ and $u_s > u_s^*$.

Based on the above considerations, we can now determine the magnitude of R_i that will avoid singularities of the face gear tooth surface σ_2. Instead, quantity R_e determines the area free of pointed teeth of the same surface σ_2. For the calculation of R_i, we must first consider the limiting point $K(u_s^*, \vartheta_s^*)$ of the line of singularities, which belongs to the addendum of the shaper. The parameter ϑ_s^* of this limiting point is determined by the following equation:

$$\vartheta_s^* = \frac{\sqrt{r_{as}^2 - r_{bs}^2}}{r_{bs}} \tag{14.36}$$

where r_{as} and r_{bs} are respectively the radii of addendum circle and base circle of the shaper. We must then take into account the following fact, which is confirmed by the investigations: to determine singularities of surface σ_2, rather than solving Eq. (14.34), just take $\Delta_2 = 0$ or $\Delta_3 = 0$. The equations $\Delta_2 = 0$ or $\Delta_3 = 0$ include partial derivatives $\partial z_s / \partial u_s$ and $\partial z_s / \partial \vartheta_s$, while equation $\Delta_1 = 0$ does not include these partial derivatives. Using, for example, the equation $\Delta_2 = 0$ as well as the meshing equation $f_{s2}(u_s, \vartheta_s, \varphi_s) = 0$, we get two equations in the two unknowns u_s and φ_s. The parameter u_s thus obtained determines the sought-for magnitude of R_i.

Fig. 14.15 Sub-area of space of parameter (u_s, θ_s) free of singularities for the face gear tooth surface σ_2 obtained by an involute shaper

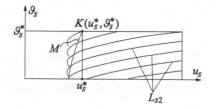

14.7 Conditions to Avoid Face Gear Pointed Teeth, and Fillet Surface

Even for face gears, it is necessary to prevent the tooth thickness on the tooth top land becoming zero, so the teeth become pointed teeth. To break the danger of the pointed teeth, it is necessary to locate the area where it can manifest, considering the intersection of the two opposite flank surfaces at the tooth top land. To locate this area, special computing programs can be used. Here we will use an approximate method that guarantees sufficiently reliable results for face gears generated by an involute shaper.

With reference to Figs. 14.14a and 14.16, we indicate with $O_m Q \equiv t_{s2}$ the instantaneous axis of rotation in the generation process, which sees the generation shaper and the to-be-generated face gear respectively rotating about their axes, z_s and z_2. Consider the two transverse sections of the shaper obtained as intersections with the two transverse planes π_1 and π_2, which are perpendicular to the z_s-axis and pass respectively through the pitch point, C, and a point, Q, located on the instantaneous axis of rotation, t_{s2} (see Fig. 14.16). We have to determine the location of π_2-plane where the two opposite tooth flanks of the face gear intersect.

Figures 14.17a and 14.17b show respectively the tooth profiles of the shaper and face gear in planes π_1 and π_2. The point of intersection of the two opposite transverse profiles of the face gear in plane π_2 is indicated by A (Fig. 14.17b). This point must be located on the addendum line of the face gear, so its location on the x'_n-axis is given by $(r_s - m)$. It is now necessary to determine the magnitude of R_e, which is defined by distance Δl between the planes π_1 and π_2 (Fig. 14.16). This last figure and Fig. 14.17b highlight the procedure for calculating Δl and R_e.

It is first necessary to determine the pressure angle α of the pointed teeth (Fig. 14.17b). For this purpose, we use the following vector equation:

$$\overline{O'_n A} = \overline{O'_n T} + \overline{TP} + \overline{PA}, \tag{14.37}$$

where P is the point of tangency of profiles of the shaper and face gear in plane π_2. Let's denote then with λ_s and $r_{bs}\vartheta_s$ the absolute values of vectors $\overline{PA}$ and $\overline{TP}$, that is $|\overline{PA}| = \lambda_s$, and $|\overline{TP}| = r_{bs}\vartheta_s$. Moreover, the absolute value of the vector $\overline{O'_n A}$ is given by (Fig. 14.17a):

Fig. 14.16 Limiting tooth dimension R_i and R_e of the face gear

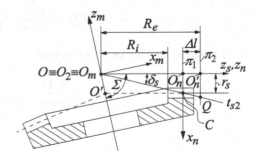

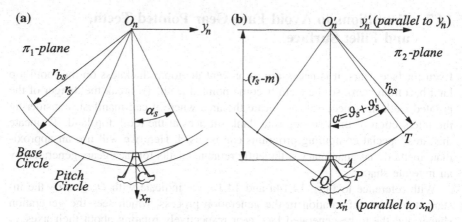

Fig. 14.17 Transverse tooth profiles of shaper and face gear: **a** in plane π_1; **b** in plane π_2

$$\overline{O'_n A} = r_s - m, \tag{14.38}$$

where m is the module.

Vector Eq. (14.37), by projections on the x'_n- and y'_n- axes, gives the following equation system, consisting of two scalar equations in the two unknowns $\alpha = \left(\vartheta_s + \vartheta'_s\right)$ and λ_s:

$$r_{bs}(\cos\alpha + \vartheta_s \sin\alpha) - \lambda_s \cos\alpha = \frac{z_s - 2}{2} m$$
$$r_{bs}(\sin\alpha - \vartheta_s \cos\alpha) - \lambda_s \sin\alpha = 0, \tag{14.39}$$

where $r_{bs} = (z_s/2)m \cos\alpha_s$ and $\vartheta_s = \alpha - \vartheta'_s$, with ϑ'_s given by Eq. (14.7).

Eliminating λ_s, from equation system (14.39) we obtain the following nonlinear equation:

$$\alpha - \sin\alpha \frac{z_s - 2}{z_s \cos\alpha_s} = \frac{\pi}{2z_s} - \mathrm{inv}\alpha_s, \tag{14.40}$$

from whose solution we get the sought-for pressure angle α. From Fig. 14.17b we then get the following relationship:

$$O'_n Q = \frac{r_{bs}}{\cos\alpha} = \frac{z_s \cos\alpha_s}{2 \cos\alpha} m. \tag{14.41}$$

Finally, from Fig. 14.16 we obtain:

$$R_e = \frac{O'_n Q}{\tan\delta_s} = \frac{z_s \cos\alpha_s}{2 \cos\alpha \tan\delta_s} m. \tag{14.42}$$

Based on the determined values of R_i and R_e, it becomes possible to design a face gear that is free from undercut and pointed teeth.

The fillet surface depends on the shape of the shaper top land, which can be configured with sharp edge tips (points A and A' in Fig. 14.8 and addendum generatrixes in Fig. 14.12a corresponding to the intersections of the addendum cylinder generatrixes with the involute opposite tooth profiles). It can be also configured with rounded tips, characterized by certain radii of curvature, which may be ugual or different from one another, as shown in the same Fig. 14.8.

Based on Fig. 14.13, we can express the addendum generatrix (see Fig. 14.12a) in coordinate system $O_s(x_s, y_s, z_s)$ by the vector function $r_s(u_s, \vartheta_s^*)$, where ϑ_s^* is given by Eq. (14.36), while the radius r_{as} of addendum circle, which appears in this last equation, is given by the following relationship:

$$r_{as} = r_s + 1.25 \, m = \frac{(z_s + 2.5)}{2} m. \tag{14.43}$$

In coordinate system $O_2(x_2, y_2, z_2)$, the fillet surface is expressed by the equation:

$$r_2(u_s, \varphi_s) = M_{2s}(\varphi_s) r_s(u_s, \vartheta_s^*), \tag{14.44}$$

where $M_{2s} = M_{2m} M_{mn} M_{ns}(\varphi_s)$.

At the end of Sect. 14.3, we emphasized the importance of an appropriate choice of the dimensionless coefficient $k = (R_e - R_i)/m = b/m$. The guidelines we had given then were based on results of investigations concerning the influence that this coefficient exerts on the structure of the face gear teeth. The above investigation assumed that the outer radius R_e was known, as previously determined by imposing conditions that could guarantee the absence of *pointing*, i.e. pointed teeth. Therefore, once R_e is known, we can eliminate the portion of the tooth where the fillet exists, by increasing the inner radius R_i (Fig. 14.16). This means that the coefficient k will have to be decreased (see Fig. 14.9). However, choosing a sufficient value of coefficient k, allows us to obtain a more uniform structure, eliminating the weaker part of the face gear tooth.

14.8 Meshing and Contact Condition Between Tooth Surfaces of a Face Gear Pair

The technological development and the improvement of the quality of a face gear pair necessarily entail an in-depth knowledge of the meshing and bearing contact between the tooth flank surfaces of the pinion and face gear, which make up a face gear pair. This in-depth knowledge can be acquired through expensive experiment tests on prototypes, or more often by computerized simulation techniques of meshing and bearing contact between the tooth surfaces, in the actual operating conditions of the gear drive under consideration.

These simulation techniques undoubtedly have the advantage of being able to be used in the design stage of the face gear pair in order to optimize design choices in relation to the goals to be achieved. To do so, Tooth Contact Analysis (TCA) programs are available, which are based on the theoretical considerations described below. With these programs, several goals can be achieved, such as:

- definition and location of the bearing contact as a series of instantaneous contact ellipses, in order to eliminate the risk of localized edge contacts;
- definition of the paths of contact on tooth flank surfaces;
- evaluation of transmission errors due to misalignment of the face gear members.

This type of analysis involves first the definition of the continuous tangency conditions between the tooth surfaces σ_1 and σ_2 of the pinion and face gear. To this end, we introduce three coordinate systems, $O_1(x_1, y_1, z_1)$, $O_2(x_2, y_2, z_2)$ and $O_f(x_f, y_f, z_f)$, which are rigidly connected with the pinion, face gear and frame of the face gear drive respectively. Coordinate systems $O_1(x_1, y_1, z_1)$ and $O_2(x_2, y_2, z_2)$ are therefore mobile systems, while coordinate system $O_f(x_f, y_f, z_f)$ is a fixed system. It should be noted that, since in this section we are interested in determining the meshing and contact conditions between tooth surfaces of a face gear pair in the actual assembly conditions, the fixed coordinate system $O_f(x_f, y_f, z_f)$ is unique.

We then introduce an additional fixed-coordinate system, $O_a(x_a, y_a, z_a)$, with which we simulate the misalignment. In addition, we introduce two more auxiliary coordinate system, $O_b(x_b, y_b, z_b)$ and $O_c(x_c, y_c, z_c)$, which are used not only to simulate the misalignment of the face gear, but also to better define the complex geometry of the problem being discussed. These six coordinate systems are shown in Fig. 14.18a–d.

All misalignments are referred to the face gear. Quantities Δa, b, and $b \cot \Sigma$ determine the position of origin O_a of the $O_a(x_a, y_a, z_a)$-system with respect to origin O_f of the $O_f(x_f, y_f, z_f)$-system (Fig. 14.18c). Quantity Δa represents the shortest distance between the axes of pinion and face gear, when they do not intersect, but are crossed axes (Fig. 14.18c); b and $b \cot \Sigma$ represent the position and orientation of the coordinate system $O_a(x_a, y_a, z_a)$ with respect to the coordinate system $O_f(x_f, y_f, z_f)$ (Fig. 14.18c). The misaligned face gear rotates about the z_c-axis (Fig. 14.18b). The position of coordinate system $O_c(x_c, y_c, z_c)$ with respect to coordinate system $O_b(x_b, y_b, z_b)$ simulates the axial displacement Δc of the face gear (Fig. 14.18d). The orientation of coordinate system $O_b(x_b, y_b, z_b)$ with respect to coordinate system $O_a(x_a, y_a, z_a)$ simulates the actual (or simulated) crossing angle $\Sigma_a = \Sigma + \Delta \Sigma$, where Σ is the shaft angle, while $\Delta \Sigma$ is its variation due to misalignment (Fig. 14.18d).

In the coordinate systems $O_1(x_1, y_1, z_1)$ and $O_2(x_2, y_2, z_2)$, the tooth flank surfaces σ_1 and σ_2 are represented by the two following vector functions of class C^2:

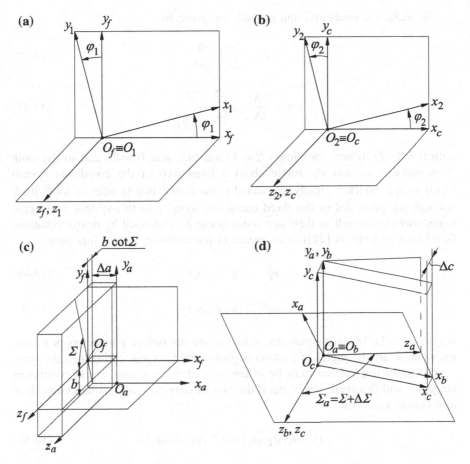

Fig. 14.18 Coordinate systems used for simulation of meshing: **a** coordinate system O_1 and O_f; **b** coordinate systems O_2 and O_c; **c** coordinate systems O_a and O_f; **d** coordinate systems O_a, O_b, and O_c

$$r_i(u_i, \vartheta_i) \in C^2, \tag{14.45}$$

with ($i = 1, 2$) and $(u_i, \vartheta_i) \in A_i$ (i.e. the elements u_i and ϑ_i belong to a set A_i), under the condition that singular points do not exist, so the following condition must be meet:

$$\frac{\partial r_i}{\partial u_i} \times \frac{\partial r_i}{\partial \vartheta_i} \neq 0. \tag{14.46}$$

This last cross vector is different from zero only when the pair of vectors $r_u = \partial r_i/\partial u_i$ and $r_\vartheta = \partial r_i/\partial \vartheta_i$, which are tangent to the u-lines and ϑ-lines respectively, are both different from zero, and are not collinear.

The surface normals and unit normals are given by:

$$N_i = \frac{\partial r_i}{\partial u_i} \times \frac{\partial r_i}{\partial \vartheta_i} \qquad (14.47)$$

$$n_i = \frac{N_i}{|N_i|} = \frac{\frac{\partial r_i}{\partial u_i} \times \frac{\partial r_i}{\partial \vartheta_i}}{\left|\frac{\partial r_i}{\partial u_i} \times \frac{\partial r_i}{\partial \vartheta_i}\right|}, \qquad (14.48)$$

with $(i = 1, 2)$. If now the pinion $(i = 1)$ and face gear $(i = 2)$, and so the tooth flank surfaces σ_1 and σ_2, rotate about a fixed axis of the coordinate system $O_f(x_f, y_f, z_f)$, which is rigidly connected to the frame, two families of tooth flank surfaces are generated in this fixed coordinate system. In this system, these generated surfaces as well as their unit normals can be expressed by matrix equations (see Litvin and Fuents [24]) or by means of the following vector functions:

$$r_f^{(i)} = (u_i, \vartheta_i, \varphi_i) \qquad (14.49)$$

$$n_f^{(i)} = (u_i, \vartheta_i, \varphi_i) \qquad (14.50)$$

with $(i = 1, 2)$. In these equations, u_i and ϑ_i are the surface parameters of σ_1 and σ_2, while φ_i are the angles of rotation of pinion and face gear during their meshing.

The contacting surface must be in tangency at every instant, i.e. in continuous tangency, and this condition is met if their position vectors and normals coincide at any instant, i.e. when:

$$r_f^{(1)}(u_1, \vartheta_1, \varphi_1) = r_f^{(2)}(u_2, \vartheta_2, \varphi_2) \qquad (14.51)$$

$$n_f^{(1)}(u_1, \vartheta_1, \varphi_1) = n_f^{(2)}(u_2, \vartheta_2, \varphi_2). \qquad (14.52)$$

Vector Eq. (14.51) gives three independent scalar equations, while vector Eq. (14.52) yields only two independent scalar equations, since the absolute values of the unit normals coincide, that is:

$$\left|n_f^{(1)}\right| = \left|n_f^{(2)}\right| = 1. \qquad (14.53)$$

Therefore, vector Eqs. (14.51) and (14.52) yield a system of five independent scalar equations in six unknowns $u_1, \vartheta_1, \varphi_1, u_2, \vartheta_2, \varphi_2$. This system of equations can be expressed as:

$$f_i(u_1, \vartheta_1, \varphi_1, u_2, \vartheta_2, \varphi_2) = 0 \qquad (14.54)$$

with $f_i \in C^1$ and $(i = 1, 2, 3, 4, 5)$.

One of the six quantities that come into play can be chosen as input parameter. Here we choose φ_1 as an example. The condition that surface σ_1 and σ_2 are in point contact leads to the following inequality:

$$Df = \frac{\partial(f_1, f_2, f_3, f_4, f_5)}{\partial(u_1, \vartheta_1, u_2, \vartheta_2, \varphi_2)} \neq 0, \tag{14.55}$$

whose left-hand side represents the *Jacobian matrix*, i.e. the matrix of all first-order partial derivatives of the functions f_i with respect to the five unknowns. In this case, the Jacobian matrix is a square matrix, for which both the matrix and its determinant are called simply *Jacobian* (see Buzano [4], Hirsch et al. [14], Tricomi [36]).

Therefore, the solution of system of Eq. (14.54) can be expressed by the functions:

$$f_i[u_1(\varphi_1), \vartheta_1(\varphi_1), u_2(\varphi_1), \vartheta_2(\varphi_1), \varphi_2(\varphi_1)] \in C^1. \tag{14.56}$$

In accordance with the theorem of implicit function system existence (see Korn and Korn [18], Miller [30]), we can state that functions (14.56) exist in the neighborhood of a point

$$P^{(0)}\left(u_1^{(0)}, \vartheta_1^{(0)}, \varphi_1^{(0)}, u_2^{(0)}, \vartheta_2^{(0)}, \varphi_2^{(0)}\right), \tag{14.57}$$

if the following requirements are met:

- functions $[f_1, f_2, f_3, f_4, f_5] \in C^1$, i.e. they are of class C^1;
- equations (14.54) are satisfied at point $P^{(0)}$;
- the following *Jacobian determinant* is different from zero, that is:

$$Df = \frac{\partial(f_1, f_2, f_3, f_4, f_5)}{\partial(u_1, \vartheta_1, u_2, \vartheta_2, \varphi_2)} = \begin{vmatrix} \frac{\partial f_1}{\partial u_1} & \frac{\partial f_1}{\partial \vartheta_1} & \frac{\partial f_1}{\partial u_2} & \frac{\partial f_1}{\partial \vartheta_2} & \frac{\partial f_1}{\partial \varphi_2} \\ \vdots & \vdots & \vdots & \vdots & \vdots \\ \frac{\partial f_5}{\partial u_1} & \frac{\partial f_5}{\partial \vartheta_1} & \frac{\partial f_5}{\partial u_2} & \frac{\partial f_5}{\partial \vartheta_2} & \frac{\partial f_5}{\partial \varphi_2} \end{vmatrix} \neq 0. \tag{14.58}$$

When this last requirement is met, functions f_i have locally, in the neighborhood of point $P^{(0)}$, their inverse functions that are differentiable if and only if the Jacobian determinant is nonzero at $P^{(0)}$.

Functions (14.56) provide complete information on the meshing condition between the pinion and face gear, which are in point contact. The function $\varphi_2 = \varphi_2(\varphi_1)$ represents the *law of motion*, as it correlates the angles of rotation of the face gear pair members about their axes. The solution of system of Eqs. (14.51) and (14.52) by means of functions (14.56) is an iterative procedure. This procedure requires, as a first step, the identification of the set of parameters characterizing the point $P^{(0)}$, and satisfying locally the system of Eqs. (14.51) and (14.52).

The aforementioned solution, by means of functions (14.56), allows us to reach important goals. The first goal is to define the law of motion, or transmission function, $\varphi_2 = \varphi_2(\varphi_1)$, and function of transmission error defined by the equation:

$$\Delta\varphi_2(\varphi_1) = \varphi_2(\varphi_1) - (z_1/z_2)\varphi_1. \qquad (14.59)$$

The second goal is to determine the paths of contact on surface σ_1 and σ_2, which are given, respectively, by the equations:

$$r_1(u_1(\varphi_1), \vartheta_1(\varphi_1)) \qquad (14.60)$$

$$r_2(u_2(\varphi_1), \vartheta_2(\varphi_1)). \qquad (14.61)$$

It is not the case here to go further into the theoretical discussion described above. Instead, we should focus on some results obtained by Litvin and Fuentes [24], who applied the aforementioned theory to a face gear generated by an involute shaper with ideal theoretical cutting profiles. These results show an important shift in bearing contact, due to alignment errors. This bearing contact is oriented across the tooth surface, and is sensitive to the variation $\Delta\Sigma$ of the shaft angle. Therefore, the risk of dangerous edge contact may occur. The sensitivity to the shaft angle variation can be compensated by an axial correction Δc (Fig. 14.18d) of the face gear during assembly. Faced with these disadvantages, the geometry of the face gear pair obtained with the above-mentioned generation process has the considerable advantage of not causing transmission errors, whatever the misalignment.

The same Litvin and Fuentes [24] have also proposed and studied other face gear pair geometries, obtained by a process that generates pinion and shaper by means of rack-cutters with a modified profile compared to a reference rack-cutter with straight-line profile. The face gear pair thus obtained are characterized by a lengthwise orientation of the bearing contact, which removes or decreases the danger of edge contact, and results in an appreciable reduction in stress. However, they have the disadvantage of being still sensitive to misalignment, which is moreover accompanied with transmission errors. For more information on this topic, we refer the reader to Litvin et al. [25, 26].

14.9   Generation and Grinding of Face Gears by Worms

14.9.1   General Considerations

In Sect. 14.2 we introduced the Litvin's method that, as we have already pointed out, can be used for generation or grinding of face gears, respectively by hobs (or generation worms) or grinding worms. We have already highlighted the advantages of this method in comparison with the conventional generation method of face gears, by means of a spur involute shaper and a manufacturing process that simulates the meshing of this shaper and face gear to be generated.

In this section we want to describe the theoretical bases of the Litvin's method [22, 23]. This method solves the following design problems of the generating and grinding worms: (i), determination of the shaft angle between axes of the imaginary shaper and actual worm, and definition of the worm thread surfaces; (ii), choices to avoid singularities of these worm surfaces, and definition of their dressing process.

The surfaces involved in the problem we are discussing are those of the imaginary shaper, worm and face gear, which are here indicated with σ_s, σ_w and σ_2 respectively. Figure 14.5 shows the simultaneous meshing of these three surfaces. We here assume that surface σ_s has been generated as the envelope to the family of an ideal rack cutter having straight tooth profiles. Therefore, surface σ_s is the involute surface of a spur gear, which is represented by the vector function $r_s(u_s, \vartheta_s)$ given by Eq. (14.8). It should be noted that if we want to take into account both the profiles shown in Fig. 14.13, in the second row of the matrix in Eq. (14.8) we must introduce the signs that appear in the second Eq. (14.6). Surfaces σ_w and σ_2 are generated as the envelope to the family of shaper surfaces σ_s.

We must first determine the shaft angle between axes of imaginary shaper and worm. In Sect. 14.2 we have highlighted that the installation of the generating or grinding worm on the face gear to be generated or grinded must take into account the lead angle, γ_w, of its thread, so the worm and face gear axes are skewed or crossed axes. Consequently, the installation of the worm with respect to the shaper must take into account this fact, so shaper and worm axes will also have crossed axes, with the same shaft angle of the gear pair formed by worm and face gear.

To determine this shaft angle between axes of shaper and worm, and to solve other problems, such as those relating to tangency of surfaces σ_s and σ_2, and generation of the face gear by the worm, we use the coordinate systems shown in Fig. 14.19, namely:

- mobile coordinate system $O_s(x_s, y_s, z_s)$, rigidly connected to the shaper, whose z_s-axis coincides with the axis of rotation of the shaper;
- mobile coordinate system $O_w(x_w, y_w, z_w)$, rigidly connected to the worm, whose z_w-axis coincides with the axis of rotation of the worm;
- fixed coordinate system $O_a(x_a, y_a, z_a)$, rigidly connected to the frame, and having the same origin and the same z-axis of the O_s-system, that is $O_a \equiv O_s$ and $z_a \equiv z_s$;
- fixed coordinate systems $O_b(x_b, y_b, z_b)$ and $O_c(x_c, y_c, z_c)$, rigidly connected to the frame, both of which have the same origin of the O_w-system (i.e. $O_b \equiv O_c \equiv O_w$), and with z_b-axis parallel to the axes z_a and z_s, and z_c-axis coinciding with the z_w-axis;
- fixed coordinate system $O_m(x_m, y_m, z_m)$ rigidly connected to the frame as defined in the Sect. 14.5 and also shown in Fig. 14.14.

Axes z_s and z_w of shaper and worm are crossed axes, and their positive directions form the shaft angle $\Sigma = (\pi/2) \pm \gamma_w$, where γ_w is the lead angle of the worm

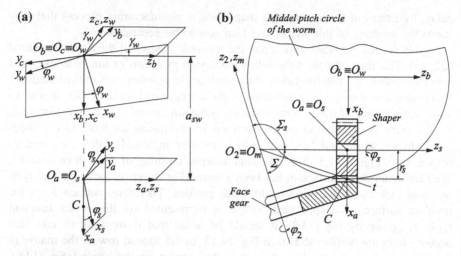

Fig. 14.19 Coordinate systems to: **a** define the position of the worm with respect to the shaper; **b** impose the simultaneous meshing of surface σ_s, σ_w and σ_2, and generate the face gear by the worm

thread, given by Eq. (14.1). Just remember the installation of the same worm on the face gear described in Sect. 14.2. It is to be noted that here we depart slightly from the definition of ISO 1122-1[15], according to which the shaft angle is the smallest angle through which one of the axes of the crossed gear pair must be swiveled so that the axes are parallel and their directions of rotation are opposite. The upper and lower sign in expression above corresponds to the application of a right-hand or left-hand worm. The center distance, a_{sw}, is the shortest distance between axes z_s and z_w.

14.9.2   Determination of the Shaft Angle

To determine the shaft angle $\Sigma = (\pi/2) \pm \gamma_w$, it is necessary to first calculate the lead angle of the worm thread, γ_w. The determination of this quantity necessarily implies that surfaces σ_s, σ_w and σ_2 are in simultaneous tangency. To simplify as much as possible this determination, in a way equivalent to simultaneous tangency of these three surfaces, we can consider the tangency of those surfaces that are equidistant to σ_s, σ_w and σ_2 and pass through point C that, in O_a-system, is expressed by the following position vector:

$$\boldsymbol{r}_a^{(C)} = [r_s \quad 0 \quad 0]^T, \tag{14.62}$$

where r_s is the radius of the pitch circle of the shaper.

Let's first consider the tangency at point C of surfaces that are equidistant to σ_s and σ_2. Axes of rotation z_s and z_2 of shaper and face gear intersect (see

Fig. 14.19b), so there is an instantaneous axis of rotation, $O_m t$, which passes through the point of intersection, $O_m \equiv O_2$, of the two axes of rotation. We choose point C on this instantaneous axis of rotation, as shown in Fig. 14.19b. The tangency of the surfaces σ_s and σ_2 is assured and proven because the normals to σ_s-surface pass through point C (Figs. 14.13 and 14.19b). Tangency of surfaces σ_s and σ_w at point C occurs if the following meshing condition is satisfied at point C:

$$N^{(s)} \cdot v^{(sw)} = 0, \tag{14.63}$$

where $N^{(s)}$ is the normal to σ_s-surface, while $v^{(sw)} = v^{(s)} - v^{(w)}$ is the relative velocity between the two surfaces at point C. Of course, $v^{(s)}$ and $v^{(w)}$ are the velocity vectors of point C of the shaper and worm. By developing Eq. (14.62), we get the following expression of the lead angle of the worm thread:

$$\gamma_w = \arcsin \frac{r_s}{z_s(a_{ws} + r_s)}, \tag{14.64}$$

where z_s and r_s are respectively the number of teeth and radius of pitch circle of the shaper, while a_{ws} is the shortest distance between axes of shaper and worm. The magnitude of the center distance, a_{sw}, affects the dimensions of the generation or grinding worm as well as the avoidance of surface singularities of the worm. Figure 14.20 shows schematically the meshing between worm and shaper.

14.9.3   Determination of σ_w-Surface

Figure 14.5 shows the imaginary pinion, worm and face gear in simultaneous meshing, during which all three surfaces σ_s, σ_w and σ_2 are involved. For our purpose, the σ_s-surface of the imaginary shaper is here considered as a given generation surface. Instead, surfaces σ_2 and σ_w are generated as the envelopes to the family of σ_s-surface of the shaper.

Now, the generation surface σ_s, whose geometry is known, is given by the vector function (14.8). Even surface σ_2 of the face gear has already been determined, simultaneously using Eq. (14.11) describing the family of shaper surfaces σ_s in the coordinate system $O_2(x_2, y_2, z_2)$, and the meshing equation given by Eq. (14.16). Therefore, only the surface σ_w of the worm remains to be determined. To determine this surface in coordinate system $O_w(x_w, y_w, z_w)$, we simultaneously use the following two equations [24, 25]:

$$r_w(u_s, \vartheta_s, \varphi_s) = M_{ws}(\varphi_s) r_s(u_s, \vartheta_s) \tag{14.65}$$

$$\left(\frac{\partial r_w}{\partial u_s} \times \frac{\partial r_w}{\partial \vartheta_s} \right) \cdot \frac{\partial r_w}{\partial \varphi_s} = f_{ws}(u_s, \vartheta_s, \varphi_s) = 0, \tag{14.66}$$

Fig. 14.20 Schematic drawing of meshing between worm and shaper

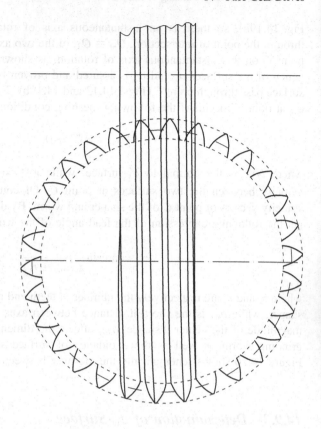

where: vector function $r_w(u_s, \vartheta_s, \varphi_s)$ is the family of surfaces σ_s of shaper expressed in coordinate system $O_w(x_w, y_w, z_w)$; matrix $M_{ws}(\varphi_s)$ provides coordinate transformation from system $O_s(x_s, y_s, z_s)$ to system $O_w(x_w, y_w, z_w)$; the dot product of cross product given by Eq. (14.66) is the meshing equation between surfaces σ_s and σ_w. The vector function $r_w(u_s, \theta_s, \varphi_s)$, given by Eq. (14.65), is expressed in three-parameter form: the first two parameters that appear in this equation designate the surface parameters of the shaper (u_s, ϑ_s), while the third parameter is the generalized parameter of motion, φ_s.

During generation of the worm by shaper, the worm and shaper rotate about their corresponding axes, z_w and z_s, which are crossed axes. Angles of rotation of the worm and shaper, φ_w and φ_s, are related by the relationship:

$$\frac{\varphi_s}{\varphi_w} = \frac{\omega_s}{\omega_w} = \frac{z_w}{z_s} \qquad (14.67)$$

where z_s is the number of teeth of the shaper, and z_w the number of threads of the worm. In accordance with the usual choices in this regard, we here assume that the worm has only one thread, so $z_w = 1$.

The meshing Eq. (14.66) can be written in the form [21, 24]:

$$N_s \cdot v_s^{(sw)} = f_{ws}(u_s, \vartheta_s, \varphi_s) = 0, \tag{14.68}$$

where N_s is the normal to any point of the shaper surface σ_s, and $v_s^{(sw)}$ is the relative velocity of a point on surface σ_s with respect to a point on surface σ_w. Of course, when these two points coincide with each other, and the two surfaces σ_s and σ_w are regular surfaces, they represent the common point of tangency of the two surfaces.

Vector function $r_w(u_s, \vartheta_s, \varphi_s)$ given by Eq. (14.65), and equation of meshing $f_{ws}(u_s, \vartheta_s, \varphi_s) = 0$, given by Eq. (14.66) or Eq. (14.68), represent the worm thread surface σ_w expressed by the three related parameters u_s, ϑ_s and φ_s.

From all this, it follows that the generation shaper surface σ_s is in line contact with the worm thread surface σ_w. Such a type of line contact is due to the fact that surface σ_w is generated as the envelope of the shaper surface σ_s. We designate by L_{sw} the lines of tangency between surfaces σ_s and σ_w in relative motion.

The surface σ_w can be also written in two-parameter form, using the theorem of implicit function system existence (see Korn and Korn [18], Zalgaller [39]). Here too, following the procedure proposed by Zalgaller and Litvin [38], we assume that the meshing equation, $f_{sw}(u_s, \vartheta_s, \varphi_s) = 0$, is satisfied at a point $P^0(u_s^0, \vartheta_s^0, \varphi_s^0)$. We also assume that at this point the following condition is satisfied:

$$\frac{\partial f_{sw}}{\partial u_s} \neq 0. \tag{14.69}$$

Thus, the meshing equation $f_{sw}(u_s, \vartheta_s, \varphi_s) = 0$ can be solved in the neighborhood of point P^0 by a function of class C^1 as:

$$\vartheta_s = \vartheta_s(u_s, \varphi_s) \in C^1 \tag{14.70}$$

and the worm thread surface, σ_w, can be determined locally as:

$$r_w(u_s, \vartheta_s(u_s, \varphi_s), \varphi_s) = R_w(u_s, \varphi_s). \tag{14.71}$$

Lines of contact L_{sw} on the worm thread surface are determined by vector function $R_w(u_s, \varphi_s)$, taking $\varphi_s = const.$

Results concerning the lines of contact, obtained using the equations described above, show that the lines L_{sw} and L_{s2} do not coincide with each other, but intersect at any point where the two surfaces σ_w and σ_2 are in meshing. Both of these lines of contact are a function of the parameter of motion, φ_s. In the plane of the parameter (u_s, ϑ_s), the qualitative results shown in Fig. 14.21 are obtained. Figure 14.21a shows that the shaper surface σ_s is in line contact with the worm surface σ_w, and that the corresponding contact lines referring to a given value of $\varphi_s^{(i)}$ are non-linear curves. Figure 14.21b shows that the shaper surface σ_s is also in line contact with

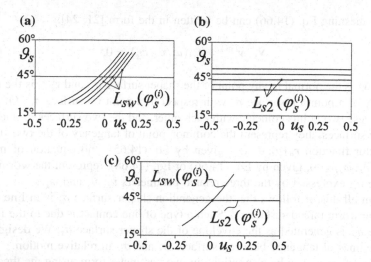

Fig. 14.21 Lines of contact: **a** lines L_{sw} between surfaces σ_s and σ_w; **b** lines L_{s2} between surfaces σ_s and σ_2; **c**, intersection of lines $L_{sw}\left(\varphi_s^{(i)}\right)$ and $L_{s2}\left(\varphi_s^{(i)}\right)$ at meshing position where surfaces σ_w and σ_2 are in point contact

the face gear surface σ_2, and that the corresponding contact lines referring to a given value of $\varphi_s^{(i)}$ are straight lines all parallel to the u_s-axis. Figure 14.21c shows that lines of contact $L_{sw}\left(\varphi_s^{(i)}\right)$ and $L_{s2}\left(\varphi_s^{(i)}\right)$ intersect with each other at a position of meshing for which $\varphi_s = \varphi_s^{(i)}$. This point of intersection corresponds to the point of tangency between surfaces σ_w and σ_2.

14.9.4 Generation of Surface σ_2 by the Worm Surface σ_w

From what we have seen in the previous section, we know that both contact between surfaces σ_s and σ_w, and between surface σ_s and σ_2, are line contacts. Instead, surfaces σ_w and σ_2 of worm and face gear are in point contact with each other at any instant. This means that neither the generation cutting nor the finishing grinding of the surface σ_2 of face gear by the worm surface σ_w can be carried out using a one-parameter enveloping process. In effect, such a process would not provide the entire required surface σ_2, but only a small strip of it.

To obtain the entire required surface σ_s, however, it is necessary to use a two-parameter enveloping process, characterized by two set of parameters, independent of each other, namely:

- a set of angles of rotation φ_w and φ_2 of the cutting (or grinding) worm and face gear to be generated (or grinded), which must be correlated by the relationship:

where

- z_2 and z_w (usually $z_w = 1$, as a single start worm is used) are respectively the number of teeth of the face gear, and the number of threads of the worm;

$$\frac{\omega_w}{\omega_2} = \frac{\varphi_w}{\varphi_2} = \frac{z_2}{z_w}, \tag{14.72}$$

- a translation motion of the worm collinear to the axis of the imaginary shaper (see Fig. 14.5), and characterized by the parameter l_w.

To obtain the surface σ_2 of the face gear generated by the worm, we use the same theoretical basis that allowed us to derive the surfaces σ_w and σ_2 as the envelopes of surface σ_s of the imaginary shaper. We also use the coordinate systems described in Sect. 14.9.1 and shown in Fig. 14.19. In this way we get the following equations, to be solved simultaneously:

$$r_2(u_s, \vartheta_s, \varphi_w, l_w) = M_{2w}(\varphi_w, l_w) r_w(u_s, \vartheta_s(u_s, \varphi_s), \varphi_s) \tag{14.73}$$

$$\left(\frac{\partial r_2}{\partial u_s} \times \frac{\partial r_2}{\partial \vartheta_s} \frac{\partial \vartheta_s}{\partial u_s}\right) \times \left(\frac{\partial r_2}{\partial \varphi_s} \times \frac{\partial r_2}{\partial \vartheta_s} \frac{\partial \vartheta_s}{\partial \varphi_s}\right) \cdot \frac{\partial r_2}{\partial \varphi_w} = 0 \tag{14.74}$$

$$\left(\frac{\partial r_2}{\partial u_s} \times \frac{\partial r_2}{\partial \vartheta_s} \frac{\partial \vartheta_s}{\partial u_s}\right) \times \left(\frac{\partial r_2}{\partial \varphi_s} \times \frac{\partial r_2}{\partial \vartheta_s} \frac{\partial \vartheta_s}{\partial \varphi_s}\right) \cdot \frac{\partial r_2}{\partial l_w} = 0, \tag{14.75}$$

where

$$r_w(u_s, \vartheta_s(u_s, \varphi_s), \varphi_s) = R_w(u_s, \varphi_s) \tag{14.76}$$

is the worm surface expressed in vector function form.

In the aforementioned equations, it should be noted that: $M_{2w}(\varphi_w, l_w)$ is the matrix of coordinate transformation from system $O_w(x_w, y_w, z_w)$ to system $O_2(x_2, y_2, z_2)$; $r_2(u_s, \vartheta_s, \varphi_w, l_w)$ is the vector function that represents the family of worm surfaces σ_w in the coordinate system $O_2(x_2, y_2, z_2)$; $\vartheta_s(u_s, \varphi_s)$ is a function obtained from the meshing Eq. (14.66) by using the theorem of implicit function system existence, already mentioned.

It is also to be noted that, since the enveloping process of generation is a two-parameter process, the problem we are dealing with is not characterized by a single meshing equation, but by two meshing equations, expressed by the relationships (14.74) and (14.75). The cross product that appears in these relationships represents the normal to the worm surface in the coordinate system $O_2(x_2, y_2, z_2)$. Finally, it is to be noted that vectors $\partial r_2/\partial \varphi_w$ and $\partial r_2/\partial l_w$ are equivalent to the relative velocities in the generation process where two set of independent motions come into play.

Two families of lines of contact can be identified on surface σ_2 of the face gear thus generated (Fig. 14.22); the first family consists of lines corresponding to

Fig. 14.22 Lines of contact during generation of the face gear by worm: **a** lines corresponding to $\varphi_w = const$ and l_w variable; **b** lines corresponding to $l_w = const$ and φ_w variable; **c**, instantaneous point of tangency between surfaces σ_w and σ_2

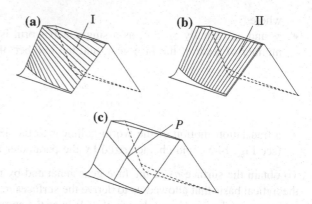

parameter $\varphi_w = const$, and variable parameter l_w (lines I in Fig. 14.22a), while the second family consists of lines corresponding to variable parameter φ_w, and parameter $l_w = const$ (lines II in Fig. 14.22b). The instantaneous point of tangency P between the worm thread surface σ_w and the face gear tooth surface σ_2 is the intersection point of lines I and II (Fig. 14.22c).

The correctness of the results obtained with the above-mentioned calculation procedure is easily controllable, because the surface σ_2 generated using the worm as a generation cutting tool must coincide with that previously obtained using the shaper as a generation cutting tool. In other words, the results obtained using Eqs. (14.73) to (14.75) must coincide with those obtained using Eqs. (14.11) and (14.16).

14.9.5   Singularities of the Worm Thread Surface

The thread surfaces σ_w of the generation worm must be designed so that they are regular surfaces, i.e. without singular points, as these points cause undercut of the face gear tooth surfaces to be generated. The procedures for determining the singularities of worm thread surfaces are the same two procedures we described in Sect. 14.6, and used to determine the singularities of face gear tooth surface.

With the first procedure, we find the singular points of the worm thread surface σ_w under the condition that the local normal N_w is equal to zero. Therefore, in the case that we are here examining, instead of Eq. (14.20) we will have:

$$\left(\frac{\partial \boldsymbol{r}_w}{\partial \vartheta_s} \times \frac{\partial \boldsymbol{r}_w}{\partial \varphi_s}\right) \frac{\partial f_{ws}}{\partial u_s} + \left(\frac{\partial \boldsymbol{r}_w}{\partial \varphi_s} \times \frac{\partial \boldsymbol{r}_w}{\partial u_s}\right) \frac{\partial f_{ws}}{\partial \vartheta_s} + \left(\frac{\partial \boldsymbol{r}_w}{\partial u_s} \times \frac{\partial \boldsymbol{r}_w}{\partial \vartheta_s}\right) \frac{\partial f_{ws}}{\partial \varphi_s} = 0, \quad (14.77)$$

where vector function $\boldsymbol{r}_w(u_s, \vartheta_s, \varphi_s)$ is determined by Eq. (14.65), and $f_{ws}(u_s, \vartheta_s, \varphi_s) = 0$ is the meshing equation, given by Eq. (14.66).

With the second procedure, we find the singular points of the worm thread surface σ_w, using the following relationships:

$$v_r^{(s)} + v_s^{(sw)} = 0 \tag{14.78}$$

$$\frac{\partial f_{ws}}{\partial u_s} \frac{du_s}{dt} + \frac{\partial f_{ws}}{\partial \vartheta_s} \frac{d\vartheta_s}{dt} = -\frac{\partial f_{ws}}{\partial \varphi_s} \frac{d\varphi_s}{dt} \tag{14.79}$$

$$M = \begin{bmatrix} \dfrac{\partial r_s}{\partial u_s} & \dfrac{\partial r_s}{\partial \vartheta_s} & -v_s^{(sw)} \\ \dfrac{\partial f_{sw}}{\partial u_s} & \dfrac{\partial f_{sw}}{\partial \vartheta_s} & -\dfrac{\partial f_{sw}}{\partial \varphi_s} \dfrac{d\varphi_s}{dt} \end{bmatrix} \tag{14.80}$$

$$f_{ws}(u_s, \vartheta_s, \varphi_s) = 0, \tag{14.81}$$

which are respectively quite similar to relationships (14.23), (14.26), (14.27) and (14.32), already obtained and discussed for find singular points on face gear tooth surface, σ_2. Quantities and expressions that appear in these relationships are: $v_r^{(s)}$, the velocity of a point that moves over surface of the imaginary shaper; $v_s^{(sw)}$, the sliding velocity at a tangency point between the meshing surface σ_s and σ_w.

As we have already seen in Sect. 14.6, it should be noted that:

- vectors that appear in Eq. (14.78) are expressed in the coordinate system $O_s(x_s, y_s, z_s)$;
- Equation (14.78) and the differential meshing Eq. (14.66), considered simultaneously, allow to obtain Eq. (14.79);
- Equations (14.78) and (14.79) result in a system of four linear equations in two unknowns, which has a certain solution if the matrix M, given by Eq. (14.80), has the rank $r = 2$;
- Equation (14.81) results from the solution of the above system of equations;
- the meshing Eqs. (14.66) and (14.81) allow to determine the sought-for line on surface of the imaginary shaper σ_s, that generates singular points on the surface of the worm σ_w.

The calculation procedure for determining and highlighting singularities on the worm surfaces consists of the following step sequence:

- By means of the meshing Eq. (14.66), we determine the lines of contact L_{sw} between shaper and worm surfaces, and represent them on the plane of parameters of the shaper (u_s, ϑ_s) as functions of the generalized parameter φ_s. In this way we can see what happens on both the side surfaces that delineate the shaper tooth space.
- With Eq. (14.81), we represent on the same plane (u_s, ϑ_s) the limiting lines of the singular points of the worm. These lines enable us to determine the maximum angle of rotation of the shaper, which is permissible for avoidance of worm singularities. It is thus possible to determine the maximum number of turns of the worm thread.
- Using Eqs. (14.66) and (14.81) as well as the equation of the shaper tooth surface σ_s, we determine the limiting lines of the regular points of the shaper, which generate the worm singularities. These limiting lines are represented on

both the side surfaces that delineate the shaper tooth space. Each of these side surfaces is characterized by two limiting lines, located close to the extreme transverse planes delimiting the face width.

- Finally, using coordinate transformation from the shaper surface σ_s to the worm thread surface σ_w, we determine the limiting regular points on surface of the shaper, and the singularity points on the worm surface that are generated by these limiting regular points of the shaper.

For more details on this subject, we refer the reader to the paper of Litvin et al. [25].

14.9.6 Dressing of the Worm

So far, we have never dealt with technological processes on how to sharp or dress the tools that are used to cut, shave or grind the gears. This is because this is a typical topic of mechanical technology, which is therefore outside the scope of this textbook or is entirely marginal. We are here to make an exception, with a short note of the sharpening and dressing problems of cutting tools (hobs) and grinding wheels used for manufacturing face gears. The reason for this exception follows from the fact that, as we have already mentioned in Sects. 14.1 and 14.2, the solution to the face-gear grinding problem was the key that opened the doors for the growing affirmation of these types of gears.

We are faced with two typical problems: those concerning the sharpening of cutting tools, and those concerning the dressing of grinding tools. From a theoretical point of view, there are no differences between the two types of problems. Therefore, for brevity, we only talk about dressing problems, with the tacit understanding that even the sharpening problems are included.

The worm dressing can be done using both planar disks, as proposed by Tan [34], and conical disks, conforming to the tooth space of the pinion, as proposed by Litvin et al. [22]. Face gears ground with any method must have tooth surfaces perfectly conjugate to the mating pinion. To achieve this goal, the installation and motion of the dressing tool relative to the worm (i.e. the grinding wheel) must meet well-defined conditions, which we here briefly describe, with reference to a dressing tool having the shape of a planar disk. As Fig. 14.5 shows, this dressing tool is a disk with a flat surface, having mathematically a plane as generation surface. In Sect. 14.2 we have already described the great advantages associated with this extremely simple geometry, so it is no longer the case to come back to the subject.

The installation and motion of the dressing planar disk relative to the grinding worm can be specified with reference to the imaginary involute pinion. The dressing disk must be positioned in such a way that it is always in tangency with the pinion tooth surface. As shown in Fig. 14.5, this positioning can be uniquely defined by the two parameters α and s, given by:

$$\alpha = \varphi_1 + \vartheta \qquad (14.82)$$

$$s = \vartheta r_{b1} \qquad (14.83)$$

where: α is the angle between the dressing surface and the shortest distance line through the axes of rotation of the grinding worm and imaginary pinion; s is the shortest distance between the dressing surface and the pinion axis; φ_1 is the angle of rotation of the pinion; ϑ is the roll angle on the pinion involute profile at point where it is in tangency with the dressing surface; r_{b1} is the base radius of the pinion.

The installation of the grinding worm on the face gear to be grind must also take into account the lead angle γ_w, given by Eq. (14.1). The motion of the dressing disk relative to the grinding worm, when it is specified with reference to the imaginary involute pinion, implies that we consider:

- the meshing between the pinion and face gear, whose rotations are governed by Eq. (14.2), which determines the transmission ratio of the gear drive;
- the grinding process, where rotation of the grinding worm is related to rotation of the face gear by Eq. (14.72).

Using simultaneously Eqs. (14.1), (14.2), (14.72), (14.82) and (14.83), we get the following relationship, which correlates the positioning parameters of the dressing planar disk and grinding worm:

$$s = r_{b1}\left(\alpha \pm \varphi_w \frac{z_w}{z_1}\right), \qquad (14.84)$$

where the plus-sign is to be used for dressing of the right side of a left-hand thread or the left side of a right-hand thread, while the minus-sign is to be used for dressing the right side of a right-hand thread or the left side of a left-hand thread.

An infinite number of pairs of parameters (α, s) satisfies Eq. (14.84), for a given value φ_w of the angular position of the grinding worm. Each of these pairs corresponds to the dressing disk in tangency condition with different points on the thread surface of the grinding worm.

If we assume that the angular velocity of the grinding worm is constant $(\omega_w = d\varphi_w/dt = const)$, differentiating Eq. (14.84) we obtain the following equation expressing the velocity v_d of the dressing disk along the direction of the normal to its flat surface (Fig. 14.5):

$$v_d = \frac{ds}{dt} = r_{b1}\left(\frac{d\alpha}{dt} \pm \omega_w \frac{z_w}{z_1}\right). \qquad (14.85)$$

Since the grinding worm is rotating, so that its angular position defined by parameter φ_w is continuously variable, the instantaneous position and orientation of the dressing disk as well as its motion are specified with two parameters (α, s). These two parameters can be varied independently, while still satisfying conditions of meshing. All dressing disk positions corresponding to different combinations of parameters (α, s), which even satisfy Eq. (14.84), constitute a family of generating planes on the grinding worm's thread surface. The final shape of the grinding worm is generated as the envelope to a family of these planar surfaces.

In gearing theory, such a generation process is called two-parameter enveloping process. Practical application of this process involves the introduction of specifically designed relationships between the two theoretically independent parameters in order to make the implementation realistically feasible.

As for the dressing process discussed here, the motion of the dressing disk is specified to be in the direction of normal to its flat surface, so $\alpha = const$. In this case, the motion of the dressing disk is a pure translation in the direction of this normal, and Eq. (14.85) becomes:

$$v_d = \pm r_{b1} \omega_w \frac{z_w}{z_1}; \qquad (14.86)$$

therefore, the magnitude of the dressing disk velocity is constant for any value of $\alpha = const$. The pure translation of the dressing disk makes it possible to avoid working into sensitive areas in the grinding worm where undercut may occur. This is one of the main advantages of the use of a dressing disk with planar surface compared to a dressing tool with an involute profile.

Another important feature of the two-parameter enveloping process is that it guarantees at every instant a point contact condition between the generation and the generated surfaces. This contact condition, combined with the use of CNC technology, allows topological control or modifications of profile of the grinding worm. These control or modifications can then be transferred on the face gear during the grinding process. The desired meshing characteristics between the pinion and face gear can be translated into specifications for modifications of the face gear teeth, and eventually into motion control in the dressing process by CNC technology. This is due to one-to-one correspondence between the position of the dressing disk, a point on the grinding worm, a point on the face gear tooth and a point on the pinion. Topological control of teeth of the face gear has important practical implications, on which we cannot dwell more extensively. In this regard, we refer the reader to the paper of Heath et al. [13]. Concluding this topic, we must highlight an elegant analytic discussion on the subject, made by Litvin et al. [25], to which we refer the reader.

References

1. Akahori H, Sato Y, Nishida Y, Kubo A (2001) Test of the durability of face gears. In: The Proceedings of the JSME International Conference on MTP, Fukuoka, Japan
2. Basstein G, Sijtstra A (1993) New developments concerning design and manufacturing of face gears. Antriebstechnik 32(11)
3. Bloomfield B (1947) Designing face gears. Mach Des 19(4):129–134

4. Buzano P (1961) Lezioni di Analisi Matematica – Parte Prima, 5th edn. Libreria Universitaria Levrotto&Bella, Torino
5. Chakraborty J, Bhadoria BS (1971) Design of face gears. Mechanisms 6:435–445
6. Chakraborty J, Bhadoria BS (1973) Some studies on hypoid face gears. Mech Mach Theory 8:339–349
7. Dudley DW, Winter H (1961) Zahnräder: Berechnung, Enfwurf und Herstellung nach amerikanischen Enfahrungen. Springer, Berlin, Göttingen, Heidelberg
8. Feng GS, Xie ZF, Zhou M (2019) Geometric design and analysis of face-gear drive with involute helical pinion. Mech Mach Theory 134:169–196
9. Francis V, Silvagi J (1950) Face Gear Design Factors. Prod Eng 21:117–121
10. Goldfarb V, Barmina N (eds) (2016) Theory and Practice of Gearing and Transmissions. In Honor of Professor Faydor L. Litvin. Springer International Publishing Switzerland, Chan Heidelberg
11. Guingand M, de Vaujany J-P, Jacquin C-Y (2005) Quasi-static analysis of a face gear under torque. Comput Methods Appl Mech Eng 194:4301–4318
12. Handschuh RF, Lewicki DG, Bossler RB (1994) Experimental testing of prototype face gears for helicopter transmissions. Proc Inst Mech Eng Part G J Aerosp Eng 208(2):129–136
13. Heath GF, Filler RR, Tan J (2002) Development of face gear technology for industrial and aerospace power transmission, NASA/CR-2002-211320, ARL-CR-0485, 1L18211-FR-01001
14. Hirsch MW, Smale S, Devaney RL (2004) Differential equations, dynamical systems, and an introduction to chaos. Elsevier Academic Press, Amsterdam
15. ISO 1122-1:1998. Vocabulary of gear terms—part 1: Definition related to geometry
16. Kawasaki K, Tsuji I, Gunbara H (2018) Geometric design of a face gear drive with a helical pinion. J Mech Sci Technol 32(4):1653–1659
17. Kissling U, Beermann S (2007) Face gears: geometry and strengh. Gear Technol 1(2):54–61
18. Korn GA, Korn TM (1968) Matematics hanbook for scientist and engineers, 2nd edn. McGraw-Hill Inc, New York
19. Lewicki DG, Handschuh RF, Heath GF, Sheth V (1999) Evaluation of carbonized face gears. In: The american helicopter society 55th annual forum, Montreal, Canada
20. Litvin FL (1992) Design and geometry of face gear drives. Trans ASME, J Mech Des 114(4): 642–647
21. Litvin FL (1998) Development of gear technology and theory of gearing, NASA reference publication 1406, ARL-TR-1500
22. Litvin FL, Chen YS, Heath GF, Sheth VJ, Chen N (2000) Apparates and method for precision grinding face gears. U.A. Patent No. 6.146.253
23. Litvin FL, Egelja A, Tan J, Chen DY-D, Heath G (2000) Handbook on face gear drives with a spur involute pinion. NASA/CR-2000-209909, ARL-CR-447
24. Litvin FL, Fuentes A (2004) Gear geometry and applied theory, 2nd edn. Cambridge University Press, Cambridge
25. Litvin FL, Fuentes A, Zanzi C, Pontiggia M (2002) Face gear drive with spur involute pinion: geometry, generation by a worm, stress analysis. NASA/CR-2002-211362. ARL-CR-491-2002
26. Litvin FL, Fuentes A, Zanzi C, Pontiggia M, Handschuh RF (2002) Face-gear drive with spur involute pinion: geometry, generation by a worm, stress analysis. Comput Methods Appl Mech Eng 191:2785–2813
27. Litvin FL, Fuentes A, Zanzi C, Pontiggia M (2002) Design, generation and stress analysis of two versions geometry of face-gear drives. Mech Mach Theory 37(23):1179–1211
28. Litvin FL, Gonzalez-Perez I, Fuentes A, Vecchiato D, Hausen BD, Binney D (2005) Design, generation and stress analysis of face-gear drive with helical pinion. Comput Methods Appl Mech Eng 194:3870–3901
29. Litvin FL, Wang JC, Bossler Jr RB, Chen Y-JD, Heath G, Lewicki DG (1992) Application of face-gear drives in helicopter transmission. In: Proceeding of the 6th international power transmission and gearing conference, Scottsdale, Arizona, AVSCOM Technical Report 91-C-036 and NASA Technical Memorandum 105655, 13–16 Sept 1992

30. Litvin FL, Zhang Y, Krenzer TJ, Goldrich RN (1989) Hypoid gear drive with face-milled teeth: condition of pinion non-undercut and fillet generation, AGMA Paper 89FTM7

31. Miller EW (1942) Hob for generation of crown gears. U.S. Patent No. 2.304.588

32. Novozhilov VV (1970) Thin shell theory. 2nd augmented and revised edition. Wolters-Noordhoff Publishing, Groningen

33. Radzevich SP (2012) Dudley's handbook of practical gear design and manufacture, 2nd edn. CRC Press, Taylor & Francis Group, Boca Raton

34. Tan J (1998) Apparates and method for improved precision grinding of face gears. U.S. Patent No. 5.823.857

35. Townsend DP (ed) (1991) Dudley's gear handbook, 2nd edn. McGraw-Hill, New York

36. Tricomi FG (1956) Lezioni di Analisi Matematica – Parte Seconda, 7th edn. CEDAM-Case Editrice Dott. Antonio Milani, Padova

37. Ventsel E, Krauthammer T (2001) Thin plates and shell: theory, analysis, and applications. Marcel Dekker Inc, New York

38. Zalgaller VA, Litvin FL (1977) Sufficient condition of existence of envelope contact lines and edge of regression on the surface of the envelope to the parametric family of surfaces represented in parametric form. Proc Univ Math 178(3):20–23 (in Russian)

39. Zalgaller VA (1975) Theory of envelopes. Publishing House Nauka, Department of Physics and Mathematics, Moscow

40. Zanzi C, Pedrero JI (2005) Application of modified geometry of face gear drive. Comput Methods Appl Mech Eng 194(27–29):3047–3066

Index of Standards

© Springer Nature Switzerland AG 2020
V. Vullo, *Gears*, Springer Series in Solid and Structural Mechanics 10,
https://doi.org/10.1007/978-3-030-36502-8

- ISO 6336-3:2006, *Calculation of load capacity of spur and helical gears—Part 3: Calculation of tooth bending strength.*
- ISO/TR 10828:1997(E), *Worm gears—Geometry of worm profiles.*
- ISO/TR 4467: *Addendum Modification of the Teeth of Cylindrical Gears for Speed- Addendum Modification Reducing and Speed-Increasing Gear Pairs.*

Name of Index

© Springer Nature Switzerland AG 2020
V. Vullo, *Gears*, Springer Series in Solid and Structural Mechanics 10,
https://doi.org/10.1007/978-3-030-36502-8

Subject Index

© Springer Nature Switzerland AG 2020
V. Vullo, *Gears*, Springer Series in Solid and Structural Mechanics 10,
https://doi.org/10.1007/978-3-030-36502-8

Printed in the United States
By Bookmasters

Printed in the United States
By Bookmasters